Lecture Notes in Computer Science 15691

Advanced Research in Computing and Software Science

Subline of Lecture Notes in Computer Science

More information about this series at https://link.springer.com/bookseries/558

Parosh Aziz Abdulla · Delia Kesner
Editors

Foundations of Software Science and Computation Structures

28th International Conference, FoSSaCS 2025
Held as Part of the International Joint Conferences on Theory and Practice of Software, ETAPS 2025
Hamilton, ON, Canada, May 3–8, 2025, Proceedings

Springer

Editors
Parosh Aziz Abdulla
Uppsala University
Uppsala, Sweden

Delia Kesner
Université Paris Cité
Paris, France

ISSN 0302-9743 ISSN 1611-3349 (electronic)
Lecture Notes in Computer Science
ISBN 978-3-031-90896-5 ISBN 978-3-031-90897-2 (eBook)
https://doi.org/10.1007/978-3-031-90897-2

This Springer imprint is published by the registered company Springer Nature Switzerland AG
The registered company address is: Gewerbestrasse 11, 6330 Cham, Switzerland

ETAPS Foreword

Welcome to the 28th ETAPS! ETAPS 2025 took place in Hamilton, Canada. It is the first time ETAPS was held outside of Europe.

ETAPS 2025 was the 28th instance of the International Joint Conferences on Theory and Practice of Software. ETAPS is an annual federated conference established in 1998, and consists of four conferences: ESOP, FASE, FoSSaCS, and TACAS. Each conference has its own Program Committee (PC) and its own Steering Committee (SC). The conferences cover various aspects of software systems, ranging from theoretical computer science to foundations of programming languages, analysis tools, and formal approaches to software engineering. Organizing these conferences in a coherent, highly synchronized conference programme enables researchers to participate in an exciting event, having the possibility to meet many colleagues working in different directions in the field, and to easily attend talks of different conferences. On the weekend before the main conference, numerous satellite workshops took place that attracted many researchers from all over the globe.

ETAPS 2025 received 329 submissions in total, 106 of which were accepted, yielding an overall acceptance rate of 32.2%. I thank all the authors for their interest in ETAPS, all the reviewers for their reviewing efforts, the PC members for their contributions, and in particular the PC (co-)chairs for their hard work in running this entire intensive process. Last but not least, my congratulations to all authors of the accepted papers!

ETAPS 2025 featured the unifying invited speakers Ina Schaefer (Karlsruhe Institute of Technology, Germany) and Matthew B. Dwyer (University of Virginia, USA), and the invited speakers Amal Ahmed (Northeastern University, USA) for ESOP and José Meseguer (University of Illinois Urbana-Champaign, USA) for FASE. Invited tutorials were provided by Suguman Bansal (Georgia Institute of Techology, USA) on reinforcement learning from logical specifications and Arun Ross (Michigan State University, USA) on biometrics.

ETAPS 2025 was organized by McMaster University. The Faculty of Engineering at McMaster University has a reputation for innovative programs, cutting-edge research, leading faculty, and aspiring students. It has earned a strong reputation as a center for academic excellence and innovation. The Faculty has approximately 180 faculty members, along with close to 4,500 undergraduate and 1,000 graduate students. The local organization team consisted of Claudio Menghi and Mark Lawford (general chairs), Melissa Alzaeim (event organizer), Alan Wassyng and Angelo Gargantini (workshop chairs), Sébastien Mosser and Matt Luckcuck (publicity chairs), Patrizio Pelliccione (sponsor chair), Silvia Bonfanti and Andrea Bombarda (web chairs), Jacques Carette and Christos Tsigkanos (local proceedings chair), Lena Liberale and Martin von Mohrenschildt (finance chairs), Damiano Torre and Lina Marsso (registration chairs), and Vera Pantelic and Denise Geiskkovitch (student volunteer chairs).

ETAPS 2025 is further supported by the following associations and societies: ETAPS e.V., EATCS (European Association for Theoretical Computer Science), EAPLS (European Association for Programming Languages and Systems), and EASST (European Association of Software Science and Technology).

The ETAPS Steering Committee consists of an Executive Board, and representatives of the individual ETAPS conferences, as well as representatives of EATCS, EAPLS, and EASST. The Executive Board consists of Marieke Huisman (Twente, chair), Andrzej Wąsowski (Copenhagen), Thomas Noll (Aachen), Jan Kofroň (Prague), Barbara König (Duisburg-Essen), Arnd Hartmanns (Twente), Caterina Urban (Inria), Jan Křetínský (Munich), Elizabeth Polgreen (Edinburgh), and Lenore Zuck (Chicago).

Other members of the steering committee are: Elvira Albert (Madrid), Maurice ter Beek (Pisa), Nathalie Bertrand (Rennes), Dirk Beyer (Munich), Artur Boronat (Leicester), Luís Caires (Lisboa), Ferruccio Damiani (Torino), Gordon Fraser (Passau), Arie Gurfinkel (Waterloo), Reiner Hähnle (Darmstadt), Reiko Heckel (Leicester), Marijn Heule (Pittsburgh), Sebastian Junges (Nijmegen), Joost-Pieter Katoen (Aachen and Twente), Guy Katz (Jerusalem), Delia Kesner (Paris), Fabrice Kordon (Paris), Robbert Krebbers (Nijmegen), Kim Guldstrand Larsen (Aalborg), Mark Lawford (Hamilton), Claudio Menghi (Hamilton and Bergamo), Stefan Milius (Erlangen-Nürnberg), Andrzej Murawski (Oxford), Corina Păsăreanu (Ames), Laure Petrucci (Paris), Peter Y.A. Ryan (Luxembourg), Don Sannella (Edinburgh), Viktor Vafeiadis (Kaiserslautern), and Anton Wijs (Eindhoven).

I would like to take this opportunity to thank all authors, keynote speakers, attendees, organizers of the satellite workshops, and Springer Nature for their support. ETAPS 2025 was also generously supported by Tourism Hamilton and the Tutte Institute for Mathematics and Computing. I hope you all enjoyed ETAPS 2025.

Finally, a big thanks to Claudio, Mark and Melissa and their local organization team for all their enormous efforts to make ETAPS a fantastic event.

May 2025

Marieke Huisman
ETAPS SC Chair
ETAPS e.V. President

Preface

This volume contains the papers presented at FoSSaCS 2025, the 28th International Conference on Foundations of Software Science and Computation Structures, held May 3–9, 2025, in Hamilton.

The conference is dedicated to foundational research that is clearly significant for software science and brings together research on theories and methods to support the analysis, integration, synthesis, transformation, and verification of programs and software systems.

There were 58 submissions, from which the committee accepted 19 papers. Each submission was assessed by three or more Program Committee members, with the help of external reviewers. The conference management system EasyChair was used to handle the submissions, conduct the electronic Program Committee discussions, and assist with the assembly of the proceedings.

We wish to thank all the authors who submitted papers for consideration, the members of the Program Committee for their conscientious work, and all additional reviewers who assisted the Program Committee in the evaluation process. We would also like to thank Andrzej Murawski, the FoSSaCS Steering Committee Chair for various pieces of advice, and the ESOP/FASE/FoSSaCS joint Artifact Evaluation Committee members for the artifact evaluation. Finally, we thank the ETAPS organization for providing an excellent environment for FoSSaCS, the other conferences and the workshops.

March 2025

Parosh Aziz Abdulla
Delia Kesner
Christos Tsigkanos
Jacques Carette

Organization

Program Committee

Parosh Aziz Abdulla	Uppsala University, Sweden
Zena M. Ariola	University of Oregon, USA
Nathalie Bertrand	Inria Rennes Bretagne-Atlantique, France
Luis Caires	Técnico ULisboa and INESC ID, Portugal
Yu-Fang Chen	Academia Sinica, Taiwan
Aiswarya Cyriac	Chennai Mathematical Institute, India
Ugo Dal Lago	Università di Bologna, Italy
Laure Daviaud	University of East Anglia, UK
Yuxin Deng	East China Normal University, China
Santiago Figueira	Universidad de Buenos Aires, Argentina
Dana Fisman	Ben-Gurion University, Israel
Ambrus Kaposi	Eötvös Loránd University, Hungary
Delia Kesner	Université Paris Cité, France
Naoki Kobayashi	The University of Tokyo, Japan
Orna Kupferman	Hebrew University of Jerusalem, Israel
Anthony W. Lin	TU Kaiserslautern, Germany
Assia Mahboubi	Inria, France and Vrije Universiteit Amsterdam, Netherlands
Dylan McDermott	Reykjavik University, Iceland
Aleksandar Nanevski	IMDEA Software Institute, Spain
Luca Padovani	Università di Camerino, Italy
Matija Pretnar	University of Ljubljana, Slovenia
Karin Quaas	University of Leipzig, Germany
Jean-Francois Raskin	Université libre de Bruxelles, Belgium
Prakash Saivasan	Institute of Mathematical Sciences, Chennai, India
Aleksy Schubert	University of Warsaw, Poland
Mahsa Shirmohammadi	CNRS, Université Paris Cité, France
Alwen Tiu	Australian National University, Australia
Patrick Totzke	University of Liverpool, UK
Caterina Urban	Inria and PSL, France
James Worrell	University of Oxford, UK

Contents

Context-Free Languages of String Diagrams

Matthew Earnshaw[1(✉)] and Mario Román[1,2]

[1] Department of Software Science, Tallinn University of Technology, Tallinn, Estonia
matt@earnshaw.org.uk

[2] Department of Computer Science, University of Oxford, Oxford, UK
mromang08@gmail.com

Abstract. We introduce context-free languages of morphisms in monoidal categories, extending recent work on the categorification of context-free languages, and regular languages of string diagrams. Context-free languages of string diagrams include classical context-free languages of words, trees, and hypergraphs, when instantiated over appropriate monoidal categories. We prove a representation theorem for context-free languages of string diagrams: every such language arises as the image under a monoidal functor of a regular language of string diagrams.

1 Introduction

Monoids are the classical algebraic home of formal languages, and a long line of research beginning in the 60s has sought to extend the tools and concepts of language theory to other algebraic structures, such as trees [24,44], traces [14], hypergraphs [12,26,45], models of algebraic theories [20,49], algebras for monads [3,4], and categories [2,50].

Categories are "monoids with many objects", and passing from the theory of context-free languages in monoids to the theory of context-free languages in categories has been the subject of recent work by Melliès and Zeilberger [35,36]. This novel structural point of view suggests a natural generalization to categories with additional structure. Here, we pursue this idea for *monoidal categories*. On the one hand, strict monoidal categories are *two-dimensional monoids*, and so a natural step from their one-dimensional counterpart. On the other hand, they have a natural graphical syntax, *string diagrams*, providing a fresh approach to languages of graphs.

A vast literature has explored language theory in various algebras of graphs, culminating in the celebrated results of Courcelle [12]. Our point of departure is the claim that many graphical notions can be naturally viewed as *morphisms in monoidal categories*; that is, monoidal categories provide a suitable algebraic framework for graphical formal languages. This manuscript pursues this idea in the context of recent work in the foundations of language theory which takes a structural approach to *context-freeness*. Ultimately, this line of work seeks to unify the various generalizations of context-free languages, and identify reusable tools for reasoning about them.

P. A. Abdulla and D. Kesner (Eds.): FoSSaCS 2025, LNCS 15691, pp. 1–23, 2025.
https://doi.org/10.1007/978-3-031-90897-2_1

1.1 Languages of string diagrams

Monoidal categories have an intuitive, sound and complete graphical syntax: *string diagrams*. String diagrams resemble graphical languages commonly found in engineering and science, and indeed, they allow us to reason about Markov kernels [23], linear algebra [6], or quantum processes [1]. In computer science, they provide foundations for visual programming [29,31].

The use of string diagrams as a syntax in these various domains suggests the need for a corresponding theory of string diagrams as a formal language. This is one aim of recent work on languages of string diagrams or *monoidal languages*, such as that elaborated by Sobociński and the first author [17,18], who introduced the class of *regular monoidal languages*. A monoidal language in this sense is simply a subset of morphisms in a strict monoidal category, just as a classical formal language is a subset of a monoid. In this work, we introduce a natural class of *context-free* monoidal languages, which capture various extended notions of context-free language found in the computer science literature.

1.2 Context-free languages over categories

Our main point of reference in this paper is the recent work of Melliès and Zeilberger [35,36]. This work is a thoroughgoing refashioning of the theory of context-free languages from a "fibrational" point of view. Melliès and Zeilberger demonstrate that it is natural and fruitful to consider context-free languages over arbitrary *categories*. They introduce an adjunction between *splicing* (introducing gaps or contexts in terms) and *contouring* (linearizing derivation trees), and use it to give a novel conceptual proof of the Chomsky-Schützenberger representation theorem: every context-free language is the image of the intersection of a regular language and the language of balanced parentheses [9].

Melliès and Zeilberger provide an ample supply of examples of context-free languages in categories, such as context-free languages of runs over an automaton, languages with an explicit end-of-input marker, multiple context-free grammars [46] and a grammar of series-parallel graphs. However, it is less clear how notions such as context-free grammars of trees and hypergraphs fit into this framework. In this paper, we show how this can be accomplished by adapting the machinery of Melliès and Zeilberger to the wider setting of monoidal categories and their string diagrams. This generalization is non-trivial, and sheds light on the intriguing differences between languages of string diagrams and classical languages. In particular, our two-dimensional version of the Chomsky-Schützenberger representation theorem says that every context-free language of string diagrams is the image under a monoidal functor of a regular language of string diagrams: no intersection of context-free and regular languages is necessary.

Related work Bruggink and König have investigated recognizable languages of morphisms in a category using a notion of automaton functor [8], which is similar to our notion of non-deterministic monoidal automaton. Similar ideas have also been investigated by Colcombet and Petrisan [11]. Griffing has also introduced a notion of recognizable set of morphisms in a category [25]. These works deal with languages over categories, rather than monoidal categories.

The representation of context-free grammars as certain morphisms of multigraphs was introduced by Walters in a short paper [51]. A similar type-theoretical version of this idea was also introduced by De Groote [13]. As discussed more extensively above, this idea was taken up and substantially refined by Melliès and Zeilberger, first in a conference paper [35] and later in an extended version [36].

A different notion of context-free families of string diagrams has been introduced by Zamdzhiev [52]. There, string diagrams are defined combinatorially as *string graphs*, and context-free families are then generated by B-edNCE graph grammars [45]. Though similar, the resulting notion is not directly comparable to ours. Here, we use the native algebra of monoidal categories and their multicategories of contexts to define and investigate languages.

Finally, Heindel's abstract [27] claims a proof of a Chomsky-Schützenberger theorem for morphisms in symmetric monoidal categories, but the work described in this abstract was never published. Our development is quite different from that outlined in Heindel's abstract. We prove a stronger representation theorem that does not require an intersection of languages; we work without the assumption of symmetry; and we generalize the categorical machinery of Melliès and Zeilberger.

Contributions We introduce context-free languages of string diagrams (Definition 4.12) and show that they include a wide variety of examples in the computer science literature including context-free languages of trees and hypergraphs. We introduce the category of raw optics (Definition 5.1) over a monoidal category, and its left adjoint, the optical contour (Definition 5.5, Theorem 5.6). We use this machinery to prove a representation theorem for context-free monoidal languages (Theorem 6.6), relating them to previous work on regular monoidal languages.

2 Preliminaries

In this paper, we define context-free grammars as particular *morphisms of multigraphs*. This point of view, while perhaps unfamiliar, is simple and powerful. It suggests natural generalizations of context-free grammars, such as we will pursue in the main body of the paper, and new conceptual tools for reasoning about them. This idea is not original to us; its roots go back to Walters [51], with recent refinement and extension by Melliès and Zeilberger [35,36].

2.1 Context-free languages in free monoids and other categories

We introduce the definition of context-free grammars as morphisms of certain *multigraphs*. Multigraphs (or *species* in the work of Melliès and Zeilberger [36]) are a kind of graph in which edges have a *list* of sources and single target. It is often helpful to think of a multigraph as a *signature*, specifying a set of typed operations. Note that this is a different use of the term *multigraph* from that specifying graphs allowing multiple parallel edges.

Definition 2.1. *A* multigraph M *is a set* S *of* sorts, *and sets* $M(X_1, ..., X_n; Y)$ *of generating* operations *(or* multimorphisms*), for each pair of a list of sorts* $X_1, ..., X_n$ *and a sort* Y. *A multigraph is* finite *if sorts and operations are finite sets. A* morphism of multigraphs *is given by a function* f *on sorts and functions* $M(X_1, ..., X_n; Y) \to M(fX_1, ..., fX_n; fY)$ *between sets of operations.*

Multigraphs freely generate multicategories, also known as *operads* (though this term sometimes refers only to the single-sorted, symmetric case). See Leinster [34] for a comprehensive reference on multicategories. The free multicategory $\mathcal{F}_\nabla M$ over a multigraph M has as multimorphisms $\mathcal{F}_\nabla M(X_1, ..., X_n; Y)$ the "trees" rooted at Y, with open leaves $X_1, ..., X_n$, that one can build by "plugging together" operations in M. We call closed trees, i.e. nullary multimorphisms $d \in \mathcal{F}_\nabla M(; Y)$, *derivations.* Every multicategory $\mathbb{M}$ has an underlying multigraph, denoted $|\mathbb{M}|$, given by forgetting identities and composition.

Every rule in a context-free grammar is of the form $R \to w_1 R_1 ... R_{n-1} w_n$, where R, R_i are non-terminals, and w_i are (possibly empty) words over an alphabet Σ. The insight of Melliès and Zeilberger [36] is that this data may be arranged as an operation $R_1, ..., R_n \to R$ in a multigraph over an n-ary operation $w_1 - ... - w_n$ called a *spliced word*: a word with n gaps, as in Figure 1. We introduce the multicategory of *spliced arrows in a category.*

Definition 2.2 (Melliès and Zeilberger [36]). *The* multicategory of spliced arrows, $\mathscr{W}\mathbb{C}$, *over a category* $\mathbb{C}$, *contains, as objects, pairs of objects of* $\mathbb{C}$, *denoted as* $\frac{A}{B}$. *Its multimorphisms are morphisms of the original category, but with n "gaps" or "holes", into which other morphisms (with holes) may be* spliced. *More precisely, the multimorphisms of* $\mathscr{W}\mathbb{C}$ *are given by:*

$$\mathscr{W}\mathbb{C}(\tfrac{A_1}{B_1}, \ldots, \tfrac{A_n}{B_n}; \tfrac{X}{Y}) := \mathbb{C}(X; A_1) \times \prod_{i=1}^{n-1} \mathbb{C}(B_i; A_{i+1}) \times \mathbb{C}(B_n; Y).$$

By convention, nullary multimorphisms are morphisms of $\mathbb{C}$, *that is* $\mathscr{W}\mathbb{C}(; \tfrac{X}{Y}) := \mathbb{C}(X; Y)$. *The identity is given by a pair of identities of the original category, multicategorical composition is derived from the composition of the original category.*

We can now present a context-free grammar in terms of a morphism of multigraphs from a multigraph of non-terminals to the underlying multigraph of spliced arrows, as in Figure 1.

Definition 2.3 (Melliès and Zeilberger [36]). *A context-free grammar of morphisms in a category* $\mathbb{C}$ *is a morphism of multigraphs* $G \to |\mathscr{W}\mathbb{C}|$ *and a sort* S *in* G *(the start symbol).*

By the free-forgetful adjunction between multicategories and multigraphs, morphisms $\phi : G \to |\mathscr{W}\mathbb{C}|$ and morphisms of multicategories (or multifunctors)

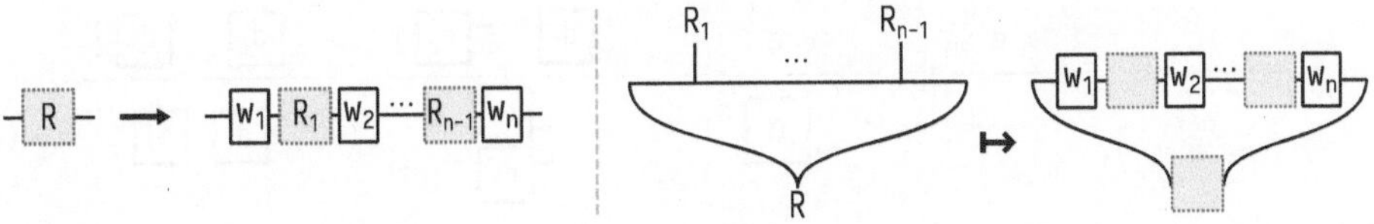

Fig. 1: (Left) Generic form of a context-free rule. (Right) Context-free rules as a morphism of multigraphs into spliced arrows; here, spliced arrows in a monoid.

$\widehat{\phi} : \mathcal{F}_\nabla G \to \mathscr{W}\mathbb{C}$ are in bijection. This allows for a slick definition of the language of a grammar.

Definition 2.4 (Melliès and Zeilberger [36]). *Let $\mathcal{G} = (\phi : G \to |\mathscr{W}\mathbb{C}|, S)$ be a context-free grammar of morphisms in $\mathbb{C}$. The language of $\mathcal{G}$ is given by the image of the set of derivations $\mathcal{F}_\nabla G(; S)$ under the multifunctor $\widehat{\phi}$.*

When $\mathbb{C}$ is a finitely generated free monoid considered as a one-object category, then context-free grammars over $\mathbb{C}$ correspond precisely to the classical context-free grammars.

An important realization of Melliès and Zeilberger is that the operation of forming the multicategory of spliced arrows in $\mathbb{C}$ has a left adjoint. That is, every multicategory gives rise to a category called the *contour* of $\mathbb{M}$, and this contouring operation is left adjoint to splicing. We refer to their paper for more details [36, Section 3.2]. Contours give a conceptual replacement for Dyck languages in the classical theory of context-free languages: they linearize derivation trees.

In Section 5, we define a new contour of multicategories which we call the *optical contour*; we shall use it to prove a representation theorem for languages of string diagrams (Theorem 6.6), inspired by generalized Chomsky-Schützenberger representation theorem proved by Melliès and Zeilberger.

2.2 Monoidal categories, their string diagrams and languages

In this paper, we will mostly be concerned with monoidal categories presented by generators and equations between the string diagrams built from these generators. Generators are given by *polygraphs*.

Definition 2.5. *A* polygraph *Γ is a set S_Γ of* sorts, *and sets $\Gamma(X_1 \otimes ... \otimes X_n; Y_1 \otimes ... \otimes Y_m)$ of* generators *for every pair of lists X_i, Y_j of sorts. A polygraph is* finite *if sorts and generators are finite sets. A* morphism of polygraphs *is a function f on sorts and functions $\Gamma(X_1 \otimes ... \otimes X_n; Y_1 \otimes ... \otimes Y_m) \to \Gamma(fX_1 \otimes ... \otimes fX_n; fY_1 \otimes ... \otimes fY_m)$ between generators.*

For a generator γ of arity $X_1 \otimes ... \otimes X_n$ and coarity $Y_1 \otimes ... \otimes Y_m$ we write $\gamma : X_1 \otimes ... \otimes X_n \to Y_1 \otimes ... \otimes Y_m$. When S is single-sorted, we use natural numbers for the arities and coarities; this case will cover most of the examples in the following. We depict generators as boxes with strings on the left and right for their arities and coarities (Figure 2).

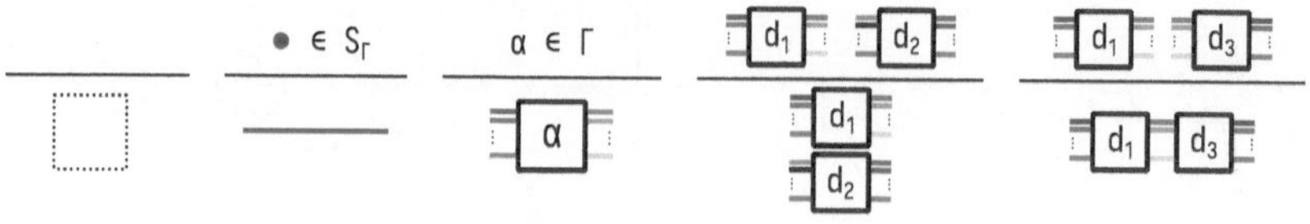

Fig. 2: The free strict monoidal category over a polygraph Γ has set of objects S_Γ^* and morphisms string diagrams given inductively over the generators of Γ as above quotiented by the equivalence relation generated by planar isotopy of diagrams, keeping left and right boundaries fixed. The leftmost rule denotes the empty diagram. We use colours to indicate sorts.

Proposition 2.6. *String diagrams with generators in a polygraph construct a monoidal category (Figure 2). The monoidal category of string diagrams over a polygraph is the* free strict monoidal category *over the polygraph [30,47]. Every monoidal category is equivalent to a strict one. In particular, string diagrams are sound and complete for monoidal categories.*

We shall need to impose equations between string diagrams, such as in defining symmetric monoidal categories, cartesian monoidal categories and hypergraph categories. To this end, we introduce the following notion of presentation.

Definition 2.7. *A finite presentation of a strict monoidal category consists of a finite polygraph of generators, $\mathcal{P}$, and a finite polygraph of equations, $\mathcal{E}$, with projections for the two sides of each equation, $l, r \colon \mathcal{E} \to |\mathcal{F}_\otimes \mathcal{P}|$. The strict monoidal category presented by $(\mathcal{P}, \mathcal{E}, l, r)$, is defined as the free strict monoidal category generated by $\mathcal{P}$ and quotiented by the equations in $\mathcal{E}$; in other words, the equalizer of the two projections $l^*, r^* \colon \mathcal{F}_\otimes \mathcal{E} \to \mathcal{F}_\otimes \mathcal{P}$.*

For the soundness and completeness of string diagrams, see Joyal and Street [30]. For a survey of string diagrams for monoidal categories, see Selinger [47].

Definition 2.8. *A* monoidal language *or* language of string diagrams *is a subset of morphisms in a strict monoidal category.*

3 Regular Monoidal Languages

Before introducing context-free monoidal languages, we introduce the *regular* case, which shall play an important role in Section 6. Regular monoidal languages were introduced by Sobociński and the first author [17,18], following earlier work of Bossut and Heindel [7,28]. They are defined by a simple automaton model, reminiscent of tree automata. In a regular monoidal language, the *alphabet* is given by a finite polygraph.

Definition 3.1. *A* non-deterministic monoidal automaton *comprises: a finite polygraph Γ (the alphabet); a finite set of states Q; for each generator $\gamma : n \to m$ in Γ, a transition function $\Delta_\gamma : Q^n \to \mathscr{P}(Q^m)$; and initial and final state vectors $i, f \in Q^*$.*

Example 3.2. Classical non-deterministic finite state automata arise as monoidal automata over single-sorted polygraphs in which every generator has arity and coarity 1. Bottom-up regular tree automata [24] arise precisely from monoidal automata over single-sorted polygraphs in which every generator has coarity 1 and arbitrary arity, with initial state the empty word and final state a singleton.

A finite state automaton over an alphabet Σ accepts elements of the free monoid Σ^*. A monoidal automaton over a polygraph Γ accepts morphisms in the free monoidal category $\mathcal{F}_{\otimes}\Gamma$ over Γ. Let us see some examples before giving the formal definition of the accepted language. We depict the transitions of a monoidal automaton as elements of a polygraph with strings labelled by states, and generators labelled by the corresponding element of Γ.

Example 3.3. Consider the following polygraph containing generators (left, below) for an opening and closing parenthesis, and the monoidal automaton over this polygraph with $Q = \{S, M\}$, $i = f = S$, and transitions shown below, centre. An accepting run over this automaton is shown below, right. The string diagram accepted by this run is what we obtain by erasing the states from this picture.

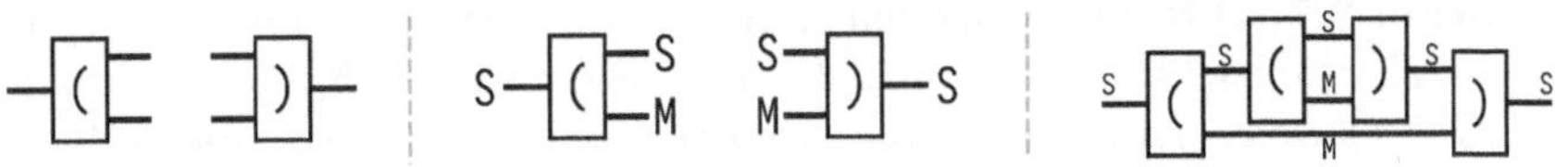

It is clear that the language accepted by this automaton is exactly the "balanced parentheses", but note that this is not a language of *words*, since we use an extra string to keep track of opening and closing parentheses. This principle will play an important role in our representation theorem in Section 6. Roughly speaking, this extra wire arises from the optical contour of a string language of balanced parentheses.

Example 3.4. In the field of DNA computing, Rothemund, Papadakis and Winfree demonstrated self-assembly of Sierpiński triangles from *DNA tiles* [43]. Sobociński and the first author [17] showed how to recast the tile model as a regular monoidal language over a polygraph containing two tile generators (white and grey), along with start and end generators, as in Figure 3. Note that the start (end) generators have arity (coarity) 0, and hence effect a transition from (to) the empty word of states.

Transitions of a non-deterministic monoidal automaton over Γ extend inductively to string diagrams in $\mathcal{F}_{\otimes}\Gamma$, giving functions $\hat{\delta}_{n,m} : Q^n \times \mathcal{F}_{\otimes}\Gamma(n,m) \to \mathscr{P}(Q^m)$.

Definition 3.5. *A string diagram $s : n \to m$ in the free monoidal category $\mathcal{F}_{\otimes}\Gamma$ over a polygraph Γ is in the language of a non-deterministic monoidal automaton $(\{\Delta_\gamma\}_{\gamma\in\Gamma}, i, f)$ if and only if the run over s reaches the final state, $f \in \hat{\delta}_{n,m}(i, s)$.*

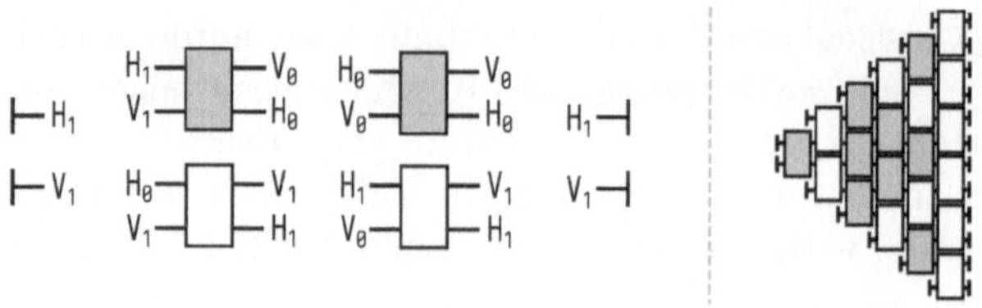

Fig. 3: Transitions for the Sierpiński monoidal automaton (left) and an element of the language (right). The initial and final states are the empty word.

Definition 3.6. *A monoidal language L is a* regular monoidal language *if and only if there exists a non-deterministic monoidal automaton accepting L.*

The data of a monoidal automaton is equivalent to a morphism of finite polygraphs, which we call a *regular monoidal grammar*, following Walters' [51] use of the term *grammar* when data is presented as *fibered* over an alphabet, and automata when the alphabet *indexes* transitions as in Definition 3.1. We shall use this convenient presentation in the following.

Definition 3.7. *A* regular monoidal grammar *is a morphism of finite polygraphs $\psi : \mathbb{Q} \to \Gamma$, equipped with finite initial and final sorts $i, f \in S^*_{\mathbb{Q}}$. The morphisms in $\mathcal{F}_\otimes\mathbb{Q}(i, f)$ are derivations in the grammar, and their image under the free monoidal functor $\mathcal{F}_\otimes\psi$ is the language of the grammar; a subset of morphisms in $\mathcal{F}_\otimes\Gamma$.*

Proposition 3.8. *For every non-deterministic monoidal automaton there is a regular monoidal grammar with the same language, and vice-versa.*

Proof (Sketch). From the transitions $\Delta_\gamma \subseteq Q^n \times Q^m$ of a monoidal automaton, we can build a polygraph $\mathbb{Q}$ by taking a generator $\gamma_i : q_1 \otimes ... \otimes q_n \to q'_1 \otimes ... \otimes q'_m$ for each $((q_1, ..., q_n), (q'_1, ..., q'_m)) \in \Delta_\gamma$. The morphism of polygraphs $\psi : \mathbb{Q} \to \Gamma$ simply maps γ_i to γ. The reverse is analogous.

Not every monoidal language is a regular monoidal language. The following is an example.

Proposition 3.9. *Let Γ be the polygraph containing two generators: one for "over-braiding"*  *and one for "under-braiding"* *. The language of* unbraids *on two strings over Γ, i.e. diagrams equivalent under planar isotopy to untangled strings, is not a regular monoidal language.*

Proof (Sketch). We can use the pumping lemma for regular monoidal languages [19, Lemma 5.1], with $k = 2$. The argument is analogous to that for classical languages of balanced parentheses: every over-braiding or under-braiding must be eventually balanced with its opposite.

In the next section, we introduce *context-free monoidal languages* and we shall see that unbraids fall in this class. In Section 6, we prove a surprising representation theorem: every context-free monoidal language is the image under a monoidal functor of a regular monoidal language.

Remark 3.10. As defined, regular monoidal languages are subsets of *free* strict monoidal categories: we shall need only this case in order to prove our main theorem. Context-free monoidal languages will be defined over *arbitrary* strict monoidal categories, so this raises the question of extending the regular case to monoidal categories that are not free. We suggest this can be done by a generalization of Melliès and Zeilberger's definition of *finite-state automata over a category* as finitary *unique lifting of factorizations* functors [36, Section 2].

4 Context-Free Monoidal Languages

We now turn our attention to context-free grammars over monoidal categories. The multicategory of spliced arrows is defined for any category. However, for categories equipped with a monoidal structure, it is natural to consider more general kinds of holes than allowed by the spliced arrows construction (Figure 4). Rather than tuples of disjoint pieces, we should allow the possibility that a hole can be surrounded by strings. The necessity of considering these more general holes is forced upon us by various examples that could not be captured using spliced arrows (e.g. Examples 4.14 and 4.15). Proofs omitted from this section may be found in the full version [15].

4.1 The symmetric multicategory of diagram contexts

Context-free monoidal grammars should contain productions from a variable to an incomplete diagram containing multiple variables or "holes". This section constructs diagram contexts over an arbitrary polygraph. Diagram contexts represent the incomplete derivation of a monoidal term: as such, they consist of string diagrams over which we add "holes". We shall notate these holes in string diagrams as pink boxes (e.g. Figure 4).

Fig. 4: (Left) A spliced arrow is a tuple of morphisms. (Right) In a monoidal category, there is the possibility of more general holes, which do not split a morphism into disjoint pieces.

Substituting another diagram context inside a hole induces a symmetric multicategorical structure on the diagrams: symmetry means that we do not distinguish the specific order in which the holes appear. This allows us to avoid

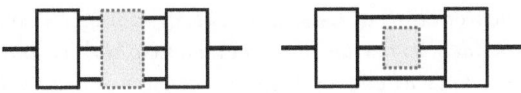

declaring a particular ordering of holes when defining a context-free monoidal grammar. We could achieve this by introducing a rule that allows us to permute contexts. However, this breaks the correspondence between terms and derivations. Instead, we shall use shufflings, inspired by the work of Shulman [48].

Definition 4.1. *A shuffling of two lists, $\Psi \in \mathrm{Shuf}(\Gamma, \Delta)$ is any list Ψ that contains the elements of both Γ and Δ in any order but preserving the relative orders of Γ and Δ.*

For instance, if $\Gamma = [\boxed{x}, \boxed{y}, \boxed{z}]$ and $\Delta = [\boxed{u}, \boxed{v}]$, a shuffling is $\Psi = [\boxed{x}, \boxed{u}, \boxed{y}, \boxed{z}, \boxed{v}]$, but not $[\boxed{y}, \boxed{u}, \boxed{z}, \boxed{x}, \boxed{v}]$. The theory of diagram contexts will introduce a shuffling every time it mixes two contexts: this way, if a term was derived by combining two contexts, we can always reorder these contexts however we want. For instance, the term $\boxed{u}, \boxed{v} \vdash \boxed{u} \mathbin{⨟} \boxed{v}$ was derived from composing the axioms $\boxed{u} \vdash \boxed{u}$ and $\boxed{v} \vdash \boxed{v}$; by choosing a different shuffling, we can also derive the term $\boxed{v}, \boxed{u} \vdash \boxed{u} \mathbin{⨟} \boxed{v}$. Let us now formally introduce the theory.

Definition 4.2. *The theory of* diagram contexts $\boxed{\mathcal{P}}$ *over a polygraph, $\mathcal{P}$, is described by the following logic. This logic contains objects ($A, B, C, ... \in \mathcal{P}^*_{obj}$) that consist of lists of types of the polygraph, $X, Y, Z, ... \in \mathcal{P}_{obj}$; it also contains contexts ($\Gamma, \Delta, \Psi, ...$) that consist of lists of pairs of objects. Apart from the single variables ($x, y, z, ..$) and the generators of the polygraph ($f, g, h, ...$); we consider fully formed terms ($t_1, t_2, ...$).*

IDENTITY
$$\frac{}{\vdash \mathrm{id} : {}^{X}_{X}}$$

GENERATOR
$$\frac{}{\vdash f : {}^{X_1,\dots,X_n}_{Y_1,\dots,Y_m}}$$

HOLE
$$\frac{}{\boxed{x} : {}^{A}_{B} \vdash \boxed{x} : {}^{A}_{B}}$$

SEQUENTIAL
$$\frac{\Gamma \vdash t_1 : {}^{A}_{B} \qquad \Delta \vdash t_2 : {}^{B}_{C} \qquad \Phi \in \mathrm{Shuf}(\Gamma; \Delta)}{\Phi \vdash t_1 \mathbin{⨟} t_2 : {}^{A}_{C}}$$

PARALLEL
$$\frac{\Gamma \vdash t_1 : {}^{A_1}_{B_1} \qquad \Delta \vdash t_2 : {}^{A_2}_{B_2} \qquad \Phi \in \mathrm{Shuf}(\Gamma; \Delta)}{\Phi \vdash t_1 \otimes t_2 : {}^{A_1 +\!\!+ A_2}_{B_1 +\!\!+ B_2}}$$

$+\!\!+$ denotes the concatenation of lists. Every term in a given context has a unique derivation. We consider terms up to α-equivalence and we impose the following equations over the terms whenever they are constructed over the same context: $(t_1 ⨟ t_2) ⨟ t_3 = t_1 ⨟ (t_2 ⨟ t_3)$; $t ⨟ \mathrm{id} = t$; $t_1 \otimes (t_2 \otimes t_3) = (t_1 \otimes t_2) \otimes t_3$; $(t_1 ⨟ t_2) \otimes (t_3 ⨟ t_4) = (t_1 \otimes t_3) ⨟ (t_2 \otimes t_4)$.

Proposition 4.3. *The multicategory of derivable sequents in the theory of diagram contexts is symmetric. In logical terms, exchange is* admissible *in the theory of diagram contexts: whenever we can prove that a diagram context exists under certain context Γ, we can prove that it exists under a permutation of Γ.*

Proposition 4.4. *Derivable sequents in the theory of diagram contexts over a polygraph $\mathcal{P}$ form the free strict monoidal category over the polygraph extended with special "hole" generators, $\mathcal{P} + \{h_{A,B} : A \to B \mid A, B \in \mathcal{P}^*_{obj}\}$. Derivable sequents over the empty context form the free strict monoidal category over the polygraph $\mathcal{P}$. Moreover, there exists a symmetric multifunctor*

$$i \colon \boxed{|\mathcal{F}_\otimes \mathcal{P}|} \to \boxed{\mathcal{P}}$$

interpreting each monoidal term as its derivable sequent.

Remark 4.5. Various notions of "holes in a monoidal category" exist in the literature, under names such as *optics*, *contexts*, or *wiring diagrams* [38,40]. Hefford and the authors [41,16] gave a universal characterization of the *produoidal* category of optics over a monoidal category. This produoidal structure is useful for describing *decompositions* of diagrams. The above logic generates a multicategory similar to the operad of directed, acyclic *wiring diagrams* introduced by Patterson, Spivak and Vagner [38]; whose operations are generic morphism shapes, rather than holes in a specific monoidal category.

Definition 4.6. *A* symmetric multigraph *is a multigraph G equipped with bijections $\sigma^* \colon G(X_1, ..., X_n; Y) \cong G(X_{\sigma(1)}, ..., X_{\sigma(n)}; Y)$ for every list $X_1, ..., X_n$ of sorts and every permutation σ, satisfying $(\sigma \cdot \tau)^* = \sigma^* \mathbin{⨾} \tau^*$ and $\mathsf{id}^* = \mathsf{id}$. A* morphism of symmetric multigraphs *is a morphism of multigraphs which commutes with the bijections.*

Definition 4.7. *Every multigraph, M, freely induces a symmetric multigraph, $\mathsf{clique}(M)$, with the same objects and, for each $f \in M(X_1, ..., X_n; Y)$, a clique of elements*

$$f_\sigma \in \mathsf{clique}(M)(X_{\sigma(1)}, ..., X_{\sigma(n)}; Y),$$

connected by symmetries, meaning that $\sigma^*(f_\tau) = f_{\sigma \cdot \tau}$. This is the left adjoint to the inclusion of symmetric multigraphs into multigraphs.

Remark 4.8. Given any symmetric multigraph G, finding a multigraph M whose clique recovers it, $\mathsf{clique}(M) = G$, amounts to choosing a representative for each one of the cliques of the multigraph. Any symmetric multigraph can be (non-uniquely) recovered in this way: for each multimorphism $f \in G(X_1, ..., X_n; Y)$, we can consider its orbit under the action of the symmetric group, $\mathsf{orb}(f) = \{\sigma^*(f) \mid \sigma \in S_n\}$ – the orbits of different elements may coincide, but each element does have one – and picking an element g_o for each orbit, $o \in \{\mathsf{orb}(f) \mid f \in G\}$, recovers a multigraph giving rise to the original symmetric multigraph.

Definition 4.9. *The theory of diagram contexts over a finitely presented monoidal category, $(\mathcal{P}, \mathcal{E}, l, r)$ (Definition 2.7), is the theory of diagram contexts over its generators, quotiented by its equations; in other words, it is the equalizer of the two projections of each equation, interpreted as derivable sequents $(\boxed{l^*} \mathbin{⨾} i), (\boxed{r^*} \mathbin{⨾} i) \colon \boxed{\mathcal{E}} \to \boxed{\mathcal{P}}$.*

Proposition 4.10. *The formation of diagram contexts in a monoidal category or polygraph extends to functors* $\boxed{\ } : \mathsf{MonCat} \to \mathsf{MultiCat}$ *and* $\boxed{\ } : \mathsf{PolyGraph} \to \mathsf{MultiGraph}$*, which moreover commute with the free multicategory* $\mathcal{F}_\triangledown$ *and free monoidal category functors* $\mathcal{F}_\otimes$*.*

At this point, the reader may doubt that the formation of diagram contexts has a left adjoint similar to the contour functor for spliced arrows. Indeed, in order to recover a left adjoint, we shall need to introduce another multicategory of diagrams which we call raw optics. This technical device will allow us to prove our main theorem (Theorem 6.6). However, let us first see the definition of context-free monoidal grammar, and some examples.

4.2 Context-Free Monoidal Grammars

We now have the ingredients for our central definition. A context-free monoidal grammar specifies a language of string diagrams by a collection of rewrites between *diagram contexts*, where the non-terminals of a context-free grammar are now (labelled) *holes* in a diagram (e.g. Figure 8). Our definition is entirely analogous to Definition 2.3, but using our new symmetric multicategory of diagram contexts in a monoidal category, instead of spliced arrows.

Definition 4.11. *A* context-free monoidal grammar *over a strict monoidal category* $(\mathbb{C}, \otimes, I)$ *is a morphism of symmetric multigraphs* $\Psi : \mathcal{G} \to |\boxed{\mathbb{C}}|$*, into the underlying multigraph of diagram contexts in* $\mathbb{C}$*, where* $\mathcal{G}$ *is finite, and a start sort* $S \in \mathcal{G}$*.*

We shall use the notation $S \sqsubset {}^{A}_{B}$ to indicate that $\Psi(S) = {}^{A}_{B}$, following the convention in the literature [36]. A morphism of symmetric multigraphs $\Psi : \mathcal{G} \to |\boxed{\mathbb{C}}|$ defining a grammar uniquely determines, via the free-forgetful adjunction, a symmetric multifunctor $\hat{\Psi} : \mathcal{F}_\triangledown \mathcal{G} \to \boxed{\mathbb{C}}$, mapping (closed) derivations to morphisms of $\mathbb{C}$. The language of a grammar is then defined analogously to Definition 2.4:

Definition 4.12. *Let* $(\Psi : \mathcal{G} \to |\boxed{\mathbb{C}}|, S \sqsubset {}^{A}_{B})$ *be a context-free monoidal grammar. The language of* Ψ *is the set of morphisms in* $\mathbb{C}(A; B)$ *given by the image under* $\hat{\Psi}$ *of the set of derivations* $\mathcal{F}_\triangledown \mathcal{G}(; S)$*. A set of morphisms* L *in* $\mathbb{C}$ *is a* context-free monoidal language *if and only if there exists a context-free monoidal grammar whose language is* L*.*

Example 4.13 (Classical context-free languages). Every context-free monoidal grammar of the following form is equivalent to a classical context-free grammar of words. Let Γ be a single-sorted finite polygraph whose generators are all of arity and coarity 1. Then context-free monoidal grammars over $\mathcal{F}_\otimes \Gamma$ with a start symbol $\phi(S) \sqsubset {}^{1}_{1}$ are context-free grammars of words over Γ. Figure 5 gives the classical example of balanced parentheses. Similarly, every context-free grammar of words may be encoded as a context-free monoidal grammar in this way.

Fig. 5: Balanced parentheses as a context-free monoidal grammar.

Example 4.14 (Context-free tree grammars). Context-free tree grammars [24,44] are defined over *ranked alphabets* of terminals and non-terminals, which amount to polygraphs in which the generators have arbitrary arity (the rank) and coarity 1. Productions have the form $A(x_1, ..., x_m) \to t$ where the left hand side is a non-terminal of rank m whose frontier is labelled by the variables x_i in order, and whose right hand side is a tree t built from terminals and non-terminals, and whose frontier is labelled by variables from the set $\{x_1, ..., x_m\}$. Note that t may use the variables non-linearly.

For example, let S be a non-terminal with arity 0, A a non-terminal with arity 2, f a terminal of arity 2, and x a terminal of arity 0 (a leaf). Then a possible rule over these generators is $A(x_1, x_2) \to f(x_1, A(x_1, x_2))$, where x_1 appears non-linearly. In order to allow such non-linear use of variables in a context-free monoidal grammar, we can consider the free *cartesian* category over Γ. In terms of string diagrams, this amounts to introducing new generators for copying (—⊂) and deleting variables (—•).

Let Γ be a polygraph in which generators have arbitrary arity, and coarity 1, as above. Context-free monoidal grammars over the free cartesian category on Γ, with a start symbol $S \sqsubset {}^{0}_{1}$ are context-free tree grammars. In Figure 6 we extend the above data to a full example. Note that by allowing start symbols $S \sqsubset {}^{0}_{n}$, we can produce forests of n trees.

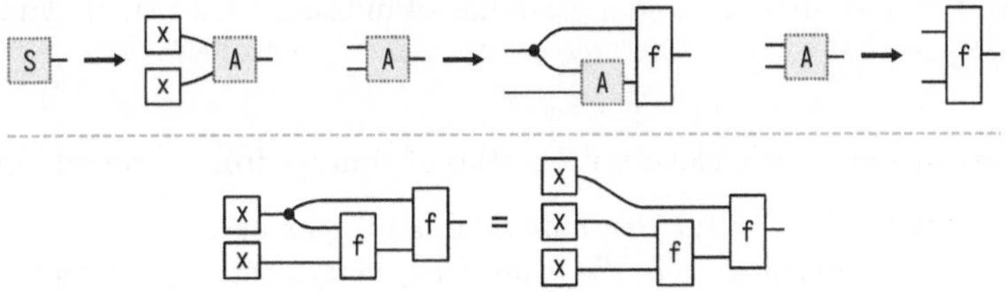

Fig. 6: Example of a context-free tree grammar as a context-free monoidal grammar. The string diagrams at the bottom are *equal* in the free *cartesian* category over the polygraph of terminals.

Example 4.15 (Hyperedge-replacement grammars). Hyperedge-replacement (HR) grammars are a kind of *context-free graph grammar* [21]. We consider HR grammars in *normal form* in the sense of Habel [26, Theorem 4.1]. A production $N \to R$ of an HR grammar has N a non-terminal with arity and coarity, and R

Fig. 7: A hypergraph grammar for simple *control flow graphs* with branching and looping, as a context-free monoidal grammar. Based on Habel [26, Example 3.3].

a hypergraph with the same arity and coarity[3], whose hyperedges are labelled by some finite set of terminals and non-terminals. Just as trees are morphisms in free *cartesian* monoidal categories (Example 4.14), hypergraphs are the morphisms of monoidal categories equipped with extra structure, known as *hypergraph categories* [42,5,22]. Generators in a polygraph are exactly directed hyperedges. The extra structure in a hypergraph category, amounts to a combinatorial encoding of patterns of wiring between nodes.

Let Γ be a polygraph of terminal hyperedges, G a multigraph of non-terminal rules, and $S \in G$ a start symbol. Then context-free monoidal grammars $(G \to |\mathsf{Hyp}[\Gamma]|, S)$ over the free hypergraph category on Γ are exactly hyperedge replacement grammars over Γ (e.g. Figure 7). A hole in a morphism in $\mathsf{Hyp}[\Gamma]$ is a placeholder for an (n, m) hyperedge, the grammar labels these holes by non-terminals, and composition corresponds to hyperedge replacement.

Example 4.16 (Unbraids). We return to the language of unbraids suggested in Proposition 3.9. Take the grammar over the over- and under-braiding polygraph depicted in Figure 8, with start symbol $S \sqsubset {}^{2}_{2}$. The language of this grammar consists of unbraids on two strings.

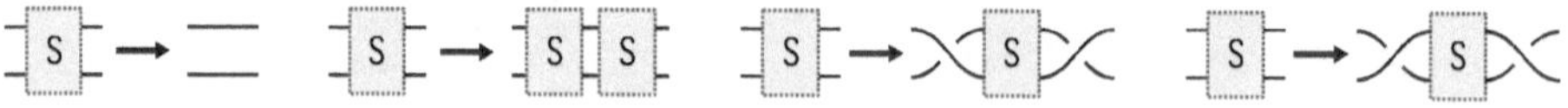

Fig. 8: A context-free monoidal grammar of unbraids, with start symbol S.

Let us record some basic closure properties of context-free monoidal languages.

Proposition 4.17. *Context-free monoidal languages over* $\mathbb{C}$ *with start symbol* $S \sqsubset {}^{X}_{Y}$ *are closed under union. The underlying morphism is given by the copairing, and start symbols can be unified by introducing a fresh symbol and productions where necessary, as in the classical case. Context-free monoidal languages are also closed under images of strict monoidal functors: the underlying morphism is given by postcomposition.*

5 Optical Contour of a Multicategory

An important realization of Melliès and Zeilberger is that the formation of spliced arrows in a category has a left adjoint, which they call the *contour* of a mul-

[3] A *multi-pointed hypergraph* in Habel's terminology.

ticategory [36, Section 3.2]. This adjunction is a key conceptual tool in their generalized version of the Chomsky-Schützenberger representation theorem, and is closely linked to the notion of *item* in LR parsing [36]. In this section, we present a similar adjunction for the monoidal setting. However, it is not clear that the formation of diagram contexts has a left adjoint. We must therefore first conduct a dissection of diagram contexts into *raw optics.*

5.1 The multicategory of raw optics

A raw optic is a tuple of morphisms obtained by cutting a diagram context into a sequence of disjoint pieces. The term *optics* refers to a notion closely related to diagram contexts, which are defined exactly as a quotient of raw optics [10,39]. In Section 5 we shall see that raw optics has a left adjoint, the optical contour, and this will be enough to prove our representation theorem (Theorem 6.6).

Definition 5.1. *The* multicategory of raw optics *over a strict monoidal category* $\mathbb{C}$*, denoted* $\mathsf{ROpt}[\mathbb{C}]$*, is defined to have, as objects, pairs* $\frac{A}{B}$ *of objects of* $\mathbb{C}$*, and, as its set of multimorphisms,* $\mathsf{ROpt}[\mathbb{C}](\frac{A_1}{B_1},\dots,\frac{A_n}{B_n};\frac{S}{T})$*, the following set, where we write* AB *for* $A \otimes B$*,*

$$\sum_{M_i,N_i \in \mathbb{C}} \mathbb{C}(S; M_1A_1N_1) \times \prod_{i=1}^{n-1} \mathbb{C}(M_iB_iN_i; M_{i+1}A_{i+1}N_{i+1}) \times \mathbb{C}(M_nB_nN_n; T).$$

As a special case, $\mathsf{ROpt}[\mathbb{C}](;\frac{S}{T}) := \mathbb{C}(S;T)$*. In other words, a multimorphism, from* $\frac{A_1}{B_1},\dots,\frac{A_n}{B_n}$ *to* $\frac{S}{T}$*, consists of two families of objects,* $M_1,\dots,M_n$ *and* $N_1,\dots,N_n$*, and a family of functions,* $(f_0,\dots,f_n)$*, with types* $f_0 \colon S \to M_1 \otimes A_1 \otimes N_1$*; with* $f_i \colon M_i \otimes B_i \otimes N_i \to M_{i+1} \otimes A_{i+1} \otimes N_{i+1}$*; for each* $1 \le i \le n-1$*; and* $f_n \colon M_n \otimes B_n \otimes N_n \to T$*. In the special nullary case, we have a single morphism* $f_0 \colon S \to T$*.*

Identities are given by pairs $(\mathsf{id}_A, \mathsf{id}_B)$*. Given two raw optics* $f = (f_0,\dots,f_n)$ *and* $g = (g_0,\dots,g_m)$*, their composition is defined by*

$$f \fatsemi_i g := (g_0,\dots,g_i \fatsemi (\mathsf{id} \otimes f_0 \otimes \mathsf{id}),\dots,\mathsf{id} \otimes f_i \otimes \mathsf{id},\dots,(\mathsf{id} \otimes f_n \otimes \mathsf{id}) \fatsemi g_{i+1},\dots,g_n).$$

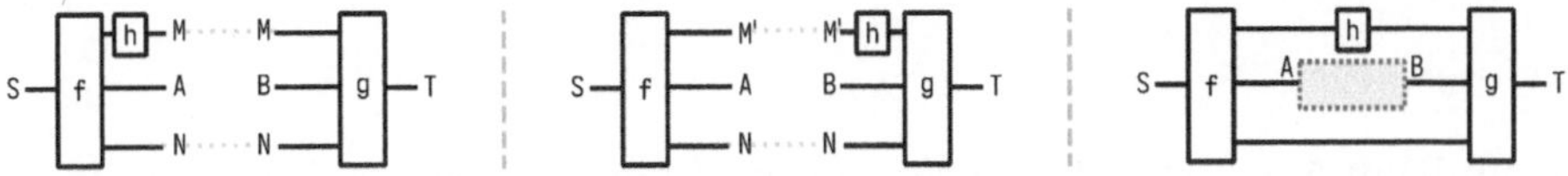

Fig. 9: Two raw optics (left, centre) in $\mathsf{ROpt}[\mathbb{C}](\frac{A}{B};\frac{S}{T})$ which quotient to the same diagram context. Note that a raw optic is not the same as a spliced arrow: the types M, N must match.

Every raw optic can be glued into a diagram context, as illustrated in Figure 9. More precisely, we have the following result.

Proposition 5.2. *There is an identity on objects multifunctor* $q : \mathsf{ROpt}[\mathbb{C}] \to \boxed{\mathbb{C}}$ *mapping each raw optic to its corresponding diagram context. Equivalently, there is an identity on objects symmetric multifunctor* $q^* \colon \mathsf{clique}(\mathsf{ROpt}[\mathbb{C}]) \to \boxed{\mathbb{C}}$*; this symmetric multifunctor is full.*

Proposition 5.3. *The construction of raw optics extends to a functor* $\mathsf{ROpt} : \mathsf{MonCat} \to \mathsf{MultiCat}$ *between the categories of strict monoidal categories and strict monoidal functors, and multicategories and multifunctors.*

Remark 5.4. We could have defined context-free monoidal grammars as morphisms into raw optics, rather than diagram contexts, but this would require an arbitrary choice of raw optic for each rule, as in Figure 9. In particular, this would force us to choose a particular ordering of the holes, since raw optics do not form a *symmetric* multicategory. On the other hand, that such a choice exists will be needed to prove our representation theorem (Section 6).

5.2 Optical contour

We now introduce the left adjoint to the formation of raw optics, which we call the *optical contour* of a multicategory. The difference from the contour of Section 2 is that additional objects M_i, N_i are introduced which keep track of strings that might surround holes. This gives rise to a strict monoidal category.

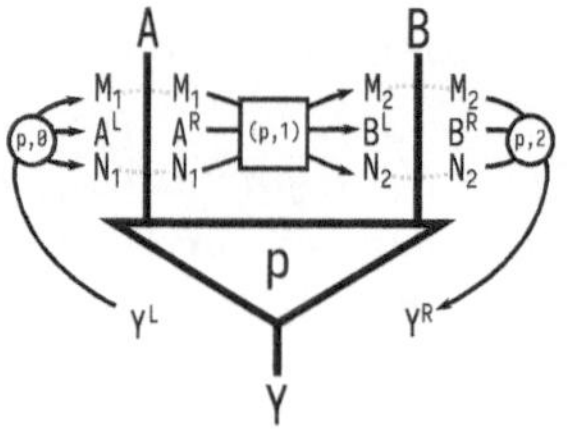

Fig. 10: A multimorphism $p \in \mathbb{M}(A, B; Y)$ and its three sectors given by optical contour: $(p, 0) : Y^L \to M_1 \otimes A^L \otimes N_1, (p, 1) : M_1 \otimes A^R \otimes N_1 \to M_2 \otimes B^L \otimes N_2, (p, 2) : M_2 \otimes B^R \otimes N_2 \to Y^R$.

Definition 5.5. *Let* $\mathbb{M}$ *be a multicategory. Its* optical contour, $\mathcal{C}\mathbb{M}$, *is the strict monoidal category presented by a polygraph whose generators are given by taking* contours *of multimorphisms in* $\mathbb{M}$. *Each multimorphism gives rise to a set of generators for the monoidal category* $\mathcal{C}\mathbb{M}$ *– its set of* sectors, *as in Figure 10.*

Explicitly, for each object $A \in \mathbb{M}$, *the optical contour* $\mathcal{C}\mathbb{M}$ *contains a left polarized,* A^L, *and a right polarized,* A^R, *version of the object. Additionally, for each multimorphism* $f \in \mathbb{M}(X_1, ..., X_n; Y)$, *there exists a family of objects* $M_1^f, ..., M_n^f, N_1^f, ..., N_n^f$, *whose superscripts we omit when they are clear from context. The morphisms are given by the following generators. For each* $f \in \mathbb{M}(X_1, ..., X_n; Y)$*, we consider the following* $n+1$ *generators:*

$$(f, 0) : Y^L \to M_1^f \otimes X_1^L \otimes N_1^f,$$
$$(f, i) : M_i^f \otimes X_i^R \otimes N_i^f \to M_{i+1}^f \otimes X_{i+1}^L \otimes N_{i+1}^f, \text{ for } 1 \leqslant i \leqslant n-1, \text{ and}$$
$$(f, n) : M_n^f \otimes X_n^R \otimes N_n^f \to Y^R.$$

In particular, for a nullary multimorphism $f \in \mathbb{M}(; Y)$*, we consider a generator* $(f, 0) : Y^L \to Y^R$*. Further, we ask for the following equations which ensure that the optical contour preserves identities and composition: for all* $x \in \mathbb{M}$*,* $(\mathsf{id}_X, 0) = \mathsf{id}_{X^L}, (\mathsf{id}_X, 1) = \mathsf{id}_{X^R}$ *with* $M_1^{\mathsf{id}_X} = N_1^{\mathsf{id}_X} = I$*; and given any* $f \in \mathbb{M}(X_1, ..., X_n; Y_i)$ *and* $g \in \mathbb{M}(Y_1, ..., Y_m; Z)$*,*

$$(f \mathbin{⨾_i} g, j) = \begin{cases} (g, j) & j < i, \text{ with } M_j^{f⨾g} = M_j^g, N_j^{f⨾g} = N_j^g \\ (g, i) ⨾ (\mathsf{id} \otimes (f, 0) \otimes \mathsf{id}) & j = i, \text{ with } M_i^{f⨾g} = M_j^g \otimes M_0^f \\ \mathsf{id}_{M_i^g} \otimes (f, j-i) \otimes \mathsf{id}_{N_i^g} & i < j < i+n, \text{ with } M_j^{f⨾g} = M_i^g \otimes M_{j-i}^f \\ (\mathsf{id} \otimes (f, n) \otimes \mathsf{id}) ⨾ (g, i+1) & j = i+n+1, \text{ with } M_j^{f⨾g} = M_i^g \otimes M_n^f \\ (g, j-n) & j > i+n+1 \text{ with } M_j^{f⨾g} = M_{j-n}^g. \end{cases}$$

In particular, when $f \in \mathbb{M}(; Y_i)$ *is nullary,* $(f \mathbin{⨾_i} g, 0) = g_i ⨾ f_0 ⨾ g_{i+1}$.

Theorem 5.6. *Optical contour is left adjoint to raw optics; there exists an adjunction* $(\mathcal{C} \dashv \mathsf{ROpt}) : \mathit{MonCat} \to \mathit{MultiCat}$.

6 A Monoidal Representation Theorem

The Chomsky-Schützenberger representation theorem says that every context-free language can be obtained as the image under a homomorphism of the intersection of a Dyck language and a regular language [9]. Melliès and Zeilberger [35] use their splicing-contour adjunction to give a novel proof of this theorem for context-free languages in categories: the classical version is recovered when the category is a free monoid. The role of the Dyck language, providing linearizations of derivation trees, is taken over by *contours* of derivations.

Monoidal categories provide a more striking case: the Dyck language is not needed because the information that parentheses encode can be carried instead by tensor products. In this section, we show that a regular monoidal language of optical contours is sufficient to reconstruct the original language. Theorem 6.6 states that *every context-free monoidal language is the image under a monoidal functor of a regular monoidal language.*

Our strategy will be to first choose a factoring of a grammar into raw optics, then use the optical contour/raw optics adjunction to produce the required monoidal functor. We must first establish that such a factoring exists. Omitted proofs may be found in the full version [15].

Lemma 6.1. *Any morphism of symmetric multigraphs underlying a context-free monoidal grammar, $\phi : G \to |\boxed{\mathbb{C}}|$, factors (non-uniquely) through the quotienting of raw optics (Proposition 5.2); meaning that there exists some multigraph G' satisfying $G = \mathsf{clique}(G')$, and some morphism $\phi_r \colon G' \to |\mathsf{ROpt}[\mathbb{C}]|$, such that $\phi = \mathsf{clique}(\phi_r) \mathbin{⨾} q^*$.*

Call the factor $\phi_r : G' \to |\mathsf{ROpt}[\mathbb{C}]|$ a *raw representative* of ϕ. It amounts to choosing a fixed ordering of the holes in a diagram context for each rule in the grammar, and a particular splicing into a raw optic.

Lemma 6.2. *Let $\mathcal{G} = (\phi, S)$ be a context-free monoidal grammar. Then the language of any raw representative ϕ_r of ϕ (with start symbol S) equals the language of $\mathcal{G}$. That is, $\phi_r[\mathcal{F}_\triangledown G'(;S)] = \phi[\mathcal{F}_\triangledown G(;S)]$.*

Lemma 6.3. *A raw representative $\phi_r : G' \to |\mathsf{ROpt}[\mathbb{C}]|$ uniquely determines a strict monoidal functor $I_\phi : \mathcal{F}_\otimes(\mathcal{C}G') \to \mathbb{C}$.*

We shall see that this monoidal functor maps the following regular monoidal language over $\mathcal{C}G$ to the language of the original context-free monoidal grammar.

Definition 6.4. *Let $\mathcal{G} = (\phi : G \to |\boxed{\mathbb{C}}|, S)$ be a context-free monoidal grammar, and ϕ_r a raw representative with domain G'. Define a* regular representative *of $\mathcal{G}$ to be the regular monoidal grammar $\mathcal{R} = (\mathsf{id} \colon \mathcal{C}G' \to \mathcal{C}G', S^L, S^R)$ over optical contours of G' whose morphism of polygraphs is the identity.*

Lemma 6.5. *Given a multigraph G, there is a bijection between derivations rooted at a sort S and optical contours from S^L to S^R, i.e.*

$$\mathcal{F}_\triangledown G(;S) \cong \mathcal{F}_\otimes(\mathcal{C}G)(S^L; S^R).$$

Theorem 6.6. *The language of a context-free monoidal grammar $\mathcal{G} = (\phi : G \to |\boxed{\mathbb{C}}|, S)$ equals the image of a regular representative under the monoidal functor I_ϕ of Lemma 6.3.*

Theorem 6.6 is at first quite surprising, since in comparison with the usual Chomsky-Schützenberger theorem and its generalization [36], one might expect to see an *intersection* of a regular monoidal language and a context-free monoidal language. Instead, this theorem tells us that regular monoidal languages are powerful enough to encode context-free monoidal languages, even while the latter is strictly more expressive than the former. Just as a context-free grammar suffices to specify a programming language which may encode instructions for arbitrary computations, regular monoidal languages can specify arbitrary context-free monoidal languages, with a monoidal functor effecting the "compilation".

7 Conclusion

There are still many avenues to explore in this structural approach to context-free languages. One obvious direction is to investigate a notion of pushdown automaton for context-free monoidal languages. In fact, it still remains to be elaborated how pushdown automata emerge for context-free languages over plain categories. Following the general principle of *parsing as a lifting problem* [36], and the duality of grammars (fibered) and automata (indexed) may provide some clue to characterizing such automata by a universal property.

The study of languages and the dependence relations that diagram contexts naturally present may be useful to the study of complexity in monoidal categories, such as the notion of "monoidal width" proposed by Di Lavore and Sobociński [32,33]. Conversely, measures of monoidal complexity may inform the cost of parsing different terms.

Finally, different types of string diagram exist for a variety of widely applied categorical structures beyond monoidal categories, such as double categories [37]. There are many opportunities to extend the general principle elaborated here to a notion of context-free language in these structures.

8 Acknowledgements

Matthew Earnshaw was supported by Estonian Research Council grant PRG1210. Mario Román was partly supported by the Air Force Office of Scientific Research (AFOSR) award number FA9550-21-1-0038 and partly supported by the Advanced Research + Invention Agency (ARIA) Safeguarded AI Programme.

References

1. Samson Abramsky and Bob Coecke. Categorical quantum mechanics. In Kurt Engesser, Dov M. Gabbay, and Daniel Lehmann, editors, *Handbook of Quantum Logic and Quantum Structures*, pages 261–323. Elsevier, Amsterdam, 2009. URL: `https://www.sciencedirect.com/science/article/pii/B9780444528698500104`, `doi:10.1016/B978-0-444-52869-8.50010-4`.
2. Jirí Adámek, Robert S.R. Myers, Henning Urbat, and Stefan Milius. Varieties of languages in a category. In *2015 30th Annual ACM/IEEE Symposium on Logic in Computer Science*, pages 414–425, 2015. `doi:10.1109/LICS.2015.46`.
3. Achim Blumensath. Algebraic language theory for Eilenberg–Moore algebras. *Logical Methods in Computer Science*, 17, 2021.
4. Mikołaj Bojańczyk, Bartek Klin, and Julian Salamanca. *Monadic Monadic Second Order Logic*, pages 701–754. Springer International Publishing, Cham, 2023. `doi:10.1007/978-3-031-24117-8_19`.
5. Filippo Bonchi, Fabio Gadducci, Aleks Kissinger, Pawel Sobocinski, and Fabio Zanasi. String diagram rewrite theory I: Rewriting with Frobenius structure. *J. ACM*, 69(2), 2022. `doi:10.1145/3502719`.

6. Filippo Bonchi, Robin Piedeleu, Pawel Sobocinski, and Fabio Zanasi. Graphical affine algebra. In *34th Annual ACM/IEEE Symposium on Logic in Computer Science, LICS 2019, Vancouver, BC, Canada, June 24-27, 2019*, pages 1–12. IEEE, 2019. doi:10.1109/LICS.2019.8785877.
7. Francis Bossut, Max Dauchet, and Bruno Warin. A Kleene theorem for a class of planar acyclic graphs. *Inf. Comput.*, 117:251–265, 03 1995. doi:10.1006/inco.1995.1043.
8. H.J. Sander Bruggink and Barbara König. Recognizable languages of arrows and cospans. *Mathematical Structures in Computer Science*, 28(8):1290–1332, 2018.
9. Noam Chomsky and Marcel-Paul Schützenberger. The algebraic theory of context-free languages. In P. Braffort and D. Hirschberg, editors, *Computer Programming and Formal Systems*, volume 35 of *Studies in Logic and the Foundations of Mathematics*, pages 118–161. Elsevier, 1963. URL: https://www.sciencedirect.com/science/article/pii/S0049237X08720238, doi:10.1016/S0049-237X(08)72023-8.
10. Bryce Clarke, Derek Elkins, Jeremy Gibbons, Fosco Loregiàn, Bartosz Milewski, Emily Pillmore, and Mario Román. Profunctor optics, a categorical update. *CoRR*, abs/2001.07488, 2020. URL: https://arxiv.org/abs/2001.07488, arXiv:2001.07488.
11. Thomas Colcombet and Daniela Petrisan. Automata minimization: a functorial approach. *Log. Methods Comput. Sci.*, 16(1), 2020.
12. Bruno Courcelle and Joost Engelfriet. *Graph Structure and Monadic Second-Order Logic: A Language-Theoretic Approach.* Encyclopedia of Mathematics and its Applications. Cambridge University Press, 2012. doi:10.1017/CBO9780511977619.
13. Philippe de Groote. Towards abstract categorial grammars. In *Proceedings of the 39th Annual Meeting of the Association for Computational Linguistics*, pages 252–259, Toulouse, France, July 2001. Association for Computational Linguistics. URL: https://aclanthology.org/P01-1033, doi:10.3115/1073012.1073045.
14. Volker Diekert and Grzegorz Rozenberg. *The Book of Traces.* World Scientific, 1995. doi:10.1142/2563.
15. Matt Earnshaw and Mario Román. Context-free languages of string diagrams. *CoRR*, abs/2404.10653, 2024. URL: https://doi.org/10.48550/arXiv.2404.10653, arXiv:2404.10653, doi:10.48550/ARXIV.2404.10653.
16. Matthew Earnshaw, James Hefford, and Mario Román. The Produoidal Algebra of Process Decomposition. In Aniello Murano and Alexandra Silva, editors, *32nd EACSL Annual Conference on Computer Science Logic (CSL 2024)*, volume 288 of *Leibniz International Proceedings in Informatics (LIPIcs)*, pages 25:1–25:19, Dagstuhl, Germany, 2024. Schloss Dagstuhl – Leibniz-Zentrum für Informatik. URL: https://drops.dagstuhl.de/entities/document/10.4230/LIPIcs.CSL.2024.25, doi:10.4230/LIPIcs.CSL.2024.25.
17. Matthew Earnshaw and Paweł Sobociński. Regular monoidal languages. In Stefan Szeider, Robert Ganian, and Alexandra Silva, editors, *47th International Symposium on Mathematical Foundations of Computer Science (MFCS 2022)*, volume 241 of *Leibniz International Proceedings in Informatics (LIPIcs)*, pages 44:1–44:14, Dagstuhl, Germany, 2022. Schloss Dagstuhl – Leibniz-Zentrum für Informatik. URL: https://drops.dagstuhl.de/opus/volltexte/2022/16842, doi:10.4230/LIPIcs.MFCS.2022.44.
18. Matthew Earnshaw and Paweł Sobociński. String Diagrammatic Trace Theory. In Jérôme Leroux, Sylvain Lombardy, and David Peleg, editors, *48th International Symposium on Mathematical Foundations of Computer Science (MFCS*

2023), volume 272 of *Leibniz International Proceedings in Informatics (LIPIcs)*, pages 43:1–43:15, Dagstuhl, Germany, 2023. Schloss Dagstuhl – Leibniz-Zentrum für Informatik. URL: https://drops.dagstuhl.de/opus/volltexte/2023/18577, doi:10.4230/LIPIcs.MFCS.2023.43.

19. Matthew Earnshaw and Pawel Sobociński. Regular planar monoidal languages. *Journal of Logical and Algebraic Methods in Programming*, 2024. In Press.
20. Samuel Eilenberg and Jesse B. Wright. Automata in general algebras. *Information and Control*, 11(4):452–470, 1967. doi:10.1016/S0019-9958(67)90670-5.
21. Joost Engelfriet. *Context-Free Graph Grammars*, pages 125–213. Springer Berlin Heidelberg, Berlin, Heidelberg, 1997. doi:10.1007/978-3-642-59126-6_3.
22. Brendan Fong and David I. Spivak. Hypergraph categories. *Journal of Pure and Applied Algebra*, 223(11):4746–4777, 2019. URL: https://www.sciencedirect.com/science/article/pii/S0022404919300489, doi:10.1016/j.jpaa.2019.02.014.
23. Tobias Fritz. A synthetic approach to Markov kernels, conditional independence, and theorems on sufficient statistics. *CoRR*, abs/1908.07021, 2019. URL: http://arxiv.org/abs/1908.07021, arXiv:1908.07021.
24. Ferenc Gécseg and Magnus Steinby. *Tree Languages*, page 1–68. Springer-Verlag, Berlin, Heidelberg, 1997.
25. Gary Griffing. Composition-representative subsets. *Theory and Applications of Categories*, 11(19):420–437, 2003.
26. Annegret Habel. *Hyperedge Replacement: Grammars and Languages*, volume 643 of *Lecture Notes in Computer Science*. Springer, 1992. URL: https://doi.org/10.1007/BFb0013875, doi:10.1007/BFB0013875.
27. Tobias Heindel. The Chomsky-Schützenberger theorem with circuit diagrams in the role of words. *Abstract*, 2017.
28. Tobias Heindel. A Myhill-Nerode theorem beyond trees and forests via finite syntactic categories internal to monoids. *Preprint*, 2017.
29. Alan Jeffrey. Premonoidal categories and a graphical view of programs. *Preprint, Dec*, pages 80688–7, 1997.
30. André Joyal and Ross Street. The geometry of tensor calculus, I. *Advances in Mathematics*, 88(1):55–112, 1991. URL: https://www.sciencedirect.com/science/article/pii/000187089190003P, doi:10.1016/0001-8708(91)90003-P.
31. Mohammad Amin Kuhail, Shahbano Farooq, Rawad Hammad, and Mohammed Bahja. Characterizing visual programming approaches for end-user developers: A systematic review. *IEEE Access*, 9:14181–14202, 2021.
32. Elena Di Lavore. *Monoidal Width.* PhD thesis, Tallinn University of Technology, 2023.
33. Elena Di Lavore and Pawel Sobocinski. Monoidal width. *Log. Methods Comput. Sci.*, 19(3), 2023. URL: https://doi.org/10.46298/lmcs-19(3:15)2023, doi:10.46298/LMCS-19(3:15)2023.
34. Tom Leinster. *Higher Operads, Higher Categories.* London Mathematical Society Lecture Note Series. Cambridge University Press, 2004. doi:10.1017/CBO9780511525896.
35. Paul-André Melliès and Noam Zeilberger. Parsing as a Lifting Problem and the Chomsky-Schützenberger Representation Theorem. In *MFPS 2022-38th conference on Mathematical Foundations for Programming Semantics*, 2022.
36. Paul-André Melliès and Noam Zeilberger. The categorical contours of the Chomsky-Schützenberger representation theorem. Preprint, December 2023. URL: https://hal.science/hal-04399404.

37. David Jaz Myers. String diagrams for double categories and equipments, 2018. arXiv:1612.02762.
38. Evan Patterson, David I. Spivak, and Dmitry Vagner. Wiring diagrams as normal forms for computing in symmetric monoidal categories. *Electronic Proceedings in Theoretical Computer Science*, 333:49–64, February 2021. URL: http://dx.doi.org/10.4204/EPTCS.333.4, doi:10.4204/eptcs.333.4.
39. Mitchell Riley. Categories of Optics. *arXiv preprint arXiv:1809.00738*, 2018.
40. Mario Román. Open diagrams via coend calculus. *Electronic Proceedings in Theoretical Computer Science*, 333:65–78, Feb 2021. URL: http://dx.doi.org/10.4204/EPTCS.333.5, doi:10.4204/eptcs.333.5.
41. Mario Román. *Monoidal Context Theory*. PhD thesis, Tallinn University of Technology, 2023.
42. Robert Rosebrugh, Nicoletta Sabadini, and Robert F.C. Walters. Generic commutative separable algebras and cospans of graphs. *Theory and applications of categories*, 15:164–177, 2005.
43. Paul W. K Rothemund, Nick Papadakis, and Erik Winfree. Algorithmic self-assembly of DNA Sierpinski triangles. *PLOS Biology*, 2(12), 12 2004. doi:10.1371/journal.pbio.0020424.
44. William C. Rounds. Context-free grammars on trees. In *Proceedings of the First Annual ACM Symposium on Theory of Computing*, STOC '69, page 143–148, New York, NY, USA, 1969. Association for Computing Machinery. doi:10.1145/800169.805428.
45. Grzegorz Rozenberg. *Handbook Of Graph Grammars And Computing By Graph Transformation, Vol 1: Foundations*. World Scientific Publishing Company, 1997. URL: https://books.google.ee/books?id=KwbtCgAAQBAJ.
46. Hiroyuki Seki, Takashi Matsumura, Mamoru Fujii, and Tadao Kasami. On multiple context-free grammars. *Theoretical Computer Science*, 88(2):191–229, 1991. URL: https://www.sciencedirect.com/science/article/pii/030439759190374B, doi:10.1016/0304-3975(91)90374-B.
47. Peter Selinger. A survey of graphical languages for monoidal categories. In Bob Coecke, editor, *New Structures for Physics*, pages 289–355. Springer Berlin Heidelberg, Berlin, Heidelberg, 2011. doi:10.1007/978-3-642-12821-9_4.
48. Michael Shulman. Categorical logic from a categorical point of view. *Available on the web*, 2016. URL: https://mikeshulman.github.io/catlog/catlog.pdf.
49. James W. Thatcher and J. B. Wright. Generalized finite automata theory with an application to a decision problem of second-order logic. *Mathematical systems theory*, 2(1):57–81, Mar 1968.
50. Bret Tilson. Categories as algebra: An essential ingredient in the theory of monoids. *Journal of Pure and Applied Algebra*, 48(1):83–198, 1987. URL: https://www.sciencedirect.com/science/article/pii/0022404987901083, doi:10.1016/0022-4049(87)90108-3.
51. Robert F.C. Walters. A note on context-free languages. *Journal of Pure and Applied Algebra*, 62(2):199–203, 1989. doi:10.1016/0022-4049(89)90151-5.
52. Vladimir Zamdzhiev. *Rewriting Context-free Families of String Diagrams*. PhD thesis, University of Oxford, 2016.

A behavioural pseudometric for continuous-time Markov processes

Linan Chen[1], Florence Clerc[2(✉)], and Prakash Panangaden[1]

[1] McGill University, Montreal, Canada
[2] Heriot-Watt University, Edinburgh, Scotland
florence.clerc@mail.mcgill.ca

Abstract. In this work, we generalize the concept of bisimulation metric in order to metrize the behaviour of continuous-time processes. Similarly to what is done for discrete-time systems, we follow two approaches and show that they coincide: as a fixpoint of a functional and through a real-valued logic.

The whole discrete-time approach relies entirely on the step-based dynamics: the process jumps from state to state. We define a behavioural pseudometric for processes that evolve continuously through time, such as Brownian motion or involve jumps or both.

1 Introduction

Bisimulation is a concept that captures behavioural equivalence of states in a variety of types of transition systems. It has been widely studied in a discrete-time setting where the notion of a step is fundamental. An important and especially useful further notion is that of bisimulation metric which quantifies "how similar two states are".

Most of the theoretical work that exists is on discrete time but a growing part of what computer science allows us to do is in real-time: robotics, self-driving cars, online machine-learning etc. A common solution is to discretize time, however it is well-known that this can lead to errors that are hopefully small but that may accumulate over time and lead to vastly different outcomes. For that reason, it is important to have a continuous-time way of quantifying the error made.

Bisimulation [21,23,25] is a fundamental concept in the theory of transition systems capturing a strong notion of behavioural equivalence. The extension to probabilistic systems is due to Larsen and Skou [20]; henceforth we will simply say "bisimulation" instead of "probabilistic bisimulation". Bisimulation has been studied for discrete-time systems where transitions happen as steps, both on discrete [20] and continuous state spaces [3,12,13]. In all these types of systems, a crucial ingredient of the definition of bisimulation is the ability to talk about *the next step*. This notion of bisimulation is characterized by a modal logic [20] even when the state space is continuous [12].

Some work had previously been done in what are called continuous-time systems, see for example [2], but even in so-called continuous-time Markov chains

P. A. Abdulla and D. Kesner (Eds.): FoSSaCS 2025, LNCS 15691, pp. 24–44, 2025.
https://doi.org/10.1007/978-3-031-90897-2_2

there is a discrete notion of time *step*; it is only that there is a real-valued duration associated with each state that leads to people calling such systems continuous-time. They are often called "jump processes" in the mathematical literature (see, for example, [24,28]), a phrase that better captures the true nature of such processes. Metrics and equivalences for such processes were studied by Gupta et al. [17,18].

The processes we consider have continuous state spaces and are governed by a continuous-time evolution, a paradigmatic example is Brownian motion. When approximating such processes by discrete-time processes, entirely new phenomena and difficulties manifest themselves in this procedure. For example, even the basic properties of trajectories of Brownian motion are vastly more complicated than the counterparts of a random walk. Basic concepts like "the time at which a process exits a given subset of the state space" becomes intricate to define. Notions like "matching transition steps" are no longer applicable as the notion of "step" does not make sense.

In [7,8,9], we proposed different notions of behavioural equivalences on continuous-time processes. We showed that there were several possible extensions of the notion of bisimulation to continuous time and that the continuous-time notions needed to involve trajectories in order to be meaningful. There were significant mathematical challenges in even proving that an equivalence relation existed. For example, obstacles occurred in establishing measurability of various functions and sets, due to the inability to countably generate the relevant σ-algebras. Those papers left completely open the question of defining a suitable pseudometric analogue, a concept that would be more useful in practice than an equivalence relation.

Previous work on discrete-time Markov processes by Desharnais et al. [14,15] extended the modal logic characterizing bisimulation to a real-valued logic that allowed to not only state if two states were "behaviourally equivalent" but, more interestingly, how similarly they behaved. This shifts the notion from a qualitative notion (an equivalence) to a quantitative one (a pseudometric).

Other work also on discrete-time Markov processes by van Breugel et al. [26] introduced a slightly different real-valued logic and compared the corresponding pseudometric to another pseudometric obtained as a terminal coalgebra of a carefully crafted functor. We also mention in this connexion the work by Ferns et al. on Markov Decision Processes and the connexion between bisimulation and optimal value functions [16].

In this work, we are looking to extend the notion of bisimulation metric to a behavioural pseudometric on continuous-time processes. Very broadly speaking, we are following a familiar path from equivalences to logics to metrics. However, it is necessary for us to redevelop the framework and the mathematical techniques from scratch. Indeed, a very important aspect in discrete-time is the fact that the process is a jump process, "hopping" from state to state. This limitation also applies to continuous-time Markov chains. In our case, we want to cover processes that evolve through time. A standard example would be Brownian motion or other diffusion processes (often described by stochastic differential equations). As one will see throughout this work, there are new mathematical challenges that need to be overcome. This means that the similarity between the pre-existing work on discrete-time and our generalization to continuous-time is only at the highest level of abstraction.

Outline of the paper[3]*:* The first two sections after the introduction are background. We will start by recalling some mathematical notions in Section 2, introducing the continuous-time processes that we will be studying in Section 3. In Section 4, we will introduce a functional $\mathcal{F}$ and define a pseudometric $\overline{\delta}$ using this functional. We will also show that the pseudometric $\overline{\delta}$ is a fixpoint of $\mathcal{F}$. In Section 5, we will show that this pseudometric is characterized by a real-valued logic. We will further emphasize the novelty of this work wrt discrete time and summarize the obstacles that we had to overcome in Section 6. We will provide some examples in Section 7. Finally we will discuss the limitations of our approach and how it relates to previous works in Section 8.

2 Mathematical background

We assume the reader to be familiar with basic measure theory and topology. Nevertheless we provide a brief review of the relevant notions and theorems. Let us start with clarifying a few notations on integrals: Given a measurable space X equipped with a measure μ and a measurable function $f : X \to \mathbb{R}$, we can write either $\int f \, \mathrm{d}\mu$ or $\int f(x) \, \mu(\mathrm{d}x)$ interchangeably. The second notation will be especially useful when considering a Markov kernel P_t for some $t \geq 0$ and $x \in X$: $\int f(y) \, P_t(x, \mathrm{d}y) = \int f \, \mathrm{d}P_t(x)$.

2.1 Couplings

Definition 1. *Let (X, Σ_X, P) and (Y, Σ_Y, Q) be two probability spaces. Then a* coupling *γ of P and Q is a probability distribution on $(X \times Y, \Sigma_X \otimes \Sigma_Y)$ such that for every $B_X \in \Sigma_X$, $\gamma(B_X \times Y) = P(B_X)$ and for every $B_Y \in \Sigma_Y$, $\gamma(X \times B_Y) = Q(B_Y)$ (P, Q are called the* marginals *of γ). We write $\Gamma(P, Q)$ for the set of couplings of P and Q.*

Lemma 1. *Given two probability measures P and Q on Polish spaces X and Y respectively, the set of couplings $\Gamma(P, Q)$ is compact under the topology of weak convergence.*

2.2 Optimal transport theory

A lot of this work is based on optimal transport theory. This whole subsection is based on [27] and will be adapted to our framework.

[3] We provide a longer version of this paper with additional appendices [11]

Consider a Polish space $\mathcal{X}$ and a lower semi-continuous cost function $c : \mathcal{X} \times \mathcal{X} \to [0, 1]$, meaning that for any for every $x_0, y_0 \in \mathcal{X}$, $\liminf_{(x,y)\to(x_0,y_0)} c(x, y) \geq c(x_0, y_0)$. Assume that c also satisfies that for every $x \in \mathcal{X}$, $c(x, x) = 0$.

For every two probability distributions μ and ν on $\mathcal{X}$, we write $W(c)(\mu, \nu)$ for the optimal transport cost from μ to ν. Adapting Theorem 5.10(iii) of [27] to our framework, we get the following statement for the Kantorovich duality:

$$W(c)(\mu, \nu) = \min_{\gamma \in \Gamma(\mu,\nu)} \int c \, \mathrm{d}\gamma = \max_{h \in \mathcal{H}(c)} \left| \int h \, \mathrm{d}\mu - \int h \, \mathrm{d}\nu \right|$$

where $\mathcal{H}(c) = \{h : \mathcal{X} \to [0, 1] \mid \forall x, y \ \ |h(x) - h(y)| \leq c(x, y)\}$.

Lemma 2. *If the cost function c is a 1-bounded pseudometric on $\mathcal{X}$, then $W(c)$ is a 1-bounded pseudometric on the space of probability distributions on $\mathcal{X}$.*

We will later need the following technical lemma. Theorem 5.20 of [27] states that a sequence $W(c_k)(P_k, Q_k)$ converges to $W(c)(P, Q)$ if c_k uniformly converges to c and P_k and Q_k converge weakly to P and Q respectively. Uniform convergence in the cost function may be too strong a condition for us, but the following lemma is enough for what we need.

Lemma 3. *Consider a Polish space $\mathcal{X}$ and a cost function $c : \mathcal{X} \times \mathcal{X} \to [0, 1]$ such that there exists an increasing ($c_{k+1} \geq c_k$ for every k) sequence of continuous cost functions $c_k : \mathcal{X} \times \mathcal{X} \to [0, 1]$ that converges to c pointwise. Then, given two probability distributions P and Q on $\mathcal{X}$,*

$$\lim_{k\to\infty} W(c_k)(P, Q) = W(c)(P, Q).$$

3 Background on continuous-time Markov processes

This work focuses on continuous-time processes that are honest (without loss of mass over time) and with additional regularity conditions. In order to define what we mean by continuous-time Markov processes here, we first define Feller-Dynkin processes. Much of this material is adapted from [24] and we use their notations. Another useful source is [4].

Let E be a locally compact, Hausdorff space with a countable base. We also equip the set E with its Borel σ-algebra $\mathcal{B}(E)$, denoted $\mathcal{E}$. The previous topological hypotheses also imply that E is σ-compact and Polish (see corollary IX.57 in [5]). We will denote Δ for the 1-bounded metric that generates the topology making E Polish.

Definition 2. *A* semigroup *of operators on any Banach space X is a family of linear continuous (bounded) operators $\mathcal{P}_t : X \to X$ indexed by $t \in \mathbb{R}_{\geq 0}$ such that*

$$\forall s, t \geq 0, \mathcal{P}_s \circ \mathcal{P}_t = \mathcal{P}_{s+t} \qquad \textit{(semigroup property)}$$

and

$$\mathcal{P}_0 = I \qquad \textit{(the identity)}.$$

Definition 3. *For X a Banach space, we say that a semigroup $\mathcal{P}_t : X \to X$ is* strongly continuous *if*

$$\forall x \in X, \lim_{t \downarrow 0} \|\mathcal{P}_t x - x\| \to 0.$$

What the semigroup property expresses is that we do not need to understand the past (what happens before time t) in order to compute the future (what happens after some additional time s, so at time $t+s$) as long as we know the present (at time t).

We say that a continuous real-valued function f on E "vanishes at infinity" if for every $\varepsilon > 0$ there is a compact subset $K \subseteq E$ such that for every $x \in E \backslash K$, we have $|f(x)| \leq \varepsilon$. To give an intuition, if E is the real line, this means that $\lim_{x \to \pm\infty} f(x) = 0$. The space $C_0(E)$ of continuous real-valued functions that vanish at infinity is a Banach space with the sup norm.

Definition 4. *A* Feller-Dynkin (FD) semigroup *is a strongly continuous semigroup $(\hat{P}_t)_{t \geq 0}$ of linear operators on $C_0(E)$ satisfying the additional condition:*

$$\forall t \geq 0 \quad \forall f \in C_0(E), \text{ if } \quad 0 \leq f \leq 1, \text{ then } \quad 0 \leq \hat{P}_t f \leq 1$$

The Riesz representation theorem can be found as Theorem II.80.3 of [24]. From it, we can derive the following important proposition which relates these FD-semigroups with Markov kernels. This allows one to see the connection with familiar probabilistic transition systems.

Proposition 1. *Given an FD-semigroup $(\hat{P}_t)_{t \geq 0}$ on $C_0(E)$, it is possible to define a unique family of sub-Markov kernels $(P_t)_{t \geq 0} : E \times \mathcal{E} \to [0,1]$ such that for all $t \geq 0$ and $f \in C_0(E)$,*

$$\hat{P}_t f(x) = \int f(y) P_t(x, \mathrm{d}y).$$

Given a time t and a state x, we will often write $P_t(x)$ for the measure $P_t(x, \cdot)$ on E. Note that since E is Polish, then $P_t(x)$ is tight.

Definition 5. *A process described by the FD-semigroup $(\hat{P}_t)_{t \geq 0}$ is* honest *if for every $x \in E$ and every time $t \geq 0$, $P_t(x, E) = 1$.*

Worded differently, a process is honest if there is no loss of mass over time. A standard example of an honest process is Brownian motion.

3.1 Observables

In previous sections, we defined Feller-Dynkin processes. In order to bring the processes more in line with the kind of transition systems that have hitherto been studied in the computer science literature, we also equip the state space E with an additional continuous function $obs : E \to [0,1]$. One should think of it as the interface between the process and the user (or an external observer): external observers won't see the exact state in which the process is at a given time, but they will see the associated *observables*. What could be a real-life example is the depth at which a diver goes: while the diver does not know precisely his location underwater, at least his watch is giving him the depth at which he is.

Note that the condition on the observable is a major difference from our previous work [7,8] since we used a countable set of atomic propositions AP and *obs* was a discrete function $E \to 2^{AP}$.

Definition 6. *In this study, a* Continuous-time Markov process *(abbreviated CTMP) is an honest FD-semigroup on* $C_0(E)$ *equipped with a continuous function obs* : $E \to [0,1]$ *and that satisfies the following additional property: if a sequence* $(x_n)_{n\in\mathbb{N}}$ *converges to* x *in* E*, then for every* t*, the sequence of measures* $(P_t(x_n))_{n\in\mathbb{N}}$ *weakly converges to the measure* $P_t(x)$*.*

Remark 1. Some properties could be relaxed. For instance, in some cases, a non honest process could be made into a CTMP by adding a state ∂. Another hypothesis that could be relaxed is the one on *obs* by imposing some stronger conditions on the FD-process.

4 Generalizing to continuous-time through a functional

We start by defining a behavioural pseudometric on our CTMPs by defining a functional $\mathcal{F}$ on the lattice of 1-bounded pseudometrics. As we will see, unlike in the discrete-time case, it is not possible to apply the Banach fixpoint theorem and get a fixpoint metric a priori: instead we need to construct a candidate and then show that it is a fixpoint of our functional. More specifically, the idea is to iteratively apply our functional to a metric and then consider the supremum of the sequence of pseudometrics. Doing so requires to first restrict the scope of our functional $\mathcal{F}$.

4.1 Lattices

At the core of this construction is the definition of a functional on the lattice of 1-bounded pseudometrics.

Let $\mathcal{M}$ be the lattice of 1-bounded pseudometrics on the state space E equipped with the order $\leq$ defined as: $m_1 \leq m_2$ if and only if for every (x, y), $m_1(x, y) \leq m_2(x, y)$. We can define a sublattice $\mathcal{P}$ of $\mathcal{M}$ by restricting to pseudometrics that are lower semi-continuous (wrt the original topology $\mathcal{O}$ on E generated by the metric Δ making the space E Polish). We will further require to define the sublattice $\mathcal{C}$ which is the set of pseudometrics $m \in \mathcal{M}$ on the state space E such that the topology generated by m on E is a subtopology of the original topology $\mathcal{O}$, *i.e.* m is a continuous function $E \times E \to [0, 1]$.

We have the following inclusion: $\mathcal{C} \subset \mathcal{P} \subset \mathcal{M}$.

Remark 2. One has to be careful here. The topology $\mathcal{O}$ on E is generated by the 1-bounded metric Δ, and hence Δ is in $\mathcal{C}$. However, we can define many pseudometrics that are not related to $\mathcal{O}$. As an example, the discrete pseudometric[4] on the real line is not related to the usual topology on $\mathbb{R}$.

4.2 Defining our functional

Throughout the rest of the paper, $(P_t)_{t>=0}$ is the family of Markov kernels associated with a CTMP. Given a discount factor $0 < c < 1$, we define the functional $\mathcal{F}_c : \mathcal{P} \to \mathcal{M}$ as follows: for every pseudometric $m \in \mathcal{P}$ and every two states x, y,

$$\mathcal{F}_c(m)(x, y) = \sup_{t \geq 0} c^t W(m)(P_t(x), P_t(y)).$$

$\mathcal{F}_c(m)(x, y)$ compares all the distributions $P_t(x)$ and $P_t(y)$ through transport theory and takes their supremum.

There are several remarks to make on this definition. First, we can only define $\mathcal{F}_c(m)$ if m is lower semi-continuous since we are using optimal transport theory which is why the domain of $\mathcal{F}_c$ is only $\mathcal{P}$.

Additionally, even if m is lower semi-continuous, $\mathcal{F}_c(m)$ may not even be measurable which means that the range of $\mathcal{F}_c$ is not the lattice $\mathcal{P}$. At least, Lemma 2 ensures that $\mathcal{F}_c(m)$ is indeed in $\mathcal{M}$, as a supremum of pseudometrics. This subtlety was not present in the work on continuous-time Markov chains in [17,18].

Second, we will use the Kantorovich duality throughout this work. It only holds for probability measures, and that is why we restrict this work to honest processes.

As a direct consequence of the definition of $\mathcal{F}_c$, we have that $\mathcal{F}_c$ is monotone: if $m_1 \leq m_2$ in $\mathcal{P}$, then $\mathcal{F}_c(m_1) \leq \mathcal{F}_c(m_2)$. It is also easy to prove the following result.

Lemma 4. *For every pseudometric m in $\mathcal{P}$, discount factor $0 < c < 1$ and pair of states x, y,*

$$m(x, y) \leq \mathcal{F}_c(m)(x, y).$$

4.3 When restricted to continuous pseudometrics

We wish to iteratively apply $\mathcal{F}_c$ in order to construct a fixpoint (in a similar fashion to the proof of the Knaster-Tarski theorem). While $\mathcal{F}_c(m)$ is a pseudometric (for $m \in \mathcal{P}$), there is no reason for it to be in $\mathcal{P}$. This means that we cannot hastily apply $\mathcal{F}_c$ iteratively to just any pseudometric in order to obtain a fixpoint.

However, if m is a pseudometric which is continuous wrt the original topology, then so is $\mathcal{F}_c(m)$.

---

[4] The discrete pseudometric is defined as $m(x, y) = 1$ if $x \neq y$ and $m(x, x) = 0$

Lemma 5. *Consider a pseudometric* $m \in \mathcal{C}$. *Then the topology generated by* $\mathcal{F}_c(m)$ *is a subtopology of the original topology* $\mathcal{O}$ *for any discount factor* $0 < c < 1$.

This is where we need that the discount factor $c < 1$. The condition that $c < 1$ enables us to maintain continuity by allowing to bound the time interval we consider. Indeed, given $T > 0$, for any time $t \geq T$ and any $x, y \in E$, we know that $c^t W(m)(P_t(x), P_t(y)) \leq c^T$.

Proof. Since c and m are fixed throughout the proof, we will omit noting them and for instance write $\mathcal{F}(x, y)$ instead of $\mathcal{F}_c(m)(x, y)$ and $W(P_t(x), P_t(y))$ instead of $W(m)(P_t(x), P_t(y))$. We will also write $\Phi(t, x, y) = c^t W(m)(P_t(x), P_t(y))$, i.e. $\mathcal{F}(x, y) = \sup_t \Phi(t, x, y)$.

It is enough to show that for a fixed state x, the map $y \mapsto \mathcal{F}(x, y)$ is continuous.

Pick $\epsilon > 0$ and a sequence of states $(y_n)_{n\in\mathbb{N}}$ converging to y. We want to show that there exists M such that for all $n \geq M$,

$$|\mathcal{F}(x, y) - \mathcal{F}(x, y_n)| \leq \epsilon. \tag{1}$$

Pick t such that $\mathcal{F}(x, y) = \sup_s \Phi(s, x, y) \leq \Phi(t, x, y) + \epsilon/4$, i.e.

$$|\Phi(t, x, y) - \mathcal{F}(x, y)| \leq \epsilon/4. \tag{2}$$

Recall that $P_t(y_n)$ converges weakly to $P_t(y)$ and hence we can apply Theorem 5.20 of [27] and get:

$$\lim_{n\to\infty} W(P_t(x), P_t(y_n)) = W(P_t(x), P_t(y)).$$

This means that there exists N' such that for all $n \geq N'$,

$$|W(P_t(x), P_t(y_n)) - W(P_t(x), P_t(y))| \leq \epsilon/4.$$

This further implies that for all $n \geq N'$,

$$|\Phi(t, x, y_n) - \Phi(t, x, y)| \leq c^t \epsilon/4 \leq \epsilon/4. \tag{3}$$

In order to show (1), it is enough to show that there exists N such that for every $n \geq N$,

$$|\Phi(t, x, y_n) - \mathcal{F}(x, y_n)| \leq \epsilon/2. \tag{4}$$

Indeed, in that case, $\forall n \geq \max\{N, N'\}$, using Equations (2) and (3),

$$\begin{aligned}
&|\mathcal{F}(x, y) - \mathcal{F}(x, y_n)| \\
&\leq |\mathcal{F}(x, y) - \Phi(t, x, y)| + |\Phi(t, x, y) - \Phi(t, x, y_n)| + |\Phi(t, x, y_n) - \mathcal{F}(x, y_n)| \\
&\leq \epsilon/4 + \epsilon/4 + \epsilon/2 = \epsilon.
\end{aligned}$$

So let us show (4). Assume it is not the case: for all N, there exists $n \geq N$ such that $|\Phi(t, x, y_n) - \mathcal{F}(x, y_n)| > \epsilon/2$, i.e.

$$\Phi(t, x, y_n) + \epsilon/2 < \mathcal{F}(x, y_n).$$

Define the sequence $(N_k)_{k \in \mathbb{N}}$ by: $N_{-1} = -1$ and if N_k is defined, define N_{k+1} to be the smallest $n \geq N_k + 1$ such that $\Phi(t, x, y_n) + \epsilon/2 < \mathcal{F}(x, y_n)$. In particular for every $k \in \mathbb{N}$, $\mathcal{F}(x, y_{N_k}) > \epsilon/2$. There exists T such that for every $s \geq T$, $c^s < \epsilon/2$. We thus have that

$$\forall k \in \mathbb{N} \quad \mathcal{F}(x, y_{N_k}) = \sup_{0 \leq s \leq T} \Phi(s, x, y_{N_k}).$$

Therefore for every $k \in \mathbb{N}$, there exists $s_k \in [0, T]$ such that

$$\mathcal{F}(x, y_{N_k}) \leq \Phi(s_k, x, y_{N_k}) + \epsilon/8. \tag{5}$$

We get a sequence $(s_k)_{k \in \mathbb{N}} \subset [0, T]$, and there is thus a subsequence $(t_k)_{k \in \mathbb{N}}$ converging to some $t' \in [0, T]$. There is a corresponding subsequence $(z_k)_{k \in \mathbb{N}}$ of the original sequence $(y_{N_k})_{k \in \mathbb{N}}$. Since $\lim_{n \to \infty} y_n = y$, $\lim_{k \to \infty} z_k = y$.

We constructed the sequence $(N_k)_{k \in \mathbb{N}}$ such that $\Phi(t, x, y_{N_k}) + \epsilon/2 < \mathcal{F}(x, y_{N_k})$. Hence by Equation (5),

$$\Phi(t, x, z_k) + \epsilon/2 < \mathcal{F}(x, y_{N_k}) \leq \Phi(t_k, x, z_k) + \epsilon/8,$$

which means that by taking the limit $k \to \infty$,

$$\Phi(t, x, y) + \epsilon/2 \leq \Phi(t', x, y) + \epsilon/8. \tag{6}$$

At the start of this proof, we picked t such that $\mathcal{F}(x, y) = \sup_s \Phi(s, x, y) \leq \Phi(t, x, y) + \epsilon/4$ which means that

$$\Phi(t', x, y) \leq \Phi(t, x, y) + \epsilon/4. \tag{7}$$

Equations (6) and (7) are incompatible which concludes the proof. □

4.4 Defining our family of pseudometrics

We are now finally able to iteratively apply our functional $\mathcal{F}_c$ on continuous pseudometrics and thus construct a sequence of increasing pseudometrics and its limit.

Since *obs* is a continuous function $E \to \mathbb{R}_{\geq 0}$ and by Lemma 5, we can define the sequence of pseudometrics in $\mathcal{C}$ for each $0 < c < 1$:

$$\delta_0^c(x, y) = |obs(x) - obs(y)|,$$
$$\delta_{n+1}^c = \mathcal{F}_c(\delta_n^c).$$

By Lemma 4, for every two states x and y, $\delta_{n+1}^c(x, y) \geq \delta_n^c(x, y)$. Define the pseudometric $\overline{\delta}^c = \sup_n \delta_n^c$ (which is also a limit since the sequence is non-decreasing).

Since it is the supremum of continuous functions, the pseudometric $\overline{\delta}^c$ is lower semi-continuous and is thus in the lattice $\mathcal{P}$ for any $0 < c < 1$.

Remark 3. Note that the lattice $\mathcal{C}$ is not complete which means that, although the metrics δ^c_n all belong to $\mathcal{C}$, $\overline{\delta}^c$ does not need to be in $\mathcal{C}$. For that reason, we cannot directly use the Knaster-Tarski theorem in this work.

4.5 Fixpoint

Even though we are not able to define the metric $\overline{\delta}^c$ as a fixpoint directly, it is actually a fixpoint.

Theorem 1. *The pseudometric $\overline{\delta}^c$ is a fixpoint for $\mathcal{F}_c$.*

The proof relies on the use of Sion's minimax theorem on the function

$$\begin{aligned} \varXi : \Gamma(P_t(x), P_t(y)) \times Y &\to [0,1] \\ (\gamma, m) &\mapsto \int m \, \mathrm{d}\gamma. \end{aligned}$$

where Y is the set of linear combinations of pseudometrics $\sum_{n\in\mathbb{N}} a_n \delta_n$ such that for every n, $a_n \geq 0$ (and finitely many are non-zero) and $\sum_{n\in\mathbb{N}} a_n = 1$.

Lemma 6. *Consider a discount factor $0 < c < 1$ and a pseudometric m in $\mathcal{P}$ such that*

1. *m is a fixpoint for $\mathcal{F}_c$,*
2. *for every two states x and y, $m(x,y) \geq |obs(x) - obs(y)|$,*

then $m \geq \overline{\delta}^c$.

The proof is done by showing that $m \geq \delta^c_n$ by induction on n.

We even have the following characterization of $\overline{\delta}^c$ using Lemma 6 and Theorem 1.

Theorem 2. *The pseudometric $\overline{\delta}^c$ is the least fixpoint of $\mathcal{F}_c$ that is greater than the pseudometric $(x,y) \mapsto |obs(x) - obs(y)|$.*

5 Corresponding real-valued logic

Similarly to what happens in the discrete-time setting, this behavioural pseudometric $\overline{\delta}^c$ can be characterized by a real-valued logic. This real-valued logic should be thought of as tests performed on the diffusion process, for instance "what is the expected value of *obs* after letting the process evolve for time t?" and it generates a pseudometric on the state space by looking at how different the process performs on those tests starting from different positions.

5.1 The logic

Definition of the logic: The logic is defined inductively and is denoted Λ:

$$f \in \Lambda := q \mid obs \mid \min\{f_1, f_2\} \mid 1 - f \mid f \ominus q \mid \langle t \rangle f$$

for all $f_1, f_2, f \in \Lambda$, $q \in [0,1] \cap \mathbb{Q}$ and $t \in \mathbb{Q}_{\geq 0}$.

This logic closely resembles the ones introduced for discrete-time systems by Desharnais et al. [14,15] and by van Breugel et al. [26]. The key difference is the term $\langle t \rangle f$ which deals with continuous-time.

Interpretation of the logics: We fix a discount factor $0 < c < 1$. The expressions in Λ are interpreted inductively as functions $E \to [0,1]$ as follows for a state $x \in E$:

$$\begin{aligned} q(x) &= q, \\ obs(x) &= obs(x), \\ (\min\{f_1, f_2\})(x) &= \min\{f_1(x), f_2(x)\}, \\ (1 - f)(x) &= 1 - f(x), \\ (f \ominus q)(x) &= \max\{0, f(x) - q\}, \\ (\langle t \rangle f)(x) &= c^t \int f(y)\ P_t(x, \mathrm{d}y) = c^t \left(\hat{P}_t f\right)(x). \end{aligned}$$

Whenever we want to emphasize the fact that the expressions are interpreted for a discount factor $0 < c < 1$, we will write Λ_c.

Remark 4. Let us clarify what the difference is between an expression in Λ and its interpretation. Expressions can be thought of as the notation $+, ^2, \times$ etc. They don't carry much meaning by themselves but one can then interpret them for a given set: $\mathbb{R}, C_0(E)$ (continuous functions $E \to \mathbb{R}$ that vanish at infinity) for instance. Combining notations, one can write expressions that can then be interpreted on a given set.

From Λ, we can also define the expression $f \oplus q = 1 - ((1 - f) \ominus q)$ which is interpreted as a function $E \to [0,1]$ as $(f \oplus q)(x) = \min\{1, f(x) + q\}$.

5.2 Definition of the pseudometric

The pseudometric we derive from the logic Λ corresponds to how different the test results are when the process starts from x compared to the case when it starts from y.

Given a fixed discount factor $0 < c < 1$, we can define the pseudometric λ^c:

$$\lambda^c(x, y) = \sup_{f \in \Lambda_c} (f(x) - f(y)) = \sup_{f \in \Lambda_c} |f(x) - f(y)|.$$

The latter equality holds since for every $f \in \Lambda_c$, Λ_c also contains $1 - f$.

5.3 Comparison to the fixpoint metric

This real-valued logic Λ_c is especially interesting as the corresponding pseudometric λ^c matches the fixpoint pseudometric $\overline{\delta}^c$ for the functional $\mathcal{F}_c$ that we defined in Section 4.4. In order to show that $\lambda^c = \overline{\delta}^c$, we establish the inequalities in both directions. One of the direction is proven by induction on the structure of the terms in our logic Λ_c:

Lemma 7. *For every f in Λ_c, there exists n such that for every x, y, $f(x) - f(y) \leq \delta_n^c(x, y)$.*

Since $\lambda^c(x, y) = \sup_{f \in \Lambda_c}(f(x) - f(y))$, we get the following corollary as an immediate consequence of Lemma 7.

Corollary 1. *For every discount factor $0 < c < 1$, $\overline{\delta}^c \geq \lambda^c$.*

We now aim at proving the reverse inequality. We will use the following result (Lemma A.7.2) from [1] which is also used in [26] for discrete-time processes.

Lemma 8. *Let X be a compact Hausdorff space. Let A be a subset of the set of continuous functions $X \to \mathbb{R}$ such that if $f, g \in A$, then $\max\{f, g\}$ and $\min\{f, g\}$ are also in A. Consider a function h that can be approximated at each pair of points by functions in A, meaning that*

$$\forall x, y \in X \ \forall \epsilon > 0 \ \exists g \in A \ |h(x) - g(x)| \leq \epsilon \text{ and } |h(y) - g(y)| \leq \epsilon$$

Then h can be approximated by functions in A, meaning that $\forall \epsilon > 0 \ \exists g \in A \ \forall x \in X \ |h(x) - g(x)| \leq \epsilon$.

In order to use this lemma, we need the following lemmas:

Lemma 9. *For every $f \in \Lambda_c$, the function $x \mapsto f(x)$ is continuous.*

Proof. This is done by induction on the structure of Λ_c. The only case that is not straightforward is when $f = \langle t \rangle g$. By induction hypothesis, g is continuous. Since the map $x \mapsto P_t(x)$ is continuous (onto the weak topology), $f = c^t \hat{P}_t g$ is continuous.

Lemma 10. *Consider a continuous function $h : E \to [0, 1]$ and two states z, z' such that there exists f in the logic Λ_c such that $|h(z) - h(z')| \leq |f(z) - f(z')|$. Then for every $\epsilon > 0$, there exists $g \in \Lambda_c$ such that $|h(z) - g(z)| \leq 2\epsilon$ and $|h(z') - g(z')| \leq 2\epsilon$.*

Proof. WLOG $h(z) \geq h(z')$ and $f(z) \geq f(z')$ (otherwise consider $1 - f$ instead of f). Pick $p, q, r \in \mathbb{Q} \cap [0, 1]$ such that

$$p \in [f(z') - \epsilon, f(z')],$$
$$q \in [h(z) - h(z') - \epsilon, h(z) - h(z')],$$
$$r \in [h(z'), h(z') + \epsilon].$$

One can then show that $g = (\min\{f \ominus p, q\}) \oplus r$ satisfies the desired property.

Corollary 2. *Consider a continuous function* $h : E \to [0,1]$ *such that for any two states* z, z' *there exists* f *in the logic* Λ_c *such that* $|h(z) - h(z')| \leq |f(z) - f(z')|$. *Then for every compact set* K *in* E, *there exists a sequence* $(g_n)_{n\in\mathbb{N}}$ *in* Λ_c *that approximates* h *on* K.

Proof. We have proven in Lemma 10 that such a function h can be approximated at pairs of states by functions in Λ_c. Now recall that all the functions in Λ_c are continuous (Lemma 9).

We can thus apply Lemma 8 on the compact set K, and we get that the function h can be approximated by functions in Λ_c.

Theorem 3. *The pseudometric* λ^c *is a fixpoint of* $\mathcal{F}_c$.

Proof. We will omit writing c as an index in this proof. We already have that $\lambda \leq \mathcal{F}(\lambda)$ (cf Lemma 4), so we only need to prove the reverse direction.

There are countably many expressions in Λ so we can number them: $\Lambda = \{f_0, f_1, ...\}$. Write $m_k(x,y) = \max_{j\leq k} |f_j(x) - f_j(y)|$. Since all the f_j are continuous (Lemma 9), the map m_k is also continuous. Furthermore, $m_k \leq m_{k+1}$ and for every two states x and y, $\lim_{k\to\infty} m_k(x,y) = \lambda(x,y)$.

Using Lemma 3, we know that for every states x, y and time t,

$$\sup_k W(m_k)(P_t(x), P_t(y)) = W(\lambda)(P_t(x), P_t(y)).$$

This implies that

$$\mathcal{F}(\lambda)(x,y) = \sup_{t\geq 0} c^t W(\lambda)(P_t(x), P_t(y)) = \sup_k \sup_{t\geq 0} c^t W(m_k)(P_t(x), P_t(y)).$$

It is therefore enough to show that for every k, every time $t \in \mathbb{Q}_{\geq 0}$ and every pair of states x, y, $c^t W(m_k)(P_t(x), P_t(y)) \leq \lambda(x,y)$. There exists $h \in \mathcal{H}(m_k)$ such that $W(m_k)(P_t(x), P_t(y)) = \left|\int h \; \mathrm{d}P_t(x) - \int h \; \mathrm{d}P_t(y)\right|$.

Since $P_t(x)$ and $P_t(y)$ are tight, for every $\epsilon > 0$, there exists a compact set $K \subset E$ such that $P_t(x, E \setminus K) \leq \epsilon/4$ and $P_t(y, E \setminus K) \leq \epsilon/4$.

By Corollary 2, there exists $(g_n)_{n\in\mathbb{N}}$ in Λ that pointwise converge to h on K. In particular, for n large enough,

$$\left|\int_K g_n \; \mathrm{d}P_t(x) - \int_K h \; \mathrm{d}P_t(x)\right| \leq \epsilon/4,$$

and similarly for $P_t(y)$. We get that for n large enough.

$$\begin{aligned} &\left|\int_E g_n \; \mathrm{d}P_t(x) - \int_E h \; \mathrm{d}P_t(x)\right| \\ &\leq \left|\int_K g_n \; \mathrm{d}P_t(x) - \int_K h \; \mathrm{d}P_t(x)\right| + \int_{E\setminus K} |g_n - h| \; \mathrm{d}P_t(x) \leq \epsilon/2, \end{aligned}$$

and similarly for $P_t(y)$. We can thus conclude that

$$\begin{aligned}
c^t W(m_k)(P_t(x), P_t(y)) &= c^t \left| \int_E h \, \mathrm{d}P_t(x) - \int_E h \, \mathrm{d}P_t(y) \right| \\
&\leq c^t \left| \int_E g_n \, \mathrm{d}P_t(x) - \int_E h \, \mathrm{d}P_t(x) \right| + c^t \left| \int_E g_n \, \mathrm{d}P_t(y) - \int_E h \, \mathrm{d}P_t(y) \right| \\
&\quad + |(\langle t \rangle g_n)(x) - (\langle t \rangle g_n)(y)| \\
&\leq \epsilon + \lambda(x, y).
\end{aligned}$$

Since ϵ is arbitrary, $c^t W(m_k)(P_t(x), P_t(y)) \leq \lambda(x, y)$. □

As a consequence of Theorem 3 and using Lemma 6, we get that $\overline{\delta}^c \leq \lambda^c$. Then with Corollary 1, we finally get:

Theorem 4. *The two pseudometrics are equal:* $\overline{\delta}^c = \lambda^c$.

6 Obstacles in continuous time

As we have pointed out throughout this work, although the overall outline is similar to that employed in the discrete case, we are forced to develop new strategies to overcome technical challenges arising from the continuum setting, where the topological properties of the time/state space become crucial elements in the study.

For example, a key obstacle we face in this work is that the fixpoint pseudometric can't be derived directly from the Banach fixed point theorem. There is no notion of step and we therefore need to consider all times in $\mathbb{R}_{\geq 0}$. In discrete-time, the counterpart of our functional $\mathcal{F}_c$ is c-Lipschitz. However, in our case, this should be thought of as c^1. Since we are forced to consider all times and since $\sup_{t \geq 0} c^t = 1$, we cannot find a constant $k < 1$ such that $\mathcal{F}_c$ is k-Lipschitz. To overcome this issue, we construct a candidate pseudometric with brute force (we first define a sequence of pseudometrics δ_n^c and then define our candidate $\overline{\delta}^c$ as their supremum) and then prove that it is indeed the greatest fixpoint.

In addition, the scope of the functional $\mathcal{F}_c$ also requires careful treatment: for measurability reasons, we have to restrict to the lattice of pseudometrics which generate a subtopology of the pre-existing one. We cannot apply the Kleene fixed point theorem either since this lattice is not complete. This whole approach toward a fixpoint pseudometric differs substantially from the long-existing methodology.

Furthermore, for some of our key results (e.g., Theorem 3 and Lemma 5), the proofs rely on the compactness argument to establish certain convergence relations in temporal and/or spatial variables and to further achieve the goals. This type of procedure in general is not required in the discrete setting.

Although from time to time, we do restrict the time variable to rational values thanks to the continuity of the FD-semigroups, this is very different from treating discrete-time models, because rational time stamps cannot be "ordered" to represent the notion of "next step" in a continuous-time setting. Therefore, we are still working with a true continuous-time dynamics, and it cannot be reduced to a discrete-time problem.

7 Examples

7.1 A toy example

Let us consider a process defined on $\{0, x, y, z, \partial\}$. Let us first give an intuition for what we are trying to model. In the states x, y, z, the process is trying to learn a value. In the state 0, the correct value has been learnt, but in the state ∂, an incorrect value has been learnt. From the three " learning" states x, y and z, the process has very different learning strategies:

- from x, the process exponentially decays to the correct value represented by the state 0,
- from y, the process is not even attempting to learn the correct value and thus remains in a learning state, and
- from z, the process slowly learns but it may either learn the correct value (0) or an incorrect one (∂).

The word "learning" is here used only to give colour to the example.

Formally, this process is described by the time-indexed kernels:

$$
\begin{array}{lll}
P_t(x, \{0\}) = 1 - e^{-\lambda t} & P_t(z, \{0\}) = \frac{1}{2}(1 - e^{-\lambda t}) & P_t(0, \{0\}) = 1 \\
P_t(x, \{x\}) = e^{-\lambda t} & P_t(z, \{\partial\}) = \frac{1}{2}(1 - e^{-\lambda t}) & P_t(y, \{y\}) = 1 \\
 & P_t(z, \{z\}) = e^{-\lambda t} & P_t(\partial, \{\partial\}) = 1
\end{array}
$$

(where $\lambda \geq 0$) and by the observable function $obs(x) = obs(y) = obs(z) = r \in (0, 1)$, $obs(\partial) = 0$ and $obs(0) = 1$.

We will pick as a discount factor $c = e^{-\lambda}$ to simplify notations. We can compute the distance $\overline{\delta}$ (we omit adding c in the notation).

We will only detail the computations for $\overline{\delta}(0, z)$ in the main section of this paper. First of all, note that $\delta_0(0, z) = 1 - r$ and that

$$
\begin{aligned}
\delta_{n+1}(0, z) &= \sup_{t \geq 0} e^{-\lambda t} \left(e^{-\lambda t} \delta_n(0, z) + \frac{1}{2}(1 - e^{-\lambda t}) \delta_n(0, \partial) \right) \\
&= \sup_{0 < \theta \leq 1} \theta \left(\theta \delta_n(0, z) + \frac{1}{2}(1 - \theta) \right) \\
&= \sup_{0 < \theta \leq 1} \theta \left[\frac{1}{2} + \theta \left(\delta_n(0, z) - \frac{1}{2} \right) \right]
\end{aligned}
$$

We are thus studying the function $\phi : \theta \mapsto \theta \left[\frac{1}{2} + \theta \left(\delta_n(0, z) - \frac{1}{2} \right) \right]$. Its derivative ϕ' has value 0 at $\theta_0 = \frac{1}{2(1 - 2\delta_n(0, z))}$. There are three distinct cases:

1. If $0 < \delta_n(0,z) \leq \frac{1}{4}$, in that case, $0 < \theta_0 \leq 1$ and $\sup_{1<\theta\leq 1} \phi(\theta)$ is attained in θ_0 and in this case, we have that $\delta_{n+1}(0,z) = \frac{1}{8(1-2\delta_n(0,z))} \leq \frac{1}{4}$. This means in particular that if $1 - r \leq \frac{1}{4}$, then $\overline{\delta}(0,z) = \frac{1}{4}$.
2. If $\frac{1}{4} \leq \delta_n(0,z) \leq \frac{1}{2}$, then $\theta_0 \geq 1$ and ϕ is increasing on $(-\infty, \theta_0]$. In that case, we therefore have that $\sup_{1<\theta\leq 1} \phi(\theta)$ is attained in 1 and therefore $\delta_{n+1}(0,z) = \delta_n(0,z)$.
3. If $\frac{1}{2} \leq \delta_n(0,z)$, then $\theta_0 \leq 0$ and ϕ is increasing on $[\theta_0, +\infty)$. In that case, we therefore have that $\sup_{1<\theta\leq 1} \phi(\theta)$ is attained in 1 and therefore $\delta_{n+1}(0,z) = \delta_n(0,z)$.

We therefore have that

$$\overline{\delta}(0,z) = \begin{cases} \frac{1}{4} & \text{if } r \geq \frac{3}{4} \\ 1 - r & \text{otherwise.} \end{cases}$$

We summarize the other cases in the following table. Note that the computation of $\overline{\delta}(x,z)$ is too involved and we therefore only provide an interval.

	x	y	z	∂	0
x	0	$\frac{1-r}{2}$	$\in \left[\frac{1}{8}, \frac{1}{4}\right]$	$\begin{cases} r & \text{if } r \geq \frac{1}{2} \\ \frac{1}{2} & \text{otherwise} \end{cases}$	$1-r$
y	$\frac{1-r}{2}$	0	$\frac{1}{4}$	r	$1-r$
z	$\in \left[\frac{1}{8}, \frac{1}{4}\right]$	$\frac{1}{4}$	0	$\begin{cases} r & \text{if } r \geq \frac{1}{4} \\ \frac{1}{4} & \text{otherwise} \end{cases}$	$\begin{cases} \frac{1}{4} & \text{if } r \geq \frac{3}{4} \\ 1-r & \text{otherwise} \end{cases}$
∂	$\begin{cases} r & \text{if } r \geq \frac{1}{2} \\ \frac{1}{2} & \text{otherwise} \end{cases}$	r	$\begin{cases} r & \text{if } r \geq \frac{1}{4} \\ \frac{1}{4} & \text{otherwise} \end{cases}$	0	1
0	$1-r$	$1-r$	$\begin{cases} \frac{1}{4} & \text{if } r \geq \frac{3}{4} \\ 1-r & \text{otherwise} \end{cases}$	1	0

Note that even though the process behaves vastly differently from x than from y, we have that $\overline{\delta}(x,0) = \overline{\delta}(y,0)$, even though for $t > 0$, we have that $\hat{P}_t obs(x) > \hat{P}_t obs(y)$. However note that x and y have different distances to other states.

This also happens in the discrete-time setting and for continuous-time Markov chains. A continuous-time Markov chain is a continuous-time type of processes but where the evolution still occurs as steps. They can be described as jump processes over continuous time. They have been studied in [18] by considering the whole trace starting from a single state.

It is possible to adapt our work to traces (called trajectories in [24]) by adding some additional regularity conditions on the processes; this can be found in [10]. However one should consider the added complexity. For instance, for Brownian motion the kernels P_t are well-known and easy to describe with a density function but that is not the case for the probability measures on trajectories.

7.2 Two examples based on Brownian motion

Previous example is a very simple example which emphasizes some of the difficulties of computing our metric. It may then be tempting to think that our approach cannot yield any result when applied to the real world. The next examples show that we can still provide some meaningful results when looking at real-life processes such as Brownian motion.

Brownian motion is a stochastic process describing the irregular motion of a particle being buffeted by invisible molecules. We refer the reader to [19] for a detailed introduction. One might recall that standard Brownian motion is a stochastic process $(B_t)_{t\geq 0}$ is defined on a probability space Ω with the properties that $B_0 = 0$ almost surely and for $0 \leq s < t$, $B_t - B_s$ is independent of $\mathcal{F}_s$ and is normally distributed with mean 0 and variance $t - s$.

In this very special process, one can start at any place, there is an overall translation symmetry which makes calculations more tractable. We denote $(B_t^x)_{t\geq 0}$ the standard Brownian motion on the real line starting from x.

First example: We can then define the hitting time of 0 or 1 of our standard Brownian motion $(B_t^x)_{t\geq 0}$: $\tau := \inf\{t \geq 0 : B_t^x = 0 \text{ or } B_t^x = 1\}$. The intuition is that this is the first time that the process hits either 0 or 1. It is a random time $\Omega \to \mathbb{R}_{\geq 0}$.

Now consider the stochastic process $(B_{t\wedge\tau}^x)_{t\geq 0}$: the intuition is that this process behaves like Brownian motion until it reaches one of the "boundaries" (either 0 or 1) where it becomes "stuck".

We now restrict the state space to the interval $[0, 1]$. For every $x \in [0, 1]$ and every $t \geq 0$, let $P_t(x, \cdot)$ be distribution of $B_{t\wedge\tau}^x$. In other words, $P_t(x, \cdot)$ is the distribution of Brownian motion starting from x running until hitting either 0 or 1 and getting trapped upon hitting a boundary. We equip the state space $[0, 1]$ with *obs* defined as $obs(x) = x$. Then, $obs(B_{t\wedge\tau}^x) = B_{t\wedge\tau}^x$ and hence for every $x \in (0, 1)$

$$\begin{aligned}
\delta_1(0, x) &= \sup_{t\geq 0} c^t W(\delta_0)(P_t(0), P_t(x)) = \sup_{t\geq 0} c^t \mathbb{E}[B_{t\wedge\tau}^x] \\
&= \sup_{t\geq 0} c^t \cdot x \text{ (because } B_{t\wedge\tau}^x \text{ is a martingale)} \\
&= x.
\end{aligned}$$

where the first equality follows from the fact that $P_t(0)$ is the dirac distribution $\eth_0$ at 0 and the second equality comes from the fact that any coupling $\gamma \in \Gamma(\eth_0, P_t(x))$ is reduced to the marginal $P_t(x)$. Since $\delta_0(0, x) = \delta_1(0, x)$, we then have $\overline{\delta}(0, x) = x$.

Similarly, $\overline{\delta}(1, y) = 1 - y$ for every $y \in [0, 1]$. It is difficult to compute $\overline{\delta}(x, y)$ for general $x, y \in [0, 1]$ though.

Second example: This example relies on stochastic differential equations (SDE). We refer the reader to [22] for a comprehensive introduction.

Let the state space and *obs* be the same as above. Let $Q_t(x, \cdot)$ be the distribution of the solution to the SDE

$$dX_t = X_t dB_t^0 + \frac{1}{2} X_t dt \text{ with } X_0 = x.$$

It can be verified that the solution to this equation is the process $X_t = xe^{B^0_t}$. In this case, we also have $Q_t(0) = \eth_0$. Again, for every $x \in [0,1]$,

$$\delta_1(0,x) = \sup_{t\geq 0} c^t W(\delta_0)(Q_t(0), Q_t(x)) = \sup_{t\geq 0} c^t \mathbb{E}[obs(xe^{B^0_t})]$$
$$= \sup_{t\geq 0} c^t \left(x\mathbb{E}[e^{B^0_t} \mid t \leq \tau'] + \mathbb{P}(t > \tau') \right),$$

where $\tau' := \inf\{s \geq 0 \mid B^0_s \geq -\ln x\}$ and $\mathbb{E}$ and $\mathbb{P}$ denote expected values and probabilities for the standard Brownian motion starting in 0. The distribution of τ', as well as the joint distribution of (B^0_t, τ'), has been determined explicitly (see Chapter 2, Section 8 in [19] for instance). Even if we only consider the second term in the expression above, we have

$$\delta_1(0,x) \geq \sup_{t\geq 0} c^t \mathbb{P}(t > \tau') = \sup_{t\geq 0} c^t \frac{2}{\sqrt{2\pi}} \int_{\frac{-\ln x}{\sqrt{t}}}^{\infty} e^{-\frac{y^2}{2}} dy.$$

It is possible for the right hand side above to be greater than x. For example, if $x = \frac{1}{e}$, then by choosing $t = 4$, we have that the right hand side above is no smaller than $c\frac{2}{\sqrt{2\pi}} \int_{\frac{1}{2}}^{\infty} e^{-\frac{y^2}{2}} dy$, which will be (strictly) greater than $\frac{1}{e}$ (i.e., x) provided that c is sufficiently close to 1. As a consequence, for this example, we have $\bar{\delta}(0,x) > x$.

These two examples demonstrate different behaviors of the two systems, while the first system "maintains" the mass at the starting point (expectation is constant x), the second system dissipates the mass to the right (which is the direction of larger values of *obs*). Therefore, processes starting from the same point x possess different distances to the static path between these two systems.

A very important observation from these examples based on Brownian motion is that even though we cannot explicitly compute values, we are still able to compare state behaviours.

8 Conclusion

In our previous work [7,8], we showed that we needed to use trajectories in order to define a meaningful notion of behavioural equivalence. However working with trajectories is extremely complex as notions do not translate easily from states to trajectories; for instance, we said that a measurable set B of trajectories is time-*obs*-closed if for every two trajectories ω, ω' such that for every time t, $obs(\omega(t)) = obs(\omega'(t))$, then $\omega \in B$ if, and only if $\omega' \in B$. The σ-algebra of all time-*obs*-closed sets cannot be simply described.

To explain why this present work does not deal with trajectories, we need to first discuss the example that lead us to trajectories in [7]: consider Brownian motion on the real line equipped with the function $obs = \mathbb{1}_{\{0\}}$. There are four *obs*-closed sets: $\emptyset$, $\{0\}$, $\mathbb{R}\backslash\{0\}$, $\mathbb{R}$ and for any $x \neq 0$ and any time t, $P_t(x, \{0\}) = 0$. This meant that one could not distinguish between the states 1 and 1000 in this specific case. Using trajectories enabled us to consider intervals of time. In this current work, we have decided to instead "smooth" the function *obs* so as to prevent singling points out without needing to deal with trajectories.

Let us go back to our examples. As shown in those examples, computing $\overline{\delta}$ is quite an involved process. It would have been interesting to adapt our example in Section 7.1 to the real-line with $obs(x) = e^{-x^2}$ and consider other processes such as Brownian motion which stops once it hits 0 or the Ornstein–Uhlenbeck process (a variation of Brownian motion which is "attracted" to 0). Having to deal with transport theory, a supremum (over time) and the inductive definition of $\overline{\delta}$ makes it virtually impossible to compute any harder example. As we have seen in the second example (Section 7.2), it may still be possible to compare the behaviours of two states by stating for instance that "the behaviour of the process starting from x is closer to that starting from y than from z". The problem of transport theory and the Kantorovich metric also exists in discrete time and there are interesting ways to deal with it, for instance through the MICo distance [6].

One of the advantages of optimal transport is that the Kantorovich duality gives us a way of computing bounds. As one notes however, there is again some difficulty with having a supremum over time in the definition of our functional $\mathcal{F}_c$. In particular, we can only provide lower bounds for $\overline{\delta}$.

One avenue of work on this could be to study replacements for this supremum, such as integrals over time for instance. Note that the real-valued logic would also need to be adapted even though it seems to generalize discrete-time really well.

This work is, as far as we know, the first behavioural metric adapted to the continuous-time case. Clearly much remains to be done, particularly the exploration of examples and connexions to broader classes of processes, such as for example those defined by stochastic differential equations.

Acknowledgments. Linan Chen and Prakash Panangaden were supported by NSERC discovery grants

Florence Clerc was supported by the EPSRC-UKRI grant EP/Y000455/1, by a CREATE grant for the project INTER-MATH-AI, by NSERC and by IVADO through the DEEL Project CRDPJ 537462-18.

Disclosure of Interests. The authors have no competing interests to declare.

References

1. R. B. Ash. *Real Analysis and Probability.* Academic Press, 1972.
2. C. Baier and J.-P. Katoen. *Principles of Model Checking.* MIT Press, 2008.
3. R. Blute, J. Desharnais, A. Edalat, and P. Panangaden. Bisimulation for labelled Markov processes. In *Proceedings of the Twelfth IEEE Symposium On Logic In Computer Science, Warsaw, Poland.*, 1997.
4. A. Bobrowski. *Functional analysis for probability and stochastic processes: an introduction.* Cambridge University Press, 2005.

5. N. Bourbaki. *General Topology Chapters 5-10.* Springer-Verlag, 1989.
6. P. S. Castro, T. Kastner, P. Panangaden, and M. Rowland. Mico: Improved representations via sampling-based state similarity for markov decision processes. *Advances in Neural Information Processing Systems*, 34:30113–30126, 2021.
7. L. Chen, F. Clerc, and P. Panangaden. Bisimulation for feller-dynkin processes. *Electronic Notes in Theoretical Computer Science*, 347:45 – 63, 2019. Proceedings of the Thirty-Fifth Conference on the Mathematical Foundations of Programming Semantics.
8. L. Chen, F. Clerc, and P. Panangaden. Towards a classification of behavioural equivalences in continuous-time Markov processes. *Electronic Notes in Theoretical Computer Science*, 2020. Proceedings of the Thirty-Sixth Conference on the Mathematical Foundations of Programming Semantics.
9. L. Chen, F. Clerc, and P. Panangaden. Behavioural equivalences for continuous-time markov processes. *Mathematical Structures in Computer Science*, page 1–37, 2023.
10. L. Chen, F. Clerc, and P. Panangaden. Behavioural pseudometrics for continuous-time diffusions. arXiv 2312.16729, 2024.
11. L. Chen, F. Clerc, and P. Panangaden. A behavioural pseudometric for continuous-time markov processes. arXiv 2501.13008, 2025.
12. J. Desharnais, A. Edalat, and P. Panangaden. A logical characterization of bisimulation for labelled Markov processes. In *proceedings of the 13th IEEE Symposium On Logic In Computer Science, Indianapolis*, pages 478–489. IEEE Press, June 1998.
13. J. Desharnais, A. Edalat, and P. Panangaden. Bisimulation for labeled Markov processes. *Information and Computation*, 179(2):163–193, Dec 2002.
14. J. Desharnais, V. Gupta, R. Jagadeesan, and P. Panangaden. Metrics for labeled Markov systems. In *Proceedings of CONCUR99*, number 1664 in Lecture Notes in Computer Science. Springer-Verlag, 1999.
15. J. Desharnais, V. Gupta, R. Jagadeesan, and P. Panangaden. A metric for labelled Markov processes. *Theoretical Computer Science*, 318(3):323–354, June 2004.
16. N. Ferns, P. Panangaden, and D. Precup. Bisimulation metrics for continuous Markov decision processes. *SIAM Journal of Computing*, 40(6):1662–1714, 2011.
17. V. Gupta, R. Jagadeesan, and P. Panangaden. Approximate reasoning for real-time probabilistic processes. In *The Quantitative Evaluation of Systems, First International Conference QEST04*, pages 304–313. IEEE Press, 2004.
18. V. Gupta, R. Jagadeesan, and P. Panangaden. Approximate reasoning for real-time probabilistic processes. *Logical Methods in Computer Science*, 2(1):paper 4, 2006.
19. I. Karatzas and S. Shreve. *Brownian motion and stochastic calculus*, volume 113. Springer Science and Business Media, 2012.
20. K. G. Larsen and A. Skou. Bisimulation through probablistic testing. *Information and Computation*, 94:1–28, 1991.
21. R. Milner. *A Calculus for Communicating Systems*, volume 92 of *Lecture Notes in Computer Science*. Springer-Verlag, 1980.
22. B. Oksendal. *Stochastic differential equations: an introduction with applications.* Springer Science & Business Media, 2013.
23. D. Park. Concurrency and automata on infinite sequences. In *Proceedings of the 5th GI Conference on Theoretical Computer Science*, number 104 in Lecture Notes In Computer Science, pages 167–183. Springer-Verlag, 1981.
24. L. C. G. Rogers and D. Williams. *Diffusions, Markov processes and martingales: Volume 1. Foundations.* Cambridge university press, 2nd edition, 2000.

25. D. Sangiorgi. On the origins of bisimulation and coinduction. *ACM Transactions on Programming Languages and Systems (TOPLAS)*, 31(4):15, 2009.
26. F. van Breugel and J. Worrell. A behavioural pseudometric for probabilistic transition systems. *Theoretical Computer Science*, 331(1):115 – 142, 2005.
27. C. Villani. *Optimal transport: old and new.* Springer-Verlag, 2008.
28. W. Whitt. *An Introduction to Stochastic-Process Limits and their Applications to Queues.* Springer Series in Operations Research. Springer-Verlag, 2002.

Idempotent Resources in Separation Logic

The Heart of core in Iris

Daniel Gratzer(✉), Mathias Adam Møller, and Lars Birkedal

Aarhus University, Aarhus, Denmark
{gratzer,birkedal}@cs.au.dk, 201704440@post.au.dk

Abstract. We revisit the foundational notion of "resources" used by separation logics from a categorical and algebraic viewpoint. In particular, we show that the *cameras* used by concurrent, higher-order, impredicative separation logics like Iris as a generalization of partial commutative monoids can be simplified and clarified and we introduce a category of cameras in which many vital cameras exhibit simple universal properties. We do this by observing that an important structure on cameras (the *core* operator) can be uniquely constrained and replaced by the property governing the idempotent elements of the camera. We verify that all cameras used in practice in Iris satisfy this property and use this insight to simplify the existing Iris formalization.

1 Introduction

Since its introduction in the late 1990s and early 2000s separation logics [22, 24], have gone through numerous conceptual revisions and alterations. Modern higher-order concurrent separation logics such as Iris [18] are now vastly more complex and built on top of several layers of abstraction in order to account for concurrency and higher-order reasoning, but the premise is still fundamentally unchanged: specifications for a program are given by pre- and post-conditions which are some form of predicate on program state and a "disjoint union" operator on program state gives rise to a separating conjunction in these specifications.

To manage this complexity, Iris attempts to realize the vast majority of the program logic purely within a flexible base logic. The result is that base Iris is a standard higher-order logic supplemented with the connectives of bunched implication logic, a handful of modalities, and a family of propositions denoting ownership. The heart of Iris is therefore in its highly abstracted definition of resource used by ownership propositions. These must be flexible enough to handle not only program state but also to encode the various features (invariants, state transition systems, etc.) that program logics utilize to verify programs.

Our goal is to revisit these abstract resources, termed *cameras*[1] in Iris, and isolate a new property of these structures. We capitalize on this new property to remove some of the complexity of cameras and obtain a simpler notion of *resource algebras* (RAs). We then construct a category of resource algebras and show that important cameras in Iris admit simple universal properties as RAs.

[1] The name derives from an older acronym CMRA which proved inaccurate for Iris 3.0.

P. A. Abdulla and D. Kesner (Eds.): FoSSaCS 2025, LNCS 15691, pp. 45–66, 2025.
https://doi.org/10.1007/978-3-031-90897-2_3

1.1 From heaps to cameras

The impetus for the more general notion of resource in Iris traces back to the advent of separation logic [24]. In particular, separation logic's eponymous separating conjunction is derived from an important piece of additional structure carried by program heaps (the most basic realization of program state): a disjoint union operation. Recall that one may represent heaps as partial maps from locations (e.g., natural numbers) to values. Given two such heaps h_1, h_2 with disjoint domains of definition, one may unambiguously form their union $h_1 \uplus h_2$. With this to hand, one defines the separating conjunction as follows:

$$(P * Q)\, h \triangleq \exists h_1, h_2.\, P\, h_1 \wedge Q\, h_2 \wedge h = h_1 \uplus h_2$$

While predicates in the original separation logic were taken over exactly the data needed to execute a program—i.e., the state of the heap—it was quickly realized that this was inadequate for realistic programs. Their correctness often depended upon invariants or conditions that might not be reflected directly in the memory. While a variety of solutions were proposed, the most elegant was to simply enlarge the definition of "program state" to include *ghost state*: objects which were not strictly necessary to describe the program execution. A simple example of ghost state is heaps with fractional permissions, which can be used to account for whether a thread has read or write ownership of a heap location [5].

The unifying feature of this more general ghost state is a partial 'union' operation. Whatever ghost state is used, it forms a set M equipped with a partial commutative and associative operation $\cdot$ (see, e.g., [6, 9, 10, 16]). Our definition of $*$ generalizes to such an M without change[2] whereby we can define the primitives of separation logic over an arbitrary *partial commutative monoid* M. This extra generality allows for a more conceptual definition of ghost state.

This story is complicated by higher-order state, concurrency, and other features. To cope with them, separation logics leverage *step-indexing* [1]. Rather than pre- and post-conditions being drawn from (monotone) predicates $M \to \mathrm{Prop}$, they are further indexed by a natural number $\mathbb{N} \times M \to \mathrm{Prop}$. The extra indexing soundly simulates arbitrary recursion in predicates.

In fact, embracing a small amount of revisionism, one can see all of this as ordinary separation logic done *internally* to a particular presheaf topos $\mathbf{PSh}(\omega)$ [3, 11, 21]. Working internally like this, it becomes natural to regard the partial commutative monoid as a "step-indexed" set with partial operations i.e., a partial commutative monoid internally to $\mathbf{PSh}(\omega)$. With this extra layer of generality, Jung et al. [18] showed that all the features of a modern concurrent separation logic could be built on top of ghost state within a higher-order logic.

With this motivation, we give a slightly imprecise definition of an Iris camera:

Definition 1. *A camera consists of a (step-indexed) set M, a (step-indexed) partial multiplication operation $(\cdot)$, and a partial operation $|-|$ which (when defined) maps m to an idempotent element $|m|$ such that $|m| \cdot m = m$.*

[2] In fact, it is now more recognizable as *Day convolution* [8]

1.2 Core and duplicable propositions

Unfortunately, the definition of camera is far more complex than the above might indicate. As a result, it is difficult to give a high-level definition of cameras without eliding important details and it is challenging to articulate why the camera axioms are in any way complete.[3] One manifestation of this poor behavior is the lack of a suitable category of cameras: there is no well-behaved notion of morphism of cameras and even simple constructions like the product of two cameras $M_1 \times M_2$ lack any universal property which could give a more conceptual description.

Some of the complexity is not, strictly speaking, mathematical. For instance, representing partial operations and step-indexing in a proof assistant is known to be arduous and the definition of camera is tuned to make it as easy as possible to realize in Coq at the cost of more uniform behavior. However, a central aspect of the generalization of cameras over partial commutative monoids is to replace the unit element ϵ with a *core operation* $|-|$ which is both complex and slightly ad-hoc. As it stands, requiring the existence of this operator is a major obstacle towards a good category of cameras.

To properly situate the definition of core and motivate our proposed replacement for it, we once more turn to traditional separation logic. In this setting, there is a *comonadic modality* $\square$ defined by:

$$(\square P)\, h \triangleq P(\emptyset)$$

The $\square$ modality occurs frequently when working within separation logic: it isolates those specifications for which separating conjunction coincides with ordinary conjunction such that, e.g., $\square P$ can be duplicated to $\square P * \square P$. In Iris, for instance, $\square$ is crucial for specifying features like *invariants* as well as for embedding ordinary propositions into the program logic. In the simple setting of traditional separation logic, $\square$ can actually be characterized in a number of distinct ways, any of which could be used to generalize $\square$ to arbitrary cameras:

- $\square$ is defined by evaluating at the *unit* of $\uplus$.
- $\square$ is the "always" comonad from Kripke models of modal logic.
- $\square$ is the unique operation preserving $\top$, $\wedge$, and $\exists$ such that $\square(\ell \hookrightarrow v) = \bot$.
- $\square$ is defined by evaluating at the unique *idempotent* with respect to $\uplus$.

In order to lift $\square$ to an arbitrary camera M, the definition of a camera more-or-less hard-codes the necessary structure for the fourth definition; the axioms of $|-|$ are exactly those needed to ensure that $P \mapsto P \circ |-|$ (modulo partiality) gives rise to a $\square$ modality.

Even this description hides some complexity. Bizjak and Birkedal [4] observed that absent step-indexing, there is a bijection between core structures and well-behaved $\square$ modalities. However, this (1) does not extend to the step-indexed case and (2) offers no insight as to which of the many $\square$ modalities should be chosen. In particular, when working with Heap there was a single clear choice of $\square$ modality with a variety of different characterizations. In contrast, for a general camera there are many distinct choices of core inducing distinct $\square$ modalities.

[3] Indeed, they differ between versions of Iris!

1.3 The maximal idempotent axiom

In this work, we simplify the definition of cameras by *removing* the core operation from their specification entirely. In its place, we instead impose a single additional property on our cameras insisting that for each $a \in M$, there exists a maximal idempotent below a or no idempotents at all. To disambiguate, we term the resulting structure a *resource algebra*. In other words, rather than requiring each resource algebra to come equipped with a partial assignment of elements to idempotents below them, we require that there is always a 'best' such assignment which picks out the largest possible idempotent. This choice in turn yields a $\Box$ modality uniquely defined as the largest modality satisfying the expected axioms.

This substantially simplifies the definition of a resource algebra: we exchange one operation and four axioms for this single new axiom. More than this, however, the removal of core makes it possible to give a satisfactory *category of resource algebras* **RA**. Given that cameras are nearly partial commutative monoids, it was long suspected that they should organize into a well-behaved category. In particular, it was conjectured that various examples of cameras used in Iris (sums, products, etc.) ought to enjoy universal properties within this category. We confirm this conjecture and observe that all of the cameras by Iris can be realized as resource algebras and that many common cameras induce resource algebras arising from adjoints or universal constructions within **RA**.

Our development of the theory of resource algebras takes place entirely within the internal language of $\mathbf{PSh}(\omega)$ and, consequently, does not mention step-indexing at all. In fact, our proofs are designed to apply also to transfinite versions of Iris (based on presheaves over larger ordinals) or separation logics without any step-indexing at all. The only cost for this generality is the need to work within a constructive setting.

Remark 1. The importance of maximum idempotent elements to isolate a universal core was noted independently by Dardinier et al. [7] in the context of automatic verification. We defer a more detailed comparison of till Section 7.

1.4 Contributions

We contribute a simpler and more abstract definition of "resources" for concurrent separation logics like Iris. We do this by *removing* a structure from Iris's definition of camera and replacing it with a new axiom (the maximal idempotent axiom), thereby obtaining the new notion of a *resource algebra*. We show the following:

- Every camera used in Iris satisfies the maximal idempotent axiom and therefore induces a resource algebra.
- Every resource algebra induces a universal $\Box$ modality.
- Resource algebras assemble into a well-behaved category **RA** in which many important resource algebras enjoy simple universal properties.

All of our results are carried out within a version of extensional type theory and apply not just to step-indexing over ω but to transfinite separation logics as well.

Finally, we show how to adapt the Iris formalization in Coq as it stands today to partially benefit from these results. In so doing, we also note various places where expediency of formalization has complicated the definition of cameras and give a precise account of the trade-offs these induce.

Acknowledgments This work was supported in part by a Villum Investigator grant (no. 25804), Center for Basic Research in Program Verification (CPV), from the VILLUM Foundation. This work was co-funded by the European Union (ERC, CHORDS, 101096090). Views and opinions expressed are however those of the authors only and do not necessarily reflect those of the European Union or the European Research Council. Neither the European Union nor the granting authority can be held responsible for them.

Data Availability Statement

An environment with the tools and data used for the experimental evaluation in the current study is available at 10.5281/zenodo.14765017 [12].

2 A brief reprise of PSh(ω)

For much of this paper (Sections 3 to 5) we work *internally* to the topos of trees **PSh**(ω). In this section, we review the basic properties of this internal language.

2.1 Type theory within PSh(ω)

To begin with, we recall that **PSh**(ω) is a *presheaf topos* and therefore supports a rich *internal language*: a model of extensional dependent type theory with all the standard connectives, including a hierarchy of universes ($\mathcal{U}_i$) [13, 14]. Among others, included in these are various inductive types such as the type of natural numbers, lists, and so on. We note that as we are dealing with extensional equality, both function extensionality (funext) and unicity of identity proofs (UIP) hold. As is standard, we largely ignore size issues and simply write $\mathcal{U}$. Also included is a type of propositions Prop closed under all the standard logical connectives, including impredicative quantification. There is a function $\mathrm{Prf} : \mathrm{Prop} \to \mathcal{U}$ which sends a proposition to the type of its proofs such that $\mathrm{Prf}(\phi)$ contains at most one element. Finally, Prop satisfies propositional extensionality: $\mathrm{Prf}(\phi) \cong \mathrm{Prf}(\psi) \iff \phi = \psi$ where $\cong$ refers to the standard notion of isomorphism of types.

Rather than explicitly adding all the necessary logical connectives to Prop, we realize them all at once through the addition of an "inverse" of sorts to Prf sending a type $A : \mathcal{U}_i$ to the propositional truncation $[A] : \mathrm{Prop}$. The defining characteristic of propositional truncation is that if B is a type with at most one element, then $A \to B \cong \mathrm{Prf}([A]) \to B$. Consequently, $\mathrm{Prf}([A]) \cong A$ if A is a type containing at most one element e.g., if $A = \mathrm{Prf}(\phi)$. One can then show, for instance, that logical entailment $\phi \to \psi$ can be defined as $[\mathrm{Prf}(\phi) \to \mathrm{Prf}(\psi)]$.

Finally, the internal language supports the axioms of guarded recursion (a later modality $\triangleright$, Löb induction, etc.). Surprisingly, this structure is not relevant to our discussion of resource algebras and so we refer the reader to Birkedal et al. [3] for details and an explanation of the semantics of the internal language.

2.2 Partial maps within $\mathbf{PSh}(\omega)$

For a first example of working within the internal language of $\mathbf{PSh}(\omega)$ and in preparation for Section 3, we define the notion of a partial map. A naïve first guess might be to say that a partial map from $A \rightharpoonup B$ consists of an ordinary function $A \to \mathbf{1} + B$. While this definition suffices in 'ordinary' external mathematics, it is too strict when working internally as we are. Indeed, this definition only captures functions whose domain of definition is *decidable* and we will frequently be interested in partial functions f where $f(a)\downarrow$ is an arbitrary proposition and therefore—since $\mathbf{PSh}(\omega)$ does not satisfy excluded middle—typically undecidable.

We instead use the *partial map classifier* following Rosolini [25]:

Definition 2. *If $A : \mathcal{U}$ then the partial map classifier $A^?$ is $\sum_{\phi:\mathrm{Prop}} \mathrm{Prf}(\phi) \to A$.*

A map $f : A \to B^?$ encodes a partial map from A to B. Note that f consists of (1) a map $f_0 : A \to \mathrm{Prop}$ indicating where f is defined and (2) a map $f_1 : (a : A) \to \mathrm{Prf}(f_0\, a) \to B$ giving the value of f when it is defined.

Definition 3. *We define the type of partial maps $A \rightharpoonup B$ to be $A \to B^?$*

Lemma 1. *$-^?$ is a monad whose unit and join operations are defined as follows:*

$$\mathsf{ret} : A \to A^?$$
$$\mathsf{ret}(x) \triangleq (\top, \lambda\_.\, x)$$
$$\mathsf{join} : A^{?^?} \to A^?$$
$$\mathsf{join}(\phi, \lambda z_1.\, (\psi(z_1), \lambda z_2.\, \tilde{a}(z_1, z_2))) \triangleq (\textstyle\sum_{\exists z_1 : \phi.\, \psi(z_1)}, \lambda\,(z_1, z_2).\tilde{a}(z_1, z_2))$$

We use standard monadic notation $(x \leftarrow a; b(x))$ to manipulate $A^?$ and write $f^? : A^? \to B^?$ for the functorial action.

Definition 4. *We define the approximation partial order $(\phi, a) \sqsubseteq (\psi, b) : \mathrm{Prop}$ on $A^?$ as $\exists f : \mathrm{Prf}(\phi) \to \mathrm{Prf}(\psi).\, (z : \mathrm{Prf}(\phi)) \to a\, z = b(f\, z)$.*

3 Resource Algebras and the Maximal Idempotent Axiom

We now analyze the abstract definition of resources using the internal language. While this definition is closely related to Iris's cameras, we avoid several maneuvers used by Iris to optimize for formalization in order to take full advantage of the internal language. A full comparison is provided in Section 6. We show how various portions of the definition may be replaced and folded into a single new axiom we term the *maximal idempotent axiom*.

3.1 Partial commutative semigroups and predicates

We begin with the basic notion of a type equipped with a partial multiplication:

Definition 5. *A partial commutative semigroup (PCS) consists of a type A and a partial map $(\cdot) : A \times A \rightharpoonup A$ satisfying the following conditions:*

1. *If $a, b : A$ then $a \cdot b = b \cdot a : A^?$*
2. *If $a, b, c : A$ then $x \leftarrow (a \cdot b); x \cdot c = x \leftarrow (b \cdot c); a \cdot x : A^?$*

In light of (1) and (2), we may unambiguously write $a \cdot b \cdot c$.

Lemma 2. *If A is a PCS then $a \leq b \triangleq a = b \vee \exists c : A.\, a \cdot c = \mathsf{ret}\, b$ is a preorder.*

As mentioned in the introduction, the semantics of separation logics like Iris can be constructed within monotone predicates on a PCS $(A, \cdot)$:

$$A \xrightarrow{\mathsf{mon}} \mathrm{Prop} \triangleq \{P : A \to \mathrm{Prop} \mid \forall a \leq b.\, P(a) \to P(b)\}$$

We will not spell out how all the connectives of higher-order logic are interpreted into $A \xrightarrow{\mathsf{mon}} \mathrm{Prop}$, but we record the definition of $*$ for intuition:

$$(P * Q)\, a \triangleq \exists a_0, a_1 : A, a_0 \cdot a_1 = \mathsf{ret}(a) \wedge P(a_0) \wedge Q(a_1)$$

3.2 Core structures

The classical definition of a camera is a PCS with a core operation:

Definition 6. *If $(A, \cdot)$ is a PCS, a core structure on A is a map $|-| : A \to \mathbf{1} + A$ satisfying the following conditions:*

1. *If $|a| = \mathsf{in}_2(a')$ then $a' \cdot a = \mathsf{ret}\, a$.*
2. *If $|a| = \mathsf{in}_2(a')$ then $|a'| = \mathsf{in}_2(a')$.*
3. *If $a \leq b$ then $|a| \leq |b|$*

For the last point, we order $\mathbf{1} + A$ such that $\mathsf{in}_1(\star)$ is the minimum element.

Definition 7. *A camera is a PCS equipped with a core structure.*[4]

As mentioned in the introduction, any core structure induces a comonadic modality $\Box$ on monotone predicates $A \xrightarrow{\mathsf{mon}} \mathrm{Prop}$:

$$\Box P \triangleq \lambda a \to \begin{cases} \bot & |a| = \mathsf{in}_1(\star) \\ P(a') & |a| = \mathsf{in}_2(a') \end{cases}$$

Lemma 3. *If $|-|$ is a core structure then the induced $\Box$ modality is monotone, comonadic ($\Box P \to P$ and $\Box\Box P \iff \Box P$), and commutes with $\exists$, $\triangleright$, and satisfies the following: $(\Box P) \wedge Q \iff (\Box P) * Q$*

[4] Jung et al. [18], required an additional axiom to force $\triangleright$ to commute with $*$. We have followed e.g. Spies et al. [26] and removed this requirement to allow for more models.

For the purposes of separation logic, it is the last property that is most vital. It allows a user of separation logic to freely exchange separating and ordinary conjunction when manipulating $\Box P$ and thereby allows one to e.g., duplicate $\Box P$ into $\Box P * \Box P$ and similar. Intuitively, $\Box P$ is the closest replacement of P which satisfies this property. Unfortunately, this idea is complicated by the fact that there are often great many distinct core structures which one may choose to equip $(A, \cdot)$ with, and they induce distinct modalities. It is thus far from clear which core structure (if any) actually forces $\Box P$ to realize the above description.

This situation is in contrast to the motivating example with monotone predicates on heaps where the $\Box$ modality could be characterized by several universal properties. Our goal is to isolate a property of core which fully constrains its definition to give the largest possible $\Box$ modality. That is, we wish to find a core structure $|-|$ so that if $|-|'$ is another core structure then $\forall P. \Box_{|-|'} P \to \Box_{|-|} P$.

We begin by massaging Definition 6 into a form which is easier to work with. If $|-|$ is a core structure and $|a| = \mathsf{in}_2(a')$ then a' is an idempotent element (by 1 and 2). In light of 3, we may therefore view $|-|$ as a monotone assignment $A \to \mathbf{1} + \mathsf{Idem}(A)$ and, more specifically, an assignment $(a : A) \to \{x : \mathbf{1} + \mathsf{Idem}(A) \mid x \leq \mathsf{in}_2(a)\}$.

Lemma 4. *A core structure on* $(A, \cdot)$ *is equivalent to a monotone map* $(a : A) \to \{x : \mathbf{1} + \mathsf{Idem}(A) \mid x \leq \mathsf{in}_2(a)\}$.

A great many such monotone maps are possible for most PCSs. For instance, the constant map $\lambda a \to \mathsf{in}_1(\star)$ is always available and if A has a global unit ϵ, then $\lambda a \to \mathsf{in}_2(\epsilon)$ is also a valid choice. As core structures are particular monotone maps between preorders, they themselves inherit a preorder defined pointwise.

Lemma 5. *The assignment of core structures to* $\Box$ *modalities is monotone: if* $|-|$ *and* $|-|'$ *are two core structures such that* $|-| \leq |-|'$ *then* $\Box_{|-|} P \to \Box_{|-|'} P$.

We note that we have have already encountered the unique minimal core structure for any PCS: the assignment $\lambda a \to \mathsf{in}_1(\star)$ which discards all information of a. Unfolding, this core structure induces the $\Box$ modality which sends every proposition to $\bot$, which is clearly the smallest possible modality. Much more useful is the *maximal* $\Box$ modality.

In order to obtain the largest possible $\Box$ modality, it suffices to construct a core structure which is maximal. In particular, it would suffice to ensure $|-|$ sends a to the largest possible idempotent element contained within a and which is undefined only if no idempotent element exists. Notice that a priori there may be distinct maximal core structures as A is merely pre-ordered, but every PCS is partially-ordered on idempotent elements, in particular:

Lemma 6. *If* i, j *are idempotents such that* $i \leq j$ *and* $j \leq i$ *then* $i = j$.

Proof. By assumption so there exists i' and j' such that $i \cdot i' = \mathsf{ret}\, j$ and $j \cdot j' = \mathsf{ret}\, i$. Since both i and j are idempotent, this tells us that $j \cdot i = \mathsf{ret}\, i$ and $i \cdot j = \mathsf{ret}\, j$. Commutativity then tells us that $i = j$.

In particular, core structures are partially-ordered and (using pointwise multiplication) form a join semi-lattice. We therefore conclude the following:

Corollary 1. *Maximal core structures are unique.*

Unfortunately, there is no guarantee that a maximal core structure need exist:

Lemma 7. *There exists a PCS for which there is no maximal core structure.*

Proof. Consider the PCS on $A = \mathbb{N}+\mathbf{1} = \{0\ldots\infty\}$ where (1) $n \cdot m = \max(n, m)$ (2) $n \cdot \infty = \infty \cdot n = \infty$ and (3) $\infty \cdot \infty$ is undefined. Note that $\mathsf{Idem}(A)$ consists of all elements of A except ∞.

To define a core structure on A, note that if $|\infty| = \mathsf{in}_2(n)$, then $|m| \leq \mathsf{in}_2(n)$ for all m by (3) and if $|\infty| = \mathsf{in}_1(\star)$ then $|m| = \mathsf{in}_1(\star)$. It follows that any core structure $|-|$ where $|\infty| = \mathsf{in}_2(n)$ must be smaller than the core structure $|-|^{n+1}$ defined by $|\infty|^{n+1} = \mathsf{in}_2(n+1)$ and $|m|^{n+1} = \mathsf{in}_2(\min(n+1, m))$. This yields an ascending chain of core structures $|-|^1 < |-|^2 \ldots$ for which there is no upper bound.

Fortunately, maximal core structures do exist in a wide variety of circumstances. For instance, it was necessary that our above counterexample involved an infinite number of idempotent elements.

Lemma 8. *If for each* $a : A$, $\{x : \mathbf{1} + \mathsf{Idem}(A) \mid x \leq \mathsf{in}_2(a)\}$ *is equivalent to a finite type* $\mathbb{N}_{\leq n}$ *for some* n*, then* A *has a maximal core structure.*

Many common PCSs have only a single idempotent element each element e.g.:

Corollary 2. Heap *has a maximal core structure:* $\lambda h \to \epsilon$.

3.3 The maximal idempotent axiom

We can reformulate the requirement that PCS possess a maximal core structure into the following property:

Definition 8 (The Maximal Idempotent Axiom). *A PCS* $(A, \cdot)$ *satisfies the maximal idempotent (MI) axiom if the following holds for all* $a : A$*:*

$$\mathsf{MI}(a) \triangleq \neg(\exists i \leq a.\, i\,\mathsf{idem}) \vee (\exists i \leq a.\, i\,\mathsf{idem} \wedge \forall j \leq a. j\,\mathsf{idem} \to j \leq i)$$

In other words, below each $a : A$ *there either exists a maximal idempotent or there are no idempotents at all.*

A priori, one may worry that the mere existence of a maximal idempotent below $a : A$ does not suffice to construct a function $|-|$ picking it out. After all, in general constructing such a function amounts to the axiom of choice and the lack of excluded middle implies the failure of the axiom of choice internal to $\mathbf{PSh}(\omega)$. However, as a topos $\mathbf{PSh}(\omega)$ does satisfy the axiom of *unique* choice: if $\forall a : A. \exists! b : B. \Phi(a, b)$ then $\exists! f : A \to B. \forall a : A. \Phi(a, f(a))$. Since maximal idempotents are unique, we obtain the following:

Theorem 1. *If $(A, \cdot)$ satisfies the MI axiom then A has a maximal core structure.*

Lemma 9. *If $\phi, \psi : \mathrm{Prop}$ such that $\phi \wedge \psi = \bot$ then $\mathrm{Prf}(\phi \vee \psi) \cong \mathrm{Prf}(\phi) + \mathrm{Prf}(\psi)$.*

Lemma 10. *If $A : \mathcal{U}$ and $\phi : A \to \mathrm{Prop}$ such that $\exists a : A.\, \phi\, a = \exists! a : A.\, \phi\, a$ then $\mathrm{Prf}(\exists a : A.\, \phi\, a) \cong \sum_{a:A} \mathrm{Prf}(\phi\, a)$.*

Proof (Theorem 1). Since the total absence of idempotents and the presence of maximal idempotent are disjoint, $\mathrm{Prf}(\mathsf{MI}(a))$ is equal to the following:

$$(\mathrm{Prf}(\exists i \leq a.\, i\, \mathsf{idem}) \to \bot) + \mathrm{Prf}(\exists i \leq a.\, i\, \mathsf{idem} \wedge \forall j \leq a. j\, \mathsf{idem} \to j \leq i)$$

Next, since maximal idempotents are unique, the second summand is equivalent to *unique* existence, whereby we may replace it with the following:

$$((i : A) \to i \leq a \to i\, \mathsf{idem} \to \bot) + \textstyle\sum_{i \leq a} i\, \mathsf{idem} \times ((j \leq a) \to j\, \mathsf{idem} \to j \leq i)$$

In other words, $\mathrm{Prf}(\forall a : A.\, \mathsf{MI}(a))$ is equivalent to a choice function sending each element of A to the maximal idempotent below A or a proof that there are no idempotents below A at all. This function is the required maximal core structure.

This addresses a rather mysterious fact about cameras: why are core structures partial in a different way than $\cdot$? In light of the above, we see that a core structure $|-|$ has decidable support because $|a|$ is undefined just when a contains *no* idempotents at all and the MI axiom guarantees that this is decidable.

We now *redefine* and simplify cameras from their formulation in Iris. Rather than insisting that a camera be a PCS equipped with a core structure, we instead ask that it satisfy the MI axiom. This has the simultaneous advantages of (1) ensuring that the induced core structure is as large as possible while also (2) merely being a property to check, not an additional structure.

Definition 9. *A resource algebra is a PCS satisfying the maximal idempotent axiom. Write* RA *for the type of all resource algebras.*

To justify this definition, we show that it is not overly restrictive (Section 4) and that it opens up more conceptual descriptions of existing cameras (Section 5).

4 Examples of cameras

If Definition 9 is to be an adequate replacement for the current definition of cameras in Iris, it must not rule out examples which are currently used in practice. Fortunately, every camera currently used in the Iris formalization in Coq satisfies the MI axiom. For reasons of space, we present only a handful of the most important resource algebras used in separation logic and illustrate the proofs that they satisfy the MI axiom. We divide these examples into two categories: basic stock cameras which are used in a wide variety of constructions and a few operations on cameras used to build up more complex examples.

4.1 Basic resource algebras

For the basic 'building block' resource algebras, Lemma 8 suffices to show that the relevant PCSs satisfy the MI axiom; in practice, they all have no idempotent elements, one idempotent element, or every element is idempotent. We illustrate this with two representative examples.

First, the *exclusive resource algebra* which is used to encode ownership of non-duplicable abstract resources:

Lemma 11. *If* $A : \mathcal{U}$*, regard* A *as a PCS* $\mathsf{Excl}(A)$ *where* $\cdot_{\mathsf{Excl}(A)}$ *is undefined everywhere.* $\mathsf{Excl}(A)$ *satisfies the MI axiom.*

Proof. Such a camera has *no* idempotents, so Lemma 8 trivially applies.

For another extreme example, we consider the *agreement* resource algebra which is used to enforce global agreement about a particular value. For this, take a type $A : \mathcal{U}$ and regard A as a PCS $\mathsf{Ag}(A)$ using the following partial multiplication operation: $a \cdot_{\mathsf{Ag}(A)} a' \triangleq (a = a', \lambda\_ \to a)$.

Lemma 12. *For any* $A : \mathcal{U}$, $\mathsf{Ag}(A)$ *satisfies the MI axiom.*

Proof. Lemma 8 applies as the only element contained in $\mathsf{Idem}(A)_{\leq a}$ is a itself.

4.2 Constructions on resource algebras

The power of resource algebras is their composability: given simple resource algebras like those outlined above, we may stitch them together into more powerful constructions like Heap. We outline the most important constructions for combining resource algebras and show that they preserve the MI axiom.

Given two resource algebras A and B we consider pointwise multiplication:

$$(a, b) \cdot_{A \times B} (a', b') \triangleq \bar{a} \leftarrow a \cdot a'; \bar{b} \leftarrow b \cdot b'; \mathsf{ret}(\bar{a}, \bar{b})$$

Lemma 13. *If* A *and* B *are resource algebras,* $A \times B$ *satisfies the MI axiom.*

Proof. Fix a pair (a, b). An idempotent below (a, b) consists of a pair (i, j) of idempotents i below a and j below b. By the MI axiom of both A and B, it is decidable whether or not there exists any idempotents below either a or b whence whether any such pair (i, j) exists. By a further application of the MI axiom, if such a pair exists, there is a pair of maximal idempotents as required.

Given a resource algebra A, we can construct the resource algebra $\mathsf{Opt}(A) = \mathbf{1} + A$ adjoining a new element to A. This element will act as the unit of multiplication, meaning the partial multiplication operation will be defined as:

$$x \cdot_{\mathsf{Opt}(A)} y \triangleq \begin{cases} \mathsf{ret}(y) & x = \mathsf{in}_1(\star) \\ \mathsf{ret}(x) & y = \mathsf{in}_1(\star) \\ \mathsf{in}_2{}^?(a \cdot_A a') & x = \mathsf{in}_2(a), y = \mathsf{in}_2(a') \end{cases}$$

Lemma 14. $\mathsf{Opt}(A)$ *satisfies the MI axiom.*

Proof. Fix $x : \mathsf{Opt}(A)$. If $x = \mathsf{in}_1(\star)$, then x itself is idempotent and thus the necessary maximal idempotent. If instead $x = \mathsf{in}_2(a)$, then the idempotents below x are either of the form $\mathsf{in}_1(\star)$ or $\mathsf{in}_2(i)$ where i is an idempotent below a. Accordingly, the maximal idempotent below $\mathsf{in}_2(a)$ is either the maximal idempotent below a or $\mathsf{in}_1(\star)$ if no such idempotent exists.

For our final example, we describe a more complex construction assembling a collection of resource algebras together. This 'direct sum' operation is used pervasively in Iris: any real use of Iris is based on this resource algebra to stitch together all of the resources required the verification of a given program. We describe it in more generality than it is presented in Iris by allowing for direct sums to be indexed by a type I with decidable equality.

Fix a family of resource algebras $A : (i : I) \to \mathsf{RA}$, let $\bigoplus_{i:I} A\,i$ denote the type of partial maps $(i : I) \to A(i)$ with decidable and non-empty, finite support. Representing these in type theory can be somewhat difficult, but our internal language is strong enough to define quotient types (using Prop). Accordingly, we realize $\bigoplus_{i:I} A\,i$ as a quotient of the following type:

$$\sum_{n:\mathbb{N}} \sum_{f:\mathbb{N}_{\leq n} \hookrightarrow I} (n : \mathbb{N}_{\leq n}) \to A(f\,n)$$

The quotienting relation identifies two triples (n, f, a) and (n', f', a') just when $n = n'$ and there is a bijection $\sigma : \mathbb{N}_{\leq n} \to \mathbb{N}_{\leq n}$ such that $f = f' \circ \sigma$ and $a = a' \circ \sigma$.

For convenience, we will write elements of $\bigoplus_{i:I} A\,i$ as $\{(i_0, a_0), \ldots, (i_n, a_n)\}$ where $i_0, \ldots, i_n$ are disjoint elements of I and $a_k : A\,i_k$.

Lemma 15. $\bigoplus_{i:I} A\,i$ *is a partial semigroup under pointwise multiplication.*

Proof (Sketch). The definition of pointwise multiplication is somewhat involved and so we only sketch it here. Fix two elements $\rho = \{(i_0, a_0), \ldots, (i_n, a_n)\}$ and $\rho' = \{(j_0, b_0), \ldots, (j_m, b_m)\}$. Rearranging the elements of ρ' if necessary, we assume that there exists ℓ such that $i_k = j_k$ if $k < \ell$ and outside of these pairs, the sequences Is are disjoint. Note that this requires I to have decidable equality.

We define the $\rho \cdot \rho'$ by multiplying together the ℓ elements overlapping between ρ and ρ' and, when these are all defined, returning the partial map with these the results and along with the elements of ρ and ρ' outside the overlap:

$$\begin{aligned}
&\rho \cdot \rho' \triangleq \\
&\quad c_0 \leftarrow a_0 \cdot_{A\,i_0} b_0; \\
&\quad \ldots \\
&\quad c_{\ell-1} \leftarrow a_{\ell-1} \cdot_{A(\ell-1)} b_{\ell-1}; \\
&\quad \mathsf{ret}\,\{(i_0, c_0), \ldots (i_{\ell-1}, c_{\ell-1}), (i_\ell, a_\ell) \ldots (i_n, a_n), (j_\ell, b_\ell) \ldots (j_m, b_m)\}
\end{aligned}$$

Since $-^?$ is a commutative monad, the order of these multiplications are irrelevant and so this process respects the equivalence relation quotienting $\bigoplus_{i:I} A\,i$.

Theorem 2. $\bigoplus_{i:I} A\,i$ *satisfies the MI axiom.*

Proof. Fix $\rho = \{(i_0, a_0), \ldots, (i_n, a_n)\}$. Since $\mathsf{MI}(\rho)$ is a proposition the quotienting is irrelevant and we ignore it. Next, note that if $\sigma \cdot \sigma' = \mathsf{ret}(\rho)$, then the support of σ and σ' must be subsets of the support of ρ. Accordingly, if σ is an idempotent within ρ, then it consists of a collection of idempotents drawn from $A\,i_k$ below the corresponding element a_k.

Since each $A\,i_k$ is a resource algebra, the MI axiom tells us that there is either (1) no idempotent below a_k or (2) a maximal idempotent a'_k below a_k. Without loss of generality, let us assume that a_k has a maximal idempotent if and only if $k \leq m$ for some m and no idempotent below it $k > m$.

The maximal idempotent below ρ is then given by $\sigma = \{(i_0, a'_0), \ldots, (i_m, a'_m)\}$. Indeed, any other idempotent σ' below ρ must have a smaller support than σ and, where both are defined, the value of σ' is smaller than that of σ by construction.

Note that the above depends critically on the fact that the existence of a maximal idempotent is *decidable*. We must be able to analyze whether the i_k entry is to be included based on whether or not a_k contains a maximal idempotent.

To conclude this section, we describe how one can combine these simpler resource algebras into more complex constructions needed for realistic program logics. For instance, the foundational resource algebra representing heaps can be constructed as follows. Writing Loc and Val to be the types of locations and (syntactic) values in a hypothetical imperative language, we define the resource algebra of heaps in this language as follows:

$$\mathsf{Heap} \triangleq \mathsf{Opt}\big(\textstyle\bigoplus_{l:\mathsf{Loc}} \mathsf{Excl}(\mathsf{Val})\big)$$

In this definition $\mathsf{in}_1(\star)$ represents the empty heap, and the exclusive resouce algebra ensures that multiplication of heaps is only defined when the locations of the relevant heaps are disjoint. In particular, there is no need to check the MI axiom here: it is automatic since all constructions involved preserve its validity.

5 The Category of Resource Algebras

We now turn to the properties of resource algebras *collectively* by structuring them into a category. In particular, we show that many resource algebras that are important in practice satisfy simple and recognizable universal properties. The proofs of these facts are not terribly difficult, but this is the point; by isolating a well-behaved definition of resource algebra these calculations are of the sort familiar to any category of algebraic structures.

To define the category of resource algebras, we must, of course, decide on what a morphism of resource algebras should be. Since a resource algebra is a PCS satisfying an additional property, a first idea is to define the category of resource algebras to be a full subcategory of partial commutative semigroups. This is analogous to how, e.g., a morphism of abelian groups is an ordinary group homomorphism whose domain and codomain happen to be abelian. When formulating morphisms of PCSs, the real wrinkle is partiality; from a categorical point of view, PCSs are commutative semigroups in the *Kleisli category* of $-^?$. But Kleisli categories for a monad are usually poorly behaved and $-^?$ is no exception. The solution is to integrate the partial ordering $\sqsubseteq$ introduced in Section 2:

Definition 10. *A morphism of resource algebras $f : (A, \cdot) \to (B, \cdot)$ is a function $f : A \to B$ such that $f^?(a_0 \cdot a_1) \sqsubseteq f(a_0) \cdot f(a_1)$ when $a_0, a_1 : A$.*

Informally: a morphism of resource algebras is a morphism between carriers which commutes with multiplication *provided the multiplication is defined in the domain.* Such morphisms are commonly referred to as *lax* morphisms since we do not, e.g., require $f(a_0) \cdot f(a_1)$ to be *undefined* if $a_0 \cdot a_1$ is undefined. As we shall see, lax morphisms enjoy better categorical properties.

Lemma 16. *Resource algebras and morphisms of such form a category* **RA**.

5.1 First steps with RA

We now record a few basic properties of **RA**:

Lemma 17. **RA** *has finite products and finite coproducts.*

Proof. Initial and terminal objects are given by **0** and **1** with the obvious multiplication maps. We have already encountered Lemma 13 and, since the construction of coproducts is unsurprising, we content ourselves with showing that $A \times B$ has the expected universal property.

To this end, observe that the projection maps $\pi_1 : A \times B \to A$ and $\pi_2 : A \times B \to B$ are both morphisms of resource algebras. Surprisingly, this is the most subtle part of the proof and it depends upon the fact that we require only that ${\pi_1}^?(p \cdot q) \sqsubseteq \pi_1\, p \cdot \pi_p\, q$ rather than equality. To see that the universal property is satisfied, we need only show that if C is a resource algebra such that $f : C \to A$ and $g : C \to B$ are morphisms, then $\lambda c.\,(f\, c, g\, c)$ is a morphism to $A \times B$. This is a routine calculation.

The universal property of the direct sum algebra is slightly longer to describe as it is not a simple limit or colimit. However, in light of the complexity of the definition of multiplication in $\bigoplus_{i:I} A\, i$, we note that even though the universal property is complex, it is significantly more conceptual than the actual definition.

Definition 11. *A compatible cocone over A is a family of morphisms $(f_i : A\, i \to C)_{i:I}$ such that for every finite family of elements $a_0 : A\, i_0, \ldots, a_n : A\, i_n$ the product $f_{i_0}\, a_0 \cdot \ldots \cdot f_{i_n}\, a_n$ is defined.*

Lemma 18. *The resource algebra $\bigoplus_{i:I} A\, i$ along with the canonical inclusions $A\, i \to \bigoplus_{i:I} A\, i$ is the universal compatible cocone over A.*

In other words, modulo partiality $\bigoplus_{i:I} A\, i$ is the coproduct of A; it is the universal way to include each $A\, i$ inside a single resource algebra in such a way that multiplications between disjoint elements are permitted.

5.2 The relation between RA and other categories

A number of the resource algebras encountered in Section 4 can be characterized as adjoints of natural functors between **RA** and other categories. For instance, Excl and Ag are both left adjoints to certain functors from **RA** to the category of small types and functions between them (denoted **TY**).

Lemma 19. Excl *is left adjoint to the forgetful functor* $\mathbf{RA} \to \mathbf{TY}$.

Proof. Examining definitions, since multiplication in Excl is always undefined, a morphism $\mathsf{Excl}(A) \to B$ is precisely a map from A to the carrier of B, as required.

A similar argument gives a universal characterization of Ag:

Lemma 20. Ag *is left adjoint to the functor* $\mathsf{Idem} : \mathbf{RA} \to \mathbf{TY}$ *that maps a resource algebra to its set of idempotents.*

The final result concerns Opt. As this resource algebra extends an existing resource algebra with a unit element (i.e., an element ϵ where $\epsilon \cdot - = \mathsf{ret}$), it is reasonable to guess that $\mathsf{Opt}(A)$ is the resource algebra "freely generated" by A along with a unit. To state this precisely, we introduce *unital* resource algebras:

Definition 12. *A* unital resource algebra *is a resource algebra* $(A, \cdot)$ *with a unit and a unital morphism a morphism of resource algebras preserving the unit.*

Definition 13. *The category of unital resource algebras is denoted* **uRA**.

Lemma 21. *There is a forgetful functor* $U : \mathbf{uRA} \to \mathbf{RA}$ *and the left adjoint to this functor is given by* Opt.

We note that while the proof of this lemma is straightforward, it is simply *false* if the definition of resource algebra included a core structure as data.

For a small example of how all of these universal properties combine, let us turn to the 'map of resource algebras' used to instantiate Iris: $\mathsf{Opt}(\bigoplus_i A\, i)$. We can give a concise description of this rather large resource algebra as the universal method of collecting together the family of resource algebras A and freely adjoining a unit.

6 From the topos of trees to the Iris formalization

Thus far, we have taken our motivation from Iris and its formalization in Coq but focused our attention on the better-behaved $\mathbf{PSh}(\omega)$ and its internal type theory. While this is ideal for working on paper, Iris and its realization in Coq must also optimize for the practicalities of formalization with Coq and therefore use only a subcategory of $\mathbf{PSh}(\omega)$ and a more ad-hoc replacement for $-^?$ in its definition of cameras. In this section, we discuss the impact of our results in Sections 3 to 5 for Iris and its formalization in Coq. Within this section we disregard the internal language of $\mathbf{PSh}(\omega)$ and work externally. For clarity, we refer to cameras as they appear in the Iris formalization as *Iris cameras* and reserve the terms camera and resource algebra for the notions discussed in Section 3.

6.1 Cameras in Iris

We begin with a brief summary of Iris. Rather than using all of $\mathbf{PSh}(\omega)$, Iris's basic types are drawn from the full subcategory of *total presheaves*:

Definition 14. *A total presheaf* $X : \mathbf{PSh}(\omega)$ *is one whose restriction maps* $X(n+1) \to X(n)$ *are surjective.*

Equivalently, a total presheaf is a set X together with a family of equivalence relations $(\equiv_0) \subseteq (\equiv_1) \ldots$ such that $X = \mathsf{lim}_n\, X/(\equiv_n)$ i.e., a COFE in Iris parlance [18]. Under this encoding, a natural transformation between total presheaves is a map of sets which respects all the equivalence relations. Total presheaves are closed under products, sums, exponentials, the partial map classifier, and various other operations in $\mathbf{PSh}(\omega)$. However, they are not closed under finite limits, do not form a locally cartesian closed category, and lack a strong internal language.

To ease formalization, Iris avoids using $-^?$ and encodes of partiality indirectly:

Definition 15. *A (total)* lifted *PCS (LPCS)* $(X, \cdot, \mathcal{V})$ *is a (total) presheaf* X *with an associative and commutative natural transformation* $\cdot : X \times X \to X$ *and a* validity predicate $\mathcal{V} : X \to \mathrm{Prop}$ *such that* $\forall x, y \in X(n).\, \mathcal{V}_n(x \cdot_n y) \subseteq \mathcal{V}_n(x)$.[5]

Intuitively, we have arranged for $\cdot$ to be total by having X also contain sentinel values. The role of $\mathcal{V}$ is then to carve out those elements of X which are genuine from those merely representing undefined multiplications.

One PCS $(M, \cdot)$ induces an LPCS by taking $(M^?, \bar{\cdot}, \pi_1)$ where $\bar{\cdot} : M^? \times M^? \to M^?$ is the Kleisli extension of $\cdot$. The reverse is also possible: given $(X, \cdot, \mathcal{V})$ one can take the subpresheaf of valid elements along with the partial map induced by $\cdot$ defined only when $x \cdot y$ is valid. The loop PCS $\to$ LPCS $\to$ PCS is the identity.

Definition 16. *A core structure on a total LPCS* $(X, \cdot, \mathcal{V})$ *is a natural transformation* $|{-}| : X \to \mathbf{1} + X$ *satisfying the following properties:*

- *If* $x \in X(n)$ *and* $|x| = \mathsf{in}_2(x')$ *then* $x' \cdot x = x$ *and* $|x'| = \mathsf{in}_2(x')$.
- $|{-}|$ *is monotone with respect to the extension order on* X.

In general, a core structure on an LPCS is distinct from a core structure on the induced PCS due to the presence of invalid elements. However, one can always extend a core structure on a PCS to a core structure on the induced LPCS.

Definition 17. *An Iris camera is a total LPCS equipped with a core structure.*

In fact, the definition is more relaxed: a total presheaf can be represented as a family of equivalence relations on a set $(X, (\equiv)_n)$ satisfying $X = \mathsf{lim}_n\, X/(\equiv)_n$. Iris currently requires only the strictly weaker condition that $(\bigcap_n(\equiv_n)) = \{(x, x) \mid x \in X\}$ of its cameras to e.g., give a particular definition of the agreement camera. We will return to this point in Section 6.3.

[5] Recall that in $\mathbf{PSh}(\omega)$ the set $\mathrm{Prop}(n)$ is given by down-closed sets of $\{0 \ldots n\}$.

6.2 Maximal idempotent axiom for Iris cameras

Despite the difference between a core structure on an Iris camera and a core structure on the induced partial commutative semigroup, one can translate the definition of the MI axiom into the language Iris cameras:

Definition 18. *An Iris camera X satisfies the maximal idempotent axiom if the following holds for every n and $x : X(n)$ such that $n \in \mathcal{V}_n(x)$:*[6]

$$(\forall y : X(0).\, y \leq x|_0 \wedge y \cdot_0 y = y \to \bot)$$
$$\vee\ (\exists y : X(n).\, y \leq x \wedge y \cdot_n y = y \wedge \forall m \leq n, z : X(m).\, z \leq x|_m \wedge z \cdot_m z = z \to z \leq y|_m)$$

We have written e.g., $x|_m$ for the functorial action $X(m \leq n)(x)$.

This definition is mechanically produced by unfolding the internal definition in $\mathbf{PSh}(\omega)$ of the maximal idempotent axiom [3]. For instance, PCS satisfies the MI axiom if and only if the induced LPCS satisfies the above proposition.

Theorem 3. *Every Iris camera in the Iris formalization satisfies the MI axiom.*

We have proven this by extending the Iris formalization with proofs of the MI axiom for every Iris camera [12]. More precisely, the accompanying Coq formalization of this paper modifies the existing Iris formalization in Coq by (1) removing the core operation from the structure defining cameras and (2) replacing it with the above translation of the MI axiom. We have then carried this change through the rest of the formalization by updating every camera used in the codebase with a proof of the MI axiom. Notably, we did not need to make significant modifications to other parts of the Iris codebase and no case studies were affected.

In total then, the theoretical work done in Section 3 has allowed us to remove explicit constructions in the Iris formalization and replace them with more conceptual arguments all without requiring any change to the complicated proofs in separation logic currently carried out within Iris.

Remark 2. As discussed in Section 3, one may as well replace $\vee$ with $+$ and $\exists$ with $\sum$. When extending the Iris formalization, we have made these substitutions as Coq's metatheory is too weak to justify either replacement without axioms.

6.3 Categories of Iris cameras

While the MI axiom can be used to save effort and complexity in the formalization, the results of Section 5 do not translate so easily. There are several incompatible possible definitions of morphisms for Iris cameras and none are fully satisfactory. For instance, morphisms may be themselves total or only defined on valid elements,

[6] The reader may be surprised to see $y : X(0)$ rather than $X(n)$. To make this predicate monotone in n, we must require that it holds whenever $y : X(m)$ for $m \leq n$ and it is therefore both necessary and sufficient to check when $m = 0$.

they may preserve validity laxly or strictly, etc. Each of these definitions gives rise to a category satisfying some, but not all, of the properties of **RA**. Fundamentally, the issue is that totalizing a PCS forces us to consider invalid elements.

There exists a well-behaved subclass of LPCSs which organize into a category: the image of the map $(M, \cdot) \mapsto (M^?, \bar{\cdot}, \pi_1)$. However, important Iris cameras do not land in this class of LPCSs because e.g., their underlying types are discrete.

For a concrete example of how this state of affairs complicates research in Iris, consider the resource algebra $\mathsf{Ag}(A)$ from Section 4. It induces an Iris camera by passing to the induced LPCS $\hat{\mathsf{Ag}}(A) = \mathsf{Ag}(A)^?$ which satisfies the desiderata of the agreement Iris camera [18, Section 4.3]. A closely related construction appeared in Jung et al. [17], but was criticized for being exceptionally difficult to define [18]; this suggests an advantage of our categorical approach which yields $\hat{\mathsf{Ag}}(A)$ as the composition of a number of simple and canonical results.

Rather than $\hat{\mathsf{Ag}}(A)$, Jung et al. [18] use a more ad-hoc construction $\mathsf{Ag}'(A)$ as it is easier to formalize. However, $\mathsf{Ag}'(A)$ is *not* a total presheaf: we do not have $\mathsf{Ag}'(A) = \mathsf{lim}_n \mathsf{Ag}'(A)/(\equiv)_n$ because of the presence of 'junk' elements which are never valid. This motivated practitioners to stop requiring that Iris cameras satisfy the completeness condition. However, these junk elements ensure that $\mathsf{Ag}'(A)$ is not the totalization of a PCS and it is therefore not among those LPCSs which organize into a well-behaved category. Consequently, we have no good universal property describing $\mathsf{Ag}'(A)$.

From our point of view, however, the issue is not with $\hat{\mathsf{Ag}}(A)$—it has the expected universal property—but with the insistence on lifted PCSs. For instance, while $\mathsf{Ag}'(A)$ is easier to deal with than $\mathsf{Ag}(A)^?$, it is still more complex than the underlying type of $\mathsf{Ag}(A)$: A! Strikingly, $\mathsf{Ag}(A)$ is simpler to define than $\mathsf{Ag}'(A)$ and the motivating advantage besides simplicity recommending $\mathsf{Ag}'(A)$ over $\hat{\mathsf{Ag}}(A)$—a technical property involving the interaction between $\mathsf{Ag}'(A)$ and the later modality—follows without any additional effort for the resource algebra $\mathsf{Ag}(A)$. More generally, if one works with resource algebras instead of Iris cameras, the underlying types are simplified, the multiplication operation is more transparent, and the results enjoy universal properties.

Fortunately, most of the payoff of our investigation of resource algebras is contained in Theorem 3; for formalization purposes, it is not necessary to have a well-behaved category of Iris cameras, even if one is useful for discovering and explaining particular examples. We therefore view Section 5 as evidence that future iterations of Iris and similar separation logics would benefit from using a (1) broader class of presheaves and (2) using PCSs over LPCSs.

7 Related Work

A great deal of effort has focused on the logical underpinnings of (concurrent) separation logics and our work fits into this tradition. The logical core of separation logic was developed under the name bunched implication logic [15, 23] and subsequently generalized by Biering et al. [2] to account for higher-order separation logics. The importance of partial commutative monoids in separation logic was apparent in these models and PCMs were subsequently featured prominently in concurrent separation logics [6, 9, 10, 16]. Early versions of Iris [19] also included partial commutative monoids, but later replaced them with cameras [17, 18, 20].

Directly related is the work by Bizjak and Birkedal [4] which analyzes the behavior of $\Box$ modalities in terms of the core structures which induce them. They show that there is a bijection between $\Box$ modalities and core structures in the non-step-indexed setting and a weaker correspondence in a constructive setting. We have used this as inspiration for our approach to uniquely *define* core as the operation inducing the largest possible $\Box$ modality. They also elucidate the connection between $\mathbf{PSh}(\omega)$ and the model of Iris [18].

Also closely related is the work of Dardinier et al. [7]. In Appendix A of this work, the authors define a subclass of partial commutative monoids wherein each element $a : M$ satisfies the additional property that $\{b \mid b \leq a \wedge b \cdot b = b\}$ is finite and non-empty. They then observe that in such cases, there is a unique largest core operation on M which they then use in their separation logic. Interestingly, their motivation for introducing this property was not to obtain a better behaved category of PCMs, but rather to facilitate automation by ensuring resources could be decomposed into duplicable and non-duplicable components. their condition is less general than the MI axiom: it applies only to discrete resources and requires the conditions of Lemma 8. However, in the future we hope to explore further how the MI axiom and the work of Dardinier et al. [7] interact, in particular, whether the latter can be extended to a step-indexed setting in this manner.

Finally, we note that our observation that total presheaves can be limiting when extending the model of Iris was also noted by Spies et al. [26] who found several technical issues generalizing total presheaves to a transfinite setting.

8 Conclusion and Future Work

A key aspect of modern separation logics is how resources are modeled. In this paper we have introduced a novel notion of *resource algebra*, which differs in a subtle but crucial way from the camera definition used in Iris, by omitting the core operation and replacing it by a property, the maximal idempotent axiom. We have demonstrated that the new definition improves upon the earlier one in the sense that it induces a universal $\Box$ modality and a well-behaved category RA of resource algebras. Moreover, we have shown that many of the known resource algebra constructions from Iris can be adapted to our new definition of resource algebra and that they satisfy universal properties. Finally, we have also shown that all the Iris cameras satisfy the maximal idempotent axiom.

Future work includes extending the Coq implementation of Iris to use the full category of presheaves rather than just a subcategory thereof, and thence to adopt our notion of resource algebra to such an implementation.

References

[1] Appel, A.W., McAllester, D.: An indexed model of recursive types for foundational proof-carrying code. ACM Transactions on Programming Languages and Systems 23(5), 657–683 (2001)

[2] Biering, B., Birkedal, L., Torp-Smith, N.: Bi-hyperdoctrines, higher-order separation logic, and abstraction. ACM Trans. Program. Lang. Syst. 29(5), 24–es (Aug 2007), https://doi.org/10.1145/1275497.1275499

[3] Birkedal, L., Møgelberg, R., Schwinghammer, J., Støvring, K.: First steps in synthetic guarded domain theory: step-indexing in the topos of trees. Logical Methods in Computer Science 8(4) (2012)

[4] Bizjak, A., Birkedal, L.: On models of higher-order separation logic. Electronic Notes in Theoretical Computer Science 336, 57–78 (04 2018), http://dx.doi.org/10.1016/j.entcs.2018.03.016

[5] Bornat, R., Calcagno, C., O'Hearn, P.W., Parkinson, M.J.: Permission accounting in separation logic. In: Palsberg, J., Abadi, M. (eds.) Proceedings of the 32nd ACM SIGPLAN-SIGACT Symposium on Principles of Programming Languages, POPL 2005, Long Beach, California, USA, January 12-14, 2005. pp. 259–270. ACM (2005), https://doi.org/10.1145/1040305.1040327

[6] Calcagno, C., O'Hearn, P.W., Yang, H.: Local action and abstract separation logic. In: 22nd IEEE Symposium on Logic in Computer Science (LICS 2007), 10-12 July 2007, Wroclaw, Poland, Proceedings. pp. 366–378. IEEE Computer Society (2007), https://doi.org/10.1109/LICS.2007.30

[7] Dardinier, T., Parthasarathy, G., Müller, P.: Verification-preserving inlining in automatic separation logic verifiers. Proceedings of the ACM on Programming Languages 7(OOPSLA1), 789–818 (Apr 2023)

[8] Day, B.: On closed categories of functors. In: MacLane, S., Applegate, H., Barr, M., Day, B., Dubuc, E., Phreilambud, Pultr, A., Street, R., Tierney, M., Swierczkowski, S. (eds.) Reports of the Midwest Category Seminar IV. pp. 1–38. Springer Berlin Heidelberg, Berlin, Heidelberg (1970)

[9] Dinsdale-Young, T., Birkedal, L., Gardner, P., Parkinson, M.J., Yang, H.: Views: compositional reasoning for concurrent programs. In: Giacobazzi, R., Cousot, R. (eds.) The 40th Annual ACM SIGPLAN-SIGACT Symposium on Principles of Programming Languages, POPL '13, Rome, Italy - January 23 - 25, 2013. pp. 287–300. ACM (2013), https://doi.org/10.1145/2429069.2429104

[10] Dockins, R., Hobor, A., Appel, A.W.: A fresh look at separation algebras and share accounting. In: Hu, Z. (ed.) Programming Languages and Systems, 7th Asian Symposium, APLAS 2009, Seoul, Korea, December 14-16, 2009. Proceedings. Lecture Notes in Computer Science, vol. 5904, pp. 161–177. Springer (2009), https://doi.org/10.1007/978-3-642-10672-9_13

[11] Dreyer, D., Ahmed, A., Birkedal, L.: Logical step-indexed logical relations. Logical Methods in Computer Science Volume 7, Issue 2 (06 2011)

[12] Gratzer, D., Møller, M.A., Birkedal, L.: Coq formalization accompanying "Idempotent Resources in Separation Logic" (2025), https://doi.org/10.5281/zenodo.14765017

[13] Hofmann, M.: Syntax and Semantics of Dependent Types. In: Pitts, A.M., Dybjer, P. (eds.) Semantics and Logics of Computation, pp. 79–130. Cambridge University Press (1997), https://www.tcs.ifi.lmu.de/mitarbeiter/martin-hofmann/pdfs/syntaxandsemanticsof-dependenttypes.pdf

[14] Hofmann, M., Streicher, T.: Lifting Grothendieck universes (1997), https://www2.mathematik.tu-darmstadt.de/~streicher/NOTES/lift.pdf, unpublished note

[15] Ishtiaq, S.S., O'Hearn, P.W.: Bi as an assertion language for mutable data structures. In: Proceedings of the 28th ACM SIGPLAN-SIGACT symposium on Principles of programming languages. POPL01, vol. 7, p. 14–26. ACM (01 2001), http://dx.doi.org/10.1145/360204.375719

[16] Jensen, J.B., Birkedal, L.: Fictional separation logic. In: Seidl, H. (ed.) Programming Languages and Systems - 21st European Symposium on Programming, ESOP 2012, Held as Part of the European Joint Conferences on Theory and Practice of Software, ETAPS 2012, Tallinn, Estonia, March 24 - April 1, 2012. Proceedings. Lecture Notes in Computer Science, vol. 7211, pp. 377–396. Springer (2012), https://doi.org/10.1007/978-3-642-28869-2_19

[17] Jung, R., Krebbers, R., Birkedal, L., Dreyer, D.: Higher-order ghost state. SIGPLAN Not. 51(9), 256–269 (09 2016)

[18] Jung, R., Krebbers, R., Jourdan, J.H., Bizjak, A., Birkedal, L., Dreyer, D.: Iris from the ground up: A modular foundation for higher-order concurrent separation logic. Journal of Functional Programming 28 (2018)

[19] Jung, R., Swasey, D., Sieczkowski, F., Svendsen, K., Turon, A., Birkedal, L., Dreyer, D.: Iris: Monoids and invariants as an orthogonal basis for concurrent reasoning. SIGPLAN Not. 50(1), 637–650 (01 2015)

[20] Krebbers, R., Jung, R., Bizjak, A., Jourdan, J.H., Dreyer, D., Birkedal, L.: The Essence of Higher-Order Concurrent Separation Logic, p. 696–723. Springer Berlin Heidelberg (2017), http://dx.doi.org/10.1007/978-3-662-54434-1_26

[21] Mac Lane, S., Moerdijk, I.: Sheaves in geometry and logic : a first introduction to topos theory. Universitext, Springer (1992)

[22] O'Hearn, P., Reynolds, J., Yang, H.: Local Reasoning about Programs that Alter Data Structures, p. 1–19. Springer Berlin Heidelberg (2001), http://dx.doi.org/10.1007/3-540-44802-0_1

[23] O'Hearn, P.W., Pym, D.J.: The logic of bunched implications. Bulletin of Symbolic Logic 5(2), 215–244 (06 1999), http://dx.doi.org/10.2307/421090

[24] Reynolds, J.: Separation logic: a logic for shared mutable data structures. In: Proceedings 17th Annual IEEE Symposium on Logic in Computer Science (2002)

[25] Rosolini, G.: Continuity and effectiveness in topoi. Ph.D. thesis, University of Oxford (1986)

[26] Spies, S., Gäher, L., Gratzer, D., Tassarotti, J., Krebbers, R., Dreyer, D., Birkedal, L.: Transfinite Iris: Resolving an Existential Dilemma of Step-Indexed Separation Logic, p. 80–95. Association for Computing Machinery, New York, NY, USA (2021), https://doi.org/10.1145/3453483.3454031

Alternative Characterizations of Hereditary History-Preserving Bisimilarity via Backward Ready Multisets

Marco Bernardo(✉), Andrea Esposito, and Claudio A. Mezzina

Dipartimento di Scienze Pure e Applicate, Università di Urbino, Urbino, Italy
marco.bernardo@uniurb.it

Abstract. We provide two alternative characterizations of hereditary history-preserving bisimilarity: a denotational one, on stable configuration structures, and an operational one, on a reversible process calculus. The characterizing equivalence is forward-reverse bisimilarity extended with a check for backward ready multiset equality. Unlike previous approaches, the focus is thus on counting identically labeled events rather than uniquely identifying them. We also investigate the relationships between event identifier logic, characterizing the former bisimilarity, and backward ready multiset logic, characterizing the latter bisimilarity.

1 Introduction

In the spectrum of truly concurrent bisimilarities [23,19,32], there are two equivalences that are particularly important: history-preserving bisimilarity [34] and hereditary history-preserving bisimilarity [6]. They are the coarsest equivalence and the finest equivalence, respectively, that are preserved under action refinement and are capable of respecting causality, branching, and their interplay while abstracting from choices between identical alternatives [23]. Moreover, hereditary history-preserving bisimilarity can be obtained as a special case of a categorical definition of bisimilarity over concurrency models [25].

History-preserving and hereditary history-preserving bisimilarities are defined over truly concurrent models such as event structures [35] or their variants, in particular configuration structures [24]. A configuration is a finite set of non-conflicting events that is downward-closed with respect to a causality relation over events. The bisimulation game compares configuration transitions. While history-preserving bisimilarity considers only outgoing transitions, hereditary history-preserving bisimilarity takes into account also incoming transitions. In other words, the former stepwise matches only forward computations, whereas the latter examines backward computations too. Both equivalences rely on ternary bisimulation relations, where the third component is a labeling- and causality-preserving bijection from the set of events executed so far in the first structure to the set of events executed so far in the second structure.

Logical characterizations of both equivalences have been provided in [33,4]. Furthermore, an axiomatization for hereditary history-preserving bisimilarity has been developed over forward-only processes in [21]. Finally, history-preserving

P. A. Abdulla and D. Kesner (Eds.): FoSSaCS 2025, LNCS 15691, pp. 67–87, 2025.
https://doi.org/10.1007/978-3-031-90897-2_4

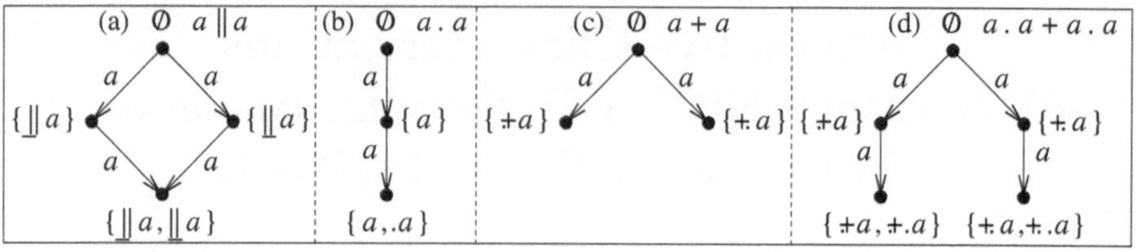

Fig. 1. Configuration graphs: autoconcurrency (a), autocausation (b), and autoconflict

bisimilarity is known to coincide with causal bisimilarity [15,16], hence the latter offers a characterization and an axiomatization [18] for the former. In this paper, we concentrate on characterizations of hereditary history-preserving bisimilarity.

The first alternative characterization of hereditary history-preserving bisimilarity has appeared in [6] for configuration graphs of prime event structures. The characterizing equivalence is called back-and-forth bisimilarity – not to be confused with the homonymous one in [17], which retrieves an interleaving semantics by constraining backward computations to take place along the corresponding forward computations even in the presence of concurrency. The main difference between hereditary history-preserving bisimilarity and back-and-forth bisimilarity is that the latter relies on binary bisimulation relations, hence no labeling- and causality-preserving bijection is stepwise built during the bisimulation game. The characterization result holds under the assumption of no autoconcurrency, i.e., the absence of configurations from which it is possible to execute two identically labeled, distinct events that are not in conflict with each other.

In Figures 1(a) and (b) we show the configuration graphs respectively associated with the following two processes for a given action a:

- Autoconcurrency on a, which is expressed as $a \parallel a$ where $\parallel$ stands for parallel composition. There are two equally labeled, non-conflicting events, denoted by $\underline{|}|a$ and $|\underline{|}a$, that can be executed in any order.
- Autocausation on a, which is expressed as $a\,.\,a$ where dot represents action prefix. There are two equally labeled, non-conflicting events, denoted by a and $.a$, such that the former has to be executed before the latter.

These two configuration graphs are back-and-forth bisimilar as witnessed by the symmetric binary relation that contains the pairs of configurations $(\emptyset, \emptyset)$, $(\{\underline{|}|a\}, \{a\})$, $(\{|\underline{|}a\}, \{a\})$, and $(\{\underline{|}|a, |\underline{|}a\}, \{a, .a\})$. However, they are not hereditary history-preserving bisimilar because, with respect to the last pair, there is no (labeling- and) causality-preserving bijection that maps the two independent events $\underline{|}|a$ and $|\underline{|}a$ to the two causally-related events a and $.a$.

The second alternative characterization of hereditary history-preserving bisimilarity has been given in [31]. The characterizing equivalence is the forward-reverse bisimilarity – very close in spirit to the back-and-forth bisimilarity of [6] – originally defined in [30] for a reversible variant of CCS [29] called CCSK. The operational semantics of CCSK produces labeled transition systems based on a forward transition relation and a backward one ensuring the loop property [13]. Each transition label comprises an action and a communication key; the latter is necessary when building backward transitions so as to know who synchronized with whom in the forward direction. In [31] forward-reverse bisimilarity has been generalized to configuration graphs of prime event structures and shown to coincide with hereditary history-preserving bisimilarity in the absence of repeated, identically labeled events along forward computations, which implies the absence of autoconcurrency (and autocausation), i.e., the assumption made in [6].

In [32] it has been shown, by working on stable configuration structures, how to relax the conditions under which the two characterization results of [6] and [31] hold. Specifically, it is sufficient to require the absence of equidepth autoconcurrency, i.e., the absence of identically labeled events occurring at the same depth within a configuration; the depth of an event is defined as the length of the longest causal chain of events up to and including the considered event.

The third alternative characterization of hereditary history-preserving bisimilarity has been provided in [3] and, unlike the previous two, does not need any restrictive assumption. Based on earlier work [2] – in which hereditary history-preserving bisimilarity was shown to coincide with back-and-forth barbed bisimulation congruence over singly-labeled processes, i.e., processes with no autoconcurrency and autoconflict (see Figure 1(c)) – it has been developed in the setting of a different reversible variant of CCS called RCCS [13,14,27]. While in CCSK all executed actions and discarded alternative subprocesses are kept within the syntax of processes so as to enable reversibility, in RCCS the same information is stored into stack-based memories attached to processes; the two approaches have been proven to be equivalent in [28]. The idea in [3] is to import hereditary history-preserving bisimilarity in the RCCS setting by encoding memories, i.e., the past behavior, as identified configuration structures. These are stable configuration structures enriched with unique event identifiers, used in transition labels and exploited when undoing synchronizations. The characterizing equivalence, called back-and-forth bisimilarity and defined over RCCS processes, relies on ternary bisimulation relations in which the third component is a bijection from the set of identifiers of the actions executed so far in the first process to the set of identifiers of the actions executed so far in the second process.

Having to reintroduce a third component in the bisimulation relations in order to exactly characterize hereditary history-preserving bisimilarity amounts to certifying that "reversibility is not just back and forth" [3], i.e., the forward and backward bisimulation games alone are not enough. The question then becomes whether and to what extent a systematic event identification is really necessary.

This question also arises from the fact that, in the aforementioned bisimulation games, CCSK transition labels such as $a[i]$ and $a[j]$ are deemed to be different if the two keys i and j are different [30] – which results in the absence of repeated, identically labeled events along forward computations [31] – while identified RCCS transition labels like $i{:}a$ and $j{:}a$ are viewed as compatible even if i and j are different [3]. On the one hand, in CCSK the two processes $a \parallel a$ and $a \,.\, a$ are told apart by forward-reverse bisimilarity because the former evolves to

$a[i] \parallel a[j]$, which can undo $a[i]$ and $a[j]$ in any order, while the latter evolves to $a[i] \,.\, a[j]$, from which only $a[j]$ can be undone, hence undoing $a[i]$ cannot be matched by undoing $a[j]$. On the other hand, in identified RCCS the same two processes are distinguished by back-and-forth bisimilarity because, although undoing $i{:}a$ can be matched by undoing $j{:}a$, it is not possible to establish a suitable bijection from a distributed memory containing $i{:}a$ in a location and $j{:}a$ in another location to a centralized memory containing $j{:}a$ on top of $i{:}a$.

In this paper we propose a totally different approach to exactly characterize hereditary history-preserving bisimilarity. Rather than the unique identification of identically labeled events, the focus is on *counting* them. Let us consider again Figures 1(a) and (b). If we look at the two top (resp. bottom) configurations, we note that the one on the left has two outgoing (resp. incoming) transitions, while the one on the right has only one. As for the top configuration on the left, in principle we may not know whether the branch is due to the fact that the two events are concurrent or conflicting. However, for the bottom configuration on the left we can certainly say that the two events are concurrent, as the models we are considering are truly concurrent and hence the configuration graph of process $a \,.\, a + a \,.\, a$ where $+$ stands for nondeterministic choice (see Figure 1(d)) cannot be isomorphic to the one of $a \parallel a$ because it must have two different bottom configurations ($\{\underset{\cdot}{+}a, \underset{\cdot}{+}.a\}$ and $\{\dot{+}a, \dot{+}.a\}$) instead of a single one.

As an extension of the notion of backward ready set exploited in [8] to axiomatize forward-reverse bisimilarity over reversible concurrent processes, we define the *backward ready multiset* of a configuration or process to be the multiset of labels of its incoming transitions. After recalling in Section 2 the definitions of stable configuration structure [24], hereditary history-preserving bisimilarity [6], and event identifier logic [33], we provide the following contributions:

- In Section 3 we exhibit a denotational characterization on stable configuration structures: hereditary history-preserving bisimilarity turns out to coincide with forward-reverse bisimilarity extended with a clause for checking the equality of the backward ready multisets of matching configurations.
- In Section 4 we exhibit an operational characterization based on a variant of the reversible process calculus of [9,8] where executed action identification is limited to synchronizations. After revising its proved operational semantics inspired by [18] so as to faithfully account for causality and concurrency, we set up a backward-ready-multiset variant of forward-reverse bisimilarity and devise a backward ready multiset logic characterizing it. Then we define a denotational semantics based on stable configuration structures in which events are formalized as proof terms [10,11], so as to import the notion of hereditary history-preserving bisimilarity. We show that the stable configuration structures associated with two processes are hereditary history-preserving bisimilar iff the two processes are equated by the backward-ready-multiset variant of forward-reverse bisimilarity.
- In Section 5 we start the investigation of the relationships between the event identifier logic of [33] and our backward ready multiset logic.

Section 6 concludes the paper with directions for future work.

2 Hereditary History-Preserving Bisimilarity

In this section we recall hereditary history-preserving bisimilarity [6] over stable configuration structures [24] along with its logical characterization based on event identifier logic [33].

In the following two definitions taken from [23], $\mathcal{P}_{\mathrm{fin}}(\mathcal{E})$ denotes the set of finite subsets of set $\mathcal{E}$ while $f \upharpoonright X$ denotes the restriction of function f to set X.

Definition 1. *A* configuration structure *is a quadruple* $\mathcal{C} = (\mathcal{E}, C, \mathcal{A}, l)$ *where:*

- $\mathcal{E}$ *is a set of* events.
- $C \subseteq \mathcal{P}_{\mathrm{fin}}(\mathcal{E})$ *is a set of* configurations.
- $\mathcal{A}$ *is a countable set of* labels.
- $l : \bigcup_{X \in C} X \to \mathcal{A}$ *is a* labeling function.

$\mathcal{C}$ *is said to be* stable *iff it is:*

- *Rooted:* $\emptyset \in C$.
- *Connected:* $\forall X \in C \setminus \{\emptyset\}. \exists e \in X. X \setminus \{e\} \in C$.
- *Closed under bounded unions and intersections:* $\forall X, Y, Z \in C. X \cup Y \subseteq Z \Longrightarrow X \cup Y, X \cap Y \in C$.

The causality relation *over* $X \in C$ *is defined by letting* $e_1 \leq_X e_2$ *for* $e_1, e_2 \in X$ *iff* $e_2 \in Y$ *implies* $e_1 \in Y$ *for all* $Y \in C$ *such that* $Y \subseteq X$*; we write* $e_1 <_X e_2$ *when* $e_1 \leq_X e_2$ *and* $e_1 \neq e_2$. *Two events* $e_1, e_2 \in X$ *are* concurrent *in* X *iff* $e_1 \not<_X e_2$ *and* $e_2 \not<_X e_1$. *We write* $X \xrightarrow{a}_{\mathcal{C}} X'$ *for* $X, X' \in C$ *and* $a \in \mathcal{A}$ *iff* $X \subseteq X'$, $X' \setminus X = \{e\}$, *and* $l(e) = a$. ■

Definition 2. *We say that two stable configuration structures* $\mathcal{C}_i = (\mathcal{E}_i, C_i, \mathcal{A}, l_i)$, $i \in \{1, 2\}$, *are* hereditary history-preserving bisimilar, *written* $\mathcal{C}_1 \sim_{\mathrm{HHPB}} \mathcal{C}_2$, *iff there exists a hereditary history-preserving bisimulation between* $\mathcal{C}_1$ *and* $\mathcal{C}_2$, *i.e., a relation* $\mathcal{B} \subseteq C_1 \times C_2 \times \mathcal{P}(\mathcal{E}_1 \times \mathcal{E}_2)$ *such that:*

- $(\emptyset, \emptyset, \emptyset) \in \mathcal{B}$.
- *Whenever* $(X_1, X_2, f) \in \mathcal{B}$ *then:*
  - $f \subseteq \mathcal{E}_1 \times \mathcal{E}_2$ *is a bijection from* $X_1 \in C_1$ *to* $X_2 \in C_2$ *that preserves:*
    - *Labeling:* $l_1(e) = l_2(f(e))$ *for all* $e \in X_1$.
    - *Causality:* $e \leq_{X_1} e' \Longleftrightarrow f(e) \leq_{X_2} f(e')$ *for all* $e, e' \in X_1$.
  - *For each* $X_1 \xrightarrow{a}_{\mathcal{C}_1} X_1'$ *there exist* $X_2 \xrightarrow{a}_{\mathcal{C}_2} X_2'$ *and* $f' \subseteq \mathcal{E}_1 \times \mathcal{E}_2$ *such that* $(X_1', X_2', f') \in \mathcal{B}$ *and* $f' \upharpoonright X_1 = f$, *and vice versa.*
  - *For each* $X_1' \xrightarrow{a}_{\mathcal{C}_1} X_1$ *there exist* $X_2' \xrightarrow{a}_{\mathcal{C}_2} X_2$ *and* $f' \subseteq \mathcal{E}_1 \times \mathcal{E}_2$ *such that* $(X_1', X_2', f') \in \mathcal{B}$ *and* $f \upharpoonright X_1' = f'$, *and vice versa.* ■

Since there is a single transition relation, similar to [17,6] in the bisimulation game above a distinction is made between the outgoing transitions of X_1 and X_2 ($X_1 \xrightarrow{a}_{\mathcal{C}_1} X_1'$ and $X_2 \xrightarrow{a}_{\mathcal{C}_2} X_2'$ in the forward direction) and their incoming transitions ($X_1' \xrightarrow{a}_{\mathcal{C}_1} X_1$ and $X_2' \xrightarrow{a}_{\mathcal{C}_2} X_2$ in the backward direction).

Hereditary history-preserving bisimilarity is characterized by *event identifier logic* [33]. The set $\mathcal{L}_{\rm EI}$ of its formulas is generated by the following syntax:

$$\phi ::= \mathsf{true} \mid \neg\phi \mid \phi \wedge \phi \mid \langle x : a \rangle\rangle \phi \mid (x : a)\phi \mid \langle\langle x \rangle \phi$$

where $a \in \mathcal{A}$ and $x \in \mathcal{I}$, with $\mathcal{I}$ being a countable set of identifiers. The unary operators $\langle x : a \rangle\rangle\_$ and $(x : a)\_$ act as binders for the identifiers inside them. Therefore, the set of identifiers that occur *free* in $\phi \in \mathcal{L}_{\rm EI}$ is defined by induction on the syntactical structure of ϕ as follows:

$$\begin{aligned} \mathit{fi}(\mathsf{true}) &= \emptyset \\ \mathit{fi}(\neg\phi) &= \mathit{fi}(\phi) \\ \mathit{fi}(\phi_1 \wedge \phi_2) &= \mathit{fi}(\phi_1) \cup \mathit{fi}(\phi_2) \\ \mathit{fi}(\langle x : a \rangle\rangle \phi) &= \mathit{fi}(\phi) \setminus \{x\} \\ \mathit{fi}((x : a)\phi) &= \mathit{fi}(\phi) \setminus \{x\} \\ \mathit{fi}(\langle\langle x \rangle \phi) &= \mathit{fi}(\phi) \cup \{x\} \end{aligned}$$

where we say that ϕ is *closed* if $\mathit{fi}(\phi) = \emptyset$, *open* otherwise.

In order to assign meaning to open formulas, environments are employed to indicate what events the free identifiers are bound to. Given a configuration structure $\mathsf{C} = (\mathcal{E}, \mathcal{C}, \mathcal{A}, l)$, an *environment* is a partial function $\rho : \mathcal{I} \rightharpoonup \mathcal{E}$. Given $X \in \mathcal{C}$ and $\phi \in \mathcal{L}_{\rm EI}$, we say that ρ is a *permissible* environment for X and ϕ iff ρ maps every free identifier in ϕ to an event in X. Denoting with $\mathit{dom}(\rho)$ the domain of ρ, $\mathit{rge}(\rho)$ the codomain of ρ, and ρ_ϕ the restriction $\rho \restriction \mathit{fi}(\phi)$, permissibility is formalized as $\mathit{fi}(\phi) \subseteq \mathit{dom}(\rho)$ and $\mathit{rge}(\rho_\phi) \subseteq X$. The set of permissible environments for X and ϕ is indicated by $\mathit{pe}(X, \phi)$.

The satisfaction relation $\models \subseteq (\mathcal{C} \times \mathcal{E}^{\mathcal{I}}) \times \mathcal{L}_{\rm EI}$, with $\mathcal{E}^{\mathcal{I}}$ being the set of functions from $\mathcal{I}$ to $\mathcal{E}$, i.e., the set of environments, is defined by induction on the syntactical structure of $\phi \in \mathcal{L}_{\rm EI}$ as follows:

$$\begin{array}{ll} X \models_\rho \mathsf{true} & \\ X \models_\rho \neg\phi' & \text{iff } X \not\models_\rho \phi' \\ X \models_\rho \phi_1 \wedge \phi_2 & \text{iff } X \models_\rho \phi_1 \text{ and } X \models_\rho \phi_2 \\ X \models_\rho \langle x : a \rangle\rangle \phi' & \text{iff there is } X \xrightarrow{l(e)}_{\mathsf{C}} X' \text{ such that } l(e) = a \text{ and } X' \models_{\rho[x \mapsto e]} \phi' \\ X \models_\rho (x : a)\phi' & \text{iff there is } e \in X \text{ such that } l(e) = a \text{ and } X \models_{\rho[x \mapsto e]} \phi' \\ X \models_\rho \langle\langle x \rangle \phi' & \text{iff there is } X' \xrightarrow{l(e)}_{\mathsf{C}} X \text{ such that } \rho(x) = e \text{ and } X' \models_\rho \phi' \end{array}$$

where it is understood that the environment in the subscript of every occurrence of $\models$ is permissible for the configuration on the left and the formula on the right. Moreover, $\rho[x \mapsto e]$ is $\rho \setminus \{(x, \rho(x))\} \cup \{(x, e)\}$ if $x \in \mathit{dom}(\rho)$, $\rho \cup \{(x, e)\}$ otherwise.

Let $\mathcal{L}_{\rm EI}^{\rm c}$ be the set of closed formulas of $\mathcal{L}_{\rm EI}$. Given $\phi \in \mathcal{L}_{\rm EI}^{\rm c}$, we write $X \models \phi$ as a shorthand for $X \models_\emptyset \phi$ and $\mathsf{C} \models \phi$ as a shorthand for $\emptyset \models \phi$. Image finiteness means no configuration has infinitely many transitions with the same label.

Theorem 1 ([33]). *Let $\mathsf{C}_i = (\mathcal{E}_i, \mathcal{C}_i, \mathcal{A}, l_i)$, $i \in \{1, 2\}$, be two image-finite stable configuration structures. Then $\mathsf{C}_1 \sim_{\rm HHPB} \mathsf{C}_2$ iff $\forall \phi \in \mathcal{L}_{\rm EI}^{\rm c}.\ \mathsf{C}_1 \models \phi \iff \mathsf{C}_2 \models \phi$.* ■

3 Characterization on Stable Configuration Structures

The first characterization that we provide for $\sim_{\rm HHPB}$ is on stable configuration structures. From a ternary bisimulation relation we move to a binary one where, instead of stepwise building a labeling- and causality-preserving bijection between the events of matching configurations – which are the events executed so far in both stable configuration structures – we just count the identically labeled incoming transitions of matching configurations. Given a configuration X, its *backward ready multiset* is defined as $brm(X) = \{\!| a \in \mathcal{A} \mid X' \xrightarrow{a}_{\mathcal{C}} X |\!\}$ where $\{\!|$ and $|\!\}$ are multiset delimiters. We thus decorate the resulting forward-reverse bisimilarity with the acronym brm, standing for backward ready multiset.

Definition 3. *We say that two stable configuration structures* $\mathcal{C}_i = (\mathcal{E}_i, C_i, \mathcal{A}, l_i)$, $i \in \{1, 2\}$, *are* brm-forward-reverse bisimilar, *written* $\mathcal{C}_1 \sim_{\mathrm{FRB:brm}} \mathcal{C}_2$, *iff there exists a brm-forward-reverse bisimulation between* $\mathcal{C}_1$ *and* $\mathcal{C}_2$, *i.e., a relation* $\mathcal{B} \subseteq C_1 \times C_2$ *such that* $(\emptyset, \emptyset) \in \mathcal{B}$ *and, whenever* $(X_1, X_2) \in \mathcal{B}$, *then:*

- *For each* $X_1 \xrightarrow{a}_{\mathcal{C}_1} X_1'$ *there exists* $X_2 \xrightarrow{a}_{\mathcal{C}_2} X_2'$ *such that* $(X_1', X_2') \in \mathcal{B}$, *and vice versa.*
- *For each* $X_1' \xrightarrow{a}_{\mathcal{C}_1} X_1$ *there exists* $X_2' \xrightarrow{a}_{\mathcal{C}_2} X_2$ *such that* $(X_1', X_2') \in \mathcal{B}$, *and vice versa.*
- $brm(X_1) = brm(X_2)$. ■

Theorem 2. *Let* $\mathcal{C}_i = (\mathcal{E}_i, C_i, \mathcal{A}, l_i)$, $i \in \{1, 2\}$, *be two stable configuration structures. Then* $\mathcal{C}_1 \sim_{\mathrm{HHPB}} \mathcal{C}_2$ *iff* $\mathcal{C}_1 \sim_{\mathrm{FRB:brm}} \mathcal{C}_2$. ■

4 Operational Characterization

The second characterization that we provide for $\sim_{\mathrm{HHPB}}$ is operational. More precisely, we present a variant of the syntax (Section 4.1) and the proved operational semantics (Section 4.2) of the reversible process calculus of [9,8], followed by a redefinition of brm-forward-reverse bisimilarity on that variant along with a modal logic characterization (Section 4.3). Then we develop a denotational semantics for the modified calculus based on stable configuration structures (Section 4.4), so as to import the notion of hereditary history-preserving bisimilarity. Finally, we prove that the stable configuration structures associated with two processes are hereditary history-preserving bisimilar iff the two processes are brm-forward-reverse bisimilar (Section 4.5).

4.1 Syntax of Reversible Concurrent Processes

In the representation of a process, we are used to describe only its future behavior. However, in order to support reversibility in the style of [30], we need to equip the syntax with information about the past, in particular the actions that have already been executed. Taking inspiration from CCS [29] and CSP [12], given a countable set $\mathcal{A}$ of actions including an unobservable action denoted by τ, we extend as follows the syntax for reversible concurrent processes of [9,8]:

$$P ::= \underline{0} \mid a \,.\, P \mid a^{\dagger\xi} . P \mid P + P \mid P \parallel_L P$$
$$\xi ::= \varepsilon \mid \langle \theta, \theta \rangle_L$$

where $a \in \mathcal{A}$, $L \subseteq \mathcal{A} \setminus \{\tau\}$, ε is the empty string, θ is a proof term (its syntax will be provided in Section 4.2), and:

- $\underline{0}$ is the terminated process.
- $a \,.\, P$ is a process that can execute action a and whose forward continuation is P (unexecuted action prefix).
- $a^{\dagger\xi} \,.\, P$ is a process that executed action a and whose forward continuation is inside P, which can undo action a after all executed actions within P have been undone (executed action prefix).
- $P_1 + P_2$ expresses a nondeterministic choice between P_1 and P_2 as far as neither has executed any action yet, otherwise only the one that was selected in the past can move (past-sensitive alternative composition).
- $P_1 \parallel_L P_2$ expresses that P_1 and P_2 proceed independently of each other on actions in $\overline{L} = \mathcal{A} \setminus L$, while they have to synchronize on every action in L (parallel composition).

We can characterize two important classes of processes via as many predicates. Firstly, we define *initial* processes, in which all actions are unexecuted and hence no †-decoration appears:

$$\begin{array}{rl} & \mathit{init}(\underline{0}) \\ \mathit{init}(a \,.\, P) & \text{if } \mathit{init}(P) \\ \mathit{init}(P_1 + P_2) & \text{if } \mathit{init}(P_1) \wedge \mathit{init}(P_2) \\ \mathit{init}(P_1 \parallel_L P_2) & \text{if } \mathit{init}(P_1) \wedge \mathit{init}(P_2) \end{array}$$

Secondly, we define *well-formed* processes, whose set we denote by $\mathcal{P}$, in which both unexecuted and executed actions can occur in certain circumstances:

$$\begin{array}{rl} \mathit{wf}(\underline{0}) & \\ \mathit{wf}(a \,.\, P) & \text{if } \mathit{init}(P) \\ \mathit{wf}(a^{\dagger\xi} \,.\, P) & \text{if } \mathit{wf}(P) \\ \mathit{wf}(P_1 + P_2) & \text{if } (\mathit{wf}(P_1) \wedge \mathit{init}(P_2)) \vee (\mathit{init}(P_1) \wedge \mathit{wf}(P_2)) \\ \mathit{wf}(P_1 \parallel_L P_2) & \text{if } \mathit{wf}(P_1) \wedge \mathit{wf}(P_2) \end{array}$$

Well formedness not only imposes that every unexecuted action is followed by an initial process, but also that in every alternative composition at least one subprocess is initial. Multiple paths may arise in the presence of both alternative and parallel compositions. However, at each occurrence of the former, only the subprocess chosen for execution can move. Although not selected, the other subprocess is kept as an initial subprocess within the overall process, in the same way as executed actions are kept inside the syntax [11,30], so as to support reversibility. As an example, in $a^{\dagger} \,.\, b \,.\, \underline{0} + c \,.\, d \,.\, \underline{0}$ the subprocess $c \,.\, d \,.\, \underline{0}$ cannot move because a was selected in the choice between a and c.

It is worth noting that:

- $\underline{0}$ is both initial and well-formed.
- Any initial process is well-formed too.
- $\mathcal{P}$ also contains processes that are not initial like, e.g., $a^{\dagger} \,.\, b \,.\, \underline{0}$, which can either do b or undo a.
- In $\mathcal{P}$ the relative positions of already executed actions and actions to be executed matter. Precisely, an action of the former kind can never occur after one of the latter kind. For instance, $a^{\dagger} \,.\, b \,.\, \underline{0} \in \mathcal{P}$ whereas $b \,.\, a^{\dagger} \,.\, \underline{0} \notin \mathcal{P}$.
- In $\mathcal{P}$ the subprocesses of an alternative composition can be both initial, but cannot be both non-initial. For example, $a \,.\, \underline{0} + b \,.\, \underline{0} \in \mathcal{P}$ while $a^{\dagger} \,.\, \underline{0} + b^{\dagger} \,.\, \underline{0} \notin \mathcal{P}$.

Sometimes we will need to bring a process back to its initial version. This is accomplished by removing all †-decorations through function $to\_init : \mathcal{P} \rightarrow \mathcal{P}_{\mathrm{init}}$ with $\mathcal{P}_{\mathrm{init}}$ being the set of initial processes of $\mathcal{P}$, which is defined as follows:

$$\begin{array}{rcll} to\_init(P) & = & P & \text{if } init(P) \\ to\_init(a^{\dagger\xi} . P') & = & a \,.\, to\_init(P') & \\ to\_init(P_1 + P_2) & = & to\_init(P_1) + to\_init(P_2) & \text{if } \neg init(P_1) \vee \neg init(P_2) \\ to\_init(P_1 \parallel_L P_2) & = & to\_init(P_1) \parallel_L to\_init(P_2) & \text{if } \neg init(P_1) \vee \neg init(P_2) \end{array}$$

4.2 Proved Operational Semantics

According to [30] dynamic operators such as action prefix and alternative composition have to be made static in the operational semantic rules, so as to retain within the syntax all the information needed to enable reversibility. Unlike [30] we do not generate a forward transition relation and a backward one, but a single transition relation that we deem to be symmetric in order to enforce the *loop property* [13]: every executed action can be undone and every undone action can be redone. A backward transition from P' to P is subsumed by the corresponding forward transition t from P to P'. As already done in Sections 2 and 3 as well as in [17,6], we will view t as an *outgoing* transition of P when going forward, while we will view t as an *incoming* transition of P' when going backward.

Following [8] we provide an operational semantics based on [18], which is very concrete as every transition is labeled with a *proof term* [10,11]. This is an action preceded by the sequence of operator symbols in the scope of which the action occurs inside the source process of the transition. In the case of a binary operator, the corresponding symbol also specifies whether the action occurs to the left or to the right. The syntax that we adopt for the set Θ of proof terms is the following where $a \in \mathcal{A}$ and $L \subseteq \mathcal{A} \setminus \{\tau\}$:

$$\theta \ ::= \ a \mid ._{a}\theta \mid \dot{+}\theta \mid +\!\!\!\cdot\,\theta \mid \rfloor\!\!\rfloor_L\theta \mid \lfloor\!\!\lfloor_L\theta \mid \langle \theta, \theta \rangle_L$$

The proved operational semantic rules are in Table 1 and generate the proved labeled transition system $(\mathcal{P}, \Theta, \longrightarrow)$ where $\longrightarrow \, \subseteq \mathcal{P} \times \Theta \times \mathcal{P}$ is the proved transition relation. We denote by $\mathbb{P} \subsetneq \mathcal{P}$ the set of processes that are *reachable* from an initial one via $\longrightarrow$. Not all well-formed processes are reachable; for example, $a^{\dagger} . \underline{0} \parallel_{\{a\}} \underline{0}$ is not reachable from $a \,.\, \underline{0} \parallel_{\{a\}} \underline{0}$ as action a on the left cannot synchronize with any action on the right. From now on we consider only $\mathbb{P}$ and denote by $\mathbb{P}_{\mathrm{init}}$ the subset of its initial processes. Every process in $\mathbb{P}$ may have several outgoing transitions and, if it is not initial, has at least one incoming transition.

The first rule for action prefix ($\textsc{Act}_{\mathrm{f}}$ where f stands for forward) applies only if P is initial and retains the executed action in the target process of the generated forward transition by decorating the action itself with †. The second rule ($\textsc{Act}_{\mathrm{p}}$ where p stands for propagation) propagates actions of inner initial subprocesses by putting an a-dot before them in the label for each outer executed a-action prefix that is encountered.

In both rules for alternative composition ($\textsc{Cho}_{\mathrm{l}}$ and $\textsc{Cho}_{\mathrm{r}}$ where l stands for left and r stands for right), the subprocess that has not been selected for

$$(\textsc{Act}_{\mathrm{f}})\ \frac{init(P)}{a\,.\,P \xrightarrow{a} a^{\dagger}\,.\,P} \qquad (\textsc{Act}_{\mathrm{p}})\ \frac{P \xrightarrow{\theta} P'}{a^{\dagger\xi}\,.\,P \xrightarrow{._a\theta} a^{\dagger\xi}\,.\,P'}$$

$$(\textsc{Cho}_{\mathrm{l}})\ \frac{P_1 \xrightarrow{\theta} P_1' \quad init(P_2)}{P_1 + P_2 \xrightarrow{\underset{.}{+}\theta} P_1' + P_2} \qquad (\textsc{Cho}_{\mathrm{r}})\ \frac{P_2 \xrightarrow{\theta} P_2' \quad init(P_1)}{P_1 + P_2 \xrightarrow{\dot{+}\theta} P_1 + P_2'}$$

$$(\textsc{Par}_{\mathrm{l}})\ \frac{P_1 \xrightarrow{\theta} P_1' \quad act(\theta) \notin L}{P_1 \parallel_L P_2 \xrightarrow{\underset{.}{\parallel}_L\theta} P_1' \parallel_L P_2} \qquad (\textsc{Par}_{\mathrm{r}})\ \frac{P_2 \xrightarrow{\theta} P_2' \quad act(\theta) \notin L}{P_1 \parallel_L P_2 \xrightarrow{\dot{\parallel}_L\theta} P_1 \parallel_L P_2'}$$

$$(\textsc{Syn})\ \frac{P_1 \xrightarrow{\theta_1} P_1' \quad P_2 \xrightarrow{\theta_2} P_2' \quad act(\theta_1) = act(\theta_2) \in L}{P_1 \parallel_L P_2 \xrightarrow{\langle\theta_1,\theta_2\rangle_L} enr(P_1' \parallel_L P_2', \langle\theta_1,\theta_2\rangle_L)}$$

Table 1. Proved operational semantic rules for reversible concurrent processes

execution is retained as an initial subprocess in the target process of the generated transition. When both subprocesses are initial, both rules for alternative composition are applicable, otherwise only one of them can be applied and in that case it is the non-initial subprocess that can move, because the other one has been discarded at the moment of the selection. The symbol $\underset{.}{+}$ or $\dot{+}$ is added at the beginning of the proof term.

Due to the †-decorations of executed actions inside the process syntax, over the set $\mathbb{P}_{\mathrm{seq}}$ of *sequential* processes – in which there are no occurrences of parallel composition – every non-initial process has exactly one incoming transition, proved labeled transition systems turn out to be trees, and well formedness coincides with reachability [9].

Example 1. The proved labeled transition system underlying the initial sequential process $a\,.\,\underline{0}$ has a single transition $a\,.\,\underline{0} \xrightarrow{a} a^{\dagger}\,.\,\underline{0}$. In contrast, the proved labeled transition system underlying the initial sequential process $a\,.\,\underline{0} + a\,.\,\underline{0}$ has the two transitions $a\,.\,\underline{0} + a\,.\,\underline{0} \xrightarrow{\underset{.}{+}a} a^{\dagger}\,.\,\underline{0} + a\,.\,\underline{0}$ and $a\,.\,\underline{0} + a\,.\,\underline{0} \xrightarrow{\dot{+}a} a\,.\,\underline{0} + a^{\dagger}\,.\,\underline{0}$. Note that the two target processes are different from each other due to the presence of action decorations, whereas a single a-transition from $a\,.\,\underline{0} + a\,.\,\underline{0}$ to $\underline{0}$ would be generated in the setting of a forward-only process calculus. ■

The three rules for parallel composition use partial function $act : \Theta \rightharpoonup \mathcal{A}$ to extract an action from a proof term θ. This function, which will be used throughout the paper, is defined by induction on the syntactical structure of θ as follows:

$$\begin{aligned}
act(a) &= a \\
act(._a\theta') &= act(\theta') \\
act(\underset{.}{+}\theta') = act(\dot{+}\theta') &= act(\theta') \\
act(\underset{.}{\parallel}_L\theta') = act(\dot{\parallel}_L\theta') &= act(\theta') \\
act(\langle\theta_1,\theta_2\rangle_L) &= \begin{cases} act(\theta_1) & \text{if } act(\theta_1) = act(\theta_2) \\ \text{undefined} & \text{otherwise} \end{cases}
\end{aligned}$$

In the first two rules ($\textsc{Par}_{\mathrm{l}}$ and $\textsc{Par}_{\mathrm{r}}$), a single subprocess proceeds by performing an action not belonging to L, with $\rfloor\!\rfloor_L$ or $\lfloor\!\lfloor_L$ being placed at the beginning of the proof term. In the third rule (SYN), both subprocesses synchronize on an action in L and the resulting proof term contains both individual proof terms. If $L = \emptyset$ or $L = \mathcal{A} \setminus \{\tau\}$, then the two subprocesses are fully independent or fully synchronized, respectively, on observable actions.

The natural target process $P'_1 \parallel_L P'_2$ of a synchronization has to be suitably manipulated in rule SYN to correctly reflect causality and concurrency. More precisely, the †-decoration of every executed action participating in the synchronization has to be enriched with a proof term of the form $\langle \theta_1, \theta_2 \rangle_L$. This is accomplished by taking $enr(P'_1 \parallel_L P'_2, \langle \theta_1, \theta_2 \rangle_L) = enr'(P'_1 \parallel_L P'_2, \langle \theta_1, \theta_2 \rangle_L, \langle \theta_1, \theta_2 \rangle_L)$ as target process, where partial function $enr' : \mathbb{P} \times \Theta \times \Theta \rightharpoonup \mathbb{P}$ is defined by induction on the syntactical structure of its first argument $P \in \mathbb{P}$ as follows:

$$
\begin{aligned}
enr'(\underline{0}, \theta, \bar{\theta}) &= \underline{0} \\
enr'(a \,.\, P', \theta, \bar{\theta}) &= \text{undefined} \\
enr'(a^{\dagger \xi} \,.\, P', \theta, \bar{\theta}) &= \begin{cases} a^{\dagger \bar{\theta}} \,.\, P' & \text{if } \theta = a \\ a^{\dagger \xi} \,.\, enr'(P', \theta', \bar{\theta}) & \text{if } \theta = {\cdot_a}\theta' \\ \text{undefined} & \text{otherwise} \end{cases} \\
enr'(P_1 + P_2, \theta, \bar{\theta}) &= \begin{cases} enr'(P_1, \theta', \bar{\theta}) + P_2 & \text{if } \theta = \dot{+}\theta' \\ P_1 + enr'(P_2, \theta', \bar{\theta}) & \text{if } \theta = +\!\!\cdot\,\theta' \\ \text{undefined} & \text{otherwise} \end{cases} \\
enr'(P_1 \parallel_L P_2, \theta, \bar{\theta}) &= \begin{cases} enr'(P_1, \theta', \bar{\theta}) \parallel_L P_2 & \text{if } \theta = \rfloor\!\rfloor_L \theta' \\ P_1 \parallel_L enr'(P_2, \theta', \bar{\theta}) & \text{if } \theta = \lfloor\!\lfloor_L \theta' \\ enr'(P_1, \theta_1, \bar{\theta}) \parallel_L enr'(P_2, \theta_2, \bar{\theta}) & \text{if } \theta = \langle \theta_1, \theta_2 \rangle_L \\ \text{undefined} & \text{otherwise} \end{cases}
\end{aligned}
$$

Example 2. The proved labeled transition system underlying the initial process $(a \,.\, \underline{0} \parallel_\emptyset a \,.\, \underline{0}) \parallel_{\{a\}} a \,.\, a \,.\, \underline{0}$, which is the synchronization of autoconcurrency with autocausation, has the following two maximal transition sequences:

- $(a \,.\, \underline{0} \parallel_\emptyset a \,.\, \underline{0}) \parallel_{\{a\}} a \,.\, a \,.\, \underline{0}$
  $\xrightarrow{\langle \rfloor\!\rfloor_\emptyset a, a \rangle_{\{a\}}} (a^{\dagger \langle \rfloor\!\rfloor_\emptyset a, a \rangle_{\{a\}}} \,.\, \underline{0} \parallel_\emptyset a \,.\, \underline{0}) \parallel_{\{a\}} a^{\dagger \langle \rfloor\!\rfloor_\emptyset a, a \rangle_{\{a\}}} \,.\, a \,.\, \underline{0}$
  $\xrightarrow{\langle \lfloor\!\lfloor_\emptyset a, \cdot_a a \rangle_{\{a\}}} (a^{\dagger \langle \rfloor\!\rfloor_\emptyset a, a \rangle_{\{a\}}} \,.\, \underline{0} \parallel_\emptyset a^{\dagger \langle \lfloor\!\lfloor_\emptyset a, \cdot_a a \rangle_{\{a\}}} \,.\, \underline{0}) \parallel_{\{a\}} a^{\dagger \langle \rfloor\!\rfloor_\emptyset a, a \rangle_{\{a\}}} \,.\, a^{\dagger \langle \lfloor\!\lfloor_\emptyset a, \cdot_a a \rangle_{\{a\}}} \,.\, \underline{0}$
- $(a \,.\, \underline{0} \parallel_\emptyset a \,.\, \underline{0}) \parallel_{\{a\}} a \,.\, a \,.\, \underline{0}$
  $\xrightarrow{\langle \lfloor\!\lfloor_\emptyset a, a \rangle_{\{a\}}} (a \,.\, \underline{0} \parallel_\emptyset a^{\dagger \langle \lfloor\!\lfloor_\emptyset a, a \rangle_{\{a\}}} \,.\, \underline{0}) \parallel_{\{a\}} a^{\dagger \langle \lfloor\!\lfloor_\emptyset a, a \rangle_{\{a\}}} \,.\, a \,.\, \underline{0}$
  $\xrightarrow{\langle \rfloor\!\rfloor_\emptyset a, \cdot_a a \rangle_{\{a\}}} (a^{\dagger \langle \rfloor\!\rfloor_\emptyset a, \cdot_a a \rangle_{\{a\}}} \,.\, \underline{0} \parallel_\emptyset a^{\dagger \langle \lfloor\!\lfloor_\emptyset a, a \rangle_{\{a\}}} \,.\, \underline{0}) \parallel_{\{a\}} a^{\dagger \langle \lfloor\!\lfloor_\emptyset a, a \rangle_{\{a\}}} \,.\, a^{\dagger \langle \rfloor\!\rfloor_\emptyset a, \cdot_a a \rangle_{\{a\}}} \,.\, \underline{0}$

Note that the target processes of the two sequences are different thanks to the different additional decorations of the pairs of synchronizing executed actions. Without those decorations, the two sequences would end up in the same process $(a^\dagger \,.\, \underline{0} \parallel_\emptyset a^\dagger \,.\, \underline{0}) \parallel_{\{a\}} a^\dagger \,.\, a^\dagger \,.\, \underline{0}$ – thus yielding a diamond-shaped transition system – which would not reflect the fact that the two executed a-actions in $a^\dagger \,.\, a^\dagger \,.\, \underline{0}$ cannot be undone in any order as the first one causes the second one. ■

Example 3. The proved labeled transition system underlying the initial process $(a \,.\, \underline{0} \parallel_\emptyset a \,.\, \underline{0}) \parallel_{\{a\}} (a \,.\, \underline{0} \parallel_\emptyset a \,.\, \underline{0})$, which is the synchronization of autoconcurrency with itself, has the following four maximal transition sequences:

– $(a\,.\,\underline{0} \parallel_{\emptyset} a\,.\,\underline{0}) \parallel_{\{a\}} (a\,.\,\underline{0} \parallel_{\emptyset} a\,.\,\underline{0})$
$$\xrightarrow{\langle \rrfloor_{\emptyset} a, \rrfloor_{\emptyset} a\rangle_{\{a\}}} (a^{\dagger\langle \rrfloor_{\emptyset} a, \rrfloor_{\emptyset} a\rangle_{\{a\}}}.\,\underline{0} \parallel_{\emptyset} a\,.\,\underline{0}) \parallel_{\{a\}} (a^{\dagger\langle \rrfloor_{\emptyset} a, \rrfloor_{\emptyset} a\rangle_{\{a\}}}.\,\underline{0} \parallel_{\emptyset} a\,.\,\underline{0})$$
$$\xrightarrow{\langle \llfloor_{\emptyset} a, \llfloor_{\emptyset} a\rangle_{\{a\}}} (a^{\dagger\langle \rrfloor_{\emptyset} a, \rrfloor_{\emptyset} a\rangle_{\{a\}}}.\,\underline{0} \parallel_{\emptyset} a^{\dagger\langle \llfloor_{\emptyset} a, \llfloor_{\emptyset} a\rangle_{\{a\}}}.\,\underline{0}) \parallel_{\{a\}} (a^{\dagger\langle \rrfloor_{\emptyset} a, \rrfloor_{\emptyset} a\rangle_{\{a\}}}.\,\underline{0} \parallel_{\emptyset} a^{\dagger\langle \llfloor_{\emptyset} a, \llfloor_{\emptyset} a\rangle_{\{a\}}}.\,\underline{0})$$

– $(a\,.\,\underline{0} \parallel_{\emptyset} a\,.\,\underline{0}) \parallel_{\{a\}} (a\,.\,\underline{0} \parallel_{\emptyset} a\,.\,\underline{0})$
$$\xrightarrow{\langle \llfloor_{\emptyset} a, \llfloor_{\emptyset} a\rangle_{\{a\}}} (a\,.\,\underline{0} \parallel_{\emptyset} a^{\dagger\langle \llfloor_{\emptyset} a, \llfloor_{\emptyset} a\rangle_{\{a\}}}.\,\underline{0}) \parallel_{\{a\}} (a\,.\,\underline{0} \parallel_{\emptyset} a^{\dagger\langle \llfloor_{\emptyset} a, \llfloor_{\emptyset} a\rangle_{\{a\}}}.\,\underline{0})$$
$$\xrightarrow{\langle \rrfloor_{\emptyset} a, \rrfloor_{\emptyset} a\rangle_{\{a\}}} (a^{\dagger\langle \rrfloor_{\emptyset} a, \rrfloor_{\emptyset} a\rangle_{\{a\}}}.\,\underline{0} \parallel_{\emptyset} a^{\dagger\langle \llfloor_{\emptyset} a, \llfloor_{\emptyset} a\rangle_{\{a\}}}.\,\underline{0}) \parallel_{\{a\}} (a^{\dagger\langle \rrfloor_{\emptyset} a, \rrfloor_{\emptyset} a\rangle_{\{a\}}}.\,\underline{0} \parallel_{\emptyset} a^{\dagger\langle \llfloor_{\emptyset} a, \llfloor_{\emptyset} a\rangle_{\{a\}}}.\,\underline{0})$$

– $(a\,.\,\underline{0} \parallel_{\emptyset} a\,.\,\underline{0}) \parallel_{\{a\}} (a\,.\,\underline{0} \parallel_{\emptyset} a\,.\,\underline{0})$
$$\xrightarrow{\langle \rrfloor_{\emptyset} a, \llfloor_{\emptyset} a\rangle_{\{a\}}} (a^{\dagger\langle \rrfloor_{\emptyset} a, \llfloor_{\emptyset} a\rangle_{\{a\}}}.\,\underline{0} \parallel_{\emptyset} a\,.\,\underline{0}) \parallel_{\{a\}} (a\,.\,\underline{0} \parallel_{\emptyset} a^{\dagger\langle \rrfloor_{\emptyset} a, \llfloor_{\emptyset} a\rangle_{\{a\}}}.\,\underline{0})$$
$$\xrightarrow{\langle \llfloor_{\emptyset} a, \rrfloor_{\emptyset} a\rangle_{\{a\}}} (a^{\dagger\langle \rrfloor_{\emptyset} a, \llfloor_{\emptyset} a\rangle_{\{a\}}}.\,\underline{0} \parallel_{\emptyset} a^{\dagger\langle \llfloor_{\emptyset} a, \rrfloor_{\emptyset} a\rangle_{\{a\}}}.\,\underline{0}) \parallel_{\{a\}} (a^{\dagger\langle \llfloor_{\emptyset} a, \rrfloor_{\emptyset} a\rangle_{\{a\}}}.\,\underline{0} \parallel_{\emptyset} a^{\dagger\langle \rrfloor_{\emptyset} a, \llfloor_{\emptyset} a\rangle_{\{a\}}}.\,\underline{0})$$

– $(a\,.\,\underline{0} \parallel_{\emptyset} a\,.\,\underline{0}) \parallel_{\{a\}} (a\,.\,\underline{0} \parallel_{\emptyset} a\,.\,\underline{0})$
$$\xrightarrow{\langle \llfloor_{\emptyset} a, \rrfloor_{\emptyset} a\rangle_{\{a\}}} (a\,.\,\underline{0} \parallel_{\emptyset} a^{\dagger\langle \llfloor_{\emptyset} a, \rrfloor_{\emptyset} a\rangle_{\{a\}}}.\,\underline{0}) \parallel_{\{a\}} (a^{\dagger\langle \llfloor_{\emptyset} a, \rrfloor_{\emptyset} a\rangle_{\{a\}}}.\,\underline{0} \parallel_{\emptyset} a\,.\,\underline{0})$$
$$\xrightarrow{\langle \rrfloor_{\emptyset} a, \llfloor_{\emptyset} a\rangle_{\{a\}}} (a^{\dagger\langle \rrfloor_{\emptyset} a, \llfloor_{\emptyset} a\rangle_{\{a\}}}.\,\underline{0} \parallel_{\emptyset} a^{\dagger\langle \llfloor_{\emptyset} a, \rrfloor_{\emptyset} a\rangle_{\{a\}}}.\,\underline{0}) \parallel_{\{a\}} (a^{\dagger\langle \llfloor_{\emptyset} a, \rrfloor_{\emptyset} a\rangle_{\{a\}}}.\,\underline{0} \parallel_{\emptyset} a^{\dagger\langle \rrfloor_{\emptyset} a, \llfloor_{\emptyset} a\rangle_{\{a\}}}.\,\underline{0})$$

While the target processes of the first (resp. last) two sequences are equal, the target process of the first two sequences is different from the one of the last two sequences due to the different additional decorations of the pairs of synchronizing executed actions. This results in a double-diamond-shaped transition system like the one of $(a\,.\,\underline{0} \parallel_{\emptyset} a\,.\,\underline{0}) + (a\,.\,\underline{0} \parallel_{\emptyset} a\,.\,\underline{0})$. Without those decorations, the four sequences would end up in the same process $(a^{\dagger}.\,\underline{0} \parallel_{\emptyset} a^{\dagger}.\,\underline{0}) \parallel_{\{a\}} (a^{\dagger}.\,\underline{0} \parallel_{\emptyset} a^{\dagger}.\,\underline{0})$, thus yielding a single-diamond-shaped transition system. ■

4.3 Forward-Reverse Bisimilarity and Backward Ready Multisets

We now redefine brm-forward-reverse bisimilarity over $\mathbb{P}$. Unlike stable configuration structures, for processes we can syntactically construct their (finite) backward ready multisets, intended as the multisets of actions occurring in the labels of their incoming transitions. In the following we use $\oplus$ for multiset union, which adds multiplicities of identical elements, and $\otimes$ for multiset intersection, which multiplies the multiplicities of those elements. The *backward ready multiset* of $P \in \mathbb{P}$ is inductively defined as follows where $\overline{L} = \mathcal{A} \setminus L$:

$$\begin{aligned}
brm(\underline{0}) &= \emptyset \\
brm(a\,.\,P') &= \emptyset \\
brm(a^{\dagger\xi}.\,P') &= \begin{cases} \{\!| \, a \, |\!\} & \text{if } init(P') \\ brm(P') & \text{if } \neg init(P') \end{cases} \\
brm(P_1 + P_2) &= \begin{cases} \emptyset & \text{if } init(P_1) \wedge init(P_2) \\ brm(P_1) & \text{if } \neg init(P_1) \wedge init(P_2) \\ brm(P_2) & \text{if } init(P_1) \wedge \neg init(P_2) \end{cases} \\
brm(P_1 \parallel_L P_2) &= (brm(P_1) \otimes \overline{L}) \oplus (brm(P_2) \otimes \overline{L}) \oplus (brm(P_1) \otimes brm(P_2) \otimes L)
\end{aligned}$$

The first two clauses stated below are the same as the ones of forward-reverse bisimilarity $\sim_{\rm FRB}$ over a single transition relation defined in [9,8]. Note the use of function *act* to abstract from operator symbols inside transition labels.

Definition 4. *We say that* $P_1, P_2 \in \mathbb{P}$ *are* brm-forward-reverse bisimilar, *written* $P_1 \sim_{\rm FRB:brm} P_2$, *iff* P_1 *and* P_2 *are related by a brm-forward-reverse bisimulation, i.e., a symmetric relation* $\mathcal{B}$ *over* $\mathbb{P}$ *such that, whenever* $(Q_1, Q_2) \in \mathcal{B}$, *then:*

- *For each* $Q_1 \xrightarrow{\theta_1} Q'_1$ *there exists* $Q_2 \xrightarrow{\theta_2} Q'_2$ *such that* $act(\theta_1) = act(\theta_2)$ *and* $(Q'_1, Q'_2) \in \mathcal{B}$.
- *For each* $Q'_1 \xrightarrow{\theta_1} Q_1$ *there exists* $Q'_2 \xrightarrow{\theta_2} Q_2$ *such that* $act(\theta_1) = act(\theta_2)$ *and* $(Q'_1, Q'_2) \in \mathcal{B}$.
- $brm(Q_1) = brm(Q_2)$. ■

Example 4. $a\,.\,\underline{0} \parallel_\emptyset a\,.\,\underline{0} \not\sim_{\rm FRB:brm} a\,.\,a\,.\,\underline{0}$ as in the forward bisimulation game they respectively reach $a^\dagger.\,\underline{0} \parallel_\emptyset a^\dagger\,.\,\underline{0}$ and $a^\dagger.\,a^\dagger.\,\underline{0}$ after performing two a-transitions, where $brm(a^\dagger.\,\underline{0} \parallel_\emptyset a^\dagger\,.\,\underline{0}) = \{\!| a, a |\!\} \neq \{\!| a |\!\} = brm(a^\dagger.\,a^\dagger.\,\underline{0})$. Likewise, $(a\,.\,\underline{0} \parallel_\emptyset a\,.\,\underline{0}) \parallel_{\{a\}} a\,.\,a\,.\,\underline{0} \not\sim_{\rm FRB:brm} (a\,.\,\underline{0} \parallel_\emptyset a\,.\,\underline{0})$. In contrast, $(a\,.\,\underline{0} \parallel_\emptyset a\,.\,\underline{0}) \parallel_{\{a\}} (a\,.\,\underline{0} \parallel_\emptyset a\,.\,\underline{0}) \sim_{\rm FRB:brm} (a\,.\,\underline{0} \parallel_\emptyset a\,.\,\underline{0}) + (a\,.\,\underline{0} \parallel_\emptyset a\,.\,\underline{0}) \sim_{\rm FRB:brm} a\,.\,\underline{0} \parallel_\emptyset a\,.\,\underline{0}$. ■

An axiomatization of $\sim_{\rm FRB:brm}$ can be derived from the one of $\sim_{\rm FRB}$ in [8] by using backward ready multisets instead of backward ready sets when extending action prefixes at process encoding time. We conclude this section by developing a modal logic characterization for $\sim_{\rm FRB:brm}$ inspired by the one of $\sim_{\rm FRB}$ in [7].

The set $\mathcal{L}_{\rm BRM}$ of formulas of the *backward ready multiset logic* is generated by the following syntax:

$$\phi ::= \mathsf{true} \mid M \mid \neg\phi \mid \phi \wedge \phi \mid \langle a \rangle \phi \mid \langle a^\dagger \rangle \phi$$

where $M : \mathcal{A} \to \mathbb{N}$ and $a \in \mathcal{A}$. The satisfaction relation $\models \subseteq \mathbb{P} \times \mathcal{L}_{\rm BRM}$ is defined by induction on the syntactical structure of $\phi \in \mathcal{L}_{\rm BRM}$ as follows:

$P \models \mathsf{true}$
$P \models M$ iff $brm(P) = M$
$P \models \neg\phi'$ iff $P \not\models \phi'$
$P \models \phi_1 \wedge \phi_2$ iff $P \models \phi_1$ and $P \models \phi_2$
$P \models \langle a \rangle \phi'$ iff there exists $P \xrightarrow{\theta} P'$ such that $act(\theta) = a$ and $P' \models \phi'$
$P \models \langle a^\dagger \rangle \phi'$ iff there exists $P' \xrightarrow{\theta} P$ such that $act(\theta) = a$ and $P' \models \phi'$

Note that every $P \in \mathbb{P}$ is image finite, i.e., it has finitely many outgoing (and incoming) transitions labeled with proof terms containing the same action.

Theorem 3. *Let* $P_1, P_2 \in \mathbb{P}$. *Then* $P_1 \sim_{\rm FRB:brm} P_2$ *iff* $\forall \phi \in \mathcal{L}_{\rm BRM}.\, P_1 \models \phi \Longleftrightarrow P_2 \models \phi$. ■

4.4 Denotational Semantics on Stable Configuration Structures

To enable a comparison between hereditary history-preserving bisimilarity over stable configuration structures and brm-forward-reverse bisimilarity over processes, we proceed with the introduction of a denotational semantics for $\mathbb{P}$ based on stable configuration structures. The first step consists of redefining the process operators of Section 4.1 over stable configuration structures. Taking inspiration from [11], we do this by using proof terms in Θ to formalize events:

- The terminated stable configuration structure N is defined as $(\emptyset, \{\emptyset\}, \mathcal{A}, \emptyset)$.
- Let $a \in \mathcal{A}$ and $\mathsf{C} = (\mathcal{E}, \mathcal{C}, \mathcal{A}, l)$ be a stable configuration structure such that $\mathcal{E} \subseteq \Theta$. The action prefix $a \,.\, \mathsf{C}$ is defined as $(\mathcal{E}', \mathcal{C}', \mathcal{A}, l')$ where:
  - $\mathcal{E}' = \{a\} \cup \{._a\theta \mid \theta \in \mathcal{E}\}$.
  - $\mathcal{C}' = \{\emptyset\} \cup \{X' \in \mathcal{P}_{\text{fin}}(\mathcal{E}') \mid \exists X \in \mathcal{C} .\, X' = \{a\} \cup \{._a\theta \mid \theta \in X\}\}$.
  - $l' = \{(a, a)\} \cup \{(._a\theta, act(._a\theta)) \mid \exists X \in \mathcal{C} .\, \theta \in X\}$.
- Let $\mathsf{C}_i = (\mathcal{E}_i, \mathcal{C}_i, \mathcal{A}, l_i)$ be a stable configuration structure such that $\mathcal{E}_i \subseteq \Theta$ for $i \in \{1, 2\}$. The alternative composition $\mathsf{C}_1 + \mathsf{C}_2$ is defined as $(\mathcal{E}, \mathcal{C}, \mathcal{A}, l)$ where:
  - $\mathcal{E} = \{{}_{\cdot}\!+\theta \mid \theta \in \mathcal{E}_1\} \cup \{+_{\cdot}\theta \mid \theta \in \mathcal{E}_2\}$.
  - $\mathcal{C} = \{X \in \mathcal{P}_{\text{fin}}(\mathcal{E}) \mid \exists X_1 \in \mathcal{C}_1 .\, X = \{{}_{\cdot}\!+\theta \mid \theta \in X_1\}\} \cup$
    $\{X \in \mathcal{P}_{\text{fin}}(\mathcal{E}) \mid \exists X_2 \in \mathcal{C}_2 .\, X = \{+_{\cdot}\theta \mid \theta \in X_2\}\}$.
  - $l = \{({}_{\cdot}\!+\theta, act({}_{\cdot}\!+\theta)) \mid \exists X_1 \in \mathcal{C}_1 .\, \theta \in X_1\} \cup$
    $\{(+_{\cdot}\theta, act(+_{\cdot}\theta)) \mid \exists X_2 \in \mathcal{C}_2 .\, \theta \in X_2\}$.
- Let $\mathsf{C}_i = (\mathcal{E}_i, \mathcal{C}_i, \mathcal{A}, l_i)$ be a stable configuration structure such that $\mathcal{E}_i \subseteq \Theta$ for $i \in \{1, 2\}$ and $L \subseteq \mathcal{A} \setminus \{\tau\}$. The parallel composition $\mathsf{C}_1 \,\|_L\, \mathsf{C}_2$ is defined as $(\mathcal{E}, \mathcal{C}, \mathcal{A}, l)$ where:
  - $\mathcal{E} = \{\rfloor\!\rfloor_L\theta \mid \theta \in \mathcal{E}_1 \wedge act(\theta) \notin L\} \cup \{\lfloor\!\lfloor_L\theta \mid \theta \in \mathcal{E}_2 \wedge act(\theta) \notin L\} \cup$
    $\{\langle\theta_1, \theta_2\rangle_L \mid \theta_1 \in \mathcal{E}_1 \wedge \theta_2 \in \mathcal{E}_2 \wedge act(\theta_1) = act(\theta_2) \in L\}$.
  - $\mathcal{C} = \{X \in \mathcal{P}_{\text{fin}}(\mathcal{E}) \mid proj_1(X) \in \mathcal{C}_1 \wedge proj_2(X) \in \mathcal{C}_2 \wedge \forall e, e' \in X.$
    $((proj_1(\{e\}) = proj_1(\{e'\}) \neq \emptyset \vee proj_2(\{e\}) = proj_2(\{e'\}) \neq \emptyset) \Longrightarrow$
    $e = e') \wedge$ [local injectivity of projections]
    $(e \neq e' \Longrightarrow$ [coincidence freeness (a single event per transition)]
    $\exists Y \subseteq X .\, (proj_1(Y) \in \mathcal{C}_1 \wedge proj_2(Y) \in \mathcal{C}_2 \wedge (e \in Y \Longleftrightarrow e' \notin Y))))\}$

    with projections being defined as follows:
    * $proj_1(X) = \{\theta_1 \in \mathcal{E}_1 \mid \rfloor\!\rfloor_L\theta_1 \in X \vee \exists \theta_2 \in \mathcal{E}_2 .\, \langle\theta_1, \theta_2\rangle_L \in X\}$.
    * $proj_2(X) = \{\theta_2 \in \mathcal{E}_2 \mid \lfloor\!\lfloor_L\theta_2 \in X \vee \exists \theta_1 \in \mathcal{E}_1 .\, \langle\theta_1, \theta_2\rangle_L \in X\}$.
  - $l = \{(\rfloor\!\rfloor_L\theta, act(\rfloor\!\rfloor_L\theta)) \mid \exists X_1 \in \mathcal{C}_1 .\, \theta \in X_1 \wedge act(\theta) \notin L\} \cup$
    $\{(\lfloor\!\lfloor_L\theta, act(\lfloor\!\lfloor_L\theta)) \mid \exists X_2 \in \mathcal{C}_2 .\, \theta \in X_2 \wedge act(\theta) \notin L\} \cup$
    $\{(\langle\theta_1, \theta_2\rangle_L, act(\langle\theta_1, \theta_2\rangle_L)) \mid \exists X_1 \in \mathcal{C}_1 .\, \exists X_2 \in \mathcal{C}_2 .$
    $\theta_1 \in X_1 \wedge \theta_2 \in X_2 \wedge act(\theta_1) = act(\theta_2) \in L\}$.

Then with each process $P \in \mathbb{P}$ we denotationally associate a stable configuration structure semantics in a way similar to [35,3], with the notable difference that we represent events via proof terms. More precisely, each process is given a pair formed by a stable configuration structure, built by using the operators above, and a configuration of that structure. The idea is that all processes reachable from the same initial process share the same configuration structure. In contrast, the designated configuration uniquely identifies the specific process through the proof terms labeling a sequence of proved transitions by means of which the considered process is reached from the initial one.

Note that such a sequence is empty if P is initial – which corresponds to the empty configuration – and unique if P is sequential. In the case that P is neither initial nor sequential, if there are several transition sequences reaching it – meaning that non-synchronizing actions of different parallel subprocesses have been executed – then they result in the same configuration [11], because independent actions can be executed in any order and the order of the elements within a configuration – which is a set – does not matter.

Definition 5. *The* stable configuration structure semantics *of* $P \in \mathbb{P}$ *is the pair* $[\![P]\!] = (C_P, X_P)$ *where:*

- $C_P = scs(to\_init(P))$*, with the stable configuration structure* $scs(Q)$ *associated with an initial process* $Q \in \mathbb{P}$ *being defined by induction on the syntactical structure of* Q *as follows:*
  - $scs(\underline{0}) = \mathsf{N}$.
  - $scs(a \,.\, Q') = a \,.\, scs(Q')$.
  - $scs(Q_1 + Q_2) = scs(Q_1) + scs(Q_2)$.
  - $scs(Q_1 \parallel_L Q_2) = scs(Q_1) \parallel_L scs(Q_2)$.
- $X_P = \emptyset$ *if* P *is initial, otherwise* $X_P = \{\theta_i \mid 1 \le i \le n\}$ *for some* $n \in \mathbb{N}_{\ge 1}$ *such that there exists* $P_{i-1} \xrightarrow{\theta_i} P_i$ *for all* $1 \le i \le n$ *with* $P_0 = to\_init(P)$ *and* $P_n = P$. ■

Example 5. $[\![a \,.\, \underline{0} \parallel_{\emptyset} a \,.\, \underline{0}]\!]$ comprises (see Figure 1(a)):

- The two events $\rfloor\!\rfloor_{\emptyset} a$ and $\lfloor\!\lfloor_{\emptyset} a$.
- The four configurations $\emptyset$, $\{\rfloor\!\rfloor_{\emptyset} a\}$, $\{\lfloor\!\lfloor_{\emptyset} a\}$, and $\{\rfloor\!\rfloor_{\emptyset} a, \lfloor\!\lfloor_{\emptyset} a\}$.
- The two maximal computations $\emptyset \xrightarrow{a}_{C_{a \,.\, \underline{0} \parallel_{\emptyset} a \,.\, \underline{0}}} \{\rfloor\!\rfloor_{\emptyset} a\} \xrightarrow{a}_{C_{a \,.\, \underline{0} \parallel_{\emptyset} a \,.\, \underline{0}}} \{\rfloor\!\rfloor_{\emptyset} a, \lfloor\!\lfloor_{\emptyset} a\}$ and $\emptyset \xrightarrow{a}_{C_{a \,.\, \underline{0} \parallel_{\emptyset} a \,.\, \underline{0}}} \{\lfloor\!\lfloor_{\emptyset} a\} \xrightarrow{a}_{C_{a \,.\, \underline{0} \parallel_{\emptyset} a \,.\, \underline{0}}} \{\rfloor\!\rfloor_{\emptyset} a, \lfloor\!\lfloor_{\emptyset} a\}$.

In contrast, $[\![a \,.\, a \,.\, \underline{0}]\!]$ comprises (see Figure 1(b)):

- The two events a and $._{a}a$.
- The three configurations $\emptyset$, $\{a\}$, and $\{a, ._{a}a\}$.
- The only maximal computation $\emptyset \xrightarrow{a}_{C_{a \,.\, a \,.\, \underline{0}}} \{a\} \xrightarrow{a}_{C_{a \,.\, a \,.\, \underline{0}}} \{a, ._{a}a\}$.

Therefore, $[\![a \,.\, \underline{0} \parallel_{\emptyset} a \,.\, \underline{0}]\!] \not\sim_{\mathrm{HHPB}} [\![a \,.\, a \,.\, \underline{0}]\!]$ because a causally precedes $._{a}a$ while $\rfloor\!\rfloor_{\emptyset} a$ and $\lfloor\!\lfloor_{\emptyset} a$ are independent of each other and hence no (labeling- and) causality-preserving bijection would relate the former two events to the latter two. ■

4.5 Operational Characterization Result

We start by establishing a connection between proved transitions of processes and transitions of stable configuration structures associated with processes.

Lemma 1. *Let* $P, P' \in \mathbb{P}$ *and* $\theta \in \Theta$*. Then* $P \xrightarrow{\theta} P'$ *iff* $X_P \xrightarrow{act(\theta)}_{C_P} X_{P'}$. ■

We are now in a position of proving our operational characterization result.

Theorem 4. *Let* $P_1, P_2 \in \mathbb{P}$*. Then* $[\![P_1]\!] \sim_{\mathrm{HHPB}} [\![P_2]\!]$ *iff* $P_1 \sim_{\mathrm{FRB:brm}} P_2$. ■

5 Relationships between $\mathcal{L}_{\mathrm{EI}}$ and $\mathcal{L}_{\mathrm{BRM}}$

From Theorems 4, 3, and 1 it follows that two processes satisfy the same formulas of $\mathcal{L}_{\mathrm{BRM}}$ iff their associated stable configuration structures satisfy the same formulas of $\mathcal{L}_{\mathrm{EI}}$. It is therefore interesting to investigate the relationships between the two logics. On the one hand, we show how to reinterpret $\mathcal{L}_{\mathrm{EI}}$ over processes (Section 5.1) and $\mathcal{L}_{\mathrm{BRM}}$ over stable configuration structures (Section 5.2). On the other hand, we discuss how to translate $\mathcal{L}_{\mathrm{BRM}}$ into $\mathcal{L}_{\mathrm{EI}}$ (Section 5.3) and vice versa (Section 5.4).

5.1 Reinterpreting $\mathcal{L}_{\mathrm{EI}}$ over $\mathbb{P}$

The only non-trivial case is the one of the binder $(x : a)$. The process analogous of an event in a configuration that is labeled with a certain action is a subprocess starting with an executed occurrence of that action. Indicating with $sp(P)$ the set of all subprocesses of P, let $apt(a^{\dagger} . P', P)$ be the proof term associated with the execution of action a of the subterm $a^{\dagger} . P'$ of P. Formally, $apt(a^{\dagger} . P', P) = \theta$ iff $a^{\dagger} . P' \in sp(P)$ and there exist $P'', P''' \in \mathbb{P}$ such that $a \,.\, to\_init(P') \in sp(P'')$, $P'' \xrightarrow{\theta} P'''$, $act(\theta) = a$, and $a^{\dagger} . \, to\_init(P') \in sp(P''')$.

The satisfaction relation $\models \subseteq (\mathbb{P} \times \Theta^{\mathcal{I}}) \times \mathcal{L}_{\mathrm{EI}}$ is defined by induction on the syntactical structure of $\phi \in \mathcal{L}_{\mathrm{EI}}$ as follows:

$P \models_{\rho} \mathsf{true}$
$P \models_{\rho} \neg\phi'$ iff $P \not\models_{\rho} \phi'$
$P \models_{\rho} \phi_1 \wedge \phi_2$ iff $P \models_{\rho} \phi_1$ and $P \models_{\rho} \phi_2$
$P \models_{\rho} \langle x : a \rangle\rangle \phi'$ iff there is $P \xrightarrow{\theta} P'$ such that $act(\theta) = a$ and $P' \models_{\rho[x \mapsto \theta]} \phi'$
$P \models_{\rho} (x : a)\phi'$ iff there is $a^{\dagger} . P' \in sp(P)$ such that $P \models_{\rho[x \mapsto apt(a^{\dagger} . P', P)]} \phi$
$P \models_{\rho} \langle\langle x \rangle \phi'$ iff there is $P' \xrightarrow{\theta} P$ such that $\rho(x) = \theta$ and $P' \models_{\rho} \phi'$

where it is understood that the environment in the subscript of every occurrence of $\models$ is permissible for the configuration identifying (in the associated denotational semantics) the process on the left – e.g., X_P in the case of process P – and the formula on the right.

Theorem 5. *Let* $P \in \mathbb{P}$. *Then* $\forall \phi \in \mathcal{L}_{\mathrm{EI}} . \forall \rho \in pe(X_P, \phi) . P \models_{\rho} \phi \iff [\![P]\!] \models_{\rho} \phi$. ∎

To prove that, consequently, $\mathcal{L}_{\mathrm{EI}}$ reinterpreted over $\mathbb{P}$ characterizes $\sim_{\mathrm{FRB:brm}}$, we follow [33] and hence first show that any substitution of the variables freely occurring in a formula requires a modification of the permissible environment.

Lemma 2. *Let* $P \in \mathbb{P}$, $\phi \in \mathcal{L}_{\mathrm{EI}}$, *and* $\rho \in pe(X_P, \phi)$. *Given a substitution* σ *that – not necessarily injectively – maps* $fi(\phi)$ *to a set of fresh identifiers that do not occur either free or bound in* ϕ, *let* $\sigma(\phi)$ *be the formula obtained from* ϕ *by replacing each occurrence of* $x \in fi(\phi)$ *with* $\sigma(x)$ *and let* $\rho^{\sigma} \in pe(X_P, \sigma(\phi))$ *be the environment obtained from* ρ *by replacing each* $x \in fi(\phi)$ *with* $\sigma(x)$. *Then* $P \models_{\rho} \phi$ *iff* $P \models_{\rho^{\sigma}} \sigma(\phi)$. ∎

Corollary 1. *Let* $P_1, P_2 \in \mathbb{P}$. *Then* $P_1 \sim_{\mathrm{FRB:brm}} P_2$ *iff* $\exists f_{1,2} . \forall \phi \in \mathcal{L}_{\mathrm{EI}} . \forall \rho \in pe(X_{P_1}, \phi) . P_1 \models_{\rho} \phi \iff P_2 \models_{f_{1,2} \circ \rho} \phi$ *where* $f_{1,2}$ *is a label-preserving bijection from* X_{P_1} *to* X_{P_2}. ∎

5.2 Reinterpreting $\mathcal{L}_{\mathrm{BRM}}$ over Stable Configuration Structures

Let us denote by $[\![\mathbb{P}]\!]$ the set of all stable configuration structures – whose events are proof terms – that turn out to be the denotational semantics of some $P \in \mathbb{P}$. Recalling that $[\![P]\!] = (C_P, X_P)$, when writing $[\![P]\!] \models \phi$ we mean $X_P \models \phi$.

The satisfaction relation $\models \subseteq [\![\mathbb{P}]\!] \times \mathcal{L}_{\mathrm{BRM}}$ is defined by induction on the syntactical structure of $\phi \in \mathcal{L}_{\mathrm{BRM}}$ as follows:

$$\begin{array}{ll}
[\![P]\!] \models \mathsf{true} & \\
[\![P]\!] \models M & \text{iff } \{\!| a \in \mathcal{A} \mid X_{P'} \xrightarrow{a}_{C_{P'}} X_P |\!\} = M \\
[\![P]\!] \models \neg\phi' & \text{iff } [\![P]\!] \not\models \phi' \\
[\![P]\!] \models \phi_1 \wedge \phi_2 & \text{iff } [\![P]\!] \models \phi_1 \text{ and } [\![P]\!] \models \phi_2 \\
[\![P]\!] \models \langle a \rangle \phi' & \text{iff there exists } X_P \xrightarrow{a}_{C_P} X_{P'} \text{ such that } [\![P']\!] \models \phi' \\
[\![P]\!] \models \langle a^\dagger \rangle \phi' & \text{iff there exists } X_{P'} \xrightarrow{a}_{C_{P'}} X_P \text{ such that } [\![P']\!] \models \phi'
\end{array}$$

Every process and its associated stable configuration structure satisfy the same formulas of $\mathcal{L}_{\mathrm{BRM}}$. As a consequence, $\mathcal{L}_{\mathrm{BRM}}$ reinterpreted over stable configuration structures characterizes $\sim_{\mathrm{HHPB}}$.

Theorem 6. *Let $P \in \mathbb{P}$. Then $\forall \phi \in \mathcal{L}_{\mathrm{BRM}}. P \models \phi \iff [\![P]\!] \models \phi$.* ∎

Corollary 2. *Let $P_1, P_2 \in \mathbb{P}$. Then $[\![P_1]\!] \sim_{\mathrm{HHPB}} [\![P_2]\!]$ iff $\forall \phi \in \mathcal{L}_{\mathrm{BRM}}. [\![P_1]\!] \models \phi \iff [\![P_2]\!] \models \phi$.* ∎

5.3 Translating $\mathcal{L}_{\mathrm{BRM}}$ into $\mathcal{L}_{\mathrm{EI}}$

The main difficulty is the encoding of multisets, as they are not present in $\mathcal{L}_{\mathrm{EI}}$. In the translation function we thus introduce two additional parameters:

- The first one is a finite set A of actions, e.g., those occurring in a process P. Since $P \models M$ iff $brm(P) = M$, the $\mathcal{L}_{\mathrm{EI}}$ formula corresponding to M has to express the fact that every action in the support of M, i.e., $supp(M) = \{a \in \mathcal{A} \mid M(a) > 0\}$, can be undone a number of times equal to its multiplicity, while any action in $A \setminus supp(M)$ cannot be undone at all. We assume that $supp(M)$ is finite to avoid infinite conjunctions in the translation.
- The second one is a finite sequence $\varrho_n : \{1, \dots, n\} \to \mathcal{I} \times \mathcal{A}$ of pairs each formed by an identifier and an action. It acts like a stack-based memory that keeps track of executed actions, bound to variables via $\langle x : a \rangle\!\rangle$ and $(x : a)$.

The translation function $\mathcal{T}_{\mathrm{BE}} : \mathcal{L}_{\mathrm{BRM}} \times \mathcal{P}_{\mathrm{fin}}(\mathcal{A}) \times \{\varrho_n \mid n \in \mathbb{N}_{\geq 1}\} \to \mathcal{L}_{\mathrm{EI}}$ is defined by induction on the syntactical structure of $\phi \in \mathcal{L}_{\mathrm{BRM}}$ as follows:

$$\begin{array}{rcl}
\mathcal{T}_{\mathrm{BE}}(\mathsf{true}, A, \varrho_n) & = & \mathsf{true} \\
\mathcal{T}_{\mathrm{BE}}(M, A, \varrho_n) & = & \displaystyle\bigwedge_{a_i \in supp(M)} \left(\bigwedge_{k=1}^{M(a_i)} \langle\!\langle x_{i,k} \rangle \mathsf{true} \wedge \bigwedge_{h=1}^{\sharp(a_i, \varrho_n) - M(a_i)} \neg \langle\!\langle z_{i,h} \rangle \mathsf{true} \right) \\
& & \displaystyle\wedge \bigwedge_{b \in A \setminus supp(M)} \neg (y : b) \langle\!\langle y \rangle \mathsf{true} \qquad \text{with } y \text{ fresh} \\
\mathcal{T}_{\mathrm{BE}}(\neg\phi', A, \varrho_n) & = & \neg \mathcal{T}_{\mathrm{BE}}(\phi', A, \varrho_n) \\
\mathcal{T}_{\mathrm{BE}}(\phi_1 \wedge \phi_2, A, \varrho_n) & = & \mathcal{T}_{\mathrm{BE}}(\phi_1, A, \varrho_n) \wedge \mathcal{T}_{\mathrm{BE}}(\phi_2, A, \varrho_n) \\
\mathcal{T}_{\mathrm{BE}}(\langle a \rangle \phi', A, \varrho_n) & = & \langle x : a \rangle\!\rangle \mathcal{T}_{\mathrm{BE}}(\phi', A, \varrho_n \cup \{(n+1, (x, a))\}) \quad \text{with } x \text{ fresh} \\
\mathcal{T}_{\mathrm{BE}}(\langle a^\dagger \rangle \phi', A, \varrho_n) & = & (x : a) \langle\!\langle x \rangle \mathcal{T}_{\mathrm{BE}}(\phi', A, \varrho_n) \qquad \text{with } x \text{ fresh}
\end{array}$$

where in the translation of M the finite sequence ϱ_n is such that:

- $(x_{i,k}, a_i) \in rge(\varrho_n)$, with $x_{i,k} \neq x_{i,k'}$ for $k \neq k'$ and all the identifiers $x_{i,k}$ being taken starting from the end of ϱ_n, i.e., the top of the stack.
- $\sharp(a_i, \varrho_n)$ is the number of occurrences of a_i in ϱ_n.
- $(z_{i,h}, a_i) \in rge(\varrho_n)$, with $z_{i,h} \notin \{x_{i,k} \mid 1 \leq k \leq M(a_i)\}$ and $z_{i,h} \neq z_{i,h'}$ for $h \neq h'$.

Theorem 7. *Let $P \in \mathbb{P}$, $\phi \in \mathcal{L}_{\text{BRM}}$, and $act(P)$ be the set of actions in P. Then $P \models \phi$ iff $\exists \varrho_n . \exists \rho \in pe(P, \mathcal{T}_{\text{BE}}(\phi, act(P), \varrho_n)). P \models_\rho \mathcal{T}_{\text{BE}}(\phi, act(P), \varrho_n)$.* ∎

5.4 Translating $\mathcal{L}_{\text{EI}}$ into $\mathcal{L}_{\text{BRM}}$

The challenge is the encoding of formulas of the form $(x : a)\phi$, because identifiers are not present in $\mathcal{L}_{\text{BRM}}$ and the satisfaction of these formulas is not necessarily related to actions to be done or undone in this moment. Rather, it is generically related to executed actions. On the other hand, it is not clear how multisets would come into play. The study of this translation is left for future work.

6 Conclusions

In this paper we have proposed an entirely new approach to characterize hereditary history-preserving bisimilarity, both denotationally and operationally, even in the presence of autoconcurrency, autocausation, and autoconflict. Unlike [3], the focus is on counting identically labeled events rather than uniquely identifying them, thus avoiding bijections between events altogether. Moreover, on the operational side, it has turned out that proof terms naturally lend themselves to identification purposes; in a reversible setting like ours, they have been used for the first time in [1]. Finally, we have logically characterized backward-ready-multiset forward-reverse bisimilarity with backward ready multiset logic and investigated the relationships of the latter with event identifier logic [33].

The operational characterization is particularly important for several reasons. Firstly, in addition to the equational characterization over forward-only processes developed in [21], hereditary history-preserving bisimilarity can be axiomatized over reversible processes by resorting to the approach of [18] as applied in [8], provided that backward ready multisets are considered in place of backward ready sets. Secondly, in addition to the logics of [33,4], hereditary history-preserving bisimilarity can be characterized also in terms of backward ready multiset logic. The latter is simpler as it does not make use of variables and binders, but the former contain fragments that have been proven to characterize various behavioral equivalences in the true concurrency spectrum [23,19,32].

As for future work, we would like to complete the investigation of the relationships between backward ready multiset logic and event identifier logic, as well as to extend it to the logic of [4]. Another direction to pursue is whether our results apply to configuration structures that are not stable, i.e., in which it is not necessarily the case that causality among events can be always represented in terms of partial orders, possibly defined locally to each configuration. Hereditary history-preserving bisimilarity has been defined over non-stable models in [22,5] and a logical characterization for it has been provided in [5], which is a conservative extension of the one in [4].

Finally, we plan to study backward-ready-multiset forward-reverse bisimilarity checking by taking into account, as far as hereditary history-preserving bisimilarity is concerned, the undecidability result over finite labeled transition systems extended with an independence relation between transitions of [26] and the polynomial-time algorithm over basic parallel processes of [20].

Acknowledgments. We are grateful to Rob van Glabbeek for bringing to our attention the process examined in Example 2 as well as the anonymous reviewers for their comments and suggestions. This research has been supported by the PRIN 2020 project *NiRvAna – Noninterference and Reversibility Analysis in Private Blockchains*, the PRIN 2022 project *DeKLA – Developing Kleene Logics and Their Applications*, and the INdAM-GNCS 2024 project *MARVEL – Modelli Composizionali per l'Analisi di Sistemi Reversibili Distribuiti*.

References

1. Aubert, C.: Concurrencies in reversible concurrent calculi. In: Proc. of the 14th Int. Conf. on Reversible Computation (RC 2022). LNCS, vol. 13354, pp. 146–163. Springer (2022)
2. Aubert, C., Cristescu, I.: Contextual equivalences in configuration structures and reversibility. Journal of Logical and Algebraic Methods in Programming **86**, 77–106 (2017)
3. Aubert, C., Cristescu, I.: How reversibility can solve traditional questions: The example of hereditary history-preserving bisimulation. In: Proc. of the 31st Int. Conf. on Concurrency Theory (CONCUR 2020). LIPIcs, vol. 171, pp. 7:1–7:23 (2020)
4. Baldan, P., Crafa, S.: A logic for true concurrency. Journal of the ACM **61**, 24:1–24:36 (2014)
5. Baldan, P., Gorla, D., Padoan, T., Salvo, I.: Behavioural logics for configuration structures. Theoretical Computer Science **913**, 94–112 (2022)
6. Bednarczyk, M.A.: Hereditary history preserving bisimulations or what is the power of the future perfect in program logics. Technical Report, Polish Academy of Sciences, Gdansk (1991)
7. Bernardo, M., Esposito, A.: Modal logic characterizations of forward, reverse, and forward-reverse bisimilarities. In: Proc. of the 14th Int. Symp. on Games, Automata, Logics, and Formal Verification (GANDALF 2023). EPTCS, vol. 390, pp. 67–81 (2023)
8. Bernardo, M., Esposito, A., Mezzina, C.A.: Expansion laws for forward-reverse, forward, and reverse bisimilarities via proved encodings. In: Proc. of the Combined 31st Int. Workshop on Expressiveness in Concurrency and 21st Workshop on Structural Operational Semantics (EXPRESS/SOS 2024). EPTCS, vol. 412, pp. 51–70 (2024)

9. Bernardo, M., Rossi, S.: Reverse bisimilarity vs. forward bisimilarity. In: Proc. of the 26th Int. Conf. on Foundations of Software Science and Computation Structures (FOSSACS 2023). LNCS, vol. 13992, pp. 265–284. Springer (2023)
10. Boudol, G., Castellani, I.: A non-interleaving semantics for CCS based on proved transitions. Fundamenta Informaticae **11**, 433–452 (1988)
11. Boudol, G., Castellani, I.: Flow models of distributed computations: Three equivalent semantics for CCS. Information and Computation **114**, 247–314 (1994)
12. Brookes, S.D., Hoare, C.A.R., Roscoe, A.W.: A theory of communicating sequential processes. Journal of the ACM **31**, 560–599 (1984)
13. Danos, V., Krivine, J.: Reversible communicating systems. In: Proc. of the 15th Int. Conf. on Concurrency Theory (CONCUR 2004). LNCS, vol. 3170, pp. 292–307. Springer (2004)
14. Danos, V., Krivine, J.: Transactions in RCCS. In: Proc. of the 16th Int. Conf. on Concurrency Theory (CONCUR 2005). LNCS, vol. 3653, pp. 398–412. Springer (2005)
15. Darondeau, P., Degano, P.: Causal trees. In: Proc. of the 16th Int. Coll. on Automata, Languages and Programming (ICALP 1989). LNCS, vol. 372, pp. 234–248. Springer (1989)
16. Darondeau, P., Degano, P.: Causal trees: Interleaving + causality. In: Proc. of the LITP Spring School on Theoretical Computer Science: Semantics of Systems of Concurrent Processes. LNCS, vol. 469, pp. 239–255. Springer (1990)
17. De Nicola, R., Montanari, U., Vaandrager, F.: Back and forth bisimulations. In: Proc. of the 1st Int. Conf. on Concurrency Theory (CONCUR 1990). LNCS, vol. 458, pp. 152–165. Springer (1990)
18. Degano, P., Priami, C.: Proved trees. In: Proc. of the 19th Int. Coll. on Automata, Languages and Programming (ICALP 1992). LNCS, vol. 623, pp. 629–640. Springer (1992)
19. Fecher, H.: A completed hierarchy of true concurrent equivalences. Information Processing Letters **89**, 261–265 (2004)
20. Fröschle, S., Jančar, P., Lasota, S., Sawa, Z.: Non-interleaving bisimulation equivalences on basic parallel processes. Information and Computation **208**, 42–62 (2010)
21. Fröschle, S., Lasota, S.: Decomposition and complexity of hereditary history preserving bisimulation on BPP. In: Proc. of the 16th Int. Conf. on Concurrency Theory (CONCUR 2005). LNCS, vol. 3653, pp. 263–277. Springer (2005)
22. van Glabbeek, R.J.: On the expressiveness of higher dimensional automata. Theoretical Computer Science **368**, 168–194 (2006)
23. van Glabbeek, R.J., Goltz, U.: Refinement of actions and equivalence notions for concurrent systems. Acta Informatica **37**, 229–327 (2001)
24. van Glabbeek, R.J., Plotkin, G.D.: Configuration structures, event structures and Petri nets. Theoretical Computer Science **410**, 4111–4159 (2009)
25. Joyal, A., Nielsen, M., Winskel, G.: Bisimulation from open maps. Information and Computation **127**, 164–185 (1996)
26. Jurdzinski, M., Nielsen, M., Srba, J.: Undecidability of domino games and hhp-bisimilarity. Information and Computation **184**, 343–368 (2003)
27. Krivine, J.: A verification technique for reversible process algebra. In: Proc. of the 4th Int. Workshop on Reversible Computation (RC 2012). LNCS, vol. 7581, pp. 204–217. Springer (2012)
28. Lanese, I., Medić, D., Mezzina, C.A.: Static versus dynamic reversibility in CCS. Acta Informatica **58**, 1–34 (2021)
29. Milner, R.: Communication and Concurrency. Prentice Hall (1989)

30. Phillips, I., Ulidowski, I.: Reversing algebraic process calculi. Journal of Logic and Algebraic Programming **73**, 70–96 (2007)
31. Phillips, I., Ulidowski, I.: Reversibility and models for concurrency. In: Proc. of the 4th Int. Workshop on Structural Operational Semantics (SOS 2007). ENTCS, vol. 192(1), pp. 93–108. Elsevier (2007)
32. Phillips, I., Ulidowski, I.: A hierarchy of reverse bisimulations on stable configuration structures. Mathematical Structures in Computer Science **22**, 333–372 (2012)
33. Phillips, I., Ulidowski, I.: Event identifier logic. Mathematical Structures in Computer Science **24(2)**, 1–51 (2014)
34. Rabinovich, A.M., Trakhtenbrot, B.A.: Behavior structures and nets. Fundamenta Informaticae **11**, 357–404 (1988)
35. Winskel, G.: Event structures. In: Advances in Petri Nets. LNCS, vol. 255, pp. 325–392. Springer (1986)

Complementation of Emerson-Lei Automata

Vojtěch Havlena, Ondřej Lengál(✉), and Barbora Šmahlíková

Faculty of Information Technology, Brno University of Technology, Brno, Czech Republic
lengal@fit.vut.cz

Abstract. We give new constructions for complementing subclasses of Emerson-Lei automata using modifications of rank-based Büchi automata complementation. In particular, we propose a specialized rank-based construction for a Boolean combination of Inf acceptance conditions, which heavily relies on a novel way of a run DAG labelling enhancing the ranking functions with models of the acceptance condition. Moreover, we propose a technique for complementing generalized Rabin automata, which are structurally as concise as general Emerson-Lei automata (but can have a larger acceptance condition). The construction is modular in the sense that it extends a given complementation algorithm for a condition φ in a way that the resulting procedure handles conditions of the form Fin $\wedge\ \varphi$. The proposed constructions give upper bounds that are exponentially better than the state of the art for some of the classes.

1 Introduction

Complementation of ω-automata is an important operation in formal verification with various applications, for example in model checking wrt expressive temporal logics such as QPTL [25] or HyperLTL [10]; testing language inclusion of ω-automata, or in decision procedures of various logics [6,21]. For Büchi automata (BAs)—i.e., ω-automata with the simplest acceptance condition—complementation has been, from the theoretical point of view, thoroughly explored, starting with constructions having the $2^{2^{O(n)}}$ state complexity [6] coming down to constructions asymptotically matching the lower bound $\Omega((0.76n)^n)$ (modulo a polynomial factor) [38,1]. Over the years, ω-automata with more complex acceptance conditions (such as generalized Büchi (GBAs), (generalized) Rabin/Streett, parity) have found uses in practice. The most general acceptance condition used is the so-called *Emerson-Lei* condition [11], which is an arbitrary Boolean formula consisting of Fin and Inf atoms. Fin(ⓒ) denotes that all transitions labeled with ⓒ must occur only finitely often in an accepting run and Inf(ⓒ) denotes that there must be a transition labeled with ⓒ occurring infinitely often. There are two main reasons for using more complex acceptance conditions: (i) more compact representation of automata and (ii) the ability to determinize (deterministic BAs are strictly less expressive than BAs).

From the theoretical point of view, precise bounds on complementation of automata with more complex acceptance condition is much less researched, demonstrated by the best upper bound for (transition-based) Emerson-Lei automata (TELAs) being $2^{2^{O(n)}}$ [37] states. Here, the O in the exponent can hide a linear (or constant) factor, which would have a doubly-exponential effect, giving little information about the actual complexity. In this paper, we present complementation algorithms for several subclasses of TELAs and thoroughly study their complexity, giving better upper bounds than the currently-best known algorithms.

P. A. Abdulla and D. Kesner (Eds.): FoSSaCS 2025, LNCS 15691, pp. 88–110, 2025.
https://doi.org/10.1007/978-3-031-90897-2_5

Our contributions can be summarized as follows:

1. We propose a rank-based complementation algorithm for Inf-TELAs, i.e., TELAs where the acceptance condition does not contain any Fin atom, with the state complexity $O(n(0.76nk)^n)$ where n is the number of states and k is the number of *minimal models* of the acceptance condition.
2. By instantiating the previously mentioned algorithm, we obtain a complementation algorithm for generalized Büchi automata with k colours constructing a BA with the state complexity $O(n(0.76nk)^n)$, which is, to the best of our knowledge, better than the best previously known algorithms.
3. We propose a modular procedure for complementing TELAs with the acceptance condition $\mathsf{Fin}(\textcircled{c}) \wedge \varphi$ given a compatible complementation procedure for formula φ.
4. Next, we instantiate the modular procedure to handle Rabin pairs ($\mathsf{Fin}(\textcircled{0}) \wedge \mathsf{Inf}(\textcircled{1})$) and, in turn, obtain an algorithm for complementing Rabin automata with k Rabin pairs with the complexity $O(n^k(0.76n)^{nk})$, which is, again, better than any other algorithm that we know of.
5. Finally, we instantiate the procedure also for generalized Rabin pairs ($\mathsf{Fin}(\textcircled{0}) \wedge \mathsf{Inf}(\textcircled{1}) \wedge \ldots \wedge \mathsf{Inf}(\textcircled{\ell})$) and obtain complementation constructions for generalized Rabin automata and TELAs with the upper bound $O(n^{2^k}(0.76nk)^{n2^k})$, which is the best upper bound for complementation of general TELAs that we are aware of.

An extended version of the paper with missing proofs can be found at [20].

2 Preliminaries

We fix a finite non-empty alphabet Σ and the first infinite ordinal ω. For $k \in \omega$, we use $\llbracket k \rrbracket$ to represent the largest even number less than or equal to k, e.g., $\llbracket 43 \rrbracket = \llbracket 42 \rrbracket = 42$. An (infinite) word w is a function $w\colon \omega \to \Sigma$ where the i-th symbol is denoted as w_i. Sometimes, we represent w as an infinite sequence $w = w_0 w_1 \ldots$ We denote the set of all infinite words over Σ as Σ^ω; an *ω-language* is a subset of Σ^ω. We use $\cdot$ for ellipsis, e.g., if interested only in the second component of a triple, we may write the triple as $(\cdot, x, \cdot)$.

2.1 Emerson-Lei Acceptance Conditions

Given a set $\Gamma = \{0, \ldots, k-1\}$ of k *colours* (often depicted as $\textcircled{0}$, $\textcircled{1}$, etc.), we define the set of *Emerson-Lei acceptance conditions* $\mathbb{EL}(\Gamma)$ as the set of formulae constructed according to the following grammar:

$$\alpha ::= tt \mid ff \mid \mathsf{Inf}(c) \mid \mathsf{Fin}(c) \mid (\alpha \wedge \alpha) \mid (\alpha \vee \alpha)$$

for $c \in \Gamma$. The *satisfaction* relation $\models$ for a set of colours $M \subseteq \Gamma$ and a condition α is defined inductively as follows (for $c \in \Gamma$):

$$\begin{array}{lll} M \models tt, & M \models \mathsf{Fin}(c) \text{ iff } c \notin M, & M \models \alpha_1 \vee \alpha_2 \text{ iff } M \models \alpha_1 \text{ or } M \models \alpha_2, \\ M \not\models ff, & M \models \mathsf{Inf}(c) \text{ iff } c \in M, & M \models \alpha_1 \wedge \alpha_2 \text{ iff } M \models \alpha_1 \text{ and } M \models \alpha_2. \end{array}$$

If $M \models \alpha$, we say that M is a *model* of α We denote by $|\alpha|$ the number of atomic conditions contained in α, where multiple occurrences of the same atomic condition are counted multiple times.

2.2 Emerson-Lei Automata

A (nondeterministic) transition-based[1] *Emerson-Lei automaton* (TELA) over Σ is a tuple $\mathcal{A} = (Q, \delta, I, \Gamma, \mathsf{p}, \mathsf{Acc})$, where Q is a finite set of *states* (we often use n to denote $|Q|$), $\delta \subseteq Q\times\Sigma\times Q$ is a set of *transitions*[2], $I \subseteq Q$ is the set of *initial* states, Γ is the set of *colours*, $\mathsf{p}\colon \delta \to 2^{\Gamma}$ is a *colouring* of transitions, and $\mathsf{Acc} \in \mathbb{EL}(\Gamma)$. We use $p \xrightarrow{a} q$ to denote that $(p, a, q) \in \delta$ and sometimes treat δ as a function $\delta\colon Q \times \Sigma \to 2^{Q}$. Moreover, we extend δ to sets of states $P \subseteq Q$ as $\delta(P, a) = \bigcup_{p\in P} \delta(p, a)$. See Fig. 1 for an example TELA $\mathcal{A}_{ex}$ over $\Sigma = \{a, b, c\}$ with 3 colours $\Gamma = \{⓪, ①, ②\}$ and the acceptance condition Inf(⓪) $\wedge$ Inf(①). We define $|\mathcal{A}| = |Q|$.

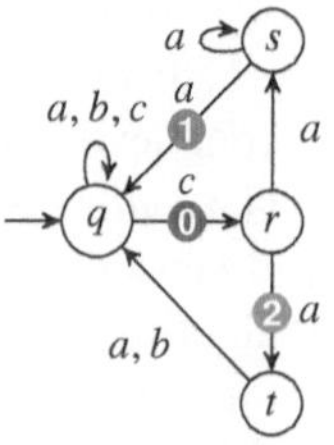

Inf(⓪) ∧ Inf(①)

Fig. 1: $\mathcal{A}_{ex}$

A *run* of $\mathcal{A}$ from $q \in Q$ on an input word w is an infinite sequence $\rho\colon \omega \to Q$ that starts in q and respects δ, i.e., $\rho(0) = q$ and $\forall i \geq 0\colon \rho(i) \xrightarrow{w_i} \rho(i+1) \in \delta$. Let $\mathit{inf}(\rho) \subseteq \delta$ denote the set of transitions occurring in ρ infinitely often and $\mathit{inf}_{\Gamma}(\rho) = \bigcup\{\mathsf{p}(x) \mid x \in \mathit{inf}(\rho)\}$ be the set of infinitely often occurring colours. A run ρ is *accepting* wrt an acceptance condition α, written as $\rho \models \alpha$, iff $\mathit{inf}_{\Gamma}(\rho) \models \alpha$ and ρ is accepting in $\mathcal{A}$ iff $\rho \models \mathsf{Acc}$. The *language* of $\mathcal{A}$, denoted as $\mathcal{L}(\mathcal{A})$, is defined as the set of words $w \in \Sigma^{\omega}$ for which there exists an accepting run in $\mathcal{A}$ starting with some state in I. Classical acceptance conditions can be in this more general framework described as follows (we only provide those used later in the paper and include their abbreviations):

- *Büchi* (BA): Acc = Inf(⓪),
- *co-Büchi* (CBA): Acc = Fin(⓪),
- *Generalized Büchi* (GBA): Acc = $\bigwedge_{0\leq j<k}$ Inf(ⓙ),
- *Generalized co-Büchi* (GCBA): Acc = $\bigvee_{0\leq j<k}$ Fin(ⓙ),
- *Rabin*: $\bigvee_{0\leq j<k} \mathrm{Fin}(B_j) \wedge \mathrm{Inf}(G_j)$,
- *Generalized Rabin*: $\bigvee_{0\leq j<k}(\mathrm{Fin}(B_j) \wedge \bigwedge_{0\leq \ell<m_j} \mathrm{Inf}(G_{j,\ell}))$, and
- *Parity*[3]: Fin(⓪) ∧ (Inf(①) ∨ (Fin(②) ∧ (Inf(③) ∨ (Fin(④) ∧ ...)))),

where $B_j, G_j, G_{j,\ell} \in \Gamma$ for all j, ℓ. Further, we use Inf-TELA to denote a TELA where the acceptance condition contains no Fin atoms. We also use the syntactic sugar $\mathcal{A} = (Q, \delta, I, F)$ to denote a (transition-based) BA that would be defined using the TELA definition above as $(Q, \delta, I, \{⓪\}, \{t \mapsto \emptyset \mid t \in \delta \setminus F\} \cup \{t \mapsto \{⓪\} \mid t \in F\}, \mathsf{Inf}(⓪))$.

2.3 Run DAGs

In this section, we recall the terminology from [19] (which is a minor modification of the terminology from [26] and [38]) used heavily in the paper. Let $\mathcal{A} = (Q, \delta, I, \Gamma, \mathsf{p}, \mathsf{Acc})$ be a TELA. We fix the definition of the *run DAG* of $\mathcal{A}$ over a word w to be a DAG (directed acyclic graph) $\mathcal{G}_w = (V, E)$ of vertices V and edges E where

---

[1] We only consider transition-based acceptance in order to avoid cluttering the paper by dealing with accepting states *and* accepting transitions. Extending our approach to state/transition-based (or just state-based) automata is straightforward.

[2] Note that there is also a more general definition of TELAs with $\delta \subseteq Q \times \Sigma \times 2^{\Gamma} \times Q$; in this paper, we use the simpler one.

[3] We consider the so-called *parity min odd* condition; any parity condition from the set $\{\min, \max\} \times \{\mathrm{even}, \mathrm{odd}\}$ can be easily translated to it.

– $V \subseteq Q \times \omega$ s.t. $(q, i) \in V$ iff there is a run ρ of $\mathcal{A}$ from I over w with $\rho_i = q$,
– $E \subseteq V \times V$ s.t. $((q, i), (q', i')) \in E$ iff $i' = i + 1$ and $q' \in \delta(q, w_i)$.

See Fig. 2 for an example of a run DAG of $\mathcal{A}_{ex}$ from Fig. 1 over the word $caa(cab)^\omega \notin \mathcal{L}(\mathcal{A}_{ex})$ (we will return to the additional labels in the figure later). Given a DAG $\mathcal{G} = (V, E)$, we often identify $\mathcal{G}$ with V, for instance, we will write $(p, i) \in \mathcal{G}$ to denote that $(p, i) \in V$. For a vertex $v \in \mathcal{G}$, we denote the set of vertices of $\mathcal{G}$ reachable from v (including v itself) as $reach_{\mathcal{G}}(v)$ or just $reach(v)$ if $\mathcal{G}$ is clear from the context. A vertex $v \in \mathcal{G}$ is *finite* iff $reach(v)$ is finite and *infinite* if it is not finite. In Fig. 2, the vertices $(s, 2), (s, 3), (s, 5), \ldots$ are finite and all other vertices are infinite. Moreover, for a colour ⓒ $\in \Gamma$, an edge $((q, i), (q', i + 1)) \in E$ is a ⓒ-edge if ⓒ $\in \mathsf{p}(q \xrightarrow{w_i} q')$ and a vertex $v \in V$ is ⓒ-*endangered* iff it cannot reach any ⓒ-edge. For a set of colours $C \subseteq \Gamma$, v is C-*endangered* iff it is ⓒ-endangered for every ⓒ $\in C$. For example, in Fig. 2, the vertices $(q, 1)$ and $(t, 2)$ are {①}-endangered but they are not {⓪, ①}-endangered. A pair of vertices $v_1, v_2 \in V$ is *converging* iff $reach(v_1) \cap reach(v_2) \neq \emptyset$ (v_1 and v_2 *converge*). A function $r : V \to \omega$ is a *run DAG ranking* if for every $v \in V$ it holds that $\forall u \in reach(v) : r(u) \leq r(v)$. We use $\max(r)$ to denote the *rank* of r, i.e., the maximum value from $\{r(u) \mid u \in V\}$ if it exists and ∞ otherwise. A ranking r of $\mathcal{G}$ is called *tight* iff there exists a level i such that (i) $m = \max\{r((q, i)) \mid q \in Q\}$ is odd and (ii) for all levels $j \geq i$ it holds that $\{1, 3, \ldots, m\} \subseteq \{r((q, j)) \mid q \in Q\}$.

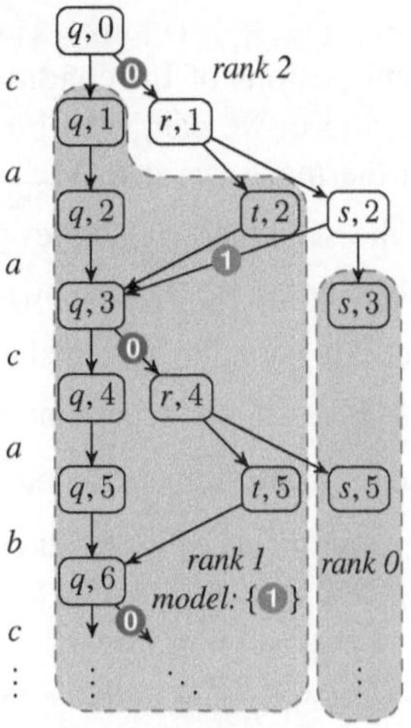

Fig. 2: A labelled run DAG of $\mathcal{A}_{ex}$ over the word $caa(cab)^\omega \notin \mathcal{L}(\mathcal{A}_{ex})$

3 Complementation of Inf-TELAs

In this section, we describe a complement construction for Inf-TELAs. Our approach is an extension of rank-based BA complementation algorithms [26,14,38,24,9,16,19], which construct a BA whose runs simulate a run DAG ranking procedure. We start with giving the run DAG ranking procedure (which extends the ranking procedure from [26] with the introduction of models) and then proceed to the complementation algorithm itself. One can see our algorithm also as an improvement of the algorithm for complementing GBAs in [28] by (i) introducing model assignments, (ii) getting better complexity through the use of tight rankings, and (iii) generalizing the construction from GBAs to arbitrary Inf-TELAs.

3.1 Inf-TELA Run DAG Labelling

Let $\mathcal{A} = (Q, \delta, I, \Gamma, \mathsf{p}, \mathsf{Acc})$ be an Inf-TELA. We use $\overline{\mathsf{Acc}}$ to denote the propositional formula obtained from Acc by replacing conjunctions by disjunctions and vice versa, and substituting atoms of the form Inf(ⓙ) by ⓙ (this can be viewed as negating Acc, transforming it into the negation normal form, substituting ¬Inf(ⓙ) by

Fin(ⓙ), and denoting each Fin(ⓙ) just by ⓙ). Let $\mathcal{M}_{\overline{\mathsf{Acc}}}$ be the set of models of $\overline{\mathsf{Acc}}$ where the colours ⓙ are interpreted as propositional variables. For example, if $\mathsf{Acc} = \mathrm{Inf}(⓿) \wedge (\mathrm{Inf}(❶) \vee \mathrm{Inf}(❷))$, then $\overline{\mathsf{Acc}} = ⓿ \vee (❶ \wedge ❷)$ and $\mathcal{M}_{\overline{\mathsf{Acc}}} = \{\{⓿\}, \{❶, ❷\}, \{⓿, ❶\}, \{⓿, ❷\}, \{⓿, ❶, ❷\}\}$ ($\mathcal{M}_{\overline{\mathsf{Acc}}}$ can be interpreted as saying which combinations of Inf-conditions need to be broken in order to break Acc; in the example above, we can, e.g., break Inf(⓿), we can break both Inf(❶) and Inf(❷), etc.). Furthermore, we use $\mathcal{M}^{\min}_{\overline{\mathsf{Acc}}}$ to denote the set of *minimal models* of $\overline{\mathsf{Acc}}$, i.e., $\mathcal{M}^{\min}_{\overline{\mathsf{Acc}}}$ is the set where (i) for every model $m \in \mathcal{M}_{\overline{\mathsf{Acc}}}$, there exists a model $m' \in \mathcal{M}^{\min}_{\overline{\mathsf{Acc}}}$ such that $m' \subseteq m$, and (ii) there are no $m, m' \in \mathcal{M}^{\min}_{\overline{\mathsf{Acc}}}$ such that $m \subset m'$. We note that $\mathcal{M}_{\overline{\mathsf{Acc}}}$ can be obtained as the upward closure of $\mathcal{M}^{\min}_{\overline{\mathsf{Acc}}}$ (and $\mathcal{M}^{\min}_{\overline{\mathsf{Acc}}}$ is an *antichain*). For the example acceptance condition above, $\mathcal{M}^{\min}_{\overline{\mathsf{Acc}}} = \{\{⓿\}, \{❶, ❷\}\}$. Moreover, we use **lex-min**($\overline{\mathsf{Acc}}$) to denote the lexicographically smallest model from $\mathcal{M}^{\min}_{\overline{\mathsf{Acc}}}$ (w.l.o.g., we assume $\mathcal{M}_{\overline{\mathsf{Acc}}} \neq \emptyset$). **lex-min**($\overline{\mathsf{Acc}}$) is used to pinpoint one model when any model can be used.

Let $\mathcal{G} = (V, E)$ be a run DAG of $\mathcal{A}$ over w. For a set of vertices $U \subseteq V$, a mapping $\eta\colon U \to \mathcal{M}^{\min}_{\overline{\mathsf{Acc}}}$ is called *endangered in* $\mathcal{G}$ if

1. U is finite and nonempty,
2. each $v \in U$ is $\eta(v)$-endangered in $\mathcal{G}$, and
3. for each pair of vertices $v_1, v_2 \in U$ converging in $\mathcal{G}$, we have $\eta(v_1) = \eta(v_2)$.

A function m with the signature $m\colon V \to \mathcal{M}^{\min}_{\overline{\mathsf{Acc}}}$ is called a *model assignment*. For instance, for $\mathcal{A}_{ex}$ in Fig. 1, we have $\mathcal{M}^{\min}_{\overline{\mathsf{Acc}}} = \{\{⓿\}, \{❶\}\}$ since $\mathcal{A}_{ex}$ is a GBA. In addition, for the run DAG in Fig. 2 and a set $\{(q, 1), (t, 2)\}$, the mapping $\{(q, 1) \mapsto \{❶\}, (t, 2) \mapsto \{❶\}\}$ is endangered in $\mathcal{G}$. On the other hand, there exists no endangered mapping for the set $\{(s, 2)\}$ in $\mathcal{G}$, as $(s, 2)$ can reach both a ⓿-edge as well as a ❶-edge.

In Algorithm 1, we give a (nondeterministic) ranking procedure that assigns ranks and minimal models of $\overline{\mathsf{Acc}}$ to each vertex of $\mathcal{G}$. Intuitively, the algorithm starts by giving all initially finite vertices the rank 0 and assigning their model to **lex-min**($\overline{\mathsf{Acc}}$) (Line 4). Then, it proceeds in iterations, each starting with the DAG $\mathcal{G}^i$ and consisting of two steps:

1. First, the algorithm tries to find a model assignment $\eta\colon U \to \mathcal{M}^{\min}_{\overline{\mathsf{Acc}}}$ for a finite nonempty set of vertices U of $\mathcal{G}^i$ s.t. for all $u \in U$, if $\eta(u) = \{\textcircled{c_1}, \ldots, \textcircled{c_\ell}\}$, then every path in $\mathcal{G}^i$ starting in u satisfies the condition $\bigwedge_{1 \le j \le \ell} \mathrm{Fin}(\textcircled{c_j})$ (the path breaks all the $\mathrm{Inf}(\textcircled{c_j})$ conditions, i.e., η is endangered). If such a model assignment exists (the choice is nondeterministic), the algorithm assigns rank $i+1$ to all vertices reachable from U and removes them from the DAG, creating DAG $\mathcal{G}^{i+1}$ (Lines 7–9).
2. Second, the algorithm assigns rank $i + 2$ to all vertices in $\mathcal{G}^{i+1}$ that became finite (after the previous step) and removes them from the DAG, creating DAG $\mathcal{G}^{i+2}$ (Lines 10–12). The counter i is incremented by two and the next iteration continues.

The iterations end when $\mathcal{G}^i$ is empty or when no suitable model assignment η is found (which happens when w is accepted by $\mathcal{A}$). Note that due to the nondeterminism within

Algorithm 1: Inf-TELA run DAG labelling

```
   Input: A run DAG G_w of A over w, acceptance condition Acc
   Output: A run DAG ranking r and a model assignment m if w ∉ L(A), else ⊥
1  i ← 0, r ← ∅, m ← ∅;          // i ∈ ω, r: V ⇀ {0,...,2|Q|}, m: V ⇀ M^min_Acc
2  G^0 = (V^0, E^0) ← G_w without finite vertices;
3  foreach v ∈ G_w s.t. v is finite do
4  |  r(v) ← 0, m(v) ← lex-min(¬Acc);
5  while G^i ≠ ∅ do
6  |  if ∃(η: U → M^min_Acc) s.t. U ⊆ V^i and η is endangered in G^i then
7  |  |  foreach v ∈ U and u ∈ reach_{G^i}(v) do
8  |  |  |  r(u) ← i + 1, m(u) ← η(v);
9  |  |  G^{i+1} ← G^i without vertices with the rank i + 1;
10 |  |  foreach v ∈ G^{i+1} s.t. v is finite in G^{i+1} do
11 |  |  |  r(v) ← i + 2, m(v) ← lex-min(¬Acc);
12 |  |  G^{i+2} ← G^{i+1} without vertices with the rank i + 2;
13 |  |  i ← i + 2;
14 |  else
15 |  |  return ⊥;
16 return (r, m);
```

the algorithm, it may be possible to obtain, in two different runs of the algorithm on the same run DAG, two different pairs (r_1, m_1) and (r_2, m_2) with $\max(r_1) \neq \max(r_2)$.

Example 1. See Fig. 2 for a possible labelling of the run DAG of $\mathcal{A}_{ex}$ over the word $caa(cab)^\omega$. The ranking procedure proceeds in the following steps:

1. $(i = 0)$ First, all finite vertices, which are in this example vertices of the form $(s, 3)$, $(s, 5), \ldots, (s, 3j + 2)$ for all $1 \leq j$, are assigned rank 0 and model **lex-min**$(\overline{\mathsf{Acc}})$, and $\mathcal{G}^0$ is set to be $\mathcal{G}^w$ without those vertices. (Lines 2–4)
2. Second, we set η_1 to the mapping $\eta_1 = \{(q, 1) \mapsto \{\textcircled{1}\}, (t, 2) \mapsto \{\textcircled{1}\}\}$. The mapping η_1 is endangered in $\mathcal{G}^0$ because the following conditions hold:
   (a) η_1 is finite and nonempty,
   (b) neither $(q, 1)$ nor $(t, 2)$ can reach a $\textcircled{1}$ transition, and
   (c) $(q, 1)$ and $(t, 2)$ converge (in $(q, 3)$) and they are both assigned the same model $(\eta_1((q, 1)) = \eta_1((t, 2)) = \{\textcircled{1}\})$.

   In particular, η_1 is the endangered mapping that gives the largest number of vertices of $\mathcal{G}^0$ rank 1. (Line 6)
3. Third, we assign every vertex in $\mathcal{G}^0$ reachable from $(q, 1)$ or $(t, 2)$ the rank 1 and model $\{\textcircled{1}\}$. (Line 7)
4. Fourth, we obtain $\mathcal{G}^1$ from $\mathcal{G}^0$ by removing vertices with rank 1. (Line 9)
5. $\mathcal{G}^1$ contains three vertices $(\{(q, 0), (r, 1), (s, 2)\})$, which all get rank 2 (Line 10) and are removed in $\mathcal{G}^2$ (Line 12). The ranking procedure finishes. □

Lemma 2. *If Algorithm 1 returns* $\perp$*, then* $w \in \mathcal{L}(\mathcal{A})$.

Proof. Let Acc' be a formula in the disjunctive normal form (DNF) equivalent to Acc, i.e., $\mathsf{Acc}' = \bigvee_{j=1}^{\ell} \varphi_j$ where $\varphi_j = \mathrm{Inf}(c_1^j) \wedge \cdots \wedge \mathrm{Inf}(c_{k_j}^j)$ for some ℓ and $k_1, \ldots, k_\ell$. Note that $\mathcal{M}_{\mathsf{Acc}}^{\min} = \mathcal{M}_{\mathsf{Acc}'}^{\min}$ contains sets of colours $M \subseteq \Gamma$, each of them with at least one colour from φ_1, at least one colour from φ_2, etc. In order for Algorithm 1 to return $\bot$, it needs to hold that there is some $i \geq 0$ such that $\mathcal{G}^i$ is nonempty and there does not exist any mapping $\eta \colon U \to \mathcal{M}_{\mathsf{Acc}}^{\min}$, with $U \subseteq V^i$, that would be endangered in $\mathcal{G}^i$. In particular, such a (nonempty) mapping η does not exist iff no vertex $v \in \mathcal{G}^i$ satisfies point (2) of the definition of an endangered mapping (i.e., when we can find an accepting path from all vertices remaining in $\mathcal{G}^i$). Therefore, it follows that no vertex $v \in \mathcal{G}^i$ is M-endangered for any $M \in \mathcal{M}_{\mathsf{Acc}}^{\min}$, i.e., in other words,

for every vertex $v \in \mathcal{G}^i$ there is some clause φ_j such that v can in $\mathcal{G}^i$ reach a c_p^j-edge for each $1 \leq p \leq k_j$. (Reach)

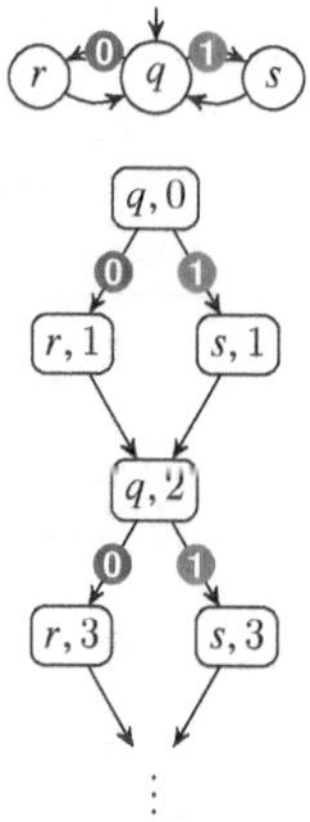

We will now construct an accepting path π in $\mathcal{G}_w$. Note that not all paths in $\mathcal{G}^i$ are necessarily accepting (consider the TELA over a unary alphabet and the run DAG in the right, with the acceptance condition $\mathrm{Inf}(⓿) \wedge \mathrm{Inf}(❶)$; there are many non-accepting paths from $(q, 0)$—e.g., a path that alternates between a q-vertex and an r-vertex and never touches any s-vertex). While constructing π, for every clause φ_j we will be tracking the information about which atom of φ_j we should see next in order to satisfy φ_j on the path. In particular, we will start from a vertex v_0 that is a root vertex of $\mathcal{G}^i$ and we will use the tuple $t_0 = (c_1^1, \ldots, c_1^\ell)$ to keep track of the colours. Using (Reach), it follows that there is a clause φ_j s.t. v_0 can reach a c_1^j-edge e_1. We will set $t_1 = (c_1^1, \ldots, c_2^j, \ldots, c_1^\ell)$ and continue in a similar way: from every vertex we encounter, we use (Reach) to obtain an edge that is a c-edge for some c in t_i. In the case we need to increment some component of t_i from $c_{k_j}^j$, we set the new value to c_1^j. The path π is then constructed as an infinite path that goes through the infinite sequence $v_0, e_1, e_2, \ldots$ Note that because the sequence $v_0, e_1, e_2 \ldots$ is infinite, due to the pigeonhole principle there will be a clause φ_j s.t. the sequence $t_0, t_1, \ldots$ infinitely often increments the j-th component and so π is accepting. From π, we can now extract the accepting run of $\mathcal{A}$ on w. □

Lemma 3. *Algorithm 1 always terminates with $i \leq 2n$.*

Proof. Consider a run DAG $\mathcal{G}_w$ for a word w. First observe that at the end of the main loop of Algorithm 1 (Line 13), $\mathcal{G}^i$ has no finite vertices (all of them were removed). Due to Line 2, $\mathcal{G}^i$ at the beginning of the main loop (Line 6) also has no finite vertices. Let $\mathcal{G}_m^i$ be the DAG $(V_m^i, E^i \cap (V_m^i \times V_m^i))$ where $V_m^i = \{(q, j) \in V^i \mid j \geq m\}$, i.e., the projection of $\mathcal{G}^i$ from level m down, and $\mathit{width}(\mathcal{G}_m^i)$ be the maximum number of vertices on any level of the run DAG below level m, formally, $\mathit{width}(\mathcal{G}_m^i) = \max\{|\{(q, j) : (q, j) \in V_m^i\}| : j \geq m\}$. From the definition of endangered mapping and the loop on Line 7, we have that if the condition on Line 6 holds, there is some $m \in \omega$ s.t. $\mathit{width}(\mathcal{G}_m^{i+1}) < \mathit{width}(\mathcal{G}_m^i)$. This holds because if the mapping η is non-empty, then there is at least one infinite path in $\mathcal{G}^i$ that is completely removed in the next step, i.e., from some level m, the width of all levels below get decreased by at least one. If the condition on Line 6 does not hold, the algorithm terminates and we are done. From the previous claim we have that in each successful iteration of the main loop, the width of $\mathcal{G}^{i+2}$ in the limit is at most the one of $\mathcal{G}^i$ minus one. Since the maximum width of $\mathcal{G}_w$ is n, then, if $w \notin \mathcal{L}(\mathcal{A})$, at latest $\mathcal{G}_m^{2n-1}$ is empty for some $m \in \omega$, and hence $\mathcal{G}^{2n}$ is empty and the algorithm terminates. □

Lemma 4. *If* $w \in \mathcal{L}(\mathcal{A})$, *then Algorithm 1 terminates with* $\perp$.

Proof. Consider some $w \in \mathcal{L}(\mathcal{A})$. Then, there is an accepting run ρ on w in $\mathcal{A}$. We have $(\rho_j, j) \in \mathcal{G}_w$ for all $j \in \omega$; we show that (ρ_j, j) is not M-endangered for every $M \in \mathcal{M}^{\min}_{\overline{\mathsf{Acc}}}$. The fact that no ρ_j is finite follows from the fact that ρ is infinite. Observe that for each $M \in \mathcal{M}^{\min}_{\overline{\mathsf{Acc}}}$, there is some $\textcircled{c} \in M$ s.t. $\textcircled{c} \in \mathrm{Inf}(\rho)$ (otherwise, w would not be accepted by $\mathcal{A}$). Therefore, (ρ_j, j) is not M-endangered. Hence, in every iteration of Algorithm 1, all vertices (ρ_j, j) stay in $\mathcal{G}^i$. From Lemma 3 we have that Algorithm 1 always terminates, but $\mathcal{G}^i \neq \emptyset$ for each i. Therefore, the algorithm terminates with $\perp$. □

Corollary 5. $w \notin \mathcal{L}(\mathcal{A})$ *iff Algorithm 1 on* $\mathcal{G}_w$ *terminates with* (r, m).

The following lemma about the ranking procedure will be useful later.

Lemma 6. *If Algorithm 1 terminates with* (r, m), *then* $\max(r) \leq 2n$ *and, moreover, either* $\max(r) = 0$ *or* r *is tight.*

Proof. The first part ($\max(r) \leq 2n$) follows directly from Lemma 3. For the second part, there are two options: either $\mathcal{G}_w$ is finite (i.e., there is no infinite run of $\mathcal{A}$ on w), in which case Algorithm 1 assigns all vertices in $\mathcal{G}_w$ rank 0 and does not even enter the loop at Line 5. In the other case ($\mathcal{G}$ is infinite), let $k = \max(r)$ if $\max(r)$ is odd and $k = \max(r) - 1$ otherwise (from the previous case, we know that $k \geq 1$). We know that for every $\ell \in \{1, 3, \ldots, k\}$, there is a vertex $v_\ell = (q_\ell, i_\ell) \in \mathcal{G}_w$ with $r(v_\ell) = \ell$ (this is because the mapping at Line 6 in the algorithm needs to be non-empty) and that such a vertex is the beginning of an infinite path of vertices with rank ℓ. Therefore, there needs to be a level i containing vertices with all ranks $\{1, 3, \ldots, k\}$. From the previous, all levels $j > i$ will also have all of the odd ranks up to k. Choosing i large enough will prevent level i having a vertex with an even rank higher than k. Therefore, r is tight. □

3.2 Inf-TELA Complement Construction

Let $\mathcal{A} = (Q, \delta, I, \Gamma, \mathsf{p}, \mathsf{Acc})$ be an Inf-TELA and $n = |Q|$. We define a *(level) ranking* to be a function $f : Q \to \{0, \ldots, 2n\}$. The *rank* of f is defined as $f = \max\{f(q) \mid q \in Q\}$. We call a mapping $\mu : Q \to \mathcal{M}^{\min}_{\overline{\mathsf{Acc}}}$ a *level model*. We say that μ is *consistent* wrt f if $\mu(q) = \textbf{lex-min}(\overline{\mathsf{Acc}})$ whenever $f(q)$ is even. We denote the set of all level models by LM. For a set of states $S \subseteq Q$ and a level model μ, f is (S, μ)*-tight* if

(i) it has an odd rank r,
(ii) $f(S) \supseteq \{1, 3, \ldots, r\}$,
(iii) $f(Q \setminus S) = \{0\}$, and
(iv) μ is consistent wrt f.

A ranking is *μ-tight* if it is (Q,μ)-tight; we use $\mathcal{T}$ to denote the set of all μ-tight rankings for some level model μ.

For two level rankings f, f' and two level models μ, μ', we say that (f',μ') is a *consistent successor* of (f,μ) over $a \in \Sigma$, denoted as $(f,\mu) \rightsquigarrow^{a}_{\delta} (f',\mu')$, iff

(i) μ and μ' are consistent wrt f and f', respectively, and
(ii) for all $q \in Q$ and $q' \in \delta(q,a)$ the following holds:
 (a) $f'(q') \leq f(q)$,
 (b) $(\mathsf{p}(q \xrightarrow{a} q') \cap \mu(q) \neq \emptyset) \Rightarrow f'(q') \leq \llbracket f(q) \rrbracket$, and
 (c) $\mu'(q') \neq \mu(q) \Rightarrow f'(q') \leq \llbracket f(q) \rrbracket$.

Intuitively, the rankings guess the ranks of states in the run DAG and the level models guess the models assigned to states in the labelling procedure from Section 3.1. Consistent successors respect the labelling procedure. On every path in a run DAG, the ranks are nonincreasing. If some odd-ranked vertex v has an outgoing ⓒ-edge to v' and ⓒ is in the model assigned to v, the vertex v' has to have a lower rank than v, because when v is removed from $\mathcal{G}_w^i$, there is no reachable ⓒ-edge in $\mathcal{G}_w^i$. Moreover, if the model is changed between v and v' and the rank is odd, then the rank also has to be decreased.

The complement of $\mathcal{A}$ is given as the BA CInfTela($\mathcal{A}$) $= (Q', \delta', I', F')$ whose components are defined as follows:

- $Q' = Q_1 \cup Q_2$ where
  - $Q_1 = 2^Q$ and
  - $Q_2 = \{(S, O, f, i, \mu) \in 2^Q \times 2^Q \times \mathcal{T} \times \{0, 2, \ldots, 2n-2\} \times \mathsf{LM} \mid f \text{ is } (S,\mu)\text{-tight}, O \subseteq S \cap f^{-1}(i)\}$,
- $I' = \{I\}$,
- $\delta' = \delta_1 \cup \delta_2 \cup \delta_3$ where
  - $\delta_1 \colon Q_1 \times \Sigma \to 2^{Q_1}$ such that $\delta_1(S,a) = \{\delta(S,a)\}$,
  - $\delta_2 \colon Q_1 \times \Sigma \to 2^{Q_2}$ s.t. $\delta_2(S,a) = \{(S', \emptyset, f, 0, \mu) \mid S' = \delta(S,a)\}$, and
  - $\delta_3 \colon Q_2 \times \Sigma \to 2^{Q_2}$ such that $(S', O', f', i', \mu') \in \delta_3((S, O, f, i, \mu), a)$ iff
    * $S' = \delta(S,a)$,
    * $(f,\mu) \rightsquigarrow^{a}_{\delta} (f',\mu')$,
    * $\mathit{rank}(f) = \mathit{rank}(f')$,
    * and
      · $i' = (i+2) \mod (\mathit{rank}(f')+1)$ and $O' = f'^{-1}(i')$ if $O = \emptyset$ or
      · $i' = i$ and $O' = \delta(O,a) \cap f'^{-1}(i)$ if $O \neq \emptyset$, and
- $F' = \{\emptyset \xrightarrow{a} \emptyset \in \delta_1 \mid a \in \Sigma\} \cup \{M_1 \xrightarrow{a} M_2 \in \delta_3 \mid M_1 = (\cdot, \emptyset, \cdot, \cdot, \cdot), a \in \Sigma\}$

Intuitively, a run of CInfTela($\mathcal{A}$) on a word w tries to construct the run DAG $\mathcal{G}_w$ of $\mathcal{A}$ on the same word, with rankings encoded within the states. The restrictions on δ_3 encode the rules from Algorithm 1. The partitioning of Q' into Q_1 and Q_2 allows us to consider only tight rankings, as in [14]. Moreover, the i-component of a macrostate allows us to further decrease the number of states in the same way as in [38] (we know that all states in O have the same rank i).

Theorem 7. *Let $\mathcal{A}$ be an Inf-TELA. Then, $\mathcal{L}(CInfTela(\mathcal{A})) = \Sigma^\omega \setminus \mathcal{L}(\mathcal{A})$.*

Proof. ($\subseteq$) We use Boolean laws and prove an equivalent statement $\mathcal{L}(\mathcal{A}) \subseteq \Sigma^\omega \setminus \mathcal{L}(\text{CInfTELA}(\mathcal{A}))$. Let $w \in \mathcal{L}(\mathcal{A})$ be a word and ρ be an accepting run of $\mathcal{A}$ on w. First, let ρ' be the run $\rho' = S_0 S_1 \dots$ with $S_0 = I$ and $S_{i+1} = \delta_1(S_i, w(i))$ for all $i \in \omega$ (i.e., ρ' stays in Q_1). The run ρ' cannot be accepting in CInfTELA($\mathcal{A}$), because $\rho(i) \in S_i$ and so $S_i \neq \emptyset$ for any $i \in \omega$ (in Q_1, the only accepting transitions are those from state $\emptyset$ to state $\emptyset$). Second, let

$$\rho'' = S_0 S_1 \dots S_p (S_{p+1}, O_{p+1}, f_{p+1}, i_{p+1}, \mu_{p+1})(S_{p+2}, O_{p+2}, f_{p+2}, i_{p+2}, \mu_{p+2}) \dots$$

be a run of CInfTELA($\mathcal{A}$) on w (ρ'' jumps to Q_2 at position p). From the construction, it holds that $(f_j, \mu_j) \twoheadrightarrow^a_\delta (f_{j+1}, \mu_{j+1})$ for all $j > p$. Since ρ is accepting in $\mathcal{A}$, eventually there will be a position $k > p$ such that $f_k(\rho(k)), f_{k+1}(\rho(k+1)), f_{k+2}(\rho(k+2)), \dots$ are all even (because there is no model satisfying ρ in $\mathcal{M}^{\min}_{\mathsf{Acc}}$, so points (iib) and (iic) from the definition of $\twoheadrightarrow^a_\delta$ will enforce this). For the sake of contradiction, assume that ρ'' is accepting. Then for some position $\ell > k$, because the i-component of a macrostate rotates over all even ranks, it holds that $i_\ell = f_\ell(\rho(\ell))$ and $\rho(\ell) \in O_\ell = f_\ell^{-1}(\rho(\ell))$. We can easily show by induction that for all $j \geq \ell$, it holds that $\rho(j) \in O_j \neq \emptyset$, which is in contradiction with the assumption that ρ'' is accepting.

($\supseteq$) Consider any word $w \notin \mathcal{L}(\mathcal{A})$. From Corollary 5 and Lemma 6 it follows that the run DAG $\mathcal{G}_w$ has a bounded rank. If all vertices of $\mathcal{G}_w$ are finite, then there is an accepting run ρ' on CInfTELA($\mathcal{A}$) where $\rho' = S_0 S_1 \dots$ with $S_0 = I$ and $S_{i+1} = \delta(S_i, w_i)$ for all $i \in \omega$. Otherwise, Algorithm 1 terminates with a tight ranking r and a model m. From the definition of $\twoheadrightarrow^a_\delta$, there is a run

$$\rho'' = S_0 S_1 \dots S_p (S_{p+1}, O_{p+1}, f_{p+1}, i_{p+1}, \mu_{p+1})(S_{p+2}, O_{p+2}, f_{p+2}, i_{p+2}, \mu_{p+2}) \dots$$

such that $f_k(q) = r((q, k))$ and $\mu_k(q) = m((q, k))$ for all $k > p$. In order to show that ρ'' is acepting, we need to show that the O-component of the macrostates on the run is empty infinitely often. Assume by contradiction that there is an index $\ell > p$ such that O_j is non-empty for all $j \geq \ell$. Then, there is a vertex $(q, \ell) \in \mathcal{G}_w$ s.t. $r((q, \ell))$ is even and there are infinitely many vertices reachable from (q, ℓ) with the same even rank, which is a contradiction with the construction of r in Algorithm 1, which would give some of the vertices odd ranks. □

For the complexity analysis, we use $tight(n)$ to denote the number of μ-tight level rankings for an automaton with n states (μ-tight rankings for Inf-TELAs correspond to *tight* rankings for BAs). It holds that $tight(n) \approx (0.76n)^n$ [14,38].

Theorem 8. *The number of states of CInfTELA($\mathcal{A}$) is in* $O(k^n \cdot tight(n+1)) = O(n(0.76nk)^n)$ *for* $k = |\mathcal{M}^{\min}_{\mathsf{Acc}}|$.

Proof. The set of macrostates Q_1 is obtained by a simple subset construction, therefore $|Q_1| \in O(2^n)$. That is much smaller than $O(k^n \cdot tight(n+1))$, so it is sufficient to count only the number of macrostates of the form (S, O, f, i, μ). For this, we uniquely encode each macrostate as a pair (h, i) where $h\colon Q \to \{-2, -1, \dots, 2n-1\} \times \mathcal{M}^{\min}_{\mathsf{Acc}}$ is defined as follows:

$$h(q) = \begin{cases} (-1, \mu) & \text{if } q \in O, \\ (-2, \mu) & \text{if } q \in Q \setminus S, \\ (f(q), \mu) & \text{otherwise.} \end{cases} \tag{1}$$

We compute the number of encodings h for a fixed i. We divide all encodings into four groups according to the set $\mathrm{img}(h)_0 \cap \{-2,-1\}$ where $\mathrm{img}(h)_0$ denotes the set of first elements of the pairs in $\mathrm{img}(h)$. We show that we can obtain the bound $O(k^n \cdot \mathit{tight}(n))$ for each of the groups. The groups are denoted by g_M with $M \subseteq \{-2,-1\}$. For $h(q) = (m,\mu)$, we use $h(q)_m$ and $h(q)_\mu$ to denote m and μ.

$g_\emptyset$: from the definition, f is μ-tight. The level model μ is of the form $\mu\colon Q \to \mathcal{M}^{\min}_{\overline{\mathsf{Acc}}}$, so there are k possible assignments for every state from Q. The number of level models is therefore k^n and $|g_\emptyset| = O(k^n \cdot \mathit{tight}(n))$.

$g_{\{-1\}}$: since there is at least one state q with $h(q)_m = -1$, this means that $q \in O$ so q has an even rank. As a consequence, at least one of the positive odd ranks of h (up to $2n-1$) will not be taken, so we can infer that $h\colon Q \to \{-1,\ldots,2n-3\} \times \mathcal{M}^{\min}_{\overline{\mathsf{Acc}}}$. We can therefore uniquely represent h by a mapping h' by incrementing all ranks of h by two, so $h'\colon Q \to \{0,\ldots,2n-1\} \times \mathcal{M}^{\min}_{\overline{\mathsf{Acc}}}$. But then $h' \in \mathcal{T}(n)$ and the number of all level models is k^n, so $|g_{\{-1\}}| \in O(k^n \cdot \mathit{tight}(n))$.

$g_{\{-2,-1\}}$: similarly as for $g_{\{-1\}}$ we get that $|g_{\{-2,-1\}}| \in O(k^n \cdot \mathit{tight}(n))$.

$g_{\{-2\}}$: the reasoning is similar to the one for $g_{\{-1\}}$, with the exception that now, we know that there is a state $q \in Q \setminus S$, which is, according to the definition of a ranking, assigned the rank 0. This means that one positive odd rank of h is, again, not taken, so we increment all non-negative ranks of h by two and map all states in $Q \setminus S$ to 1, obtaining a tight ranking $h' \in \mathcal{T}(n)$. The number of level models is k^n, therefore, $|g_{\{-2\}}| \in O(k^n \cdot \mathit{tight}(n))$.

Since the size of all groups is bounded by $O(k^n \cdot \mathit{tight}(n))$, for a fixed i, the total number of these encodings is still $O(k^n \cdot \mathit{tight}(n))$. When we sum the encodings for all i's, we obtain that the number is bounded by $O(k^n \cdot \mathit{tight}(n+1))$, since $O(n \cdot \mathit{tight}(n)) = O(\mathit{tight}(n+1))$ [38]. The rest follows from the approximation of $\mathit{tight}(n)$. □

Corollary 9. *Let $\mathcal{A}$ be an Inf-TELA with n states and k colours Γ. The number of states of* CInfTela$(\mathcal{A})$ *is in* $O(\binom{k}{\lfloor k/2 \rfloor}^n \cdot \mathit{tight}(n+1)) = O(n \cdot (\binom{k}{\lfloor k/2 \rfloor} \cdot 0.76n)^n) \subseteq O(n(2^k \cdot 0.76n)^n)$.

Proof. The proof of the more precise bound follows directly from Theorem 8 and the fact that the size of $\mathcal{M}^{\min}_{\overline{\mathsf{Acc}}}$ is bounded by the size of the largest antichain in 2^Γ, which is at most $\binom{k}{\lfloor k/2 \rfloor}$ by Sperner's theorem. □

Corollary 10. *Let $\mathcal{A}$ be a GBA with n states and k colours. Then the number of states of* CInfTela$(\mathcal{A})$ *is in* $O(k^n \cdot \mathit{tight}(n+1)) = O(n(0.76nk)^n)$.

Proof. The proof follows directly from Theorem 8. For a GBA it holds that $\overline{\mathsf{Acc}} = \bigvee_{0 \le j < k} \textcircled{j}$. The formula is in DNF, hence $\mathcal{M}^{\min}_{\overline{\mathsf{Acc}}} = \{\{\textcircled{j}\} \mid 0 \le j < k\}$ and $|\mathcal{M}^{\min}_{\overline{\mathsf{Acc}}}| = k$. The number of all level models is k^n. □

We note that to the best of our knowledge, our bound on the complementation of GBAs is better than other bounds in the literature. In particular, it is clearly better than the bound $O(k^n(2n+1)^n)$ from [28], which is the best upper bound for complementing GBAs that we are aware of. It is also better than an approach that would go through determinization by using the procedure in [39], which outputs a deterministic Rabin automaton with the number states bounded by $(1.47nk)^n$ for large k and $2^n - 1$ accepting pairs, which can be complemented easily into a Streett automaton.

4 Modular Complementation of $\mathrm{Fin}(\textcircled{c}) \land \varphi$ TELAs

In this section, we propose a modular algorithm $\mathsf{FinCompl}$ for complementation of TELAs with the acceptance condition $\mathrm{Fin}(\textcircled{c}) \land \varphi$ for any φ, parameterized by an algorithm for complementing TELAs with the condition φ. In Section 5, we will then instantiate the algorithm for some common acceptance conditions, eventually obtaining an efficient complementation algorithm for general TELAs.

Let us fix a TELA $\mathcal{A} = (Q, \delta, I, \Gamma, \mathrm{p}, \mathrm{Fin}(\textcircled{c}) \land \varphi)$ and let Δ be δ without transitions whose label contains $\textcircled{c}$. For a word $w \in \Sigma^\omega$, we define a *relaxed run DAG* (RRDAG) over w, denoted by $\mathcal{G}_w^\Delta$, as any sequence of states $\mathcal{G}_w^\Delta = (S_0, S_1, \dots)$ where $S_i \subseteq Q$ and $\Delta(S_i, w_i) \subseteq S_{i+1}$. Intuitively, an RRDAG over a word may contain more states on each level than it is necessary from the reachability of Δ. Note that this definition of RRDAGs is equivalent to having vertices of the form (q, i), where $q \in S_i$ with edges given implicitly by Δ. We use these definitions interchangeably. Clearly, there may be multiple RRDAGs over a single word, they are all, however, subgraphs of the (standard) run DAG $\mathcal{G}_w$. We say that $\mathcal{G}_w^\Delta = (S_0, S_1, \dots)$ is *accepting* wrt φ, written as $\mathcal{G}_w^\Delta \models \varphi$, if there is a run $\rho = q_k q_{k+1} \dots$ for $k \geq 0$ in Δ such that for every $i \geq k$ it holds that $q_i \in S_i$ and $q_{i+1} \in \Delta(q_i, w_i)$, and, moreover, $\rho \models \varphi$ (i.e., the accepting run does not need to start at the beginning of $\mathcal{G}_w^\Delta$). The reason for introducing RRDAGs is that the algorithm for condition φ will construct a BA that runs over RRDAGs constructed using the restricted transition relation Δ. The relaxation allows us to introduce new vertices (not connected to the root of the RRDAG) at any level that represent runs that have seen finitely many times a $\textcircled{c}$ transition in δ.

Our definition of the modular procedure $\mathsf{FinCompl}$ for $\mathrm{Fin}(\textcircled{c}) \land \varphi$ is given wrt a *subprocedure* for complementing a TELA with condition φ. The subprocedure is given as a tuple $\mathbb{S}_\Delta^\varphi = (\mathcal{M}, \mathcal{M}_0, \mathsf{SuccAct}_\Delta, \mathsf{SuccTrack}_\Delta, \mathsf{EmptyBreak})$, where

(i) $\mathcal{M}$ is a set of *macrostates*,
(ii) $\mathcal{M}_0 \subseteq \mathcal{M}$ is a set of *initial macrostates*,
(iii) $\mathsf{SuccAct}_\Delta \colon 2^Q \times \Sigma \times \mathcal{M} \to 2^{\mathcal{M}}$ is an *active transition function*,
(iv) $\mathsf{SuccTrack}_\Delta \colon 2^Q \times \Sigma \times \mathcal{M} \to 2^{\mathcal{M}}$ is a *tracking transition function*, and
(v) $\mathsf{EmptyBreak} \subseteq \mathcal{M}$ is an *empty-breakpoint* predicate.

We use Succ_Δ to denote $\mathsf{SuccAct}_\Delta \cup \mathsf{SuccTrack}_\Delta$ (when treated as relations). Intuitively, $\mathcal{M}$ is a set of macrostates given by the subprocedure for φ. $\mathsf{EmptyBreak}$ is a condition that has to hold for a macrostate to be accepting in $\mathbb{S}_\Delta^\varphi$. The transitions between macrostates of $\mathcal{M}$ are described using transition functions $\mathsf{SuccAct}_\Delta$ and $\mathsf{SuccTrack}_\Delta$. In particular, $\mathsf{M}' \in \mathsf{Succ}_\Delta(P', a, \mathsf{M})$ is computed by taking the successor of the macrostate M over a, but also while taking into account the set P' of states (M corresponds to index i of the run while M' and P' correspond to index $i+1$) provided by $\mathsf{FinCompl}$, which represent breaking the $\mathrm{Fin}(\textcircled{c})$ condition. The reason for using two transition functions ($\mathsf{SuccAct}_\Delta$ and $\mathsf{SuccTrack}_\Delta$) is that some subprocedures that we will introduce later will use two types of macrostates: active and tracking. For instance, if $\mathbb{S}_\Delta^\varphi$ is a rank-based procedure (cf. Section 5.2), active macrostates will contain breakpoints, which the construction will try to empty, and once a breakpoint is seen, FinCompl will add some more runs to the rank-based algorithm. The new runs might not be tight at the given point, so we switch into the tracking mode and wait for newly added runs to become tight before switching into the active mode again.

Let w be a word and $\mathcal{G}_w^\Delta = (S_0, S_1, \dots)$ be an RRDAG over w. A *Fin-run* R of $\mathbb{S}_\Delta^\varphi$ over $\mathcal{G}_w^\Delta$ is a sequence $(\mathsf{M}_0, \mathsf{M}_1, \dots)$ s.t. $\mathsf{M}_0 \in \mathcal{M}_0$ and $\mathsf{M}_{i+1} \in \mathsf{Succ}_\Delta(S_{i+1}, w_i, \mathsf{M}_i)$ for all $i \geq 0$. R is *accepting* if $\mathsf{EmptyBreak}(\mathsf{M}_i)$ holds for infinitely many i's. We say that the subprocedure $\mathbb{S}_\Delta^\varphi$ is *correct for* φ if for each word w and every RRDAG $\mathcal{G}_w^\Delta$ over w it holds that $\mathcal{G}_w^\Delta$ is not accepting wrt φ iff there is an accepting Fin-run R of $\mathbb{S}_\Delta^\varphi$ over $\mathcal{G}_w^\Delta$.

Let us now move to the definition of FinCompl. For subprocedure $\mathbb{S}_\Delta^\varphi$ and TELA $\mathcal{A}$ given above, the algorithm will construct the BA $\mathsf{FinCompl}(\mathbb{S}_\Delta^\varphi, \mathcal{A}) = (Q', I', \delta', F')$ defined as follows:

- $Q' = \{(S, P, \mathsf{M}) \in 2^Q \times 2^Q \times \mathcal{M}\}$,
- $I' = \{(I, I, \mathsf{M}_0) \mid \mathsf{M}_0 \in \mathcal{M}_0\}$,
- $\delta' = \delta_1 \cup \delta_2$ where
  - $\delta_1 \colon Q' \times \Sigma \to 2^{Q'}$ such that $(S', P', \mathsf{M}') \in \delta_1((S, P, \mathsf{M}), a)$ iff
    * $S' = \delta(S, a)$,
    * if $\mathsf{EmptyBreak}(\mathsf{M})$: $P' = S'$,
    * if $\neg\mathsf{EmptyBreak}(\mathsf{M})$: $P' = \Delta(P, a)$,
    * $\mathsf{M}' \in \mathsf{SuccAct}_\Delta(P', a, \mathsf{M})$,
  - $\delta_2 \colon Q' \times \Sigma \to 2^{Q'}$ such that $(S', P', \mathsf{M}') \in \delta_2((S, P, \mathsf{M}), a)$ iff
    * $S' = \delta(S, a)$,
    * $P' = \Delta(P, a)$,
    * $\mathsf{M}' \in \mathsf{SuccTrack}_\Delta(P', a, \mathsf{M})$, and
- $F' = \{(S, P, \mathsf{M}) \xrightarrow{a} (S', P', \mathsf{M}') \in \delta' \mid a \in \Sigma, \mathsf{EmptyBreak}(\mathsf{M}')\}$.

Intuitively, the construction executes $\mathbb{S}_\Delta^\varphi$ on the restricted transition relation Δ, while also keeping track of all runs (in S) and runs that either need to terminate or see a ⓪-transition (in P). Whenever $\mathbb{S}_\Delta^\varphi$ clears its breakpoint, P is re-sampled (and some new runs can be added to $\mathbb{S}_\Delta^\varphi$).

Theorem 11. *For a correct subprocedure* $\mathbb{S}_\Delta^\varphi$, $\mathcal{L}(\mathsf{FinCompl}(\mathbb{S}_\Delta^\varphi, \mathcal{A})) = \Sigma^\omega \setminus \mathcal{L}(\mathcal{A})$.

The overhead of the procedure over the subprocedure $\mathbb{S}_\Delta^\varphi$ is at most 3^n-times.

Theorem 12. *Suppose* $\mathbb{S}_\Delta^\varphi = (\mathcal{M}, \cdot, \cdot, \cdot)$. *Then* $|\mathsf{FinCompl}(\mathbb{S}_\Delta^\varphi, \mathcal{A})| \in \mathcal{O}(3^n \cdot |\mathcal{M}|)$.

Proof. Since in (S, P, M), it always holds that $P \subseteq S$, each state of $\mathcal{A}$ can be in one of the three following sets: (i) $Q \setminus S$, (ii) $S \cap P$, and (iii) $S \setminus P$. □

5 Complementation of TELAs and their Subclasses

We proceed by instantiating the modular algorithm FinCompl from the previous section for several common automata classes—co-Büchi automata, Rabin automata, parity automata, generalized Rabin automata, and, eventually, TELAs.

5.1 Co-Büchi Automata

As a simple demonstration of instantiation of $\mathsf{FinCompl}$, we use it to create a complementation algorithm for co-Büchi automata. The acceptance condition for co-Büchi automata is Fin(⓿) = Fin(⓿) ∧ tt, we therefore need to provide a trivial subprocedure $\mathbb{S}^{tt} = (\mathcal{M}^{tt}, \mathcal{M}_0^{tt}, \mathsf{SuccAct}_\Delta^{tt}, \emptyset, \mathsf{EmptyBreak}^{tt})$ that is correct for tt (notice that $\mathsf{SuccTrack}_\Delta^{tt}$ is empty). In the subprocedure, $\mathcal{M}^{tt} = 2^Q$, $\mathcal{M}_0^{tt} = \{I\}$, and the remaining components are given as follows:

$$\mathsf{SuccAct}_\Delta^{tt}(P, a, S) = \{P\} \qquad \text{and} \qquad \mathsf{EmptyBreak}^{tt}(P) \iff P = \emptyset.$$

Intuitively, the instantiated procedure works with macrostates (S, P, P) (i.e., to adhere to the formal definition of $\mathsf{FinCompl}$, P is there twice) where S tracks all runs and P is a breakpoint that contains runs that yet need to either terminate or see ⓿. To accept, P needs to be emptied infinitely often. One can observe that $\mathsf{FinCompl}(\mathbb{S}^{tt}, \mathcal{A})$ resembles the well-known Miyano-Hayashi construction [34] for complementation of co-Büchi automata.

Lemma 13. *The subprocedure $\mathbb{S}^{tt}$ is correct for the acceptance condition tt.*

Corollary 14. *For a co-Büchi automaton $\mathcal{A}$, $\mathcal{L}(\mathsf{FinCompl}(\mathbb{S}^{tt}, \mathcal{A})) = \Sigma^\omega \setminus \mathcal{L}(\mathcal{A})$.*

Proof. Follows from Lemma 13 and Theorem 11. □

Since the result of the construction can be mapped to the Miyano-Hayashi's algorithm [34], the complexities also match.

Corollary 15. $|\mathsf{FinCompl}(\mathbb{S}^{tt}, \mathcal{A})| \in \mathcal{O}(3^n)$.

5.2 Rabin Automata

In this section, we give an instantiation of $\mathsf{FinCompl}$ with subprocedure $\mathbb{S}^{\mathsf{inf}} = (\mathcal{M}^{\mathsf{inf}}, \mathcal{M}_0^{\mathsf{inf}}, \mathsf{SuccAct}_\Delta^{\mathsf{inf}}, \mathsf{SuccTrack}_\Delta^{\mathsf{inf}}, \mathsf{EmptyBreak}^{\mathsf{inf}})$ for Inf(❶), which will allow us to complement TELAs where the acceptance condition is a single Rabin pair. The algorithm is based on the optimal rank-based BA complementation algorithm from [38] adjusted to the needs of the modular construction. The macrostates of the instantiation are given as

$$\mathcal{M}^{\mathsf{inf}} = \overbrace{2^Q \cup (\mathcal{T} \times 2^Q \times \{0, 2, \ldots, 2n-2\})}^{\mathcal{M}^{\mathsf{inf}}_{\mathsf{Act}}} \cup \overbrace{(\mathcal{T} \times \{0, 2, \ldots, 2n-2\})}^{\mathcal{M}^{\mathsf{inf}}_{\mathsf{Track}}}$$

where $\mathcal{M}_0^{\mathsf{inf}} = \{I\}$. Notice that *active macrostates* ($\mathcal{M}^{\mathsf{inf}}_{\mathsf{Act}}$) are either sets of states (from 2^Q, just keeping track of all runs) or states of the form (f, O, i) (representing tight runs). On the other hand, *tracking macrostates* ($\mathcal{M}^{\mathsf{inf}}_{\mathsf{Track}}$) are of the form (f, i); these are used to wait for newly arrived runs to become tight. The remaining components are then defined as follows:

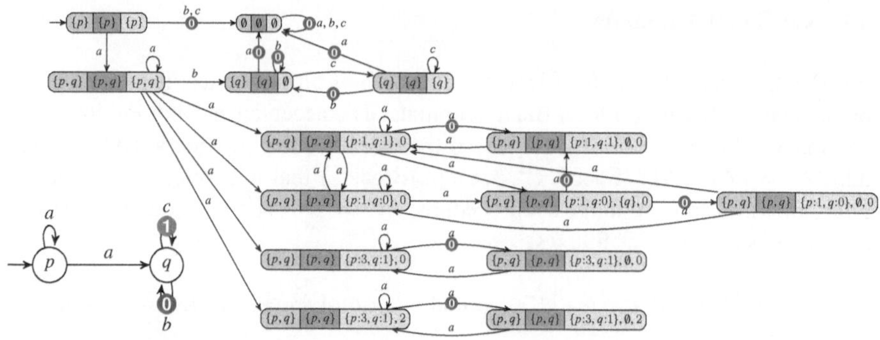

(a) Example of a Rabin automaton with the acceptance condition Fin(⓪) ∧ Inf(❶).

(b) The resulting complement automaton with the acceptance condition Inf(⓪). The macrostates are depicted in the form S (grey), P (blue), M (green).

Fig. 3: Example of FinCompl instantiated with $\mathbb{S}^{\mathsf{inf}}$ for complementation of automata with the acceptance condition containing a single Rabin pair.

- $(f', O', i') \in \mathsf{SuccAct}^{\mathsf{inf}}_{\Delta}(P, a, (f, O, i))$ iff
  - $f \sqsubseteq^a_\Delta f'$ and $rank(f) = rank(f')$,
  - $\mathrm{dom}(f') = P$,
  - $O \neq \emptyset$,
  - $i' = i$,
  - $O' = \Delta(O, a) \cap f'^{-1}(i)$
- $(f', i') \in \mathsf{SuccAct}^{\mathsf{inf}}_{\Delta}(P, a, (f, O, i))$ iff
  - $f \sqsubseteq^a_\Delta f'$ and $rank(f) = rank(f')$,
  - $O = \emptyset$,
  - $i' = (i + 2) \mod (rank(f') + 1)$
- $P' \in \mathsf{SuccAct}^{\mathsf{inf}}_{\Delta}(P, a, P)$ iff
  - $P' = P$
- $(f', i') \in \mathsf{SuccTrack}^{\mathsf{inf}}_{\Delta}(P, a, P)$ iff
  - f' is P-tight
  - $i' = 0$
- $\{(f', i'), (f', O', i')\} \subseteq \mathsf{SuccTrack}^{\mathsf{inf}}_{\Delta}(P, a, (f, i))$ iff
  - $f \sqsubseteq^a_\Delta f'$ and $rank(f) = rank(f')$,
  - $O' = f'^{-1}(i')$,
  - $i' = i$
- $\mathsf{EmptyBreak}^{\mathsf{inf}}((f, O, i)) \iff O = \emptyset$
- $\mathsf{EmptyBreak}^{\mathsf{inf}}(P) \iff P = \emptyset$
- $\mathsf{EmptyBreak}^{\mathsf{inf}}((f, i)) \iff \mathit{false}$

An example of the construction is shown in Fig. 3. The correctness of the instantiation is then summarized by the following lemma.

Lemma 16. *The subprocedure $\mathbb{S}^{\mathsf{inf}}$ is correct for the acceptance condition Inf(❶).*

Proof (Sketch). In order to show that the subprocedure $\mathbb{S}^{\mathsf{inf}}$ is correct, we need to show that for each word w and every RRDAG $\mathcal{G}^\Delta_w$ it holds that $\mathcal{G}^\Delta_w$ is not accepting wrt Inf(❶) iff there is an accepting Fin-run of $\mathbb{S}^{\mathsf{inf}}$ over $\mathcal{G}^\Delta_w$. We begin with the proof of the statement from left to right. Assume that $\mathcal{G}^\Delta_w$ is not accepting wrt Inf(❶). There is either no run of $\mathcal{G}^\Delta_w$ on w at all or all runs do not satisfy the formula. If there is no run of $\mathcal{G}^\Delta_w$ on w, then there is a sequence $(M_0, M_1, \ldots)$ where $M_0 = I$ and $M_{j+1} = \Delta(M_j, a)$ for all $j \geq 0$ such that there is some $i \geq 0$ such that $M_l = \emptyset$ for all $l \geq i$. The predicate $\mathsf{EmptyBreak}(M_l)$ is true for all $l \geq i$, so it holds infinitely often, and there therefore exists an accepting run of $\mathbb{S}^{\mathsf{inf}}$ over $\mathcal{G}^\Delta_w$. Now assume that there is a run of $\mathcal{G}^\Delta_w$ on w. Then, no matter from which point there are no transitions from ❶, the condition Inf(❶) does not hold for the particular run. With every transition $(f, i) \to (f', O', i')$ we sample all currently reachable states and then check that all runs from these states contain transitions from ❶ only finitely often by modified Schewe's rank-based algorithm. The O-component is emptied infinitely often and so there is an accepting run of $\mathbb{S}^{\mathsf{inf}}$ over $\mathcal{G}^\Delta_w$.

Now we prove the equivalence in the opposite direction. Assume that there is an accepting run of $\mathbb{S}^{\mathsf{inf}}$ over $\mathcal{G}^\Delta_w$. There is therefore a run where the $\mathsf{EmptyBreak}$ predicate is true infinitely often. The first possible option is that $\mathsf{EmptyBreak}(P)$ is true infinitely many times. That can happen only if there is no run on w and $\mathcal{G}_w^{\Delta}$ is finite. If there is no such run, the formula is not satisfied and $\mathcal{G}_w^{\Delta}$ is not accepting. The second option is that $\mathsf{EmptyBreak}((f, O, i))$ is true infinitely many times. That means that the formula Inf(❶) does not hold for any run, no matter when the run stops containing transitions from ⓒ. The formula is therefore not satisfied in any run and $\mathcal{G}_w^{\Delta}$ is not accepting. □

The following lemma shows that using our approach, handling the Fin(ⓒ) condition is "*for free*," i.e., the asymptotical complexity stays the same as for the optimal algorithm for BA complementation from [38].

Lemma 17. $|\mathsf{FinCompl}(\mathbb{S}^{\mathsf{inf}}, \mathcal{A})| \in O(tight(n+1))$.

Proof. It suffices to count the number of macrostates of the form (S, P, f, O, i). Consider a macrostate (S, P, f, O, i). We uniquely encode the macrostate as (h, i) where $h\colon Q \to \{-3, \dots, 2n-1\}$ is defined as follows:

$$h(q) = \begin{cases} -1 & \text{if } q \in O, \\ -2 & \text{if } q \in Q \setminus S, \\ -3 & \text{if } q \in S \setminus P, \text{ and} \\ f(q) & \text{otherwise.} \end{cases} \tag{2}$$

For a fixed i we compute the number of such encodings h. First we divide all encodings into groups according to the set $\mathrm{img}(h) \cap \{-3, -2, -1\}$ (8 groups at most) and we will show for each of the groups how we can "*shuffle*" the ranks in h to obtain the bound $O(tight(n))$ for each of the groups. We will denote each of the groups by g_M with $M \subseteq \{-3, -2, -1\}$.

$g_\emptyset$: from the definition, f is tight so $|g_\emptyset| = O(tight(n))$

$g_{\{-1\}}$: since there is at least one state q with $h(q) = -1$, this means that $q \in O$ so q has an even rank. As a consequence, at least one of the positive odd ranks of h will not be taken, so we can infer that $h\colon Q \to \{-1, \dots, 2n-3\}$. We can therefore uniquely map h to a mapping h' by incrementing all ranks of h by two, so $h'\colon Q \to \{1, \dots, 2n-1\}$. But then $h' \in \mathcal{T}(n)$, so $|g_{\{-1\}}| \in O(tight(n))$.

$g_{\{-2,-1\}}$: via the same reasoning as for $g_{\{-1\}}$ we get that $|g_{\{-2,-1\}}| \in O(tight(n))$.

$g_{\{-2\}}$: the reasoning is similar to the one for $g_{\{-1\}}$, with the exception that now, we know that there is a state $q \in Q \setminus S$, which is, according to the definition of a ranking, assigned the rank 0. This means that one positive odd rank of h is, again, not taken, so we increment all non-negative ranks of h by two and map all states in $Q \setminus S$ to 1, obtaining a tight ranking $h' \in \mathcal{T}(n)$. Therefore, $|g_{\{-2\}}| \in O(tight(n))$.

$g_{\{-3\}}$: the reasoning is, again, similar to the one for $g_{\{-1\}}$, with the exception that now, we know that there is a state $q \in S \setminus P$ such that its rank is, according to the definition 0. Therefore, we increment all non-negative ranks of h by two and map the states in $S \setminus P$ to 1, obtaining a tight ranking $h' \in \mathcal{T}(n)$; therefore, $|g_{\{-3\}}| \in O(tight(n))$.

$g_{\{-3,-2\}}, g_{\{-3,-1\}}$: similarly as for $g_{\{-2\}}$, we increment all non-negative ranks of h by two and set $h'(q) = 0$ if $h(q) = -3$ and $h'(q) = 1$ if $h(q) = -2$ (resp. if $h(q) = -1$). Then $h' \in \mathcal{T}(n)$ and so $|g_{\{-3,-2\}}| = O(tight(n))$ and $|g_{\{-3,-1\}}| \in O(tight(n))$.

$g_{\{-3,-2,-1\}}$: in this case, we know that there is at least one state $q_1 \in O$ and at least one state $q_2 \in Q \setminus S$. Therefore, there will be at least two odd positions not taken in h, so we can infer that h: $\{-3, \ldots, 2n-5\}$. We create h' by incrementing all ranks in h by *four*; in this way, we obtain a tight ranking $h' : Q \to \{0, \ldots, 2n-1\}$, so $|g_{\{-3,-2,-1\}}| \in O(tight(n))$.

Since the size of all groups is bounded by $O(tight(n))$, for a fixed i, the total number of these encodings is still $O(tight(n))$. When we sum the encodings for all possible i's, we obtain that the number is bounded by $O(tight(n+1))$, since $O(n \cdot tight(n)) = O(tight(n+1))$ [38]. □

The modular construction instantiated with $\mathbb{S}^{\mathsf{inf}}$ gives us a procedure for complementing Rabin automata with a single pair. To get a procedure for general Rabin automata, we construct a complement automaton for each Rabin pair, make a product of these automata, and obtain a GBA accepting the complement of the original automaton. The complexity reasoning is straightforward and is summarized by the following corollary.

Corollary 18. *Let $\mathcal{A}$ be a Rabin automaton with k Rabin pairs. Then we can construct a GBA accepting the complement of the language of $\mathcal{A}$ with $O(tight(n+1)^k) = O(n^k(0.76n)^{nk})$ states and k colours.*

Proof. $O(tight(n+1)^k) = O((n \cdot tight(n))^k) = O((n(0.76n)^n)^k) = O(n^k(0.76n)^{nk})$ □

To the best of our knowledge, the state complexity of our procedure is better than the complexity of other approaches for complementing Rabin automata (even if we require the output to be a BA and not a GBA—the BA would have $O(k \cdot tight(n+1)^k) = O(kn^k(0.76n)^{nk})$ states). In particular, it is better than the complexity $O(k \cdot 3^n \cdot (2n+1)^{nk})$ of [27]. Comparing the two techniques, the main difference is that our modular approach allows us to use tight rankings (and the optimal construction of, e.g., Schewe [38]), which are a significant factor in decreasing the size of the complement (both in theory and in practice). On the other hand, [27] does not use tight rankings (their run DAG ranking procedure does not allow it since ranks can change arbitrarily when a Fin state is encountered), however, it performs the complementation for the k Rabin pairs at once and avoids performing the product. The complexity of our approach is better; combining the two approaches to get an even better complexity is future work.

The complexity of our approach is also better than the complexity of a procedure that would first transform the input Rabin automaton into a BA with $m = nk$ states and run the optimal BA complementation with complexity $O(m(0.76m)^m) = O(nk(0.76nk)^{nk})$ [38], as shown by the following lemmas.

Lemma 19. $O(n^k(0.76n)^{nk}) \subset O(k \cdot 3^n \cdot (2n+1)^{nk})$

Proof. $n^k(0.76n)^{nk} = (\sqrt[n]{n} \cdot 0.76n)^{nk}$. The global maximum of the function $\sqrt[n]{n}$ is less than 1.5, so $(\sqrt[n]{n} \cdot 0.76n)^{nk} < (1.14n)^{nk} < (2n+1)^{nk}$ for $n \geq 1$. □

Lemma 20. $O(n^k(0.76n)^{nk}) \subset O(nk(0.76nk)^{nk})$

Proof. Similar reasoning as in the proof of Lemma 19. □

5.3 Parity Automata

Since the parity condition is a special case of the Rabin condition [15], we can easily give an upper bound on the complementation of parity automata.

Lemma 21. *For a parity automaton $\mathcal{A}$ with index k, there is a GBA for the complement of $\mathcal{L}(\mathcal{A})$ with $\frac{k}{2}$ colours and $O(\mathit{tight}(n+1)^{\frac{k}{2}}) = O(n^{\frac{k}{2}}(0.76n)^{\frac{nk}{2}})$ states.*

Proof. The min-odd parity acceptance condition is of the form Acc = Fin(⓿) ∧ (Inf(❶) ∨ (Fin(❷) ∧ (Inf(❸) ∨ (Fin(❹) ∧ ...)))). If we transform the acceptance condition into the DNF, we obtain Acc′ = (Fin(⓿) ∧ Inf(❶)) ∨ (Fin(⓿+❷) ∧ Inf(❶+❸)) ∨ (Fin(⓿+❷+❹) ∧ Inf(❶+❸+❺)) ∨ ... which is a Rabin acceptance condition with $\frac{k}{2}$ Rabin pairs. In the condition, e.g., ⓿ + ❷ denotes *union* of colours ⓿ and ❷, obtained by changing all occurrences of ⓿ and ❷ in $\mathcal{A}$'s colouring function p to the new colour ⓿ + ❷. Note that we can use a new colour for each union of colours and we obtain the same number of colours as in Acc. According to Corollary 18, the parity automaton $\mathcal{A}$ can be complemented into a GBA with $O(\mathit{tight}(n+1)^{\frac{k}{2}})$ states. □

We note that the complexity obtained by our general procedure is worse than the best one we are aware of, which is $2^{O(n \log n)}$ [7].

5.4 Generalized Rabin Automata

Recall that the generalized Rabin pair is of the form Fin(⓿) ∧ $\bigwedge_{j=1}^{n}$ Inf(ⓙ). We can now easily combine the procedure for (standard) Rabin automata from the previous section and the procedure for Inf-TELA from Section 3.2 to construct the subprocedure $\mathbb{S}^{\wedge \mathsf{inf}}$ for $\bigwedge_{j=1}^{n}$ Inf(ⓙ). The set of macrostates will be

$$\mathcal{M}^{\wedge \mathsf{inf}} = 2^Q \cup (\mathcal{T} \times 2^Q \times \{0, 2, \ldots, 2n-2\} \times \mathsf{LM}) \cup (\mathcal{T} \times \{0, 2, \ldots, 2n-2\} \times \mathsf{LM})$$

Details are given in [20]. Similarly to Sections 3.2 and 5.2, one can then obtain the following bound on the size of the complement.

Lemma 22. *Let $\mathcal{A}$ be a generalized Rabin automaton with one generalized Rabin pair with ℓ Infs. Then, there exists a BA accepting the complement of $\mathcal{A}$ with $O(\ell^n \mathit{tight}(n+1)) = O(n\ell^n(0.76n)^n)$ states.*

Theorem 23. *Let $\mathcal{A}$ be a generalized Rabin automaton with k generalized Rabin pairs, each with at most ℓ Infs. Then, there exists a GBA with k colours and $O(\ell^{nk} \mathit{tight}(n+1)^k) = O(n^k(0.76\ell n)^{nk})$ states accepting $\Sigma^\omega \setminus \mathcal{L}(\mathcal{A})$.*

There is not much work on the complementation of generalized Rabin automata or general TELAs (we are only aware of the upper bound $2^{2^{O(n)}}$ from [37])). One could approach the complementation by translation of the generalized Rabin automaton into a GBA using the technique from [22]. The technique first performs Fin-removal, i.e., it makes k copies of $\mathcal{A}$, each with the corresponding Fin-transitions removed, obtaining a GBA with $n(k+1)$ states and ℓ colours (one can share colours across the independent copies). After that, we could use our GBA complementation algorithm from Section 3, which would give us a BA with $O(n(k+1)(0.76\ell n(k+1))^{n(k+1)})$ states, which is worse.

Lemma 24. $O(n^k(0.76\ell n)^{nk}) \subset O(n(k+1)(0.76\ell n(k+1))^{n(k+1)})$

Proof (Idea). Let us observe the behaviour of the fraction with a simplified right-hand side: $\frac{nk(0.76\ell nk)^{nk}}{n^k(0.76\ell n)^{nk}} = \frac{nk^{nk+1}}{n^k}$. There are two options:

(i) $n \geq k$: in this case, $k^{nk} \gg n^k$ and the claim holds.
(ii) $k \geq n$: in this case, $k^k \gg n^k$ and the claim holds. □

5.5 General TELAs

For complementation of general TELAs, we use the fact that any TELA can be converted into a generalized Rabin automaton with the same structure by modifying the acceptance condition into the DNF form (and not touching the structure of the automaton). For a TELA with k colours, the DNF will have at most 2^k clauses (i.e., generalized Rabin pairs), each one with at most k literals.

Theorem 25. *Let $\mathcal{A}$ be a TELA with k colours. Then, there exists a GBA with* 2^k *colours and* $O(k^{n2^k} \mathit{tight}(n+1)^{2^k}) = O(n^{2^k}(0.76nk)^{n2^k})$ *states accepting* $\Sigma^\omega \setminus \mathcal{L}(\mathcal{A})$.

Proof. By substituting to Theorem 23. □

6 Related Work

Lower bounds for complementation of classes of ω-automata using the full automata technique were established in [45] (improving the previous $\Omega(n!)$ lower bound of Michel [33]). The technique was later generalized to improve the lower bound of Rabin automata complementation [8]. A double exponential lower bound for complementation of general Emerson-Lei automata was given in [37]. See the survey in [4] for more details.

Simultaneously to establishing the lower bound, there emerged algorithms for determinizing and complementing various classes of ω-automata. The optimal determinization approach for GBAs introduced in [39] yields a deterministic Rabin automaton with the number of states bounded by $(1.47nk)^n$ for large k and $2^n - 1$ Rabin pairs. In [13], the Miyano-Hayashi construction [34] is used within Büchi determinization. Rank-based complementation of GBAs was proposed in [28]. Furthermore, there are approaches for semideterminization-based complementation of GBAs [3] with double exponential complexity. Regarding other acceptance conditions, determinization of parity automata based on root history trees was proposed in [40]. A rank-based complementation of Streett and Rabin automata was introduced in [27] and later improved by tree structures in [7]. Tight determinization of Streett automata was presented in [43]. A tight complementation technique for parity automata based on flattened nested history trees was then proposed in [41]. A lot of effort has been put into complementation of Büchi automata leading to algorithms roughly divided into several groups: Ramsey-based [5,6,42], rank-based [16,19,18,44,26,38], determinization-based [36,35,30], slice-based [23], and others [1,17,31]. There are specialized more efficient algorithms for subclasses of BAs, such as inherently-weak [34], deterministic [29], semideterministic [2], elevator [19,17], or unambiguous [32,12] BAs.

Acknowledgments

We thank the anonymous reviewers for their insightful comments that helped improve the quality of the paper. This work has been supported by the Czech Ministry of Education, Youth and Sports ERC.CZ project LL1908, the Czech Science Foundation project 25-18318S, and the FIT BUT internal project FIT-S-23-8151.

References

1. Joël D. Allred and Ulrich Ultes-Nitsche. A simple and optimal complementation algorithm for Büchi automata. In Anuj Dawar and Erich Grädel, editors, *Proceedings of the 33rd Annual ACM/IEEE Symposium on Logic in Computer Science, LICS 2018, Oxford, UK, July 09-12, 2018*, pages 46–55. ACM, 2018. doi:10.1145/3209108.3209138.
2. Frantisek Blahoudek, Matthias Heizmann, Sven Schewe, Jan Strejcek, and Ming-Hsien Tsai. Complementing semi-deterministic Büchi automata. In Marsha Chechik and Jean-François Raskin, editors, *Tools and Algorithms for the Construction and Analysis of Systems - 22nd International Conference, TACAS 2016, Held as Part of the European Joint Conferences on Theory and Practice of Software, ETAPS 2016, Eindhoven, The Netherlands, April 2-8, 2016, Proceedings*, volume 9636 of *Lecture Notes in Computer Science*, pages 770–787. Springer, 2016. doi:10.1007/978-3-662-49674-9_49.
3. František Blahoudek, Alexandre Duret-Lutz, and Jan Strejček. Seminator 2 can complement generalized Büchi automata via improved semi-determinization. In *Proceedings of the 32nd International Conference on Computer-Aided Verification (CAV'20)*, volume 12225 of *Lecture Notes in Computer Science*, pages 15–27. Springer, July 2020. doi:10.1007/978-3-030-53291-8_2.
4. Udi Boker. Why these automata types? In *LPAR-22. 22nd International Conference on Logic for Programming, Artificial Intelligence and Reasoning, Awassa, Ethiopia, 16-21 November 2018*, volume 57 of *EPiC Series in Computing*, pages 143–163. EasyChair, 2018.
5. Stefan Breuers, Christof Löding, and Jörg Olschewski. Improved Ramsey-based Büchi complementation. In *Proc. of FOSSACS'12*, pages 150–164. Springer, 2012.
6. J. Richard Büchi. On a decision method in restricted second order arithmetic. In *Proc. of International Congress on Logic, Method, and Philosophy of Science 1960*. Stanford Univ. Press, Stanford, 1962.
7. Yang Cai and Ting Zhang. Tight upper bounds for Streett and parity complementation. In Marc Bezem, editor, *Computer Science Logic, 25th International Workshop / 20th Annual Conference of the EACSL, CSL 2011, September 12-15, 2011, Bergen, Norway, Proceedings*, volume 12 of *LIPIcs*, pages 112–128. Schloss Dagstuhl - Leibniz-Zentrum für Informatik, 2011. doi:10.4230/LIPIcs.CSL.2011.112.
8. Yang Cai, Ting Zhang, and Haifeng Luo. An improved lower bound for the complementation of Rabin automata. In *Proceedings of the 2009 24th Annual IEEE Symposium on Logic In Computer Science*, LICS '09, page 167–176, USA, 2009. IEEE Computer Society. doi:10.1109/LICS.2009.13.
9. Yu-Fang Chen, Vojtech Havlena, and Ondrej Lengál. Simulations in rank-based Büchi automata complementation. In Anthony Widjaja Lin, editor, *Programming Languages and Systems - 17th Asian Symposium, APLAS 2019, Nusa Dua, Bali, Indonesia, December 1-4, 2019, Proceedings*, volume 11893 of *Lecture Notes in Computer Science*, pages 447–467. Springer, 2019. doi:10.1007/978-3-030-34175-6_23.
10. Michael R. Clarkson, Bernd Finkbeiner, Masoud Koleini, Kristopher K. Micinski, Markus N. Rabe, and César Sánchez. Temporal logics for hyperproperties. In Martín Abadi and Steve Kremer, editors, *Principles of Security and Trust - Third International Conference, POST 2014, Held as Part of the European Joint Conferences on Theory and Practice of Software, ETAPS 2014, Grenoble, France, April 5-13, 2014, Proceedings*, volume 8414 of *Lecture Notes in Computer Science*, pages 265–284. Springer, 2014. doi:10.1007/978-3-642-54792-8_15.
11. E. Allen Emerson and Chin-Laung Lei. Modalities for model checking: Branching time logic strikes back. *Sci. Comput. Program.*, 8(3):275–306, 1987. doi:10.1016/0167-6423(87)90036-0.

12. Weizhi Feng, Yong Li, Andrea Turrini, Moshe Y. Vardi, and Lijun Zhang. On the power of finite ambiguity in Büchi complementation. *Inf. Comput.*, 292:105032, 2023. URL: https://doi.org/10.1016/j.ic.2023.105032, doi:10.1016/J.IC.2023.105032.
13. Dana Fisman and Yoad Lustig. A modular approach for Büchi determinization. In Luca Aceto and David de Frutos-Escrig, editors, *26th International Conference on Concurrency Theory, CONCUR 2015, Madrid, Spain, September 1.4, 2015*, volume 42 of *LIPIcs*, pages 368–382. Schloss Dagstuhl - Leibniz-Zentrum für Informatik, 2015. URL: https://doi.org/10.4230/LIPIcs.CONCUR.2015.368, doi:10.4230/LIPICS.CONCUR.2015.368.
14. Ehud Friedgut, Orna Kupferman, and Moshe Y. Vardi. Büchi complementation made tighter. *Int. J. Found. Comput. Sci.*, 17(4):851–868, 2006. doi:10.1142/S0129054106004145.
15. Erich Grädel, Wolfgang Thomas, and Thomas Wilke, editors. *Automata, Logics, and Infinite Games: A Guide to Current Research [outcome of a Dagstuhl seminar, February 2001]*, volume 2500 of *Lecture Notes in Computer Science*. Springer, 2002. doi:10.1007/3-540-36387-4.
16. Vojtech Havlena and Ondrej Lengál. Reducing (to) the ranks: Efficient rank-based Büchi automata complementation. In Serge Haddad and Daniele Varacca, editors, *32nd International Conference on Concurrency Theory, CONCUR 2021, August 24-27, 2021, Virtual Conference*, volume 203 of *LIPIcs*, pages 2:1–2:19. Schloss Dagstuhl - Leibniz-Zentrum für Informatik, 2021. URL: https://doi.org/10.4230/LIPIcs.CONCUR.2021.2, doi:10.4230/LIPICS.CONCUR.2021.2.
17. Vojtěch Havlena, Ondřej Lengál, Yong Li, Barbora Šmahlíková, and Andrea Turrini. Modular mix-and-match complementation of Büchi automata. In *Tools and Algorithms for the Construction and Analysis of Systems - 28th International Conference, TACAS 2023, Held as Part of the European Joint Conferences on Theory and Practice of Software, ETAPS 2023, Paris, France*, Lecture Notes in Computer Science. Springer, 2023.
18. Vojtěch Havlena, Ondřej Lengál, and Barbora Šmahlíková. Complementing Büchi automata with ranker. In Sharon Shoham and Yakir Vizel, editors, *Computer Aided Verification - 34th International Conference, CAV 2022, Haifa, Israel, August 7-10, 2022, Proceedings, Part II*, volume 13372 of *Lecture Notes in Computer Science*, pages 188–201. Springer, 2022. doi:10.1007/978-3-031-13188-2_10.
19. Vojtěch Havlena, Ondřej Lengál, and Barbora Šmahlíková. Sky is not the limit: Tighter rank bounds for elevator automata in Büchi automata complementation. In Dana Fisman and Grigore Rosu, editors, *Tools and Algorithms for the Construction and Analysis of Systems - 28th International Conference, TACAS 2022, Held as Part of the European Joint Conferences on Theory and Practice of Software, ETAPS 2022, Munich, Germany, April 2-7, 2022, Proceedings, Part II*, volume 13244 of *Lecture Notes in Computer Science*, pages 118–136. Springer, 2022. doi:10.1007/978-3-030-99527-0_7.
20. Vojtěch Havlena, Ondřej Lengál, and Barbora Šmahlíková. Complementation of Emerson-Lei automata (technical report). *CoRR*, abs/2410.11644, 2024. URL: https://arxiv.org/abs/2410.11644, arXiv:2410.11644.
21. Philipp Hieronymi, Dun Ma, Reed Oei, Luke Schaeffer, Christian Schulz, and Jeffrey O. Shallit. Decidability for Sturmian words. *Log. Methods Comput. Sci.*, 20(3), 2024. URL: https://doi.org/10.46298/lmcs-20(3:12)2024, doi:10.46298/LMCS-20(3:12)2024.
22. Tobias John, Simon Jantsch, Christel Baier, and Sascha Klüppelholz. From Emerson-Lei automata to deterministic, limit-deterministic or good-for-MDP automata. *Innov. Syst. Softw. Eng.*, 18(3):385–403, 2022. doi:10.1007/s11334-022-00445-7.
23. Detlef Kähler and Thomas Wilke. Complementation, disambiguation, and determinization of Büchi automata unified. In *Proc. of ICALP'08*, pages 724–735. Springer, 2008.
24. Hrishikesh Karmarkar and Supratik Chakraborty. On minimal odd rankings for Büchi complementation. In Zhiming Liu and Anders P. Ravn, editors, *Automated Technology for*

Verification and Analysis, 7th International Symposium, ATVA 2009, Macao, China, October 14-16, 2009. Proceedings, volume 5799 of *Lecture Notes in Computer Science*, pages 228–243. Springer, 2009. doi:10.1007/978-3-642-04761-9_18.
25. Yonit Kesten and Amir Pnueli. A complete proof systems for QPTL. In *Proceedings, 10th Annual IEEE Symposium on Logic in Computer Science, San Diego, California, USA, June 26-29, 1995*, pages 2–12. IEEE Computer Society, 1995. doi:10.1109/LICS.1995.523239.
26. Orna Kupferman and Moshe Y. Vardi. Weak alternating automata are not that weak. *ACM Trans. Comput. Log.*, 2(3):408–429, 2001. doi:10.1145/377978.377993.
27. Orna Kupferman and Moshe Y. Vardi. Complementation constructions for nondeterministic automata on infinite words. In Nicolas Halbwachs and Lenore D. Zuck, editors, *Tools and Algorithms for the Construction and Analysis of Systems, 11th International Conference, TACAS 2005, Held as Part of the Joint European Conferences on Theory and Practice of Software, ETAPS 2005, Edinburgh, UK, April 4-8, 2005, Proceedings*, volume 3440 of *Lecture Notes in Computer Science*, pages 206–221. Springer, 2005. doi:10.1007/978-3-540-31980-1_14.
28. Orna Kupferman and Moshe Y. Vardi. From complementation to certification. *Theor. Comput. Sci.*, 345(1):83–100, 2005. doi:10.1016/j.tcs.2005.07.021.
29. Robert P. Kurshan. Complementing deterministic Büchi automata in polynomial time. *J. Comput. Syst. Sci.*, 35(1):59–71, 1987. doi:10.1016/0022-0000(87)90036-5.
30. Yong Li, Andrea Turrini, Weizhi Feng, Moshe Y. Vardi, and Lijun Zhang. Divide-and-conquer determinization of Büchi automata based on SCC decomposition. In Sharon Shoham and Yakir Vizel, editors, *Computer Aided Verification - 34th International Conference, CAV 2022, Haifa, Israel, August 7-10, 2022, Proceedings, Part II*, volume 13372 of *Lecture Notes in Computer Science*, pages 152–173. Springer, 2022. doi:10.1007/978-3-031-13188-2_8.
31. Yong Li, Andrea Turrini, Lijun Zhang, and Sven Schewe. Learning to complement Büchi automata. In *Proc. of VMCAI'18*, pages 313–335. Springer, 2018.
32. Yong Li, Moshe Y. Vardi, and Lijun Zhang. On the power of unambiguity in Büchi complementation. In Jean-Francois Raskin and Davide Bresolin, editors, Proceedings 11th International Symposium on *Games, Automata, Logics, and Formal Verification,* Brussels, Belgium, September 21-22, 2020, volume 326 of *Electronic Proceedings in Theoretical Computer Science*, pages 182–198. Open Publishing Association, 2020. doi:10.4204/EPTCS.326.12.
33. Max Michel. Complementation is more difficult with automata on infinite words. *CNET, Paris*, 15, 1988.
34. Satoru Miyano and Takeshi Hayashi. Alternating finite automata on omega-words. In Bruno Courcelle, editor, *CAAP'84, 9th Colloquium on Trees in Algebra and Programming, Bordeaux, France, March 5-7, 1984, Proceedings*, pages 195–210. Cambridge University Press, 1984.
35. Nir Piterman. From nondeterministic Büchi and Streett automata to deterministic parity automata. In *Proc. of LICS'06*, pages 255–264. IEEE, 2006.
36. Shmuel Safra. On the complexity of ω-automata. In *Proc. of FOCS'88*, pages 319–327. IEEE, 1988.
37. Shmuel Safra and Moshe Y. Vardi. On ω-automata and temporal logic (preliminary report). In David S. Johnson, editor, *Proceedings of the 21st Annual ACM Symposium on Theory of Computing, May 14-17, 1989, Seattle, Washington, USA*, pages 127–137. ACM, 1989. doi:10.1145/73007.73019.
38. Sven Schewe. Büchi complementation made tight. In Susanne Albers and Jean-Yves Marion, editors, *26th International Symposium on Theoretical Aspects of Computer Science, STACS 2009, February 26-28, 2009, Freiburg, Germany, Proceedings*, volume 3 of *LIPIcs*, pages

661–672. Schloss Dagstuhl - Leibniz-Zentrum fuer Informatik, Germany, 2009. doi:10.4230/LIPIcs.STACS.2009.1854.
39. Sven Schewe and Thomas Varghese. Tight bounds for the determinisation and complementation of generalised Büchi automata. In Supratik Chakraborty and Madhavan Mukund, editors, *Automated Technology for Verification and Analysis - 10th International Symposium, ATVA 2012, Thiruvananthapuram, India, October 3-6, 2012. Proceedings*, volume 7561 of *Lecture Notes in Computer Science*, pages 42–56. Springer, 2012. doi:10.1007/978-3-642-33386-6_5.
40. Sven Schewe and Thomas Varghese. Determinising parity automata. In Erzsébet Csuhaj-Varjú, Martin Dietzfelbinger, and Zoltán Ésik, editors, *Mathematical Foundations of Computer Science 2014 - 39th International Symposium, MFCS 2014, Budapest, Hungary, August 25-29, 2014. Proceedings, Part I*, volume 8634 of *Lecture Notes in Computer Science*, pages 486–498. Springer, 2014. doi:10.1007/978-3-662-44522-8_41.
41. Sven Schewe and Thomas Varghese. Tight bounds for complementing parity automata. In Erzsébet Csuhaj-Varjú, Martin Dietzfelbinger, and Zoltán Ésik, editors, *Mathematical Foundations of Computer Science 2014 - 39th International Symposium, MFCS 2014, Budapest, Hungary, August 25-29, 2014. Proceedings, Part I*, volume 8634 of *Lecture Notes in Computer Science*, pages 499–510. Springer, 2014. doi:10.1007/978-3-662-44522-8_42.
42. A. Prasad Sistla, Moshe Y. Vardi, and Pierre Wolper. The Complementation Problem for Büchi Automata with Applications to Temporal Logic. *Theoretical Computer Science*, 49(2-3):217–237, 1987.
43. Cong Tian, Wensheng Wang, and Zhenhua Duan. Making streett determinization tight. In *Proceedings of the 35th Annual ACM/IEEE Symposium on Logic in Computer Science*, LICS '20, page 859–872, New York, NY, USA, 2020. Association for Computing Machinery. doi:10.1145/3373718.3394757.
44. Moshe Y. Vardi. The Büchi complementation saga. In *Proc. of STACS'07*, pages 12–22. Springer, 2007.
45. Qiqi Yan. Lower bounds for complementation of ω-automata via the full automata technique. In Michele Bugliesi, Bart Preneel, Vladimiro Sassone, and Ingo Wegener, editors, *Automata, Languages and Programming*, pages 589–600, Berlin, Heidelberg, 2006. Springer Berlin Heidelberg.

Relational Connectors and Heterogeneous Simulations

Pedro Nora[2], Jurriaan Rot[2], Lutz Schröder[1(✉)], and Paul Wild[1]

[1] Friedrich-Alexander-Universität Erlangen-Nürnberg, Erlangen, Germany
lutz.schroeder@fau.de

[2] Radboud University, Nijmegen, The Netherlands

Abstract. While behavioural equivalences among systems of the same type, such as Park/Milner bisimilarity of labelled transition systems, are an established notion, a systematic treatment of relationships between systems of different types is currently missing. We provide such a treatment in the framework of universal coalgebra, in which the type of a system (nondeterministic, probabilistic, weighted, game-based etc.) is abstracted as a set functor: We introduce *relational connectors* among set functors, which induce notions of heterogeneous (bi)simulation among coalgebras of the respective types. We give a number of constructions on relational connectors. In particular, we identify composition and converse operations on relational connectors; we construct corresponding identity relational connectors, showing that the latter generalize the standard Barr extension of weak-pullback-preserving functors; and we introduce a Kantorovich construction in which relational connectors are induced from relations between modalities. For Kantorovich relational connectors, one has a notion of dual-purpose modal logic interpreted over both system types, and we prove a corresponding Hennessy-Milner-type theorem stating that generalized (bi)similarity coincides with theory inclusion on finitely-branching systems. We apply these results to a number of example scenarios involving labelled transition systems with different label alphabets, probabilistic systems, and input/output conformances.

1 Introduction

Notions of simulation and bisimulation are pervasive in the specification and verification of reactive systems (e.g. [31]). For instance, they appear in state space reduction (e.g. [6]), they are used to specify concrete systems in terms of abstract systems (e.g. in connection with the analysis of ePassport protocols [22]), and, classically, they relate tightly to indistinguishability in modal logic [19]. Originally introduced for (labelled) transition systems, notions of (bi)simulation have been extended to a wide range of system types, e.g. probabilistic systems [27,10], weighted systems [9], or monotone neighbourhood frames [34,17]. They have

J. Rot—Funded by the Dutch Research Council (NWO) – project number VI.Vidi.223.096

L. Schröder—Funded by the Deutsche Forschungsgemeinschaft (DFG, German Research Foundation) – project number 531706730

P. A. Abdulla and D. Kesner (Eds.): FoSSaCS 2025, LNCS 15691, pp. 111–132, 2025.
https://doi.org/10.1007/978-3-031-90897-2_6

received a uniform treatment in the framework of universal coalgebra [37]. However, so far, notions of (bi)simulation have typically been confined to settings where the two systems being compared are of the same type in a strict sense, e.g. labelled transition systems (LTS) over the same alphabet. In the present paper, we introduce a principled approach to comparing behaviour across different system types by means of *heterogeneous (bi)simulations.*

To this end, we encapsulate system types as set functors in the paradigm of universal coalgebra, and introduce *(relational) connectors* between system types. The latter generalize *lax extensions*, which induce notions of (bi)simulation on a single system type [29,30]. A connector between functors F and G induces a notion of (bi)simulation between F-coalgebras and G-coalgebras, i.e. between the systems of the types represented by F and G, respectively, for instance between nondeterministic and probabilistic systems. We give a range of constructions of connectors, such as converse, composition, and pulling back along natural transformations. Notably, we show that the composition of relational connectors admits identities. Identity relational connectors satisfy a minimality condition, and form smallest lax extensions of functors; for weak-pullback-preserving functors, they coincide with the *Barr extension* [5], which instantiates, e.g., to the well-known Egli-Milner relation lifting for the powerset functor. We use these constructions to cover a number of application scenarios, e.g. transferring bisimilarity among LTS over different alphabets; sharing of infinite traces among LTS; nondeterministic abstractions of probabilistic LTS; and input-output conformances (*ioco*) [8].

We go on to give a construction of relational connectors based on relating modalities, modelled as predicate liftings in the style of coalgebraic logic [33,38]. In reference to constructions of behavioural metrics (on a single system type) from modalities [3,47], we call such relational connectors *Kantorovich.* Many of our running examples turn out to be Kantorovich. We then prove a Hennessy-Milner-type result for Kantorovich connectors, showing that on finitely branching systems, the induced similarity coincides with theory inclusion in a generic dual-purpose modal logic that can be interpreted over both of the involved system types. The generic theorem instantiates to logical characterizations of bisimulation between LTS with different alphabets, trace sharing between LTS, nondeterministic abstraction of probabilistic LTS, and ioco compatibility.

Proofs are often omitted or only sketched; details can be found in the full version [32].

Related work Relational connectors generalize lax extensions [40,39,29,30], which belong to an extended strand of work on extending set functors to act on relations (e.g. [2,42,21,28]). The Kantorovich construction similarly generalizes constructions of functor liftings and lax extensions in both two-valued and quantitative settings [29,30,16,3,47,13]. Our heterogeneous Hennessy-Milner theorem generalizes (the monotone case of) coalgebraic Hennessy-Milner theorems for behavioural equivalence [33,38] and behavioural preorders [23,46]. A different generalization of notions of bisimulation occurs via functor liftings along fibrations [20,18], which have also been connected to modal logics [25,24]. The

Kantorovich construction is generalized there by the so-called codensity lifting [41]. Heterogeneous notions of bisimulation have not been considered there.

2 Preliminaries: Coalgebras and Lax Extensions

We assume basic familiarity with category theory (e.g. [1]). We proceed to recall requisite background on relations, coalgebras, and lax extensions.

Relations A *relation* from a set X to a set Y is a subset $r \subseteq X \times Y$, denoted $r\colon X \nrightarrow Y$; we write $x\ r\ y$ for $(x, y) \in r$. Given $r\colon X \nrightarrow Y$ and $s\colon Y \nrightarrow Z$, we write $s \cdot r$ for the applicative-order relational composite of r and s, i.e.

$$s \cdot r = \{(x, z) \mid \exists y \in Y.\, x\ r\ y\ s\ z\}.$$

The *join* of a family of relations is just its union. Relational composition is *join continuous* in both arguments, i.e. we have $(\bigvee_{i \in I} s_i) \cdot r = \bigvee_{i \in I}(s_i \cdot r)$ and $s \cdot (\bigvee_{i \in I} r_i) = \bigvee_{i \in I}(s \cdot r_i)$. We define the *relational converse* $r^\circ\colon Y \nrightarrow X$ by $r^\circ = \{(y, x) \mid (x, y) \in r\}$. We identify a function $f\colon X \to Y$ with its graph, i.e. the relation $\{(x, f(x)) \mid x \in X\}$. For clarity, we sometimes write $\Delta_X = \{(x, x) \mid x \in X\}$ for the diagonal relation on X, i.e. the graph of the identity function on X, which is neutral for relational composition. Functions $f\colon X \to Y$ are characterized by the inequalities

$$\Delta_X \subseteq f^\circ \cdot f \quad \textit{(totality)} \qquad f \cdot f^\circ \subseteq \Delta_Y \quad \textit{(univalence)}.$$

Given a subset $A \subseteq X$ and a relation $r\colon X \nrightarrow Y$, we write $r[A] = \{y \in Y \mid \exists x \in A.\, x\ r\ y\}$ for the *relational image* of A under r. We say that r is *right total* if $r[X] = Y$, and *left total* if $r^\circ[Y] = X$.

Universal coalgebra State-based systems of a wide range of transition types can be usefully abstracted as coalgebras for a given functor encapsulating the system type [37]. We work more specifically over the category of sets, and thus model a system type as a functor $F\colon \mathsf{Set} \to \mathsf{Set}$. Then, an F-*coalgebra* is a pair (C, γ) consisting of a set C of *states* and a *transition map* $\gamma\colon C \to FC$. Following tradition in algebra, we often just write C for the coalgebra (C, γ). We think of C as a set of *states*, and of γ as assigning to each state $c \in C$ a collection $\gamma(c)$ of successor states, structured according to F. For instance, if $F = \mathcal{P}$ is the usual (covariant) powerset functor, then γ assigns to each state a *set* of successors, so a $\mathcal{P}$-coalgebra is just a standard relational transition system. More generally, given a set $\mathcal{A}$ of *labels*, F-coalgebras for the functor $F = \mathcal{P}(\mathcal{A} \times (-))$ are $\mathcal{A}$-labelled transition systems ($\mathcal{A}$-LTS). On the other hand, we write $\mathcal{D}$ for the (discrete) *distribution functor*, which assigns to a set X the set of discrete probability distributions on X (which may be represented as functions $\alpha\colon X \to [0, 1]$ such that $\sum_{x \in X} \alpha(x) = 1$, extended to subsets $A \subseteq X$ by $\alpha(A) = \sum_{x \in A} \alpha(x)$) and acts on maps by taking direct images. Then, $\mathcal{D}$-coalgebras are probabilistic transition systems, or Markov chains, while $\mathcal{D}(\mathcal{A} \times (-))$-coalgebras are probabilistic $\mathcal{A}$-labelled transition systems (probabilistic $\mathcal{A}$-LTS). We assume w.l.o.g. that functors preserve injective maps [5], and then in fact that subset inclusions are preserved.

A *morphism* $f\colon C \to D$ of F-coalgebras (C, γ), (D, δ) is a map $f\colon C \to D$ such that $Ff \cdot \gamma = \delta \cdot f$. States $c \in C$, $d \in D$ in F-coalgebras C, D are *behaviourally equivalent* if there exist an F-coalgebra (E, ϵ) and morphisms $f\colon C \to E$, $g\colon D \to E$ such that $f(x) = g(y)$. For instance, morphisms of $\mathcal{P}(\mathcal{A} \times (-))$-coalgebras are bounded morphisms of $\mathcal{A}$-LTS in the usual sense (i.e. functional bisimulations), and behavioural equivalence instantiates to the usual notion of (strong) bisimilarity on LTS.

Lax extensions As indicated in the introduction, relational connectors are largely intended as a generalization of lax extensions, which extend a single functor to act also on relations, to settings where relations need to connect elements of different functors. A *lax extension* L (references are in Section 1) of a set functor F assigns to each relation $r\colon X \nrightarrow Y$ a relation $Lr\colon FX \nrightarrow FY$ such that

(L1)	$r_1 \subseteq r_2 \to Lr_1 \subseteq Lr_2$	*(monotonicity)*
(L2)	$Ls \cdot Lr \subseteq L(s \cdot r)$	*(lax functoriality)*
(L3)	$Ff \subseteq Lf$ and $(Ff)^\circ \subseteq L(f^\circ)$	

for all sets X, Y, Z, and $r, r_1, r_2\colon X \nrightarrow Y$, $s\colon Y \nrightarrow Z$, $f\colon X \to Y$. These conditions imply *naturality* [40,30]:

$$L(g^\circ \cdot r \cdot f) = (Fg)^\circ \cdot Lr \cdot Ff$$

for $r\colon X \nrightarrow Y$ and maps $f\colon X' \to X$, $g\colon Y' \to Y$. We say that L *preserves diagonals* if $L\Delta_X \subseteq \Delta_{FX}$ for all X, equivalently, $Lf \subseteq Ff$ for all maps f. Moreover, L *preserves converse* if $L(r^\circ) = (Lr)^\circ$ for all r. (Indeed, this property is often included in the definition of lax extension [30].)

Lax extensions induce notions of *(bi)simulation*, that is, of relations that witness behavioural equivalence in the sense recalled above. Given a lax extension L of a functor F, a relation $r\colon C \nrightarrow D$ between F-coalgebras (C, γ), (D, δ) is an *L-simulation* if $\delta \cdot r \leq Lr \cdot \gamma$; that is, whenever $c\ r\ d$, then $\gamma(c)\ Lr\ \delta(d)$. Two states $c \in C$, $d \in D$ are *L-similar* if there exists an L-simulation $r\colon C \nrightarrow D$ such that $c\ r\ d$. If L preserves converse, then the converse r° of an L-simulation r is also an L-simulation and, hence, L-similarity is symmetric; one thus speaks more appropriately of *L-bisimulations* and *L-bisimilarity*. Notably, if L preserves converse and diagonals, then L-bisimilarity coincides with behavioural equivalence [29,30]. Every lax extension can be induced from a choice of modalities [29,30]; we return to this point in Section 5. We recall only the most basic example:

Example 2.1. Let $\mathcal{A}$ be a set of labels, and let $F = \mathcal{P}(\mathcal{A} \times (-))$ be the functor modelling $\mathcal{A}$-LTS as recalled above. We have a converse- and diagonal-preserving lax extension L of F given by $S\ Lr\ T$ iff (i) for all $(l,x) \in S$, there is $(l,y) \in T$ such that $x\ r\ y$ ('*forth*'), and (ii) for all $(l,y) \in T$, there is $(l,x) \in S$ such that $x\ r\ y$ ('*back*'). Indeed, L is even a strict extension, i.e. condition (L2) holds in the stronger form $Ls \cdot Lr = L(s \cdot r)$ for composable s, r (such strict extensions exist, and then are unique, iff the underlying functor preserves weak pullbacks [4,44]). L-bisimulations in the sense recalled above are precisely (strong) bisimulations of LTS in the standard sense.

Remark 2.2 (Barr extension). The above-mentioned strict extension L of a weak-pullback-preserving functor F, often called the *Barr extension*, is described as follows [4]: A relation $r\colon X \nrightarrow Y$ itself forms a set (a subset of $X \times Y$), and as such comes with two projection maps $\pi_1\colon r \to X$, $\pi_2\colon r \to Y$. Then, $Lr = F\pi_2 \cdot (F\pi_1)^\circ$. A slightly simpler example than Example 2.1 is the Barr extension L of the powerset functor $\mathcal{P}$, which coincides with the well-known Egli-Milner extension: For $r\colon X \nrightarrow Y$ and $S \in \mathcal{P}(X)$, $T \in \mathcal{P}(Y)$, we have $S\ Lr\ T$ iff for every $x \in S$ there is $y \in T$ such that $x\ r\ y$ and symmetrically.

3 Relational Connectors

We proceed to introduce relational connectors and associated constructions.

3.1 Axiomatics

The main idea is that while a lax extension of a functor F (Section 2) lifts relations between sets X and Y to relations between FX and FY, a relational connector between functors F and G lifts relations between sets X and Y to relations between FX and GY. The axiomatics of relational connectors is inspired by that of lax extensions, but forcibly deviates in some respects:

Definition 3.1 (Relational connector). Let F, G be set functors. A *relational connector* (or occasionally just a *connector*) $L\colon F \to G$ assigns to each relation $r\colon X \nrightarrow Y$ a relation

$$Lr\colon FX \nrightarrow GY$$

such that the following conditions hold:

1. Whenever $r_1 \subseteq r_2$ for $r_1, r_2\colon X \nrightarrow Y$, then $Lr_1 \subseteq Lr_2$ (*monotonicity*).
2. Whenever $f\colon X' \to X$, $g\colon Y' \to Y$, and $r\colon X \nrightarrow Y$, then

$$L(g^\circ \cdot r \cdot f) = (Gg)^\circ \cdot Lr \cdot Ff \qquad (\textit{naturality}).$$

We define an ordering on connectors $F \to G$ by $L \leq K$ iff $Lr \subseteq Kr$ for all r.

In pointful notation,.naturality says that for data as above and $a \in FX'$, $b \in GY'$, we have

$$Ff(a)\ Lr\ Gg(b) \quad \text{iff} \quad a\ L(g^\circ \cdot r \cdot f)\ b. \tag{3.1}$$

Example 3.2. Let $F = \mathcal{P}(\mathcal{A}\times(-))$, $G = \mathcal{P}(\mathcal{B}\times(-))$ be the functors determining $\mathcal{A}$-LTS and $\mathcal{B}$-LTS as their coalgebras, respectively (Section 2). For $R \subseteq \mathcal{A} \times \mathcal{B}$, we define a relational connector $L_R \colon F \to G$ by

$$S\ L_R r\ T \iff \forall (l,m) \in R.\, \forall (l,x) \in S.\, \exists (m,y) \in T.\, x\ r\ y\ \wedge \\ \forall (m,y) \in T.\, \exists (l,x) \in S.\, x\ r\ y$$

for $r \colon X \nrightarrow Y$. We will later use instances of this type of relational connector to transfer bisimilarity between $\mathcal{A}$-LTS and $\mathcal{B}$-LTS.

Of course, every lax extension of F is a relational connector $F \to F$. In the axiomatics of relational connectors, notable omissions in comparison to lax extensions include (L2) and (L3), both of which in general just fail to type for relational connectors. We will later discuss these conditions and further ones as properties that a relational connector may or may not have, if applicable. Note that we do retain an important consequence of these properties, viz., naturality.

3.2 Constructions

Our perspective on relational connectors is partly driven by constructions enabled by the axiomatics; maybe the most central ones among these are composition and identities, introduced next.

Definition 3.3 (Composition of relational connectors). Given relational connectors $K \colon F \to G$, $L \colon G \to H$, we define the *composite* $L \cdot K \colon F \to H$ by

$$(L \cdot K)r = \bigvee_{r = s \cdot t} Ls \cdot Kt \quad \text{for } r \colon X \nrightarrow Z, \tag{3.2}$$

where the join is over all $t \colon X \nrightarrow Y$, $s \colon Y \nrightarrow Z$ such that $s \cdot t = r$, with Y ranging over all sets (see however Theorem 3.8 and Lemma 3.5).

Lemma 3.4. *Given relational connectors $K \colon F \to G$, $L \colon G \to H$, the composite $L \cdot K \colon F \to H$ is a relational connector.*

Proof (sketch). Monotonicity: Let $r \subseteq r' \colon X \nrightarrow Z$. If $a\ (L \cdot K)r\ c$ is witnessed by a factorization $r = s \cdot t$ where $t \colon X \nrightarrow Y$, $s \colon Y \nrightarrow Z$, then $a\ (L \cdot K)r'\ c$ is witnessed by the factorization $r' = s' \cdot t'$ where $t' \colon X \nrightarrow Y'$, $s' \colon Y' \nrightarrow Z$ with $Y' = Y \cup (r' \setminus r)$ (w.l.o.g. a disjoint union) and

$$t' = t \cup \{(x,(x,z)) \mid (x,z) \in r' \setminus r\} \qquad s' = s \cup \{((x,z),z) \mid (x,z) \in r' \setminus r\}.$$

Remarkably, the further proof uses naturality (w.r.t. $Y \hookrightarrow Y'$) but not monotonicity of K and L.

Naturality: $(L \cdot K)(g^\circ \cdot r \cdot f) = (Hg)^\circ \cdot (L \cdot K)r \cdot Ff$ is shown using naturality and monotonicity of K and L, monotonicity of $L \cdot K$, and totality and univalence of f and g. □

As an immediate consequence of monotonicity of composite relational connectors, we have the following alternative description of composition:

Lemma 3.5. *Given relational connectors $K\colon F \to G$, $L\colon G \to H$, we have*

$$(L \cdot K)r = \bigvee_{r \supseteq s \cdot t} Ls \cdot Kt \quad \text{for } r\colon X \nrightarrow Z$$

where the join is over all $t\colon X \nrightarrow Y$, $s\colon Y \nrightarrow Z$ such that $r \supseteq s \cdot t$, with Y ranging over all sets.

In order to compute composites of relational connectors, the following observation is sometimes useful.

Definition 3.6. The *couniversal factorization* $r = s \cdot t$ of a relation $r\colon X \nrightarrow Z$ is given by

$$\begin{aligned} Y &= \{(A, B) \in \mathcal{P}(X) \times \mathcal{P}(Z) \mid A \times B \subseteq r\} \\ t &= \{(x, (A, B)) \mid x \in A\}\colon X \nrightarrow Y \\ s &= \{((A, B), z) \mid z \in B\}\colon Y \nrightarrow Z. \end{aligned}$$

Lemma 3.7. *Let $s\colon Y \nrightarrow Z$, $t\colon X \nrightarrow Y$ be the couniversal factorization of $r\colon X \nrightarrow Z$. Then indeed $r = s \cdot t$, and for every factorization $r = s' \cdot t'$ of r into $s'\colon Y' \nrightarrow Z$, $t'\colon X \nrightarrow Y'$, there is a map $f\colon Y' \to Y$ such that $s' = s \cdot f$ and $t' = f^{\circ} \cdot t$.*

Theorem 3.8. *Let $K\colon F \to G$, $L\colon G \to H$ be relational connectors, and let $r = s \cdot t$ be the couniversal factorization of $r\colon X \nrightarrow Z$. Then*

$$(L \cdot K)r = Ls \cdot Kt.$$

We proceed to establish that the composition operation defined above equips relational connectors with the structure of a quasicategory (i.e. overlarge category). We first check associativity:

Lemma 3.9. *Let $K\colon F \to G$, $L\colon G \to H$, and $M\colon H \to V$ be relational connectors. Then $(M \cdot K) \cdot L = M \cdot (K \cdot L)$.*

The straightforward proof uses join continuity of relational composition. We next construct identities:

Definition 3.10 (Identity relational connectors). The *identity relational connector* $\mathsf{Id}^{\mathsf{c}}_F\colon F \to F$ on a set functor F is defined as follows. For $r\colon X \nrightarrow Y$, $b \in FX$, and $c \in FY$, we put $b\ \mathsf{Id}^{\mathsf{c}}_F r\ c$ iff for all set functors G, all relational connectors $L\colon G \to F$, all $s\colon Z \nrightarrow X$, and all $a \in GZ$,

$$a\ Ls\ b \quad \text{implies} \quad a\ L(r \cdot s)\ c.$$

(This definition is highly impredicative, but we will later give a characterization of $\mathsf{Id}^{\mathsf{c}}_F$ that eliminates quantification over relational connectors.) We will show that $\mathsf{Id}^{\mathsf{c}}_F$ is neutral w.r.t. composition of relational connectors. We first note that, as an immediate consequence of the definition,

$$\Delta_{FX} \subseteq \mathsf{Id}^{\mathsf{c}}_F \Delta_X \qquad \text{for all } X. \tag{3.3}$$

Lemma 3.11. *For each functor F, $\mathsf{Id}^{\mathsf{c}}_F$ is a relational connector.*

The proof of naturality relies in particular on monotonicity of relational connectors in combination with totality and univalence of maps. We show next that identity connectors do actually act as identities under composition:

Lemma 3.12. *For each $L\colon G \to F$, we have $L = \mathsf{Id}^{\mathsf{c}}_F \cdot L = L \cdot \mathsf{Id}^{\mathsf{c}}_G$.*

Proof (sketch). One shows, using (3.3) inter alia, that $\mathsf{Id}^{\mathsf{c}}_F$ is a left identity ($L = \mathsf{Id}^{\mathsf{c}}_F \cdot L$). By a symmetric argument, composition of relational connectors also has right identities, and then the left and right identities are necessarily equal. □

Relational connectors admit a natural notion of converse:

Definition 3.13 (Converse, meet and product of relational connectors). The *converse* $L^\circ\colon G \to F$ of a relational connector $L\colon F \to G$ is given by

$$L^\circ r = (Lr^\circ)^\circ\colon GX \nrightarrow FY$$

for $r\colon X \nrightarrow Y$. The *meet* $L \cap K$ of relational connectors $L, K\colon F \to G$ is their componentwise intersection ($(L \cap K)r = Lr \cap Kr$). For relational connectors $L_1\colon F_1 \to G_1$ and $L_2\colon F_2 \to G_2$, their *product* $L_1 \times L_2\colon F_1 \times F_2 \to G_1 \times G_2$ is given by

$$(a, b)\ (L_1 \times L_2)r\ (c, d) \iff a\ L_1 r\ c \text{ and } b\ L_2 r\ d.$$

Lemma 3.14. *The converse, meet and product of relational connectors are again relational connectors.*

We record some expected properties of converse:

Lemma 3.15. *Converse is involutive ($(L^\circ)^\circ = L$) and monotone. Moreover, for relational connectors $K\colon F \to G$ and $L\colon G \to H$, we have*

$$(L \cdot K)^\circ = K^\circ \cdot L^\circ.$$

Remark 3.16. In view of the above properties, one may ask whether relational connectors form an overlarge allegory [12]. We leave this question open for the moment; specifically, it is not clear that relational connectors satisfy the *modular law* $(L \cdot K) \cap M \leq L \cdot (K \cap (L^\circ \cdot M))$.

Example 3.17 (Constructions of relational connectors). We can decompose the connector $L_R\colon \mathcal{P}(\mathcal{A} \times (-)) \to \mathcal{P}(\mathcal{B} \times (-))$ from Example 3.2 as follows. Define a further relational connector $K_R\colon \mathcal{P}(\mathcal{A} \times (-)) \to \mathcal{P}(\mathcal{B} \times (-))$ similarly as L_R but omit one of the directions, putting $S\ K_R r\ T$ (for $S \in \mathcal{P}(\mathcal{A} \times X)$, $T \in \mathcal{P}(\mathcal{B} \times Y)$, and $r\colon X \nrightarrow Y$) iff for all $(l, m) \in R$ and $(l, x) \in S$, there is $(m, y) \in T$ such that $x\ r\ y$. While L_R has the feel of inducing a notion of heterogeneous bisimilarity (this will be made formal in Section 4), K_R has a flavour of similarity, including as it does only a 'forth'-type condition. Clearly, we have

$$L_R = K_R \cap K^\circ_{R^\circ}.$$

Given a further set $\mathcal{C}$ of labels and a relation $Q \subseteq \mathcal{B} \times \mathcal{C}$, we have

$$K_Q \cdot K_R = K_{Q \cdot R} \quad \text{and} \quad L_Q \cdot L_R \leq L_{Q \cdot R}.$$

It is a fairly typical phenomenon in describing composites of relational connectors that upper bounds such as the above are often straightforward, while the converse inequalities are more elusive or fail to hold. When showing $K_{Q \cdot R}\, r \subseteq (K_Q \cdot K_R)\, r$ for $r\colon X \nrightarrow Z$, one gets away with using the trivial factorization $r = s \cdot t$ given by $s = r$, $t = \Delta_X$, while for a full description of $L_Q \cdot L_R$, we need to use Theorem 3.8. Specifically, for $S \in \mathcal{P}(\mathcal{A} \times X)$, $U \in \mathcal{P}(\mathcal{B} \times X)$, we have $S\ (L_Q \cdot L_R)r\ U$ iff S and U satisfy conditions *forth* and *back*, where *forth* is given as follows and *back* is given symmetrically: Whenever $(l, m) \in R$ and $(l, x) \in S$, then there are $A \in \mathcal{P}(X)$, $B \in \mathcal{P}(Z)$ such that $A \times B \subseteq r$ and $x \in A$, and moreover (i) for all $(l', m) \in R$, there is $x' \in A$ such that $(l', x') \in S$, and (ii) for all $(m, p) \in Q$, there is $z \in B$ such that $(p, z) \in U$.

A further straightforward way to obtain relational connectors is to pull them back along natural transformations:

Lemma and Definition 3.18. *Let $L\colon F \to G$ be a relational connector, and let $\alpha\colon F' \Rightarrow F$, $\beta\colon G' \Rightarrow G$ be natural transformations. Then we have relational connectors $L \bullet \alpha\colon F' \to G$, $\beta^\circ \bullet L\colon F \to G'$ defined on $r\colon X \nrightarrow Y$ by $(L \bullet \alpha)r = Lr \cdot \alpha_X$ and $(\beta^\circ \bullet L)r = (\beta_Y)^\circ \cdot Lr$, respectively.*

In particular, from $\alpha\colon F \to G$, we always obtain a relational connector $\alpha \bullet \mathsf{Id}^{\mathsf{c}}_G\colon F \to G$, which plays a distinguished role:

Definition 3.19. A relational connector $L\colon F \to G$ *extends* a natural transformation $\alpha\colon F \to G$ if $\alpha_X \leq L\Delta_X$ for all X.

(In particular, $L\colon F \to F$ extends F iff L extends id_F.)

Theorem 3.20. *Let $\alpha\colon F \to G$ be a natural transformation. The relational connector $\mathsf{Id}^{\mathsf{c}}_G \bullet \alpha$ is the least relational connector that extends α. In particular, $\mathsf{Id}^{\mathsf{c}}_G$ is the least relational connector that extends G.*

Example 3.21. We have a variant L_{f} of the Barr extension of the functor $F = \mathcal{P}(\mathcal{A} \times (-))$ modelling $\mathcal{A}$-LTS (Example 2.1) given by including only the *forth* condition: For $r\colon X \nrightarrow Y$, $S \in FX$, $T \in FY$, we put $S\ L_{\mathsf{f}}r\ T$ iff for all $(l, x) \in S$, there is $(l, y) \in T$ such that $x\ r\ y$. Now let $\iota\colon \mathcal{A} \times (-) \Rightarrow F$ be the inclusion natural transformation. Then we have a relational connector $L_{\mathsf{t}} = L_{\mathsf{f}} \bullet \iota\colon \mathcal{A} \times (-) \to F$; explicitly, for $r\colon X \nrightarrow Y$, $(l, x) \in \mathcal{A} \times X$, and $T \in FY$, we have $(l, x)\ L_{\mathsf{t}}r\ T$ iff there exists $(l, y) \in T$ such that $x\ r\ y$. By itself, L_{t} is not yet very interesting, but we can build further relational connectors using the constructions introduced above; for instance, we have a relational connector $L_{\mathsf{t}} \cdot L_{\mathsf{t}}^\circ\colon F \to F$, described by $S\ (L_{\mathsf{t}} \cdot L_{\mathsf{t}}^\circ)r\ T$ iff there exist $(a, x) \in S$, $(a, y) \in T$ such that $x\ r\ y$; this connector is symmetric and extends F but fails to be transitive, hence is not a lax extension. We will later employ $L_{\mathsf{t}} \cdot L_{\mathsf{t}}^\circ$ to relate LTS that share an infinite trace (Example 4.8).

Example 3.22. Consider again the functors $F = \mathcal{P}(\mathcal{A}\times(-))$ and $G = \mathcal{P}(\mathcal{B}\times(-))$ together with a fixed relation on labels $R \subseteq \mathcal{A} \times \mathcal{B}$. Note that, for every set X, the elements of FX and GX can be interpreted as relations $\mathcal{A} \nrightarrow X$ and $\mathcal{B} \nrightarrow X$, respectively. Define the natural transformation $\alpha\colon F \Rightarrow G$ by $\alpha_X(S) = S \cdot R^\circ$. Let $L^G_{\mathsf{f}}\colon G \to G$ be the 'forth' relational connector from Example 3.21 instantiated to G, and consider the relational connector $L^G_{\mathsf{f}} \bullet \alpha$. For $S \in FX$, $T \in GY$ and $r\colon X \nrightarrow Y$, we have $S\ (L^G_{\mathsf{f}} \bullet \alpha)r\ T$ iff $(S \cdot R^\circ)\ L^G_{\mathsf{f}} r\ T$. Explicitly, the latter means that if $(l,m) \in R$ and $(l,x) \in S$, then there is $y \in Y$ such that $(m,y) \in T$ and $x\ r\ y$. This coincides with the relational connector K_R from Example 3.17, which is hence induced by a natural transformation and a lax extension. (It does not seem to be the case that L_R as per Example 3.2/Example 3.17 is induced in this way.)

We can instead compose with a natural transformation on the other side. Let $\beta\colon G \Rightarrow F$ be given by $\beta_X(T) = T \cdot R$, and let $L^F_{\mathsf{f}} : G \to G$ be the connector L^F_{f} from Example 3.21, instantiated to F. The connector $\beta^\circ \bullet L^F_{\mathsf{f}} : F \to G$ is given, for $S \in FX$, $T \in GY$ and $r\colon X \nrightarrow Y$, by $S\ (\beta^\circ \bullet L^F_{\mathsf{f}})r\ T$ iff $S\ L^F_{\mathsf{f}} r\ (T \cdot R)$. Hence,

$$S\ (\beta^\circ \bullet L^F_{\mathsf{f}})r\ T \iff \forall (l,x) \in S.\ \exists (l,m) \in R.\ (m,y) \in T \text{ and } x\ r\ y,$$

which differs from K_R in that here, the quantification over R is existential.

Remark 3.23. Analogously to the fact that lax extensions of a functor F are equivalent to certain liftings of F to the category of preordered sets [13], relational connectors $F \to G$ can be identified with certain liftings of $F \times G\colon \mathsf{Set}^2 \to \mathsf{Set}^2$ to the category of binary relations and relation-preserving pairs of functions. Indeed, this category is a fibration over Set^2, and the relational connectors are precisely the liftings that preserve cartesian morphisms; a condition that has featured in situations where liftings of a functor F are used to derive notions of "behavioural conformance" for F-coalgebras (e.g. [3,18,11,45]).

3.3 Lax Extensions as Relational Connectors

For context, we briefly discuss how the additional properties of lax extensions are phrased in terms of the constructions from Section 3, and in particular how lax extensions relate to identity relational connectors.

Definition 3.24. A relational connector $L\colon F \to F$ is *transitive* if $L \cdot L \leq L$, and *symmetric* if $L^\circ \leq L$. Moreover, L *extends* F if $\Delta_{FX} \subseteq L\Delta_X$ for all X.

The following observations are straightforward.

Lemma 3.25. *Let $L\colon F \to F$ be a relational connector. Then L is symmetric iff $L^\circ = L$ iff $L \leq L^\circ$.*

Lemma 3.26. *Let $L\colon F \to F$ be a relational connector. Then the following hold.*

1. *L satisfies condition (L2) in the definition of lax extension iff L is transitive.*
2. *L satisfies condition (L3) in the definition of lax extension iff L extends F.*

3. *L preserves converse iff L is symmetric.*
4. *L is a lax extension of F iff L is transitive and extends F.*
5. *If L extends F, then $L \subseteq L \cdot L$.*
6. *If L is a lax extension, then L is idempotent, i.e. $L \cdot L = L$.*

Since lax extensions satisfy naturality, this implies

Theorem 3.27. *The lax extensions of a set functor F are precisely the transitive relational connectors that extend F.*

As indicated above, a special role is played by identity relational connectors:

Theorem 3.28. *Let F be a set functor. Then, $\mathsf{Id}^{\mathsf{c}}_F$ is a symmetric lax extension of F. Moreover, F has a diagonal-preserving lax extension iff $\mathsf{Id}^{\mathsf{c}}_F$ preserves diagonals.*

Proof (sketch). Most subclaims are obvious by Lemma 3.26 and (3.3). To see that $\mathsf{Id}^{\mathsf{c}}_F$ is symmetric, show that $(\mathsf{Id}^{\mathsf{c}}_F)^\circ$ is a right identity: For $L\colon F \to G$, we have $L \cdot (\mathsf{Id}^{\mathsf{c}}_F)^\circ = (\mathsf{Id}^{\mathsf{c}}_F \cdot L^\circ)^\circ = (L^\circ)^\circ = L$ (using Lemma 3.15). □

In connection with Theorem 3.20, we obtain moreover:

Corollary 3.29. *The identity relational connector $\mathsf{Id}^{\mathsf{c}}_F$ is both the smallest lax extension and the smallest symmetric lax extension of a set functor F.*

Example 3.30. If F preserves weak pullbacks, then $\mathsf{Id}^{\mathsf{c}}_F$ is the Barr extension of F (cf. Remark 2.2); this is immediate from Theorem 3.28, as one shows easily that the Barr extension is below every converse-preserving lax extension. For instance, the standard Egli-Milner lifting is an identity relational connector.

4 Heterogeneous (Bi)simulations

We proceed to introduce a notion of heterogeneous (bi)simulations relating systems of different type; we induce such notions from relational connectors.

Definition 4.1. Let $L\colon F \to G$ be a relational connector. A relation $r\colon C \nrightarrow D$ is an *L-simulation* between an F-coalgebra (C, γ) and a G-coalgebra (D, δ) if

$$\text{whenever } x\ r\ y, \text{ then } \gamma(x)\ Lr\ \delta(y);$$

in pointfree notation, this means that $r \subseteq \delta^\circ \cdot Lr \cdot \gamma$, equivalently $\delta \cdot r \subseteq Lr \cdot \gamma$. States $x \in C$, $y \in D$ are *L-similar* if there exists an L-simulation r such that $x\ r\ y$, in which case we write $x \preceq_L y$. Occasionally, we will designate the ambient coalgebras C, D explicitly by writing $x \preceq_L^{C,D} y$; thus, $\preceq_L^{C,D}$ is a relation $C \nrightarrow D$.

In case $F = G$, r is an *L-bisimulation* if r and r° are L-simulations. Correspondingly, states $x \in C$, $y \in D$ are *L-bisimilar* if there exists an L-bisimulation r such that $x\ r\ y$, in which case we write $x \simeq_L y$ or, more explicitly, $x \simeq_L^{C,D} y$.

We note that in case L is a lax extension, these definitions match existing terminology (e.g. [30]). Monotonicity of relational connectors ensures that by the Knaster-Tarski theorem, $\preceq_L$ is the greatest fixpoint of the map taking r to $\delta^\circ \cdot Lr \cdot \gamma$, and in particular is itself an L-simulation, correspondingly for $\simeq_L$. We note that L-similarity is invariant under coalgebra morphisms (Section 2), a key fact that hinges on monotonicity and naturality of relational connectors, lending further support to our choice of axiomatics:

Lemma 4.2. *Let $L\colon F \to G$ be a connector, let $r\colon C \nrightarrow D$ be an L-simulation between an F-coalgebra (C,γ) and a G-coalgebra (D,δ), and let $f\colon (C',\gamma') \to (C,\gamma)$, $g\colon (C,\gamma) \to (C'',\gamma'')$ be F-coalgebra morphisms. Then $r \cdot f$ and $r \cdot g^\circ$ are L-simulations. Symmetric properties hold for G-coalgebra morphisms. Thus, L-similarity is closed under behavioural equivalence (Section 2) on both sides.*

Notions of (bi)simulation interact well with composition and converse of relational connectors:

Lemma 4.3 (Composites of simulations). *Let $K\colon F \to G$ and $L\colon G \to H$ be relational connectors, and let (C,γ) be an F-coalgebra, (D,δ) a G-coalgebra, and (E,ε) an H-coalgebra. Then the composite $s \cdot r\colon C \nrightarrow E$ of a K-simulation $r\colon C \nrightarrow D$ and an L-simulation $s\colon D \nrightarrow E$ is an $L \cdot K$-simulation. Thus,*

$$\preceq_L^{D,E} \cdot \preceq_K^{C,D} \subseteq \preceq_{L\cdot K}^{C,E} \quad \text{and (if } F = G\text{)} \quad \simeq_L^{D,E} \cdot \simeq_K^{C,D} \subseteq \simeq_{L\cdot K}^{C,E}.$$

Lemma 4.4 (Converses of simulations). *Let $L\colon F \to G$ be a relational connector, let (C,γ) be an F-coalgebra, and let (D,δ) be a G-coalgebra. If $r\colon C \nrightarrow D$ is an L-simulation, then $r^\circ\colon D \nrightarrow C$ is an L°-simulation. Thus,*

$$\preceq_{L^\circ}^{C,D} = (\preceq_L^{D,C})^\circ \quad \text{and (if } F = G\text{)} \quad \simeq_{L^\circ}^{C,D} = (\simeq_L^{D,C})^\circ$$

It follows that notions of (bi)similarity inherit properties expressed in terms of converse and composition from the inducing lax extensions; for instance:

Lemma 4.5. *Let $L\colon F \to F$ be a relational connector. Then the following hold.*

1. *If L is transitive, then $\preceq_L$ and $\simeq_L$ are transitive.*
2. *If L is symmetric, then $\simeq_L$ is symmetric. Moreover, every L-simulation is an L-bisimulation, so $\preceq_L = \simeq_L$.*
3. *If L extends F, then $\preceq_L$ and $\simeq_L$ are reflexive.*

As a further immediate consequence of Lemma 4.3 and Lemma 4.4, we have the following criterion for preservation of (bi)similarity under relational connectors:

Theorem 4.6 (Transfer of bisimilarity). *Let $K\colon F \to F$, $L\colon F \to G$, $H\colon G \to G$ be relational connectors such that $L \cdot K \cdot L^\circ \leq H$. Then $\preceq_L \cdot \preceq_K \cdot \preceq_L^\circ \subseteq \preceq_H$ and $\simeq_L \cdot \simeq_K \cdot \simeq_L^\circ \subseteq \simeq_H$.*

Example 4.7 (Transfer of bisimilarity between LTS of different type). Recall the relational connector $L_R\colon \mathcal{P}(\mathcal{A} \times (-)) \to \mathcal{P}(\mathcal{B} \times (-))$ induced from a relation $R\colon \mathcal{A} \nrightarrow \mathcal{B}$ as per Example 3.2. We note that $L_{R^\circ} = (L_R)^\circ$. This implies that for every L-simulation r, r° is an L_{R°-simulation, so we suggestively write $\simeq_R$ for $\preceq_{L_R}$ and speak of L_R-*bisimilarity.*

Recall that the usual notion of bisimilarity on LTS is captured by the identity relational connectors on F and G, respectively (Example 2.1, Example 3.30). It is straightforward to check that if R is right total, then

$$L_R \cdot id_F \cdot L_R^\circ = L_R \cdot L_R^\circ \leq id_G,$$

so that by Theorem 4.6, $\simeq_R$ transfers bisimilarity from F-coalgebras to G-coalgebras. In elementwise notation, this is phrased as follows: Let c, c' be states in an F-coalgebra C, and let d, d' be states in a G-coalgebra D such that $c' \simeq_R d'$, $c \simeq_R d$, and $c \simeq_F c'$. Then $d \simeq_G d'$. Similarly, if R is left total, then $\simeq_R$ transfers bisimilarity from G-coalgebras to F-coalgebras, so of course if R is left and right total, then it transfers bisimilarity in both directions. A similar principle is under the hood of the proof of the operational equivalence of the standard λ-calculus and a variable-free variant called the algebraic λ-calculus in recent work on higher-order mathematical operational semantics [15].

Example 4.8 (Shared traces). Recall the symmetric relational connector $L_t \cdot L_t^\circ \colon F \nrightarrow F$ from Example 3.21, where $F = \mathcal{P}(\mathcal{A} \times (-))$ is the functor modelling $\mathcal{A}$-LTS. States x, y in $\mathcal{A}$-LTS are $L_t \cdot L_t^\circ$-bisimilar iff x and y have a common infinite trace. We may view x as specifying a set of bad infinite traces; then x and y are *not* $L_t \cdot L_t^\circ$-bisimilar iff y does *not* have a bad infinite trace.

Example 4.9 (Weak simulation). Let $\mathcal{A}$ be a set of labels, with $\tau \in \mathcal{A}$ a distinguished label for "internal" steps. Let $\mathcal{A}^*$ be the set of words over $\mathcal{A}$, with the empty word denoted by ε, $F = \mathcal{P}(\mathcal{A} \times (-))$ and $G = \mathcal{P}(\mathcal{A}^* \times (-))$. We define a relational connector $L \colon F \nrightarrow G$ by instantiating (the second half of) Example 3.22 to $R \subseteq \mathcal{A} \times \mathcal{A}^*$ given by $R = \{(l, \tau^i l \tau^j) \mid l \in \mathcal{A}, i, j \geq 0\} \cup \{(\tau, \varepsilon)\}$. In the particular case where the transitions in the G-coalgebra (D, δ) at hand arise by composing transitions from an F-coalgebra (D, δ_0), L-simulations from an F-coalgebra (C, γ) to (D, δ) are precisely *weak simulations* between the $\mathcal{A}$-LTS (C, γ) and (D, δ_0).

Example 4.10 (Conformance testing). In model-based testing, a *specification* is compared to an *implementation.* Typically, both specifications and implementations are modelled as transition systems, and a given notion of *conformance* stipulates when an implementation is correct w.r.t. a specification. In the case of the *ioco* (input/output conformance) relation [43], the specification is an LTS over a set of input and output labels. The implementation is an LTS as well, but is required to be *input-enabled,* meaning that for every state and every input label there is an outgoing transition with that label. We focus on the deterministic case, which enables a coinductive formulation of ioco conformance [8]. This example has been cast in a general coalgebraic framework [36], in which however the distinction between the type of specification and implementation cannot be made (and in fact, they are assumed to have the same state space).

We write $X \to Y$ and $X \rightharpoonup Y$ for the sets of total and partial functions from X to Y, respectively. We denote the domain of $f \colon X \rightharpoonup Y$ by $\mathrm{dom}(f) \subseteq X$, and put $X \rightharpoonup_{\mathrm{ne}} Y = \{f: X \rightharpoonup Y \mid \mathrm{dom}(f) \neq \emptyset\}$. Now let I, O be input and output alphabets, respectively. Define the functor F by $F(X) = (I \rightharpoonup X) \times (O \rightharpoonup_{\mathrm{ne}} X)$, and the functor G by $G(X) = (I \to X) \times (O \rightharpoonup_{\mathrm{ne}} X)$. An F-coalgebra is a *suspension automaton*, which is *non-blocking* (there is always at least one output-labelled transition from every state). A G-coalgebra is an *input-enabled* suspension automaton.

Define $L: F \to G$ on $r: X \nrightarrow Y$ by

$$(\delta_I, \delta_O)\; Lr\; (\tau_I, \tau_O) \iff \begin{array}{l} \forall i \in \mathrm{dom}(\delta_I).\ \delta_I(i)\ r\ \tau_I(i), \quad \text{and} \\ \forall o \in \mathrm{dom}(\tau_O).\ o \in \mathrm{dom}(\delta_O) \text{ and } \delta_O(o)\ r\ \tau_O(o). \end{array}$$

This is a relational connector, and L-simulations capture precisely the ioco-relation on suspension automata, in the coinductive formulation given in [8].

The composite relational connector $L^\circ \cdot L: F \to F$ is described as follows:

$$(\delta_I, \delta_O)\; (L^\circ \cdot L)r\; (\delta'_I, \delta'_O) \iff \begin{array}{l} \forall i \in \mathrm{dom}(\delta_I) \cap \mathrm{dom}(\delta'_I).\ \delta_I(i)\ r\ \delta'_I(i), \quad \text{and} \\ \exists o \in \mathrm{dom}(\delta_I) \cap \mathrm{dom}(\delta'_I).\ \delta_O(o)\ r\ \delta'_O(o). \end{array}$$

The existential quantification on outputs arises in this factorization due to the non-emptyness of the domain of partial functions $O \rightharpoonup_{\mathrm{ne}} X$. Simulations for this composite relational connector are precisely the *ioco compatibility* relations between specifications [8], generalized to a coalgebraic setting in [36].

5 Kantorovich Relational Connectors

We next present a construction of relational connectors from relations between modalities for the given functors; in honour of the formal analogy with the classical Kantorovich metric and its coalgebraic generalizations [3,47,41], we refer to the arising connectors as *Kantorovich relational connectors.*

In this context, modalities are understood as induced by predicate liftings in the style of coalgebraic logic [33,38], and indeed we use the terms *modality* and *predicate lifting* interchangeably. Recall that an n-ary *predicate lifting* for a functor F is a natural transformation λ with components

$$\lambda_X: (2^X)^n \to 2^{FX}$$

(or just λ) where $2^{(-)}$ denotes the *contravariant powerset functor*; that is, 2^X is the powerset of a set X, and $2^f: 2^Y \to 2^X$ takes preimages under a map $f: X \to Y$. The naturality condition thus says explicitly that, for $a \in FX$ $f: X \to Y$, and $A_1, \ldots, A_n \in 2^Y$, we have $Ff(a) \in \lambda_Y(A_1, \ldots, A_n)$ iff $a \in \lambda_X(f^{-1}[A_1], \ldots, f^{-1}[A_n])$. We say that λ is *monotone* if $\lambda(A_1, \ldots, A_n) \subseteq \lambda(B_1, \ldots, B_n)$ whenever $A_i \subseteq B_i$ for $i = 1, \ldots, n$. The *dual* $\overline{\lambda}$ of λ is the predicate lifting defined by $\overline{\lambda}_X(A_1, \ldots, A_n) = FX \setminus \lambda_X(X \setminus A_1, \ldots, X \setminus A_n)$.

In logical syntax, we abuse λ as an n-ary modality: If $\phi_1, \ldots, \phi_n$ are formulae in some modal logic equipped with a satisfaction relation $\models$ between states in F-coalgebras and formulae, with extensions $[\![\phi_i]\!] = \{x \in C \mid x \models \phi_i\} \in 2^C$ in a given F-coalgebra (C, γ), then the semantics of the modalized formula

$\lambda(\phi_1, \ldots, \phi_n)$ is given by $x \models \lambda(\phi_1, \ldots, \phi_n)$ iff $\gamma(x) \in \lambda_C([\![\phi_1]\!], \ldots, [\![\phi_n]\!])$. For instance, the unary predicate lifting $\Diamond$ for the powerset functor $\mathcal{P}$ given by $\Diamond_X(A) = \{S \in \mathcal{P}(X) \mid S \cap A \neq \emptyset\}$ captures precisely the usual diamond modality on Kripke frames ('there exists some successor such that').

A set Λ of monotone predicate liftings for F induces a lax extension L_Λ of F defined for $r \colon X \nrightarrow Y$, $a \in FX$, and $b \in FY$ by $a\ L_\Lambda r\ b$ iff whenever $a \in \lambda_X(A_1, \ldots, A_n)$ for n-ary $\lambda \in \Lambda$ and $A_1, \ldots, A_n$, then $b \in \lambda_Y(r[A_1], \ldots, r[A_n])$ (cf. [29,30,16]). We show that more generally, one can induce relational connectors from *relations* between predicate liftings:

Definition 5.1 (Kantorovich connectors). For a functor F, we write $\mathsf{PL}(F)$ for the set of monotone predicate liftings for F. Now let F, G be functors, and let Λ be a relation $\Lambda \colon \mathsf{PL}(F) \nrightarrow \mathsf{PL}(G)$ that *preserves arity*; that is, if $(\lambda, \mu) \in \Lambda$, then λ and μ have the same arity, which we then view as the *arity* of (λ, μ). We define a relational connector $L_\Lambda \colon F \to G$ for $r \colon X \nrightarrow Y$, $a \in FX$, and $b \in GY$ by $a\ L_\Lambda r\ b$ iff whenever $(\lambda, \mu) \in \Lambda$ is n-ary and $A_1, \ldots, A_n \in 2^X$, then

$$a \in \lambda_X(A_1, \ldots, A_n) \quad \text{implies} \quad b \in \mu_Y(r[A_1], \ldots, r[A_n]).$$

We briefly refer to L_Λ-similarity as *Λ-similarity*, and write $\preceq_\Lambda$ for $\preceq_{L_\Lambda}$. A relational connector L is *Kantorovich* if it has the form $L = L_\Lambda$ for a suitable Λ as above. We write $\overline{\Lambda} = \{(\overline{\lambda}, \overline{\mu}) \mid (\lambda, \mu) \in \Lambda\}$.

Theorem 5.2. *Under Definition 5.1, L_Λ is indeed a relational connector.*

Example 5.3. 1. For every $l \in \mathcal{A}$, we have a predicate lifting $\Diamond_l$ for $\mathcal{P}(\mathcal{A} \times (-))$ given by $\Diamond_l(A) = \{S \in \mathcal{P}(\mathcal{A} \times X) \mid \exists x \in A.\ (l, x) \in S\}$. The arising modality is the usual diamond modality of Hennessy-Milner logic, and the dual of $\Diamond_l$ is the usual box modality $\Box_l$. The connectors $K_R, L_R \colon \mathcal{P}(\mathcal{A} \times (-)) \to \mathcal{P}(\mathcal{B} \times (-))$ from Example 3.17 are Kantorovich: We have $K_R = L_\Lambda$ and $L_R = L_{\Lambda \cup \overline{\Lambda}}$ for $\Lambda = \{(\Diamond_l, \Diamond_m) \mid (l, m) \in R\}$.

2. We can restrict the predicate lifting $\Diamond_l$ from the previous item to a predicate lifting $\Diamond_l$ for $\mathcal{A} \times (-)$ (so $\Diamond_l(A) = \{(l, x) \mid x \in A\}$). The relational connector $L_\mathsf{t} = L_\mathsf{f} \bullet \iota \colon \mathcal{A} \times (-) \to \mathcal{P}(\mathcal{A} \times (-))$ from Example 3.21 is Kantorovich for $\Lambda = \{(\Diamond_l, \Diamond_l) \mid l \in \mathcal{A}\}$. We will later give a Kantorovich description of the composite connector $L_\mathsf{t} \cdot L_\mathsf{t}^\circ$ (Example 5.7).

3. Given a label $l \in \mathcal{A}$, define the predicate lifting $\Diamond_l$ for $\mathcal{A} \rightharpoonup (-)$ by $\Diamond_l(A) = \{\delta \colon \mathcal{A} \rightharpoonup X \mid l \in \mathrm{dom}(\delta) \text{ and } \delta(l) \in A\}$ for $A \in 2^X$. Its dual is given by $\Box_l(A) = \{\delta \mid l \in \mathrm{dom}(\delta) \text{ implies } \delta(l) \in A\}$. Further, we define a 0-ary modality $\downarrow_l = \{\delta \mid l \notin \mathrm{dom}(\delta)\}$. These modalities allow us to capture the *ioco* connector $L \colon F \to G$ from Example 4.10. First, assuming that I and O are disjoint, the modalities $\Diamond_l, \Box_l, \downarrow_l$ for $i \in I \cup O$ can be extended to F and G in the obvious way by projection (and they are extended to total functions and partial functions with a non-empty domain without change). Now, L is Kantorovich for $\Lambda = \{(\Diamond_i, \Diamond_i) \mid i \in I\} \cup \{(\Box_o, \Box_o) \mid o \in O\} \cup \{(\downarrow_o, \downarrow_o) \mid o \in O\}$.

4. Given $\epsilon \in [0,1]$, we have predicate liftings $L_{\epsilon,l}$ (for $l \in \mathcal{A}$) for the functor $\mathcal{D}(\mathcal{A} \times (-))$ modelling probabilistic LTS, given by $L_{\epsilon,l}(A) = \{\alpha \in \mathcal{D}(\mathcal{A} \times X) \mid$

$\alpha(\{l\} \times A) \geq \epsilon\}$ for $A \in 2^X$. Putting $\Lambda = \{(\Diamond_l, L_{\epsilon,l}) \mid l \in \mathcal{A}\}$, we obtain relational connectors $L_\Lambda, L_{\overline{\Lambda}}, L_{\Lambda\cup\overline{\Lambda}} \colon \mathcal{P}(\mathcal{A} \times (-)) \to \mathcal{D}(\mathcal{A} \times (-))$. Explicitly, for $r \colon X \nrightarrow Y$, $S \in \mathcal{P}(\mathcal{A} \times X)$, and $\alpha \in \mathcal{D}(\mathcal{A} \times Y)$, we have (i) $S\ L_\Lambda r\ \alpha$ iff whenever $(l,x) \in S$, then $\alpha(\{l\} \times r[\{x\}]) \geq \epsilon$; (ii) $S\ L_{\overline{\Lambda}} r\ \alpha$ iff whenever $\alpha(\{l\} \times B) \geq \epsilon$, then there are $(l,x) \in S$ and $y \in B$ such that $x\ r\ y$; and (iii) $S\ L_{\Lambda\cup\overline{\Lambda}} r\ \alpha$ iff both (i) and (ii) hold. Roughly speaking, similarity w.r.t. these connectors between an $\mathcal{A}$-LTS C and probabilistic $\mathcal{A}$-LTS D specifies what may happen in D with non-negligible probability, where ϵ specifies the negligibility threshold. For instance, an L_Λ-simulation $r \colon C \to D$ witnesses that behaviour embodied in C is enabled with non-negligible probability in D, while an $L_{\overline{\Lambda}}$-simulation $r \colon C \to D$ witnesses that things that can happen with non-negligible probability in D are foreseen in C.

We record basic facts on the interaction of the Kantorovich construction with composition and converse of relational connectors:

Theorem 5.4. *Let $\Lambda \colon \mathsf{PL}(F) \nrightarrow \mathsf{PL}(G)$ and $\Theta \colon \mathsf{PL}(G) \nrightarrow \mathsf{PL}(H)$. Then*

1. $L_\Theta \cdot L_\Lambda \leq L_{\Theta\cdot\Lambda}$
2. $(L_\Lambda)^\circ = L_{(\overline{\Lambda})^\circ}$.

(Recall that $\overline{\Lambda}$ dualizes all modalities.) Specializing to relational connectors $F \to F$, we thus recover the standard way of inducing lax extensions from predicate liftings [30,14] as described above:

Corollary 5.5. *Let F be a functor, and let $\Lambda \colon \mathsf{PL}(F) \nrightarrow \mathsf{PL}(F)$.*

1. *If $\Lambda \cdot \Lambda \subseteq \Lambda$, then L_Λ is transitive.*
2. *If Λ is closed under duals, i.e. $\overline{\Lambda} \subseteq \Lambda$ (equivalently $\overline{\Lambda} = \Lambda$), and symmetric, then L_Λ is symmetric.*
3. *If $\Lambda \subseteq id$, then L_Λ is a lax extension of F.*
4. *If $\Lambda \subseteq id$ and the set $\{\lambda \mid (\lambda,\lambda) \in \Lambda\}$ of predicate liftings is separating, then L_Λ is a normal lax extension of F.*

Remark 5.6 (Composing Kantororovich connectors). The upper bound $L_\Theta \cdot L_\Lambda \leq L_{\Theta\cdot\Lambda}$ on composites of Kantorovich connectors L_Θ, L_Λ given in Theorem 5.4 is not always tight. In the simple case of the connectors $K_R \colon \mathcal{P}(\mathcal{A} \times (-)) \to \mathcal{P}(\mathcal{B} \times (-))$ induced by relations $R \colon \mathcal{A} \nrightarrow \mathcal{B}$ (Examples 3.17 and 5.3), we do indeed have exact equality (Example 3.17). For the general case, one can improve the upper bound (by including *more* pairs of modalities) in at least two ways. First, in the composite $\Theta \cdot \Lambda$ of the relations $\Lambda \colon \mathsf{PL}(F) \nrightarrow \mathsf{PL}(G)$ and $\Theta \colon \mathsf{PL}(G) \nrightarrow \mathsf{PL}(H)$, one can include weakening in the middle step. Formally, we write $\leq$ for the pointwise inclusion order on predicate liftings, and put $\Theta \blacktriangleright \Lambda = \{(\lambda,\pi) \mid \exists(\lambda,\mu) \in \Lambda, (\mu',\pi) \in \Theta \mid \mu \leq \mu'\}$. Then $L_\Theta \cdot L_\Lambda \leq L_{\Theta\blacktriangleright\Lambda}$. Moreover, monotone predicate liftings are closed under taking positive Boolean combinations both above and below; e.g. if λ and μ are unary monotone predicate liftings, then the transformation π taking predicates A, B to $\lambda(A \vee B) \wedge \mu(A \wedge B)$ is a binary monotone predicate lifting. We write Λ^{pos} for the closure of Λ under componentwise positive Boolean combinations in this sense; e.g. if

$(\lambda_1, \lambda_2), (\mu_1, \mu_2) \in \Lambda$, then $(\pi_1, \pi_2) \in \Lambda^{\mathsf{pos}}$ where $\pi_i(A, B) = \lambda_i(A \vee B) \wedge \mu_i(A \wedge B)$. One checks easily that $L_\Lambda = L_{\Lambda^{\mathsf{pos}}}$, so overall we have

$$L_\Theta \cdot L_\Lambda \leq L_{\Theta^{\mathsf{pos}} \blacktriangleright \Lambda^{\mathsf{pos}}}. \tag{5.1}$$

We next give an example where one actually has equality; we leave it as an open problem whether equality holds in general.

Example 5.7. Recall from Example 5.3.2 that the connector $L_{\mathsf{t}} \colon \mathcal{A} \times (-) \to \mathcal{P}(\mathcal{A} \times (-))$ equals L_Λ where $\Lambda = \{(\Diamond_l, \Diamond_l) \mid l \in \mathcal{A}\}$; thus, $L_{\mathsf{t}}^\circ = L_{(\overline{\Lambda})^\circ}$ by Theorem 5.4. Assume for simplicity that $\mathcal{A}$ is finite. Note that we have

$$(\textstyle\bigwedge_{l \in \mathcal{A}} \Box_l(-)_l, \bigvee_{l \in \mathcal{A}} \Diamond_l(-)_l) \in \Lambda^{\mathsf{pos}} \blacktriangleright (\overline{\Lambda}^{\mathsf{pos}})^\circ,$$

where $\bigwedge_{l \in \mathcal{A}} \Box_l(-)_l$ takes an $\mathcal{A}$-indexed family of predicates A_l to $\bigcap_{l \in \mathcal{A}} \Box_l A_l$, correspondingly for $\bigvee_{l \in \mathcal{A}} \Diamond_l(-)_l$, since this pair represents a valid implication over $\mathcal{A} \times (-)$. From this observation, one easily concludes that $L_{\mathsf{t}} \cdot L_{\mathsf{t}}^\circ = L_\Lambda \cdot L_{(\overline{\Lambda})^\circ} = L_{\Lambda^{\mathsf{pos}} \blacktriangleright (\overline{\Lambda}^{\mathsf{pos}})^\circ}$, i.e. we have equality in the applicable instance of (5.1). We will use this fact to obtain a logical characterization of $L_{\mathsf{t}} \cdot L_{\mathsf{t}}^\circ$-bisimilarity (i.e. of sharing an infinite trace) in Example 6.5.

Remark 5.8. Every lax extension of a finitary functor is induced by monotone predicate liftings as described above [26,29,30]. We leave it as an open problem whether every relational connector among finitary functors is Kantorovich.

6 Expressiveness

We now go on to establish an expressiveness theorem in the style of the classical Hennessy-Milner theorem, which states that two states in finitely branching LTS are bisimilar iff they satisfy the same formulae of Hennessy-Milner logic. Our present version subsumes the classical theorem and coalgebraic generalizations, but also variants for asymmetric comparisons such as simulation, and hence instead works with forward preservation of formula satisfaction in a logic with only positive Boolean combinations, introduced next:

Definition 6.1. Let $\Lambda \colon \mathsf{PL}(F) \nrightarrow \mathsf{PL}(G)$. Then the set $\mathcal{F}(\Lambda)$ of *Λ-formulae* ϕ, ψ is given by the grammar

$$\mathcal{F}(\Lambda) \ni \phi, \psi ::= \bot \mid \top \mid \phi \wedge \psi \mid \phi \vee \psi \mid \langle \lambda, \mu \rangle \phi \qquad ((\lambda, \mu) \in \Lambda).$$

We interpret Λ-formulae over both F-coalgebras and G-coalgebras. For a state x in an F-coalgebra (C, γ) and a Λ-formula ϕ, we write $x \models_F \phi$, or just $x \models \phi$, to indicate that x *satisfies* ϕ; similarly, we write $y \models_G \phi$ or just $y \models \phi$ to indicate that a state y in a G-coalgebra (D, δ) *satisfies* ϕ. We denote the *extension* of ϕ in C by $[\![\phi]\!]_C = \{x \in C \mid x \models C\}$, similarly for D. The satisfaction relations are then defined by the usual clauses for the Boolean operators, and by

$$x \models_F \langle \lambda, \mu \rangle \phi \text{ iff } \gamma(x) \in \lambda([\![\phi]\!]_C), \qquad y \models_G \langle \lambda, \mu \rangle \phi \text{ iff } \delta(y) \in \mu([\![\phi]\!]_D).$$

We refer to the modal logic thus defined as $\mathcal{L}(\Lambda)$.

One shows easily that the logic $\mathcal{L}(\Lambda)$ is preserved under L_Λ-similarity:

Proposition 6.2. *Let $\Lambda\colon \mathsf{PL}(F) \nrightarrow \mathsf{PL}(G)$, and let ϕ be a Λ-formula. Whenever $x \preceq_\Lambda y$ and $x \models_F \phi$, then $y \models_G \phi$.*

The converse is less straightforward, and (like the classical Hennessy-Milner theorem) depends on finite branching. For brevity, we say that an F-coalgebra (C, γ) is *finitely branching* if for every $x \in C$, there exists a finite subset $C' \subseteq C$ such that $\gamma(x) \in FC' \subseteq FC$ (cf. assumptions made in Section 2).

Theorem 6.3 (Expressiveness). *Let $\Lambda\colon \mathsf{PL}(F) \nrightarrow \mathsf{PL}(G)$. Then Λ-similarity coincides with theory inclusion in $\mathcal{L}(\Lambda)$ on finitely branching coalgebras; that is, for states $x \in C$, $y \in D$ in finitely branching coalgebras $(C, \gamma\colon C \to FC)$ and $(D, \delta\colon D \to GD)$, we have $x \preceq_\Lambda y$ iff for every Λ-formula ϕ, whenever $x \models_F \phi$, then $y \models_G \phi$.*

Proof (sketch). Show that theory inclusion $r = \{(x, y) \in C \times D \mid \forall \phi \in \mathcal{F}(\Lambda).\, x \models_F \phi \implies y \models_G \phi\}$ is an L_Λ-simulation. □

Remark 6.4. From Theorem 6.3, we recover in particular the coalgebraic generalization of the Hennessy-Milner theorem for behavioural equivalence [33,38], restricted to monotone modalities, by instantiating to $\Lambda \subseteq id$ satisfying the usual separation condition (cf. Corollary 5.5). This theorem applies to a logic with full Boolean propositional base; note here that when Λ is closed under duals, our logic admits an encoding of negation via negation normal forms. Also, Theorem 6.3 subsumes coalgebraic Hennessy-Milner theorems for behavioural preorders such as simulation [23,46]. Our main interest is in heterogeneous examples, listed next.

Example 6.5. 1. From the Kantorovich description of the relational connectors $K_R, L_R\colon \mathcal{P}(\mathcal{A} \times (-)) \to \mathcal{P}(\mathcal{B} \times (-))$ induced from $R\colon \mathcal{A} \nrightarrow \mathcal{B}$ (Example 5.3.1), we obtain logical characterizations of K_R-similarity and L_R-(bi)similarity on finitely branching $\mathcal{A}$-LTS and $\mathcal{B}$-LTS. For instance, states $x \in C$, $y \in D$ in an $\mathcal{A}$-LTS C and a $\mathcal{B}$-LTS D, both finitely branching, are L_R-bisimilar iff x and y satisfy the same formulae in a modal logic with modalities $\langle \Diamond_l, \Diamond_m \rangle$ and $\langle \Box_l, \Box_m \rangle$ for $(l, m) \in R$.

2. In Example 5.3.3, a Kantorovich description is given for *ioco* simulation, yielding a logical characterization by Theorem 6.3. The logic features the modalities $\Diamond_i$ for inputs $i \in I$, $\Box_o$ for outputs $o \in O$ and "undefinedness" modalities $\downarrow_o$, which hold at a state iff there is no outgoing o-transition from that state.

3. The Kantorovich definition of the relational connector $L_\Lambda\colon \mathcal{P}(\mathcal{A} \times (-)) \to \mathcal{P}(\mathcal{B} \times (-))$, for $\Lambda = \{(\Diamond_l, L_{l,\epsilon}) \mid l \in \mathcal{A}\}$ as per Example 5.3.4, implies a logical characterization of L_Λ-simulation: Given states $x \in C$, $y \in D$ in an finitely branching $\mathcal{A}$-LTS C and a finitely branching probabilistic $\mathcal{A}$-LTS D, we have $x \preceq_\Lambda y$ iff whenever x satisfies a formula ϕ in the positive fragment of Hennessy-Milner logic with only diamond modalities $\Diamond_l$, then y satisfies the probabilistic modal formula obtained from ϕ by replacing $\Diamond_l$ with $L_{l,\epsilon}$ throughout. Corresponding characterizations hold for $L_{\overline{\Lambda}}$-similarity and $L_{\Lambda \cup \overline{\Lambda}}$-similarity.

4. From the Kantorovich description of the connector $L_t \cdot L_t^\circ : \mathcal{P}(\mathcal{A} \times (-)) \to \mathcal{P}(\mathcal{A} \times (-))$, we obtain a logical characterization of $L_t \cdot L_t^\circ$-bisimilarity, i.e. of sharing an infinite trace: States x, y in finitely branching $\mathcal{A}$-LTS are $L_t \cdot L_t^\circ$-bisimilar iff whenever x satisfies a formula ϕ in a positive modal logic with $|\mathcal{A}|$-ary modalities $\bigwedge_{l \in \mathcal{A}} \Box_l(-)_l$ as per Example 5.7, then y satisfies the formula obtained from ϕ by replacing $\bigwedge_{l \in \mathcal{A}} \Box_l(-)_l$ with $\bigvee_{l \in \mathcal{A}} \Diamond_l(-)_l$ throughout. We note that in a scenario where we view x as specifying a set of bad traces, this means that the fact that y does *not* have a bad trace can be witnessed by a single counterexample formula ϕ.

7 Conclusions

We have presented a systematic approach to comparing systems of different transition types, abstracted as set functors in the paradigm of universal coalgebra [37]: We induce notions of *heterogeneous (bi)simulation* from *relational connectors* among set functors. We have exhibited a number of key constructions of relational connectors, including composition, converse, identity, and a Kantorovich construction in which a connector is induced from a relation between modalities. Building on the latter, we have proved a Hennessy-Milner type theorem that characterizes heterogeneous (bi)similarity in terms of theory inclusion in a flavour of positive coalgebraic modal logic [23] that is interpretable over both of the involved system types. One instance of this result asserts that absence of a shared trace between LTS can be witnessed by a pair of modal formulae in Hennessy-Milner logic.

We leave quite a few problems open for further investigation, maybe most notably including the question whether every relational connector among finitary functors is Kantorovich (this holds for lax extensions [26,29,30], which form a special case of relational connectors, and generalizes to arbitrary functors when infinitary modalities are allowed [38,14]). More specifically, one would be interested in a logical descriptions of composites of Kantorovich connectors, working from Remark 5.6. A further open question is under what conditions similarity for a composite relational connector $L \cdot K$ equals the composite of the similarity relations for L and K respectively—currently, we only have one inclusion (Lemma 4.3). This is of particular interest for the example on ioco conformance (Example 4.10), where two specifications are known to be compatible iff they have a common conforming implementation [8], a result that has been recovered in a coalgebraic setting [36]. A further issue for future research is to develop the coinductive up-to techniques [35,7] for relational connectors.

References

1. Adámek, J., Herrlich, H., Strecker, G.E.: Abstract and concrete categories: The joy of cats. John Wiley & Sons Inc. (1990), http://tac.mta.ca/tac/reprints/articles/17/tr17abs.html, republished in: Reprints in Theory and Applications of Categories, No. 17 (2006) pp. 1–507

2. Backhouse, R.C., de Bruin, P.J., Hoogendijk, P.F., Malcolm, G., Voermans, E., van der Woude, J.: Polynomial relators (extended abstract). In: Nivat, M., Rattray, C., Rus, T., Scollo, G. (eds.) Algebraic Methodology and Software Technology, AMAST 1991. pp. 303–326. Workshops in Computing, Springer (1991)
3. Baldan, P., Bonchi, F., Kerstan, H., König, B.: Coalgebraic behavioral metrics. Log. Methods Comput. Sci. **14**(3) (2018). https://doi.org/10.23638/LMCS-14(3:20)2018
4. Barr, M.: Relational algebras. In: Reports of the Midwest Category Seminar IV. pp. 39–55. No. 137 in Lect. Notes Math., Springer (1970)
5. Barr, M.: Terminal coalgebras in well-founded set theory. Theor. Comput. Sci. **114**(2), 299–315 (1993). https://doi.org/10.1016/0304-3975(93)90076-6
6. Blom, S., Orzan, S.: A distributed algorithm for strong bisimulation reduction of state spaces. Int. J. Softw. Tools Technol. Transf. **7**(1), 74–86 (2005). https://doi.org/10.1007/S10009-004-0159-4
7. Bonchi, F., Petrisan, D., Pous, D., Rot, J.: A general account of coinduction up-to. Acta Informatica **54**(2), 127–190 (2017)
8. van den Bos, P., Janssen, R., Moerman, J.: n-complete test suites for IOCO. Softw. Qual. J. **27**(2), 563–588 (2019). https://doi.org/10.1007/S11219-018-9422-X
9. Buchholz, P.: Bisimulation relations for weighted automata. Theor. Comput. Sci. **393**(1-3), 109–123 (2008). https://doi.org/10.1016/J.TCS.2007.11.018
10. Desharnais, J., Edalat, A., Panangaden, P.: Bisimulation for labelled Markov processes. Inf. Comput. **179**(2), 163–193 (2002). https://doi.org/10.1006/INCO.2001.2962
11. Forster, J., Goncharov, S., Hofmann, D., Nora, P., Schröder, L., Wild, P.: Quantitative Hennessy-Milner theorems via notions of density. In: Computer Science Logic, CSL 2025. LIPIcs, vol. 252, pp. 22:1–22:20. Schloss Dagstuhl – Leibniz-Zentrum für Informatik (2023)
12. Freyd, P.J., Scedrov, A.: Categories, allegories, North-Holland mathematical library, vol. 39. North-Holland (1990)
13. Goncharov, S., Hofmann, D., Nora, P., Schröder, L., Wild, P.: Kantorovich functors and characteristic logics for behavioural distances. In: Kupferman, O., Sobocinski, P. (eds.) Foundations of Software Science and Computation Structures, FoSSaCS 2023. LNCS, vol. 13992, pp. 46–67. Springer (2023). https://doi.org/10.1007/978-3-031-30829-1_3
14. Goncharov, S., Hofmann, D., Nora, P., Schröder, L., Wild, P.: A point-free perspective on lax extensions and predicate liftings. Math. Struct. Comput. Sci. **34**(2), 98–127 (2024). https://doi.org/10.1017/S096012952300035X
15. Goncharov, S., Milius, S., Schröder, L., Tsampas, S., Urbat, H.: Higher-order mathematical operational semantics. CoRR **abs/2405.16708** (2024). https://doi.org/10.48550/ARXIV.2405.16708, earlier version in Amal Ahmed, ed., Principles of Programming Languages, POPL 2024, Proc. ACM Prog.Lang. 7: 632-658 (2023)
16. Gorín, D., Schröder, L.: Simulations and bisimulations for coalgebraic modal logics. In: Heckel, R., Milius, S. (eds.) Algebra and Coalgebra in Computer Science, CALCO 2013. LNCS, vol. 8089, pp. 253–266. Springer (2013). https://doi.org/10.1007/978-3-642-40206-7_19
17. Hansen, H.H., Kupke, C.: A coalgebraic perspective on monotone modal logic. In: Adámek, J., Milius, S. (eds.) Coalgebraic Methods in Computer Science, CMCS 2004. ENTCS, vol. 106, pp. 121–143. Elsevier (2004). https://doi.org/10.1016/J.ENTCS.2004.02.028
18. Hasuo, I., Kataoka, T., Cho, K.: Coinductive predicates and final sequences in a fibration. Math. Struct. Comput. Sci. **28**(4), 562–611 (2018). https://doi.org/10.1017/S0960129517000056

19. Hennessy, M., Milner, R.: Algebraic laws for non-determinism and concurrency. J. ACM **32**, 137–161 (1985). https://doi.org/10.1145/2455.2460
20. Hermida, C., Jacobs, B.: Structural induction and coinduction in a fibrational setting. Inf. Comput. **145**(2), 107–152 (1998). https://doi.org/10.1006/INCO.1998.2725
21. Hesselink, W.H., Thijs, A.: Fixpoint semantics and simulation. Theor. Comput. Sci. **238**(1-2), 275–311 (2000). https://doi.org/10.1016/S0304-3975(98)00176-5
22. Horne, R., Mauw, S.: Discovering ePassport vulnerabilities using bisimilarity. Log. Methods Comput. Sci. **17**(2), 24 (2021). https://doi.org/10.23638/LMCS-17(2:24)2021
23. Kapulkin, K., Kurz, A., Velebil, J.: Expressiveness of positive coalgebraic logic. In: Bolander, T., Braüner, T., Ghilardi, S., Moss, L.S. (eds.) Advances in Modal Logic, AiML 2021. pp. 368–385. College Publications (2012), http://www.aiml.net/volumes/volume9/Kapulkin-Kurz-Velebil.pdf
24. Komorida, Y., Katsumata, S., Kupke, C., Rot, J., Hasuo, I.: Expressivity of quantitative modal logics : Categorical foundations via codensity and approximation. In: Logic in Computer Science, LICS 2021. pp. 1–14. IEEE (2021). https://doi.org/10.1109/LICS52264.2021.9470656
25. Kupke, C., Rot, J.: Expressive logics for coinductive predicates. Log. Methods Comput. Sci. **17**(4) (2021). https://doi.org/10.46298/LMCS-17(4:19)2021
26. Kurz, A., Leal, R.A.: Modalities in the stone age: A comparison of coalgebraic logics. Theor. Comput. Sci. **430**, 88–116 (2012). https://doi.org/10.1016/J.TCS.2012.03.027
27. Larsen, K.G., Skou, A.: Bisimulation through probabilistic testing. Inf. Comput. **94**(1), 1–28 (1991). https://doi.org/10.1016/0890-5401(91)90030-6
28. Levy, P.B.: Similarity quotients as final coalgebras. In: Hofmann, M. (ed.) Foundations of Software Science and Computational Structures, FOSSACS 2011. LNCS, vol. 6604, pp. 27–41. Springer (2011). https://doi.org/10.1007/978-3-642-19805-2_3
29. Marti, J., Venema, Y.: Lax extensions of coalgebra functors. In: Pattinson, D., Schröder, L. (eds.) Coalgebraic Methods in Computer Science, CMCS 2012. LNCS, vol. 7399, pp. 150–169. Springer (2012). https://doi.org/10.1007/978-3-642-32784-1_9
30. Marti, J., Venema, Y.: Lax extensions of coalgebra functors and their logic. J. Comput. Syst. Sci. **81**(5), 880–900 (2015). https://doi.org/10.1016/J.JCSS.2014.12.006
31. Milner, R.: Communication and Concurrency. Prentice Hall (1989)
32. Nora, P., Rot, J., Schröder, L., Wild, P.: Relational connectors and heterogeneous bisimulations. CoRR **abs/2410.14460** (2024). https://doi.org/10.48550/ARXIV.2410.14460
33. Pattinson, D.: Expressive logics for coalgebras via terminal sequence induction. Notre Dame J. Formal Log. **45**(1), 19–33 (2004). https://doi.org/10.1305/NDJFL/1094155277
34. Pauly, M.: Bisimulation for general non-normal modal logic (1999), manuscript
35. Pous, D., Sangiorgi, D.: Enhancements of the bisimulation proof method. In: Advanced Topics in Bisimulation and Coinduction, Cambridge tracts in theoretical computer science, vol. 52, pp. 233–289. Cambridge University Press (2012)
36. Rot, J., Wißmann, T.: Bisimilar states in uncertain structures. In: Baldan, P., de Paiva, V. (eds.) Algebra and Coalgebra in Computer Science, CALCO 2023. LIPIcs, vol. 270, pp. 12:1–12:17. Schloss Dagstuhl – Leibniz-Zentrum für Informatik (2023). https://doi.org/10.4230/LIPICS.CALCO.2023.12
37. Rutten, J.J.M.M.: Universal coalgebra: a theory of systems. Theor. Comput. Sci. **249**(1), 3–80 (2000). https://doi.org/10.1016/S0304-3975(00)00056-6

38. Schröder, L.: Expressivity of coalgebraic modal logic: The limits and beyond. Theor. Comput. Sci. **390**(2-3), 230–247 (2008). https://doi.org/10.1016/j.tcs.2007.09.023
39. Schubert, C., Seal, G.J.: Extensions in the theory of lax algebras. Theory and Applications of Categories **21**(7), 118–151 (2008)
40. Seal, G.J.: Canonical and op-canonical lax algebras. Theory and Applications of Categories **14**(10), 221–243 (2005), http://www.tac.mta.ca/tac/volumes/14/10/14-10abs.html
41. Sprunger, D., Katsumata, S., Dubut, J., Hasuo, I.: Fibrational bisimulations and quantitative reasoning: Extended version. J. Log. Comput. **31**(6), 1526–1559 (2021). https://doi.org/10.1093/LOGCOM/EXAB051
42. Thijs, A.: Simulation and fixpoint semantics. Ph.D. thesis, University of Groningen (1996)
43. Tretmans, J.: Model based testing with labelled transition systems. In: Hierons, R.M., Bowen, J.P., Harman, M. (eds.) Formal Methods and Testing. LNCS, vol. 4949, pp. 1–38. Springer (2008). https://doi.org/10.1007/978-3-540-78917-8_1
44. Trnková, V.: General theory of relational automata. Fund. Inform. **3**(2), 189–233 (1980)
45. Turkenburg, R., Beohar, H., Kupke, C., Rot, J.: Forward and backward steps in a fibration. In: Baldan, P., de Paiva, V. (eds.) Algebra and Coalgebra in Computer Science, CALCO 2023. LIPIcs, vol. 270, pp. 6:1–6:18. Schloss Dagstuhl – Leibniz-Zentrum für Informatik (2023). https://doi.org/10.4230/LIPICS.CALCO.2023.6
46. Wild, P., Schröder, L.: A quantified coalgebraic van Benthem theorem. In: Kiefer, S., Tasson, C. (eds.) Foundations of Software Science and Computation Structures, FOSSACS 2021. LNCS, vol. 12650, pp. 551–571. Springer (2021). https://doi.org/10.1007/978-3-030-71995-1_28
47. Wild, P., Schröder, L.: Characteristic logics for behavioural hemimetrics via fuzzy lax extensions. Log. Methods Comput. Sci. **18**(2) (2022). https://doi.org/10.46298/LMCS-18(2:19)2022

On the cut-elimination of the modal μ-calculus: Linear Logic to the rescue

Esaïe Bauer(✉) and Alexis Saurin(✉)

Université Paris Cité & CNRS & INRIA, Pl. Aurélie Nemours, 75013 Paris, France
{esaie.bauer,alexis.saurin}@irif.fr

Abstract. This paper presents a proof-theoretic analysis of the modal μ-calculus. More precisely, we prove a syntactic cut-elimination for the non-wellfounded modal μ-calculus, using methods from linear logic. and its exponential modalities. To achieve this, we introduce a new system, $\mu\mathsf{LL}_{\square}^{\infty}$, which is a linear version of the modal μ-calculus, intertwining the modalities from the modal μ-calculus with the exponential modalities from linear logic. Our strategy for proving cut-elimination involves (i) proving cut-elimination for $\mu\mathsf{LL}_{\square}^{\infty}$ and (ii) translating proofs of the modal mu-calculus into this new system via a "linear translation", allowing us to extract the cut-elimination result.

1 Introduction

Eliminability of cuts and the modal μ-calculus. Since Kozen's seminal work on the *modal μ-calculus* [17], this logic extending basic modal logic with least and greatest fixed-points has been extremely fruitful for the study of computational systems, especially reactive systems. In addition to its wide expressive power, its deep roots in logic also allow for a number of fruitful approaches, be they model-theoretic, proof-theoretic or automata-theoretic. Still, *cut-elimination* – a cornerstone of modern proof-theory – has only received partial solutions [1,6,21,22,24]:

– Either as cut-admissibility statements which are noneffective possibly using infinitary proof-systems such as (i) infinitely branching proof systems, allowing the ω-rule [16] or (ii) non-wellfounded or circular proof systems [1,24], allowing proof-trees with infinitely long branches;
– Or as syntactic cut-elimination results capturing only a *fragment* of the calculus in systems with the ω-rule [21,22]. A challenge in describing a *syntactic cut-elimination* for such systems is that the number of applications of a μ-rule must sometimes be determined before knowing how many are needed to match each premises of a ν-rule. In [6], the authors discuss a specific example where syntactic cut-elimination fails. While there are syntactic cut-elimination results in systems based on the ω-rule [6,21,22], they capture only strict fragments of the modal μ-calculus.

In fact, there is no syntactic cut-elimination theorem for the full modal μ-calculus. The present work establishes such a syntactic cut-elimination theorem for the modal μ-calculus in the setting of non-wellfounded sequent calculus.

P. A. Abdulla and D. Kesner (Eds.): FoSSaCS 2025, LNCS 15691, pp. 133–154, 2025.
https://doi.org/10.1007/978-3-031-90897-2_7

On unity and diversity in computational logic. Logic presents at the same time a deep unity and a wide diversity. Miller [20] argues that the *"universal character [of logic] has been badly fractured in the past few decades"*, due to the wide range of its applications, the various families of logics that have emerged and the different computational tools that are in use, often with little relationship. Miller thus proposes the following questions as the first of a list of "challenges":

Challenge 1: *Unify a wide range of logical features into a single framework. How best can we explain the many enhancements that have been designed for logic: for example, classical / intuitionistic / linear, fixed points, first-order / higher-order quantification, modalities, and temporal operators? (...)*

In the present paper, we partially address Miller's first challenge, providing a common framework for two of the main logics that emerged in the 1980s, Kozen's modal μ-calculus [17] and Girard's linear logic (LL) [14]. Working in the setting of circular and non-wellfounded proof systems for the above logics, we propose a so-called *linear decomposition* of the modal μ-calculus in linear logic with fixed-points. This proof-theoretic analysis of the modal μ-calculus allows us to do a finer-grained treatment of syntactic cut-elimination.

Cut-admissibility vs. cut-elimination. The treatment of the cut-inference in sequent-based proof-systems follows two main traditions: (i) one can consider cut-free proofs as the primitive proof-objects and establish that the cut inference is *admissible* (according to this tradition, the cut-inference essentially lives at the metalevel, ensuring compositionality of the logic) or (ii) one can consider that the cut inference lives at the object-level and is a fundamental piece of proofs: one thus establishes that the cut inference is *eliminable*, ensuring the sub-formula property (and its numerous important consequences, ranging from consistency to interpolation properties).

This second tradition may use similar techniques as the first tradition, but it also permits the investigation of a syntactic, or effective, approach to cut-elimination, consisting of a cut-reduction relation on proofs, shown to be (at least) weakly normalizing, with the normal forms being cut-free proofs. An advantage of such syntactic cut-elimination results is that, in many settings (most notably LJ and LL [14]), such cut-reductions induce an interesting relation on proofs and have a computational interpretation that is the starting point of the Curry-Howard correspondence built upon sequent calculus [9].

Linear Logic. Linear logic (LL) is often described as a resource-sensitive logic. It is more accurate, though, to view it as a logic designed for analyzing cut-elimination itself. Indeed, LL comes from an analysis of structural rules, aiming at controlling them rather than weakening them as in substructural logics. This solves some fundamental drawbacks of cut-elimination in classical logic, such as its non-termination or non-confluence. For instance, LL permits the decomposition of both intuitionistic and classical logic, in a structured and fine-grained manner allowing the refinement of the cut-elimination of those logics as well as their notion of model (allowing the building of a non-trivial denotational model of classical proofs); the prototypical example of such a linear decomposition consists in decomposing the usual *intuitionistic* arrow (*i.e.*, the function type of the λ-calculus), $A \Rightarrow B$, into a replication operator and a linear implication: $!A \multimap B$ [10,14]. Further analyses on these exponential modalities led to the discovery of *light logics*, where the complexity of cut-elimination is tamed in a flexible way, usually by considering alternative, weaker exponential modalities.

The proof theory of LL was extended to $\mu\mathsf{LL}^\infty$, that is LL with fixed-points in the finitary and non-wellfounded setting [2,12,26,27,28] and $\mu\mathsf{LL}^\infty$ allowed for the same kind of linear decomposition for (the non-wellfounded version of) $\mu\mathsf{LJ}$ and $\mu\mathsf{LK}$. A natural question is whether the extensions of LL with fixed-points can also help us achieve syntactic cut-elimination for the modal μ-calculus.

Contributions. The discussion of the above paragraph suggests a first question: what would be a linear decomposition of the modal μ-calculus? The first contribution of this paper is to provide such a linear decomposition of the modal μ-calculus which is compatible with circular and non-wellfounded proof theory, $\mu\mathsf{LL}_\square^\infty$. This *linear-logical modal μ-calculus* will allow us to complete the analysis of cut-elimination for the modal μ-calculus, proving the first syntactic cut-elimination theorem for the full modal μ-calculus (in the non-wellfounded setting). We therefore adopt the following roadmap in the body of the paper.

In Section 2, we recall the necessary technical background about $\mu\mathsf{LL}^\infty$ and $\mu\mathsf{LK}_\square^\infty$ proof theory. In Section 3, we motivate and introduce $\mu\mathsf{LL}_\square^\infty$ and prove its cut-elimination in Section 4. We then define the linear decomposition of $\mu\mathsf{LK}_\square^\infty$ into $\mu\mathsf{LL}_\square^\infty$ from which we conclude to $\mu\mathsf{LK}_\square^\infty$ cut-elimination theorem in Section 5 in the form of an infinitary weak-normalizing cut-reduction system.

2 Circular and (non-)wellfounded proof systems

We recall here some basic definitions of both wellfounded & non-wellfounded systems. Some standard definitions of proof theory such as *derivation rules*, *active formula* or *principal formula* are not recalled and can be found in [7].

2.1 The Modal μ-calculus

Formulas Let $\mathcal{V}$ and $\mathcal{A}$ be two disjoint sets of *fixed-point variables* and of *atoms* respectively. We define the pre-formulas of the modal μ-calculus, $\mu\mathsf{LK}_\square^\infty$, as: $F, G \ ::= \ a \mid X \mid \mu X.F \mid \nu X.F \mid \square F \mid \lozenge F \mid F^\perp \mid F \to G \mid F \vee G \mid F \wedge G \mid \mathsf{F} \mid \mathsf{T}$ ($a \in \mathcal{A}, X \in \mathcal{V}$). Knaster-Tarski's theorem guarantees the existence of extremal fixed-points of monotonic functions on complete lattices; monotonicity is reflected syntactically as a *positivity condition* on variables, defined as:

Definition 1 (Positive & negative occurrence of fixed-point variables) *Let $X \in \mathcal{V}$ be a fixed-point variable, one defines the fact, for X, to occur positively (resp. negatively) in a pre-formula by induction on the structure of pre-formulas:*

- *X occurs positively in X.*
- *X occurs positively (resp. negatively) in $c(F_1, \ldots, F_n)$, if there is some $1 \leq i \leq n$ such that X occurs positively (resp. negatively) in F_i for $c \in \{\Box, \Diamond, \vee, \wedge\}$.*
- *X occurs positively (resp. negatively) in $F^\perp$ if it occurs negatively (resp. positively) in F.*
- *X occurs positively (resp. negatively) in $F \to G$ if X occurs either positively (resp. negatively) in G or negatively (resp. positively) in F.*
- *X occurs positively (resp. negatively) in $\delta Y.G$ (with $Y \neq X$) if it occurs positively (resp. negatively) in G (for $\delta \in \{\mu, \nu\}$).*

We can now define the formulas of $\mu\mathsf{LK}_\Box^\infty$:

Definition 2 (Formulas) *A $\mu\mathsf{LK}_\Box^\infty$* formula *$F$ is a closed pre-formula such that for any sub-pre-formula of F of the form $\delta X.G$ (with $\delta \in \{\mu, \nu\}$), X does not occur negatively in G.*

By considering the $\{\mu, \nu, X\}$-free formulas of this system, we get $\mathsf{LK}_\Box$. By considering the $\{\Box, \Diamond\}$-free formulas of $\mu\mathsf{LK}_\Box^\infty$, we get the μ-calculus. Finally, the intersection of these two systems, is propositional classical logic.

Sequent calculus We define, here, the sequents, rules and proofs for $\mu\mathsf{LK}_\Box^\infty$.

Definition 3 (Sequent) *A* sequent *is an ordered pair of two lists of formulas Γ, the* antecedent, *and Δ, the* succedent, *that we write $\Gamma \vdash \Delta$.*

Remark 1 (Derivation rules & ancestor relation) *In the structural proof theory literature, inference rules are defined together with an* ancestor relation *(or sub-occurrence relation) between formulas of the conclusion and formulas of the premises of the rule. While this relation is often overlooked we provide some details here. Sequent being lists, we define the ancestor relation, to be a relation from the positions of the formula in the conclusion, to the positions of the formula in the premises. Those ancestor relations will be dealt graphically as in Figures 1 and 2, by drawing the ancestor relation on sequents when needed and leaving it implicit when unambiguous.*

We define the inference rules for $\mathsf{LK}_\Box$ in Figure 1. Rules for LK will be the $\{\Box, \Diamond\}$-free rules of $\mathsf{LK}_\Box$. We add rules of Figure 2 to LK, (resp. $\mathsf{LK}_\Box$) to get the fixed-point version $\mu\mathsf{LK}^\infty$ (resp. $\mu\mathsf{LK}_\Box^\infty$) of this system. The two *exchange rules* (ex_l) and (ex_r) from Figures 1 and 4 allows one to derive the rule $\dfrac{\sigma'(\Gamma) \vdash \sigma(\Delta)}{\Gamma \vdash \Delta}\,\mathrm{ex}(\sigma', \sigma)$ for any permutations σ and σ' of $\{1, \ldots, \#(\Delta)\}$ and $\{1, \ldots, \#(\Gamma)\}$ respectively, where $\sigma(\Delta)$ designates the action of σ on the list Δ, with the induced ancestor relation. In the rest of the article, we will intentionally treat the exchange rule implicitly: the reader can consider that each of our rules are preceded and followed by a finite number of rule (ex).

Proofs of non fixed-point systems, LK, $\mathsf{LK}_\Box$ are the trees inductively generated by the corresponding set of rules of each of these systems. We can define a first notion of infinite derivations, pre-proofs, that will soon be refined:

$$\frac{}{F \vdash F}\,\mathrm{ax} \qquad \frac{\Gamma_1 \vdash F, \Delta_1 \quad \Gamma_2, F \vdash \Delta_2}{\Gamma_1, \Gamma_2 \vdash \Delta_1, \Delta_2}\,\mathrm{cut} \qquad \frac{\Gamma \vdash F, \Delta}{\Box\Gamma \vdash \Box F, \Diamond\Delta}\,\Box_{\mathrm{p}} \qquad \frac{\Gamma, F \vdash \Delta}{\Box\Gamma, \Diamond F \vdash \Diamond\Delta}\,\Diamond_{\mathrm{p}}$$

$$\frac{\Gamma \vdash F_1, \Delta}{\Gamma \vdash F_1 \vee F_2, \Delta}\,\vee_r^1 \qquad \frac{\Gamma \vdash F_2, \Delta}{\Gamma \vdash F_1 \vee F_2, \Delta}\,\vee_r^2 \qquad \frac{\Gamma, F_1 \vdash \Delta \quad \Gamma, F_2 \vdash \Delta}{\Gamma, F_1 \vee F_2 \vdash \Delta}\,\vee_l \qquad \frac{}{\Gamma, \mathsf{F} \vdash \Delta}\,\mathsf{F}$$

$$\frac{\Gamma, F_1 \vdash \Delta}{\Gamma, F_1 \wedge F_2 \vdash \Delta}\,\wedge_l^1 \qquad \frac{\Gamma, F_2 \vdash \Delta}{\Gamma, F_1 \wedge F_2 \vdash \Delta}\,\wedge_l^2 \qquad \frac{\Gamma \vdash F_1, \Delta \quad \Gamma \vdash F_2, \Delta}{\Gamma \vdash F_1 \wedge F_2, \Delta}\,\wedge_r \qquad \frac{}{\Gamma \vdash \mathsf{T}, \Delta}\,\mathsf{T}$$

$$\frac{\Gamma, A \vdash B, \Delta}{\Gamma \vdash A \to B, \Delta}\,\to_r \qquad \frac{\Gamma_1, B \vdash \Delta_1 \quad \Gamma_2 \vdash A, \Delta_2}{\Gamma_1, \Gamma_2, A \to B \vdash \Delta_1, \Delta_2}\,\to_l \qquad \frac{\Gamma, A \vdash \Delta}{\Gamma \vdash A^\perp, \Delta}\,(-)_r^\perp \qquad \frac{\Gamma \vdash A, \Delta}{\Gamma, A^\perp \vdash \Delta}\,(-)_l^\perp$$

$$\frac{\Gamma \vdash \Delta}{\Gamma \vdash F, \Delta}\,\mathrm{w}_r \quad \frac{\Gamma \vdash \Delta}{\Gamma, F \vdash \Delta}\,\mathrm{w}_l \quad \frac{\Gamma, F, F \vdash \Delta}{\Gamma, F \vdash \Delta}\,\mathrm{c}_l \quad \frac{\Gamma \vdash F, F, \Delta}{\Gamma \vdash F, \Delta}\,\mathrm{c}_r \quad \frac{\Gamma_1, G, F, \Gamma_2 \vdash \Delta}{\Gamma_1, F, G, \Gamma_2 \vdash \Delta}\,\mathrm{ex}_l \quad \frac{\Gamma \vdash \Delta_1, G, F, \Delta_2}{\Gamma \vdash \Delta_1, F, G, \Delta_2}\,\mathrm{ex}_r$$

Fig. 1: Rules of $\mathsf{LK}_\Box$

$$\frac{\Gamma, F[X := \mu X.F] \vdash \Delta}{\Gamma, \mu X.F \vdash \Delta}\,\mu_l \qquad \frac{\Gamma \vdash F[X := \mu X.F], \Delta}{\Gamma \vdash \mu X.F, \Delta}\,\mu_r \qquad \frac{\Gamma, F[X := \nu X.F] \vdash \Delta}{\Gamma, \nu X.F \vdash \Delta}\,\nu_l \qquad \frac{\Gamma \vdash F[X := \nu X.F], \Delta}{\Gamma \vdash \nu X.F, \Delta}\,\nu_r$$

Fig. 2: Rules for the fixed-point fragment

Definition 4 (Pre-proofs) *Given a set of derivation rules, a* pre-proof *is a tree co-inductively generated using the rules of the system.*

Example 1 (Regular pre-proof) *Regular pre-proofs are those pre-proofs having a finite number of distinct sub-proofs. We represent them with back-edges.*

Taking $F := \nu X.\Diamond X$, we have:

$$\dfrac{\dfrac{\dfrac{F \vdash F}{\Diamond F \vdash \Diamond F}\,\Diamond_p}{F \vdash F}\,\nu_l, \nu_r \qquad \dfrac{\dfrac{F \vdash F}{\Diamond F \vdash \Diamond F}\,\Diamond_p}{F \vdash F}\,\nu_l, \nu_r}{F \vdash F}\,cut$$

Remark 2 *The pre-proofs define an inconsistent system. In fact, any sequent is provable:*

$$\dfrac{\dfrac{\Gamma \vdash \nu X.X}{\Gamma \vdash \nu X.X}\,\nu_r \qquad \dfrac{\nu X.X \vdash \Delta}{\nu X.X \vdash \Delta}\,\nu_l}{\Gamma \vdash \Delta}\,cut$$

Proofs are defined as those pre-proofs satisfying a correctness condition:

Definition 5 (Validity and proofs) *Let $b = (s_i)_{i\in\omega}$ be a consecutive sequence of sequents defining an infinite branch in a pre-proof π. A* thread *of b is a sequence $(F_i \in s_i)_{i>n}$, for some $n \in \omega$, of occurrences such that for each j, F_j and F_{j+1} satisfy the ancestor relation. We say that a thread of b is* valid *if the minimal recurring formula of this sequence, for the sub-formula ordering, exists and is (i) either a ν-formula appearing infinitely often on the succedent of its sequent or a μ-formula appearing infinitely often in the antecedent of its sequent and (ii) the thread is infinitely often principal (there are an infinite number of principal formulas in it). A branch b is* valid *if there is a valid thread of b.*

A pre-proof is valid *and is a* proof *if each of its infinite branches is valid.*

The least (μ) and greatest (ν) fixed-point constructors have the same (local) derivation rules: they are distinguished by the (global) validity condition, which is a parity condition akin to parity games for the μ-calculus.

$$\pi_0 := \cfrac{\cfrac{\cfrac{}{\vdash T}\; \top}{\vdash T \lor \mathrm{Nat}}\; \lor_r^1}{\vdash \mathrm{Nat}}\; \mu_r \qquad \pi_{n+1} := \cfrac{\cfrac{\cfrac{\pi_n}{\vdash \mathrm{Nat}}}{\vdash T \lor \mathrm{Nat}}\; \lor_r^2}{\vdash \mathrm{Nat}}\; \mu_r$$

$$\pi_\infty := \cfrac{\vdash \mathrm{Nat}}{\vdash \mathrm{Nat}}\; \mu_r, \lor_r^2$$

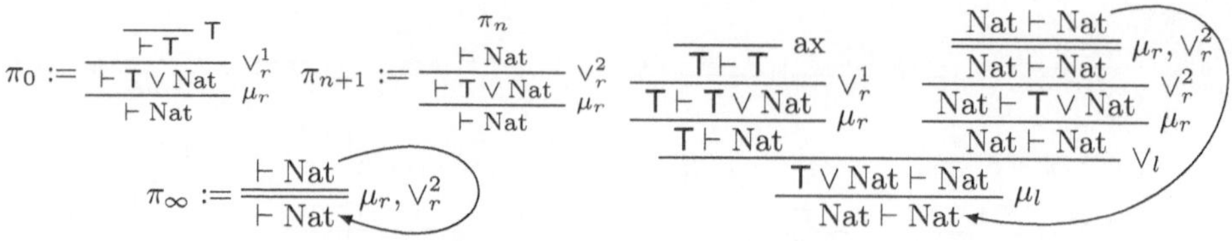

(a) Cut-free pre-proofs of Nat

$$\cfrac{\cfrac{\cfrac{\cfrac{}{T \vdash T}\; \mathrm{ax}}{T \vdash T \lor \mathrm{Nat}}\; \lor_r^1}{T \vdash \mathrm{Nat}}\; \mu_r \qquad \cfrac{\cfrac{\cfrac{\cfrac{\mathrm{Nat} \vdash \mathrm{Nat}}{\mathrm{Nat} \vdash \mathrm{Nat}}\; \mu_r, \lor_r^2}{\mathrm{Nat} \vdash T \lor \mathrm{Nat}}\; \lor_r^2}{\mathrm{Nat} \vdash \mathrm{Nat}}\; \lor_l}{T \lor \mathrm{Nat} \vdash \mathrm{Nat}}\; \mu_l}{\mathrm{Nat} \vdash \mathrm{Nat}}$$

(b) Function double

Fig. 3: Valid and invalid pre-proofs

Example 2 (Valid and invalid pre-proofs) *Here, we give some examples of infinite proofs. The pre-proof of Example 1 is valid, that of Remark 2 is invalid.*

We use the notation Nat $:= \mu X.T \lor X$, *representing the type of natural numbers. We can represent any natural number n by a finite valid proof π_n defined in Figure 3a. The infinite pre-proof π_∞, defined in Figure 3a, of $\vdash$ Nat is not valid: the infinite branch in π_∞ is supported by only one thread, which is not valid as the minimal formula is a μ-formula appearing only on the right of the proof. This is coherent with the interpretation that μ is a least fixed-point: the system would reject the infinite proof of Nat. Note that the same kind of proof with* coNat $:= \nu X.T \lor X$ *would have given a valid proof.*

The pre-proof of Figure 3b (representing the double *function) is valid: The only infinite branch in it is the one going infinitely to the right at the application of the $(\lor_l)$-rule. This branch is supported by the infinite thread in the antecedent of each sequent which has a μ-formula as its minimal formula.*

2.2   Linear Logic with fixed-points

The main difference between LK (or LJ) and LL lies in the fact that formulas are not always erasable nor duplicable. Hence, the sequent $A, B \vdash A$ is not always provable, neither is $A \vdash A \otimes A$ (a sequent similar to $A \vdash A \land A$ in LK). This restriction allows LL to interpret programs with finer resource control than LK (or LJ). Here we recall the usual definitions of both the wellfounded and non-wellfounded systems of LL, following the definitions of the previous section.

Formulas As for $\mu\mathsf{LK}_\ominus^\infty$, let us set $\mathcal{V}$ and $\mathcal{A}$ as two disjoint sets. The pre-formulas of the non-well-founded linear logic, $\mu\mathsf{LL}^\infty$ are: $F, G ::= a \mid X \mid \mu X.F \mid \nu X.F \mid F^\perp \mid F \multimap G \mid F \,\bindnasrepma\, G \mid F \otimes G \mid \perp \mid 1 \mid F \oplus G \mid F \,\&\, G \mid 0 \mid \top \mid ?F \mid !F$, with $a \in \mathcal{A}, X \in \mathcal{V}$. Positivity of pre-formulas and $\mu\mathsf{LL}^\infty$ formulas are defined in the same way as for $\mu\mathsf{LK}_\ominus^\infty$. The $\{!, ?\}$-free formulas of $\mu\mathsf{LL}^\infty$ are the formulas of $\mu\mathsf{MALL}^\infty$. The $\{\mu, \nu, X\}$-free fragment of formulas of $\mu\mathsf{LL}^\infty$ are the formulas of linear logic LL. The intersection of these two fragments is MALL.

Sequent calculus The definition of sequent is the same as for $\mu\mathsf{LK}_\ominus^\infty$. The rules of MALL are given by Figure 4, the rules of LL are the rules of MALL together with the rules of Figure 5. We add the rules of Figure 2 to MALL (resp. LL) to obtain the rules of $\mu\mathsf{MALL}^\infty$ (resp. $\mu\mathsf{LL}^\infty$). Pre-proofs as well as validity are

$$\frac{}{F \vdash F}\,\text{ax} \qquad \frac{\Gamma_1 \vdash F, \Delta_1 \quad \Gamma_2, F \vdash \Delta_2}{\Gamma_1, \Gamma_2 \vdash \Delta_1, \Delta_2}\,\text{cut} \qquad \frac{\Gamma \vdash \Delta_1, G, F, \Delta_2}{\Gamma \vdash \Delta_1, F, G, \Delta_2}\,\text{ex}_r \qquad \frac{\Gamma_1, G, F, \Gamma_2 \vdash \Delta}{\Gamma_1, F, G, \Gamma_2 \vdash \Delta}\,\text{ex}_l$$

$$\frac{\Gamma \vdash F, G, \Delta}{\Gamma \vdash F \parr G, \Delta}\,\parr_r \qquad \frac{\Gamma_1, F \vdash \Delta_1 \quad \Gamma_2, G \vdash \Delta_2}{\Gamma_1, \Gamma_2, F \parr G \vdash \Delta_1, \Delta_2}\,\parr_l \qquad \frac{\Gamma, F, G \vdash \Delta}{\Gamma, F \otimes G \vdash \Delta}\,\otimes_l \qquad \frac{\Gamma_1 \vdash F, \Delta_1 \quad \Gamma_2 \vdash G, \Delta_2}{\Gamma_1, \Gamma_2 \vdash F \otimes G, \Delta_1, \Delta_2}\,\otimes_r$$

$$\frac{\Gamma, A \vdash B, \Delta}{\Gamma \vdash A \multimap B, \Delta}\,\multimap_r \qquad \frac{\Gamma_1, B \vdash \Delta_1 \quad \Gamma_2 \vdash A, \Delta_2}{\Gamma_1, \Gamma_2, A \multimap B \vdash \Delta_1, \Delta_2}\,\multimap_l \qquad \frac{\Gamma, A \vdash \Delta}{\Gamma \vdash A^\perp, \Delta}\,(-)^\perp_r \qquad \frac{\Gamma \vdash A, \Delta}{\Gamma, A^\perp \vdash \Delta}\,(-)^\perp_l$$

$$\frac{\Gamma \vdash F_1, \Delta}{\Gamma \vdash F_1 \oplus F_2, \Delta}\,\oplus^1_r \qquad \frac{\Gamma \vdash F_2, \Delta}{\Gamma \vdash F_1 \oplus F_2, \Delta}\,\oplus^2_r \qquad \frac{\Gamma, F_1 \vdash \Delta \quad \Gamma, F_2 \vdash \Delta}{\Gamma, F_1 \oplus F_2 \vdash \Delta}\,\oplus_l$$

$$\frac{\Gamma, F_1 \vdash \Delta}{\Gamma, F_1 \,\&\, F_2 \vdash \Delta}\,\&^1_l \qquad \frac{\Gamma, F_2 \vdash \Delta}{\Gamma, F_1 \,\&\, F_2 \vdash \Delta}\,\&^2_l \qquad \frac{\Gamma \vdash F_1, \Delta \quad \Gamma \vdash F_2, \Delta}{\Gamma \vdash F_1 \,\&\, F_2, \Delta}\,\&_r$$

$$\frac{}{\vdash 1}\,1_r \qquad \frac{\Gamma \vdash \Delta}{\Gamma, 1 \vdash \Delta}\,1_l \qquad \frac{}{\bot \vdash}\,\bot_l \qquad \frac{\Gamma \vdash \Delta}{\Gamma \vdash \bot, \Delta}\,\bot_r \qquad \frac{}{\Gamma \vdash \top, \Delta}\,\top \qquad \frac{}{\Gamma, 0 \vdash \Delta}\,0$$

Fig. 4: Rules of multiplicative and additive linear logic

$$\frac{\Gamma \vdash \Delta}{\Gamma \vdash ?F, \Delta}\,?_\mathrm{w} \qquad \frac{\Gamma \vdash \Delta}{\Gamma, !F \vdash \Delta}\,!_\mathrm{w} \qquad \frac{\Gamma \vdash ?F, ?F, \Delta}{\Gamma \vdash ?F, \Delta}\,?_\mathrm{c} \qquad \frac{\Gamma, !F, !F \vdash \Delta}{\Gamma, !F \vdash \Delta}\,!_\mathrm{c}$$

$$\frac{\Gamma \vdash F, \Delta}{\Gamma \vdash ?F, \Delta}\,?_\mathrm{d} \qquad \frac{\Gamma, F \vdash \Delta}{\Gamma, !F \vdash \Delta}\,!_\mathrm{d} \qquad \frac{!\Gamma \vdash F, ?\Delta}{!\Gamma \vdash !F, ?\Delta}\,!_\mathrm{p} \qquad \frac{!\Gamma, F \vdash ?\Delta}{!\Gamma, ?F \vdash ?\Delta}\,?_\mathrm{p}$$

Fig. 5: Exponential fragment of LL

defined as in the previous section.

In linear logic, duplicability and erasability are controlled by ? (*why not*) and ! (*of course*) modalities. The ($!_\mathrm{c}$), ($!_\mathrm{w}$), and ($!_\mathrm{d}$) rules allow duplication, erasure, and use of !-prefixed antecedents, respectively. The ($!_\mathrm{p}$) rule enables using an !-prefixed succedent when all antecedents are !-prefixed. This controlled approach to contraction and weakening sequentializes certain reductions in cut-elimination, leading to strong normalization for LL [25]. However, the good normalization properties of LL can be recovered by using a linear translation from LK to LL, similar to the double negation translations from LK to LJ. Indeed, every formula, every sequent, and every proof of LK can be translated into a proof in LL by adding ? and ! modalities. For instance in Figure 6 we show an example where each connective $c(A_1, \ldots, A_n)$ can be translated as $!(c(?A_1, \ldots, ?A_n))$ adding a ? on the formula from the succedent, the additional rules being shown in blue. By taking any maximal sequence of cut reduction on such proofs in LL and using the strong normalization property, we find a cut-free proof of the same sequent in LL. This projects to an LK proof of the original sequent, by simply *forgetting* superfluous modalities. In Section 5, we will use this technique to prove the cut elimination of the modal μ-calculus.

Cut-elimination for $\mu\mathsf{LL}^\infty$ We postpone to Section 4 the discussion of cut-elimination reductions (Definition 8, 10 and 11) and theorems (Theorem 2): they are directly introduced for $\mu\mathsf{LL}^\infty_\square$ that we shall consider in the next section.

$$\dfrac{\dfrac{\dfrac{}{A \vdash A}\,\text{ax}}{\vdash A^{\perp}, A}\,(-)^{\perp}_{r} \qquad \dfrac{}{A \vdash A}\,\text{ax}}{\dfrac{A^{\perp} \to A \vdash A, A}{A^{\perp} \to A \vdash A}\,\text{c}}\,\to_l \quad \rightsquigarrow \quad \dfrac{\dfrac{\dfrac{\dfrac{\dfrac{}{A \vdash A}\,\text{ax}}{?A \vdash ?A}\,?_{\text{p}}, ?_{\text{d}}}{\vdash (?A)^{\perp}, ?A}\,(-)^{\perp}_{r}}{\vdash ?!(?A)^{\perp}, ?A}\,?_{\text{d}}, !_{\text{p}} \qquad \dfrac{\dfrac{}{A \vdash A}\,\text{ax}}{?A \vdash ?A}\,?_{\text{p}}, ?_{\text{d}}}{\dfrac{\dfrac{?!(?A)^{\perp} \multimap ?A \vdash ?A, ?A}{!(?!(?A)^{\perp} \multimap ?A) \vdash ?A, ?A}\,!_{\text{d}}}{!(?!(?A)^{\perp} \multimap ?A) \vdash ?A}\,?_{\text{c}}}\,\multimap_l$$

Fig. 6: Example of a linear translation

3 A linear-logical modal μ-calculus: $\mu\mathsf{LL}^{\infty}_{\Box}$

We now introduce $\mu\mathsf{LL}^{\infty}_{\Box}$, which can be viewed from two perspectives:

- as an extension of $\mu\mathsf{LL}^{\infty}$ with modalities akin to the modal μ-calculus;
- as a linearization of the modal μ-calculus (where *linear* refers to linear logic – by default, any assumption must be used exactly once in a proof – and not to the structures of its models, as in LTL or linear-time μ-calculus [30]).

In designing the $\mu\mathsf{LL}^{\infty}_{\Box}$ sequent calculus, the primary challenge lies in understanding the interaction between the $\Box$ and $\Diamond$ modalities and LL exponentials, while being compatible with the fixed-point inferences. A natural requirement for such a sequent calculus is to allow us to extend the linear decomposition of $\mu\mathsf{LK}^{\infty}$ [28] to $\mu\mathsf{LK}^{\infty}_{\Box}$. The constraints imposed by such a linear decomposition will therefore guide us in defining the exponential and modal rules of $\mu\mathsf{LL}^{\infty}_{\Box}$. The following discussion illustrates these requirements through examples and will lead us to the desired sequent calculus defined in Definition 6 below.

A first approach would be to simply extend $\mu\mathsf{LL}^{\infty}$ with the usual inferences for $\Box$ and $\Diamond$, extending the translation $(-)^{\bullet}$ from [28] on modalities as in Figure 6 setting $(\Diamond A)^{\bullet} := !\Diamond ?(A)^{\bullet}$ and $(\Box A)^{\bullet} := !\Box ?(A)^{\bullet}$. This approach is too simplistic, though, and fails. Consider indeed an instance of the modal rule $\Box_{\text{p}}$ (defined in Figure 1) with $\Gamma = []$ and $\Delta = [B]$: $\dfrac{\vdash A, B}{\vdash \Box A, \Diamond B}\,\Box_{\text{p}}$. Following the standard sequent translation from LK to LL, we need to derive $\vdash ?!\Box ?A^{\bullet}, ?!\Diamond ?B^{\bullet}$ from the premise $\vdash ?A^{\bullet}, ?B^{\bullet}$. Reasoning bottom-up: to conclude $\vdash ?A^{\bullet}, ?B^{\bullet}$, we must remove both the $\Diamond$ and the $\Box$ with a modal rule. However, regardless of the sequence of rules used, we always end up with a sequent containing an ! together with either $\Box$ or $\Diamond$. An example of such a derivation is shown in Figure 7a. In our attempt to translate this rule, we are left with an unprovable sequent: indeed, the top sequent cannot be the conclusion of any rule in the system (except for cut and exchange, of course): $!\Box ?A^{\bullet}$ cannot be principal since there is a $\Diamond$-formula in the context, and $\Diamond ?B^{\bullet}$ cannot be principal since there is an !-formula.

$$\dfrac{\dfrac{\dfrac{\vdash\, !\Box ?A^\bullet, \Diamond ?B^\bullet}{\vdash ?!\Box ?A^\bullet, \Diamond ?B^\bullet}\,?_{\mathrm{d}}}{\vdash ?!\Box ?A^\bullet, !\Diamond ?B^\bullet}\,!_{\mathrm{p}}}{\vdash ?!\Box ?A^\bullet, ?!\Diamond ?B^\bullet}\,?_{\mathrm{d}}$$

(a) Failed linear translation of the □ rule

$$\dfrac{\dfrac{\dfrac{\dfrac{\dfrac{\vdash ?A^\bullet, ?B^\bullet}{\vdash \Box ?A^\bullet, \Diamond ?B^\bullet}\,\Box_{\mathrm{p}}}{\vdash\, !\Box ?A^\bullet, \Diamond ?B^\bullet}\,!^{\Diamond}_{\mathrm{p}}}{\vdash ?!\Box ?A^\bullet, \Diamond ?B^\bullet}\,?_{\mathrm{d}}}{\vdash ?!\Box ?A^\bullet, !\Diamond ?B^\bullet}\,!^{\Diamond}_{\mathrm{p}}}{\vdash ?!\Box ?A^\bullet, ?!\Diamond ?B^\bullet}\,?_{\mathrm{d}}$$

(b) Successful translation of the □ rule

$$\dfrac{\Box\Gamma', !\Gamma \vdash A, ?\Delta, \Diamond\Delta'}{\Box\Gamma', !\Gamma \vdash\, !A, ?\Delta, \Diamond\Delta'}\,!^{\Diamond}_{\mathrm{p}}$$

(c) New !-rule

Fig. 7: Translating the modal rule in a linear framework

$$\dfrac{\Gamma \vdash A, \Delta}{\Box\Gamma \vdash \Box A, \Diamond\Delta}\,\Box_{\mathrm{p}}, \quad \dfrac{\Gamma \vdash \Diamond A, \Diamond A, \Delta}{\Gamma \vdash \Diamond A, \Delta}\,\Diamond_{\mathrm{c}}, \quad \dfrac{\Gamma \vdash \Delta}{\Gamma \vdash \Diamond A, \Delta}\,\Diamond_{\mathrm{w}}, \quad \dfrac{\Box\Gamma', !\Gamma \vdash A, ?\Delta, \Diamond\Delta'}{\Box\Gamma', !\Gamma \vdash\, !A, ?\Delta, \Diamond\Delta'}\,!^{\Diamond}_{\mathrm{p}}$$

$$\dfrac{\Gamma, A \vdash \Delta}{\Box\Gamma, \Diamond A \vdash \Diamond\Delta}\,\Diamond_{\mathrm{p}}, \quad \dfrac{\Gamma, \Box A, \Box A \vdash \Delta}{\Gamma, \Box A \vdash \Delta}\,\Box_{\mathrm{c}}, \quad \dfrac{\Gamma \vdash \Delta}{\Gamma, \Box A \vdash \Delta}\,\Box_{\mathrm{w}}, \quad \dfrac{\Box\Gamma', !\Gamma, A \vdash ?\Delta, \Diamond\Delta'}{\Box\Gamma', !\Gamma, ?A \vdash ?\Delta, \Diamond\Delta'}\,?^{\Box}_{\mathrm{p}}$$

Fig. 8: Rules involving modalities for $\mu\mathsf{LL}^{\infty}_{\Box}$

A natural solution is to allow !-promotion in contexts containing !, □-formulas on the left and ?, ◇-formulas on the right as in Figure 7c.[1] The derivation of the translation of our ($\Box_{\mathrm{p}}$) instance can be completed as in Figure 7b.

Allowing □/◇-formulas in the context of a promotion rule has implications for the system's robustness to cut-elimination. For instance, taking the ($?_{\mathrm{p}}/?_{\mathrm{w}}$) principal case and adding modal formulas to the context naturally requires being able to weaken □ and ◇-formulas in the antecedent (resp. succedent):

$$\dfrac{\dfrac{\Gamma \vdash \Sigma}{\Gamma \vdash ?C, \Sigma}\,?_{\mathrm{w}} \quad \dfrac{!\Delta, \Box\Lambda, C \vdash ?\Phi, \Diamond\Psi}{!\Delta, \Box\Lambda, ?C \vdash ?\Phi, \Diamond\Psi}\,?_{\mathrm{p}}}{\Gamma, !\Delta, \Box\Lambda \vdash \Sigma, ?\Phi, \Diamond\Psi}\,\mathrm{cut} \quad \rightsquigarrow \quad \dfrac{\Gamma \vdash \Sigma}{\Gamma, !\Delta, \Box\Lambda \vdash \Sigma, ?\Phi, \Diamond\Psi}\,?_{\mathrm{w}}, !_{\mathrm{w}}, \Box_{\mathrm{w}}, \Diamond_{\mathrm{w}}$$

Similarly, the ($?_{\mathrm{c}}/?_{\mathrm{p}}$) key-case requires the ability to contract □/◇-formulas.

We now give a formal definition of the linear-logical modal μ-calculus:

Definition 6 (Linear-logical modal μ-calculus) *Pre-formulas of $\mu\mathsf{LL}^{\infty}_{\Box}$ are defined as:*

$$F, G \ ::= \ a \in \mathcal{A} \mid X \in \mathcal{V} \mid \mu X.F \mid \nu X.F \mid F^{\perp} \mid F \multimap G \mid F \parr G \mid F \otimes G \mid \perp \mid 1 \mid F \oplus G \mid F \,\&\, G \mid 0 \mid \top \mid ?F \mid !F \mid \Diamond F \mid \Box F.$$

We obtain positivity of an occurrence and the definition of formulas of $\mu\mathsf{LL}^{\infty}_{\Box}$ in the same way as for $\mu\mathsf{LL}^{\infty}$ and $\mu\mathsf{LK}^{\infty}_{\Box}$. Rules for $\mu\mathsf{LL}^{\infty}_{\Box}$ are the rules for $\mu\mathsf{LL}^{\infty}$ together with those of Figure 8. Note that ($!_p$) (resp. ($?_p$)) is an instance of ($!^{\Diamond}_p$) (resp. ($?^{\Box}_p$)). We define the sequents, pre-proofs, and proofs as in Section 2.

To justify the relevance of $\mu\mathsf{LL}^{\infty}_{\Box}$ we prove that there is a forgetful operation from $\mu\mathsf{LL}^{\infty}_{\Box}$ proofs to $\mu\mathsf{LK}^{\infty}_{\Box}$, the *skeleton*. We define a translation SK(−) from $\mu\mathsf{LL}^{\infty}_{\Box}$ formulas, rules and pre-proofs to $\mu\mathsf{LK}^{\infty}_{\Box}$:

[1] Note that allowing modal rules in contexts containing ? on the right and ! on the left would also solve our problem. However, forgetting the linear information of such a rule would not give us a rule derivable in $\mu\mathsf{LK}^{\infty}_{\Box}$.

Definition 7 **($\mu\mathsf{LK}^{\infty}_{\square}$-Skeleton)** *We define the skeleton of formulas inductively:*

$$
\begin{array}{lll}
\mathrm{SK}(1) = \mathsf{T} & \mathrm{SK}(F \otimes G) = \mathrm{SK}(F) \wedge \mathrm{SK}(G) & \mathrm{SK}(F^{\perp}) = \mathrm{SK}(F)^{\perp} \\
\mathrm{SK}(\perp) = \mathsf{F} & \mathrm{SK}(F \parr G) = \mathrm{SK}(F) \vee \mathrm{SK}(G) & \mathrm{SK}(\Diamond F) = \Diamond \mathrm{SK}(F) \\
\mathrm{SK}(\top) = \mathsf{T} & \mathrm{SK}(F \,\&\, G) = \mathrm{SK}(F) \wedge \mathrm{SK}(G) & \mathrm{SK}(\Box F) = \Box \mathrm{SK}(F) \\
\mathrm{SK}(0) = \mathsf{F} & \mathrm{SK}(F \oplus G) = \mathrm{SK}(F) \vee \mathrm{SK}(G) & \mathrm{SK}(?F) = \mathrm{SK}(F) \\
\mathrm{SK}(a) = a & \mathrm{SK}(F \multimap G) = \mathrm{SK}(F) \to \mathrm{SK}(G) & \mathrm{SK}(!F) = \mathrm{SK}(F) \\
\mathrm{SK}(X) = X & \mathrm{SK}(\mu X.F) = \mu X.\mathrm{SK}(F) & \mathrm{SK}(\nu X.F) = \nu X.\mathrm{SK}(F)
\end{array}
$$

Sequents of $\mu\mathsf{LL}^{\infty}_{\square}$ are translated to sequents of skeletons of these formulas. Translation of rules are standard. Translations of pre-proofs are obtained co-inductively by applying rule translations.

The following ensures that validity is preserved both ways by SK(−):

Proposition 1 (Robustness of the skeleton to validity) *If π is a valid pre-proof, then $\mathrm{SK}(\pi)$ is a $\mu\mathsf{LK}^{\infty}_{\square}$ valid pre-proof, and vice versa.*

4 Cut-elimination for $\mu\mathsf{LL}^{\infty}_{\square}$

To eliminate cuts in $\mu\mathsf{LL}^{\infty}_{\square}$, we employ a generalization of the cut inference called multicuts, as done in previous works on similar non-wellfounded proof systems [3,4,12,28].

Definition 8 (Multicut rule) *The multicut rule is the following rule:*

$$
\frac{\Gamma_1 \vdash \Delta_1 \qquad \ldots \qquad \Gamma_n \vdash \Delta_n}{\Gamma \vdash \Delta}\ \mathsf{mcut}(\iota, \perp\!\!\!\perp)
$$

It can have any number of premises. The ancestor relation ι maps one formula of the conclusion to exactly one formula of the premises, while the $\perp\!\!\!\perp$-relation links cut-formulas together, subject to acyclicity and connectedness conditions.

Remark 3 *The idea behind the multicut is to abstract a finite tree of binary cuts quotiented by the cut-commutation rule. We provide an example of a multicut rule and represent graphically ι in red and $\perp\!\!\!\perp$ in blue.*

$$
\frac{\vdash A, B \qquad B \vdash C \qquad C \vdash D}{\vdash A, D}\ \mathsf{mcut}(\iota, \perp\!\!\!\perp)
$$

The multicut rule should be seen as a tree of binary cuts, cf (cut/mcut)-case:

$$
\frac{\mathcal{C} \qquad \dfrac{\Gamma_1 \vdash F, \Delta_1 \quad \Gamma_2, F \vdash \Delta_2}{\Gamma_1, \Gamma_2 \vdash \Delta_1, \Delta_2}\ \mathit{cut}}{\Gamma \vdash \Delta}\ \mathsf{mcut}(\iota, \perp\!\!\!\perp) \quad \rightsquigarrow \quad \frac{\mathcal{C} \quad \Gamma_1 \vdash F, \Delta_1 \quad \Gamma_2, F \vdash \Delta_2}{\Gamma \vdash \Delta}\ \mathsf{mcut}(\iota', \perp\!\!\!\perp')
$$

Here, $\iota'(p) = \iota(p)$ for positions of formulas sent to $\mathcal{C}$, and uses the ancestor relation of the cut-rule together with ι to determine the image of the other positions. The relation $\perp\!\!\!\perp'$ is obtained from $\perp\!\!\!\perp$ by adding $p \perp\!\!\!\perp p'$, where p and p' are the positions of the two cut-formulas F.

$$\dfrac{\dfrac{}{\Gamma' \vdash \top, \Delta'}\top_r \quad \mathcal{C}}{\Gamma \vdash \top, \Delta}\,\mathsf{mcut}(\iota, \perp\!\!\!\perp) \quad \rightsquigarrow \quad \dfrac{}{\Gamma \vdash \top, \Delta}\top_r$$

$$\dfrac{\dfrac{}{A \vdash A}\,\mathsf{ax}}{A \vdash A}\,\mathsf{mcut}(\iota, \perp\!\!\!\perp) \quad \rightsquigarrow \quad \dfrac{}{A \vdash A}\,\mathsf{ax}$$

$$\dfrac{\dfrac{\Gamma_1, F[X := \nu X.F] \vdash \Delta_1}{\Gamma_1, \nu X.F \vdash \Delta_1}\nu_l \quad \dfrac{\Gamma_2 \vdash F[X := \nu X.F], \Delta_2}{\Gamma_2 \vdash \nu X.F, \Delta_2}\nu_r \quad \mathcal{C}}{\Gamma \vdash \Delta}\,\mathsf{mcut}(\iota, \perp\!\!\!\perp)$$

$$\rightsquigarrow \quad \dfrac{\Gamma_1, F[X := \nu X.F] \vdash \Delta_1 \quad \Gamma_2 \vdash F[X := \nu X.F], \Delta_2 \quad \mathcal{C}}{\Gamma \vdash \Delta}\,\mathsf{mcut}(\iota, \perp\!\!\!\perp)$$

$$\dfrac{\dfrac{\Gamma_1 \vdash A_1, \Delta_1 \quad \Gamma_2 \vdash A_2, \Delta_2}{\Gamma_1, \Gamma_2 \vdash A_1 \otimes A_2, \Delta_1, \Delta_2}\otimes_r \quad \dfrac{\Gamma_3, A_1, A_2 \vdash \Delta_3}{\Gamma_3, A_1 \otimes A_2 \vdash \Delta_3}\otimes_l \quad \mathcal{C}}{\Gamma \vdash \Delta}\,\mathsf{mcut}(\iota, \perp\!\!\!\perp)$$

$$\rightsquigarrow \quad \dfrac{\Gamma_1 \vdash A_1, \Delta_1 \quad \Gamma_2 \vdash A_2, \Delta_2 \quad \Gamma_3, A_1, A_2 \vdash \Delta_3 \quad \mathcal{C}}{\Gamma \vdash \Delta}\,\mathsf{mcut}(\iota', \perp\!\!\!\perp')$$

$$\dfrac{\dfrac{\Gamma' \vdash A_1, \Delta' \quad \Gamma' \vdash A_2, \Delta'}{\Gamma' \vdash A_1 \,\&\, A_2, \Delta'}\&_r \quad \mathcal{C}}{\Gamma \vdash A_1 \,\&\, A_2, \Delta}\,\mathsf{mcut}(\iota, \perp\!\!\!\perp)$$

$$\rightsquigarrow \quad \dfrac{\dfrac{\Gamma' \vdash A_1, \Delta' \quad \mathcal{C}}{\Gamma \vdash A_1, \Delta}\,\mathsf{mcut}(\iota, \perp\!\!\!\perp) \quad \dfrac{\Gamma' \vdash A_2, \Delta' \quad \mathcal{C}}{\Gamma \vdash A_2, \Delta}\,\mathsf{mcut}(\iota, \perp\!\!\!\perp)}{\Gamma \vdash A_1 \,\&\, A_2, \Delta}\&_r$$

Fig. 9: Examples of (mcut)-step of $\mu\mathsf{MALL}^\infty$

4.1 The (mcut) reduction steps

As we use a multicut reduction strategy, we first describe the steps of reduction. To describe these mcut-steps of reduction, we will use a notation and a definition:

Notation 1 ((!)-contexts) *If $\mathcal{C}$ is a list of $\mu\mathsf{LL}_\square^\infty$-proofs, we denote by $\mathcal{C}^{!/\square}$ the list of proofs obtained by applying one of $(!_p^\Diamond)$, $(?_p^\square)$, $(\square_p)$ or $(\Diamond_p)$ rules to each proof of $\mathcal{C}$. If $\mathcal{C}$ is a list of $\mu\mathsf{LL}_\square^\infty$-proofs, we denote by $\mathcal{C}^\square$ the list of proofs obtained by applying one of $(\square_p)$ or $(\Diamond_p)$ rules to each proof of $\mathcal{C}$.*

Definition 9 (Restriction of a mcut context) *Let $\dfrac{\mathcal{C}}{s}\,\mathsf{mcut}(\iota, \perp\!\!\!\perp)$ be a multicut occurrence such that $\mathcal{C} = s_1 \ \ldots \ s_n$ with $s_i := F_1, \ldots, F_{k_i} \vdash G_1, \ldots, G_{r_i}$. We define $\mathcal{C}_{F_j}$ (resp. $\mathcal{C}_{G_j}$) to be the sequents $\perp\!\!\!\perp$-connected to the formula F_j (resp. G_j). This is extended to contexts of formulas with $\mathcal{C}_\Gamma := \cup_{F \in \Gamma} \mathcal{C}_F$.*

The mcut-reduction steps of $\mu\mathsf{MALL}^\infty$ can be found in [4,28]; we provide examples of these steps in Figure 9. The reduction steps for the exponential and modal fragment of $\mu\mathsf{LL}_\square^\infty$ are presented in Figures 10 and 11, which include the exponential steps of $\mu\mathsf{LL}^\infty$. For commutative steps, we present only the

$$\dfrac{\dfrac{\overset{\pi}{!\Gamma_1, \Box\Gamma_2 \vdash A, ?\Delta_1, \Diamond\Delta_2}}{!\Gamma_1, \Box\Gamma_2 \vdash !A, ?\Delta_1, \Diamond\Delta_2}\,!^{\Diamond}_{\mathrm{p}} \qquad \mathcal{C}^{!/\Box}}{!\Gamma', \Box\Gamma \vdash !A, ?\Delta, \Diamond\Delta'}\,\mathsf{mcut}(\iota, \perp\!\!\!\perp)$$

$$\rightsquigarrow \quad \dfrac{\dfrac{\overset{\pi}{!\Gamma_1, \Box\Gamma_2 \vdash A, ?\Delta_1, \Diamond\Delta_2} \qquad \mathcal{C}^{!/\Box}}{!\Gamma', \Box\Gamma \vdash A, ?\Delta, \Diamond\Delta'}\,\mathsf{mcut}(\iota, \perp\!\!\!\perp)}{!\Gamma', \Box\Gamma \vdash !A, ?\Delta, \Diamond\Delta'}\,!^{\Diamond}_{\mathrm{p}}$$

$$\dfrac{\dfrac{\overset{\pi}{\Gamma \vdash A, \Delta}}{\Box\Gamma \vdash \Box A, \Diamond\Delta}\,\Box_{\mathrm{p}} \qquad \mathcal{C}^{\Box}}{\Box\Gamma' \vdash \Box A, \Diamond\Delta'}\,\mathsf{mcut}(\iota, \perp\!\!\!\perp) \quad \rightsquigarrow \quad \dfrac{\dfrac{\overset{\pi}{\Gamma \vdash A, \Delta} \qquad \mathcal{C}}{\Gamma' \vdash A, \Delta'}\,\mathsf{mcut}(\iota, \perp\!\!\!\perp)}{\Box\Gamma' \vdash \Box A, \Diamond\Delta'}\,\Box_{\mathrm{p}}$$

$$\dfrac{\dfrac{\overset{\pi}{\Gamma \vdash \Delta}}{\Gamma \vdash \delta A, \Delta}\,\delta_{\mathrm{w}} \qquad \mathcal{C}}{\Gamma' \vdash \delta A, \Delta'}\,\mathsf{mcut}(\iota, \perp\!\!\!\perp) \quad \rightsquigarrow \quad \dfrac{\dfrac{\overset{\pi}{\Gamma \vdash \Delta} \qquad \mathcal{C}}{\Gamma' \vdash \Delta'}\,\mathsf{mcut}(\iota', \perp\!\!\!\perp')}{\Gamma' \vdash \delta A, \Delta'}\,\delta_{\mathrm{w}}$$

$$\dfrac{\dfrac{\overset{\pi}{\Gamma \vdash \delta A, \delta A, \Delta}}{\Gamma \vdash \delta A, \Delta}\,\delta_{\mathrm{c}} \qquad \mathcal{C}}{\Gamma' \vdash \delta A, \Delta'}\,\mathsf{mcut}(\iota, \perp\!\!\!\perp) \quad \rightsquigarrow \quad \dfrac{\dfrac{\overset{\pi}{\Gamma \vdash \delta A, \delta A, \Delta} \qquad \mathcal{C}}{\Gamma' \vdash \delta A, \delta A, \Delta'}\,\mathsf{mcut}(\iota', \perp\!\!\!\perp')}{\Gamma' \vdash \delta A, \Delta'}\,\delta_{\mathrm{c}}$$

$$\dfrac{\dfrac{\overset{\pi}{\Gamma \vdash A, \Delta}}{\Gamma \vdash ?A, \Delta}\,?_{\mathrm{d}} \qquad \mathcal{C}}{\Gamma' \vdash ?A, \Delta'}\,\mathsf{mcut}(\iota, \perp\!\!\!\perp) \quad \rightsquigarrow \quad \dfrac{\dfrac{\overset{\pi}{\Gamma \vdash A, \Delta} \qquad \mathcal{C}}{\Gamma' \vdash A, \Delta'}\,\mathsf{mcut}(\iota', \perp\!\!\!\perp')}{\Gamma' \vdash ?A, \Delta'}\,?_{\mathrm{d}}$$

Fig. 10: $\mu\mathsf{LL}^{\infty}_{\Box}$ exponential & modal commutative cut-elimination steps (commutation with right rules) – $\delta \in \{?, \Diamond\}$

cases where the principal formula of the rule being commuted appears in the succedent of the sequent. For principal steps, we present only the cases where the cut-formula is a ?- or $\Diamond$-formula. All other cases can be derived by duality.

Definition 10 (Reduction sequence) *A reduction sequence* $(\pi_i)_{i \in 1+\lambda}$ *(*$\lambda \in \omega + 1$*) is a* $\rightsquigarrow$ *sequence s.t.* π_0 *contains at most one* (mcut) *rule per branch.*

We aim to prove that each reduction sequence converges to a cut-free proof. However, this theorem is not true as stated, even for infinite reduction sequences: one could apply infinitely many reductions only on some part of the proof, without reducing cuts in another part of the proof. Therefore, we need to be more precise, which motivates the following definition, directly borrowed from [3,4] (the notion of residual is the usual one from rewriting):

Definition 11 (Fair reduction sequences [3,4]) *A mcut-reduction sequence* $(\pi_i)_{i \in \omega}$ *is* fair *if, for each* π_i *such that there is a reduction* $\mathcal{R}$ *to a proof* π'*, there exists a* $j > i$ *such that* π_j *does not contain any residual of* $\mathcal{R}$*.*

$$\dfrac{\mathcal{C}_{\Gamma,\Delta} \qquad \dfrac{\dfrac{\pi}{\Gamma \vdash \delta A, \delta A, \Delta}}{\Gamma \vdash \delta A, \Delta}\,\delta_{\mathrm{c}} \qquad \mathcal{C}^{!/\Box}_{\delta A}}{!\Gamma_1, \Box\Gamma_2, \Gamma_3 \vdash \Delta_1, ?\Delta_2, \Diamond\Delta_3}\,\mathsf{mcut}(\iota, \perp\!\!\!\perp)$$

$$\leadsto \dfrac{\dfrac{\dfrac{\mathcal{C}_{\Gamma,\Delta} \qquad \dfrac{\pi}{\Gamma \vdash \delta A, \delta A, \Delta} \qquad \mathcal{C}^{!/\Box}_{\delta A} \qquad \mathcal{C}^{!/\Box}_{\delta A}}{!\Gamma_1, \Box\Gamma_2, \quad !\Gamma_1, \Box\Gamma_2, \quad \Gamma_3 \vdash \Delta_1, \quad ?\Delta_2, \Diamond\Delta_3, \quad ?\Delta_2, \Diamond\Delta_3}\,\mathsf{mcut}(\iota', \perp\!\!\!\perp')}{!\Gamma_1, \Box\Gamma_2, \Box\Gamma_2, \Gamma_3 \vdash \Delta_1, ?\Delta_2, \Diamond\Delta_3, \Diamond\Delta_3}\,?_{\mathrm{c}}, !_{\mathrm{c}}}{!\Gamma_1, \Box\Gamma_2, \Gamma_3 \vdash \Delta_1, ?\Delta_2, \Diamond\Delta_3}\,\Diamond_{\mathrm{c}}, \Box_{\mathrm{c}}$$

$$\dfrac{\mathcal{C}_{\Gamma,\Delta} \qquad \dfrac{\Gamma \vdash \Delta}{\Gamma \vdash \delta A, \Delta}\,\delta_{\mathrm{w}} \qquad \mathcal{C}^{!/\Box}_{\delta A}}{!\Gamma_1, \Box\Gamma_2, \Gamma_3 \vdash \Delta_1, ?\Delta_2, \Diamond\Delta_3}\,\mathsf{mcut}(\iota, \perp\!\!\!\perp) \quad \leadsto \quad \dfrac{\dfrac{\dfrac{\mathcal{C}_{\Gamma,\Delta} \qquad \Gamma \vdash \Delta}{\Gamma_3 \vdash \Delta_3}\,\mathsf{mcut}(\iota', \perp\!\!\!\perp')}{!\Gamma_1, \Gamma_3 \vdash \Delta_1, ?\Delta_2}\,?_{\mathrm{w}}, !_{\mathrm{w}}}{!\Gamma_1, \Box\Gamma_2, \Gamma_3 \vdash \Delta_1, ?\Delta_2, \Diamond\Delta_3}\,\Diamond_{\mathrm{w}}, \Box_{\mathrm{w}}$$

$$\dfrac{\dfrac{\Gamma_1 \vdash A, \Delta_1}{\Gamma_1 \vdash ?A, \Delta_1}\,?_{\mathrm{d}} \qquad \dfrac{!\Gamma_2, \Box\Gamma_3, A \vdash ?\Delta_2, \Diamond\Delta_3}{!\Gamma_2, \Box\Gamma_3, ?A \vdash ?\Delta_2, \Diamond\Delta_3}\,?^{\Box}_{\mathrm{p}} \qquad \mathcal{C}}{\Gamma \vdash \Delta}\,\mathsf{mcut}(\iota, \perp\!\!\!\perp)$$

$$\leadsto \dfrac{\Gamma_1 \vdash A, \Delta_1 \qquad !\Gamma_2, \Box\Gamma_3, A \vdash ?\Delta_2, \Diamond\Delta_3 \qquad \mathcal{C}}{\Gamma \vdash \Delta}\,\mathsf{mcut}(\iota, \perp\!\!\!\perp)$$

Fig. 11: One side of the $\mu\mathsf{LL}^{\infty}_{\Box}$ exponential & modal principal cut-elimination steps – in all these proofs, $\delta \in \{?, \Diamond\}$

We can now state our cut-elimination theorem:

Theorem 1 (Cut-elimination for $\mu\mathsf{LL}^{\infty}_{\Box}$) *Each fair* (mcut)*-reduction sequence of $\mu\mathsf{LL}^{\infty}_{\Box}$ valid proofs converges to a cut-free valid proof.*

To prove it, we will translate formulas, proofs, and (mcut)-steps of $\mu\mathsf{LL}^{\infty}_{\Box}$ into $\mu\mathsf{LL}^{\infty}$ and use the following cut-elimination result from [28]:

Theorem 2 (Cut-elimination for $\mu\mathsf{LL}^{\infty}$ [28]) *Every fair (mcut)-reduction sequence of $\mu\mathsf{LL}^{\infty}$ valid proofs converges to a cut-free valid proof.*

Remark 4 *In [28], exponential formulas, proofs, and cut-steps are encoded into $\mu\mathsf{MALL}^{\infty}$. We could have encoded $\mu\mathsf{LL}^{\infty}_{\Box}$ modalities into $\mu\mathsf{MALL}^{\infty}$, replaying the proof of [28] for cut-elimination. However, using the $\mu\mathsf{LL}^{\infty}$ cut-elimination theorem makes the result more modular and adaptable to future extensions of $\mu\mathsf{LL}^{\infty}$ validity conditions or cut-elimination variants.*

4.2 Translation of $\mu\mathsf{LL}^{\infty}_{\Box}$ into $\mu\mathsf{LL}^{\infty}$

We provide a translation of $\mu\mathsf{LL}^{\infty}_{\Box}$ into $\mu\mathsf{LL}^{\infty}$:

Definition 12 (Translation of $\mu\mathsf{LL}^{\infty}_{\Box}$ into $\mu\mathsf{LL}^{\infty}$) *The translation of formulas is defined inductively:*

- *Translations of $\Diamond$ and $\Box$-formulas:* $(\Diamond A)^\circ := ?A^\circ$ *and* $(\Box A)^\circ := !A^\circ$.
- *Translations of atomic and unit formulas and variables f:* $f^\circ := f$.
- *Translations of fixed-point:* $(\delta X.F)^\circ := \delta X.F^\circ$ *(with $\delta \in \{\mu, \nu\}$).*
- *Translations of other connectives:* $c(A_1, \dots, A_n)^\circ := c(A_1^\circ, \dots, A_n^\circ)$.

Translations of structural rules for modalities, $(\Diamond_c)$, $(\Diamond_w)$, $(\Box_c)$ and $(\Box_w)$ are respectively $(?_c)$, $(?_w)$, $(!_c)$ and $(!_w)$. Translations for the promotions $(!_p^\Diamond)$ and $(?_p^\Box)$ are respectively $(!_p)$ and $(?_p)$. Translations of the modal rules are given by:

$$\dfrac{\Gamma \vdash A, \Delta}{\Box\Gamma \vdash \Box A, \Diamond\Delta}\,\Box_p \quad\rightsquigarrow\quad \dfrac{\dfrac{\Gamma^\circ \vdash A^\circ, \Delta^\circ}{!\Gamma^\circ \vdash A^\circ, ?\Delta^\circ}\,!_d^{\#(\Gamma)}, ?_d^{\#(\Delta)}}{!\Gamma^\circ \vdash !A^\circ, ?\Delta^\circ}\,!_p$$

$$\dfrac{\Gamma, A \vdash \Delta}{\Box\Gamma, \Diamond A \vdash \Diamond\Delta}\,\Diamond_p \quad\rightsquigarrow\quad \dfrac{\dfrac{\Gamma^\circ, A^\circ \vdash \Delta^\circ}{!\Gamma^\circ, A^\circ \vdash ?\Delta^\circ}\,!_d, ?_d}{!\Gamma^\circ, ?A^\circ \vdash ?\Delta^\circ}\,!_p$$

Translations of other inference rules (r) are (r) themselves. Translations of pre-proofs are defined co-inductively using translations of rules.

The translation preserves the validity of pre-proof in both directions:

Lemma 1 (Validity robustness to $(-)^\circ$ translation) *Let π be a $\mu\mathsf{LL}_\Box^\infty$ pre-proof. The proof π is valid if and only if the proof π° is valid.*

Proof. Let B be a branch of π. We have that B is validated by a thread (A_i) if and only if B° is validated by (A_i°), as the minimal recurring fixed point formula is a ν on the right (resp. μ on the left) in (A_i) if and only if it is in (A_i°).

Finally, we need to ensure that (mcut)-reduction sequences are robust under this translation. In our proof of the final theorem, we also need one-step reduction rules to be simulated by a finite number of reduction steps in the translation.

Lemma 2 *Consider a $\mu\mathsf{LL}_\Box^\infty$ reduction step $\pi_0 \rightsquigarrow \pi_1$. There exist a finite number of $\mu\mathsf{LL}^\infty$ proofs $\theta_0, \dots, \theta_n$ such that: $\pi_0^\circ = \theta_0 \to \theta_1 \to \dots \to \theta_{n-1} \to \theta_n = \pi_1^\circ$.*

Proof (Proof sketch). Reductions from the non-exponential part of $\mu\mathsf{LL}_\Box^\infty$ translate easily to one step of reduction in $\mu\mathsf{LL}^\infty$. The same is true for the exponential part, except for the commutation of the modal rule. The translation of the left proof of this step is of the form (we only consider the case of $\Diamond_\mathrm{p}$; $\Box_\mathrm{p}$ is similar):

$$\dfrac{\dfrac{\dfrac{\overset{\pi_i^\circ}{!\Gamma_i^\circ \vdash A_i^\circ, \Delta_i^\circ}}{!\Gamma_i^\circ \vdash A_i^\circ, ?\Delta_i^\circ}\,!_\mathrm{d}, ?_\mathrm{d}}{!\Gamma_i^\circ \vdash !A_i^\circ, ?\Delta_i^\circ}\,!_\mathrm{p} \qquad \dfrac{\dfrac{\overset{\pi_j^\circ}{!\Gamma_j^\circ, A_j^\circ \vdash \Delta_j^\circ}}{!\Gamma_j^\circ, A_j^\circ \vdash ?\Delta_j^\circ}\,!_\mathrm{d}, ?_\mathrm{d}}{!\Gamma_j^\circ, ?A_j^\circ \vdash ?\Delta_j^\circ}\,?_\mathrm{p} \qquad \begin{array}{c}\text{with } 1 \le i \le n\ \&\\ n+1 \le j \le n+m\end{array}}{\vdash !A^\circ, ?\Gamma^\circ}\,\mathsf{mcut}(\iota, \perp\!\!\!\perp)$$

Here, we notice that for each dereliction on a cut-formula, there exists a corresponding promotion that will be erased by a dereliction/promotion key-case. The first promotion will therefore commute under the cut, and then each dereliction on formulas of the conclusion will commute as well. Each dereliction and each promotion on cut-formulas will be erased, giving us the correct translation.

Now that we know that a step of (mcut)-reduction in $\mu\mathsf{LL}_\Box^\infty$ translates to one or more $\mu\mathsf{LL}^\infty$ (mcut)-reduction steps, it is easy to translate each reduction sequence of $\mu\mathsf{LL}_\Box^\infty$ into a reduction sequence of $\mu\mathsf{LL}^\infty$. However, to use the cut-elimination theorem of $\mu\mathsf{LL}^\infty$, we need the reduction sequence to be fair. The purpose of the following lemma is to control the fairness of the translated reduction sequence:

Lemma 3 (Completeness of the (mcut)-reduction system) *Let π and π' be two $\mu\mathsf{LL}_\Box^\infty$ proofs. If there is a $\mu\mathsf{LL}^\infty$-redex $\mathcal{R}$ sending π° to π', then there is also a $\mu\mathsf{LL}_\Box^\infty$-redex $\mathcal{R}'$ sending π to a proof π'', such that in the translation of $\mathcal{R}'$, $\mathcal{R}$ is reduced.*

Proof. We only prove the exponential cases, as the non-exponential cases are immediate. We have several cases:

- If the case is the commutative step of a weakening (resp. contraction, resp. dereliction) (r), it necessarily means that (r) is the translation of a rule (r') being a contraction (resp. weakening, resp. dereliction) which is also on top of a (mcut) in π. We can take $\mathcal{R}'$ as the step commuting (r') under the cut.
- If it is a principal case on a contraction or a weakening (r) on a formula $?A$ (resp. $!A$), it means that each proof cut-connected to $?A$ (resp. $!A$) ends with a promotion. As π° is the translation of a $\mu\mathsf{LL}_\Box^\infty$-proof, it means that (r) is the translation of a weakening or contraction rule (r') on a formula $?A'$ (resp. $!A'$) or $\Diamond A'$ (resp. $\Box A'$) on top of a (mcut). It also means that all the proofs cut-connected to these formulas are promotions or modal rules (no other rules than a modal rule or a promotion in $\mu\mathsf{LL}_\Box^\infty$ translates to a derivation ending with a promotion). Therefore, $\mathcal{R}'$ is the principal case on (r').
- If it is a dereliction/promotion key case, the dereliction (resp. promotion) is the translation of a dereliction (resp. promotion). We take $\mathcal{R}'$ to be the redex formed by these two rules.
- If it is the commutative step of a promotion (r), it means that all the proofs of the contexts of the (mcut) are promotion rules. This means that all these rules come from the translation of promotion or modal rules. We need to ensure that each (mcut) with a context full of promotions or modal rules is covered by the (mcut)-reductions of $\mu\mathsf{LL}_\Box^\infty$:
  - The commutation of $(!_{\mathrm{p}}^{\Diamond})$ (or $(?_{\mathrm{p}}^{\Box})$) is covered by the first commutative case in Figure 10.
  - If it is a modal rule that is ready to be commuted, then other rules are necessarily modal rules and therefore is covered by the second commutative case in Figure 10.

We use the two previous lemmas, to prove the following:

Corollary 1 *If $\lambda \in \omega + 1$ & $(\pi_i)_{i\in 1+\lambda}$ is a fair $\mu\mathsf{LL}_\Box^\infty$ reduction sequence, then:*

- *there exists a fair $\mu\mathsf{LL}^\infty$ reduction sequence $(\theta_i)_{i\in 1+\lambda'}$ with $\lambda' \in \omega + 1$;*
- *there exists a sequence of strictly increasing $(\varphi(i))_{i\in 1+\lambda}$ of elements of $1+\lambda'$;*

$$\dfrac{\dfrac{\dfrac{?\Gamma^\bullet \vdash ?A^\bullet, ?\Delta^\bullet}{\Box ?\Gamma^\bullet \vdash \Box ?A^\bullet, \Diamond ?\Delta^\bullet}\;\Box_{\mathrm{p}}}{\Box ?\Gamma^\bullet \vdash ?!\Box ?A^\bullet, \Diamond ?\Delta^\bullet}\;?_{\mathrm{d}}, !^{\Diamond}_{\mathrm{p}}}{!\Box ?\Gamma^\bullet \vdash ?!\Box ?A^\bullet, ?!\Diamond ?\Delta^\bullet}\;?_{\mathrm{d}}, !^{\Diamond}_{\mathrm{p}}, !_{\mathrm{d}}$$

(a) Failed linear translation of the □ rule with non-empty antecedent

$$\dfrac{\dfrac{\dfrac{\dfrac{\dfrac{\Gamma^\bullet \vdash ?A^\bullet, ?\Delta^\bullet}{!?\Gamma^\bullet \vdash ?A^\bullet, ?\Delta^\bullet}\;!_{\mathrm{d}}, ?_{\mathrm{p}}}{!?\Gamma^\bullet \vdash !?A^\bullet, ?\Delta^\bullet}\;!^{\Diamond}_{\mathrm{p}}}{\Box !?\Gamma^\bullet \vdash \Box !?A^\bullet, \Diamond ?\Delta^\bullet}\;\Box_{\mathrm{p}}}{\Box !?\Gamma^\bullet \vdash ?!\Box !?A^\bullet, \Diamond ?\Delta^\bullet}\;?_{\mathrm{d}}, !^{\Diamond}_{\mathrm{p}}}{!\Box !?\Gamma^\bullet \vdash ?!\Box !?A^\bullet, ?!\Diamond ?\Delta^\bullet}\;?_{\mathrm{d}}, !^{\Diamond}_{\mathrm{p}}, !_{\mathrm{d}}$$

(b) Successful translation of the □ rule with non-empty antecedent

Fig. 12: Translating the Modal Rule with Non-Empty Antecedent

– *such that for each* i, $\theta_{\varphi(i)} = \pi_i^\circ$.

Proof. We construct the sequence by induction on the steps of $(\pi_i)_{i \in 1+\lambda}$.

– For $i = 0$: $\theta_0 = \pi_0^\circ$ and $\varphi(0) = 0$:
– For $i + 1$, suppose we have defined both sequences up to rank i. We use Lemma 2 on the step $\pi_i \rightsquigarrow \pi_{i+1}$ to get a finite sequence of reduction $\pi_i = \theta'_0 \rightsquigarrow \cdots \rightsquigarrow \theta'_n = \pi_{i+1}$. We then construct both sequences by setting $\varphi(i + 1) := \varphi(i) + n, \quad \theta_{\varphi(i)+j} := \theta'_j$ (for $j \in [\![0, n]\!]$).

Fairness of $(\theta_i)_{i \in 1+\lambda'}$ follows from Lemma 3 and from the fact that after the translation of an (mcut)-step, $\pi^\circ \rightsquigarrow {\pi'}^\circ$, each residual of a redex $\mathcal{R}$ of π°, is contained in the translations of residuals of the associated redex $\mathcal{R}'$ of Lemma 3.

4.3 Cut-elimination for $\mu\mathsf{LL}^\infty_\Box$

Finally, we can prove the main theorem of the section:

Proof (of Theorem 1). Let $(\pi_i)_{i \in 1+\lambda}$ be a fair $\mu\mathsf{LL}^\infty_\Box$ reduction sequence. We use Corollary 1 and get a fair $\mu\mathsf{LL}^\infty$ reduction sequence $(\theta_i)_{i \in 1+\lambda'}$ and a sequence $(\varphi(i))_{i \in 1+\lambda}$ of natural numbers. By Theorem 2, we know that $(\theta_i)_{i \in \omega}$ converges to a cut-free proof θ of $\mu\mathsf{LL}^\infty$. Now suppose for the sake of contradiction that $(\pi_i)_{i \in 1+\lambda}$ does not converge to an (mcut)-free pre-proof. There is a j and a path p such that for each proof $\pi_{j'}$, with $j' \geq j$, there is an (mcut)-rule at the end of path p. This means that the translation of p leads to an (mcut) for each proof $\theta_{j'}$ with $j' \geq \varphi(j)$, contradicting the convergence of (θ_j) to a cut-free proof. Moreover (π_i) converges to a pre-proof π such that $\pi^\circ = \theta$, since $\theta_{\varphi(j)}$ equals π_j° under multicuts. Since θ is cut-free, π is both valid and cut-free by Lemma 1.

5 Cut-elimination of $\mu\mathsf{LK}^\infty_\Box$

We extend the linear translation of $\mu\mathsf{LK}^\infty$ from [28] to a translation from $\mu\mathsf{LK}^\infty_\Box$ into $\mu\mathsf{LL}^\infty_\Box$. The tentative translation of the modal formulas studied at the beginning of Section 3 actually only works with an empty antecedent. Adapting our

$$\dfrac{\Delta \vdash F, \Gamma}{\Box\Delta \vdash \Box F, \Diamond\Gamma}\,\Box_p \quad \rightsquigarrow$$

$$\dfrac{\dfrac{\dfrac{\dfrac{\dfrac{\Delta^\bullet \vdash ?F^\bullet, ?\Gamma^\bullet}{!?\Delta^\bullet \vdash ?F^\bullet, ?\Gamma^\bullet}\,(!_d, ?_p)^{\#(\Delta)}}{!?\Delta^\bullet \vdash !?F^\bullet, ?\Gamma^\bullet}\,!_p}{\Box!?\Delta^\bullet \vdash \Box!?F^\bullet, \Diamond?\Gamma^\bullet}\,\Box_p}{!\Box!?\Delta^\bullet \vdash \Box!?F^\bullet, \Diamond?\Gamma^\bullet}\,!_d^{\#(\Delta)}}{!\Box!?\Delta^\bullet \vdash ?!\Box!?F^\bullet, ?!\Diamond?\Gamma^\bullet}\,(?_d, !_p^\Diamond)^{(\#(\Gamma)+1)}$$

$$\dfrac{\Delta, F \vdash \Gamma}{\Box\Delta, \Diamond F \vdash \Diamond\Gamma}\,\Diamond_p \quad \rightsquigarrow$$

$$\dfrac{\dfrac{\dfrac{\dfrac{\dfrac{\Delta^\bullet, F^\bullet \vdash ?\Gamma^\bullet}{!?\Delta^\bullet, F^\bullet \vdash ?\Gamma^\bullet}\,(!_d, ?_p)^{\#(\Delta)}}{!?\Delta^\bullet, ?F^\bullet \vdash ?\Gamma^\bullet}\,?_p}{\Box!?\Delta^\bullet, \Diamond?F^\bullet \vdash \Diamond?\Gamma^\bullet}\,\Diamond_p}{!\Box!?\Delta^\bullet, !\Diamond?F^\bullet \vdash \Diamond?\Gamma^\bullet}\,!_d^{\#(\Delta)+1}}{!\Box!?\Delta^\bullet, !\Diamond?F^\bullet \vdash ?!\Diamond?\Gamma^\bullet}\,(?_d, !_p^\Diamond)^{\#(\Gamma)}$$

Fig. 13: Linear translation of the modal rules

example by adding an antecedent to it: $\dfrac{\Gamma \vdash F, \Delta}{\Box\Gamma \vdash \Box F, \Diamond\Delta}\,\Box_p$ results in a derivation attempt as shown in Figure 12a. If Γ contains more than two formulas, $(?_p)$ cannot be applied. However, adding an !-connective in the translation of $\Box$-formulas as $\Box A^\bullet := !\Box!?A^\bullet$, allows us to complete the derivation (see Figure 12b).

Based on this, we define a translation from $\mu\mathsf{LK}_\Box^\infty$ into $\mu\mathsf{LL}_\Box^\infty$:

Definition 13 (Linear translation of $\mu\mathsf{LK}_\Box^\infty$) *The translation $(-)^\bullet$ from $\mu\mathsf{LK}_\Box^\infty$ formulas to $\mu\mathsf{LL}_\Box^\infty$ formulas is defined by induction on formulas:*

$$\begin{array}{rlrlrl}
(A_1 \to A_2)^\bullet & := !(?A_1^\bullet \multimap ?A_2^\bullet) & X^\bullet & := !X & (\mu X.A)^\bullet & := !\mu X.?A^\bullet \\
(A_1 \land A_2)^\bullet & := !(?A_1^\bullet \,\&\, ?A_2^\bullet) & \mathsf{T}^\bullet & := !\top & (\nu X.A)^\bullet & := !\nu X.?A^\bullet \\
(A_1 \lor A_2)^\bullet & := !(?A_1^\bullet \oplus ?A_2^\bullet) & \mathsf{F}^\bullet & := !0 & (\Diamond A)^\bullet & := !\Diamond?A^\bullet \\
(A^\perp)^\bullet & := !(?A^\bullet)^\perp & a^\bullet & := !a & (\Box A)^\bullet & := !\Box!?A^\bullet
\end{array}$$

Sequents are translated as $(\Gamma \vdash \Delta)^\bullet := \Gamma^\bullet \vdash ?\Delta^\bullet$.

The following property must be kept in mind when defining rule translations and is proved by an induction the formulas of $\mu\mathsf{LK}_\Box^\infty$:

Proposition 2 *For any $\mu\mathsf{LK}_\Box^\infty$ formula A, $A^\bullet$ is an !-formula.*

We provide the translation of modal rules in Figure 13. We define translations of proofs coinductively using rule translations. As the smallest formula of a totally ordered set of translations is the translation of the smallest formula, and branches of $\pi^\bullet$ contain translations of threads from π and vice-versa, we have:

Lemma 4 (Robustness of $(-)^\bullet$ to validity) *A pre-proof π is valid if and only if $\pi^\bullet$ is valid.*

Lemma 5 (Composition of SK$(-)$ and of $(-)^\bullet$) *For any $\mu\mathsf{LK}_\Box^\infty$ pre-proof π, SK$(\pi^\bullet)$ is equal to π.*

We define our rewriting system using the SK$(-)$ translation:

Definition 14 ((mcut)-rewriting system of $\mu\mathsf{LK}_\Box^\infty$) *The* (mcut)*-rewriting system of $\mu\mathsf{LK}_\Box^\infty$ is defined as the (mcut)-system obtained from the $\mu\mathsf{LL}_\Box^\infty$ (mcut)-system by discarding the linear information of proofs in this system.*

Finally, we have the following theorem:

Theorem 3 *The reduction system of $\mu\mathsf{LK}^{\infty}_{\square}$ is infinitary weakly-normalizing.*

Proof. Consider a $\mu\mathsf{LK}^{\infty}_{\square}$ proof π and a fair reduction sequence σ_{L} from $\pi^{\bullet}$. By Theorem 1, σ_{L} converges to a cut-free $\mu\mathsf{LL}^{\infty}_{\square}$ proof. By applying SK$(-)$ to each proof in the sequence, we obtain a sequence of valid $\mu\mathsf{LK}^{\infty}_{\square}$ proofs such that either $\mathrm{SK}(\pi_i) = \mathrm{SK}(\pi_{i+1})$ or $\mathrm{SK}(\pi_i)$ reduces to $\mathrm{SK}(\pi_{i+1})$ with one step of $\mu\mathsf{LK}^{\infty}_{\square}$ mcut-reduction. We obtain a $\mu\mathsf{LK}^{\infty}_{\square}$ cut-reduction sequence σ_{K} that is infinite and converges to a valid, cut-free $\mu\mathsf{LK}^{\infty}_{\square}$ proof.

6 Conclusion

We have introduced $\mu\mathsf{LL}^{\infty}_{\square}$, a linear version of the modal μ-calculus, along with its circular and non-wellfounded proof systems. We have proved a cut-elimination theorem for fair cut-elimination reduction sequences, generalizing previous results on the non-wellfounded proof theory of linear logic. Through a linear translation of the circular and non-wellfounded proof systems for the modal μ-calculus ($\mu\mathsf{LK}^{\infty}_{\square}$) to $\mu\mathsf{LL}^{\infty}_{\square}$, we have obtained a cut-elimination theorem for the non-wellfounded sequent calculus of the modal μ-calculus.

In our opinion, this work presents a new and interesting application of linear logic to the modal μ-calculus, developing proof theories in both domains and highlighting the potential for cross-fertilization for the two communities. Indeed, this constitutes the first full syntactic cut-elimination theorem for a proof system modeling the full modal μ-calculus.

Furthermore, the fine-grained cut-elimination inherited from linear logic opens up the possibility of developing a non-trivial cut-elimination equivalence on $\mu\mathsf{LK}^{\infty}_{\square}$ proofs. This, in turn, could lead to the design of a denotational semantics for proofs of modal μ-calculus, a question that was previously out of reach due to both the lack of a syntactic cut-elimination theorem and the absence of structure in proofs of the modal μ-calculus.

However, this is not the first work studying both linear logic and modal logic. Notably, studies on modal logic S4 through linear logic have addressed both its intuitionistic fragment, proving cut-admissibility [13], and its classical version [19,29]. Our work differs from these studies in two key aspects: (i) We employ a non-wellfounded setting. (ii) The logic S4 treats modalities as linear logic exponentials with a standard promotion and dereliction, making it a sub-system of the subexponentials [23]. Our work is different because we use a functorial promotion for modal rules, which requires us to examine the interaction between functorial and non-functorial promotion.

An important direction for future work is to explore whether our linear-logical modal mu-calculus can be adapted to wellfounded proof-systems of linear logic with fixed-points in a $\mu\mathsf{LL}_{\square}$ sequent calculus. We aim to investigate if our methodology can be extended to obtain a cut-elimination theorem for the finitary modal μ-sequent calculus via a linear translation from $\mu\mathsf{LK}_\Box$ to $\mu\mathsf{LL}_\Box$, building upon Baelde's results [2]. This question presents significant challenges due to the complex structure of fixed-point rules in finitary $\mu\mathsf{LL}$. While omitting linear information from $\mu\mathsf{LL}_\Box$ proofs should yield $\mu\mathsf{LK}_\Box$ proofs that simulate cut-elimination, designing a linear translation from $\mu\mathsf{LK}_\Box$ to $\mu\mathsf{LL}_\Box$ that commutes with cut-elimination, as shown in this paper, remains a non-trivial task.

From the linear logic-theoretic point of view, our system $\mu\mathsf{LL}_\Box^\infty$ can be viewed as a linear logic with two sets of exponential modalities satisfying different structural rules and exponentials. This is akin to so-called *light logics* [15,18], that are variants of linear logics developed by taming the power of exponential modalities in order to control the complexity of cut-elimination (for instance constraining the ?-context of a promotion to be immediately derelicted after a promotion ensures that typable programs have at most elementary complexity [15]).

Building on these ideas, we propose to develop a uniform framework for proving cut elimination in wellfounded proof systems of linear logic fragments that encompass both light logics and modal logics. This framework could be inspired by the subexponential system of Nigam & Miller [23] and the first author with Laurent [5], but extended to include functorial promotion. Such a system would allow us to treat light logics and our linear-logical modal mu-calculus as instances of a more general linear logic system.

In a similar direction, we shall investigate whether our approach of reducing cut-elimination to that of simpler systems can be useful for the circular approaches to session types [11,31] as well as to the study of productivity and normalization properties in functional reactive programming [8].

Funding. This work was partially funded by the ANR project RECIPROG, project reference ANR-21-CE48-019-01.

Acknowledgments. The authors would like to thank the anonymous reviewers for their constructive and valuable comments, as well as Anupam Das, Abhishek De, Farzad Jafarrahmani, Johannes Kloibhofer, Paul-André Melliès and Bahareh Afshari for fruitful discussions on this work.

References

1. Afshari, B., Leigh, G.E.: Cut-free completeness for modal mu-calculus. In: 2017 32nd Annual ACM/IEEE Symposium on Logic in Computer Science (LICS), pp. 1–12 (2017). https://doi.org/10.1109/LICS.2017.8005088
2. Baelde, D.: Least and greatest fixed points in linear logic. ACM Trans. Comput. Log. **13**(1), 2:1–2:44 (2012). https://doi.org/10.1145/2071368.2071370. URL https://doi.org/10.1145/2071368.2071370
3. Baelde, D., Doumane, A., Kuperberg, D., Saurin, A.: Bouncing threads for circular and non-wellfounded proofs: Towards compositionality with circular proofs. In: Proceedings of the 37th Annual ACM/IEEE Symposium on Logic in Computer Science, LICS '22. Association for Computing Machinery, New York, NY, USA

(2022). https://doi.org/10.1145/3531130.3533375. URL https://doi.org/10.1145/3531130.3533375

4. Baelde, D., Doumane, A., Saurin, A.: Infinitary Proof Theory: the Multiplicative Additive Case. In: CSL 2016, *LIPIcs*, vol. 62, pp. 42:1–42:17 (2016). https://doi.org/10.4230/LIPIcs.CSL.2016.42. URL http://drops.dagstuhl.de/opus/volltexte/2016/6582
5. Bauer, E., Laurent, O.: Super exponentials in linear logic. Electronic Proceedings in Theoretical Computer Science **353**, 50–73 (2021). https://doi.org/10.4204/eptcs.353.3. URL http://dx.doi.org/10.4204/EPTCS.353.3
6. Brünnler, K., Studer, T.: Syntactic cut-elimination for a fragment of the modal mu-calculus. Annals of Pure and Applied Logic **163**(12), 1838–1853 (2012). https://doi.org/https://doi.org/10.1016/j.apal.2012.04.006. URL https://www.sciencedirect.com/science/article/pii/S0168007212000760
7. Buss, S.R. (ed.): Handbook of Proof Theory. Elsevier, New York (1998)
8. Cave, A., Ferreira, F., Panangaden, P., Pientka, B.: Fair reactive programming. In: S. Jagannathan, P. Sewell (eds.) The 41st Annual ACM SIGPLAN-SIGACT Symposium on Principles of Programming Languages, POPL '14, San Diego, CA, USA, January 20-21, 2014, pp. 361–372. ACM (2014). https://doi.org/10.1145/2535838.2535881. URL https://doi.org/10.1145/2535838.2535881
9. Curien, P., Herbelin, H.: The duality of computation. In: ICFP 2000, pp. 233–243. ACM (2000). https://doi.org/10.1145/351240.351262. URL https://doi.org/10.1145/351240.351262
10. Danos, V., Joinet, J., Schellinx, H.: A new deconstructive logic: Linear logic. J. Symb. Log. **62**(3), 755–807 (1997). https://doi.org/10.2307/2275572. URL https://doi.org/10.2307/2275572
11. Derakhshan, F., Pfenning, F.: Circular proofs as session-typed processes: A local validity condition. Logical Methods in Computer Science **Volume 18, Issue 2**, 8 (2022). https://doi.org/10.46298/lmcs-18(2:8)2022. URL https://lmcs.episciences.org/5675
12. Doumane, A.: On the infinitary proof theory of logics with fixed points. Phd thesis, Paris Diderot University (2017)
13. Fukuda, Y., Yoshimizu, A.: A linear-logical reconstruction of intuitionistic modal logic S4. CoRR **abs/1904.10605** (2019). URL http://arxiv.org/abs/1904.10605
14. Girard, J.: Linear logic. Theor. Comput. Sci. **50**, 1–102 (1987). https://doi.org/10.1016/0304-3975(87)90045-4. URL https://doi.org/10.1016/0304-3975(87)90045-4
15. Girard, J.Y.: Light linear logic **143**(2), 175–204 (1998). https://doi.org/10.1006/inco.1998.2700
16. Jäger, G., Kretz, M., Studer, T.: Canonical completeness of infinitary μ. The Journal of Logic and Algebraic Programming **76**(2), 270–292 (2008). https://doi.org/https://doi.org/10.1016/j.jlap.2008.02.005. URL https://www.sciencedirect.com/science/article/pii/S1567832608000209. Logic and Information: From Logic to Constructive Reasoning
17. Kozen, D.: Results on the propositional mu-calculus. Theor. Comput. Sci. **27**, 333–354 (1983). https://doi.org/10.1016/0304-3975(82)90125-6. URL https://doi.org/10.1016/0304-3975(82)90125-6
18. Lafont, Y.: Soft linear logic and polynomial time. Theor. Comput. Sci. **318**(1–2), 163–180 (2004). https://doi.org/10.1016/j.tcs.2003.10.018
19. Martini, S., Masini, A.: A modal view of linear logic. Journal of Symbolic Logic **59**(3), 888–899 (1994). https://doi.org/10.2307/2275915

20. Miller, D.: Finding unity in computational logic. In: Proceedings of the 2010 ACM-BCS Visions of Computer Science Conference, ACM-BCS '10. BCS Learning & Development Ltd., Swindon, GBR (2010)
21. Mints, G.: Effective cut-elimination for a fragment of modal mu-calculus. Studia Logica: An International Journal for Symbolic Logic **100**(1/2), 279–287 (2012). URL `http://www.jstor.org/stable/41475226`
22. Mints, G., Studer, T.: Cut-elimination for the mu-calculus with one variable. Electronic Proceedings in Theoretical Computer Science **77**, 47–54 (2012). `https://doi.org/10.4204/eptcs.77.7`. URL `http://dx.doi.org/10.4204/EPTCS.77.7`
23. Nigam, V., Miller, D.: Algorithmic specifications in linear logic with subexponentials. In: PPDP 2009, p. 129–140. ACM, New York, NY, USA (2009). `https://doi.org/10.1145/1599410.1599427`. URL `https://doi.org/10.1145/1599410.1599427`
24. Niwiński, D., Walukiewicz, I.: Games for the μ-calculus. Theoretical Computer Science **163**(1), 99–116 (1996). `https://doi.org/https://doi.org/10.1016/0304-3975(95)00136-0`. URL `https://www.sciencedirect.com/science/article/pii/0304397595001360`
25. Pagani, M., Tortora de Falco, L.: Strong normalization property for second order linear logic. Theoretical Computer Science **411**(2), 410–444 (2010)
26. Santocanale, L.: A calculus of circular proofs and its categorical semantics. In: M. Nielsen, U. Engberg (eds.) Foundations of Software Science and Computation Structures, 5th International Conference, FOSSACS 2002. Held as Part of the Joint European Conferences on Theory and Practice of Software, ETAPS 2002 Grenoble, France, April 8-12, 2002, Proceedings, *Lecture Notes in Computer Science*, vol. 2303, pp. 357–371. Springer (2002). `https://doi.org/10.1007/3-540-45931-6_25`. URL `https://doi.org/10.1007/3-540-45931-6_25`
27. Santocanale, L.H.M.l., Fortier, J.: Cuts for circular proofs: semantics and cut-elimination. In: Computer Science Logic 2013, *LIPIcs*, vol. 23, pp. 248–262. Schloss Dagstuhl - Leibniz-Zentrum fuer Informatik, Torino, Italy (2013). `https://doi.org/10.4230/LIPIcs.CSL.2013.248`. URL `https://hal.science/hal-01260986`
28. Saurin, A.: A linear perspective on cut-elimination for non-wellfounded sequent calculi with least and greatest fixed points. TABLEAUX '23. Springer (2023). URL `https://hal.science/hal-04169137`
29. Schellinx, H.: A Linear Approach to Modal Proof Theory, pp. 33–43. Springer Netherlands, Dordrecht (1996). `https://doi.org/10.1007/978-94-017-2798-3_3`. URL `https://doi.org/10.1007/978-94-017-2798-3_3`
30. Stirling, C.: Modal and Temporal Logics. In: Handbook of Logic in Computer Science. Oxford University Press (1992). `https://doi.org/10.1093/oso/9780198537618.003.0005`. URL `https://doi.org/10.1093/oso/9780198537618.003.0005`
31. Toninho, B., Caires, L., Pfenning, F.: Corecursion and non-divergence in session-typed processes. In: M. Maffei, E. Tuosto (eds.) Trustworthy Global Computing, pp. 159–175. Springer Berlin Heidelberg, Berlin, Heidelberg (2014)

Combining quantum and classical control: syntax, semantics and adequacy

Kinnari Dave[1,2], Louis Lemonnier[3(✉)], Romain Péchoux[2], and Vladimir Zamdzhiev[1]

[1] Université Paris-Saclay, CNRS, ENS Paris-Saclay, Inria, Laboratoire Méthodes Formelles, 91190 Gif-sur-Yvette, France
[2] Université de Lorraine, CNRS, Inria, LORIA, 54000 Nancy, France
[3] University of Edinburgh, Edinburgh, UK
louis.lemonnier@ed.ac.uk

Abstract. The two main notions of control in quantum programming languages are often referred to as "quantum" control and "classical" control. With the latter, the control flow is based on classical information, potentially resulting from a quantum measurement, and this paradigm is well-suited to mixed state quantum computation. Whereas with quantum control, we are primarily focused on pure quantum computation and there the "control" is based on superposition. The two paradigms have not mixed well traditionally and they are almost always treated separately. In this work, we show that the paradigms may be combined within the same system. The key ingredients for achieving this are: (1) syntactically: a modality for incorporating pure quantum types into a mixed state quantum type system; (2) operationally: an adaptation of the notion of "quantum configuration" from quantum lambda-calculi, where the quantum data is replaced with pure quantum primitives; (3) denotationally: suitable (sub)categories of Hilbert spaces, for pure computation and von Neumann algebras, for mixed state computation in the Heisenberg picture of quantum mechanics.

1 Introduction

There are two important paradigms in the design of quantum programming languages – "classical control" and "quantum control". In the classical control approach (e.g., [19,21,20,6,17,13]) the control flow of a program is conditioned on classical information that may result from quantum measurements. Type systems for quantum programming languages that are based on classical control are able to represent a variety of effects, e.g., quantum state preparation, state evolution through the application of unitary operators, and probabilistic effects induced by quantum measurements. Because of this, it is natural to conceptualise such lambda-calculi using quantum structures and models that are suited to describing mixed-state quantum computation and information (e.g., density matrices, superoperators). In the quantum control approach (e.g., [18,10,11,1,23]), one usually places an emphasis on pure state quantum computation and more

Supplementary Information The online version contains supplementary material available at https://doi.org/10.1007/978-3-031-90897-2_8.

P. A. Abdulla and D. Kesner (Eds.): FoSSaCS 2025, LNCS 15691, pp. 155–175, 2025.
https://doi.org/10.1007/978-3-031-90897-2_8

$$\begin{aligned} &\lambda y.\mathit{let}\ \mathcal{B}(x \otimes xs) = \mathit{Tof}(y) \\ &\quad \mathit{in\ case\ meas}(x)\ \ \mathit{of} \\ &\{\mathit{inl}(*) \to \mathcal{B}(\mathrm{CNOT})(xs), \\ &\quad \mathit{inr}(*) \to xs\} \end{aligned}$$

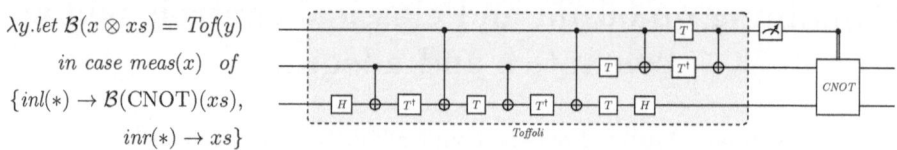

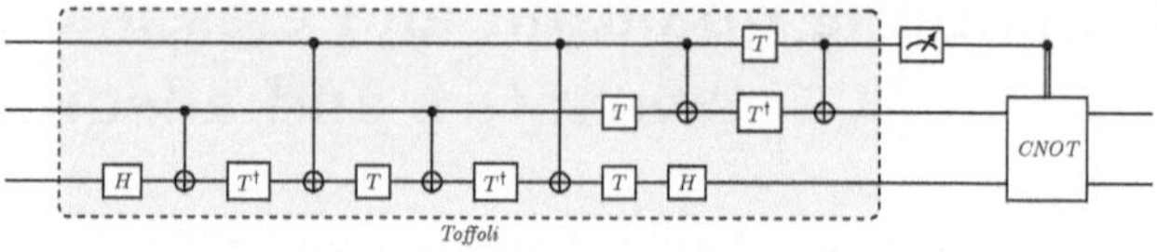

Fig. 1: A program and the corresponding circuit

specifically on superposition of terms, pure quantum states and unitary operators. In approaches that utilise classical control, one often starts with a selection of constants that represent some unitary operators (e.g., Hadamard, CNOT) and more complex unitary operations can be described in a circuit-like fashion by composing the atomic unitaries in a suitable way. Whereas in the quantum control approach, unitary operators are defined through more fundamental primitives that do not require the programmer to specify a circuit-like decomposition of the unitary operation. Because of this, a considerable economy in terms of syntax can be achieved with quantum control. For example, consider the circuit in Figure 1. It uses the well-known *Toffoli* gate, which would have to be decomposed as a large circuit like the one in the figure. However, the program representing this circuit can instead be written as a simple lambda term in our language given in Figure 1, where $\mathit{Tof} \triangleq \mathcal{B}(\text{ctrl CNOT})$. To highlight this further note that the controlled Hadamard can be defined similarly to the Toffoli in our language, but a typical classically controlled quantum programming language would require two H gates, six T gates and one $CNOT$ gate. Thus, our language offers an ease of writing programs by abstracting away quantum circuits. We aim to shift focus from drawing quantum circuits, which is a more *low-level* approach to writing code, to writing programs directly using syntax, which is a more *high-level* approach. In classical (non-quantum) programming, we can use high-level languages to write algorithms without having to specify boolean circuits or to encode integers using tuples of bits/booleans. With our paper, we are trying to advance quantum programming abstractions towards such a direction. For example, an individual qnat $|n\rangle$ can be used instead of a tensor of qubits $|q_1 q_2 ... q_k\rangle$ whose binary encoding corresponds to n. We view boolean circuits and binary encodings of integers as low-level and likewise for their quantum counterparts. Our aim is to develop higher-level quantum abstractions.

Our Contributions Because of the potential presence of quantum entanglement, it is impossible (in general) to decompose a quantum state into a non-trivial tensor product of two smaller quantum states. Indeed, in the quantum lambda-calculus (QLC) [15], which is based on classical control, the type **qbit** of qubits is an opaque type in the sense that there are no *closed* values of such a type. Program states may be described via *quantum configurations* which are triples $(|\psi\rangle, \ell, M)$, where $|\psi\rangle \in \mathbb{C}^{2^n}$ is a *pure* quantum state (possibly entangled); M is a program possibly containing free variables of type **qbit**; ℓ is a *linking* function that maps the free quantum variables of M to appropriate components of the state $|\psi\rangle$. It is possible to also view ℓ as a unitary permutation acting on $|\psi\rangle$ and this view is important for our development. Such configurations, even though they are not part of the user-facing syntax, allow us to reason about entangled states and the operational semantics are described via a relation $(|\psi\rangle, \ell, M) \rightarrow_p (|\psi'\rangle, \ell', M')$, with p the probability of reduction.

The main idea that allows us to combine both quantum and classical control is to modify the quantum configurations described above by replacing the quantum state $|\psi\rangle$ with a suitable *pure quantum term* $\boldsymbol{t}$ and to replace the linking function ℓ with a suitable *unitary permutation* u_σ, where both $\boldsymbol{t}$ and u_σ are syntactic constructs from our pure quantum language that may be assigned types.

In order to accurately model the situation with the QLC, the term $\boldsymbol{t}$ has to correspond to a *normalised* vector and we replace the unitary constants from the QLC with programmable unitary constructs, so the pure subsystem also has to ensure their *unitarity*. Our pure subsystem is based on ideas described in [3,18], but we further build on this by introducing an *equational theory* for our pure subsystem, which has unique normal forms and we describe a *denotational semantics* for it that is *sound and complete* with respect to the equational theory.

Next, we integrate the pure quantum subsystem into our main calculus, which allows us to describe both quantum and classical information and, therefore also classical control. This is a variant of the QLC with the addition of a modality $\mathcal{B}(\boldsymbol{Q})$ which allows us to view pure quantum types $\boldsymbol{Q}$ as types of mixed state quantum operations in the Heisenberg picture. The intuition behind this is the following: mixed states in the Schrödinger picture can be seen as CPTP maps[4] $(1 \mapsto \rho)\colon \mathbb{C} \rightarrow \mathcal{T}(H)$ whereas mixed states in the Heisenberg picture are given by functionals of the form $\mathrm{tr}(\rho -)\colon \mathcal{B}(H) \rightarrow \mathbb{C}$, both of which are determined by a choice of density operator $\rho\colon H \rightarrow H$ (see [5, Section 7] for detail on this matter). This modality allows us to replace the *constants* for state preparation and unitaries acting on qubits in the QLC with *terms and expressions* from the pure subsystem acting on more general quantum types beyond qubits. The operational semantics are described via a suitable adaptation of the aforementioned quantum configurations. The denotational semantics follows previous work on quantum programming semantics based on von Neumann algebras [6,17,13] and we show, in addition, that the assignment $\mathcal{B}(-)$ may be extended to a strict monoidal functor (between the relevant categories of Hilbert spaces and von Neumann algebras) that is crucial for our denotational semantics.

Overall, the two main contributions of our work are: (1) we show that quantum and classical control can be combined in a syntactic, operational and denotational sense by integrating a pure quantum control subsystem as part of the meta-theory and syntax of a quantum lambda-calculus; (2) an equational theory for a (sub)system for quantum control with unique normal forms and a sound and complete denotational semantics. Some supplementary material including extra examples, proofs, and complete figures for formation rule predicates are provided in Appendix.

Related Work We begin by commenting on semantic approaches to classical control. In [15], the authors describe a QLC where the higher-order primitives are interpreted using techniques from the semantics of quantitative models of linear

[4] CPTP stands for 'completely positive trace preserving' and $\mathcal{T}(H)$ and $\mathcal{B}(H)$ stand for the trace-class and bounded linear operators, respectively, on a Hilbert space H.

	Our work	Qunity	λ-$\mathcal{S}_1$	Sym	Quipper	QWIRE	Q.λ-calculus		
		[25]	[11]	[18]	[12]	[16]	[15]	[8,7,24]	[22]
Quantum control	✓	✓	✓	✓	✗	✗	✗	✗	✗
Classical control	✓	✓	✓	✗	✓	✓	✓	✓	✓
Infinite dimensional quantum type	✓	✗	✗	✓	✗	✗	✓	✓	✓
Unitary function spaces	✓	✗	✓	✓	✗	✓	✗	✗	✗
Normalised quantum terms	✓	✗	✓	✗	✓	✓	✓	✓	✓
Denotational semantics	✓	✓	✓	✗	✗	✓	✓	✓	✓
Adequacy	✓	✗	✓	✗	✗	✓	✓	✓	✓
Termination	✓	✗	✓	✓	✗*	✓	✗*	✗*	✗*

Fig. 2: A comparison of some quantum programming languages.

logic. In [8,7,24], the author(s) work with QLCs whose denotational semantics is given via game semantics. More recently, in [22], the authors have shown how to interpret a QLC with recursive types using presheaves that are enriched over a suitable base category and they also prove full abstraction. The approach that we use in this paper is based on von Neumann algebras which have been previously used for the semantics of quantum programming languages [6,13,17]. In particular, if we disregard the interaction between the pure subsystem and the main calculus, then our denotational model for the main calculus coincides with the one in [6]. The reason that we choose von Neumann algebras over the other approaches is because we think this gives the model which is the closest to mathematical physics out of the ones mentioned. Furthermore, our primary focus is not on developing the semantics of the classical paradigm; instead it is on the *interaction* between the pure quantum control subsystem and the main calculus based on classical control. Another distinguishing feature of our system, compared to the ones above, is the addition of types such as $\mathcal{B}(\mathbf{qnat})$ which allow us to describe states in infinite-dimensional Hilbert spaces, such as $\ell^2(\mathbb{N})$, and which we do not think can be easily included using the other approaches.

For the quantum control paradigm, relevant work includes [10,11,1,9,2]. What all these approaches have in common is that they have specific terms in their syntax for representing superposition. However, these approaches do not ensure unitarity of the relevant function spaces, in general. Our approach, instead, is based on related work first described in [18] and then later in [3]. One of the most distinctive features of this approach to quantum control is that the terms that introduce superposition are subject to strict formation conditions that ensure (linear algebraic) normalisation of the corresponding vectors. Furthermore, this approach to quantum control also imposes strict admissibility criteria on the unitary function spaces that can be formed through the type system and this ensures unitarity of the relevant expressions. Because of these considerations, we choose this approach to model quantum control in our pure subsystem. However, we have made some changes, in particular, we replace the operational semantics from these works with an equational theory instead. This works better for

* These languages admit general recursion.

our approach to *combining* quantum and classical control within the quantum configurations (in QLCs the quantum data is considered modulo equality). We continue building on this development by describing a denotational semantics that is sound and complete with respect to the equational theory. This is the first such result for languages with quantum control.

Another paper which combines quantum and classical control is Qunity [25]. However, compared to our work, Qunity does not have an operational semantics, but instead there is a compilation procedure to quantum circuits. Furthermore, the denotational semantics of Qunity: (1) does not ensure unitarity and normalisation of the pure primitives in its language, whereas ours does; (2) does not ensure trace-preservation of the mixed primitives in the Schrödinger picture, whereas ours does indeed ensure unitality of the mixed primitives in the Heisenberg picture (which corresponds to trace-preservation under the Heisenberg-Schrödinger duality). Other differences include that Qunity is restricted to finite-dimensional quantum data and they do not have higher-order lambda abstractions for dealing with mixed-state primitives. Instead, Qunity relies on a try-catch mechanism to combine quantum and classical control, whereas our approach is based on quantum configurations. The table in Figure 2 gives a comparison of our work with other quantum programming languages. *Normalised quantum terms* refers to the property that terms representing pure/mixed quantum states always correspond to complex vectors with norm one/density operators with trace one. *Adequacy* refers to the property that two observationally distinct programs have distinct denotational interpretations.

2 Quantum Control

This part is adapted from one of the authors' PhD thesis [14, Chapter 3], to which we refer the readers for full detail and proofs.

We write **Unitary** $\subseteq$ **Isometry** $\subseteq$ **Hilb** for the (sub)categories whose objects are the Hilbert spaces and whose morphisms are the unitaries (resp. isometries, bounded linear maps) between them.

2.1 Pure Quantum Syntax

We begin by describing a subsystem for quantum control. The syntax is based on symmetric pattern-matching [18]. The *pure quantum types* are given by the following grammar: $\boldsymbol{Q} ::= \boldsymbol{I} \mid \boldsymbol{Q}_1 \otimes \boldsymbol{Q}_2 \mid \boldsymbol{Q}_1 \oplus \boldsymbol{Q}_2 \mid \mathbf{qnat}$.

We interpret every quantum type as a separable Hilbert space: the unit type is interpreted as $[\![\boldsymbol{I}]\!] \triangleq \mathbb{C}$; pair types are interpreted as tensor products; (quantum) sum types are interpreted as direct sums; finally, $[\![\mathbf{qnat}]\!] \triangleq \ell^2(\mathbb{N})$. The type of qubits may be defined as $\mathbf{qbit} \triangleq \boldsymbol{I} \oplus \boldsymbol{I}$. We write $\mathbf{qbit}^n$ for the n-fold tensor product $\mathbf{qbit} \otimes \cdots \otimes \mathbf{qbit}$. The Hilbert space $\ell^2(\mathbb{N})$ allows us to form superpositions with respect to a countable orthonormal basis (e.g., $1/\sqrt{2}\,|3\rangle + 1/\sqrt{2}\,|7\rangle$). We may think of $\ell^2(\mathbb{N})$ as a quantum analogue of the natural numbers, to abstract from qubit-level computation. The addition of **qnat** allows us to reason about pure quantum computation while abstracting away from quantum circuits.

$$\boldsymbol{b} \ ::= \ \boldsymbol{x} \mid * \mid \mathbf{inl}\,\boldsymbol{b} \mid \mathbf{inr}\,\boldsymbol{b} \mid \boldsymbol{b}_1 \otimes \boldsymbol{b}_2 \mid \mathbf{0} \mid \mathbf{S}\,\boldsymbol{b}$$

$$\boldsymbol{v} \ ::= \ \textstyle\sum_{i \in \mathcal{I}} \boldsymbol{\alpha}_i \cdot \boldsymbol{b}_i$$

$$u \ ::= \ \{ \ \mid \boldsymbol{b}_1 \mapsto \boldsymbol{v}_1 \ \ldots \mid \boldsymbol{b}_n \mapsto \boldsymbol{v}_n \ \} \mid u_1 \oplus u_2 \mid u_1 \otimes u_2 \mid u_2 \circ u_1 \mid u^* \mid \mathrm{ctrl}\ u$$

$$\boldsymbol{t} \ ::= \ \boldsymbol{x} \mid * \mid \mathbf{inl}\,\boldsymbol{t} \mid \mathbf{inr}\,\boldsymbol{t} \mid \boldsymbol{t}_1 \otimes \boldsymbol{t}_2 \mid \mathbf{0} \mid \mathbf{S}\,\boldsymbol{t} \mid u\,\boldsymbol{t} \mid \textstyle\sum_{i \in \mathcal{I}} \boldsymbol{\alpha}_i \cdot \boldsymbol{t}_i$$

Fig. 3: Term grammar for the pure quantum subsystem.

The *unitary types*, written as $U(\boldsymbol{Q}_1, \boldsymbol{Q}_2)$, represent the unitary function spaces of our language. They are not interpreted as Hilbert spaces, but as sets of unitary operators: $[\![U(\boldsymbol{Q}_1, \boldsymbol{Q}_2)]\!] \triangleq \{u \colon [\![\boldsymbol{Q}_1]\!] \to [\![\boldsymbol{Q}_2]\!] \mid u \text{ is a unitary operator}\}$.

We now describe the language of the pure quantum subsystem whose grammar is given in Figure 3. (Pure) ***Terms*** $\boldsymbol{t}$ are written in bold notation in order to distinguish with other syntactic constructs. The symbols $\boldsymbol{x}, \boldsymbol{y}, \boldsymbol{z}$ denote pure quantum variables. We write $\mathbf{inl}_{\boldsymbol{Q}_1 \oplus \boldsymbol{Q}_2}\, \boldsymbol{t}$ and $\mathbf{inr}_{\boldsymbol{Q}_1 \oplus \boldsymbol{Q}_2}\, \boldsymbol{t}$ for sum type injections, and, for brevity, we often omit writing the subscripts. The term $\boldsymbol{t}_1 \otimes \boldsymbol{t}_2$ is used for forming pairs of terms. The terms $\mathbf{0}$ and $\mathbf{S}\,\boldsymbol{t}$ represent the qnat $|0\rangle$ and the qnat $|n+1\rangle$, provided that $\boldsymbol{t}$ represents the qnat $|n\rangle$. We define $|\underline{\mathbf{n}}\rangle \triangleq \mathbf{S}^{\mathbf{n}}\ \mathbf{0}$ and $|\underline{\boldsymbol{t}+\mathbf{n}}\rangle \triangleq \mathbf{S}^{\mathbf{n}}\ \boldsymbol{t}$, for $\mathbf{n} \in \mathbb{N}$. The term $u\,\boldsymbol{t}$ represents the application of a unitary u to a term $\boldsymbol{t}$. Superpositions are represented by the term $\sum_{i \in \mathcal{I}} \boldsymbol{\alpha}_i \cdot \boldsymbol{t}_i$. We impose strict conditions on these terms that ensure normalisation in §2.2.

In Figure 3, the ***Basis Terms*** (ranged over by symbols $\boldsymbol{b}, \boldsymbol{b}_i$) are the simplest kind of pure terms. The (pure) ***Values***, ranged over by $\boldsymbol{v}$ and defined in Figure 3, are the normal forms in this subsystem. They are of the form $\sum_{i \in \mathcal{I}} \boldsymbol{\alpha}_i \cdot \boldsymbol{b}_i$, where $\boldsymbol{\alpha}_i \in \mathbb{C}$, $\boldsymbol{b}_i$ are basis terms, and $\mathcal{I}$ is a finite index set. We consider $\mathcal{I}$ as a totally ordered set so that we can work with unique normal forms. We also write $\boldsymbol{\alpha}_1 \cdot \boldsymbol{b}_1 + \cdots + \boldsymbol{\alpha}_n \cdot \boldsymbol{b}_n$ for $\sum_{i \in \{1,\ldots,n\}} \boldsymbol{\alpha}_i \cdot \boldsymbol{b}_i$. The formation conditions (§2.2) ensure that all terms in the sums are pairwise distinct, orthogonal, and of the same type, and thus well-formed values correspond to *normalised* vectors. In our equational theory, every quantum term $\boldsymbol{t}$ can be rewritten to a normalised pure value $\boldsymbol{v}$ (Proposition 3). The values are the canonical forms of this subsystem.

Example 1. The basis term $* \colon \boldsymbol{I}$ represents the complex scalar $1 \in \mathbb{C}$. The basis terms that represent the computational basis for qubits can be defined by $|\mathbf{0}\rangle \triangleq \mathbf{inl}\ * \colon \mathbf{qbit}$ and $|\mathbf{1}\rangle \triangleq \mathbf{inr}\ * \colon \mathbf{qbit}$. Other basis terms that can be defined include $|\mathbf{00}\rangle \triangleq |\mathbf{0}\rangle \otimes |\mathbf{0}\rangle \colon \mathbf{qbit}^2$ and $|\mathbf{11}\rangle \triangleq |\mathbf{1}\rangle \otimes |\mathbf{1}\rangle \colon \mathbf{qbit}^2$. If $\boldsymbol{b}$ is a basis term, we often write it as $|\boldsymbol{b}\rangle$ to distinguish it from other (non-basis) terms, e.g., $|\underline{\mathbf{2}}\rangle \colon \mathbf{qnat}$ represents the qnat $|2\rangle \in \ell^2(\mathbb{N})$. Single-qubit states may be defined by $|\pm\rangle \triangleq (^1/\sqrt{2} \cdot |\mathbf{0}\rangle \pm {}^1/\sqrt{2} \cdot |\mathbf{1}\rangle) : \mathbf{qbit}$. (Entangled) quantum states can be defined, e.g., $\mathbf{Bell} \triangleq (^1/\sqrt{2} \cdot |\mathbf{00}\rangle + {}^1/\sqrt{2} \cdot |\mathbf{11}\rangle) : \mathbf{qbit}^2$. Linear combinations of qnats can also be written in the language, *i.e.* $(^1/\sqrt{2} \cdot |\underline{\mathbf{0}}\rangle + {}^1/\sqrt{6} \cdot |\underline{\mathbf{1}}\rangle + {}^1/\sqrt{3} \cdot |\underline{\mathbf{2}}\rangle) : \mathbf{qnat}$.

Next, we describe the ***Unitaries*** u of Figure 3. *Unitary pattern-matching* $\{ \ \mid \boldsymbol{b}_1 \mapsto \boldsymbol{v}_1 \ \ldots \mid \boldsymbol{b}_n \mapsto \boldsymbol{v}_n \ \}$ allows us to build unitary maps out of basis terms and values: every *closed* basis term $\boldsymbol{b}_i$ is mapped to the closed value $\boldsymbol{v}_i$; basis terms with free variables $\boldsymbol{b}_i$ determine a mapping through the use of substitution with respect to the corresponding value $\boldsymbol{v}_i$, wherein both $\boldsymbol{b}_i$ and $\boldsymbol{v}_i$ have the same free variables occurring within them, and the induced mapping is determined by performing all possible substitutions of the free variables on both sides with the same closed basis terms. The formation conditions (§2.2) ensure that the basis terms $\boldsymbol{b}_i$ (resp. the values $\boldsymbol{v}_i$) determine an ONB for the type $\boldsymbol{Q}_1$ (resp. $\boldsymbol{Q}_2$) and therefore $\{ \mid \boldsymbol{b}_1 \mapsto \boldsymbol{v}_1 \ \ldots \mid \boldsymbol{b}_n \mapsto \boldsymbol{v}_n \}$ determines a unitary map. We introduce syntactic sugar for quantum if statements, written **qif**. If u_1 is of the form $\{ \mid \boldsymbol{b}_1 \mapsto \boldsymbol{v}_1 \ \ldots \mid \boldsymbol{b}_n \mapsto \boldsymbol{v}_n \}$ and u_2 of the form $\{ \mid \boldsymbol{b}'_1 \mapsto \boldsymbol{v}'_1 \ \ldots \mid \boldsymbol{b}'_m \mapsto \boldsymbol{v}'_m \}$ then **qif** $\boldsymbol{x}$ **then** $\boldsymbol{x} \otimes u_2$ **else** $\boldsymbol{x} \otimes u_1$ is short for the unitary:

$$\left\{ \begin{matrix} |\mathbf{0}\rangle \otimes \boldsymbol{b}_1 \mapsto |\mathbf{0}\rangle \otimes \boldsymbol{v}_1 \\ \vdots \\ |\mathbf{0}\rangle \otimes \boldsymbol{b}_n \mapsto |\mathbf{0}\rangle \otimes \boldsymbol{v}_n \end{matrix} , \begin{matrix} |\mathbf{1}\rangle \otimes \boldsymbol{b}'_1 \mapsto |\mathbf{1}\rangle \otimes \boldsymbol{v}'_1 \\ \vdots \\ |\mathbf{1}\rangle \otimes \boldsymbol{b}'_m \mapsto |\mathbf{1}\rangle \otimes \boldsymbol{v}'_m \end{matrix} \right\}$$

Example 2. The unitaries Had and CNOT below encode the standard Hadamard and CNOT gates, respectively. CNOT makes use of some simple pattern-matching on its last line: it can match either with $|\mathbf{0}\rangle \otimes |\mathbf{0}\rangle$ or with $|\mathbf{0}\rangle \otimes |\mathbf{1}\rangle$. The unitary $\mathrm{Had}_{\mathbf{qnat}}$ defines an extension of the Hadamard gate on the space $\ell^2(\mathbb{N})$.

$$\mathrm{Had} : U(\mathbf{qbit}, \mathbf{qbit})$$

$$\mathrm{Had} \triangleq \left\{ \begin{matrix} \mid |\mathbf{0}\rangle \mapsto 1/\sqrt{2} \cdot |\mathbf{0}\rangle + 1/\sqrt{2} \cdot |\mathbf{1}\rangle \\ \mid |\mathbf{1}\rangle \mapsto 1/\sqrt{2} \cdot |\mathbf{0}\rangle - 1/\sqrt{2} \cdot |\mathbf{1}\rangle \end{matrix} \right\}$$

$$\mathrm{CNOT} : U(\mathbf{qbit}^2, \mathbf{qbit}^2)$$

$$\mathrm{CNOT} \triangleq \left\{ \begin{matrix} \mid |\mathbf{1}\rangle \otimes |\mathbf{0}\rangle \mapsto |\mathbf{1}\rangle \otimes |\mathbf{1}\rangle \\ \mid |\mathbf{1}\rangle \otimes |\mathbf{1}\rangle \mapsto |\mathbf{1}\rangle \otimes |\mathbf{0}\rangle \\ \mid |\mathbf{0}\rangle \otimes |\boldsymbol{x}\rangle \mapsto |\mathbf{0}\rangle \otimes |\boldsymbol{x}\rangle \end{matrix} \right\}$$

$$\mathrm{Had}_{\mathbf{qnat}} : U(\mathbf{qnat}, \mathbf{qnat})$$

$$\mathrm{Had}_{\mathbf{qnat}} \triangleq \left\{ \begin{matrix} \mid |\underline{\mathbf{0}}\rangle & \mapsto 1/\sqrt{2} \cdot |\underline{\mathbf{0}}\rangle + 1/\sqrt{2} \cdot |\underline{\mathbf{1}}\rangle \\ \mid |\underline{\mathbf{1}}\rangle & \mapsto 1/\sqrt{2} \cdot |\underline{\mathbf{0}}\rangle - 1/\sqrt{2} \cdot |\underline{\mathbf{1}}\rangle \\ \mid |\underline{\boldsymbol{x}+\mathbf{2}}\rangle & \mapsto |\underline{\boldsymbol{x}+\mathbf{2}}\rangle \end{matrix} \right\}$$

The remaining expressions for forming unitaries are easy to understand: $u_2 \circ u_1$ represents composition of unitaries, u^* represents the adjoint of a unitary, $u_1 \otimes u_2$ and $u_1 \oplus u_2$ represent tensor products and direct sums of unitaries, and finally, ctrl u represents a qubit-controlled unitary operator. For simplicity, we sometimes write $\{\ldots \mid \boldsymbol{b}_i \mapsto \boldsymbol{v}_i \ldots\}$ as a shorthand notation for the unitary $\{ \mid \boldsymbol{b}_1 \mapsto \boldsymbol{v}_1 \ \ldots \mid \boldsymbol{b}_n \mapsto \boldsymbol{v}_n \}$, when n is clear from context or unimportant. We also point out that unitaries depend only on values and basis terms and that general pure terms cannot be used for the construction of unitaries.

2.2 Formation Conditions

In order to guarantee (linear-algebraic) normalisation for quantum terms and unitarity for functions, we rely on formulating: (1) a suitable *orthogonality relation* $\perp \subseteq \boldsymbol{Terms} \times \boldsymbol{Terms}$, which is used as part of the formation condition of some terms; (2) additional predicates $\mathrm{ONB}_{\boldsymbol{Q}}$ and $\mathrm{ONB}^{val}_{\boldsymbol{Q}}$ which ensure that a set of basis terms/values of type $\boldsymbol{Q}$ determines an ONB which is used as part of the formation conditions for the atomic introduction rule of unitaries.

The relation $\perp$ is defined as the smallest binary relation $\perp \subseteq \boldsymbol{Terms} \times \boldsymbol{Terms}$ that is closed under the rules in Figure 4. Quantum terms that are orthogonal via this relation are orthogonal in a denotational sense as well (§2.3).

$$\frac{}{\mathbf{inl}\ \boldsymbol{t}_1 \perp \mathbf{inr}\ \boldsymbol{t}_2} \qquad \frac{}{\mathbf{0} \perp \mathbf{S}\ \boldsymbol{t}} \qquad \frac{\boldsymbol{t}_1 \perp \boldsymbol{t}_2}{u\,\boldsymbol{t}_1 \perp u\,\boldsymbol{t}_2} \qquad \frac{j \neq k.\ \boldsymbol{t}_j \perp \boldsymbol{t}_k \qquad \sum_{i \in \mathcal{J} \cup \mathcal{K}} \overline{\boldsymbol{\alpha}_i} \boldsymbol{\beta}_i = 0}{\sum_{j \in \mathcal{J}} (\boldsymbol{\alpha}_j \cdot \boldsymbol{t}_j) \perp \sum_{k \in \mathcal{K}} (\boldsymbol{\beta}_k \cdot \boldsymbol{t}_k)}$$

Fig. 4: Rules for the orthogonality relation (excerpt).

$$\frac{}{\mathrm{ONB}_{\boldsymbol{Q}}(\{\boldsymbol{x}\})} \qquad \frac{}{\mathrm{ONB}_{\boldsymbol{I}}(\{*\})} \qquad \frac{\mathrm{ONB}_{\boldsymbol{Q}_1}(S) \qquad \mathrm{ONB}_{\boldsymbol{Q}_2}(T)}{\mathrm{ONB}_{\boldsymbol{Q}_1 \oplus \boldsymbol{Q}_2}(\{\mathbf{inl}\ \boldsymbol{b} \mid \boldsymbol{b} \in S\} \cup \{\mathbf{inr}\ \boldsymbol{b} \mid \boldsymbol{b} \in T\})}$$

$$\frac{\mathrm{ONB}_{\boldsymbol{Q}_i}(\pi_i(S)) \qquad \forall \boldsymbol{b} \in \pi_i(S),\ \mathrm{ONB}_{\boldsymbol{Q}_j}(S^i_{\boldsymbol{b}})}{\mathrm{ONB}_{\boldsymbol{Q}_1 \otimes \boldsymbol{Q}_2}(S)}\ i, j \in \{1, 2\}, i \neq j$$

$$\frac{\mathrm{ONB}_{\mathbf{qnat}}(S)}{\mathrm{ONB}_{\mathbf{qnat}}(\{\mathbf{0}\} \cup \{\mathbf{S}\ \boldsymbol{b} \mid \boldsymbol{b} \in S\})} \qquad \frac{(\boldsymbol{\alpha}_{\boldsymbol{b},\boldsymbol{b}'})_{(\boldsymbol{b},\boldsymbol{b}') \in S \times S} \text{ is a unitary matrix} \quad \mathrm{ONB}_{\boldsymbol{Q}}(S)}{\mathrm{ONB}^{val}_{\boldsymbol{Q}}(\{\sum^*_{\boldsymbol{b}' \in S} \boldsymbol{\alpha}_{\boldsymbol{b},\boldsymbol{b}'} \cdot \boldsymbol{b}' \mid \boldsymbol{b} \in S\})}$$

Fig. 5: Orthonormal basis predicates.

This relation allows us to enforce the orthogonality of terms appearing within a superposition $\sum_{i \in \mathcal{I}} (\boldsymbol{\alpha}_i \cdot \boldsymbol{t}_i)$. Under the assumption that all terms $\boldsymbol{t}_i$ represent normalised vectors that are pairwise orthogonal, the norm of the superposition is easy to calculate (statically), and it is given by $\sum_{i \in \mathcal{I}} |\boldsymbol{\alpha}_i|^2$. This justifies the formulation of the introduction rule for superposition (Figure 6) and guarantees that the vector represented by this superposition of terms is normalised.

Example 3. The rules for the orthogonality relation are not restricted to closed quantum terms, e.g., $\mathbf{0} \perp \mathbf{S}\ \boldsymbol{x}$ and $\mathbf{inl}\ \boldsymbol{x} \perp \mathbf{inr}\ \boldsymbol{y}$. One can also show that $|+\rangle \perp |-\rangle$ (defined in Example 1). However, we cannot show that Had $|\mathbf{0}\rangle \perp |-\rangle$ even though Had $|\mathbf{0}\rangle$ is equal to $|+\rangle$ with respect to our equational logic and also with respect to our denotational semantics (both introduced later). The reason for this is that we wish to define the orthogonality relation *statically.*

In order to determine whether a finite set S of (potentially open) basis terms determines an ONB for a type $\boldsymbol{Q}$, we introduce the *orthonormal basis predicates* $\mathrm{ONB}_{\boldsymbol{Q}}(S)$ in Figure 5, originally known as *orthogonal decomposition* [18]. Note that there is a mistake with the rule for the tensor product in the original paper, *e.g.* it does not allow for the typing of a CNOT.

Given a type $\boldsymbol{Q}$ and a finite set of basis terms S, we say that S determines an *orthonormal basis of* $\boldsymbol{Q}$, which we write as $\mathrm{ONB}_{\boldsymbol{Q}}(S)$, if this can be derived via the rules in Figure 5. We now explain some of the notation used therein. If S is a set of the form $\{\boldsymbol{b}_1 \otimes \boldsymbol{b}'_1, \ldots, \boldsymbol{b}_n \otimes \boldsymbol{b}'_m\}$ then we define $\pi_1(S) \triangleq \{\boldsymbol{b}_1, \ldots, \boldsymbol{b}_n\}$ and $\pi_2(S) \triangleq \{\boldsymbol{b}'_1, \ldots, \boldsymbol{b}'_m\}$. In this situation, we also define $S^1_{\boldsymbol{b}} \triangleq \{\boldsymbol{b}' \mid \boldsymbol{b} \otimes \boldsymbol{b}' \in S\}$ and $S^2_{\boldsymbol{b}'} \triangleq \{\boldsymbol{b} \mid \boldsymbol{b} \otimes \boldsymbol{b}' \in S\}$. Furthermore, the ONB predicate implies pairwise orthogonality of the associated values. Note that the predicate $\mathrm{ONB}_{\boldsymbol{Q}}(S)$ is defined for a finite set S consisting of *basis* terms. But, in order to be able to define unitaries that make use of superposition, we also work with other ONBs. Given a finite set S consisting of *values*, we say that S determines an ONB of type $\boldsymbol{Q}$, and we write $\mathrm{ONB}^{val}_{\boldsymbol{Q}}(S)$ to indicate this, whenever this can be derived via the rules in Figure 5. We also introduce syntactic sugar: the value $\sum^*_{i \in \mathcal{I}} \boldsymbol{\alpha}_i \cdot \boldsymbol{b}_i$ is

$$\frac{\Vdash u\colon U(\boldsymbol{Q}_1,\boldsymbol{Q}_2) \quad \boldsymbol{\Gamma}\vdash \boldsymbol{t}\colon \boldsymbol{Q}_1}{\boldsymbol{\Gamma}\vdash u\,\boldsymbol{t}\colon \boldsymbol{Q}_2} \qquad \frac{\forall k,\ \boldsymbol{\Gamma}\vdash \boldsymbol{t}_k\colon \boldsymbol{Q} \quad \forall k\neq j,\ \boldsymbol{t}_k \perp \boldsymbol{t}_j \quad \sum_{i\in\mathcal{I}}|\boldsymbol{\alpha}_i|^2=1}{\boldsymbol{\Gamma}\vdash \sum_{i\in\mathcal{I}}\boldsymbol{\alpha}_i\cdot \boldsymbol{t}_i\colon \boldsymbol{Q}}$$

$$\frac{\begin{array}{ll}\forall i,\ \boldsymbol{\Gamma}_i\vdash \boldsymbol{b}_i\colon \boldsymbol{Q}_1 & \mathrm{ONB}_{\boldsymbol{Q}_1}\{\boldsymbol{b}_1,\ldots,\boldsymbol{b}_n\}\\ \forall i,\ \boldsymbol{\Gamma}_i\vdash \boldsymbol{v}_i\colon \boldsymbol{Q}_2 & \mathrm{ONB}^{val}_{\boldsymbol{Q}_2}\{\boldsymbol{v}_1,\ldots,\boldsymbol{v}_n\}\end{array}}{\Vdash \{\ \mid \boldsymbol{b}_1\mapsto \boldsymbol{v}_1\ \ldots \mid \boldsymbol{b}_n\mapsto \boldsymbol{v}_n\ \}\colon U(\boldsymbol{Q}_1,\boldsymbol{Q}_2)} \qquad \frac{\Vdash u_1\colon U(\boldsymbol{Q}_1,\boldsymbol{Q}_2) \quad \Vdash u_2\colon U(\boldsymbol{Q}_3,\boldsymbol{Q}_4)}{\Vdash u_1\oplus u_2\colon U(\boldsymbol{Q}_1\oplus \boldsymbol{Q}_3,\boldsymbol{Q}_2\oplus \boldsymbol{Q}_4)}$$

$$\frac{\Vdash u_1\colon U(\boldsymbol{Q}_1,\boldsymbol{Q}_2) \quad \Vdash u_2\colon U(\boldsymbol{Q}_3,\boldsymbol{Q}_4)}{\Vdash u_1\otimes u_2\colon U(\boldsymbol{Q}_1\otimes \boldsymbol{Q}_3,\boldsymbol{Q}_2\otimes \boldsymbol{Q}_4)} \qquad \frac{\Vdash u_1\colon U(\boldsymbol{Q}_1,\boldsymbol{Q}_2) \quad \Vdash u_2\colon U(\boldsymbol{Q}_2,\boldsymbol{Q}_3)}{\Vdash u_2\circ u_1\colon U(\boldsymbol{Q}_1,\boldsymbol{Q}_3)}$$

$$\frac{\Vdash u\colon U(\boldsymbol{Q}_1,\boldsymbol{Q}_2)}{\Vdash u^*\colon U(\boldsymbol{Q}_2,\boldsymbol{Q}_1)} \qquad \frac{\Vdash u\colon U(\boldsymbol{Q},\boldsymbol{Q})}{\Vdash \mathrm{ctrl}\ u\colon U(\mathbf{qbit}\otimes \boldsymbol{Q},\mathbf{qbit}\otimes \boldsymbol{Q})}$$

Fig. 6: Formation rules for terms and unitaries (excerpt).

the finite linear combination where elements with scalar 0 are omitted. In other words, the indices $i\in\mathcal{I}$ such that $\boldsymbol{\alpha}_i=0$ are ignored, e.g., if $\boldsymbol{\alpha}_1=1$ and $\boldsymbol{\alpha}_2=0$, $\sum^*_{i\in\{1,2\}}\boldsymbol{\alpha}_i\cdot\boldsymbol{b}_i$ is the value $\sum_{i\in\{*\}}1\cdot\boldsymbol{b}_1$.

Example 4. For **qbit**, the rules from Figure 5 show that $\mathrm{ONB}_{\mathbf{qbit}}(\{|\mathbf{0}\rangle,|\mathbf{1}\rangle\})$ and that $\mathrm{ONB}^{val}_{\mathbf{qbit}}(\{|+\rangle,|-\rangle\})$ hold (with $|\pm\rangle$ defined in Example 1). We also have that $\mathrm{ONB}_{\mathbf{qbit}}(\{\boldsymbol{x}\})$, where $\boldsymbol{x}$ is a variable. In other words, we determine an ONB for the type **qbit** by substituting $\boldsymbol{x}$ with all possible closed computational *basis* terms of type **qbit** (see Figure 3). For **qnat**, any set S with the property that $\mathrm{ONB}_{\mathbf{qnat}}(S)$ holds, must contain a term that is not closed. We have that $\mathrm{ONB}_{\mathbf{qnat}}(\{|\underline{\mathbf{x}}\rangle\})$, $\mathrm{ONB}_{\mathbf{qnat}}(\{|\underline{\mathbf{0}}\rangle,|\underline{\boldsymbol{x}+\mathbf{1}}\rangle\})$, and $\mathrm{ONB}_{\mathbf{qnat}}(\{|\underline{\mathbf{0}}\rangle,|\underline{\mathbf{1}}\rangle,|\underline{\boldsymbol{x}+\mathbf{2}}\rangle\})$ hold true and determine the same ONB for the type **qnat** (it represents the computational basis of $\ell^2(\mathbb{N})$). The set used in the last predicate allows us to conclude that $\mathrm{ONB}^{val}_{\mathbf{qnat}}(\{1/\sqrt{2}\cdot|\underline{\mathbf{0}}\rangle+1/\sqrt{2}\cdot|\underline{\mathbf{1}}\rangle, 1/\sqrt{2}\cdot|\underline{\mathbf{0}}\rangle-1/\sqrt{2}\cdot|\underline{\mathbf{1}}\rangle, |\underline{\boldsymbol{x}+\mathbf{2}}\rangle\})$ holds. This last set of values is used to type $\mathrm{Had}_{\mathbf{qnat}}$ from Example 2.

Judgements for pure quantum terms have the form $\boldsymbol{\Gamma}\vdash\boldsymbol{t}\colon\boldsymbol{Q}$ where $\boldsymbol{\Gamma}$ stands for a linear context of the form $\boldsymbol{\Gamma}\triangleq\boldsymbol{x}_1\colon\boldsymbol{Q}_1,\ldots,\boldsymbol{x}_n\colon\boldsymbol{Q}_n$. The formation rules for terms are presented in Figure 6; the first two lines are standard for a linearly-typed term calculus. The rule for superposition depends on the orthogonality relation $\perp$, and the premise ensures that the resulting superposition corresponds to a normalised vector. This is made precise later (Theorem 1). Finally, the formation rule for $u\,\boldsymbol{t}$, i.e. the application of a unitary u to a term $\boldsymbol{t}$, is akin to function application in lambda-calculi. However, the complexity is primarily relegated to the judgement $\Vdash u\colon U(\boldsymbol{Q}_1,\boldsymbol{Q}_2)$, which ensures the unitarity of u.

The formation rules for unitaries are presented in Figure 6. In the introduction rule for atomic unitaries $\{\ \mid\boldsymbol{b}_1\mapsto\boldsymbol{v}_1\ \ldots\mid\boldsymbol{b}_n\mapsto\boldsymbol{v}_n\ \}$, we have that the potentially open basis terms $\{\boldsymbol{b}_1,\ldots,\boldsymbol{b}_n\}$ (resp. values $\{\boldsymbol{v}_1,\ldots,\boldsymbol{v}_n\}$) determine an ONB for $\boldsymbol{Q}_1$ (resp. $\boldsymbol{Q}_2$) in the sense explained above. Then the resulting bijective mapping that associates $\boldsymbol{b}_i$ to $\boldsymbol{v}_i$ uniquely determines a unitary. The formation rules for the remaining unitaries are straightforward, and they align easily with the mathematical intuition that motivates them. We also wish to note that unitaries do not depend on general terms (see Figure 3).

2.3 Denotational Semantics for Pure Quantum System

Recall (§2.1) that the interpretation of a pure quantum type $\boldsymbol{Q}$ is a Hilbert space $[\![\boldsymbol{Q}]\!]$, i.e., an object of the category **Isometry**, and the interpretation of a unitary type $U(\boldsymbol{Q}_1, \boldsymbol{Q}_2)$ is the set $[\![U(\boldsymbol{Q}_1, \boldsymbol{Q}_2)]\!]$ of unitaries in $[\![\boldsymbol{Q}_1]\!] \to [\![\boldsymbol{Q}_2]\!]$.

A linear context is interpreted in the usual way $[\![\boldsymbol{\Gamma}]\!] \triangleq [\![\boldsymbol{Q}_1]\!] \otimes \cdots \otimes [\![\boldsymbol{Q}_n]\!]$. The interpretation of a judgement $\boldsymbol{\Gamma} \vdash \boldsymbol{t} \colon \boldsymbol{Q}$ is given by a morphism $[\![\boldsymbol{\Gamma} \vdash \boldsymbol{t} \colon \boldsymbol{Q}]\!] \in$ **Isometry**$([\![\boldsymbol{\Gamma}]\!], [\![\boldsymbol{Q}]\!])$. We also sometimes abuse notation and write $[\![\boldsymbol{t}]\!]$ for brevity instead of $[\![\boldsymbol{\Gamma} \vdash \boldsymbol{t} \colon \boldsymbol{Q}]\!]$. For closed quantum terms $\cdot \vdash \boldsymbol{t} \colon \boldsymbol{Q}$, we interpret the empty context as $\mathbb{C}$ (the tensor unit in **Isometry**) and so we can identify $[\![\cdot \vdash \boldsymbol{t} \colon \boldsymbol{Q}]\!](1)$ with a normalised vector.

Every value $\boldsymbol{v}$ is of the form $\sum_{i \in \mathcal{I}} \alpha_i \cdot \boldsymbol{b}_i$, but the sum is subject to strict admissibility conditions imposed by the orthogonality relation $\perp$ from §2.2. The denotational significance of this relation is provided by the statement: given $\boldsymbol{\Gamma}_1 \vdash \boldsymbol{t}_1 \colon \boldsymbol{Q}$ and $\boldsymbol{\Gamma}_2 \vdash \boldsymbol{t}_2 \colon \boldsymbol{Q}$, if $\boldsymbol{t}_1 \perp \boldsymbol{t}_2$ then $[\![\boldsymbol{t}_1]\!]^* [\![\boldsymbol{t}_2]\!] = 0_{[\![\boldsymbol{\Gamma}_2]\!], [\![\boldsymbol{\Gamma}_1]\!]}$; which is central in proving that the interpretation of values is well-defined as isometries.

The most challenging unitaries to interpret are the unitary pattern-matchings, constructed through the use of basis terms and values (but not general terms). Another key role is played by the $\mathrm{ONB}_{\boldsymbol{Q}}$ and $\mathrm{ONB}_{\boldsymbol{Q}}^{val}$ predicates: they can be understood as determining suitable ONBs of the corresponding types. The denotational significance of these predicates is that they give us a resolution of the identity at type $\boldsymbol{Q}$, as made precise by the following proposition.

Proposition 1. *Let* $\boldsymbol{\Gamma}_1 \vdash \boldsymbol{v}_1 \colon \boldsymbol{Q}, \ldots, \boldsymbol{\Gamma}_n \vdash \boldsymbol{v}_n \colon \boldsymbol{Q}$ *be well-formed values. If* $\mathrm{ONB}_{\boldsymbol{Q}}^{\mathrm{val}}\{\boldsymbol{v}_1, \ldots, \boldsymbol{v}_n\}$, *then* $\mathrm{id}_{[\![\boldsymbol{Q}]\!]} = \sum_{i=1}^{n} [\![\boldsymbol{v}_i]\!] \circ [\![\boldsymbol{v}_i]\!]^*$.

Pattern-matching is then interpreted by $[\![\Vdash \{\ldots \mid \boldsymbol{b}_i \mapsto \boldsymbol{v}_i \ldots\}]\!] \triangleq \sum_i [\![\boldsymbol{v}_i]\!] \circ [\![\boldsymbol{b}_i]\!]^*$, and the previous proposition ensures it is unitary. Finally, we can show that the interpretation of all pure quantum terms is well-defined as isometries.

Theorem 1. *If* $\boldsymbol{\Gamma} \vdash \boldsymbol{t} \colon \boldsymbol{Q}$, *then* $[\![\boldsymbol{t}]\!] \in$ **Isometry**$([\![\boldsymbol{\Gamma}]\!], [\![\boldsymbol{Q}]\!])$.

Moreover, any canonical symmetric monoidal isomorphism can be represented by the unitaries of our language. The canonical symmetric monoidal isomorphisms are simply those that can be defined using the structure of a symmetric monoidal category only, *i.e.* isomorphisms that are given by: left and right unitors, symmetric swaps, associators, closed under composition and tensor products. We use this result later for quantum configurations.

Proposition 2. *Let* $\boldsymbol{Q}_1$ *and* $\boldsymbol{Q}_2$ *be two pure types, and let* $f \colon [\![\boldsymbol{Q}_1]\!] \to [\![\boldsymbol{Q}_2]\!]$ *be a canonical symmetric monoidal isomorphism. Then, there exists a unitary* $\Vdash u \colon U(\boldsymbol{Q}_1, \boldsymbol{Q}_2)$ *such that* $[\![u]\!] = f$.

2.4 Equational Theory

We now describe the equational theory of the pure quantum subsystem. Given two pure quantum terms $\boldsymbol{t}_1$ and $\boldsymbol{t}_2$, we write $\boldsymbol{\Gamma} \vdash \boldsymbol{t}_1 = \boldsymbol{t}_2 \colon \boldsymbol{Q}$ to indicate that $\boldsymbol{t}_1$ equals $\boldsymbol{t}_2$ as terms of type $\boldsymbol{Q}$ in context $\boldsymbol{\Gamma}$. As usual if $\boldsymbol{\Gamma} \vdash \boldsymbol{t}_1 = \boldsymbol{t}_2 \colon \boldsymbol{Q}$, the rules imply that $\boldsymbol{\Gamma} \vdash \boldsymbol{t}_1 \colon \boldsymbol{Q}$ and $\boldsymbol{\Gamma} \vdash \boldsymbol{t}_2 \colon \boldsymbol{Q}$. We now explain some of the more interesting rules in greater detail. The rules of the equational theory are divided into four groups. The first one is concerned with the usual rules for reflexivity, symmetry and transitivity. Secondly, we provide rules to handle linear algebraic identities, such as exchanging sums and the linearity of injections and unitaries; this also contains the rule $\boldsymbol{\Gamma} \vdash \sum_{i=1}^{1} 1 \cdot \boldsymbol{t} = \boldsymbol{t} \colon \boldsymbol{Q}$. A small excerpt is found below.

$$\frac{\boldsymbol{\Gamma} \vdash \sum_i \boldsymbol{\alpha}_i \cdot \left(\sum_j \boldsymbol{\beta}_{i,j} \cdot \boldsymbol{t}_j\right) : \boldsymbol{Q}}{\boldsymbol{\Gamma} \vdash \sum_i \boldsymbol{\alpha}_i \cdot \left(\sum_j \boldsymbol{\beta}_{i,j} \cdot \boldsymbol{t}_j\right) = \sum_j \left(\sum_i \boldsymbol{\alpha}_i \boldsymbol{\beta}_{i,j}\right) \cdot \boldsymbol{t}_j : \boldsymbol{Q}} \qquad \frac{\Vdash u \colon U(\boldsymbol{Q}_1, \boldsymbol{Q}_2) \qquad \boldsymbol{\Gamma} \vdash \sum_i \boldsymbol{\alpha}_i \cdot \boldsymbol{t}_i \colon \boldsymbol{Q}_1}{\boldsymbol{\Gamma} \vdash u\ (\sum_i \boldsymbol{\alpha}_i \cdot \boldsymbol{t}_i) = \sum_i \boldsymbol{\alpha}_i \cdot (u\, \boldsymbol{t}_i) \colon \boldsymbol{Q}_2}$$

Congruence rules comprise the third group, e.g., $\boldsymbol{\Gamma} \vdash \mathbf{S}\, \boldsymbol{t}_1 = \mathbf{S}\, \boldsymbol{t}_2 \colon \mathbf{qnat}$ given that $\boldsymbol{\Gamma} \vdash \boldsymbol{t}_1 = \boldsymbol{t}_2 \colon \mathbf{qnat}$. Last but not least are the *computational* rules, which, read left to right, can be seen as providing an operational semantics. The most important rule and our version of β-reduction, is the rule for unitary pattern-matching applied to a basis term.

$$\frac{\cdot \vdash \boldsymbol{b}' \colon \boldsymbol{Q}_1 \qquad \Vdash \{\ \mid \boldsymbol{b}_1 \mapsto \boldsymbol{v}_1\ \ldots \mid \boldsymbol{b}_n \mapsto \boldsymbol{v}_n\ \} : U(\boldsymbol{Q}_1, \boldsymbol{Q}_2) \qquad \mathtt{s}(\boldsymbol{b}_j) = \boldsymbol{b}'}{\cdot \vdash \{\ \mid \boldsymbol{b}_1 \mapsto \boldsymbol{v}_1\ \ldots \mid \boldsymbol{b}_n \mapsto \boldsymbol{v}_n\ \}\, \boldsymbol{b}' = \mathtt{s}(\boldsymbol{v}_j) \colon \boldsymbol{Q}_2}$$

The formation conditions on unitary pattern-matching ensure that $\mathrm{ONB}_{\boldsymbol{Q}_1}(\{\boldsymbol{b}_i\}_i)$. With the latter, and given $\cdot \vdash \boldsymbol{b}' \colon \boldsymbol{Q}_1$, there exists a unique substitution $\mathtt{s}$ and a unique $\boldsymbol{b}_j$ such that $\mathtt{s}(\boldsymbol{b}_j) = \boldsymbol{b}'$. The resulting term of the rule is then the same substitution applied to the corresponding output $\boldsymbol{v}_j$. Such substitutions were first defined in [18], and can be deduced via an inference system.

Proposition 3. *Given* $\cdot \vdash \boldsymbol{t} \colon \boldsymbol{Q}$*, there is a unique value* $\boldsymbol{v}$ *s.t.* $\cdot \vdash \boldsymbol{t} = \boldsymbol{v} \colon \boldsymbol{Q}$*.*

This proposition shows that pure terms have a unique *normal form.* It is crucial for the proof of the completeness theorem, and comes in handy in the rest of the paper, especially to define the reduction system for the main calculus (§3).

Theorem 2 (Completeness). $\cdot \vdash \boldsymbol{t}_1 = \boldsymbol{t}_2 \colon \boldsymbol{Q}$ *iff* $[\![\cdot \vdash \boldsymbol{t}_1 \colon \boldsymbol{Q}]\!] = [\![\cdot \vdash \boldsymbol{t}_2 \colon \boldsymbol{Q}]\!]$.

3 Combining Classical and Quantum Control

This section lays down the syntax and semantics of the main calculus, which combines the quantum control fragment with a classically controlled layer, a variant of a call-by-value linear lambda-calculus. The subsystem which we presented in the previous section is designed to represent pure quantum information and computation, whereas in this section, we are concerned with mixed quantum information and computation. Our denotational approach towards this uses von Neumann algebras, which provide an appropriate mathematical setting to achieve this treatment. Operationally, this is achieved by adapting the configurations (also called closures) $(|\psi\rangle, \ell, M)$ from an effectful quantum lambda-calculus, e.g., [6,21], by replacing the quantum state $|\psi\rangle$ with a pure term $\boldsymbol{t}$, replacing the linking function ℓ with a unitary u that can be described via our syntax and which represents a suitable permutation, and finally by defining suitable formation conditions for the configuration.

$$
\begin{array}{lll}
(\textsc{Vars}) & x, y, z & \\
(\textsc{Types}) & \mathrm{A}, \mathrm{B} & ::= \mathrm{I} \mid \mathrm{A} + \mathrm{B} \mid \mathrm{A} \otimes \mathrm{B} \mid !\mathrm{A} \mid \mathrm{A} \multimap \mathrm{B} \mid \mathrm{Nat} \mid \mathcal{B}(\boldsymbol{Q}) \\
(\textsc{Terms}) & L, M, N & ::= * \mid x \mid \mathit{inl}(M) \mid \mathit{inr}(N) \mid \mathit{case}\, L\, \mathit{of}\{\mathit{inl}(x) \to M,\ \mathit{inr}(y) \to N\} \\
& & \mid M \otimes N \mid \mathit{let}\ x \otimes y = M\ \mathit{in}\ N \mid \mathit{lift}(M) \mid \mathit{force}(M) \\
& & \mid \lambda x.M \mid M\, N \\
& & \mid \mathit{zero} \mid \mathit{succ}(M) \mid \mathit{match}\, L\, \mathit{with}\{\mathit{zero} \to M,\ \mathit{succ}(x) \to N\} \\
& & \mid \mathit{pure}(\boldsymbol{t}) \mid \mathit{meas}(M) \mid \mathcal{B}(u)(M) \\
& & \mid \mathit{let}\ \mathcal{B}(z) = M\ \mathit{in}\ N \mid \mathit{let}\ \mathcal{B}(x \otimes y) = M\ \mathit{in}\ N \\
(\textsc{Values}) & V, W & ::= * \mid x \mid \mathit{inl}(V) \mid \mathit{inr}(W) \mid V \otimes W \mid \mathit{lift}(M) \mid \lambda x.M \\
& & \mid \mathit{zero} \mid \mathit{succ}(V)
\end{array}
$$

Fig. 7: Syntax of the classically controlled system.

3.1 Syntax and Typing Rules

The *types* of the main calculus, ranged over by capital Latin letters (e.g., A, B) are given in Figure 7. All types except Nat and $\mathcal{B}(\boldsymbol{Q})$ are standard in a system for a classical call-by-value linear lambda-calculus. Nat is a ground type representing classical natural numbers and $\mathcal{B}(\boldsymbol{Q})$ is the type that represents *mixed state* quantum computation on the Hilbert space determined by the type $\boldsymbol{Q}$. As we see later, this type plays an important role by introducing quantum and probabilistic effects into the system. The modality $\mathcal{B}$ is inspired from the denotational model, where $\mathcal{B}(H)$ represents the von Neumann algebra of bounded operators on the Hilbert space H.

Terms (and values) of the calculus are described in Figure 7. The term $\mathit{pure}(\boldsymbol{t})$ models the preparation of a pure quantum state represented by the quantum term $\boldsymbol{t}$. Measurement is modelled by the term $\mathit{meas}(M)$. The term $\mathcal{B}(u)(M)$ represents the application of the unitary u to a term M. Additionally, we have two syntactic constructs to model the isomorphism between the types $\mathcal{B}(\boldsymbol{Q}_1 \otimes \boldsymbol{Q}_2)$ and $\mathcal{B}(\boldsymbol{Q}_1) \otimes \mathcal{B}(\boldsymbol{Q}_2)$: the term $\mathit{let}\ \mathcal{B}(z) = M\ \mathit{in}\ N$ allows one to construct a term of type $\mathcal{B}(\boldsymbol{Q}_1 \otimes \boldsymbol{Q}_2)$ from a term of type $\mathcal{B}(\boldsymbol{Q}_1) \otimes \mathcal{B}(\boldsymbol{Q}_2)$ and the term $\mathit{let}\,\mathcal{B}(x \otimes y) = M\ \mathit{in}\ N$ deals with the opposite direction. This isomorphism arises because the functor $\mathcal{B}(\cdot)$ used in the denotational model is *strict monoidal.* Both constructs are necessary to allow the language to manipulate composed quantum systems and also terms possibly representing *entangled states.*

Typing contexts (ranged over by symbols $\Delta, \Sigma_1, \Sigma_2$) are finite sequences $\Delta \triangleq x_1 : \mathrm{A}_1, \ldots, x_n : \mathrm{A}_n$ mapping variables to types. We use the notation $!\Delta$ for contexts of the form $x_1 : !\mathrm{A}_1, \ldots, x_n : !\mathrm{A}_n$. *Typing judgements* have the form $\Delta \vdash M : \mathrm{A}$ and follow a linear typing discipline to deal with quantum data. The typing rules are given in Figure 8 where, as usual, Σ_1, Σ_2 denotes the disjoint union of Σ_1 and Σ_2. Classical bits correspond to the type $\mathit{bit} \triangleq I + I$ and mixed state qubits have type $\mathit{qbit} \triangleq \mathcal{B}(\mathbf{qbit})$. The rule for the term $\mathit{pure}(\boldsymbol{t})$ introduces a state $\boldsymbol{t}$ from the quantum control fragment as a term of type $\mathcal{B}(\boldsymbol{Q})$ into the main calculus. The typing rule for measurement maps a term M of quantum type $\mathcal{B}(\boldsymbol{Q})$ to a term $\mathit{meas}(M)$ of classical type $\overline{\boldsymbol{Q}}$, where the type $\overline{\boldsymbol{Q}}$ is defined inductively on the structure of the pure quantum type $\boldsymbol{Q}$:

$$\overline{\boldsymbol{I}} \triangleq I, \quad \overline{\boldsymbol{Q}_1 \otimes \boldsymbol{Q}_2} \triangleq \overline{\boldsymbol{Q}_1} \otimes \overline{\boldsymbol{Q}_2}, \quad \overline{\boldsymbol{Q}_1 \oplus \boldsymbol{Q}_2} \triangleq \overline{\boldsymbol{Q}_1} + \overline{\boldsymbol{Q}_2}, \quad \overline{\mathbf{qnat}} \triangleq \mathrm{Nat}.$$

$$\frac{!\Delta, \Sigma_1 \vdash L : \mathrm{A}+\mathrm{B} \qquad !\Delta, \Sigma_2, x : \mathrm{A} \vdash M : \mathrm{C} \qquad !\Delta, \Sigma_2, y : \mathrm{B} \vdash N : \mathrm{C}}{!\Delta, \Sigma_1, \Sigma_2 \vdash \mathit{case}\, L\, \mathit{of}\{\mathit{inl}(x) \to M,\ \mathit{inr}(y) \to N\} : \mathrm{C}}$$

$$\frac{\Delta, x : \mathrm{A} \vdash M : \mathrm{B}}{\Delta \vdash \lambda x.M : \mathrm{A} \multimap \mathrm{B}} \qquad \frac{!\Delta, \Sigma_1 \vdash M : \mathrm{A} \multimap \mathrm{B} \quad !\Delta, \Sigma_2 \vdash N : \mathrm{A}}{!\Delta, \Sigma_1, \Sigma_2 \vdash M\, N : \mathrm{B}}$$

$$\frac{!\Delta, \Sigma_1 \vdash L : \mathrm{Nat} \quad !\Delta, \Sigma_2 \vdash M : \mathrm{A} \quad !\Delta, \Sigma_2, x : \mathrm{Nat} \vdash N : \mathrm{A}}{!\Delta, \Sigma_1, \Sigma_2 \vdash \mathit{match}\, L\, \mathit{with}\{\mathit{zero} \to M,\ \mathit{succ}(x) \to N\} : \mathrm{A}} \qquad \frac{\cdot \vdash \boldsymbol{t} : \boldsymbol{Q}}{!\Delta \vdash \mathit{pure}(\boldsymbol{t}) : \mathcal{B}(\boldsymbol{Q})}$$

$$\frac{\Delta \vdash M : \mathcal{B}(\boldsymbol{Q})}{\Delta \vdash \mathit{meas}(M) : \overline{\boldsymbol{Q}}} \qquad \frac{\Vdash u : U(\boldsymbol{Q}_1, \boldsymbol{Q}_2) \qquad !\Delta, \Sigma_1 \vdash M : \mathcal{B}(\boldsymbol{Q}_1)}{!\Delta, \Sigma_1 \vdash \mathcal{B}(u)(M) : \mathcal{B}(\boldsymbol{Q}_2)}$$

$$\frac{!\Delta, \Sigma_1 \vdash M : \mathcal{B}(\boldsymbol{Q}_1) \otimes \mathcal{B}(\boldsymbol{Q}_2) \qquad !\Delta, \Sigma_2, z : \mathcal{B}(\boldsymbol{Q}_1 \otimes \boldsymbol{Q}_2) \vdash N : \mathrm{A}}{!\Delta, \Sigma_1, \Sigma_2 \vdash \mathit{let}\ \mathcal{B}(z) = M\ \mathit{in}\ N : \mathrm{A}}$$

$$\frac{!\Delta, \Sigma_1 \vdash M : \mathcal{B}(\boldsymbol{Q}_1 \otimes \boldsymbol{Q}_2) \qquad !\Delta, \Sigma_2, x : \mathcal{B}(\boldsymbol{Q}_1), y : \mathcal{B}(\boldsymbol{Q}_2) \vdash N : \mathrm{A}}{!\Delta, \Sigma_1, \Sigma_2 \vdash \mathit{let}\ \mathcal{B}(x \otimes y) = M\ \mathit{in}\ N : \mathrm{A}}$$

Fig. 8: Typing rules for terms (excerpt).

This gives the classical analogue of the type $\boldsymbol{Q}$, e.g., $\overline{\mathbf{qbit}} = \mathit{bit}$. For a quantum basis term $\boldsymbol{b}$ of type $\boldsymbol{Q}$, $\overline{\boldsymbol{b}}$ (defined below) denotes its translation to a classical value of type $\overline{\boldsymbol{Q}}$, which we use to represent measurement outcomes.

$$\overline{*} \triangleq *, \quad \overline{\boldsymbol{b}_1 \otimes \boldsymbol{b}_2} \triangleq \overline{\boldsymbol{b}_1} \otimes \overline{\boldsymbol{b}_2}, \quad \overline{\mathbf{inl}\, \boldsymbol{b}} \triangleq \mathit{inl}(\overline{\boldsymbol{b}}), \quad \overline{\mathbf{0}} \triangleq \mathit{zero}, \quad \overline{\mathbf{S}\, \boldsymbol{b}} \triangleq \mathit{succ}(\overline{\boldsymbol{b}}).$$

3.2 Operational Semantics

Quantum configurations. We lay down the operational semantics of the calculus using an adaptation of the notion of a *quantum configuration.*

Definition 1 (Quantum Configuration). *Let* $\mathtt{Conf}$ *be the set of* quantum configurations $\mathcal{C} \triangleq (\boldsymbol{t}, u_\sigma, M)$, *where:*

- $\boldsymbol{t} \in \boldsymbol{Terms}$ *represents a* pure quantum state*;*
- $u_\sigma \in \boldsymbol{Unitaries}$ *represents a symmetric monoidal isomorphism;*
- $M \in \mathrm{TERMS}$ *is a term from the main calculus.*

In the above definition, we can think of the quantum term $\boldsymbol{t}$ as representing a quantum state, because the quantum control fragment ensures that $\boldsymbol{t}$ can be rewritten (in its equational theory) to a unique quantum value (Proposition 3). Next, u_σ permutes the components of $\boldsymbol{t}$ so that they appear in the same order as variables that represent them in M. We slightly abuse terminology and refer to u_σ as a permutation, but it is meant to be understood as a canonical symmetric monoidal isomorphism (see Proposition 2). If we fix the domain and codomain of u_σ, then the essential data is given by a choice of permutation σ and this makes it easier to understand how each unitary acts on the coordinates, and the contextual rules formulation becomes precise. Note that, thanks to Proposition 2, the pure quantum syntax is expressive enough to represent all the monoidal permutations that we need in the sequel.

Example 5. Consider the quantum configuration $(\boldsymbol{t}, u_{id}, x \otimes y)$, where we have $\boldsymbol{t} \triangleq * \otimes ({}^1\!/\!\sqrt{2} \cdot |\mathbf{0}\rangle \otimes |\mathbf{0}\rangle + {}^1\!/\!\sqrt{2} \cdot |\mathbf{1}\rangle \otimes |\mathbf{1}\rangle)$ and $u_{id} \triangleq \{|\ * \otimes \boldsymbol{w} \otimes \boldsymbol{z} \mapsto \boldsymbol{w} \otimes \boldsymbol{z}\}$

is a unitary with type $\Vdash u_{id}\colon U(\boldsymbol{I} \otimes \mathbf{qbit} \otimes \mathbf{qbit}, \mathbf{qbit} \otimes \mathbf{qbit})$. It conveys the information that the variable x in $x \otimes y$ keeps track of the first qubit in $\boldsymbol{t}$ and that the variable y keeps track of the second one. Whereas in the configuration $(\boldsymbol{t}, u_{swap}, x \otimes y)$ with $u_{swap} \triangleq \{|\ * \otimes \boldsymbol{w} \otimes \boldsymbol{z} \mapsto \boldsymbol{z} \otimes \boldsymbol{w}\}$, the variable x keeps track of the second qubit and y of the first one. Note that, due to quantum entanglement, the term $\boldsymbol{t}$ cannot be decomposed into a non-trivial tensor product, and the variables x and y should be thought of as identifying qubit components of $\boldsymbol{t}$, instead of storing quantum data themselves.

A quantum configuration $\mathcal{V} = (\boldsymbol{v}, u_\sigma, V)$, when both the quantum term $\boldsymbol{v}$ and term V are values, is called a *value configuration.* For example, $(|0\rangle, u_{id}, x)$ is a value configuration. A quantum configuration $\mathcal{SV} = (\boldsymbol{v}, u_\sigma, M)$, where $\boldsymbol{v}$ is a quantum value, is a *semi-value configuration.* We write VConf (resp. SVConf) for the set of (resp. semi-)value configurations. The reason we introduce semi-value configurations, is because they allow us to define our operational semantics for the main calculus modulo equality of the pure quantum terms $\boldsymbol{t}$ (analogous to the quantum lambda-calculus). We already proved that values $\boldsymbol{v}$ are normal forms for the pure terms (Proposition 3), so this makes them a natural choice.

Definition 2. *A quantum configuration* $(\boldsymbol{t}, u_\sigma, M)$ *is* well-formed *with type* A, *written as* $(\boldsymbol{t}, u_\sigma, M)\colon \mathrm{A}$ *if the following judgments can be derived:*

- $\cdot \vdash \boldsymbol{t}\colon \boldsymbol{Q}_1 \otimes \cdots \otimes \boldsymbol{Q}_n$;
- $\cdot \Vdash u_\sigma\colon U(\boldsymbol{Q}_1 \otimes \cdots \otimes \boldsymbol{Q}_n, \boldsymbol{Q}'_1 \otimes \cdots \otimes \boldsymbol{Q}'_m)$;
- $x_1 : \mathcal{B}(\boldsymbol{Q}'_1), \ldots, x_m : \mathcal{B}(\boldsymbol{Q}'_m) \vdash M : \mathrm{A}$.

We write $\mathrm{WF}_X(A)$ for the set of well-formed configurations in X with type A, where $X \in \{\mathtt{Conf}, \mathtt{VConf}, \mathtt{SVConf}\}$.

Example 6. The configuration $((1/\sqrt{2} \cdot |\mathbf{00}\rangle + 1/\sqrt{2} \cdot |\mathbf{11}\rangle) \otimes |\mathbf{0}\rangle, u_{\mathrm{id}}, x \otimes y)$ with:

- $\cdot \vdash (1/\sqrt{2} \cdot |\mathbf{00}\rangle + 1/\sqrt{2} \cdot |\mathbf{11}\rangle) \otimes |\mathbf{0}\rangle \ : \mathbf{qbit}^3$;
- $\cdot \Vdash u_{\mathrm{id}}\colon U(\mathbf{qbit}^3, \mathbf{qbit}^3)$;
- $x : \mathcal{B}(\mathbf{qbit}^2), y : \mathit{qbit} \vdash x \otimes y : \mathcal{B}(\mathbf{qbit}^2) \otimes \mathit{qbit}$.

In this configuration, u_{id} is just the identity, x points to the Bell state in $\boldsymbol{t}$ whereas y points to $|\mathbf{0}\rangle$. We have $n = 3, m = 2$, with $\boldsymbol{Q}_1 = \boldsymbol{Q}_2 = \boldsymbol{Q}_3 = \mathbf{qbit}$, $\boldsymbol{Q}'_1 = \mathbf{qbit}^2$ and $\boldsymbol{Q}'_2 = \mathbf{qbit}$. The configuration is well-formed.

The above example highlights the need for m and n (from Definition 2) to be different values. The intuition behind this difference is that the unitary u_σ partitions the n quantum types into m blocks, where each block is represented by a variable in M. Furthermore, in situations where we wish to combine two such blocks which are not next to each other in the tensor expression of the n types, we allow the possibility of permuting these blocks and merging them so that now two blocks that were initially represented by two variables are represented by one. This is formally expressed in the reduction rules we define next. The role of u_σ in a well-formed configuration is illustrated in Figure 9. The *small-step reduction* $\cdot \to_\cdot \cdot \subseteq \mathtt{Conf} \times [0,1] \times \mathtt{Conf}$ is the relation defined by the rules of Figure 10: the reduction $\mathcal{C} \to_p \mathcal{C}'$ holds when the quantum configuration $\mathcal{C}$ reduces to $\mathcal{C}'$ with probability $p \in [0,1]$.

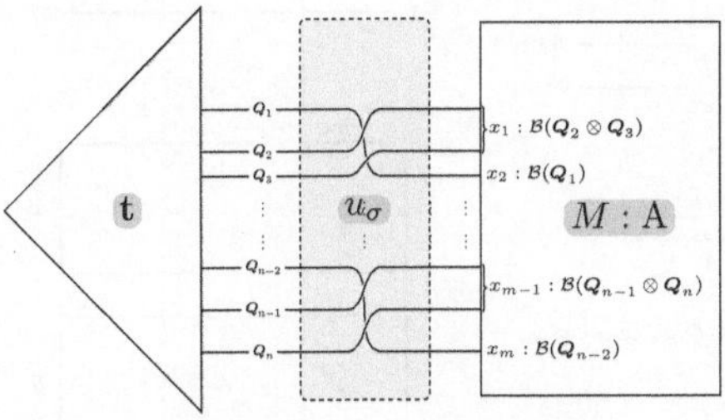

Fig. 9: Graphical representation of a well-formed quantum configuration.

The reduction rules along with the unitaries required to state them are given in Figure 10. $\{\boldsymbol{b}_{ij}\}_{i,j}$ stand for basis terms and s in the reduction rule for $meas(x)$ corresponds to the position of $\boldsymbol{b}_i$ in the tensor product appearing in the quantum term. All sums are finite. The unitary $u_{\sigma_{gather}}$ appearing in the reduction rule for the term $let\, \mathcal{B}(z) = x \otimes y \ in\ N$ merges two consecutive variables into one block, so that now a single variable can represent it. Similarly, in the reduction rule for the term $let\ \mathcal{B}(x \otimes y) = z\ in\ N$, the unitary $u_{\sigma_{divide}}$ partitions a block into two, so that two variables can represent them. The two rules are symmetric as expected, since these terms only allow switching between types $\mathcal{B}(\boldsymbol{Q}_1 \otimes \boldsymbol{Q}_2)$ and $\mathcal{B}(\boldsymbol{Q}_1) \otimes \mathcal{B}(\boldsymbol{Q}_2)$. The last rule of Figure 10 is a *contextual rule* which holds for any *evaluation context* E. Evaluation contexts are defined by the following grammar:

$$\begin{aligned} E := {} & [.] \mid E \otimes N \mid V \otimes E \mid let\ x \otimes y = E\ in\ N \mid inl(E) \mid inr(E) \\ & \mid case\, E\ of\{inl(x) \to M,\ inr(y) \to N\} \mid force(E) \mid succ(E) \mid E\, N \\ & \mid V\, E \mid meas(E) \mid \mathcal{B}(u)(E) \mid let\ \mathcal{B}(z) = E\ in\ N \mid let\ \mathcal{B}(x \otimes y) = E\ in\ N \end{aligned}$$

One of the premises of the contextual rule states that if σ is a permutation which does not act on the same set of indices as σ_1, σ_2, then we can form a "union" of these permutations by composing an *extension* of the two. For a permutation $\sigma : S \to S$, we can define its extension $\sigma^{ext} : S \sqcup S' \to S \sqcup S'$ by $\sigma^{ext}(s) \triangleq \sigma(s)$, if $s \in S$, $\sigma^{ext}(s') \triangleq s'$, if $s' \in S'$. We now introduce the reduction relation we use to define the operational semantics.

Definition 3. *The* reduction relation $\cdot \rightsquigarrow . \cdot \subseteq \mathit{SVConf} \times [0,1] \times \mathit{SVConf}$ *is defined in the following way: we write* $(\boldsymbol{v}, u_\sigma, M) \rightsquigarrow_p (\boldsymbol{v}', u_{\sigma'}, M')$ *whenever*

$$\frac{\cdot \vdash \boldsymbol{v} = \boldsymbol{t} : \boldsymbol{Q} \qquad (\boldsymbol{t}, u_\sigma, M) \to_p (\boldsymbol{t}', u_{\sigma'}, M') \qquad \cdot \vdash \boldsymbol{t}' = \boldsymbol{v}' : \boldsymbol{Q}'}{(\boldsymbol{v}, u_\sigma, M) \rightsquigarrow_p (\boldsymbol{v}', u_{\sigma'}, M')}$$

The intuition behind the above definition is that it allows us to reason modulo equality of the pure terms $\boldsymbol{t}$, as a consequence of Proposition 3.

3.3 Main properties

We show that well-formedness is preserved by the reduction relation $\cdot \rightsquigarrow . \cdot$.

Theorem 3 (Subject Reduction). *For a configuration* $\mathcal{C}_1 \in \mathrm{WF}_{\mathit{SVConf}}(A)$, *if* $\mathcal{C}_1 \rightsquigarrow_p \mathcal{C}_2$ *for some probability* $p \in [0,1]$, *then* $\mathcal{C}_2 \in \mathrm{WF}_{\mathit{SVConf}}(A)$.

Next, we show that progress and strong normalisation hold. More specifically, a well-formed semi-value configuration $\mathcal{C}$ is either a value configuration or it reduces to a finite number of semi-value configurations with total probability 1, *i.e.* if we have not reached a normal form then the probability to be stuck is 0.

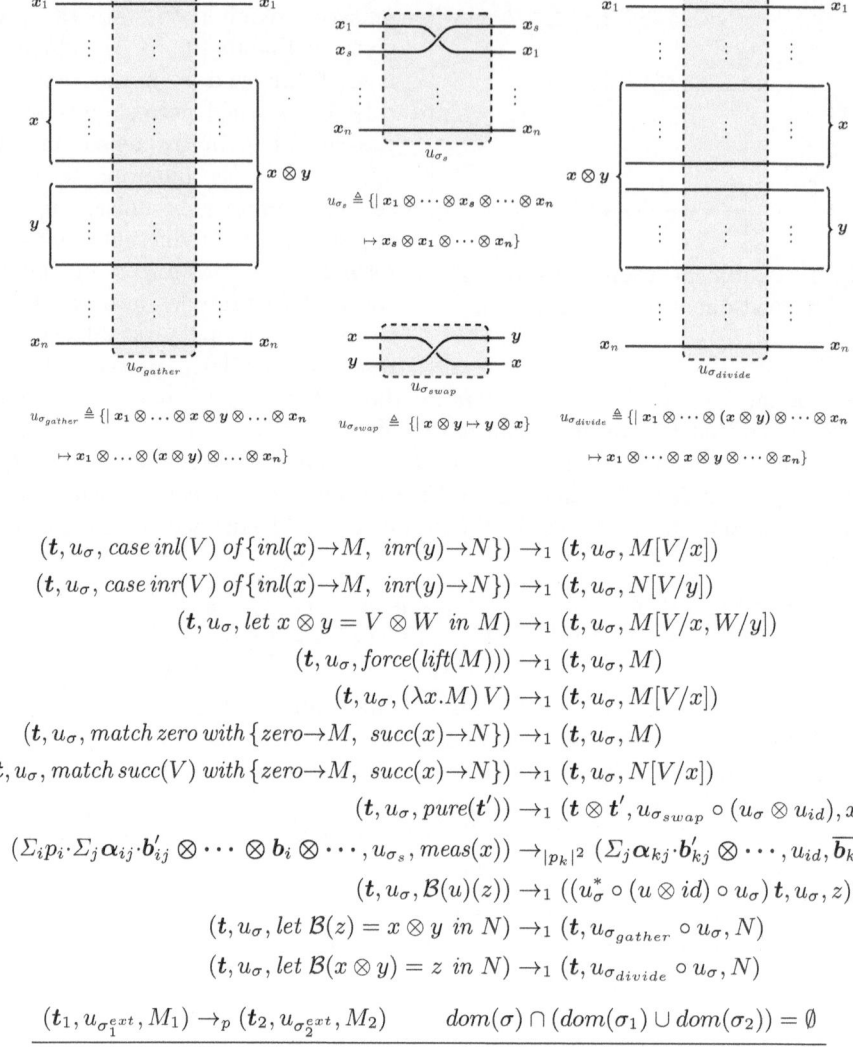

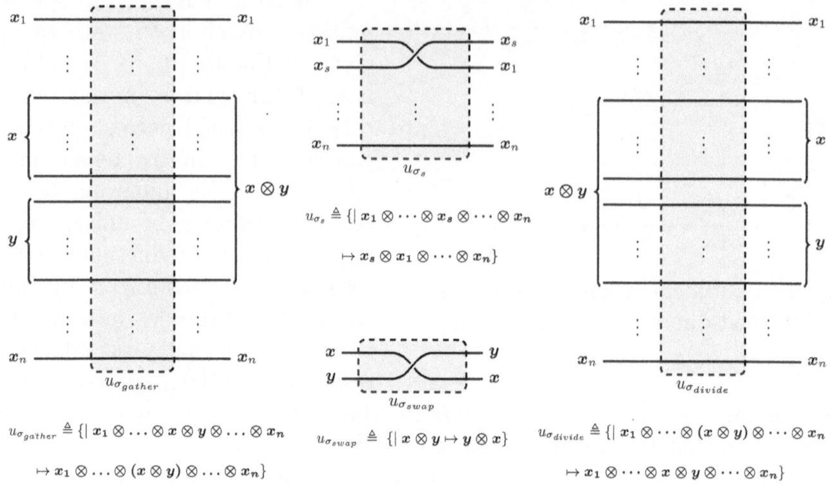

$$
\begin{aligned}
(\boldsymbol{t}, u_\sigma, \mathit{case\, inl}(V)\ \mathit{of}\{\mathit{inl}(x)\to M,\ \mathit{inr}(y)\to N\}) &\to_1 (\boldsymbol{t}, u_\sigma, M[V/x])\\
(\boldsymbol{t}, u_\sigma, \mathit{case\, inr}(V)\ \mathit{of}\{\mathit{inl}(x)\to M,\ \mathit{inr}(y)\to N\}) &\to_1 (\boldsymbol{t}, u_\sigma, N[V/y])\\
(\boldsymbol{t}, u_\sigma, \mathit{let}\ x\otimes y = V\otimes W\ \mathit{in}\ M) &\to_1 (\boldsymbol{t}, u_\sigma, M[V/x, W/y])\\
(\boldsymbol{t}, u_\sigma, \mathit{force}(\mathit{lift}(M))) &\to_1 (\boldsymbol{t}, u_\sigma, M)\\
(\boldsymbol{t}, u_\sigma, (\lambda x.M)\,V) &\to_1 (\boldsymbol{t}, u_\sigma, M[V/x])\\
(\boldsymbol{t}, u_\sigma, \mathit{match\, zero\, with}\{\mathit{zero}\to M,\ \mathit{succ}(x)\to N\}) &\to_1 (\boldsymbol{t}, u_\sigma, M)\\
(\boldsymbol{t}, u_\sigma, \mathit{match\, succ}(V)\ \mathit{with}\{\mathit{zero}\to M,\ \mathit{succ}(x)\to N\}) &\to_1 (\boldsymbol{t}, u_\sigma, N[V/x])\\
(\boldsymbol{t}, u_\sigma, \mathit{pure}(\boldsymbol{t}')) &\to_1 (\boldsymbol{t}\otimes\boldsymbol{t}', u_{\sigma_{swap}}\circ(u_\sigma\otimes u_{id}), x)\\
(\Sigma_i p_i\cdot\Sigma_j \boldsymbol{\alpha}_{ij}\cdot\boldsymbol{b}'_{ij}\otimes\cdots\otimes\boldsymbol{b}_i\otimes\cdots, u_{\sigma_s}, \mathit{meas}(x)) &\to_{|p_k|^2} (\Sigma_j \boldsymbol{\alpha}_{kj}\cdot\boldsymbol{b}'_{kj}\otimes\cdots, u_{id}, \overline{\boldsymbol{b}_k})\\
(\boldsymbol{t}, u_\sigma, \mathcal{B}(u)(z)) &\to_1 ((u_\sigma^*\circ(u\otimes id)\circ u_\sigma)\,\boldsymbol{t}, u_\sigma, z)\\
(\boldsymbol{t}, u_\sigma, \mathit{let}\ \mathcal{B}(z) = x\otimes y\ \mathit{in}\ N) &\to_1 (\boldsymbol{t}, u_{\sigma_{gather}}\circ u_\sigma, N)\\
(\boldsymbol{t}, u_\sigma, \mathit{let}\ \mathcal{B}(x\otimes y) = z\ \mathit{in}\ N) &\to_1 (\boldsymbol{t}, u_{\sigma_{divide}}\circ u_\sigma, N)
\end{aligned}
$$

$$
\frac{(\boldsymbol{t}_1, u_{\sigma_1^{ext}}, M_1) \to_p (\boldsymbol{t}_2, u_{\sigma_2^{ext}}, M_2) \qquad \mathit{dom}(\sigma)\cap(\mathit{dom}(\sigma_1)\cup\mathit{dom}(\sigma_2)) = \emptyset}{(\boldsymbol{t}_1, u_{\sigma^{ext}\circ\sigma_1^{ext}}, E[M_1]) \to_p (\boldsymbol{t}_2, u_{\sigma^{ext}\circ\sigma_2^{ext}}, E[M_2])}
$$

Fig. 10: Reduction rules and their unitary permutations.

Theorem 4 (Progress). *If $\mathcal{C} \in \mathrm{WF}_{\mathit{SVConf}}(A)$, then either $\mathcal{C} \in \mathit{VConf}$, or there exists $\mathcal{C}_i \in \mathrm{WF}_{\mathit{SVConf}}(A)$ such that $\mathcal{C} \rightsquigarrow_{p_i} \mathcal{C}_i$. Moreover, if $\{\mathcal{C}_i\}_{i\in I}$ is the set of all such distinct (not α-equivalent) configurations, then I is finite, and $\sum_i p_i = 1$.*

Theorem 5 (Strong Normalisation). *For a configuration $\mathcal{C} \in \mathrm{WF}_{\mathit{SVConf}}(A)$, there is no infinite sequence of reductions $\mathcal{C} \rightsquigarrow_{p_0} \mathcal{C}_1 \rightsquigarrow_{p_1} \mathcal{C}_2 \rightsquigarrow_{p_2} \cdots$.*

3.4 Illustrating Example: Quantum Walk Search

Given a graph with some marked vertices, a quantum walk search is the quantum analogue of a random walk [4], which looks for these marked vertices. Each node in the graph represents the state of two quantum registers. A condition on the application of a coin operator on the first register decides the direction in which a walker should take the next step. A step in this direction is represented by the application of a unitary on the second register. After a given number of steps, measurement is performed on the second register to reveal whether the walker has arrived on a marked node. Here, we present a k-step quantum walk on a cycle with 10 nodes for simplicity in our language. Representing the second register with qubits, in this case, would require 4 qubits, whereas we represent the state of the second register with a single **qnat**. In the unitary $\text{CNOT}_{\mathbf{qnat}}$ below, we use some (hopefully obvious) syntactic sugar with the expressions involving $\mathbf{n}$ in order to avoid writing all the cases; the letter $\mathbf{n}$ here should not be seen as a variable.

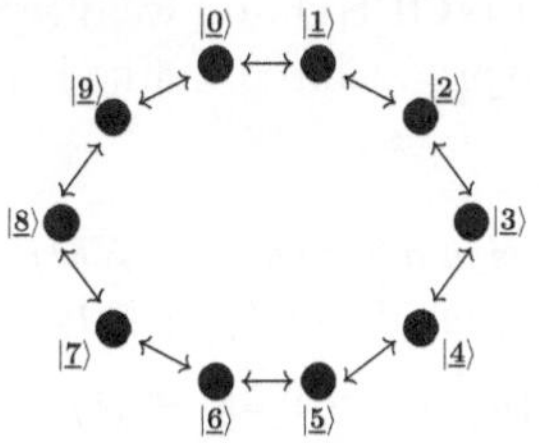

$$u_1 \triangleq \left\{ \begin{array}{lll} \mid \underline{|\mathbf{2n}\rangle} & \mapsto \underline{|\mathbf{2n{+}1}\rangle}, & \mathbf{n} \leq 4 \\ \mid \underline{|\mathbf{2n{+}1}\rangle} & \mapsto \underline{|\mathbf{2n}\rangle}, & \mathbf{n} \leq 4 \\ \mid \underline{|\mathbf{y{+}10}\rangle} & \mapsto \underline{|\mathbf{y{+}10}\rangle} & \end{array} \right\} \qquad u_2 \triangleq \left\{ \begin{array}{lll} \mid \underline{|\mathbf{0}\rangle} & \mapsto \underline{|\mathbf{9}\rangle} & \\ \mid \underline{|\mathbf{2n{+}2}\rangle} & \mapsto \underline{|\mathbf{2n{+}1}\rangle}, & \mathbf{n} \leq 3 \\ \mid \underline{|\mathbf{2n{+}1}\rangle} & \mapsto \underline{|\mathbf{2n{+}2}\rangle}, & \mathbf{n} \leq 3 \\ \mid \underline{|\mathbf{9}\rangle} & \mapsto \underline{|\mathbf{0}\rangle} & \\ \mid \underline{|\mathbf{y{+}10}\rangle} & \mapsto \underline{|\mathbf{y{+}10}\rangle} & \end{array} \right\}$$

$$\text{CNOT}_{\mathbf{qnat}} : U(\mathbf{qbit} \otimes \mathbf{qnat}, \mathbf{qbit} \otimes \mathbf{qnat})$$
$$\text{CNOT}_{\mathbf{qnat}} \triangleq \mathbf{qif}\ x\ \mathbf{then}\ x \otimes u_1\ \mathbf{else}\ x \otimes u_2$$

We define the unitary S corresponding to one step of the walk as $S \triangleq \text{CNOT}_{\mathbf{qnat}} \circ (\text{Had} \otimes \text{Id})$. Finally, the k-step walk can be represented in our language as follows, where S^k represents k compositions of S with itself:

$$\text{walk} : (\mathcal{B}(\mathbf{qbit} \otimes \mathbf{qnat}) \multimap \mathit{qbit} \otimes \text{Nat})$$
$$\text{walk} \triangleq \lambda z.\mathit{let}\ \mathcal{B}(x_1 \otimes x_2) = \mathcal{B}(S^k)(z)\ \mathit{in}\ x_1 \otimes \mathit{meas}(x_2)$$

3.5 Denotational Semantics: Soundness and Adequacy

In this section we describe the denotational semantics of the main calculus. Figure 11 succinctly describes the interaction between the pure fragment and the mixed fragment. We present the functor which facilitates the main contribution of the paper, i.e combining quantum and classical control.

The $\mathcal{B}$ functor. In order to model the interaction between the pure quantum subsystem and the main calculus, we define the functor $\mathcal{B} : \mathbf{Isometry} \to \mathbf{NCPSU}$, mapping a Hilbert space H to $\mathcal{B}(H)$. For an isometry $f \colon H_1 \to H_2$, the map $(f^* \circ (-) \circ f)$ is an NCPSU morphism $B(H_2) \to B(H_1)$, so by defining $\mathcal{B}(f) \triangleq (f^* \circ (-) \circ f)^{op}$, we get that $\mathcal{B}(f) \colon \mathcal{B}(H_1) \to \mathcal{B}(H_2)$ in **NCPSU**. We remark that this functor restricts to a functor $\mathcal{B} \colon \mathbf{Unitary} \to \mathbf{NMIU}$ and we show that the functor $\mathcal{B} : \mathbf{Isometry} \to \mathbf{NCPSU}$ is strict monoidal.

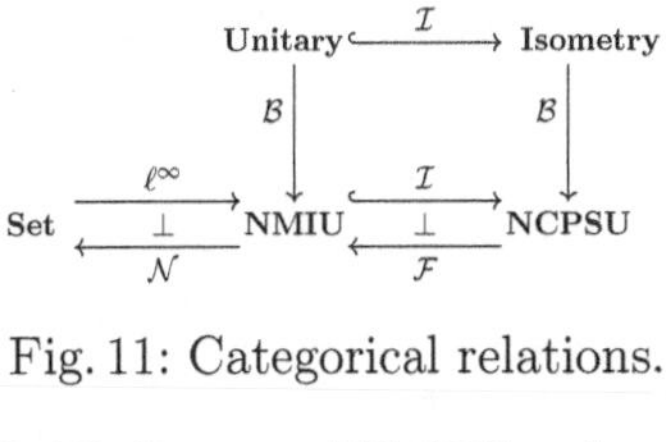

Fig. 11: Categorical relations.

$$[\![!\Delta \vdash pure(\boldsymbol{t}) : \mathcal{B}(\boldsymbol{Q})]\!] \triangleq \mathcal{B}[\![\boldsymbol{t}]\!] \circ \diamond_{!\Delta}$$
$$[\![\Delta \vdash meas(M) : \overline{\boldsymbol{Q}}]\!] \triangleq m^{\boldsymbol{Q}} \circ [\![\Delta \vdash M : \mathcal{B}(\boldsymbol{Q})]\!]$$
$$[\![\Delta \vdash \mathcal{B}(u)(M) : \mathcal{B}(\boldsymbol{Q}_2)]\!] \triangleq \mathcal{B}[\![u]\!] \circ [\![\Delta \vdash M : \mathcal{B}(\boldsymbol{Q}_1)]\!]$$

Fig. 12: Interpretation of typing judgements (an excerpt).

Interpretation of Types. We interpret types as von Neumann algebras. The types corresponding to a linear call-by-value lambda-calculus are interpreted in the standard way: $[\![\mathrm{I}]\!] \triangleq \mathbb{C}$, $[\![\mathrm{A}+\mathrm{B}]\!] \triangleq [\![\mathrm{A}]\!] \oplus [\![\mathrm{B}]\!]$, $[\![\mathrm{A} \otimes \mathrm{B}]\!] \triangleq [\![\mathrm{A}]\!] \otimes [\![\mathrm{B}]\!]$, $[\![!\mathrm{A}]\!] \triangleq ![\![\mathrm{A}]\!]$ and $[\![\mathrm{A} \multimap \mathrm{B}]\!] \triangleq [\![\mathrm{A}]\!] \multimap [\![\mathrm{B}]\!]$. The type for natural numbers is interpreted as the commutative von Neumann algebra $\ell^\infty(\mathbb{N})$, i.e $[\![\mathrm{Nat}]\!] \triangleq \ell^\infty(\mathbb{N})$. Finally, the interpretation of the type $\mathcal{B}(\boldsymbol{Q})$ uses the $\mathcal{B}$ functor that allows us to incorporate pure quantum primitives into our semantics, i.e $[\![\mathcal{B}(\boldsymbol{Q})]\!] \triangleq \mathcal{B}[\![\boldsymbol{Q}]\!]$.

Interpretation of Typing judgements. A typing judgment of the form $\Delta \vdash M \colon \mathrm{A}$ is interpreted as a morphism $[\![\Delta \vdash M \colon \mathrm{A}]\!] \colon [\![\Delta]\!] \to [\![\mathrm{A}]\!]$ in **NCPSU** and we often abbreviate this by writing $[\![M]\!]$. An excerpt of the interpretation is defined in Figure 12.

Theorem 6. *For every pure quantum type $\boldsymbol{Q}$, there exist a set X, an NMIU isomorphism $\alpha_{\overline{\boldsymbol{Q}}} : [\![\overline{\boldsymbol{Q}}]\!] \cong \ell^\infty(X)$, and an isometric isomorphism $\beta_{\boldsymbol{Q}} : [\![\boldsymbol{Q}]\!] \cong \ell^2(X)$. Moreover, $\forall \boldsymbol{b} \in$ **Basis Terms**, $\exists x \in X$ s.t. $\beta_{\boldsymbol{Q}}([\![\boldsymbol{b}]\!]) = \alpha_{\overline{\boldsymbol{Q}}}([\![\overline{\boldsymbol{b}}]\!]) = |x\rangle$.*

Using the above theorem, measurement is interpreted as the map $m^{\boldsymbol{Q}} : \mathcal{B}([\![\boldsymbol{Q}]\!]) \to [\![\overline{\boldsymbol{Q}}]\!]$ defined as $m^{\boldsymbol{Q}} \triangleq \alpha^{op}_{\overline{\boldsymbol{Q}}} \circ m^{op}_X \circ \mathcal{B}(\beta_{\boldsymbol{Q}})$.

Interpretation of Configurations. A well-formed configuration $(\boldsymbol{t}, u_\sigma, M)$ with $\cdot \vdash \boldsymbol{t} \colon \boldsymbol{Q}$, $\Vdash u_\sigma \colon (\boldsymbol{Q}, \boldsymbol{Q}'_1 \otimes \cdots \otimes \boldsymbol{Q}'_m)$, and $x_1 \colon \mathcal{B}(\boldsymbol{Q}'_1), \ldots, x_m \colon \mathcal{B}(\boldsymbol{Q}'_m) \vdash M \colon \mathrm{A}$ is interpreted as the NCPSU morphism

$$[\![(\boldsymbol{t}, u_\sigma, M) : \mathrm{A}]\!] \triangleq \left(\mathbb{C} \xrightarrow{\mathcal{B}([\![u_\sigma \boldsymbol{t}]\!])} \begin{array}{c} \mathcal{B}(\boldsymbol{Q}'_1 \otimes \cdots \otimes \boldsymbol{Q}'_m) \\ \shortparallel \\ \mathcal{B}(\boldsymbol{Q}'_1) \otimes \cdots \otimes \mathcal{B}(\boldsymbol{Q}'_m) \end{array} \xrightarrow{[\![M]\!]} \mathrm{A} \right).$$

We can now show that our semantic interpretation is sound with respect to single-step reduction.

Theorem 7 (Soundness). *For $\mathcal{C} \in \mathrm{WF}_{\mathcal{SVConf}}(A)$, if $\mathcal{C} \notin \mathcal{VConf}$, then*

$$[\![\mathcal{C} : \mathrm{A}]\!] = \sum_{\mathcal{C} \rightsquigarrow_p \mathcal{C}'} p[\![\mathcal{C}' : \mathrm{A}]\!],$$

where the sum ranges over all possible reducts $\mathcal{C} \rightsquigarrow_p \mathcal{C}'$.

Next, we can show that our interpretation is sound in a big-step sense as well, which follows easily using the previous result and strong normalisation.

Theorem 8 (Strong Adequacy). *For $\mathcal{C} \in \mathrm{WF}_{\mathcal{SVConf}}(A)$,*

$$[\![\mathcal{C}]\!] = \sum_{\mathcal{V} \in \mathcal{VConf}} P(\mathcal{C} \to_* \mathcal{V})[\![\mathcal{V}]\!],$$

where $P(\mathcal{C} \to_* \mathcal{V})$ indicates the overall probability that $\mathcal{C}$ reduces to $\mathcal{V}$.

This implies that, for any well-formed configuration $\mathcal{C}$, its interpretation $[\![\mathcal{C}]\!]$ is an NCPU map (i.e., it is not merely subunital) and therefore it corresponds to a quantum channel in the Heisenberg picture of quantum mechanics.

4 Conclusion and Perspectives

We described a programming language which has support for both pure state quantum computation and mixed state quantum computation. We began by describing the pure quantum subsystem (§2), for which we introduced an equational theory and proved its completeness w.r.t. the denotational semantics (§2.4). Then, we described the main calculus (§3.1) which uses a new adaptation of quantum configurations (§3.2), which we used to define the operational semantics of the language. Here, the interaction between the pure quantum subsystem and the main calculus becomes apparent. We showed that our denotational semantics (§3.5) has a clear and appropriate interpretation as channels in the Heisenberg picture of quantum mechanics and we showed it is sound and adequate with respect to the operational semantics (§3.5).

Although our language has support for Nat, an inductive type, an obvious extension would be to include more general inductive types. Additionally, support for higher-order *pure* quantum computation is another feature that would be interesting to add in a future work.

Another matter that could be tackled as future work is recursion. In the pure system, adding general recursion is an open and difficult problem [14, Section 5.2]. In the main calculus, our denotational model supports recursion for first-order quantum functions [17,13]. However, our model does not support general recursion when quantum lambda abstractions are involved. There are other denotational models that support general recursion and quantum lambda abstractions, but they do not support types like qnat that correspond to $\ell^2(\mathbb{N})$, instead they support recursive types that intuitively correspond to infinite classical combinations of spaces involving finite-dimensional Hilbert spaces, which is not the same as $\ell^2(\mathbb{N})$. A model for quantum lambda abstractions, qnat and general recursion is an open problem to the best of our knowledge.

Acknowledgements

The authors would like to thank Benoît Valiron for helpful comments and discussions. Louis Lemonnier's research was funded by the Engineering and Physical Sciences Research Council (EPSRC) under project EP/X025551/1 "Rubber DUQ: Flexible Dynamic Universal Quantum programming". This work is also supported by the HORIZON 2020 project NEASQC, by the Inria associate team TC(Pro)^3, by the PEPR integrated project EPiQ ANR-22-PETQ-0007, and by the HQI initiative ANR-22-PNCQ-0002.

References

1. Andrés-Martínez, P.: Unbounded loops in quantum programs: categories and weak-while loops. Ph.D. thesis, Laboratory for Foundations of Computer Science, School of Informatics, University of Edinburgh (2022). https://doi.org/https://doi.org/10.48550/arXiv.2212.05371
2. Arrighi, P., Dowek, G.: Lineal: A linear-algebraic lambda-calculus. Log. Methods Comput. Sci. **13**(1) (2017). https://doi.org/10.23638/LMCS-13(1:8)2017, https://doi.org/10.23638/LMCS-13(1:8)2017
3. Chardonnet, K.: Towards a Curry-Howard Correspondence for Quantum Computation. (Vers une correspondance de Curry-Howard pour le calcul quantique). Ph.D. thesis, Université Paris-Saclay, France (2023), https://tel.archives-ouvertes.fr/tel-03959403
4. Childs, A.M., Cleve, R., Deotto, E., Farhi, E., Gutmann, S., Spielman, D.A.: Exponential algorithmic speedup by a quantum walk. In: Proceedings of the thirty-fifth annual ACM symposium on Theory of computing. pp. 59–68 (2003)
5. Cho, K.: Semantics for a quantum programming language by operator algebras. Electronic Proceedings in Theoretical Computer Science **172**, 165–190 (Dec 2014). https://doi.org/10.4204/eptcs.172.12, http://dx.doi.org/10.4204/EPTCS.172.12
6. Cho, K., Westerbaan, A.: Von Neumann Algebras form a Model for the Quantum Lambda Calculus (2016). https://doi.org/https://doi.org/10.48550/arXiv.1603.02133
7. Clairambault, P., de Visme, M.: Full abstraction for the quantum lambda-calculus. Proc. ACM Program. Lang. **4**(POPL), 63:1–63:28 (2020). https://doi.org/10.1145/3371131, https://doi.org/10.1145/3371131
8. Clairambault, P., de Visme, M., Winskel, G.: Game semantics for quantum programming. PACMPL **3**(POPL), 32:1–32:29 (2019). https://doi.org/10.1145/3290345
9. Díaz-Caro, A., Malherbe, O.: Semimodules and the (syntactically-)linear lambda calculus. CoRR **abs/2205.02142** (2022). https://doi.org/10.48550/ARXIV.2205.02142, https://doi.org/10.48550/arXiv.2205.02142
10. Díaz-Caro, A., Guillermo, M., Miquel, A., Valiron, B.: Realizability in the unitary sphere. In: LICS 2019. IEEE (jun 2019). https://doi.org/10.1109/lics.2019.8785834, https://doi.org/10.1109%2Flics.2019.8785834
11. Díaz-Caro, A., Malherbe, O.: Quantum control in the unitary sphere: Lambda-s1 and its categorical model. Logical Methods in Computer Science **Volume 18, Issue 3** (2022). https://doi.org/10.46298/lmcs-18(3:32)2022, https://doi.org/10.46298%2Flmcs-18%283%3A32%292022
12. Green, A.S., Lumsdaine, P.L., Ross, N.J., Selinger, P., Valiron, B.: Quipper: a scalable quantum programming language. In: Proceedings of the 34th ACM SIGPLAN conference on Programming language design and implementation. pp. 333–342 (2013)
13. Jia, X., Kornell, A., Lindenhovius, B., Mislove, M.W., Zamdzhiev, V.: Semantics for variational quantum programming. In: POPL 2022. vol. 6, pp. 1–31. ACM (2022). https://doi.org/10.1145/3498687
14. Lemonnier, L.: The Semantics of Effects : Centrality, Quantum Control and Reversible Recursion. Theses, Université Paris-Saclay (Jun 2024), https://theses.hal.science/tel-04625771

15. Pagani, M., Selinger, P., Valiron, B.: Applying quantitative semantics to higher-order quantum computing. In: POPL 2014. pp. 647–658. ACM (2014). https://doi.org/10.1145/2535838.2535879
16. Paykin, J., Rand, R., Zdancewic, S.: Qwire: a core language for quantum circuits. ACM SIGPLAN Notices **52**(1), 846–858 (2017)
17. Péchoux, R., Perdrix, S., Rennela, M., Zamdzhiev, V.: Quantum programming with inductive datatypes: Causality and affine type theory. In: FoSSaCS 2020. Lecture Notes in Computer Science, vol. 12077, pp. 562–581. Springer (2020). https://doi.org/10.1007/978-3-030-45231-5_29
18. Sabry, A., Valiron, B., Vizzotto, J.K.: From symmetric pattern-matching to quantum control. In: FoSSaCS 2018). Lecture Notes in Computer Science, vol. 10803, pp. 348–364. Springer (2018). https://doi.org/10.1007/978-3-319-89366-2_19
19. Selinger, P.: Towards a quantum programming language. Mathematical Structures in Computer Science **14**(4), 527–586 (2004). https://doi.org/https://doi.org/10.1017/S0960129504004256
20. Selinger, P., Valiron, B.: A lambda calculus for quantum computation with classical control. Mathematical Structures in Computer Science **16**(3), 527–552 (2006). https://doi.org/10.1017/S0960129506005238
21. Selinger, P., Valiron, B., et al.: Quantum lambda calculus. Semantic techniques in quantum computation pp. 135–172 (2009)
22. Tsukada, T., Asada, K.: Enriched presheaf model of quantum FPC. Proc. ACM Program. Lang. **8**(POPL), 362–392 (2024). https://doi.org/10.1145/3632855, `https://doi.org/10.1145/3632855`
23. Valiron, B.: Semantics of quantum programming languages: Classical control, quantum control. Journal of Logical and Algebraic Methods in Programming **128** (2022). https://doi.org/https://doi.org/10.1016/j.jlamp.2022.100790
24. de Visme, M.: Quantum Game Semantics. (Sémantique des Jeux Quantique). Ph.D. thesis, University of Lyon, France (2020), `https://tel.archives-ouvertes.fr/tel-03045844`
25. Voichick, F., Li, L., Rand, R., Hicks, M.: Qunity: A unified language for quantum and classical computing. Proc. ACM Program. Lang. **7**(POPL), 921–951 (2023). https://doi.org/10.1145/3571225, `https://doi.org/10.1145/3571225`

Quantifier Elimination and Craig Interpolation: The Quantitative Way*

Kevin Batz[1,2(✉)], Joost-Pieter Katoen[1], and Nora Orhan[1]

[1] RWTH Aachen University, Aachen, Germany
{kevin.batz,katoen}@cs.rwth-aachen.de, nora.hiseni@rwth-aachen.de
[2] University College London, London, UK

Abstract. Quantifier elimination (QE) and Craig interpolation (CI) are central to various state-of-the-art automated approaches to hardware- and software verification. They are rooted in the Boolean setting and are successful for, e.g., first-order theories such as linear rational arithmetic. What about their applicability in the quantitative setting where formulae evaluate to numbers and quantitative supremum/infimum quantifiers are the natural pendant to traditional Boolean quantifiers? Applications include establishing quantitative properties of programs such as bounds on expected outcomes of probabilistic programs featuring unbounded nondeterminism and analyzing the flow of information through programs. In this paper, we present the — to the best of our knowledge — first QE algorithm for possibly unbounded, ∞- or $(-\infty)$-valued, or discontinuous *piecewise linear quantities*. They are the quantitative counterpart to linear rational arithmetic, and are a popular quantitative assertion language for probabilistic program verification. We provide rigorous soundness proofs as well as upper space complexity bounds. Moreover, our algorithm yields a quantitative CI theorem: Given arbitrary piecewise linear quantities f, g with $f \models g$, both the strongest and the weakest Craig interpolant of f and g are quantifier-free and effectively constructible.

Keywords: Quantifier Elimination · Craig Interpolation · Quantitative Program Verification

1 Introduction

Quantifier elimination algorithms take as input a first-order formula φ over some background theory $\mathcal{T}$ and output a quantifier-free formula $\mathsf{QE}(\varphi)$ equivalent to φ modulo $\mathcal{T}$. Prime examples include Fourier-Motzkin variable elimination [45,25] and virtual substitution [40] for linear rational arithmetic, Cooper's method [18] for linear integer arithmetic, and Cylindrical Algebraic Decomposition [16] for non-linear real arithmetic. Quantifier elimination is extensively leveraged by automatic hard- and software verification techniques for, e.g., computing images

* This research was supported by the ERC AdG project 787914 FRAPPANT.

P. A. Abdulla and D. Kesner (Eds.): FoSSaCS 2025, LNCS 15691, pp. 176–197, 2025.
https://doi.org/10.1007/978-3-031-90897-2_9

of state sets [35,36], or for synthesizing loop invariants either from templates [27,17,31] or by solving abduction problems [22].

Craig interpolation [19] is vital to various automatic hard- and software verification techniques. A first-order theory $\mathcal{T}$ is called *(quantifier-free) interpolating* [32], if for all formulae φ, ψ with $\varphi \models_{\mathcal{T}} \psi$, there is an effectively constructible (quantifier-free) formula ϑ — called *Craig interpolant* of (φ, ψ) — with $\varphi \models_{\mathcal{T}} \vartheta \models_{\mathcal{T}} \psi$ and such that all free variables occurring free in ϑ also occur free in *both* φ and ψ. Intuitively, $\varphi \models_{\mathcal{T}} \psi$ encodes some (desirable or undesirable) reachability information and ϑ is a concise explanation of this fact, abstracting away irrelevant details. The computation of (quantifier-free) Craig interpolants is a vivid area of research [1,26,15,57,13] with applications ranging from symbolic finite-state model checking [43,56,39] over computing transition power abstractions [12] to automatic infinite-state software verification [28,11,29,52,53,44,2].

Quantitative program verification includes reasoning about expected outcomes of probabilistic programs via *weakest pre-expectations* [37,38,42,30], reasoning about the quantitative flow of information via *quantitative strongest post* [59], and reasoning about competitive ratios of online algorithms via *weighted programming* [7]. Quantitative reasoning requires a shift: Predicates, i.e., maps from program states to truth values in $\{\mathsf{true}, \mathsf{false}\}$, are replaced by *quantities*[3], which map program states to *extended reals* in $\mathbb{R} \cup \{-\infty, \infty\}$.

The classical quantifiers "there exists" $\exists$ and "for all" $\forall$ from predicate logic are replaced by *quantitative* supremum S and infimum J quantifiers [8]. These quantifiers naturally occur when reasoning with quantitative program logics: Very much like *classical* strongest post-*conditions* introduce an $\exists$-quantifier for assignments [21], *quantitative* strongest post introduces a S-quantifier (cf. [59, Table 2]). Similarly, whereas *classical* weakest pre-*conditions* introduce a $\forall$-quantifier for demonically resolving unbounded non-determinism of the form $x :\approx \mathbb{Q}$ (read: assign to x an *arbitrary* rational number) [20,47], *quantitative* weakest pre-*expectations* introduce an J-quantifier [9,51].

Example 1. In this paper, we focus on *piecewise linear quantities* such as

$$g \;=\; \underbrace{[\underbrace{y_1 \geq z \longrightarrow (x - 2 < y_1 \land -x \geq y_3 \land x \geq y_2)}_{=\varphi}] \cdot (\underbrace{2 \cdot x + z}_{=\tilde{a}})}_{\text{evaluate to } \tilde{a} \text{ on variable valuation } \sigma \text{ if } \sigma \models \varphi \text{, and to } 0 \text{ otherwise}} \;,$$

where $x, y_1, \ldots$ are $\mathbb{Q}$-valued variables. We can think of g as a formula that evaluates to extended rationals from $\mathbb{Q} \cup \{-\infty, \infty\}$ instead of truth values. By prefixing g with, e.g., a supremum quantifier, we obtain a new piecewise linear quantity $\mathsf{S}x\colon g$, which, on variable valuation σ, evaluates to the (variable valuation-dependent) supremum of g under all possible values for x, i.e.,

$${}^{\sigma}[\![\mathsf{S}x\colon g]\!] \;=\; \sup\left\{{}^{\sigma[x \mapsto q]}[\![g]\!] \mid q \in \mathbb{Q}\right\} . \qquad \triangle$$

[3] We adopt this term from Zhang and Kaminski [59]. In the realm of weakest pre-expectations, quantitative assertions are usually referred to as *expectations* [42]. In weighted programming, they are called *weightings* [7].

Our Contribution: Quantitative Quantifier Elimination and Craig Interpolation. Piecewise linear quantities over $\mathbb{Q}$-valued variables are to quantitative probabilistic program verification what first-order linear rational arithmetic is to classical program verification: Their entailment problem, i.e.,

$$\textit{Given piecewise linear quantities } f \textit{ and } f', \quad \textit{does} \quad \underbrace{f \models f'}_{\text{for all variable valuations } \sigma:\ {}^{\sigma}[\![f]\!] \leq {}^{\sigma}[\![f']\!]} \quad \textit{hold?}$$

is decidable [33,6], they are effectively closed under, e.g., weakest pre-expectations of loop-free linear probabilistic programs [6], and they have been shown to be sufficiently expressive for the verification of various probabilistic programs [5,6,58,49,14]. These facts render piecewise linear quantities one of the most prevalent quantitative assertion languages in automatic reasoning.

Reasoning with piecewise linear quantities containing the quantitative S or J quantifiers has, however, received scarce attention, let alone algorithmically. Similarly, the field of *quantitative Craig interpolation* is rather unexplored, as well. The goal of this paper is to lay the foundations for (i) developing quantitative quantifier elimination- and Craig interpolation-based approaches to automatic quantitative program verification and (ii) for simplifying the reasoning with quantitative assertions involving quantitative quantifiers. Towards this end:

1. We contribute the — to the best of our knowledge — first quantitative quantifier elimination algorithm for *arbitrary, possibly unbounded,* ∞ *or* $-(\infty)$*-valued, or discontinuous* piecewise linear quantities. Put more formally, given an *arbitrary* piecewise linear quantity f possibly containing quantitative quantifiers, our algorithm computes a *quantifier-free equivalent of* f. For $\mathsf{S}x\colon g$ from Example 1, our algorithm yields (after simplification)

$$\begin{aligned}
&[y_1 < z] \cdot \infty \\
&+ [y_1 \geq z \wedge y_2 < y_1 + 2 \wedge y_2 \leq -y_3 \wedge y_1 + 2 \leq y_3] \cdot (2 \cdot y_1 + z + 4) \\
&+ [y_1 \geq z \wedge y_2 < y_1 + 2 \wedge y_2 \leq -y_3 \wedge y_1 + 2 > y_3] \cdot (-y_3 + z)\ .
\end{aligned}$$

2. We provide rigorous soundness proofs, illustrative examples, and upper space-complexity bounds on our algorithm.
3. We contribute the — to the best of our knowledge — first *Craig interpolation theorem* for piecewise linear quantities: Using our quantifier elimination algorithm, we prove that for two *arbitrary* piecewise linear quantities f, f' with $f \models f'$, both the strongest and the weakest g such that

$$f \models g \models f' \qquad \text{and} \qquad \text{the free variables in } g \text{ are free in } \textit{both } f \text{ and } f'$$

are quantifier-free and effectively constructible.

Example 2. Consider the following piecewise linear quantities:

$$f \;=\; [x \geq 0] \cdot x + [x \geq 0 \wedge y \leq x] \cdot y$$

$$f' = [x \geq 0 \wedge z \geq x] \cdot (2 \cdot x + z + 1) + [z < x] \cdot \infty$$

We have $f \models f'$. Using our quantifier elimination technique, we effectively construct both the strongest and the weakest[4] Craig interpolants of (f, f') given by

$$\underbrace{[x \geq 0] \cdot 2 \cdot x}_{\text{strongest Craig interpolant of } (f, f')} \qquad \text{and} \qquad \underbrace{[x \geq 0] \cdot (3 \cdot x + 1)}_{\text{weakest Craig interpolant of } (f, f')} \quad . \qquad \triangle$$

Remark 1. When applying *classical* Craig interpolation for a first-order theory $\mathcal{T}$ to, e.g., loop invariant generation, "simple" Craig interpolants, i.e., interpolants that lie strictly "between" (w.r.t. $\models_{\mathcal{T}}$) the strongest and the weakest ones, are often very useful [1]. Our *quantitative* Craig interpolation technique presented in this paper does *not* aim for obtaining such "simple" interpolants. Rather, our goal is to prove that quantitative Craig interpolants at all exist and that they are effectively constructible. We discuss possible directions for obtaining simpler quantitative Craig interpolants in Section 5.

Related Work. Our quantifier elimination algorithm is based on ideas related to Fourier-Motzkin variable elimination [45,25]. Most closely related is the work by Zamani, Sanner, and Fang on symbolic dynamic programming [50]. They introduce the symbolic $\max_x$ operator on piecewise defined functions of type $\mathbb{R}^n \to \mathbb{R}$, which also exploits the partitioning property (similar to Theorem 1) and disjunctive normal forms (similar to Theorem 2). We identify the following key differences: the functions considered in [50] must be (i) continuous, (ii) bounded (so that all suprema are actually maxima), and (iii) they must not contain ∞ or $-\infty$. We do not impose these restrictions since they do not apply to piecewise defined functions obtained from, e.g., applying the program logics mentioned in Section 1. [50], on the other hand, also considers piecewise quadratic functions, whereas we focus on piecewise linear functions. Extending our approach to piecewise quadratic functions is an interesting direction for future work. Finally, we provide a rigorous formalization and soundness proofs alongside upper space complexity bounds. Quantitative quantifier elimination is moreover related to *parametric programming* [54]. We are, however, not aware of an approach which tackles the computational problem we investigate as it is required from the perspective of a quantitative quantifier elimination problem.

Khatami, Pourmahdian, and Tavana [55] investigate a Craig interpolation property of first-order Gödel logic, where formulae evaluate to real numbers in the unit interval $[0, 1]$. Apart from the more restrictive semantic codomain, [55] operates in an uninterpreted setting whereas we operate within linear rational arithmetic extended by ∞ and $-\infty$. Baaz and Veith [4] investigate quantifier elimination of first-order logic over fuzzy algebras over the same semantic codomain. Teige and Fränzle [41] investigate Craig interpolation for stochastic Boolean satisfiability problems, where formulae also evaluate to numbers instead of truth values. Quantified variables are assumed to range over a finite domain.

[4] We say that g is *stronger* (resp. *weaker*) than g', if $g \models g'$ (resp. $g' \models g$).

Outline. In Section 2, we introduce piecewise linear quantities. We present our quantifier elimination algorithm alongside essential theorems and a complexity analysis in Section 3. Our quantitative Craig interpolation results are presented in Section 4. Finally, we discuss potential applications in Section 5.

2 Piecewise Linear Quantities

Throughout this paper, we fix a finite non-empty set $\mathsf{Vars} = \{x, y, z, \ldots\}$ of variables. We denote by $\mathbb{N}_0$ the set of natural numbers including 0 and let $\mathbb{N} = \mathbb{N}_0 \setminus 0$. The set of rationals (resp. reals) is denoted by $\mathbb{Q}$ (resp. $\mathbb{R}$) and we denote by $\mathbb{Q}^{\pm\infty} = \mathbb{Q} \cup \{-\infty, \infty\}$ (resp. $\mathbb{R}^{\pm\infty} = \mathbb{R} \cup \{-\infty, \infty\}$) the set of *extended* rationals (resp. reals). A *(variable) valuation* $\sigma\colon \mathsf{Vars} \to \mathbb{Q}$ assigns a rational number to each variable. The set of all valuations is denoted by Σ.

Towards defining piecewise linear quantities and their semantics, we briefly recap linear arithmetic and Boolean expressions.

Definition 1 (Linear Arithmetic Expressions). *The set* LinAX *of* linear arithmetic expressions *consists of all expressions* a *of the form*

$$a \quad = \quad q_0 + \sum_{i=1}^{|\mathsf{Vars}|} q_i \cdot x_i \ ,$$

where $q_0, \ldots, q_{|\mathsf{Vars}|} \subset \mathbb{Q}$ *and* $x_1, \ldots, x_{|\mathsf{Vars}|} \subset \mathsf{Vars}$*. Moreover, we define the set* $\mathsf{LinAX}^{\pm\infty}$ *of* extended linear arithmetic expressions *as*

$$\mathsf{LinAX}^{\pm\infty} \quad = \quad \mathsf{LinAX} \cup \{-\infty, \infty\} \ . \qquad \triangle$$

Notice that every arithmetic expression is normalized in the sense that every variable occurs exactly once. We often omit summands $q_i \cdot x_i$ (resp. q_0) with $q_i = 0$ (resp. $q_0 = 0$) for the sake of readability. Given a as above, we denote by

$$\mathsf{FV}(a) \quad = \quad \{x_i \in \mathsf{Vars} \mid q_i \neq 0\}$$

the set of (necessarily free) variables occurring in a. For $\tilde{a} = \infty$ or $\tilde{a} = -\infty$, we define $\mathsf{FV}(\tilde{a}) = \emptyset$. Given $\tilde{a} \in \mathsf{LinAX}^{\pm\infty}$ and $x_j \in \mathsf{Vars}$, we say that

$$\begin{cases} x_j \text{ occurs positively in } \tilde{a} & \text{if } \tilde{a} = q_0 + \sum_{i=1}^{|\mathsf{Vars}|} q_i \cdot x_i \text{ and } q_j > 0 \\ x_j \text{ occurs negatively in } \tilde{a} & \text{if } \tilde{a} = q_0 + \sum_{i=1}^{|\mathsf{Vars}|} q_i \cdot x_i \text{ and } q_j < 0 \ . \end{cases}$$

Finally, given $\sigma \in \Sigma$, the *semantics* ${}^{\sigma}[\![\tilde{a}]\!] \in \mathbb{Q}^{\pm\infty}$ of $\tilde{a}$ under σ is defined as

$${}^{\sigma}[\![\tilde{a}]\!] \quad = \quad \begin{cases} q_0 + \sum_{i=1}^{|\mathsf{Vars}|} q_i \cdot \sigma(x_i) & \text{if } \tilde{a} = q_0 + \sum_{i=1}^{|\mathsf{Vars}|} q_i \cdot x_i \\ -\infty & \text{if } \tilde{a} = -\infty \\ \infty & \text{if } \tilde{a} = \infty \ . \end{cases}$$

Definition 2 (Boolean Expressions). Boolean expressions φ *in the set* Bool *adhere to the following grammar, where* $\tilde{a} \in \mathsf{LinAX}^{\pm\infty}$*:*

$$\begin{aligned}
\varphi \quad \longrightarrow \quad & \tilde{a} < \tilde{a} \mid \tilde{a} \leq \tilde{a} \mid \tilde{a} > \tilde{a} \mid \tilde{a} \geq \tilde{a} && \text{(linear inequalities)} \\
& \mid \neg\varphi && \text{(negation)} \\
& \mid \varphi \wedge \varphi && \text{(conjunction)} \ \triangle
\end{aligned}$$

The Boolean constants $\mathsf{true}, \mathsf{false}$ and the Boolean connectives $\vee$ and $\longrightarrow$ are syntactic sugar. We assume that $\neg$ binds stronger than $\wedge$ binds stronger than $\vee$, and we use parentheses to resolve ambiguities. The set $\mathsf{FV}(\varphi)$ of (necessarily free) variables in φ is defined as usual. Given a valuation σ, we write $\sigma \models \varphi$ if σ satisfies φ and $\sigma \not\models \varphi$ otherwise, which is defined in the standard way. Finally, if $\sigma \not\models \varphi$ for all $\sigma \in \Sigma$, then we say that φ is *unsatisfiable*[5].

Definition 3 (Piecewise Linear Quantities (adapted from [8,33])). *The set* $\mathsf{LinQuant}$ *of* (piecewise linear) quantities *consists of all expressions*

$$f \quad = \quad Q_1 x_1 \ldots Q_k x_k \colon \sum_{i=1}^{n} [\varphi_i] \cdot \tilde{a}_i \ ,$$

where $k \in \mathbb{N}_0$, $n \in \mathbb{N}$, *and where*

1. $Q_j \in \{\reflectbox{$\mathsf{S}$}, \reflectbox{$\mathsf{L}$}\}$ *and* $x_j \in \mathsf{Vars}$ *for all* $j \in \{1, \ldots, k\}$,
2. $\varphi_i \in \mathsf{Bool}$ *and* $\tilde{a}_i \in \mathsf{LinAX}^{\pm\infty}$ *for all* $i \in \{1, \ldots, n\}$,
3. *for all* $\sigma \in \Sigma$ *and all* $i, j \in \{1, \ldots, n\}$ *with* $i \neq j$, *we have*[6]

$$\sigma \models \varphi_i \text{ and } \sigma \models \varphi_j \qquad \text{implies} \qquad \tilde{a}_i \neq \infty \text{ or } \tilde{a}_j \neq -\infty \ . \qquad \triangle$$

Here $[\varphi]$ is the *Iverson bracket* [34] of the Boolean expression φ, which evaluates to 1 under valuation σ if $\sigma \models \varphi$, and to 0 otherwise. $\reflectbox{$\mathsf{S}$}$ is called the *supremum quantifier* and $\reflectbox{$\mathsf{L}$}$ is the *infimum quantifier*. The quantitative quantifiers take over the role of the classical $\exists$- and $\forall$-quantifiers from first-order predicate logic. Their semantics is detailed further below. If $k = 0$, then we call f *quantifier-free*. Given f as above, the set of *free variables* in f is

$$\mathsf{FV}(f) \quad = \quad \bigcup_{j=1}^{n} \left(\mathsf{FV}(\varphi_j) \cup \mathsf{FV}(\tilde{a}_j) \right) \setminus \{x_1, \ldots, x_k\} \ .$$

For quantifier-free f, we introduce the shorthand $[\varphi] \cdot f = \sum_{i=1}^{n} [\varphi \wedge \varphi_i] \cdot \tilde{a}$.

Towards defining the semantics of quantities, we use the following notions: Given a valuation $\sigma \in \Sigma$, a variable $x \in \mathsf{Vars}$, and $q \in \mathbb{Q}$, we define the valuation obtained from updating the value of x under σ to q as

$$\sigma[x \mapsto q](y) \quad = \quad \begin{cases} q & \text{if } y = x \\ \sigma(y) & \text{otherwise} \ . \end{cases}$$

As is standard [3] in the realm of extended reals, we define for all $r \in \mathbb{R}$:

[5] Unsatisfiability of Boolean expressions is decidable by SMT solving over linear rational arithmetic (LRA) as is implemented, e.g., by the solver Z3 [46].

[6] This is decidable by SMT solving over LRA. Hence, the set $\mathsf{LinQuant}$ is computable.

1. $r + \infty = \infty + r = \infty$
2. $r + (-\infty) = -\infty + r = -\infty$
3. $\infty + \infty = \infty$
4. $-\infty + (-\infty) = -\infty$
5. $-\infty \cdot 0 = 0 \cdot (-\infty) = 0 = 0 \cdot \infty = \infty \cdot 0$
6. if $r > 0$, then $r \cdot \infty = \infty \cdot r = \infty$
7. if $r > 0$, then $r \cdot (-\infty) = -\infty \cdot r = -\infty$
8. if $r < 0$, then $r \cdot \infty = \infty \cdot r = -\infty$
9. if $r < 0$, then $r \cdot (-\infty) = -\infty \cdot r = \infty$

The terms $\infty + (-\infty)$ and $-\infty + \infty$ are undefined. The condition from Definition 3.3 ensures that we never encounter such undefined terms, which yields the semantics of piecewise linear quantities to be well-defined:

Definition 4 (Semantics of Piecewise Linear Quantities). *Let* $f \in \mathsf{LinQuant}$ *and* $\sigma \in \Sigma$. *The* semantics[7] ${}^{\sigma}[\![f]\!] \in \mathbb{R}^{\pm\infty}$ *of* f *under* σ *is defined inductively:*

$$
\begin{aligned}
{}^{\sigma}[\![\sum_{i=1}^{n} [\varphi_i] \cdot \tilde{a}_i]\!] &= \sum_{i=1}^{n} \begin{cases} {}^{\sigma}[\![\tilde{a}_i]\!] & \text{if } \sigma \models \varphi_i \\ 0 & \text{if } \sigma \not\models \varphi_i \end{cases} \\
{}^{\sigma}[\![\reflectbox{S} x \colon f']\!] &= \sup\left\{ {}^{\sigma[x \mapsto q]}[\![f']\!] \mid q \in \mathbb{Q} \right\} \\
{}^{\sigma}[\![\reflectbox{L} x \colon f']\!] &= \inf\left\{ {}^{\sigma[x \mapsto q]}[\![f']\!] \mid q \in \mathbb{Q} \right\}
\end{aligned}
\qquad \triangle
$$

In words: if f is quantifier-free, then ${}^{\sigma}[\![f]\!]$ evaluates to the sum of all extended arithmetic expressions $\tilde{a}_j$ for which $\sigma \models \varphi_j$. The semantics of $\reflectbox{$\mathsf{S}$}$ and $\reflectbox{$\mathsf{L}$}$ makes it evident that the quantitative quantifiers generalize the classical quantifiers: Whereas $\exists$ maximizes — so to speak — a truth value, the $\reflectbox{$\mathsf{S}$}$-quantifier maximizes a quantity by evaluating to the supremum obtained from evaluating f under all possible values for x. $\reflectbox{$\mathsf{L}$}$ behaves analogously by evaluating to an infimum.

Finally, we say that two piecewise linear quantities $f, f' \in \mathsf{LinQuant}$ are *(semantically) equivalent*, denoted by $f \equiv f'$, if ${}^{\sigma}[\![f]\!] = {}^{\sigma}[\![f']\!]$ for all $\sigma \in \Sigma$.

3 Quantitative Quantifier Elimination

In this section, we detail our quantifier elimination procedure alongside illustrative examples. Given an *arbitrary* piecewise linear quantity

$$
f \quad = \quad Q_1 x_1 \ldots Q_k x_k \colon \sum_{i=1}^{n} [\varphi_i] \cdot \tilde{a}_i \;\in\; \mathsf{LinQuant}\,,
$$

we aim to automatically compute some $\mathsf{QE}(f) \in \mathsf{LinQuant}$ satisfying

$\mathsf{QE}(f)$ is quantifier-free and $\mathsf{QE}(f)$ is equivalent to f, i.e., $f \equiv \mathsf{QE}(f)$.

[7] It follows from the soundness of our quantifier elimination algorithm (Theorem 5) that all $f \in \mathsf{LinQuant}$ evaluate to extended rationals in $\mathbb{Q}^{\pm\infty}$.

As with quantifier elimination for (theories of) classical first-order predicate logic, it suffices being able to eliminate piecewise linear quantities containing *a single* quantifier, which then enables to process quantities containing an *arbitrary* number of quantifiers in an inner- to outermost fashion, i.e.,

$$\mathsf{QE}(f) \;=\; \mathsf{QE}(Q_1 x_1\colon \mathsf{QE}(\ldots \mathsf{QE}(Q_k x_k\colon \sum_{i=1}^{n} [\varphi_i] \cdot \tilde{a}_i))) \;.$$

Throughout the next sections, we thus fix an f of the form

$$f \;=\; Qx\colon \sum_{i=1}^{n} [\varphi_i] \cdot \tilde{a}_i \;, \tag{1}$$

where $Q \in \{\reflectbox{S}, \reflectbox{L}\}$. We proceed by means of a 3-level divide-and-conquer approach. We describe each of the involved stages in Sections 3.1-3.3. In Section 3.4, we summarize our approach by providing an algorithm.

3.1 Stage 1: Exploiting the Guarded Normal Form

Our first step is to transform the input f into a normal form (an extension of [33, Section 5.1]), which enables us to subdivide the quantifier elimination problem into simpler sub-problems. Intuitively, this normal form enforces a more explicit representation of the $\mathbb{R}^{\pm\infty}$-valued function a piecewise linear quantity represents.

Definition 5 (Guarded Normal Form). *Let $g \in \mathsf{LinQuant}$ be given by*

$$g \;=\; Q_1 x_1 \ldots Q_k x_k\colon \sum_{i=1}^{n} [\varphi_i] \cdot \tilde{a}_i$$

and fix some variable $x \in \mathsf{Vars}$. We say that g is in guarded normal form w.r.t. *x, denoted by $g \in \mathsf{GNF}_x$, if all of the following conditions hold:*

1. *the φ_i partition the set Σ of valuations, i.e., for all $\sigma \in \Sigma$ there exists exactly one $i \in \{1, \ldots, n\}$ such that $\sigma \models \varphi_i$,*
2. *the φ_i are in disjunctive normal form (DNF), i.e.,*

$$\forall i \in \{1, \ldots, n\}\colon \quad \varphi_i \textit{ is of the from } \bigvee_j \bigwedge_{j'} L_{j,j'} \;,$$

where each $L_{j,j'} \in \mathsf{LinAX}^{\pm\infty}$ is a (strict or non-strict) linear inequality,

3. *for each linear inequality L in g, it holds that if $x \in \mathsf{FV}(L)$, then*

$$L \;=\; x \sim \tilde{b}$$

for some $\tilde{b} \in \mathsf{LinAX}^{\pm\infty}$ with $x \notin \mathsf{FV}\left(\tilde{b}\right)$ and $\sim \in \{>, \geq, <, \leq\}$. △

If Condition 5.1 holds, then we say that g is *partitioning*. Speaking of a *normal form* is justified by the fact that every piecewise linear quantity $g \in \mathsf{LinQuant}$ can effectively be transformed into a semantically equivalent $g' \in \mathsf{LinQuant}$ in guarded normal form with respect to variable $x \in \mathsf{Vars}$, i.e., such that $g' \in \mathsf{GNF}_x$ and $g \equiv g'$. To see this, let g be given as above. Towards obtaining g', we first establish the partitioning property by enumerating the possible assignments of truth values to the φ_i. Put more formally, we construct

$$\sum_{\left((\rho_1,\tilde{e}_1),\ldots,(\rho_n,\tilde{e}_n)\right) \in \times_{i=1}^{n}\left\{(\varphi_i,\tilde{a}_i),(\neg\varphi_i,0)\right\}} \begin{cases} \epsilon & \text{if } \bigwedge_{i=1}^{n}\rho_i \text{ is unsat.} \\ \left[\bigwedge_{i=1}^{n}\rho_i\right]\cdot\sum_{i=1}^{n}\tilde{e}_i & \text{otherwise ,} \end{cases}$$

where we let $\epsilon + \tilde{e} = \tilde{e} = \tilde{e} + \epsilon$ for all $\tilde{e} \in \mathsf{LinAX}^{\pm\infty}$ and obey the rules for treating ∞ and $-\infty$, respectively, as given in Section 2. We then obtain g' by transforming the so-obtained Boolean expressions into DNF and isolating x in every inequality where x occurs. Notice that if g satisfies the conditions from Definition 3, then so does g'. In particular, when constructing sums of the form $\sum_{i=1}^{n}\tilde{e}_i$, we never encounter expressions of the form $\infty + (-\infty)$ or $-\infty + \infty$.

Example 3. Recall the piecewise linear quantity from Example 1 given by

$$\mathsf{S}x\colon\ [y_1 \geq z \longrightarrow (x - 2 < y_1 \wedge -x \geq y_3 \wedge x \geq y_2)] \cdot (2 \cdot x + z)\ ,$$

which is *not* in GNF_x. Applying the construction from above yields

$$\begin{aligned} \mathsf{S}x\colon\ & [y_1 < z \vee (x < y_1 + 2 \wedge x \leq -y_3 \wedge x \geq y_2)] \cdot (2 \cdot x + z) \\ & + [(y_1 \geq z \wedge x \geq y_1 + 2) \vee (y_1 \geq z \wedge x > -y_3) \vee (y_1 \geq z \wedge x < y_2)] \cdot 0 \end{aligned}$$

which *is* in GNF_x and will serve us as a running example. △

Now assume w.l.o.g. that the input quantity f is in GNF_x. Each of the Conditions 5.1-3 is essential to our approach. We will now exploit that f is partitioning. Given $\varphi \in \mathsf{Bool}$ and $\tilde{a} \in \mathsf{LinAX}^{\pm\infty}$, we define the shorthands

$$\varphi \searrow \tilde{a} \;=\; [\varphi]\cdot\tilde{a} + [\neg\varphi]\cdot(-\infty) \qquad \text{and} \qquad \varphi \nearrow \tilde{a} \;=\; [\varphi]\cdot\tilde{a} + [\neg\varphi]\cdot\infty\ .$$

Notice that these quantities are always partitioning. Now consider the following:

Theorem 1. *Let* $x \in \mathsf{Vars}$ *and let* $\sum_{i=1}^{n}[\varphi_i]\cdot\tilde{a}_i \in \mathsf{GNF}_x$. *We have for all* $\sigma \in \Sigma$:

1. $^{\sigma}[\![\mathsf{S}x\colon \sum_{i=1}^{n}[\varphi_i]\cdot\tilde{a}_i]\!] = \max\{^{\sigma}[\![\mathsf{S}x\colon (\varphi_i \searrow \tilde{a}_i)]\!] \mid i \in \{1,\ldots,n\}\}$
2. $^{\sigma}[\![\mathsf{J}x\colon \sum_{i=1}^{n}[\varphi_i]\cdot\tilde{a}_i]\!] = \min\{^{\sigma}[\![\mathsf{J}x\colon (\varphi_i \nearrow \tilde{a}_i)]\!] \mid i \in \{1,\ldots,n\}\}$

Proof. This is a consequence of the fact that the quantity is partitioning and that $-\infty$ (resp. ∞) are neutral wr.t. max (resp. min). See [10, App. A.1] for details. □

We may thus consider each summand of the input quantity f separately. Assuming we can compute $\mathsf{QE}(\mathcal{S} x\colon \varphi_i \searrow \tilde{a}_i)$ and $\mathsf{QE}(\mathcal{L} x\colon \varphi_i \nearrow \tilde{a}_i)$, we obtain the sought-after quantifier-free equivalent of f by effectively constructing valuation-wise minima and maxima of finite sets of partitioning quantities as follows:

Lemma 1. *Let $M = \{h_1, \ldots, h_n\} \subseteq \mathsf{LinQuant}$ for some $n \geq 1$, where each*

$$h_i \quad = \quad \sum_{j=1}^{m_i} [\varphi_{i,j}] \cdot \tilde{a}_{i,j}$$

is partitioning. Then:

1. *There is an effectively constructible $\mathsf{MAX}(M) \in \mathsf{LinQuant}$ such that*

$$\forall \sigma \in \Sigma\colon \quad {}^{\sigma}[\![\mathsf{MAX}(M)]\!] \quad = \quad \max\{{}^{\sigma}[\![h_i]\!] \mid i \in \{1, \ldots, n\}\} \;.$$

2. *There is an effectively constructible $\mathsf{MIN}(M) \in \mathsf{LinQuant}$ such that*

$$\forall \sigma \in \Sigma\colon \quad {}^{\sigma}[\![\mathsf{MIN}(M)]\!] \quad = \quad \min\{{}^{\sigma}[\![h_i]\!] \mid i \in \{1, \ldots, n\}\} \;.$$

Moreover, both $\mathsf{MAX}(M)$ and $\mathsf{MIN}(M)$ are partitioning.

Proof. Write $\underline{m}_i = \{1, \ldots, m_i\}$. We construct[8]

$$\mathsf{MAX}(M) = \sum_{(j_1, \ldots, j_n) \in \underline{m}_1 \times \ldots \times \underline{m}_n} \; \sum_{i=1}^{n} \Big[\underbrace{\bigwedge_{k=1}^{n} \varphi_{k,j_k}}_{h_k \text{ evaluates to } \tilde{a}_{k,j_k}} \wedge \underbrace{\bigwedge_{k=1}^{i-1} \tilde{a}_{i,j_i} > \tilde{a}_{k,j_k} \wedge \bigwedge_{k=i+1}^{n} \tilde{a}_{i,j_i} \geq \tilde{a}_{k,j_k}}_{\substack{h_i \text{ is the quantity with smallest index} \\ \text{evaluating to the sought-after maximum}}} \Big] \cdot \tilde{a}_{i,j_i} \;.$$

$\mathsf{MAX}(M)$ iterates over all combinations of summands, which determine the value each of the h_i evaluate to (first summand). For each of these combinations, we check, for each $i \in \{1, \ldots, n\}$, whether h_i evaluates to the sought-after maximum (second summand). $\mathsf{MAX}(M)$ is partitioning since the h_i are and due to the fact that $\mathsf{MAX}(M)$ selects the maximizing quantity with the *smallest index*. The construction of $\mathsf{MIN}(M)$ is analogous and provided in [10, App. B.1]. □

Combining Theorem 1 and Lemma 1 thus yields:

1. $\mathsf{QE}(\mathcal{S} x\colon \sum_{i=1}^{n} [\varphi_i] \cdot \tilde{a}_i) \;=\; \mathsf{MAX}\left(\{\mathsf{QE}(\mathcal{S} x\colon (\varphi_i \searrow \tilde{a}_i)) \mid i \in \{1, \ldots, n\}\}\right)$
2. $\mathsf{QE}(\mathcal{L} x\colon \sum_{i=1}^{n} [\varphi_i] \cdot \tilde{a}_i) \;=\; \mathsf{MIN}\left(\{\mathsf{QE}(\mathcal{L} x\colon (\varphi_i \nearrow \tilde{a}_i)) \mid i \in \{1, \ldots, n\}\}\right)$

Example 4. Continuing Example 3, we have for every $\sigma \in \Sigma$,

$$\begin{aligned} \mathsf{QE}(f) \;=\; \mathsf{MAX}(\{&\mathsf{QE}(\mathcal{S} x\colon (y_1 < z \vee (x < y_1 + 2 \wedge x \geq y_2)) \searrow 2 \cdot x + z), \\ &\mathsf{QE}(\mathcal{S} x\colon ((y_1 \geq z \wedge x \geq y_1 + 2) \vee \ldots) \searrow 0)\}) \;. \end{aligned}$$

△

[8] As usual, the empty conjunction is equivalent to true.

3.2 Stage 2: Exploiting the Disjunctive Normal Form

In this stage, we aim to eliminate the quantifiers from the simpler quantities

$$\text{Ƨ} x\colon (\varphi \searrow \tilde{a}) \qquad \text{or} \qquad \text{Ɩ} x\colon (\varphi \nearrow \tilde{a}) \,.$$

Recall that we assume the input quantity f to be in guarded normal form w.r.t. x, which yields the Boolean expression φ to be in DNF (cf. Definition 5.2). Exploiting the disjunctive shape of φ yields the following:

Theorem 2. *Let $\tilde{a} \in \mathsf{LinAX}^{\pm\infty}$ be an extended arithmetic expression and let*

$$\varphi \quad = \quad \bigvee_{i=1}^{n} D_i \;\in\; \mathsf{Bool}$$

be a Boolean expression in DNF for some $n \geq 1$. We have for all $\sigma \in \Sigma$:

1. ${}^{\sigma}[\![\text{Ƨ} x\colon (\varphi \searrow \tilde{a})]\!] = \max\{{}^{\sigma}[\![\text{Ƨ} x\colon (D_i \searrow \tilde{a})]\!] \mid i \in \{1,\ldots,n\}\}$
2. ${}^{\sigma}[\![\text{Ɩ} x\colon (\varphi \nearrow \tilde{a})]\!] = \min\{{}^{\sigma}[\![\text{Ɩ} x\colon (D_i \nearrow \tilde{a})]\!] \mid i \in \{1,\ldots,n\}\}$

Proof. We first observe that

$${}^{\sigma}[\![\text{Ƨ} x\colon (\varphi \searrow \tilde{a})]\!] \quad = \quad \sup\Bigl(\bigcup_{i=1}^{n}\Bigl\{{}^{\sigma[x\mapsto q]}[\![[D_i]\cdot \tilde{a}]\!] \mid q \in \mathbb{Q} \text{ and } \sigma[x \mapsto q] \models D_i\Bigr\}\Bigr)$$

and then make use of the fact that the supremum of a finite union of extended reals is the maximum of the individual suprema, i.e., the above is equal to

$$\begin{aligned}
& \max\Bigl(\bigcup_{i=1}^{n}\Bigl\{\sup\Bigl\{{}^{\sigma[x\mapsto q]}[\![D_i \searrow \tilde{a}]\!] \mid q \in \mathbb{Q}\Bigr\}\Bigr\}\Bigr) && (-\infty \text{ is neutral w.r.t. sup})\\
=\; & \max\Bigl(\bigcup_{i=1}^{n}\{{}^{\sigma}[\![\text{Ƨ} x\colon (D_i \searrow \tilde{a})]\!]\}\Bigr) && \text{(Definition 4)}\\
=\; & \max\{{}^{\sigma}[\![\text{Ƨ} x\colon (D_i \searrow \tilde{a})]\!] \mid i \in \{1,\ldots,n\}\} \,. && \text{(rewrite set)}
\end{aligned}$$

See [10, App. A.2] for a detailed proof. The reasoning for Ɩ is analogous. □

Hence, combining Theorem 2 with Lemma 1 reduces our problem further to eliminating quantifiers from the above simpler quantities. Put formally:

1. $\mathsf{QE}(\text{Ƨ} x\colon (\varphi \searrow \tilde{a})) \;=\; \mathsf{MAX}(\{\mathsf{QE}(\text{Ƨ} x\colon (D_i \searrow \tilde{a})) \mid i \in \{1,\ldots,n\}\})$
2. $\mathsf{QE}(\text{Ɩ} x\colon (\varphi \nearrow \tilde{a})) \;=\; \mathsf{MIN}(\{\mathsf{QE}(\text{Ɩ} x\colon (D_i \nearrow \tilde{a})) \mid i \in \{1,\ldots,n\}\})$

Example 5. Continuing Example 4, we have

$$\begin{aligned}
& \mathsf{QE}(\text{Ƨ} x\colon (y_1 < z \vee (x < y_1 + 2 \wedge x \geq y_2)) \searrow 2\cdot x + z)\\
=\; & \mathsf{MAX}(\{\mathsf{QE}(\text{Ƨ} x\colon y_1 < z \searrow 2\cdot x + z),\\
& \qquad\qquad \mathsf{QE}(\text{Ƨ} x\colon (x < y_1 + 2 \wedge x \geq y_2) \searrow 2\cdot x + z)\}) \,.
\end{aligned}$$

The second argument of MAX from Example 4 is treated analogously. △

3.3 Stage 3: Computing Valuation-Dependent Suprema and Infima

This is the most involved stage since we need to operate on the atomic level of the given expressions. We aim to eliminate the quantifiers from quantities of the form

$$\mathsf{S} x\colon (\bigwedge_{i=1}^{n} L_i \searrow \tilde{a}) \quad \text{or} \quad \mathsf{J} x\colon (\bigwedge_{i=1}^{n} L_i \nearrow \tilde{a})\ ,$$

where each L_i is a linear inequality. We start with an example.

Example 6. Continuing Example 5, we perform quantifier elimination on

$$g \;=\; \mathsf{S} x\colon \underbrace{(x < y_1 + 2 \wedge x \leq -y_3 \wedge x \geq y_2)}_{=D} \searrow \underbrace{2 \cdot x + z}_{=\tilde{a}}\ .$$

Fix some valuation σ. First observe that if there is no $q \in \mathbb{Q}$ such that $\sigma[x \mapsto q] \models D$ — or, phrased in predicate logic, if $\sigma \not\models \exists x\colon D$ —, then ${}^{\sigma}[\![g]\!]$ evaluates to $-\infty$. Otherwise, we need to inspect D and $\tilde{a}$ closer in order to determine ${}^{\sigma}[\![g]\!]$. Hence, eliminating the $\mathsf{S} x$-quantifier involves characterizing whether $\sigma \models \exists x\colon D$ holds *without referring to* x. This boils down to performing *classical* quantifier elimination on the formula $\exists x\colon D$. We leverage classical Fourier-Motzkin variable elimination: Compare the bounds D imposes on x and encode whether they are consistent. Towards this end, we construct

$$\varphi_{\exists}(D, x) \quad = \quad \underbrace{y_2 < y_1 + 2 \wedge y_2 \leq -y_3}_{\text{equivalent to } \exists x\colon D} \quad \in \quad \mathsf{Bool}\ .$$

Now, how can we characterize ${}^{\sigma}[\![g]\!]$ in case $\sigma \models \varphi_{\exists}(D, x)$? We first observe that x occurs positively in $\tilde{a}$. Therefore, intuitively, the $\mathsf{S} x$-quantifier aims to maximize the value of x under all possible assignments satisfying D. Since we isolate x in every inequality where x occurs, we can readily read off from D that x's maximal (or, in fact, *supremal*) value is given by the *minimum* of $\sigma(y_1) + 2$ and $-\sigma(y_3)$ — the least of all upper bounds imposed on x. Overall, we get

$$\begin{aligned} g \;\equiv\; & [\varphi_{\exists}(D, x)] \cdot \big([y_1 + 2 \leq -y_3] \cdot (2 \cdot y_1 + z + 4) \\ & + [y_1 + 2 > -y_3] \cdot (-2 \cdot y_3 + z) \big) + [\neg \varphi_{\exists}(D, x)] \cdot (-\infty)\ . \end{aligned}$$

The above quantifier-free equivalent of g indeed evaluates to $-\infty$ if $\sigma \not\models \varphi_{\exists}(D, x)$ and, otherwise, performs a case distinction on said least upper bounds on x.

Finally, consider the quantity

$$g' \;=\; \mathsf{S} x\colon \underbrace{y_1 < z}_{=D'} \searrow \underbrace{2 \cdot x + z}_{=\tilde{a}'}$$

and observe that D' does not impose any bound on x whatsoever. This highlights the need for a careful treatment of ∞ (or, in dual situations, $-\infty$): Since x occurs positively in $\tilde{a}$, we have ${}^{\sigma}[\![g']\!] = \infty$ whenever $\sigma \models y_1 < z$, and ${}^{\sigma}[\![g']\!] = -\infty$ otherwise. We thus have

$$g' \;\equiv\; [y_1 < z] \cdot \infty + [y_1 \geq z] \cdot (-\infty)\ .$$

When considering $\reflectbox{$\mathsf{J}$}x$-quantifiers or when x occurs negatively in $\tilde{a}$, the above described observations need to be dualized, which we detail further below. $\triangle$

We condense the following steps for eliminating the $\reflectbox{$\mathsf{S}$}x$- or $\reflectbox{$\mathsf{J}$}x$-quantifiers:

1. Extract lower and upper bounds on x imposed by the Boolean expression D.
2. Construct the Boolean expression $\varphi_{\exists}(D,x)$ via classical Fourier-Motzkin.
3. Characterize least upper- and greatest lower bounds on x admitted by D.
4. Eliminate the $\reflectbox{$\mathsf{S}$}x$- or $\reflectbox{$\mathsf{J}$}x$-quantifiers by gluing the above concepts together.

We detail these steps in the subsequent paragraphs. Fix $x \in \mathsf{Vars}$, $n \geq 1$, and

$$D \quad = \quad \bigwedge_{i=1}^{n} L_i \ .$$

Extracting Lower and Upper Bounds. Given $\sim \in \{>, \geq, <, \leq\}$, we define

$$\begin{aligned}&\mathsf{Bnd}_{x\sim\cdot}(D)\\ =\ &\begin{cases}\{\tilde{a} \in \mathsf{LinAX}^{\pm\infty} \mid \exists i \in \{1,\ldots,n\}\colon L_i = x \sim \tilde{a}\} & \text{if } \sim \in \{>,<\}\\ \{\tilde{a} \in \mathsf{LinAX}^{\pm\infty} \mid \exists i \in \{1,\ldots,n\}\colon L_i = x \sim \tilde{a}\} \cup \{-\infty\} & \text{if } \sim = \geq\\ \{\tilde{a} \in \mathsf{LinAX}^{\pm\infty} \mid \exists i \in \{1,\ldots,n\}\colon L_i = x \sim \tilde{a}\} \cup \{\infty\} & \text{if } \sim = \leq\end{cases}\end{aligned}$$

and let $\mathsf{UBnd}_x = \mathsf{Bnd}_{x<\cdot}(D) \cup \mathsf{Bnd}_{x\leq\cdot}(D)$ and $\mathsf{LBnd}_x = \mathsf{Bnd}_{x>\cdot}(D) \cup \mathsf{Bnd}_{x\geq\cdot}(D)$. Including ∞ and $-\infty$, respectively, by default will be convenient when characterizing least upper- and greatest lower bounds on x admitted by D: If there is no upper (resp. lower) bound on x whatsoever imposed by D, our construction will automatically default to ∞ (resp. $-\infty$).

Classical Fourier-Motzkin Quantifier Elimination with Infinity. We define

$$\begin{aligned}\varphi_{\exists}(D,x) \quad = \quad &\bigwedge_{\substack{\tilde{a}\in\mathsf{Bnd}_{x\geq\cdot}(D),\\ \tilde{b}\in\mathsf{Bnd}_{x\leq\cdot}(D)}} \tilde{a} \leq \tilde{b} \wedge \bigwedge_{\substack{\tilde{a}\in\mathsf{Bnd}_{x\geq\cdot}(D),\\ \tilde{b}\in\mathsf{Bnd}_{x<\cdot}(D)}} \tilde{a} < \tilde{b} \wedge \bigwedge_{\substack{\tilde{a}\in\mathsf{Bnd}_{x>\cdot}(D),\\ \tilde{b}\in\mathsf{Bnd}_{x\leq\cdot}(D)}} \tilde{a} < \tilde{b}\\ &\wedge \bigwedge_{\substack{\tilde{a}\in\mathsf{Bnd}_{x>\cdot}(D),\\ \tilde{b}\in\mathsf{Bnd}_{x<\cdot}(D)}} \tilde{a} < \tilde{b} \wedge \bigwedge_{\substack{i\in\{1,\ldots,n\},\\ x\notin\mathsf{FV}(L_i)}} L_i\end{aligned}$$

as is standard in Fourier-Motzkin variable elimination. The soundness of this construction generalizes to Boolean expressions involving ∞ or $-\infty$:

Lemma 2 ([25,45]). *For all $\sigma \in \Sigma$, we have*

$$\sigma \models \varphi_{\exists}(D,x) \qquad \textit{iff} \qquad \{q \in \mathbb{Q} \mid \sigma[x \mapsto q] \models D\} \neq \emptyset \ .$$

Characterizing Suprema and Infima. Fix some ordering on the bounds on x, i.e., let $\mathsf{UBnd}_x = \{\tilde{u}_1, \ldots, \tilde{u}_{m_1}\}$ and $\mathsf{LBnd}_x = \{\tilde{\ell}_1, \ldots, \tilde{\ell}_{m_2}\}$. We define:

1. $\varphi_{\mathrm{sup}}(D, x, \tilde{u}_i) = \bigwedge_{k=1}^{i-1} \tilde{u}_i < \tilde{u}_k \wedge \bigwedge_{k=i+1}^{m_1} \tilde{u}_i \leq \tilde{u}_k$
2. $\varphi_{\mathrm{inf}}(D, x, \tilde{\ell}_i) = \bigwedge_{k=1}^{i-1} \tilde{u}_i > \tilde{u}_k \wedge \bigwedge_{k=i+1}^{m_2} \tilde{u}_i \geq \tilde{u}_k$

Intuitively, $\varphi_{\mathrm{sup}}(D, x, \tilde{u}_i)$ evaluates to true under valuation σ if ${}^{\sigma}[\![\tilde{u}_i]\!]$ evaluates to the *least upper bound* on x admitted by D under σ with *minimal* i. The intuition for $\varphi_{\mathrm{inf}}(D, x, \tilde{\ell}_i)$ is analogous. Put more formally:

Theorem 3. *Let $\sigma \in \Sigma$ such that*

$$\{q \in \mathbb{Q} \mid \sigma[x \mapsto q] \models D\} \neq \emptyset .$$

Then all of the following statements hold:

1. *There is exactly one $\tilde{u} \in \mathsf{UBnd}_x$ such that*

$$\sigma \models \varphi_{\mathrm{sup}}(D, x, \tilde{u}) .$$

2. *If $\tilde{u} \in \mathsf{UBnd}_x$ and $\sigma \models \varphi_{\mathrm{sup}}(D, x, \tilde{u})$, then*

$${}^{\sigma}[\![\tilde{u}]\!] = \sup\{q \in \mathbb{Q} \mid \sigma[x \mapsto q] \models D\} .$$

3. *There is exactly one $\tilde{\ell} \in \mathsf{LBnd}_x$ such that*

$$\sigma \models \varphi_{\mathrm{inf}}(D, x, \tilde{\ell}) .$$

4. *If $\tilde{\ell} \in \mathsf{LBnd}_x$ and $\sigma \models \varphi_{\mathrm{inf}}(D, x, \tilde{\ell})$, then*

$${}^{\sigma}[\![\tilde{\ell}]\!] = \inf\{q \in \mathbb{Q} \mid \sigma[x \mapsto q] \models D\} .$$

Proof. See [10, App. A.3].

An immediate consequence of Theorem 3 is that, for every $\sigma \models \varphi_{\exists}(D, x)$,

$$\begin{aligned} {}^{\sigma}[\![\sum_{i=1}^{m_1} [\varphi_{\mathrm{sup}}(D, x, \tilde{u}_i)] \cdot \tilde{u}_i]\!] &= \sup\{q \in \mathbb{Q} \mid \sigma[x \mapsto q] \models D\} \\ {}^{\sigma}[\![\sum_{i=1}^{m_2} \left[\varphi_{\mathrm{inf}}(D, x, \tilde{\ell}_i)\right] \cdot \tilde{\ell}_i]\!] &= \inf\{q \in \mathbb{Q} \mid \sigma[x \mapsto q] \models D\} \end{aligned}$$

It is in this sense that φ_{sup} and φ_{inf} characterize least upper- and greatest lower bounds on x admitted by D.

Eliminating the Quantitative Quantifiers Equipped with the preceding prerequisites, we formalize our construction and prove it sound. Given extended arithmetic expressions $\tilde{a}, \tilde{e} \in \mathsf{LinAX}^{\pm\infty}$ and a variable $x \in \mathsf{Vars}$, we define

$$\tilde{e}(x,\tilde{a}) = \begin{cases} \infty & \text{if } x \text{ occurs positively in } \tilde{e} \text{ and } \tilde{a} = \infty \text{ or} \\ & \quad x \text{ occurs negatively in } \tilde{e} \text{ and } \tilde{a} = -\infty \\ -\infty & \text{if } x \text{ occurs positively in } \tilde{e} \text{ and } \tilde{a} = -\infty \text{ or} \\ & \quad x \text{ occurs negatively in } \tilde{e} \text{ and } \tilde{a} = \infty \\ \tilde{e} & \text{if } x \notin \mathsf{FV}(\tilde{e}) \\ \tilde{e}[x/\tilde{a}] & \text{otherwise}, \end{cases}$$

where in the last case we have $\tilde{a} \in \mathsf{LinAX}$ so $\tilde{e}[x/\tilde{a}]$ is the standard syntactic replacement[9] of x by $\tilde{a}$ in $\tilde{e}$. Our sought-after quantifier-free equivalents are[10]

$$\begin{aligned} \mathsf{QE}(\mathsf{S}x\colon (D \searrow \tilde{e})) &= [\neg\varphi_{\exists}(D,x)] \cdot (-\infty) \\ &+ [\varphi_{\exists}(D,x)] \cdot \begin{cases} \sum_{i=1}^{m_1} [\varphi_{\sup}(D,x,\tilde{u}_i)] \cdot \tilde{e}(x,\tilde{u}_i) & \text{if } x \text{ occurs positively in } \tilde{e} \\ \sum_{i=1}^{m_2} \left[\varphi_{\inf}(D,x,\tilde{\ell}_i)\right] \cdot \tilde{e}(x,\tilde{\ell}_i) & \text{if } x \text{ occurs negatively in } \tilde{e} \\ \tilde{e} & \text{if } x \notin \mathsf{FV}(\tilde{e}) \end{cases} \end{aligned} \tag{2}$$

and

$$\begin{aligned} \mathsf{QE}(\mathsf{L}x\colon (D \nearrow \tilde{e})) &= [\neg\varphi_{\exists}(D,x)] \cdot \infty \\ &+ [\varphi_{\exists}(D,x)] \cdot \begin{cases} \sum_{i=1}^{m_2} \left[\varphi_{\inf}(D,x,\tilde{\ell}_i)\right] \cdot \tilde{e}(x,\tilde{\ell}_i) & \text{if } x \text{ occurs positively in } \tilde{e} \\ \sum_{i=1}^{m_1} [\varphi_{\sup}(D,x,\tilde{u}_i)] \cdot \tilde{e}(x,\tilde{u}_i) & \text{if } x \text{ occurs negatively in } \tilde{e} \\ \tilde{e} & \text{if } x \notin \mathsf{FV}(\tilde{e}) . \end{cases} \end{aligned} \tag{3}$$

These constructions comply with our intuition from Example 6. We apply classical Fourier-Motzkin variable elimination to check whether the respective supremum (infimum) evaluates to $-\infty$ (reps. ∞). If $\varphi_{\exists}(D,x)$ is satisfied, then we inspect $\tilde{e}$ closer, select the right bound $\tilde{u}$ (resp. $\tilde{\ell}$) on x in D via $\varphi_{\sup}$ (resp. $\varphi_{\inf}$), and substitute x in $\tilde{e}$ by $\tilde{u}$ ($\tilde{\ell}$) while obeying the arithmetic laws for the extended reals given in Section 2. The resulting quantities are partitioning and:

Theorem 4. *Let $x \in \mathsf{Vars}$ and $\tilde{e} \in \mathsf{LinAX}^{\pm\infty}$. We have:*

1. $\mathsf{S}x\colon (D \searrow \tilde{e}) \equiv \mathsf{QE}(\mathsf{S}x\colon (D \searrow \tilde{e}))$
2. $\mathsf{L}x\colon (D \nearrow \tilde{e}) \equiv \mathsf{QE}(\mathsf{L}x\colon (D \nearrow \tilde{e}))$

Proof. Fixing some $\sigma \in \Sigma$, we first distinguish the cases $\sigma \models \varphi_{\exists}(D,x)$ and $\sigma \not\models \varphi_{\exists}(D,x)$. For $\sigma \models \varphi_{\exists}(D,x)$, we then distinguish the cases $x \notin \mathsf{FV}(\tilde{e})$ and

[9] provided in [10, App. B.2]

[10] recall that $\mathsf{UBnd}_x = \{\tilde{u}_1, \ldots, \tilde{u}_{m_1}\}$ and $\mathsf{LBnd}_x = \{\tilde{\ell}_1, \ldots, \tilde{\ell}_{m_2}\}$.

Algorithm 1: $\mathsf{Elim}(\cdot)$ — Quantitative Quantifier Elimination

Input: partitioning $f \in \mathsf{LinQuant}$
Output: quantifier-free partitioning $\mathsf{Elim}(f) \in \mathsf{LinQuant}$ with $\mathsf{Elim}(f) \equiv f$
if f *is quantifier-free* **then**
  return f
else if f *is of the form* $Qx\colon g$ **then**
  $g \leftarrow \mathsf{Elim}(g)$;
  transform g into GNF_x;
  `// let` $g = \sum_{i=1}^{n} \left[\bigvee_{j=1}^{m_i} D_{i,j}\right] \cdot \tilde{a}_{i,j}$
  if $Q = \reflectbox{$\mathsf{S}$}$ **then**
    return $\mathsf{MAX}\big(\bigcup_{i=1}^{n} \bigcup_{j=1}^{m_i} \{ \underbrace{\mathsf{QE}(\reflectbox{S} x\colon D_{i,j} \searrow \tilde{a}_{i,j})}_{\text{given by Equation (2) on page 190}} \}\big)$
  else if $Q = \reflectbox{$\mathsf{L}$}$ **then**
    return $\mathsf{MIN}\big(\bigcup_{i=1}^{n} \bigcup_{j=1}^{m_i} \{ \underbrace{\mathsf{QE}(\reflectbox{L} x\colon D_{i,j} \nearrow \tilde{a}_{i,j})}_{\text{given by Equation (3) on page 190}} \}\big)$

$x \in \mathsf{FV}(\tilde{e})$, the latter case being the most interesting. The key insight is that if $\tilde{e}$ is of the form $q_0 + \sum_{y \in \mathsf{Vars}} q_y \cdot y$ and $\sigma \models \varphi_\exists(D, x)$, then

$$
\begin{aligned}
&{}^{\sigma}[\![\reflectbox{S} x\colon (D \searrow \tilde{e})]\!] \\
&= q_0 + \Big(\sum_{y \in \mathsf{Vars} \setminus \{x\}} q_y \cdot \sigma(y)\Big) + q_x \cdot \begin{cases} {}^{\sigma}[\![\sum_{i=1}^{m_1} [\varphi_{\sup}(D, x, \tilde{u}_i)] \cdot \tilde{u}_i]\!] & \text{if } q_x > 0 \\ {}^{\sigma}[\![\sum_{i=1}^{m_2} \big[\varphi_{\inf}(D, x, \tilde{\ell}_i)\big] \cdot \tilde{\ell}_i]\!] & \text{if } q_x < 0 \end{cases}
\end{aligned}
$$

by Theorem 3. See [10, App. A.4] for a detailed proof. □

3.4 Algorithmically Eliminating Quantitative Quantifiers

We summarize our quantifier elimination technique in Algorithm 1, which takes as input a partitioning $f \in \mathsf{LinQuant}$ and computes a quantifier-free equivalent $\mathsf{Elim}(f)$ of f by proceeding in a recursive inner- to outermost fashion. Since f is partitioning and since both MAX and MIN always return partitioning quantities, the transformation of g into GNF_x only involves transforming every Boolean expression into DNF and isolating x in every inequality where x occurs. The soundness of Algorithm 1 is an immediate consequence of our observations from the preceding sections. Moreover, the algorithm terminates because the number of recursive invocations equals the number of quantifiers occurring in f.

In order to upper-bound the space complexity of Algorithm 1, we agree on the following: The size $|\varphi|$ of a Boolean expression φ is the number of (not necessarily distinct) inequalities it contains. The *width* $|f|_\rightarrow$ of $f \in \mathsf{LinQuant}$ is its number of summands, and its *depth* $|f|_\downarrow$ is the maximum of the sizes of the Boolean expressions f contains.

Theorem 5. *Algorithm 1 is sound and terminates. Moreover, for partitioning*[11] $f \in \mathsf{LinQuant}$ *with* $|f|_{\rightarrow} = n$ *and* $|f|_{\downarrow} = m$ *containing exactly one quantifier,*

$$|\mathsf{Elim}(f)|_{\rightarrow} \leq n \cdot 2^m \cdot (m+2)^{n \cdot 2^m} \quad \text{and} \quad |\mathsf{Elim}(f)|_{\downarrow} \leq n \cdot 2^m \cdot \left((m+2/2)^2 + m + 1\right) .$$

Proof. We exploit that (i) transforming a Boolean expression φ of size l into DNF produces at most 2^l disjuncts, each consisting of at most l linear inequalities and (ii) if D is of size l, then $|\varphi_{\exists}(D, x)| \leq (l+2/2)^2$. See [10, App. A.6] for details. □

Fixing m and l as above, the resulting upper bounds for quantities containing k quantifiers are thus non-elementary in k. Investigating *lower* space complexity bounds of Algorithm 1 or the computational complexity of the quantitative quantifier elimination problem is left for future work.

4 Quantitative Craig Interpolation

We now employ our quantifier elimination procedure Elim from Algorithm 1 to derive a quantitative Craig interpolation theorem. Let us first agree on a notion of quantitative Craig interpolants, which is a quantitative analogue of the notion from [32]. Given $f, f' \in \mathsf{LinQuant}$, we say that f (quantitatively) *entails* f', denoted by $f \models f'$, if $\forall \sigma \in \Sigma \colon {}^{\sigma}[\![f]\!] \leq {}^{\sigma}[\![f']\!]$.

Definition 6 (Quantitative Craig Interpolant). *Given* $f, f', g \in \mathsf{LinQuant}$ *with* $f \models f'$*, we say that* g *is a* (quantitative) Craig interpolant *of* (f, f')*, if*

$$f \models g \text{ and } g \models f' \qquad \text{and} \qquad \mathsf{FV}(g) \subseteq \mathsf{FV}(f) \cap \mathsf{FV}(f') . \qquad \triangle$$

In words, g sits between f and f' and the free variables occurring in g also occur free in *both* f and f'. We will now see that piecewise linear quantities enjoy the property of being *quantifier-free interpolating* [32]: For all $f, f' \in \mathsf{LinQuant}$ with $f \models f'$, there exists a quantifier-free Craig interpolant of (f, f'). More precisely, we prove that both the *strongest* and the *weakest* Craig interpolants of (f, f') are *quantifier-free and effectively constructible.* Our construction is inspired by existing techniques for constructing *classical* Craig interpolants via *classical* quantifier elimination [23,24]: By "projecting-out" the free variables in f which are *not* free in f' via $\mathcal{S}$, we obtain the *strongest* Craig interpolant of (f, f'). Dually, by "projecting-out" the free variables in f' which are *not* free in f via $\mathcal{L}$, we obtain the *weakest* Craig interpolant of (f, f'). Put formally:

Theorem 6. *Let* $f, f' \in \mathsf{LinQuant}$ *with* $f \models f'$*. We have:*

1. *For* $\{x_1, \ldots, x_n\} = \mathsf{FV}(f) \setminus \mathsf{FV}(f')$,

$$g \;=\; \mathsf{Elim}(\mathcal{S} x_1 \ldots \mathcal{S} x_n \colon f)$$

is the strongest quantitative Craig interpolant of (f, f')*, i.e.,*

$$\forall \text{ Craig interpolants } g' \text{ of } (f, f') \colon \quad g \models g' .$$

[11] If f needs to be pre-processed to make it partitioning via the construction from Section 3.1, then n is to be substituted by 2^n and m is to be substituted by $n \cdot m$.

2. *For* $\{y_1, \ldots, y_m\} = \mathsf{FV}(f') \setminus \mathsf{FV}(f)$,

$$g = \mathsf{Elim}(\reflectbox{\textsf{L}}\, y_1 \ldots \reflectbox{\textsf{L}}\, y_m \colon f')$$

is the weakest quantitative Craig interpolant of (f, f'), *i.e.,*

$$\forall \text{ Craig interpolants } g' \text{ of } (f, f') \colon \quad g' \models g \ .$$

Proof. See [10, App. A.5].

Example 7. Consider the following quantities f, f' which satisfy $f \models f'$:

$$\begin{aligned} f &= [x \geq 0] \cdot x + [x \geq 0 \wedge y \leq x] \cdot y \\ f' &= [x \geq 0 \wedge z \geq x] \cdot (2 \cdot x + z + 1) + [z < x] \cdot \infty \end{aligned}$$

Pre-processing f and f' to make them partitioning and simplifying yields

$$\underbrace{\mathsf{Elim}(\reflectbox{\textsf{S}}\, y \colon f) = [x \geq 0] \cdot 2 \cdot x}_{\text{strongest Craig interpolant}} \quad \text{and} \quad \underbrace{\mathsf{Elim}(\reflectbox{\textsf{L}}\, z \colon f') = [x \geq 0] \cdot (3 \cdot x + 1)}_{\text{weakest Craig interpolant}} \ .$$

△

5 Conclusion

We have investigated both quantitative quantifier elimination and quantitative Craig interpolation for piecewise linear quantities — an assertion language in automatic quantitative software verification. We have provided a sound and complete quantifier elimination algorithm, proved it sound, and analyzed upper space-complexity bounds. Using our algorithm, we have derived a quantitative Craig interpolation theorem for arbitrary piecewise linear quantities.

We see ample space for future work. First, we could investigate alternative quantifier elimination procedures: Our algorithm can be understood as a quantitative generalization of Fourier-Motzkin variable elimination [45,25]. It would be interesting to apply, e.g., virtual substitution [40] in the quantitative setting and to compare the so-obtained approaches — both empirically and theoretically. We might also benefit from improvements of Fourier-Motzkin variable elimination such as FMPLEX [48] to improve the practical feasability of our approach. Moreover, we have focussed on $\mathbb{Q}$-valued variables. We plan to investigate techniques which apply to integer-valued variables using, e.g., Cooper's method [18] and in how far our results can be generalized to a non-linear setting.

Finally, we plan to investigate potential applications of our techniques:

1. Dillig et al. [22] present a quantifier elimination-based algorithm for generating inductive loop invariants of classical programs abductively. Generalizing this algorithm to the probabilistic setting, where weakest pre*conditions* are replaced by weakest pre*expectations*, might yield a promising application of our quantifier elimination algorithm.

2. We are currently investigating the applicability of McMillan's interpolation and SAT-based model checking algorithm [43] to *probabilistic* program verification. One of the major challenges is to obtain suitable quantitative interpolants and we hope that our results on the existence of interpolants spark the development of suitable techniques.
3. In the light of the above application and Remark 1, we plan to adapt Albarghouthi's and McMillan's technique for computing [1] "simpler" interpolants.

References

1. Albarghouthi, A., McMillan, K.L.: Beautiful interpolants. In: CAV. Lecture Notes in Computer Science, vol. 8044, pp. 313–329. Springer (2013)
2. Alberti, F., Bruttomesso, R., Ghilardi, S., Ranise, S., Sharygina, N.: Lazy abstraction with interpolants for arrays. In: LPAR. Lecture Notes in Computer Science, vol. 7180, pp. 46–61. Springer (2012)
3. Aliprantis, C., Burkinshaw, O.: Principles of Real Analysis. North Holland (1981)
4. Baaz, M., Veith, H.: Quantifier elimination in fuzzy logic. In: CSL. Lecture Notes in Computer Science, vol. 1584, pp. 399–414. Springer (1998)
5. Batz, K., Chen, M., Junges, S., Kaminski, B.L., Katoen, J., Matheja, C.: Probabilistic program verification via inductive synthesis of inductive invariants. In: TACAS (2). Lecture Notes in Computer Science, vol. 13994, pp. 410–429. Springer (2023)
6. Batz, K., Chen, M., Kaminski, B.L., Katoen, J., Matheja, C., Schröer, P.: Latticed k-induction with an application to probabilistic programs. In: CAV (2). Lecture Notes in Computer Science, vol. 12760, pp. 524–549. Springer (2021)
7. Batz, K., Gallus, A., Kaminski, B.L., Katoen, J.P., Winkler, T.: Weighted programming: a programming paradigm for specifying mathematical models. Proc. ACM Program. Lang. **6**(OOPSLA1) (2022)
8. Batz, K., Kaminski, B.L., Katoen, J., Matheja, C.: Relatively complete verification of probabilistic programs: an expressive language for expectation-based reasoning. Proc. ACM Program. Lang. **5**(POPL), 1–30 (2021)
9. Batz, K., Kaminski, B.L., Katoen, J.P., Matheja, C., Noll, T.: Quantitative separation logic: A logic for reasoning about probabilistic pointer programs. Proc. ACM Program. Lang. **3**(POPL), 34:1–34:29 (2019)
10. Batz, K., Katoen, J.P., Orhan, N.: Quantifier elimination and craig interpolation: The quantitative way (technical report) (2025), `https://arxiv.org/abs/2501.15156`
11. Beyer, D., Lee, N., Wendler, P.: Interpolation and sat-based model checking revisited: Adoption to software verification. CoRR **abs/2208.05046** (2022)
12. Britikov, K., Blicha, M., Sharygina, N., Fedyukovich, G.: Reachability analysis for multiloop programs using transition power abstraction. In: FM (1). Lecture Notes in Computer Science, vol. 14933, pp. 558–576. Springer (2024)
13. ten Cate, B., Comer, J.: Craig interpolation for decidable first-order fragments. In: FoSSaCS (2). Lecture Notes in Computer Science, vol. 14575, pp. 137–159. Springer (2024)
14. Chakarov, A., Sankaranarayanan, S.: Probabilistic program analysis with martingales. In: CAV. Lecture Notes in Computer Science, vol. 8044, pp. 511–526. Springer (2013)

15. Chen, M., Wang, J., An, J., Zhan, B., Kapur, D., Zhan, N.: NIL: learning nonlinear interpolants. In: CADE. Lecture Notes in Computer Science, vol. 11716, pp. 178–196. Springer (2019)
16. Collins, G.E.: Quantifier elimination for real closed fields by cylindrical algebraic decomposition. In: Automata Theory and Formal Languages. Lecture Notes in Computer Science, vol. 33, pp. 134–183. Springer (1975)
17. Colón, M., Sankaranarayanan, S., Sipma, H.: Linear invariant generation using non-linear constraint solving. In: CAV. Lecture Notes in Computer Science, vol. 2725, pp. 420–432. Springer (2003)
18. Cooper, D.: Theorem proving in arithmetic without multiplication. Machine Intelligence (1972)
19. Craig, W.: Three uses of the Herbrand-Gentzen theorem in relating model theory and proof theory. Journal of Symbolic Logic **22**(3), 269â€"285 (1957)
20. Dijkstra, E.W.: A Discipline of Programming. Prentice-Hall (1976)
21. Dijkstra, E.W., Scholten, C.S.: Predicate Calculus and Program Semantics. Texts and Monographs in Computer Science, Springer (1990)
22. Dillig, I., Dillig, T., Li, B., McMillan, K.L.: Inductive invariant generation via abductive inference. In: OOPSLA. pp. 443–456. ACM (2013)
23. Esparza, J., Kiefer, S., Schwoon, S.: Abstraction refinement with Craig interpolation and symbolic pushdown systems. In: TACAS. Lecture Notes in Computer Science, vol. 3920, pp. 489–503. Springer (2006)
24. Esparza, J., Kiefer, S., Schwoon, S.: Abstraction refinement with Craig interpolation and symbolic pushdown systems. J. Satisf. Boolean Model. Comput. **5**(1-4), 27–56 (2008)
25. Fourier, J.: Analyse des travaux de l'Académie royale des sciences pendant l'année 1824: rapport lu dans la séance publique de l'Institut le 24 avril 1825. Partie mathématique. Institut (Paris), Institut royal de France (1825)
26. Gan, T., Dai, L., Xia, B., Zhan, N., Kapur, D., Chen, M.: Interpolant synthesis for quadratic polynomial inequalities and combination with EUF. In: IJCAR. Lecture Notes in Computer Science, vol. 9706, pp. 195–212. Springer (2016)
27. Gretz, F., Katoen, J., McIver, A.: Prinsys - on a quest for probabilistic loop invariants. In: QEST. Lecture Notes in Computer Science, vol. 8054, pp. 193–208. Springer (2013)
28. Henzinger, T.A., Jhala, R., Majumdar, R., McMillan, K.L.: Abstractions from proofs. In: POPL. pp. 232–244. ACM (2004)
29. Jhala, R., McMillan, K.L.: A practical and complete approach to predicate refinement. In: TACAS. Lecture Notes in Computer Science, vol. 3920, pp. 459–473. Springer (2006)
30. Kaminski, B.L.: Advanced Weakest Precondition Calculi for Probabilistic Programs. Ph.D. thesis, RWTH Aachen University, Germany (2019)
31. Kapur, D.: Automatically generating loop invariants using quantifier elimination. In: Deduction and Applications (2005)
32. Kapur, D., Majumdar, R., Zarba, C.G.: Interpolation for data structures. p. 105â€"116. SIGSOFT '06/FSE-14, ACM (2006)
33. Katoen, J.P., McIver, A., Meinicke, L., Morgan, C.C.: Linear-invariant generation for probabilistic programs: - automated support for proof-based methods. In: SAS. Lecture Notes in Computer Science, vol. 6337, pp. 390–406. Springer (2010)
34. Knuth, D.E.: Two notes on notation. The American Mathematical Monthly **99**(5), 403–422 (1992)

35. Komuravelli, A., Gurfinkel, A., Chaki, S.: SMT-based model checking for recursive programs. In: CAV. Lecture Notes in Computer Science, vol. 8559, pp. 17–34. Springer (2014)
36. Komuravelli, A., Gurfinkel, A., Chaki, S.: SMT-based model checking for recursive programs. Formal Methods Syst. Des. **48**(3), 175–205 (2016)
37. Kozen, D.: A probabilistic PDL. In: STOC. pp. 291–297. ACM (1983)
38. Kozen, D.: A probabilistic PDL. J. Comput. Syst. Sci. **30**(2), 162–178 (1985)
39. Krishnan, H.G.V., Vizel, Y., Ganesh, V., Gurfinkel, A.: Interpolating strong induction. In: CAV (2). Lecture Notes in Computer Science, vol. 11562, pp. 367–385. Springer (2019)
40. Loos, R., Weispfenning, V.: Applying linear quantifier elimination. Comput. J. **36**(5), 450–462 (1993)
41. Mahdi, A., Fränzle, M.: Generalized Craig interpolation for stochastic satisfiability modulo theory problems. In: RP. Lecture Notes in Computer Science, vol. 8762, pp. 203–215. Springer (2014)
42. McIver, A., Morgan, C.: Abstraction, Refinement and Proof for Probabilistic Systems. Monographs in Computer Science, Springer (2005)
43. McMillan, K.L.: Interpolation and sat-based model checking. In: CAV. Lecture Notes in Computer Science, vol. 2725, pp. 1–13. Springer (2003)
44. McMillan, K.L.: Lazy abstraction with interpolants. In: CAV. Lecture Notes in Computer Science, vol. 4144, pp. 123–136. Springer (2006)
45. Motzkin, T.: Beitraege zur Theorie der linearen Ungleichungen. Universitaet Basel (1936)
46. de Moura, L.M., Bjørner, N.S.: Z3: an efficient SMT solver. In: TACAS. Lecture Notes in Computer Science, vol. 4963, pp. 337–340. Springer (2008)
47. Müller, P.: Building Deductive Program Verifiers - Lecture Notes. Engineering Secure and Dependable Software Systems (2019)
48. Nalbach, J., Promies, V., Ábrahám, E., Kobialka, P.: Fmplex: A novel method for solving linear real arithmetic problems. In: GandALF. EPTCS, vol. 390, pp. 16–32 (2023)
49. Ngo, V.C., Carbonneaux, Q., Hoffmann, J.: Bounded expectations: resource analysis for probabilistic programs. In: PLDI. pp. 496–512. ACM (2018)
50. Sanner, S., Delgado, K.V., de Barros, L.N.: Symbolic dynamic programming for discrete and continuous state MDPs. In: UAI. pp. 643–652. AUAI Press (2011)
51. Schröer, P., Batz, K., Kaminski, B.L., Katoen, J.P., Matheja, C.: A deductive verification infrastructure for probabilistic programs. Proc. ACM Program. Lang. **7**(OOPSLA2), 2052–2082 (2023)
52. Sery, O., Fedyukovich, G., Sharygina, N.: Interpolation-based function summaries in bounded model checking. In: Haifa Verification Conference. Lecture Notes in Computer Science, vol. 7261, pp. 160–175. Springer (2011)
53. Sery, O., Fedyukovich, G., Sharygina, N.: Incremental upgrade checking by means of interpolation-based function summaries. In: FMCAD. pp. 114–121. IEEE (2012)
54. Still, G.: Lectures on parametric optimization: An introduction (2018)
55. Tavana, N., Pourmahdian, M., Khatami, S.A.: The Craig interpolation property in first-order Gödel logic. Fuzzy Sets Syst. **485**, 108958 (2024)
56. Vizel, Y., Gurfinkel, A.: Interpolating property directed reachability. In: CAV. Lecture Notes in Computer Science, vol. 8559, pp. 260–276. Springer (2014)
57. Wu, H., Wang, J., Xia, B., Li, X., Zhan, N., Gan, T.: Nonlinear Craig interpolant generation over unbounded domains by separating semialgebraic sets. In: FM (1). Lecture Notes in Computer Science, vol. 14933, pp. 92–110. Springer (2024)

58. Yang, T., Fu, H., Ke, J., Zhan, N., Wu, S.: Piecewise linear expectation analysis via k-induction for probabilistic programs. CoRR **abs/2403.17567** (2024)
59. Zhang, L., Kaminski, B.L.: Quantitative strongest post: a calculus for reasoning about the flow of quantitative information. Proc. ACM Program. Lang. **6**(OOPSLA1) (2022)

Complete Test Suites for Automata in Monoidal Closed Categories*

Bálint Kocsis(✉) and Jurriaan Rot

Radboud University, Nijmegen, The Netherlands
{balint.kocsis,jurriaan.rot}@ru.nl

Abstract Conformance testing of automata is about checking the equivalence of a known specification and a black-box implementation. An important notion in conformance testing is that of a complete test suite, which guarantees that if an implementation satisfying certain conditions passes all tests, then it is equivalent to the specification.
We introduce a framework for proving completeness of test suites at the general level of automata in monoidal closed categories. Moreover, we provide a generalization of a classical conformance testing technique, the W-method. We demonstrate the applicability of our results by recovering the W-method for deterministic finite automata, Moore machines, and Mealy machines, and by deriving new instances of complete test suites for weighted automata and deterministic nominal automata.

Keywords: Coalgebra · Conformance testing · Monoidal categories

1 Introduction

In *conformance testing* of deterministic automata, the problem is to check the equivalence of a known specification and a black-box implementation [15]. This form of testing is widely used, for instance, in practical implementations of *automata learning*, where one aims to infer an automaton model by systematically interacting with a black-box system [27,28,50]. In particular, conformance testing is a standard technique to evaluate whether a learned model is correct. In Angluin's [4] minimally adequate teacher (MAT) framework for active automata learning, this is referred to as an *equivalence query* — which, when the system under learning really is a black-box, can only be implemented through testing.

In a black-box setting, testing can never be exhaustive [42]: for any (finite) collection of tests, we can construct a faulty implementation that passes all the tests. However, under certain assumptions on the implementation, it is possible to construct test suites which guarantee equivalence. These test suites are called *complete*. In particular, an *m-complete* test suite is complete with respect to all implementations with at most m states. The W-method [19,52] is a classical construction of an m-complete test suite, which has been extended and improved in numerous ways (see [21,35,40] for an overview). State-of-the-art libraries for

* This research is supported by the NWO grant No. VI.Vidi.223.096.

P. A. Abdulla and D. Kesner (Eds.): FoSSaCS 2025, LNCS 15691, pp. 198–219, 2025.
https://doi.org/10.1007/978-3-031-90897-2_10

automata learning [29,43] implement various (randomised) versions of these test suites to evaluate the equivalence query.

Active automata learning has been extended from deterministic finite automata [4] to a wide range of system types, such as Mealy machines [39], nondeterministic automata [12], register automata [1,13,18], nominal automata [41], weighted automata [8,10,25], automata over infinite words [5,38], and probabilistic models [48]. Moreover, automata learning algorithms have been generalized to the abstract level of category theory [9,20,26,49]. However, most variants of complete test suites are only developed for Mealy machines and DFAs. While the abstract frameworks for automata learning rely on equivalence queries, a general theory of evaluating these in practice through conformance testing is missing. Developing such a theory is relevant, as evaluating the equivalence query is often the bottleneck in automata learning [7].

In this paper, we study conformance testing at the generality of *machines in a category*, more specifically in monoidal closed categories. This setting goes back to classical work by Arbib and Manes [6], Goguen [24], and Adámek and Trnková [3]. This setting allows us to conveniently define reachability and language equivalence, by viewing automata both as algebras and as coalgebras.

Our main contribution is to define the W-method and to provide a framework for proving (m)-completeness of test suites at the general level of automata in monoidal closed categories. To this end, we generalize part of the bisimulation-based proof presented in [34] for Mealy machines, which ultimately goes back to [19,52]. In particular, our main result allows us to prove completeness under the assumption that the we can reach all states in the implementation, a property guaranteed by the W-method in concrete cases. We instantiate our approach to recover known complete test suites for deterministic finite automata, Mealy machines, and Moore machines. We go on to instantiate it to weighted automata [22] and deterministic nominal automata [11], providing, to the best of our knowledge, the first account of complete test suites for these systems.

Related work. Our approach is inspired by the categorical automata learning framework of Urbat and Schröder [49], which similarly relies on the possibility of viewing automata as both algebras and coalgebras in the more general setting of adjoint automata. We address conformance testing instead of automata learning.

There is extensive literature on advanced complete test suites for concrete models, in particular for DFAs and Mealy machines (but also for e.g. NFAs [44] and other conformance relations such as ioco [14]). In this paper, we focus on the W-method, which conceptually underlies many of these techniques, as a first step towards a categorical theory of conformance testing.

In [31], a coalgebraic conformance testing theory is developed for Mealy machines with monadic effects. The focus in *op. cit.* is on modelling test purposes and execution at that level, but not on m-completeness.

Outline. In Section 2, we provide background on conformance testing and an overview of our approach. Then, in Section 3, we set the scene for the generalized test suite construction by recalling automata in monoidal closed categories and related concepts such as words and languages. We introduce our framework for proving completeness of test suites in Section 4. In Section 5, we use our results to obtain complete test suites for all the abovementioned classes of automata. Finally, Section 6 summarizes our results and discusses further work.

We assume familiarity with basic category theory [37]. Some notions we rely on are recalled in the appendix [33]. Most proofs are also deferred to the appendix.

Notation. We write gf or $g \circ f$ for the composition of morphisms g and f. We denote the product of X and Y by $X \times Y$. For $f\colon Z \to X$ and $g\colon Z \to Y$, we let $\langle f, g\rangle\colon Z \to X \times Y$ denote the tupling of f and g. The coproduct of X and Y is written as $X+Y$. For $f\colon X \to Z$ and $g\colon Y \to Z$, we denote by $[f,g]\colon X+Y \to Z$ the cotupling of f and g. The coproduct of a family of objects $(X_i)_{i\in I}$ is written as $\sum_{i\in I} X_i$. For a family of morphisms $f_i\colon X_i \to Y$ indexed by $i \in I$, we write $[f_i]_{i\in I}\colon \sum_{i\in I} X_i \to Y$ for the cotupling of the family. Furthermore, we write $\kappa_i\colon X_i \to \sum_{i\in I} X_i$ for the coproduct injections.

2 Overview

In this section, we review some background on conformance testing and the W-method in particular, and we explain our approach to generalizing the test suite construction and the proof of its completeness.

Complete Test Suites. Let us introduce some terminology for conformance testing in the case of deterministic finite automata. For this section, we fix a set Σ of input symbols. A *deterministic finite automaton* (DFA) [46] is a tuple (Q, q_0, δ, F), where Q is a finite set of *states*, $q_0 \in Q$ is an *initial state*, $\delta\colon Q \times \Sigma \to Q$ is a *transition function*, and $F \subseteq Q$ is a set of *final states*. The *size* of a DFA $\mathcal{A} = (Q, q_0, \delta, F)$ is $|\mathcal{A}| = |Q|$. The transition function δ of $\mathcal{A}$ extends to words over Σ as a function $\delta^*\colon Q \times \Sigma^* \to Q$ as usual. We write $q \in \mathcal{A}$ to mean $q \in Q$.

Definition 2.1. *Let $\mathcal{A} = (Q, q_0, \delta, F)$ and $\mathcal{B} = (Q', q_0', \delta', F')$ be two DFAs.*

(i) The accepted language *of $q \in Q$ is $L_{\mathcal{A}}(q) = \{w \in \Sigma^* \mid \delta^*(q, w) \in F\}$. The* accepted language *of $\mathcal{A}$ is $L_{\mathcal{A}} = L_{\mathcal{A}}(q_0)$.*
(ii) Given a set $L \subseteq \Sigma^*$ of words, we say that two states $p \in Q$ and $q \in Q'$ are L-equivalent, *denoted by $p \sim_L q$, if $L_{\mathcal{A}}(p) \cap L = L_{\mathcal{B}}(q) \cap L$. In the case $L = \Sigma^*$, we say that p and q are* equivalent, *denoted by $p \sim q$. We write $\mathcal{A} \sim_L \mathcal{B}$ for $q_0 \sim_L q_0'$. We say that $\mathcal{A}$ and $\mathcal{B}$ are* equivalent, *denoted by $\mathcal{A} \sim \mathcal{B}$, if $q_0 \sim q_0'$.*
(iii) We say that $\mathcal{A}$ is minimal *if for all $p, q \in Q$, $p \sim q$ implies $p = q$.*

We now define the vocabulary for conformance testing. A *test suite* is a finite set $T \subseteq \Sigma^*$ of words. In the following, we use the symbol $\mathcal{S}$ for the specification and the symbol $\mathcal{M}$ for the implementation. We say that $\mathcal{S}$ and $\mathcal{M}$ *agree on* a test suite T if $\mathcal{S} \sim_T \mathcal{M}$.

Since for any test suite, there exists a faulty implementation that agrees on all the tests, we need to restrict the space of possible faulty implementations to achieve completeness. This motivates the definition of a fault domain.

Definition 2.2. *(i) A* fault domain *is a collection of DFAs.*
(ii) Let $\mathcal{S}$ be an automaton. A test suite $T \subseteq \Sigma^*$ is complete *for $\mathcal{S}$ with respect to a fault domain $\mathcal{U}$ if for all DFAs $\mathcal{M} \in \mathcal{U}$, $\mathcal{S} \sim_T \mathcal{M}$ implies $\mathcal{S} \sim \mathcal{M}$.*

Intuitively, a test suite is complete if it contains a failing test case for every inequivalent DFA in the given fault domain.

A fault domain that has often been considered in the literature [19,21,40,52] is the collection $\mathcal{U}_m$ of all DFAs that have at most m states, for some $m \in \mathbb{N}$. Test suites that are complete with respect to this fault domain are called *m-complete.*

The W-method. The method of Vasilevskii and Chow, termed the *W-method,* constructs a test suite based on the specification using sets of words with special properties, called a *state cover* and a *characterization set.* A state cover P is a set of words containing sequences with which we can cover the whole state space, and a characterization set contains words that distinguish inequivalent states.

Definition 2.3. *(i) A* state cover *for a DFA $\mathcal{S} = (Q, q_0, \delta, F)$ is a finite set $P \subseteq \Sigma^*$ of words that contains the empty word and contains, for each state $q \in Q$, an* access sequence *for q, i.e., a word $w \in \Sigma^*$ such that $\delta^*(q_0, w) = q$.*
(ii) A characterization set *for a DFA $\mathcal{S}$ is a finite set $W \subseteq \Sigma^*$ of words such that $\varepsilon \in W$ and for all states $p, q \in \mathcal{S}$, $p \sim_W q$ implies $p \sim q$.*

Intuitively, a characterization set contains a so-called *distinguishing sequence* for any pair of inequivalent states p and q, i.e. a word $w \in \Sigma^*$ such that $w \in L_{\mathcal{S}}(p)$ and $w \notin L_{\mathcal{S}}(q)$ or vice-versa. Note that, by definition, such a distinguishing sequence always exists for inequivalent states.

We are now in the position to define the W-method. In the following definition, the symbol $\cdot$ denotes concatenation of languages.

Definition 2.4. *Let $\mathcal{S}$ be a DFA. Let P be a state cover for $\mathcal{S}$, let W be a characterization set for $\mathcal{S}$, and let k be a natural number. Then the* W test suite *of order k associated to P and W is defined as $T^k_{P,W} = P \cdot \Sigma^{\leq k+1} \cdot W$.*

The state cover P makes sure that we reach all states in the specification. The role of the infixes $\Sigma^{\leq k+1}$ is to reach states in the implementation. Finally, the characterization set W is used to distinguish the states reached in the specification and the implementation after reading a word from $P \cdot \Sigma^{\leq k+1}$.

It is a classical result [19,52] that the W-method of order k is $n+k$-complete, where n is the number of states of the specification. This can be proven in two steps; this proof strategy is also used in [40] and [34]. The first step is to construct a state cover for the implementation from a state cover of the specification.

Lemma 2.5. *Let $\mathcal{S}$ be a DFA with $n = |\mathcal{S}|$, and let $\mathcal{M} \in \mathcal{U}_{n+k}$ for some $k \in \mathbb{N}$. Suppose that $P \subseteq \Sigma^*$ is a state cover for $\mathcal{S}$ and $W \subseteq \Sigma^*$ is a characterization set for $\mathcal{S}$. Suppose furthermore that $\mathcal{S}$ is minimal and $\mathcal{S} \sim_{P \cdot W} \mathcal{M}$. Then $P \cdot \Sigma^{\leq k}$ is a state cover for $\mathcal{M}$.*

Second, we prove equivalence of two DFAs using a state cover for the implementation and the assumption that the two machines agree on suitable tests.

Lemma 2.6. *Let $\mathcal{S}$ and $\mathcal{M}$ be two DFAs, and suppose $C \subseteq \Sigma^*$ is a state cover for $\mathcal{M}$ and $W \subseteq \Sigma^*$ is a characterization set for $\mathcal{S}$. Assume that $\mathcal{S}$ is minimal, and let $T = C \cdot \Sigma^{\leq 1} \cdot W$. Then $\mathcal{S} \sim_T \mathcal{M}$ implies $\mathcal{S} \sim \mathcal{M}$.*

For a proof of the above two lemmas in the case of Mealy machines, we refer the reader to [34]. In both lemmas, we assumed minimality of the specification: this is fine, as we can apply any minimization algorithm to a potential non-minimal specification. Furthermore, the hypotheses produced in active learning algorithms are generally minimal.

Altogether, we obtain $n + k$-completeness of the W-method.

Corollary 2.7. *Let $\mathcal{S}$ be a minimal DFA with $n = |\mathcal{S}|$. Then for all state covers $P \subseteq \Sigma^*$ for $\mathcal{S}$, for all characterization sets $W \subseteq \Sigma^*$ for $\mathcal{S}$, and for all $k \in \mathbb{N}$, the test suite $T^k_{P,W}$ is $n + k$-complete for $\mathcal{S}$.*

Proof. Let $\mathcal{M} \in \mathcal{U}_{n+k}$, and suppose $\mathcal{S} \sim_{T^k_{P,W}} \mathcal{M}$. Since $P \cdot W \subseteq T^k_{P,W}$, we also have $\mathcal{S} \sim_{P \cdot W} \mathcal{M}$. Hence, by Lemma 2.5, $P \cdot \Sigma^{\leq k}$ is a state cover for $\mathcal{M}$. Then, since $T^k_{P,W} = (P \cdot \Sigma^{\leq k}) \cdot \Sigma^{\leq 1} \cdot W$, $\mathcal{S} \sim \mathcal{M}$ follows by Lemma 2.6. □

Let us illustrate the W-method on a toy example. (This example is based on a Mealy machine in [34].) Suppose we have a coffee machine that can dispense coffee or espresso. Coffee costs 1 coin and espresso costs 2 coins. Suppose that the machine breaks whenever we try to order something without enough money, or when we insert more than 2 coins. Then the DFA $\mathcal{S}$ over the alphabet $\Sigma = \{c, e, 1\}$ depicted in Figure 1 accepts precisely those interaction sequences that do not break the coffee machine.

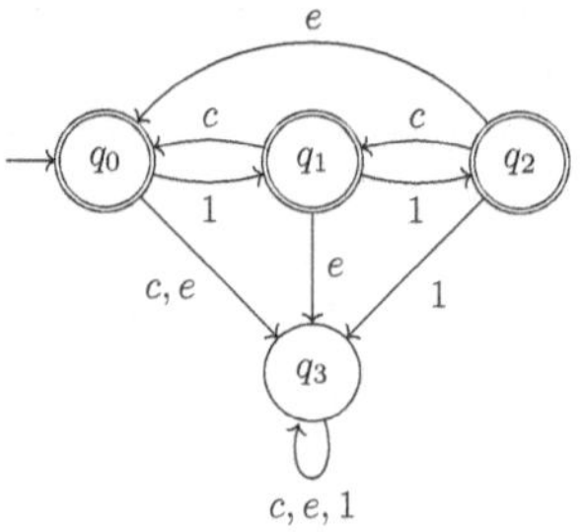

Figure 1. A DFA $\mathcal{S}$ for a coffee machine

By taking the shortest access sequences for all the states, we see that a state cover for $\mathcal{S}$ is given by $P = \{\varepsilon, c, 1, 11\}$. The input sequence c distinguishes q_0 and q_1, as well as q_2 and q_3, and the input sequence 1 distinguishes q_0 and q_2, as well as q_1 and q_2, q_0 and q_3, and q_1 and q_3. Hence, a characterization set for this DFA is $W = \{\varepsilon, c, 1\}$. Thus, setting $k = 0$, the W-method gives the test suite $T^0_{P,W} = P \cdot \Sigma^{\leq 1} \cdot W$.

By Corollary 2.7, $T^0_{P,W}$ is 4-complete for $\mathcal{S}$. For instance, consider the implementation $\mathcal{M}_1$ that is the same as $\mathcal{S}$ except that the c-transition of q_2 goes to q_2. Then the test case $11c1$ (and only this) detects this fault. Similarly, the faulty implementation $\mathcal{M}_2$ that is the same as $\mathcal{S}$ except that the e-transition of q_2 goes to q_1 is rejected by the test case $11ec$ (and only this).

Generalizing the W-method. We wish to generalize the W-method to a categorical setting. The first hurdle towards achieving this goal is giving definitions of a state cover and of a characterization set. This is challenging because the nature of the two sets is fundamentally different: the state cover concerns *reachability* of states, whereas the characterization set is about *equivalence* of states. These two notions are *dual* to each other [6]: the former is described using free algebras, while the latter by cofree algebras.

We observe that the setting of automata in monoidal closed categories (called *machines* by Goguen [24] and *sequential Σ-automata* by Adámek and Trnková [3] in the Cartesian monoidal case) provides an adequate framework for our goals. This setting defines an abstract notion of automaton which can be viewed both as an algebra and as a coalgebra for suitable endofunctors, giving a straightforward way to define reachability and equivalence of states. It covers many kinds of automata, including DFAs, Moore machines, Mealy machines, linear weighted automata, and deterministic nominal automata.

This setting is particularly nice since it allows us to work with concrete descriptions of the initial algebra and of the final coalgebra for the functors describing the algebraic and coalgebraic presentations of automata, respectively. Additionally, these descriptions are direct generalizations of the corresponding and well-known [30,47] concepts in the case of DFAs: the initial algebra arises as the object of words, and the final coalgebra as the object of languages. Crucially, this provides a link between these two objects, making it more or less straightforward to define generalizations of a state cover and of a characterization set.

In this categorical setting, we provide a generalization of Lemma 2.6. We do not generalize Lemma 2.5, as it relies on notions of size that are better handled on a case-by-case basis. As applications, we will be able to derive complete test suites for all the abovementioned classes of of automata.

3 Words, Languages, and Automata

In this section, we define the ingredients necessary to carry out our generalized test suite construction. In particular, we define words, languages, and automata in monoidal closed categories, and operations on these objects. We recall that words form an initial algebra [24] and that languages form a final coalgebra [49] of certain endofunctors. We show how to give definitions in such a way that the resulting algebraic and coalgebraic language semantics for automata coincide.

For the remainder of this paper, we fix a monoidal closed category $\mathcal{C}$ with tensor product $- \otimes -$, tensor unit I, associator α, left unitor λ, and right unitor ϱ. The internal hom of $\mathcal{C}$ is denoted by $[-,-]$. We denote by $\tau_X(-)$ the natural bijection $\mathcal{C}(Z \otimes X, Y) \xrightarrow{\cong} \mathcal{C}(Z, [X, Y])$ arising from the adjunction $- \otimes X \dashv [X, -]$. We denote the counit of same adjunction by $\varepsilon_{X,Y} \colon [X, Y] \otimes X \to Y$. Furthermore, we fix two objects $\Sigma, O \in \mathcal{C}$. We think of Σ as the *alphabet object* and of O as the *output object*. The alphabet object generalizes the set of input symbols for DFAs. In the case of DFAs, the output of a state is simply a Boolean value stating whether the state is accepting or not. The output object generalizes this concept, so that the output of a state can be an arbitrary value. We assume that $\mathcal{C}$ has binary products and countable coproducts. Finally, we assume that for any $X \in \mathcal{C}$, the functor $X \otimes -$ preserves countable coproducts. This assumption is satisfied, for instance, if the functor $X \otimes -$ also has a right adjoint, as is the case in any braided monoidal closed category.

We note that, intuitively, the internal hom $[-,-]$ internalizes the homsets of the category $\mathcal{C}$ as objects of $\mathcal{C}$, in the sense that for any pair of objects X and Y, there is a natural bijection $\mathcal{C}(X,Y) \cong \mathcal{C}(I,[X,Y])$.

3.1 Words

We begin by defining words and operations on words.

Definition 3.1. *(i) For an object $X \in \mathcal{C}$, define its n-fold tensor product by recursion on n: $X^0 = I$ and $X^{n+1} = X^n \otimes X$.*
(ii) For a natural number $k \in \mathbb{N}$, let $X^{\leq k} = \sum_{n \leq k} X^n$.
(iii) Finally, let $X^* = \sum_{n \in N} X^n$, and define $j_k \colon X^{\leq k} \to X^*$ as $[\kappa_n]_{n \leq k}$.

We think of Σ^* as the *object of words* over Σ.

At this point, we note that both functors $- \otimes X$ and $X \otimes -$ preserve countable coproducts (the former by having a right adjoint $[X,-]$, and the latter by assumption). In particular, for any object X, we have isomorphisms

$$[\kappa_n \otimes 1_X]_{n\in\mathbb{N}} \colon \sum_{n \in N} \Sigma^n \otimes X \xrightarrow{\cong} \Sigma^* \otimes X, \quad [1_X \otimes \kappa_n]_{n\in\mathbb{N}} \colon \sum_{n \in N} X \otimes \Sigma^n \xrightarrow{\cong} X \otimes \Sigma^*.$$

Combining these yields the isomorphism

$$[\kappa_m \otimes \kappa_n]_{m,n\in\mathbb{N}} \colon \sum_{m,n\in\mathbb{N}} \Sigma^m \otimes \Sigma^n \xrightarrow{\cong} \Sigma^* \otimes \Sigma^*.$$

We define two operations on words, cons and snoc. Intuitively, cons prepends a letter to the beginning of a word, and snoc appends a letter to the end.

Definition 3.2. *(i) We define a family of morphisms* $\mathrm{cons}_n \colon \Sigma \otimes \Sigma^n \to \Sigma^{n+1}$ *by recursion on n:* $\mathrm{cons}_0 = \lambda_\Sigma^{-1} \circ \varrho_\Sigma$ *and* $\mathrm{cons}_{n+1} = (\mathrm{cons}_n \otimes 1_\Sigma) \circ \alpha^{-1}_{\Sigma,\Sigma^n,\Sigma}$.
(ii) Let $\mathrm{cons} \colon \Sigma \otimes \Sigma^* \to \Sigma^*$ *be defined as the composite*

$$\Sigma \otimes \Sigma^* \xrightarrow{\cong} \sum_{n\in\mathbb{N}} \Sigma \otimes \Sigma^n \xrightarrow{\sum_{n\in\mathbb{N}} \mathrm{cons}_n} \sum_{n\in\mathbb{N}} \Sigma^{n+1} \xrightarrow{[\kappa_{n+1}]_{n\in\mathbb{N}}} \Sigma^*.$$

(iii) The map $\mathrm{snoc} \colon \Sigma^* \otimes \Sigma \to \Sigma^*$ *is defined as the composition of the isomorphism* $\Sigma^* \otimes \Sigma \xrightarrow{\cong} \sum_{n\in\mathbb{N}} \Sigma^n \otimes \Sigma$ *and* $[\kappa_{n+1}]_{n\in\mathbb{N}} \colon \sum_{n\in\mathbb{N}} \Sigma^{n+1} \to \Sigma^*$.

We can also define concatenation of words.

Definition 3.3. *(i) We define a family of maps* $\mu_{m,n} \colon \Sigma^m \otimes \Sigma^n \to \Sigma^{m+n}$ *by recursion on n:* $\mu_{m,0} = \varrho_{\Sigma^m}$ *and* $\mu_{m,n+1} = (\mu_{m,n} \otimes 1_\Sigma) \circ \alpha^{-1}_{\Sigma^m,\Sigma^n,\Sigma}$.

(ii) We define the map $\mu\colon \Sigma^* \otimes \Sigma^* \to \Sigma^*$ *as the composite*

$$\Sigma^* \otimes \Sigma^* \xrightarrow{\cong} \sum_{m,n\in\mathbb{N}} \Sigma^m \otimes \Sigma^n \xrightarrow{\sum_{m,n\in\mathbb{N}} \mu_{m,n}} \sum_{m,n\in\mathbb{N}} \Sigma^{m+n} \xrightarrow{[\kappa_{m+n}]_{m,n\in\mathbb{N}}} \Sigma^*.$$

In the next section, we will use the following concatenation operation on morphisms into Σ^*, which generalizes concatenation of languages.

Definition 3.4. *Let* $a\colon X \to \Sigma^*$, $b\colon Y \to \Sigma^*$ *be two morphisms. The* concatenation *of* a *and* b, *denoted by* $a \cdot b$, *is defined as the composite*

$$X \otimes Y \xrightarrow{a\otimes b} \Sigma^* \otimes \Sigma^* \xrightarrow{\mu} \Sigma^*.$$

By convention, $\cdot$ associates to the left.

We can characterize Σ^* as the initial algebra of a certain endofunctor on $\mathcal{C}$. From now on, we fix the functor $F\colon \mathcal{C} \to \mathcal{C}$ given on objects by $FX = I + X \otimes \Sigma$.

Proposition 3.5. *Let* $\alpha = [\kappa_0, \mathrm{snoc}]\colon I + \Sigma^* \otimes \Sigma \to \Sigma^*$. *Then the algebra* (Σ^*, α) *is an initial* F*-algebra.*

For an F-algebra (A, α_A), we denote the unique F-algebra homomorphism from (Σ^*, α) to (A, α_A) by $r_{\alpha_A}\colon \Sigma^* \to A$.

3.2 Languages

We now turn to languages. In the set-theoretical case, languages can be defined as functions $\Sigma^* \to 2$. Replacing the output set 2 by an arbitrary object yields the following definition.

Definition 3.6. *A* language *is a morphism* $\Sigma^* \to O$.

Internalizing the above notion, we may think of $[\Sigma^*, O]$ as the *object of languages.*

We define two operations on languages. The first one computes the output for the empty word. The second one corresponds to the derivative of languages [16].

Definition 3.7. *(i) Define the morphism* $\mathrm{ev}_0\colon [\Sigma^*, O] \to O$ *as the composite* $\mathrm{ev}_0 = \varepsilon_{\Sigma^*,O} \circ (1_{[\Sigma^*,O]} \otimes \kappa_0) \circ \varrho^{-1}_{[\Sigma^*,O]}$.

(ii) Define the morphism $d\colon [\Sigma^*, O] \to [\Sigma, [\Sigma^*, O]]$ *as* $\tau_\Sigma(\tau_{\Sigma^*}(d'))$, *where* d' *is the composite* $d' = \varepsilon_{\Sigma^*,O} \circ (1_{[\Sigma^*,O]} \otimes \mathrm{cons}) \circ \alpha_{[\Sigma^*,O],\Sigma,\Sigma^*}$.

We can characterize $[\Sigma^*, O]$ as the final coalgebra for a certain endofunctor. From now on, we fix the functor $G\colon \mathcal{C} \to \mathcal{C}$ given on objects by $GX = O \times [\Sigma, X]$.

Proposition 3.8. *Let* $\gamma = \langle \mathrm{ev}_0, d\rangle\colon [\Sigma^*, O] \to O \times [\Sigma, [\Sigma^*, O]]$. *Then the coalgebra* $([\Sigma^*, O], \gamma)$ *is a final* G*-coalgebra.*

For a G-coalgebra (C, γ_C), we denote the unique G-coalgebra homomorphism from (C, γ_C) to $([\Sigma^*, O], \gamma)$ by $l_{\gamma_C}\colon C \to [\Sigma^*, O]$.

3.3 Automata

We conclude this section with a discussion of automata in monoidal closed categories. These are very similar to classical DFAs. The difference is that we replace Cartesian product with tensor product, and we generalize the initial state to an initial state morphism and the set of final states to an output morphism.

Definition 3.9. *An* automaton *over* Σ *in* $\mathcal{C}$ *is a tuple* (Q, i_Q, δ_Q, f_Q), *where* $Q \in \mathcal{C}$ *and* $i_Q : I \to Q$, $\delta_Q : Q \otimes \Sigma \to Q$, $f_Q : Q \to O$ *are morphisms in* $\mathcal{C}$.

We think of Q as an object of *states*, of i_Q as selecting an *initial state*, of δ_Q as a *transition morphism*, and of f_Q as an *output morphism.*

We can view any automaton $\mathcal{A} = (Q, i_Q, \delta_Q, f_Q)$ as an F-algebra $(Q, [i_Q, \delta_Q])$ equipped with an output morphism f_Q. Hence, by the initiality of (Σ^*, α), we get the unique F-algebra homomorphism $r_{[i_Q,\delta_Q]} : \Sigma^* \to Q$. We denote this morphism by $r_{\mathcal{A}}$. Intuitively, the morphism $r_{\mathcal{A}}$ maps a word w to the state of $\mathcal{A}$ reached from the initial state upon reading the input w.

Definition 3.10. *(i) Given an automaton* $\mathcal{A} = (Q, i_Q, \delta_Q, f_Q)$, *the* recognized language *of* $\mathcal{A}$ *is defined as* $L_{\mathcal{A}} = \Sigma^* \xrightarrow{r_{\mathcal{A}}} Q \xrightarrow{f_Q} O$.
(ii) We say that two automata $\mathcal{A}$ *and* $\mathcal{B}$ *are* equivalent, *denoted by* $\mathcal{A} \sim \mathcal{B}$, *if they recognize the same language, i.e.* $L_{\mathcal{A}} = L_{\mathcal{B}}$.

The recognized language of an automaton generalizes the usual concept of accepted language of DFAs: the language assigns to a given word the output of the state reached from the initial state upon reading that word.

We can also view any automaton $\mathcal{A} = (Q, i_Q, \delta_Q, f_Q)$ as a G-coalgebra $(Q, \langle f_Q, \tau_\Sigma(\delta_Q)\rangle)$ equipped with an initial state i_Q. By the finality of $([\Sigma^*, O], \gamma)$, we get the unique G-coalgebra homomorphism $l_{\langle f_Q, \tau_\Sigma(\delta_Q)\rangle} : Q \to [\Sigma^*, O]$. We denote this morphism by $l_{\mathcal{A}}$. Intuitively, the morphism $l_{\mathcal{A}}$ maps a state of $\mathcal{A}$ to the language accepted from that state.

Note that the definition of recognized language relies on the initial algebra morphism $r_{\mathcal{A}}$. Alternatively, it can be defined coalgebraically as the morphism

$$L'_{\mathcal{A}} = I \xrightarrow{i_Q} Q \xrightarrow{l_{\mathcal{A}}} [\Sigma^*, O].$$

The following proposition asserts that these two presentations are equivalent.

Proposition 3.11. *The bijection* $\mathcal{C}(\Sigma^*, O) \cong \mathcal{C}(I, [\Sigma^*, O])$ *sends* $L_{\mathcal{A}}$ *to* $L'_{\mathcal{A}}$.

Note that, by definition, a DFA $\mathcal{A}$ is minimal if and only if the function $q \mapsto L_{\mathcal{A}}(q)$ is injective. As a direct generalization, we define an automaton $\mathcal{A}$ to be *minimal* if $l_{\mathcal{A}}$ is a monomorphism. (Goguen [24] calls such an automaton *reduced*, while Arbib and Manes [6] call such an automaton *observable*.)

4 Test Suites and the Generalized W-method

We now generalize the test suite construction to our categorical setting. We continue to use the symbol $\mathcal{S}$ for the specification and $\mathcal{M}$ for the implementation.

Since in the case of DFAs, test suites, state covers, and characterization sets are sets of words, it would make sense to define the generalizations of these as subobjects of Σ^*. However, it turns out that the theory becomes simpler if we consider arbitrary morphisms instead of monos. In particular, this allows us to bypass the need for factorization systems (cf. Remark 4.14).

We define a *test suite* to be a morphism $t\colon T \to \Sigma^*$. We say that two automata $\mathcal{S}$ and $\mathcal{M}$ *agree on* t, denoted by $\mathcal{S} \sim_t \mathcal{M}$, if $L_\mathcal{S} \circ t = L_\mathcal{M} \circ t$. Note that we removed the condition of finiteness from our generalized test suite definition. Similarly, we will drop the assumption of finiteness in the definitions of the generalizations of state cover and characterization set. We choose to do this in order to make the theory simpler. In practice, all these components will still be finitary in some sense: for instance, a finite-dimensional vector space or an orbit-finite nominal set.

Definition 4.1. *(i) A* fault domain *is a collection of automata (in $\mathcal{C}$).*
(ii) A test suite $t\colon T \to \Sigma^*$ is complete *for an automaton $\mathcal{S}$ with respect to a fault domain $\mathcal{U}$ if for all automata $\mathcal{M} \in \mathcal{U}$, $\mathcal{S} \sim_t \mathcal{M}$ implies $\mathcal{S} \sim \mathcal{M}$.*

We now define the generalization of state covers. Before we do so, we introduce an auxiliary notion.

Definition 4.2. *We say that a morphism $a\colon A \to \Sigma^*$* contains the empty word *if the morphism $\kappa_0\colon I \to \Sigma^*$ factors through a, i.e. there exists a morphism $i_A\colon I \to A$ such that $\kappa_0 = a i_A$.*

Definition 4.3. *Let $\mathcal{S} = (S, i_S, \delta_S, f_S)$ be an automaton. A* state cover *for $\mathcal{S}$ is a morphism $p\colon P \to \Sigma^*$ containing the empty word such that the composite $P \xrightarrow{p} \Sigma^* \xrightarrow{r_S} S$ is a split epi.*

Intuitively, a state cover for $\mathcal{S}$ is a collection of words such that we can effectively assign an access sequence to each state in $\mathcal{S}$.

Unfortunately, being a state cover is a relatively strong condition which cannot be satisfied in some of our examples of interest, such as ordered automata (cf. Remark 4.5). It turns out that in order to prove the generalization of Lemma 2.6, a weaker albeit more complicated condition suffices.

Definition 4.4. *Let $\mathcal{S} = (S, i_S, \delta_S, f_S)$ be an automaton. A* weak state cover *for $\mathcal{S}$ is a morphism $p\colon P \to \Sigma^*$ containing the empty word together with a map $\delta_P\colon P \otimes \Sigma \to P$ such that the following diagram commutes.*

$$\begin{array}{ccccc} P \otimes \Sigma & \xrightarrow{p \otimes 1_\Sigma} & \Sigma^* \otimes \Sigma & \xrightarrow{\text{snoc}} & \Sigma^* \\ {\scriptstyle \delta_P}\downarrow & & & & \downarrow{\scriptstyle r_S} \\ P & \xrightarrow[p]{} & \Sigma^* & \xrightarrow[r_S]{} & S \end{array}$$

Intuitively, a weak state cover assigns to each word w in P and symbol a in Σ a word $v = \delta_P(w, a)$ in P such that we reach the same state upon reading wa and v. Thus, it provides a transition structure on a restricted collection of words $p\colon P \to \Sigma^*$ that follows the transitions of the automaton $\mathcal{S}$.

Remark 4.5. For motivation of why we need the weaker definition of state covers, consider the example of ordered automata [32]. These are automata in the category **Pos** of posets and monotone maps equipped with the Cartesian monoidal structure. In this example, we choose the alphabet Σ to be a discrete finite poset. Hence, the object of words Σ^* is also a discrete poset. This implies that, apart from degenerate cases, a state cover $p\colon P \to \Sigma^*$ cannot exist. For, suppose we have an ordered automaton $\mathcal{S}$ with state space S and a morphism $p\colon P \to \Sigma^*$. For p to be a state cover, we need to find a right inverse $s\colon S \to P$ of $r_\mathcal{S} \circ p$. Such a morphism must map related elements of S to related elements of P. But if two elements of P are related, then they are necessarily sent by p to the same word due to the discrete order on Σ^*. Hence, the composite $r_\mathcal{S} \circ p \circ s$ sends every connected subset of S (viewed as an undirected graph) to the same state. Thus, s is a right inverse only if each connected component of S has size 1, i.e. if S is a discrete poset.

Proposition 4.6. *Let $\mathcal{S}$ be an automaton. Then every state cover for $\mathcal{S}$ is a weak state cover for $\mathcal{S}$.*

Proof. Let $p\colon P \to \Sigma^*$ be a state cover for $\mathcal{S} = (S, i_S, \delta_S, f_S)$. Then p contains the empty word. Since $r_\mathcal{S} \circ p$ is a split epi, it has a right inverse $s\colon S \to P$. Defining δ_P as $s \circ r_\mathcal{S} \circ \mathrm{snoc} \circ (p \otimes 1_\Sigma)$ makes the required diagram commute. □

Characterization sets are generalized as follows.

Definition 4.7. *Let $\mathcal{S} = (S, i_S, \delta_S, f_S)$ be an automaton. A* characterization morphism *for $\mathcal{S}$ is a morphism $w\colon W \to \Sigma^*$ containing the empty word such that for all $f, g\colon X \to S$, if $[w, 1_O] \circ l_\mathcal{S} \circ f = [w, 1_O] \circ l_\mathcal{S} \circ g\colon X \to [W, O]$, then $l_\mathcal{S} \circ f = l_\mathcal{S} \circ g\colon X \to [\Sigma^*, O]$.*

Remark 4.8. For motivation of the definition of characterization morphisms, note that, given an automaton $\mathcal{S} = (S, i_S, \delta_S, f_S)$, the categorical counterpart of the relation $\sim$ for $\mathcal{S}$ from Section 2 is the kernel pair (e_1, e_2) of $l_\mathcal{S}\colon S \to [\Sigma^*, O]$, and that the counterpart of the relation $\sim_W$ for a morphism $w\colon W \to \Sigma^*$ is the kernel pair (e_1^w, e_2^w) of $[w, 1_O] \circ l_\mathcal{S}\colon S \to [W, O]$. Thus, the condition $p \sim_W q \implies p \sim q$ can be expressed by saying that $\langle e_1^w, e_2^w \rangle$ factors through $\langle e_1, e_2 \rangle$. This is equivalent to the definition given above.

We now state our main theorem, which is a generalization of Lemma 2.6.

Theorem 4.9. *Let $\mathcal{S}$ and $\mathcal{M}$ be two automata. Suppose (c, δ_C) is a weak state cover for $\mathcal{M}$ and w is a characterization morphism for $\mathcal{S}$. Assume that $\mathcal{S}$ is minimal and let $t = c \cdot j_1 \cdot w$. Then $\mathcal{S} \sim_t \mathcal{M}$ implies $\mathcal{S} \sim \mathcal{M}$.*

To state completeness of the generalized test suite, we introduce an abstract fault domain, which is in a sense the most general fault domain with respect to which the test suite is complete.

Definition 4.10. *For a morphism* $c\colon C \to \Sigma^*$*, define the fault domain* $\mathcal{U}_c$ *as*

$$\mathcal{U}_c = \{\mathcal{M} \mid \exists \delta_C.\ (c, \delta_C) \text{ is a weak state cover for } \mathcal{M}\}\,.$$

Corollary 4.11. *Let* $\mathcal{S}$ *be a minimal automaton. Then for all* $c\colon C \to \Sigma^*$ *and for all characterization morphisms* $w\colon W \to \Sigma^*$ *for* $\mathcal{S}$*, the test suite* $t = c \cdot j_1 \cdot w$ *is complete for* $\mathcal{S}$ *with respect to* $\mathcal{U}_c$*.*

Proof. Let $\mathcal{M} \in \mathcal{U}_c$, and assume $\mathcal{S} \sim_t \mathcal{M}$. Then there exists a δ_C such that (c, δ_C) is a weak state cover for $\mathcal{M}$. Thus, $\mathcal{S} \sim \mathcal{M}$ follows by Theorem 4.9. □

We now turn to the generalized W-method.

Definition 4.12. *Let* $p\colon P \to \Sigma^*$ *and* $w\colon W \to \Sigma^*$ *be two morphisms, and let* $k \in \mathbb{N}$*. Then the* W test suite *of order* k *associated to* p *and* w *is defined as the morphism* $t^k_{p,w} = p \cdot j_{k+1} \cdot w$*.*

We state completeness of the generalized W-method with respect to a fault domain which is a special case of the one given in Definition 4.10, namely, $\mathcal{U}^k_p = \mathcal{U}_{p \cdot j_k}$. Completeness with respect to this fault domain in the case of Mealy machines has already been considered by Maarse [36]. The related notion of k-A-completeness for Mealy machines has been introduced by Vaandrager, Fiterău-Broştean, and Melse [51].

Corollary 4.13. *Let* $\mathcal{S}$ *be a minimal automaton. Then for all* $p\colon P \to \Sigma^*$*, for all characterization morphisms* $w\colon W \to \Sigma^*$ *for* $\mathcal{S}$*, and for all* $k \in \mathbb{N}$*, the test suite* $t^k_{p,w}$ *is complete for* $\mathcal{S}$ *with respect to* $\mathcal{U}^k_p$*.*

Remark 4.14. In order to simplify the theory, we have opted for defining test suites (as well as state covers and characterization morphisms) as arbitrary morphisms instead of subobjects. However, we note that our theory would also work out if we defined test suites to be $\mathcal{M}$-morphisms for some factorization system $(\mathcal{E}, \mathcal{M})$ such that all morphisms in $\mathcal{E}$ are epimorphisms. To obtain the adjusted test suite, we simply need to take the $(\mathcal{E}, \mathcal{M})$-factorization of the original test suite. Whether we restrict test suites to $\mathcal{M}$-morphisms or not does not influence the completeness of test suites due to the following proposition.

Proposition 4.15. *Let* $(\mathcal{E}, \mathcal{M})$ *be a factorization system on* $\mathcal{C}$ *such that all morphisms in* $\mathcal{E}$ *are epimorphisms. Let* $t\colon T \to \Sigma^*$ *be a test suite and let* $T \xrightarrow{e} T' \xrightarrow{t'} \Sigma^*$ *be its* $(\mathcal{E}, \mathcal{M})$*-factorization. Then for all automata* $\mathcal{S}$ *and* $\mathcal{M}$*,* $\mathcal{S} \sim_t \mathcal{M}$ *if and only if* $\mathcal{S} \sim_{t'} \mathcal{M}$*.*

Proof. Follows immediately from the definition of $\mathcal{S} \sim_t \mathcal{M}$ and the fact that $e \in \mathcal{E}$ is an epimorphism. □

Corollary 4.16. *Let t and t' be defined as in the previous proposition. Then for all automata $\mathcal{S}$ and for all fault domains $\mathcal{U}$, t is complete for $\mathcal{S}$ with respect to $\mathcal{U}$ if and only if t' is complete for $\mathcal{S}$ with respect to $\mathcal{U}$.*

In view of Corollary 4.16, we may replace the original test suite t with the factorized test suite t' when dealing with completeness. We will use this fact when considering instances of the general framework in the next section.

5 Applications

In this section, we use our general framework to derive complete test suites for various kinds of automata, including weighted automata and deterministic nominal automata. Details for the examples can be found in the appendix [33].

5.1 DFAs, Moore Machines, and Mealy Machines

DFAs are covered by our framework: the categorical definitions of Section 3 and Section 4 specialize to the familiar ones in Section 2 if we take as $\mathcal{C}$ the category **Set** of sets and functions with the cartesian product as monoidal structure and if we set $O = 2$. A difference is that, as mentioned in Section 4, the specialized notions of test suite, state cover, and characterization morphism are functions into Σ^* rather than mere subsets. We recover the original definitions by taking the images of the functions (cf. Remark 4.14). We also recover the W-method and Lemma 2.6.

Moore machines are essentially the same as DFAs, except that the set of final states $F \subseteq Q$ is generalized to an output function $f\colon Q \to O$ for some output set O. In the abstract framework, this amounts to changing the output object from 2 to O. Also in this case, we recover the W-method and an analogue of Lemma 2.6. As an example, take the Moore machine $\mathcal{S}$ with states and transitions identical to those of the DFA in Figure 1, and with the output function $f\colon Q \to \mathbb{N} \cup \{-1\}$ that assigns to each state the amount of coins stored in the coffee machine in that state: $f(q_0) = 0$, $f(q_1) = 1$, $f(q_2) = 2$, and $f(q_3) = -1$, where -1 is used as a value indicating an error. Then $P = \{\varepsilon, c, 1, 11\}$ is still a state cover for $\mathcal{S}$, $W = \{\varepsilon, c, 1\}$ is still a characterization set for $\mathcal{S}$, and the test suite $T^0_{P,W}$ is still 4-complete for $\mathcal{S}$. A difference in this example is that the input sequence 1 also distinguishes q_0 and q_1. Hence, we could also use the alternative and smaller characterization set $W' = \{\varepsilon, 1\}$. This results in the smaller 4-complete test suite $T^0_{P,W'} = P \cdot \Sigma^{\leq 1} \cdot W'$.

Mealy machines are similar to Moore machines, but they assign an output to each transition instead of each state. Formally, this amounts to replacing the output morphism $f\colon Q \to O$ by $\lambda\colon Q \times \Sigma \to O$. Mealy machines are covered by our framework by considering the category **Set** of sets and functions with the cartesian monoidal structure and by taking the output object to be the set O^Σ. In particular, we recover [34, Lemma 3.8].

We mention a subtle difference between our presentation of the W-method and the usual one found in the literature. Traditionally, the output function of a Mealy machine is extended to words as a function $\lambda^*\colon Q \times \Sigma^* \to O^*$ that records all the outputs encountered during a run of the machine. In contrast, the function $L_{\mathcal{A}}\colon \Sigma^* \to O^{\Sigma}$ for a Mealy machine $\mathcal{A}$ maps a word w and an input symbol a to the output of the *last* transition during the run of the input word wa. Hence, the assumption $\mathcal{S} \sim_T \mathcal{M}$ only guarantees that the last transitions of $\mathcal{S}$ and $\mathcal{M}$ match on input words from T. We can recover the stronger guarantee that $\lambda^*(q_0, w) = \lambda^*(q_0', w)$ for all $w \in T$ by making sure that the test suite is prefix-closed (i.e. any prefix of any word in T is also in T).

5.2 Weighted Automata

Weighted automata [22] are generalizations of DFAs where a weight is assigned to each transition, expressing e.g. the cost or reliability of its execution. In this subsection, we fix a field $\mathbb{K}$ and a finite set Σ of input symbols.

Definition 5.1. *A* weighted automaton *(WA) is a tuple* (Q, s_0, δ, f)*, where* Q *is a finite set of* states, $s_0\colon Q \to \mathbb{K}$ *is an* input weight function, $\delta\colon Q \times \Sigma \times Q \to \mathbb{K}$ *is a* transition weight function, *and* $f\colon Q \to \mathbb{K}$ *is an* output weight function.

The transition function δ of a WA (Q, s_0, δ, f) can be conveniently represented as a Σ-indexed family of $\mathbb{K}$-valued matrices $M_a^{\delta} \in \mathbb{K}^{Q \times Q}$ $(a \in \Sigma)$ defined as $M_a^{\delta}(p, q) = \delta(q, a, p)$. Note that the column index specifies the source state and the row index the target state. Similarly, the input and output weight functions s_0 and f can be thought of as $\mathbb{K}$-valued column vectors of cardinality Q.

WAs are covered by our framework: letting Σ' be the free vector space over Σ, WAs correspond to automata over Σ' in the category $\mathbf{Vect}_{\mathbb{K}}$ of vector spaces over $\mathbb{K}$ and linear maps equipped with the tensor product of vector spaces as monoidal structure, taking the output object $O = \mathbb{K}$. The corresponding automaton in $\mathbf{Vect}_{\mathbb{K}}$ has state space $\mathbb{K}^Q$. The correspondence is spelled out in the appendix [33]. In the remainder of the section, we identify WAs and automata in $\mathbf{Vect}_{\mathbb{K}}$ via this correspondence.

We recall that WAs recognize *weighted languages*, i.e. functions $\Sigma^* \to \mathbb{K}$. For a WA $\mathcal{A} = (Q, s_0, \delta, f)$ and a word $w \in \Sigma^*$, define the matrix M_w^{δ} by recursion on w: $M_{\varepsilon}^{\delta} = I$ (where I denotes the unit matrix) and $M_{wa}^{\delta} = M_a^{\delta} M_w^{\delta}$. The *recognized language* $L_{\mathcal{A}}(s)$ of an input weight vector $s \in \mathbb{K}^Q$ is then defined as $L_{\mathcal{A}}(s)(w) = f^t M_w^{\delta} s$, where f^t denotes the transpose of f. The *recognized language* of $\mathcal{A}$ is $L_{\mathcal{A}} = L_{\mathcal{A}}(s_0)$. We note furthermore that the function $s \mapsto L_{\mathcal{A}}(s)$ is a linear map $\mathbb{K}^Q \to \mathbb{K}^{\Sigma^*}$, which follows directly from the definition of $L_{\mathcal{A}}(s)$.

As an example, consider the WA $\mathcal{S}$ over the alphabet $\{a, b\}$ and field $\mathbb{R}$ depicted in Figure 2. (This example is taken from [22].) The input, output, and transition weights not shown are equal to 0. This automaton recognizes the weighted language $L\colon \Sigma^* \to \mathbb{R}$ that maps a word w to the decimal value of w if understood as a binary number with a standing for the digit 0 and b for the digit 1.

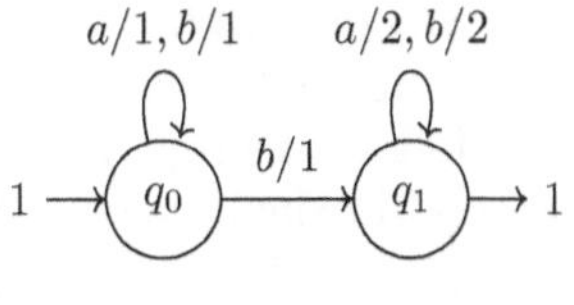

Figure 2. A WA $\mathcal{S}$ for computing the decimal value of a binary number

Let us instantiate the abstract framework to the case of WAs. We start by noting that if the vector spaces V and W have generating sets B and C, respectively, then $V \otimes W$ is generated by $\{a \otimes b \mid a \in A, b \in B\} \cong B \times C$. Furthermore, if V_i is a vector space with generating set B_i for all $i \in I$, then the direct sum $\bigoplus_{i \in I} V_i$ is generated by $\bigcup_{i \in I} \{\kappa_i(b) \mid b \in B_i\} \cong \coprod_{i \in I} B_i$ (where $\coprod$ denotes the disjoint union of sets). Hence, the vector space $(\Sigma')^n$ (where Σ' is the free vector space over Σ) is generated by the set Σ^n, $(\Sigma')^{\leq k}$ by $\Sigma^{\leq k}$, and $(\Sigma')^*$ by Σ^*. Furthermore, the map $j_k\colon (\Sigma')^{\leq k} \to (\Sigma')^*$ is the inclusion of the subspace $(\Sigma')^{\leq k}$. Since a linear map $(\Sigma')^* \to \mathbb{K}$ is completely determined by its values on the basis vectors, the categorical notion of language is in bijective correspondence with weighted languages. The following proposition states how various notions from Section 3 specialize to the case of WAs.

Proposition 5.2. *Let $\mathcal{A} = (Q, s_0, \delta, f)$ be a WA.*

(i) The morphism $\mu\colon (\Sigma')^* \otimes (\Sigma')^* \to (\Sigma')^*$ corresponds to the bilinear map $(\Sigma')^* \times (\Sigma')^* \to (\Sigma')^*$ defined on basis elements by $(u, v) \mapsto uv$.
(ii) The morphism $r_{\mathcal{A}}\colon (\Sigma')^* \to \mathbb{K}^Q$ is the linear extension of the map $\Sigma^* \to \mathbb{K}^Q$ sending w to $M_w^\delta s_0$.
(iii) The recognized language of $\mathcal{A}$ in the categorical sense is the linear extension of the recognized language of $\mathcal{A}$.
*(iv) The morphism $l_{\mathcal{A}}\colon \mathbb{K}^Q \to$ **Vect**$_{\mathbb{K}}((\Sigma')^*, \mathbb{K})$ sends an input weight vector $s \in \mathbb{K}^Q$ to the linear extension of the weighted language $L_{\mathcal{A}}(s)$.*
(v) The automaton $\mathcal{A}$ is minimal iff for all $s \in \mathbb{K}^Q$, $L_{\mathcal{A}}(s) = 0$ implies $s = 0$.

We now turn to the discussion of test suites. For the remainder of this section, we fix a specification WA $\mathcal{S} = (Q, s_0, \delta, f)$. Test suites, state covers, and characterization morphisms specialize to linear maps with codomain $(\Sigma')^*$. We focus on the case where these morphisms are actually subspace inclusions. Hence, we say that a subspace *is* a test suite, and we may refer to a certain subspace as a state cover or as a characterization *space*, the meaning being that the corresponding subspace inclusion is a state cover or a characterization morphism. We concentrate on those subspaces that are generated by subsets of Σ^*.

Proposition 5.3. *Let $\mathcal{M}$ be a WA, let $T \subseteq \Sigma^*$, and let $t\colon \mathrm{span}(T) \to (\Sigma')^*$ be the subspace inclusion. Then $\mathcal{S} \sim_t \mathcal{M}$ if and only if $L_{\mathcal{S}}|_T = L_{\mathcal{M}}|_T$.*

Thus, we obtain a notion of completeness of test suites for WAs: a test suite $T \subseteq \Sigma^*$ is complete for $\mathcal{S}$ with respect to a fault domain $\mathcal{U}$ if and only if $L_{\mathcal{S}}|_T = L_{\mathcal{M}}|_T$ implies $L_{\mathcal{S}} = L_{\mathcal{M}}$ for all $\mathcal{M} \in \mathcal{U}$. This notion is completely analogous to that for DFAs.

The following proposition characterizes the key components of the generalized W-method in the case of WAs in terms of subsets of Σ^*.

Proposition 5.4. *(i) Let $P \subseteq \Sigma^*$. Then $\mathrm{span}(P)$ is a state cover for $\mathcal{S}$ if and only if $\varepsilon \in P$ and $\mathbb{K}^Q$ is generated by $\{M_w^\delta s_0 \mid w \in P\}$.*
(ii) Let $W \subseteq \Sigma^*$. Then $\mathrm{span}(W)$ is a characterization space for $\mathcal{S}$ if and only if $\varepsilon \in W$ and for all $s \in \mathbb{K}^Q$, $L_{\mathcal{S}}(s)|_W = 0$ implies $L_{\mathcal{S}}(s) = 0$.

Motivated by the previous proposition, if $P \subseteq \Sigma^*$ is such that $\mathrm{span}(P)$ is a state cover for $\mathcal{S}$, then we also call P itself a state cover. Similarly, if $\mathrm{span}(W)$ is a characterization space, we call W a characterization *set*. Notice the similarity of the characterizations with the corresponding definitions in Section 2 for DFAs.

We are now ready to derive a specialized W-method for WAs. It turns out that, under the identification of subspaces of $(\Sigma')^*$ and subsets of Σ^*, it coincides with the classical W-method (Definition 2.4). Furthermore, it is characterized as the image of the generalized W test suite from Definition 4.12.

Lemma 5.5. *Let $P, W \subseteq \Sigma^*$, let $k \in \mathbb{N}$, and let $p\colon \mathrm{span}(P) \to (\Sigma')^*$ and $w\colon \mathrm{span}(W) \to (\Sigma')^*$ denote the inclusions. Then $\mathrm{span}(T^k_{P,W}) = \mathrm{im}(t^k_{p,w})$.*

To state completeness of the specialized W-method, we define an appropriate fault domain, which is a restriction of the fault domain $\mathcal{U}^k_p$.

Definition 5.6. *For a set $P \subseteq \Sigma^*$ and natural number k, define*

$$\mathcal{U}^k_P = \{\mathcal{M} \mid P \cdot \Sigma^{\leq k} \textit{ is a state cover for } \mathcal{M}\}\,.$$

The following theorem asserts completeness of the W-method for WAs.

Theorem 5.7. *Suppose $\mathcal{S}$ is minimal. Then for all $P \subseteq \Sigma^*$, for all characterization sets $W \subseteq \Sigma^*$ for $\mathcal{S}$, and for all $k \in \mathbb{N}$, the test suite $T^k_{P,W}$ (Definition 2.4) is complete for $\mathcal{S}$ with respect to $\mathcal{U}^k_P$.*

Continuing our previous example, we derive a complete test suite for the WA $\mathcal{S}$ of Figure 2. We have $M^\delta_\varepsilon s_0 = M^\delta_\varepsilon q_0 = q_0$ and $M^\delta_b s_0 - M^\delta_\varepsilon s_0 = M^\delta_b q_0 - M^\delta_\varepsilon q_0 = (q_0 + q_1) - q_0 = q_1$. Since $\{q_0, q_1\}$ is a basis for $\mathbb{K}^Q$, by Proposition 5.4 (i), $P = \{\varepsilon, b\}$ is a state cover for $\mathcal{S}$.

Next, we show that $W = \{\varepsilon, b\}$ is a characterization set for $\mathcal{S}$. For this, we recall that $L_\mathcal{S}(q_0) = L_\mathcal{S}$ maps a word w to the decimal value of w understood as a binary number with a standing for 0 and b for 1. Furthermore, $L_\mathcal{S}(q_1)$ maps a word w to $2^{|w|}$. Thus, if $s = k_0 q_0 + k_1 q_1$ such that $L_\mathcal{S}(s)|_W = 0$, then $L_\mathcal{S}(s)(\varepsilon) = k_0 L_\mathcal{S}(q_0)(\varepsilon) + k_1 L_\mathcal{S}(q_1)(\varepsilon) = k_1 = 0$. Moreover, we have $L_\mathcal{S}(s)(b) = k_0 L_\mathcal{S}(q_0)(b) + k_1 L_\mathcal{S}(q_1)(b) = k_0 + 2k_1 = k_0 = 0$. Thus, $L_\mathcal{S}(s)|_W = 0$ implies $s = 0$. Hence, by Proposition 5.4 (ii), $W = \{\varepsilon, b\}$ is a characterization set for $\mathcal{S}$.

By Theorem 5.7, we conclude that $T^1_{P,W} = P \cdot \Sigma^{\leq 2} \cdot W$ is a complete test suite for $\mathcal{S}$ with respect to $\mathcal{U}^1_P$. See the appendix [33] for an example of how a possible faulty implementation is rejected.

5.3 Nominal Automata

Nominal automata [11] are a model of computation over potentially infinite alphabets. They are based on the notion of *nominal sets*, which are essentially sets with certain symmetries. The idea is that we demand that the transition function of a nominal automaton respect the symmetries of the state space, so that we can represent it via finite manners.

In this subsection, we fix a countable set $\mathbb{A}$ of *atoms*. We denote by $\mathrm{Perm}(\mathbb{A})$ the group of *finite permutations* over $\mathbb{A}$ (i.e. bijections $\pi\colon \mathbb{A} \to \mathbb{A}$ such that $\pi(a) = a$ for all but finitely many $a \in \mathbb{A}$). For a $\mathrm{Perm}(\mathbb{A})$-set X, we denote the action of π on an element $x \in X$ by $\pi \cdot x$. We recall that an *equivariant function* $f\colon X \to Y$ between $\mathrm{Perm}(\mathbb{A})$-sets satisfies $f(\pi \cdot x) = \pi \cdot f(x)$, an *equivariant subset* $A \subseteq X$ is such that $x \in A \implies \pi \cdot x \in A$, and an *equivariant element* $a \in X$ is such that $\{a\}$ is an equivariant subset. For background on nominal sets in general, we refer to [45]. Finally, we fix an orbit-finite nominal set Σ acting as the alphabet.

Definition 5.8. *A* deterministic nominal automaton *(or DNA for short) is a tuple (Q, q_0, δ, F), where Q is an orbit-finite nominal set of* states, *$q_0 \in Q$ is an equivariant* initial state, *$\delta\colon Q \times \Sigma \to Q$ is an equivariant* transition map, *and $F \subseteq Q$ is an equivariant subset of* final states.

For DNA $\mathcal{A} = (Q, q_0, \delta, F)$, we write $q \in \mathcal{A}$ for $q \in Q$.

DNAs are covered by our framework, since they correspond precisely to automata over Σ in the category **Nom** of nominal sets and equivariant maps with the cartesian product as monoidal structure, taking the output object $O = 2$, the discrete nominal set with 2 elements. To see this, note that, given a nominal set Q, an equivariant element $q_0 \in Q$ corresponds to an equivariant map $1 \to Q$ (where 1 is the discrete nominal set with 1 element), and that an equivariant subset $F \subseteq Q$ corresponds to an equivariant map $Q \to 2$. We identify DNAs with automata in **Nom** via this correspondence.

DNAs recognize *nominal languages*: these are equivariant subsets of Σ^*, the nominal set of words over Σ with the pointwise group action. The transition map δ of a nominal automaton $\mathcal{A} = (Q, q_0, \delta, F)$ can be extended to an equivariant map $\delta^*\colon Q \times \Sigma^* \to Q$ just as in the case of DFAs. The *accepted language* $L_{\mathcal{A}}(q)$ of a state $q \in Q$ is also defined analogously to DFAs as the set $L_{\mathcal{A}}(q) = \{w \in \Sigma^* \mid \delta^*(q, w) \in F\}$, and the accepted language $L_{\mathcal{A}}$ of $\mathcal{A}$ is $L_{\mathcal{A}}(q_0)$.

As an example, consider the DNA $\mathcal{S}$ over the alphabet $\mathbb{A}$ of atoms (regarded as a nominal set with group action $\pi \cdot a = \pi(a)$) depicted in Figure 3. (This example is taken from [40].) The automaton has state space $Q = \{q_0, q_2, q_3\} \cup \{q_{1,a} \mid a \in \mathbb{A}\}$; the infinite amount of states $\{q_{1,a} \mid a \in \mathbb{A}\}$ is depicted with one symbolic state $q_{1,a}$. The transitions $q_0 \xrightarrow{a} q_{1,a}$ and $q_{1,a} \xrightarrow{a} q_2$ are to be understood as one transition for each $a \in \mathbb{A}$. Furthermore, transitions of the form $p \xrightarrow{A} q$ for $A \subseteq \mathbb{A}$ stand for a number of transitions, one for each $a \in A$. The automaton $\mathcal{S}$ accepts the nominal language $L = \{aa \mid a \in \mathbb{A}\}$.

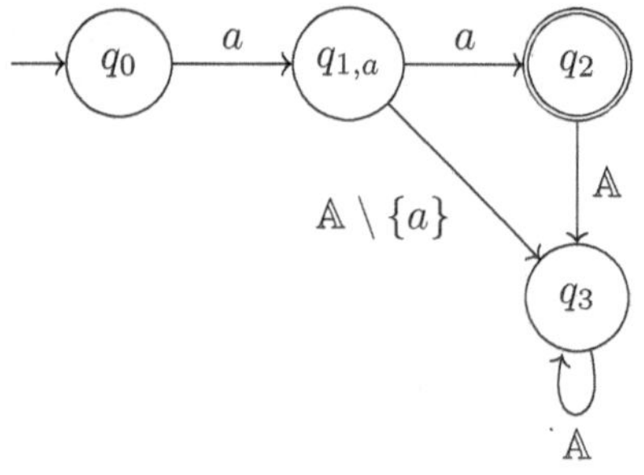

Figure 3. An example DNA $\mathcal{S}$ accepting the language $L = \{aa \mid a \in \mathbb{A}\}$

We now instantiate the abstract framework to DNAs. We first note that the categorical construction of the nominal set Σ^* coincides with the description given above. Thus, languages $\Sigma^* \to 2$ in the categorical sense correspond precisely to nominal languages. Furthermore, the map $j_k \colon \Sigma^{\leq k} \to \Sigma^*$ is simply the inclusion of a subset. The map $\mu \colon \Sigma^* \times \Sigma^* \to \Sigma^*$ is given by concatenation of words. For any DNA $\mathcal{A}$, the morphism $r_{\mathcal{A}}$ maps a word $w \in \Sigma^*$ to $\delta^*(q_0, w)$. Thus, the categorical notion of recognized language corresponds to the accepted language of $\mathcal{A}$ as defined above. Moreover, the morphism $l_{\mathcal{A}}$ maps a state $q \in \mathcal{A}$ to $L_{\mathcal{A}}(q)$. Finally, an automaton $\mathcal{A}$ is minimal (in the categorical sense) if for all $p, q \in \mathcal{A}$, $L_{\mathcal{A}}(p) = L_{\mathcal{A}}(q)$ implies $p = q$.

For the remainder of this section, we fix a specification DNA $\mathcal{S} = (Q, q_0, \delta, F)$. Similarly to the case of WAs, we focus on the case where test suites, state covers, and characterization morphisms are actually subset inclusions. A subset inclusion is equivariant if and only if the subset is equivariant. Thus, we say that an equivariant subset of Σ^* *is* a test suite, state cover, or characterization *set*. If $T \subseteq \Sigma^*$ is a test suite, and $t \colon T \to \Sigma^*$ is the corresponding inclusion, then $\mathcal{S} \sim_t \mathcal{M}$ if and only if $L_{\mathcal{S}} \cap T = L_{\mathcal{M}} \cap T$ for any DNA $\mathcal{M}$. Hence, the notion of completeness of test suites for DNAs coincides with that for DFAs. A set $W \subseteq \Sigma^*$ is a characterization set for $\mathcal{S}$ if and only if $\varepsilon \in W$ and for all $p, q \in Q$, $L_{\mathcal{S}}(p) \cap W = L_{\mathcal{S}}(q) \cap W$ implies $L_{\mathcal{S}}(p) = L_{\mathcal{S}}(q)$. Thus, characterization sets are exactly the same as for DFAs.

The specialized W-method for DNAs coincides with the W-method for DFAs. Furthermore, just as for WAs, it is characterized as the image of the generalized W test suite. Namely, if $P, W \subseteq \Sigma^*$, $p \colon P \to \Sigma^*$ and $w \colon W \to \Sigma^*$ are the subset inclusions, and $k \in \mathbb{N}$, then $T^k_{P,W} = \mathrm{im}(t^k_{p,w})$.

The following theorem asserts the completeness of the W-method for DNAs, and it is an immediate consequence of Corollary 4.13 and Corollary 4.16.

Theorem 5.9. *Suppose $\mathcal{S}$ is minimal. Then for all $P \subseteq \Sigma^*$, for all characterization sets $W \subseteq \Sigma^*$ for $\mathcal{S}$, and for all $k \in \mathbb{N}$, the test suite $T^k_{P,W}$ is complete for $\mathcal{S}$ with respect to $\mathcal{U}^k_p$ (where $p \colon P \to \Sigma^*$ is the inclusion).*

Continuing the previous example, we derive a complete test suite for the DNA $\mathcal{S}$. We first show that $P = \{\varepsilon\} \cup \{a \mid a \in \mathbb{A}\} \cup \{aa \mid a \in \mathbb{A}\} \cup \{aab \mid a, b \in \mathbb{A}\}$ is a weak state cover for $\mathcal{S}$. Define the map $\delta_P \colon P \times \Sigma \to P$ as

$$\delta_P(w, c) = \begin{cases} c & \text{if } w = \varepsilon \\ aa & \text{if } w = a \text{ for } a \in \mathbb{A} \text{ and } a = c \\ aac & \text{if } w = a \text{ for } a \in \mathbb{A} \text{ and } a \neq c \,. \\ aac & \text{if } w = aa \text{ for } a \in \mathbb{A} \\ aab & \text{if } w = aab \text{ for } a, b \in \mathbb{A} \end{cases}$$

It can be checked that δ_P is equivariant. Furthermore, we have $\delta^*(q_0, \delta_P(w, c)) = \delta^*(q_0, wc)$, as can readily be verified.

Next, we note that $W = \{\varepsilon\} \cup \{a \mid a \in \mathbb{A}\} \cup \{aa \mid a \in \mathbb{A}\}$ is a characterization set for $\mathcal{S}$. Hence, we obtain the complete test suite $T^0_{P,W} = P \cdot \mathbb{A}^{\leq 1} \cdot W$. We remark that $T^0_{P,W}$ is an infinite but orbit-finite subset of Σ^*.

6 Conclusion and Future Work

We introduced a general framework for studying the W-method and proving completeness of test suites, at the categorical level of automata in monoidal closed categories. Besides recovering basic instances such as DFAs, Mealy machines, and Moore machines, our framework instantiates to complete test suites for the new cases of weighted automata and deterministic nominal automata.

There are several open questions and avenues for future work. We have focused on the classical W-method, but it would be useful to also cover refinements thereof such as Wp [23], where distinguishing sequences depend on the state reached, or Hybrid ADS [40]. The latter requires an understanding of adaptive distinguishing sequences at an abstract level, which is of interest on its own. A further direction is to include in our framework a notion of *finiteness* of test suites using, e.g., finitely presentable objects [2]. Moreover, a key aspect of the W-method is that the size of the state cover and characterization set are polynomial in the size of the specification, and it would be useful to capture this at a more general level.

A possible direction for future work is to move from weighted automata over a field to weighted automata over a *semiring*, and complement for instance the learning algorithms for weighted automata over the integers proposed in [17,25]. Instantiating to the Boolean semiring could then connect to more concrete work on complete test suites for non-deterministic automata [44].

Our results assume a state cover and a characterization set as input. We have not touched upon the topic on how to acquire these sets from the specification. It would be of practical relevance to develop algorithms for computing state covers and characterization sets for weighted automata and for deterministic nominal automata.

Disclosure of Interests. The authors have no competing interests to declare that are relevant to the content of this article.

References

1. Aarts, F., Heidarian, F., Kuppens, H., Olsen, P., Vaandrager, F.W.: Automata learning through counterexample guided abstraction refinement. In: FM. Lecture Notes in Computer Science, vol. 7436, pp. 10–27. Springer (2012)
2. Adámek, J., Rosický, J.: Locally presentable and accessible categories, London Mathematical Society Lecture Note Series, vol. 189. Cambridge University Press (1994)
3. Adámek, J., Trnková, V.: Automata and algebras in categories, Mathematics and Its Applications, vol. 37. Springer Netherlands (1990)
4. Angluin, D.: Learning regular sets from queries and counterexamples. Inf. Comput. **75**(2), 87–106 (1987)
5. Angluin, D., Antonopoulos, T., Fisman, D.: Query learning of derived ω-tree languages in polynomial time. Log. Methods Comput. Sci. **15**(3) (2019)
6. Arbib, M.A., Manes, E.G.: Adjoint machines, state-behavior machines, and duality. Journal of Pure and Applied Algebra **6**(3), 313–344 (1975)

7. Aslam, K., Cleophas, L., Schiffelers, R.R.H., van den Brand, M.: Interface protocol inference to aid understanding legacy software components. Softw. Syst. Model. **19**(6), 1519–1540 (2020), https://doi.org/10.1007/s10270-020-00809-2
8. Balle, B., Mohri, M.: Learning weighted automata. In: CAI. Lecture Notes in Computer Science, vol. 9270, pp. 1–21. Springer (2015)
9. Barlocco, S., Kupke, C., Rot, J.: Coalgebra learning via duality. In: FoSSaCS. Lecture Notes in Computer Science, vol. 11425, pp. 62–79. Springer (2019)
10. Bergadano, F., Varricchio, S.: Learning behaviors of automata from multiplicity and equivalence queries. SIAM J. Comput. **25**(6), 1268–1280 (1996). https://doi.org/10.1137/S009753979326091X
11. Bojańczyk, M., Klin, B., Lasota, S.: Automata theory in nominal sets. Log. Methods Comput. Sci. **10**(3) (2014)
12. Bollig, B., Habermehl, P., Kern, C., Leucker, M.: Angluin-style learning of NFA. In: IJCAI. pp. 1004–1009 (2009)
13. Bollig, B., Habermehl, P., Leucker, M., Monmege, B.: A fresh approach to learning register automata. In: Developments in Language Theory. Lecture Notes in Computer Science, vol. 7907, pp. 118–130. Springer (2013)
14. van den Bos, P., Janssen, R., Moerman, J.: n-Complete test suites for IOCO. Softw. Qual. J. **27**(2), 563–588 (2019)
15. Broy, M., Jonsson, B., Katoen, J.P., Leucker, M., Pretschner, A. (eds.): Model-based testing of reactive systems, Lecture Notes in Computer Science, vol. 3472. Springer (2005)
16. Brzozowski, J.A.: Derivatives of regular expressions. J. ACM **11**(4), 481–494 (1964)
17. Buna-Marginean, A., Cheval, V., Shirmohammadi, M., Worrell, J.: On learning polynomial recursive programs. Proc. ACM Program. Lang. **8**(POPL), 1001–1027 (2024)
18. Cassel, S., Howar, F., Jonsson, B., Steffen, B.: Active learning for extended finite state machines. Formal Aspects Comput. **28**(2), 233–263 (2016)
19. Chow, T.S.: Testing software design modeled by finite-state machines. IEEE Trans. Software Eng. **4**(3), 178–187 (1978)
20. Colcombet, T., Petrisan, D., Stabile, R.: Learning automata and transducers: A categorical approach. In: CSL. LIPIcs, vol. 183, pp. 15:1–15:17. Schloss Dagstuhl - Leibniz-Zentrum für Informatik (2021)
21. Dorofeeva, R., El-Fakih, K., Maag, S., Cavalli, A.R., Yevtushenko, N.: FSM-based conformance testing methods: A survey annotated with experimental evaluation. Inf. Softw. Technol. **52**(12), 1286–1297 (2010)
22. Droste, M., Kuske, D.: Weighted automata. In: Handbook of Automata Theory (I.), pp. 113–150. European Mathematical Society Publishing House, Zürich, Switzerland (2021)
23. Fujiwara, S., von Bochmann, G., Khendek, F., Amalou, M., Ghedamsi, A.: Test selection based on finite state models. IEEE Trans. Software Eng. **17**(6), 591–603 (1991)
24. Goguen, J.A.: Discrete-time machines in closed monoidal categories I. J. Comput. Syst. Sci. **10**(1), 1–43 (1975)
25. van Heerdt, G., Kupke, C., Rot, J., Silva, A.: Learning weighted automata over principal ideal domains. In: FoSSaCS. Lecture Notes in Computer Science, vol. 12077, pp. 602–621. Springer (2020)
26. van Heerdt, G., Sammartino, M., Silva, A.: Learning automata with side-effects. In: CMCS. Lecture Notes in Computer Science, vol. 12094, pp. 68–89. Springer (2020)

27. Howar, F., Steffen, B.: Active automata learning in practice: An annotated bibliography of the years 2011 to 2016. In: Machine Learning for Dynamic Software Analysis. Lecture Notes in Computer Science, vol. 11026, pp. 123–148. Springer (2018)
28. Isberner, M.: Foundations of active automata learning: an algorithmic perspective. Ph.D. thesis, Technical University Dortmund, Germany (2015)
29. Isberner, M., Howar, F., Steffen, B.: The open-source LearnLib: A framework for active automata learning. In: CAV (1). Lecture Notes in Computer Science, vol. 9206, pp. 487–495. Springer (2015)
30. Jacobs, B., Rutten, J.J.M.M.: A tutorial on (co)algebras and (co)induction. Bulletin-European Association for Theoretical Computer Science **62**, 222–259 (1997)
31. Kanso, B., Aiguier, M., Boulanger, F., Touil, A.: Testing of abstract components. In: ICTAC. Lecture Notes in Computer Science, vol. 6255, pp. 184–198. Springer (2010)
32. Klíma, O., Polák, L.: On varieties of ordered automata. In: LATA. Lecture Notes in Computer Science, vol. 11417, pp. 108–120. Springer (2019)
33. Kocsis, B., Rot, J.: Complete test suites for automata in monoidal closed categories. CoRR **abs/2411.13412** (2024), `https://arxiv.org/abs/2411.13412`, extended version
34. Kruger, L., Junges, S., Rot, J.: Small test suites for active automata learning. In: TACAS (2). Lecture Notes in Computer Science, vol. 14571, pp. 109–129. Springer (2024)
35. Lee, D., Yannakakis, M.: Principles and methods of testing finite state machines – a survey. Proc. IEEE **84**(8), 1090–1123 (1996)
36. Maarse, T.: Active Mealy machine learning using action refinements. Master's thesis, Radboud University Nijmegen, The Netherlands (2020)
37. Mac Lane, S.: Categories for the working mathematician, Graduate Texts in Mathematics, vol. 5. Springer New York (2013)
38. Maler, O., Pnueli, A.: On the learnability of infinitary regular sets. Inf. Comput. **118**(2), 316–326 (1995), `https://doi.org/10.1006/inco.1995.1070`
39. Margaria, T., Niese, O., Raffelt, H., Steffen, B.: Efficient test-based model generation for legacy reactive systems. In: HLDVT. pp. 95–100. IEEE Computer Society (2004)
40. Moerman, J.S.: Nominal techniques and black box testing for automata learning. Ph.D. thesis, Radboud University Nijmegen, The Netherlands (2019)
41. Moerman, J.S., Sammartino, M., Silva, A., Klin, B., Szynwelski, M.: Learning nominal automata. In: POPL. pp. 613–625. ACM (2017)
42. Moore, E.F.: Gedanken-experiments on sequential machines. In: Automata Studies, pp. 129–153. Princeton University Press (1956)
43. Muskardin, E., Aichernig, B.K., Pill, I., Pferscher, A., Tappler, M.: AALpy: An active automata learning library. In: ATVA. Lecture Notes in Computer Science, vol. 12971, pp. 67–73. Springer (2021)
44. Petrenko, A., Yevtushenko, N.: Adaptive testing of nondeterministic systems with FSM. In: HASE. pp. 224–228. IEEE Computer Society (2014)
45. Pitts, A.M.: Nominal sets: Names and symmetry in computer science, Cambridge Tracts in Theoretical Computer Science, vol. 57. Cambridge University Press (2013)
46. Rabin, M.O., Scott, D.S.: Finite automata and their decision problems. IBM J. Res. Dev. **3**(2), 114–125 (1959)

47. Rutten, J.J.M.M.: Universal coalgebra: a theory of systems. Theor. Comput. Sci. **249**(1), 3–80 (2000)
48. Tappler, M., Aichernig, B.K., Bacci, G., Eichlseder, M., Larsen, K.G.: L*-based learning of Markov decision processes (extended version). Formal Aspects Comput. **33**(4), 575–615 (2021)
49. Urbat, H., Schröder, L.: Automata learning: An algebraic approach. In: LICS. pp. 900–914. ACM (2020)
50. Vaandrager, F.W.: Model learning. Commun. ACM **60**(2), 86–95 (2017)
51. Vaandrager, F.W., Fiterău-Broştean, P., Melse, I.: Completeness of FSM test suites reconsidered (2024), `https://arxiv.org/abs/2410.19405`
52. Vasilevskii, M.P.: Failure diagnosis of automata. Cybernetics **9**(4), 653–665 (1973)

Temporal Hyperproperties for Population Protocols

Nicolas Waldburger[1], Chana Weil-Kennedy[2(✉)], Pierre Ganty[2], and César Sánchez[2]

[1] Université de Rennes, IRISA, INRIA, Rennes, France
nicolas.waldburger@irisa.fr
[2] IMDEA Software Institute, Pozuelo de Alarcón, Madrid, Spain
{chana.weilkennedy,pierre.ganty,cesar.sanchez}@imdea.org

Abstract. Hyperproperties are properties over sets of traces (or runs) of a system, as opposed to properties of just one trace. They were introduced in 2010 and have been much studied since, in particular via an extension of the temporal logic LTL called HyperLTL. Most verification efforts for HyperLTL are restricted to finite-state systems, usually defined as Kripke structures. In this paper we study hyperproperties for an important class of infinite-state systems. We consider population protocols, a popular distributed computing model in which arbitrarily many identical finite-state agents interact in pairs. Population protocols are a good candidate for studying hyperproperties because the main decidable verification problem, well-specification, is a hyperproperty. We first show that even for simple (monadic) formulas, HyperLTL verification for population protocols is undecidable. We then turn our attention to immediate observation population protocols, a simpler and well-studied subclass of population protocols. We show that verification of monadic HyperLTL formulas without the next operator is decidable in 2-EXPSPACE, but that all extensions make the problem undecidable.

1 Introduction

Hyperproperties are properties that allow to relate multiple traces (also called runs) of a system simultaneously [12]. They generalize regular run properties to properties of sets of runs, and formalize a wide range of important properties such as information-flow security policies like noninterference [30,36] and observational determinism [48], consistency models in concurrent computing [10], and robustness models in cyber-physical systems [47,9].

HyperLTL [11] was introduced as an extension of LTL (linear temporal logic) with quantification over runs which can then be related across time. HyperLTL enjoys a decidable model-checking problem for finite-state systems, expressed as Kripke structures. Other logics for hyperproperties were later introduced, like HyperCTL* [27], HyperQPTL [40,13], and HyperPDL-Δ [31] which extend CTL*, QPTL [44], and PDL [28] respectively. These logics also enjoy decidable model-checking problems for finite-state systems.

P. A. Abdulla and D. Kesner (Eds.): FoSSaCS 2025, LNCS 15691, pp. 220–242, 2025.
https://doi.org/10.1007/978-3-031-90897-2_11

Most algorithmic verification results for verifying hyperproperties of temporal logics are restricted to finite-state systems. In the case of software verification, which is inherently infinite-state, the analysis of hyperproperties [7,26,43,45,46] has been limited to the class of k-safety properties — which only allow to establish the absence of a bad interaction between any k runs — and do not extend to a temporal logic for hyperproperties. A notable exception is [7], but the logic used (OHyperLTL) is a simple asynchronous logic for hyperproperties and it requires restrictions on the underlying theories of the data used in the program.

In this paper we focus on the verification of HyperLTL for an important class of infinite-state systems. We consider population protocols (PP) [2], an extensively studied (see e.g. [1,18,19]) model of distributed computation in which anonymous finite-state agents interact pairwise to change their states, following a common protocol. In a *well-specified* PP, the agents compute a predicate: the input is the initial configuration of the agents' states, and the agents interact in pairs to eventually reach a consensus opinion corresponding to the evaluation of the predicate (for *any* number of agents). Interactions are selected at random, which is modelled by considering only *fair* runs. LTL verification has been investigated for PPs in [20]. The authors consider LTL over actions, where formulas are evaluated over fair runs. They show that it is decidable, given a PP and an LTL formula, to check if all fair runs from initial configurations of the protocol verify the formula. Another related work on LTL verification for infinite-state systems is [29], where the authors consider stuttering-invariant LTL verification over shared-memory pushdown systems.

We consider PPs because, though they are infinite-state, they enjoy several decidable problems. In particular, the central verification problem checking whether a protocol is well-specified is decidable [21] and has a hyperproperty "flavor". A PP is well-specified if for every initial configuration γ_0, every fair run starting in γ_0 stabilizes to the same opinion. A run stabilizes to an opinion $b \in \{0, 1\}$ if from some position onwards it visits no configuration with an agent whose opinion differs from b. With $\mathcal{I}$ the set of initial configurations and $\mathsf{FRuns}(\gamma)$ the set of fair runs starting in γ, well-specification can be expressed as:

$$\forall \gamma_0 \in \mathcal{I},\ \mathsf{FRuns}(\gamma_0) \models \forall \rho_1 . \forall \rho_2 . \bigvee_{b \in \{0,1\}} (\mathsf{FG}(\rho_1 \text{ sees } b) \wedge \mathsf{FG}(\rho_2 \text{ sees } b))$$

where "ρ sees b" means that the run takes a transition that puts agents into states with opinion b. Then $\mathsf{FG}(\rho_i \text{ sees } b)$ ensures that ρ_i converges to b. The formula above is a proper relational hyperproperty and cannot be expressed in LTL because it requires to quantify over two traces ρ_1 and ρ_2 or quantify within the scope of an outer conjunction.

We show that for the general PP model, HyperLTL verification is already undecidable for simple (*monadic*) formulas which can be decomposed into formulas referring to only one run each (Section 3). We turn our attention to *immediate observation population protocols* (IOPP), a subclass of PP [3]. We show that HyperLTL verification over IOPP is a problem decidable in 2-EXPSPACE when the formula is monadic and does not use the temporal operator X (the formula is then *stuttering-invariant*). This result delineates the decidability frontier for verification in PP: non-monadic or non-stuttering-invariant HyperLTL verification over IOPP is undecidable (Section 4). The decidability result for HyperLTL verification of IOPP is the most technical result of the paper. In particular, the technical results of Section 5 reason on the flow of agents in runs of an IOPP in conjunction with reading the transitions in a Rabin automaton. Note that if one fixed the initial configuration (focusing on non-global model checking) the verification of monadic HyperLTL\X can be performed by a Boolean combination of LTL verification problems. However, monadic HyperLTL\X is strictly more expressive than LTL even for the non-global case (including properties like termination-sensitive non-interference [41]).

2 Preliminaries

A *finite multiset* over a finite set S is a mapping $\mu \colon S \to \mathbb{N}$ such that for each $s \in S$, $\mu(s)$ denotes the number of occurrences of element s in μ. Given a set S, $\mathcal{M}(S)$ denotes the set of finite multisets over S. Given $s \in S$, we denote by $\vec{s}$ the multiset μ such that $\mu(s) = 1$ and $\mu(s') = 0$ for all $s' \neq s$. Given $\mu, \mu' \in \mathcal{M}(S)$, the multiset $\mu + \mu'$ is defined by $(\mu + \mu')(s) = \mu(s) + \mu'(s)$ for all $s \in S$. We let $\mu \leqslant \mu'$ when $\mu(s) \leqslant \mu'(s)$ for all $s \in S$. When $\mu' \leqslant \mu$, we let $\mu - \mu'$ be the multiset such that $(\mu - \mu')(s) = \mu(s) - \mu'(s)$ for all $s \in S$. We call $|\mu| = \sum_{s \in S} \mu(s)$ the *size* of μ. A set $\mathcal{S} \subseteq \mathcal{M}(S)$ is *Presburger* if it can be written as a formula in Presburger arithmetic, *i.e.*, in $FO(\mathbb{N}, +)$.

A *strongly connected component* (SCC) in a graph is a non-empty maximal set of mutually reachable vertices. A SCC is *bottom* if no path leaves it.

2.1 Population Protocols

A *population protocol* (PP) is a tuple $\mathcal{P} = (Q, \Delta, I)$ where Q is a finite set of *states*, $\Delta \subseteq Q^2 \times Q^2$ is a set of *transitions* and $I \subseteq Q$ is the set of *initial states*. A transition $t = ((q_1, q_2), (q_3, q_4)) \in \Delta$ is denoted $(q_1, q_2) \xrightarrow{t} (q_3, q_4)$. We let $|\mathcal{P}| := |Q| + |\Delta|$ denote the *size* of $\mathcal{P}$. A *configuration* of $\mathcal{P}$ is a multiset over Q. We denote by $\Gamma := \{\mu \in \mathcal{M}(Q) \mid |\mu| > 2\}$ the set of configurations; configurations must have at least 2 agents. We note $\mathcal{I} := \{\gamma \in \Gamma \mid \forall q \notin I, \gamma(q) = 0\}$ the set of *initial configurations*. Given $\gamma, \gamma' \in \Gamma$ and $(q_1, q_2) \xrightarrow{t} (q_3, q_4) \in \Delta$, there is a *step* $\gamma \xrightarrow{t} \gamma'$ if $\gamma \geqslant \vec{q_1} + \vec{q_2}$ and $\gamma' = \gamma - \vec{q_1} - \vec{q_2} + \vec{q_3} + \vec{q_4}$. A transition $(q_1, q_2) \to (q_3, q_4)$ is *activated* at γ if $\gamma \geqslant \vec{q_1} + \vec{q_2}$, *i.e.*, if there is an agent in q_1 and an agent in q_2 (or two agents in q_1 if $q_1 = q_2$). Henceforth, we assume that for every $q_1, q_2 \in Q$, there exist $q_3, q_4 \in Q$ such that $(q_1, q_2) \to (q_3, q_4) \in \Delta$, so that there is always an activated transition. This can be done by adding self-loops $(q_1, q_2) \to (q_1, q_2)$.

A *finite run* is a sequence $\gamma_0, t_0, \gamma_1, \ldots, t_{k-1}, \gamma_k$ where $\gamma_i \xrightarrow{t_i} \gamma_{i+1}$ for all $i \leqslant k - 1$; we say t_i is *fired* at γ_i. We write $\gamma \xrightarrow{*} \gamma'$ if there exists a finite run from γ to γ', and we say γ' is *reachable* from γ. Given $\mathcal{S} \subseteq \Gamma$, let $\mathsf{post}^*(\mathcal{S})$ be the set of configurations reachable from $\mathcal{S}$, *i.e.*, $\mathsf{post}^*(\mathcal{S}) := \{\gamma \mid \exists \gamma' \in \mathcal{S} . \gamma' \xrightarrow{*} \gamma\}$. Similarly, let $\mathsf{pre}^*(\mathcal{S}) := \{\gamma \mid \exists \gamma' \in \mathcal{S} . \gamma \xrightarrow{*} \gamma'\}$.

An *infinite run* is an infinite sequence $\rho = \gamma_0, t_0, \gamma_1, t_1, \ldots$ with $\gamma_i \xrightarrow{t_i} \gamma_{i+1}$ for all $i \in \mathbb{N}$. A configuration γ is *visited* in ρ when there is i such that $\gamma_i = \gamma$; it is *visited infinitely often* when there are infinitely many such i. Similarly, $t \in \Delta$ is *fired infinitely often* in ρ where there are infinitely many i such that $t_i = t$. A finite run $\gamma'_0, t'_0, \gamma'_1, \ldots, t'_{k-1}, \gamma'_k$ *appears infinitely often* in ρ when there are infinitely many i such that $\gamma_{i+j} = \gamma'_j$ for all $j \in [0, k]$ and $t_{i+j} = t'_j$ for all $j \in [0, k-1]$. Also, ρ is *strongly fair* when, for every finite run ρ', by letting γ'_0 the first configuration in ρ', if γ'_0 is visited infinitely often in ρ then ρ' appears infinitely often in ρ. Given a configuration γ_0, the set of strongly fair runs from γ_0 is denoted $\mathsf{FRuns}(\gamma_0)$. Note that this notion of fairness differs from the one usually used for PPs. We will discuss this choice in Section 2.4.

2.2 LTL and HyperLTL

Linear temporal logic [39] (LTL) extends propositional logic with modalities to relate different positions in a run, allowing to define temporal properties of systems. HyperLTL [11] is an extension of LTL for hyperproperties, with explicit quantification over runs. We here define LTL and HyperLTL for population protocols. Let $\mathcal{P} = (Q, \Delta, I)$ be a PP. Our atomic propositions are the transitions of the run(s); we discuss this choice at the end of this section.

LTL. The syntax of LTL over $\mathcal{P}$ is:

$$\varphi ::= t \mid \varphi \vee \varphi \mid \neg\varphi \mid \mathsf{X}\varphi \mid \varphi\,\mathcal{U}\,\varphi \qquad \text{where } t \in \Delta\ .$$

The operators X (next) and $\mathcal{U}$ (until) are the temporal modalities. We use the usual additional operators: $\mathit{true} = t \vee \neg t$, $\mathit{false} = \neg \mathit{true}$, $\varphi \wedge \varphi = \neg(\neg\varphi \vee \neg\varphi)$, $\mathsf{F}\varphi = \mathit{true}\,\mathcal{U}\varphi$ and $\mathsf{G}\varphi = \neg\mathsf{F}\neg\varphi$. The *size* $|\varphi|$ of an LTL formula φ is the number of (temporal and Boolean) operators of φ. The semantics of LTL is defined over runs in the usual way (*e.g.*, [4]) over Δ^ω. An infinite run $\rho = \gamma_0, t_0, \gamma_1, t_1, \ldots$ *satisfies* an LTL formula φ, denoted $\rho \models \varphi$, when $w \models \varphi$ where $w = t_0 t_1 t_2 \cdots \in \Delta^\omega$. A configuration γ satisfies an LTL formula φ, denoted $\gamma \models \varphi$, when $\rho \models \varphi$ for all $\rho \in \mathsf{FRuns}(\gamma)$, *i.e.*, when *all* strongly fair runs starting from γ satisfy φ.

HyperLTL. The syntax of HyperLTL over $\mathcal{P}$ is:

$$\psi ::= \exists\rho.\psi \mid \forall\rho.\psi \mid \varphi \qquad\qquad \varphi ::= t_\rho \mid \varphi \vee \varphi \mid \neg\varphi \mid \mathsf{X}\varphi \mid \varphi\,\mathcal{U}\,\varphi$$

where $t \in \Delta$ and ρ is a *run variable*. Note that φ is an LTL formula with, as atomic propositions, the transitions of the run variables. A HyperLTL formula ψ must additionally be well-formed: all appearing variables are quantified and no variable is quantified twice. The size $|\psi|$ of an HyperLTL formula ψ is the number of (temporal and Boolean) operators and quantifiers of ψ. HyperLTL formulas are interpreted over strongly fair runs starting from a configuration as follows:

a configuration γ satisfies a HyperLTL formula ψ, denoted $\gamma \models \psi$, whenever $\mathsf{FRuns}(\gamma) \models \psi$. See the appendix for a formal definition of the semantics. Notice that, given a configuration γ and an LTL formula φ, $\gamma \models \varphi$ if and only if $\gamma \models \forall \rho.\varphi_\rho$ where φ_ρ is equal to φ where t is replaced by t_ρ for all $t \in \Delta$.

Example 1. Suppose that $\Delta = \{s, t\}$. Let $\psi := \forall \rho_1.\exists \rho_2.\mathsf{FG}((s_{\rho_1} \wedge t_{\rho_2}) \vee (t_{\rho_1} \wedge s_{\rho_2}))$. Given $\gamma \in \Gamma$, we have that $\gamma \models \psi$ if and only if, for every strongly fair run $\rho_1 \in \mathsf{FRuns}(\gamma)$ from γ, there is a strongly fair run $\rho_2 \in \mathsf{FRuns}(\gamma)$ from γ such that, always after some point, ρ_1 fires s whenever ρ_2 fires t and vice versa.

A HyperLTL formula $\psi : Q_1\rho_1 \dots Q_k\rho_k.\varphi$ is *monadic* if φ has a *decomposition* as a Boolean combination of temporal formulas $\varphi_1, \dots, \varphi_n$, each of which refer to exactly one run variable. We assume that a monadic formula is always given by its decomposition, *i.e.*, by giving φ_1 to φ_n and the Boolean combination.

Verification Problems. Given a PP $\mathcal{P} = (Q, \Delta, I)$ and an LTL formula φ (resp. a HyperLTL formula ψ), we denote $\mathcal{P} \models^\forall \varphi$ when $\gamma_0 \models \varphi$ (resp. $\gamma_0 \models \psi$) for all $\gamma_0 \in \mathcal{I}$. Dually, we let $\mathcal{P} \models^\exists \varphi$ (resp. $\mathcal{P} \models^\exists \psi$) when there is $\gamma_0 \in \mathcal{I}$ such that $\gamma_0 \models \varphi$ (resp. $\gamma_0 \models \psi$).

The *LTL verification problem for population protocols* consists on determining, given $\mathcal{P}$ and an LTL formula φ, whether $\mathcal{P} \models^\forall \varphi$, *i.e.*, whether all strongly fair runs from all initial configurations satisfy φ. We also consider a variant problem, the *existential LTL verification problem*, that asks whether $\mathcal{P} \models^\exists \varphi$, *i.e.*, whether there is an initial configuration from which all strongly fair runs satisfy φ. Given a HyperLTL formula ψ, the *HyperLTL verification problem for population protocols* consists on determining whether $\mathcal{P} \models^\forall \psi$; again, the existential variant consists in asking whether $\mathcal{P} \models^\exists \psi$.

Example 2. A PP $\mathcal{P} = (Q, \Delta, I)$ equipped with an *opinion* function $O\colon Q \to \{0, 1\}$ is *well-specified* if for every $\gamma_0 \in \mathcal{I}$, every run in $\mathsf{FRuns}(\gamma_0)$ eventually visits only configurations where either all agents are in states $O^{-1}(0)$ or all agents are in states $O^{-1}(1)$. Let Δ_b be the set of transitions $(q_1, q_2) \to (q_3, q_4) \in \Delta$ such that $O(q_3) = O(q_4) = b$. Well-specification of $\mathcal{P} = (Q, \Delta, I)$ with opinion function O corresponds to the HyperLTL verification problem over a monadic formula:

$$\mathcal{P} \models^\forall \forall \rho_1, \rho_2. \bigvee_{b \in \{0,1\}} \mathsf{FG}(\bigvee_{t \in \Delta_b} t_{\rho_1}) \wedge \mathsf{FG}(\bigvee_{t \in \Delta_b} t_{\rho_2}) \ .$$

LTL over transitions and LTL over states. Our LTL formulas are *over transitions*, *i.e.*, their atomic propositions are the transitions of the run. In [20, Theorems 9 and 10], the LTL verification problem defined above is proven to be decidable, although as hard as reachability for Petri nets and therefore Ackermann-complete [35,14]. The authors of [20] also show that *LTL over states*, where the atomic predicates indicate whether or not a state is currently visited by an agent, is undecidable. A slight difference between their model and ours is that their initial configurations are given by a Presburger set; however, their undecidability proof, which relies on 2-counter machines, can easily be translated to our setting. In the rest of the paper we consider only (Hyper)LTL over transitions.

Proposition 3. *The LTL over states verification problem for PP is undecidable.*

2.3 Rabin Automata and LTL

Let Σ be a finite set. The set of finite words (resp. infinite words) over Σ is denoted Σ^* (resp. Σ^ω). A *deterministic Rabin automaton* over Σ is a tuple $\mathcal{A} = (\mathcal{L}, T, \ell_0, \mathcal{W})$, where $\mathcal{L}$ is a finite set of states, $\ell_0 \in \mathcal{L}$ is the initial state, $T : \mathcal{L} \times \Sigma \to \mathcal{L}$ is the transition function and $\mathcal{W} \subseteq 2^{\mathcal{L}} \times 2^{\mathcal{L}}$ is a finite set of *Rabin pairs*. An infinite word $w \in \Sigma^\omega$ is *accepted* if there exists $(F, G) \in \mathcal{W}$ such that the run of $\mathcal{A}$ reading w visits F finitely often and G infinitely often.

Theorem 4 ([23]). *Given Σ a finite set and φ an LTL formula over Σ, one can compute, in time doubly-exponential in $|\varphi|$, a deterministic Rabin automaton $\mathcal{A}_\varphi$ over Σ, of doubly-exponential size, that recognizes (the language of) φ.*

2.4 Why Strong Fairness?

Usually, fairness in population protocols is either of the form "all configurations reachable infinitely often are reached infinitely often" [3,21], or "all steps possible infinitely often are taken infinitely often" [20]. Our notion of fairness, dubbed strong fairness, is more restrictive. A sanity check is that a (reasonable) stochastic scheduler yields a strongly fair run with probability 1. This alone does not justify using a new notion of fairness different from the literature and in particular from the prior work on LTL verification [20]. The authors motivate their choice of fairness by claiming that there is a fair run satisfying an LTL formula φ if and only if, under a stochastic scheduler, φ is satisfied with non-zero probability [20, Proposition 7]. However, we show that this claim is incorrect.

The intuition is that a (not strongly) fair run may exhibit infinite regular patterns. Consider three configurations $\gamma_1, \gamma_2, \gamma_3$ and three transitions a, b, c such that $\gamma_1 \xrightarrow{a} \gamma_2$, $\gamma_2 \xrightarrow{b} \gamma_1$, $\gamma_1 \xrightarrow{c} \gamma_3$ and $\gamma_3 \xrightarrow{d} \gamma_1$, and these are the only steps possible from each of the configurations. Consider $\varphi = \neg\mathsf{F}(a \wedge (\mathsf{X}b) \wedge (\mathsf{X}^2 a) \wedge (\mathsf{X}^3 b))$, which expresses that the sequence of transitions $abab$ does not appear. Under a stochastic scheduler, φ is satisfied with probability 0 from γ_1. However, the run which repeats sequence $abcd$ satisfies φ, and it is fair. This proves that [20, Proposition 7] does not hold; we explain the mistake in detail in the appendix.

The run from this counterexample is not strongly fair. We show that strong fairness does in fact allow the desired equivalence with stochastic schedulers. As in [20], fix a stochastic scheduler, assumed to be memoryless and guaranteeing non-zero probability for every activated transition; $\mathsf{Pr}[\gamma \models \varphi]$ denotes the probability that a run from γ satisfies φ.

Proposition 5. *Given an LTL formula φ and $\gamma_0 \in \Gamma$, $\mathsf{Pr}[\gamma_0 \models \varphi] = 1$ if and only if, for all $\rho \in \mathsf{FRuns}(\gamma_0)$, $\rho \models \varphi$.*

Proof (sketch). Let $\mathcal{A}_\varphi = (\mathcal{L}, T, \ell_0, \mathcal{W})$ be a Rabin automaton that recognizes φ (see Theorem 4). Like in the proof of [20, Proposition 7], we consider a Petri net obtained by combining $\mathcal{P}$ and $\mathcal{A}_\varphi$. We reason on the bottom SCCs of the configuration graph of this Petri net. An SCC is *winning* where there is $(F, G) \in \mathcal{W}$ such that S has some configuration with Rabin state in G but none with Rabin state in F. By [20, Proposition 6], we have $\mathsf{Pr}[\gamma_0 \models \varphi] = 1$ iff all bottom SCC S reachable from (γ_0, ℓ_0) are winning. We show that strong fairness guarantees that a run (1) always reaches a bottom SCC and (2) visits all the configurations of this bottom SCC infinitely often. The proofs of the two statements are similar; we explain here the proof of (2). Configurations of the Petri net are of the form (γ, ℓ) with γ a configuration of $\mathcal{P}$ and $\ell \in \mathcal{L}$. Consider (γ, ℓ) in a bottom SCC and $t \in \Delta$ activated from γ. Since strong fairness is only related to $\mathcal{P}$, it does not guarantee that t is eventually fired from (γ, ℓ). We circumvent this difficulty by constructing a sequence of transitions σ that, when fired from any (γ, ℓ') in the SCC, makes us go through (γ, ℓ) and fire t. Strong fairness ensures σ is fired from some (γ, ℓ'), which proves the result. □

This therefore justifies our choice to consider strong fairness for LTL verification. In particular, all results from [20] hold if strong fairness is considered instead of the usual fairness. An alternative to strong fairness for (non-Hyper)LTL verification would be to work directly with a stochastic scheduler. However, HyperLTL requires quantification over a subset of the set of runs; we make the choice to consider, for this subset, the set of strongly fair runs.

3 Undecidability of HyperLTL

One can show that verification of HyperLTL over transitions is undecidable for PP, using a proof with counter machines similar to the one for undecidability of LTL over states [20]. Intuitively, HyperLTL can be used to express whether a transition is activated at some point in the run, and hence encode zero-tests[3]. We show an even stronger undecidability result: verification of monadic HyperLTL formulas over two runs using only FG as temporal operator is undecidable.

Theorem 6. *Verification of monadic HyperLTL for PP is undecidable. If fact, it is already undecidable for formulas of the form:*

$$\forall \rho_1 . \exists \rho_2 . \neg(\mathsf{GF}\, a_{\rho_1}) \vee (\mathsf{GF}\, b_{\rho_2}) \qquad \textit{where } a, b \in \Delta\ .$$

This verification problem asks whether, for all $\gamma_0 \in \mathcal{I}$, for all $\rho_1 \in \mathsf{FRuns}(\gamma_0)$, there is $\rho_2 \in \mathsf{FRuns}(\gamma_0)$ such that if ρ_1 fires a infinitely often then ρ_2 fires b infinitely often. We first observe that the $\forall$-$\exists$ sequence of quantifiers is reminiscent of inclusion problems. Since the population protocol model is close to Petri nets, it is natural to look for undecidable inclusion-like problems for that model. Indeed, undecidability was shown multiple times [5,32] for the problem asking whether the set of reachable markings of a Petri net is included in the set of reachable marking of another Petri net with equally many places. We call this problem the *reachability set inclusion problem.* Our attempts at reducing the reachability set inclusion problem to the above problem faced a major obstacle: Petri nets allow the creation/destruction of tokens while in PPs the number of agents remains the same. We sidestepped this obstacle by looking at a particular

[3] See the proof of Theorem 11 for an illustration of this.

proof of undecidability for the reachability set inclusion problem which leverages Hilbert's Tenth Problem (shown to be undecidable by Matijasevic in the seventies). We thus obtain a reduction from Hilbert's Tenth Problem to the above problem for PPs. Our reduction uses PPs to "compute" the value of polynomials while keeping the number of agents constant during the computation.

The detailed proof is given in the appendix and we only provide here the statement of the variant of Hilbert's Tenth Problem used in the reduction.

Proposition 7 ([32]). *The following problem is undecidable:*
Input: *two polynomials* $\mathsf{P}_1(\mathsf{x}_1,\ldots,\mathsf{x}_r), \mathsf{P}_2(\mathsf{x}_1,\ldots,\mathsf{x}_r)$ *with natural coefficients*
Question: *Does it hold that, for all* $\mathsf{x}_1,\ldots,\mathsf{x}_r \in \mathbb{N}$, $\mathsf{P}_1(\mathsf{x}_1,\ldots,\mathsf{x}_r) \leqslant \mathsf{P}_2(\mathsf{x}_1,\ldots,\mathsf{x}_r)$*?*

4 Verification of HyperLTL for IOPP

Section 3 showed that verification of HyperLTL in PPs is undecidable, even when the formulas are monadic and have a simple shape. We thus turn to a subclass of PPs called *immediate observation population protocols* (IOPP) [3] that has been studied extensively (see e.g. [24,33,8,6]).

4.1 Immediate Observation PP and Preliminary Results

Definition 8. *An* immediate observation population protocol *(*IOPP*) is a population protocol where all transitions are of the form* $(q_1, q_2) \to (q_3, q_2)$*.*

We denote a transition $(q_1, q_2) \to (q_3, q_2)$ as $q_1 \xrightarrow{q_2} q_3$. Intuitively, when two agents interact, one remains in its state, as if it was observed by the other agent.

The IOPP model tends to be simpler to verify than standard PP [24], notably because it enjoys a convenient monotonicity property: whenever an agent observes an agent in q_3 and goes from q_1 to q_2, another agent in q_1 may do the same "for free". This property is however broken by the X operator of LTL. In fact, under LTL, IOPP has similar power to regular PP. Indeed, consider a PP transition $t : (q_1, q_2) \to (q_3, q_4)$. One may split this transition into immediate observation transitions $t_1 : q_1 \xrightarrow{q_2} q_3$ and $t_2 : q_2 \xrightarrow{q_3} q_4$. Using an LTL formula with the X operator, one can enforce that, whenever t_1 is fired, t_2 must be fired directly after.

We start by establishing that verification of LTL for IOPP is as hard as its counterpart for PP:

Proposition 9. *Verification of LTL for IOPP is Ackermann-complete.*

Remark 10. The fragment of LTL with no X operator is equivalent to stutter-invariant LTL [38,25]. Let φ be an LTL$\setminus$X formula φ, let $t_1, t_2, \ldots \in \Delta$ and $k_1, k_2, \ldots \geqslant 1$. This means that we have $t_1^{k_1}\, t_2^{k_2} \ldots \models \varphi$ if and only if $t_1\, t_2 \ldots \models \varphi$.

Below, we consider the fragment LTL$\setminus$X as done in prior work [29] in which the systems under study feature monotonicity due to non-atomic writes: stuttering-invariance is a natural choice for systems with monotonicity properties. We show that, even then, verification of HyperLTL$\setminus$X formulas for IOPP is undecidable.

Theorem 11. *Verification of HyperLTL$\backslash$X is undecidable for IOPP.*

Proof (sketch). We proceed by reducing from the halting problem for 2-counter machines with zero-tests, an undecidable problem [37]. A *2-counter machine* consists in two counters c_1, c_2 and a list of instructions $l_1, \ldots, l_n, \mathsf{halt}$. An instruction l_i can increment a counter, decrement a counter, or test whether a counter's value is zero. Given a 2-counter machine $\mathcal{M}$, we build an IOPP $\mathcal{P}$, with the goal of simulating executions of $\mathcal{M}$ faithfully using runs of $\mathcal{P}$. We introduce gadgets to simulate the instructions; to ensure that the gadgets are used correctly we add *bad* transitions which are activated when the simulation "cheats". A bad transition is activated in a run ρ if there exists another run which takes all the same transitions as ρ until it takes a bad transition b: $\psi_{\mathcal{B}}(\rho) = \exists \rho'.(\bigvee_{t \in \Delta} t_\rho \wedge t_{\rho'})\,\mathcal{U}\,(\bigvee_{b \in \mathcal{B}} b_{\rho'})$, where $\mathcal{B}$ is the set of bad transitions. We define the HyperLTL$\backslash$X formula $\psi = \forall \rho. \neg(\mathsf{F}\ \mathsf{halt}_\rho) \vee \psi_{\mathcal{B}}(\rho)$. Then $\mathcal{P} \models^{\forall} \psi$ if and only if $\mathcal{M}$ does not halt, and we are done. $\square$

However, we will show that the monadic HyperLTL$\backslash$X case is decidable and in 2-EXPSPACE.

4.2 Product Systems

Our approach consists, as in the proof of Proposition 5, to define *product systems* that combine the IOPP with a Rabin automaton recognizing an LTL formula. We will then characterize the set of configurations of the IOPP that satisfy the formula using reachability sets of the product system.

Definition 12. *A* product system *is a pair $\mathcal{PS} = (\mathcal{P}, \mathcal{A})$ where:*
- $\mathcal{P} = (Q, \Delta, I)$ *is an IOPP,*
- $\mathcal{A} = (\mathcal{L}, T, \ell_0, \mathcal{W})$ *is a deterministic Rabin automaton over* Δ.

We refer to the part with the Rabin automaton as the *control part.* There are two distinct notions of size for a product system: the *protocol size* $|\mathcal{PS}|_{\text{prot}} := |Q|$ and the *control size* $|\mathcal{PS}|_{\text{cont}} := |\mathcal{L}|$. The reason for this distinction is that the control size is typically exponential in the size of the LTL formulas, so that keeping track of the two sizes separately will later improve our complexity analysis.

Semantics of Product Systems. A configuration of $\mathcal{PS}$ is an element of $\mathcal{C} := \mathcal{M}(Q) \times \mathcal{L}$. Moreover, we let $\mathcal{C}_0 := \{(\gamma, \ell_0) \mid \gamma \in \mathcal{I}\}$ be the set of initial configurations of the product system. In product systems, unlike in the proof of Proposition 5, the semantics in the PP is modified to match the monotonicity properties of the system. More precisely, we rely on *accelerated semantics* for the IOPP: in $\mathcal{P}$, there is an *accelerated step* from γ to γ' with transition $t \in \Delta$ when there is $k \geqslant 1$ such that $\gamma \xrightarrow{t^k} \gamma'$. Given two configurations $c = (\gamma, \ell), c' = (\gamma', \ell') \in \mathcal{C}$ and transition $t \in \Delta$, we let $c \xrightarrow{t} c'$ when there is $k \geqslant 1$ such that $\gamma \xrightarrow{t^k} \gamma'$ in $\mathcal{P}$ and $\Delta(\ell, t) = \ell'$. A step in the product system corresponds to an accelerated step in $\mathcal{P}$ whose transition is read by $\mathcal{A}$. Note that there is no communication from the control part to the IOPP. In product systems, runs and operators $\mathsf{pre}^*(\cdot)$, $\mathsf{post}^*(\cdot)$ are defined as expected.

4.3 Satisfiability as a Reachability Problem

We fix $\mathcal{P}$ an IOPP, φ an LTL\X formula, $\mathcal{A} = (\mathcal{L}, T, \ell_0, \mathcal{W})$ a deterministic Rabin automaton recognizing φ obtained using Theorem 4, and we let $\mathcal{PS} = (\mathcal{P}, \mathcal{A})$.

Recall that, in $\mathcal{P}$, there is an *accelerated step* from γ to γ' using t when there are $k \geqslant 1$ and $t \in \Delta$ such that $\gamma \xrightarrow{t^k} \gamma'$. A (finite) *accelerated run* is a sequence $\gamma_0, t_1, \gamma_1, \ldots, t_m$ such that, for all $i \in [1, m]$, there is an accelerated step from γ_{i-1} to γ_i using t_i. We similarly define infinite accelerated runs. We extend the notion of strong fairness: an infinite accelerated run α is *strongly fair* when, for every finite accelerated run α', if the first configuration of α' is visited infinitely often in α then α' appears infinitely often in α. A run ρ of $\mathcal{PS}$ can be projected onto $\mathcal{P}$ to obtain an accelerated run of $\mathcal{P}$, denoted $\mathsf{pr}(\rho)$; ρ is called *protocol-fair* when the accelerated run $\mathsf{pr}(\rho)$ is strongly fair. Given an accelerated run $\alpha = \gamma_0, t_1, \gamma_1, t_2, \ldots$, we let $\alpha \models \varphi$ when $t_1 t_2 \ldots \models \varphi$. An accelerated infinite run $\alpha = \gamma_0, t_1, \gamma_1, t_2, \ldots$ is an *acceleration* of an infinite run ρ when there are $k_1, k_2, \ldots \geqslant 1$ such that ρ is of the form $\gamma_0, t_1^{k_1}, \gamma_1, t_2^{k_2}, \gamma_2, \ldots$

Lemma 13. *Given a strongly fair accelerated run α, there is a strongly fair run ρ such that α is an acceleration of ρ. Conversely, given a strongly fair run ρ, there is a strongly fair acceleration α of ρ.*

Proof (sketch). Given a strongly fair accelerated run α, we build ρ by choosing, for each accelerated step, the minimal number of repetitions of the transition. Any finite run ρ' available infinitely often in ρ can be seen as an accelerated finite run α'; it is easy to prove that α' is available infinitely often in α, and thus appears infinitely often in α. By minimality of the number of repetitions, ρ' appears infinitely often in ρ, so that ρ is strongly fair. For the other implication, let ρ be a strongly fair run of $\mathcal{P}$; we build an acceleration of ρ using randomization. To do so, we iteratively pick at random $m \in \mathbb{N}$ and we group the next m transitions if possible; if not, we leave the next transition without accelerating it. Assuming that the probability distribution for m gives non-zero probability for all integers, the obtained run is strongly fair with probability 1. □

For $L \subseteq \mathcal{L}$, we write $\mathcal{C}_L := \Gamma \times L \subseteq \mathcal{C}$; also, for $\mathcal{S} \subseteq \mathcal{C}$, we write $\overline{\mathcal{S}} := \mathcal{C} \setminus \mathcal{S}$. We define $[\![\exists\rho.\,\varphi]\!] := \{\gamma \in \Gamma \mid \exists\rho \in \mathsf{FRuns}(\gamma),\, \rho \models \varphi\}$ the set of configurations γ of $\mathcal{P}$ such that there exists a strongly fair run from γ satisfying formula φ. Similarly, we define $[\![\forall\rho.\,\varphi]\!] := \{\gamma \in \Gamma \mid \forall\rho \in \mathsf{FRuns}(\gamma),\, \rho \models \varphi\} = \Gamma \setminus [\![\exists\rho.\,\neg\varphi]\!]$ the set of configurations γ of $\mathcal{P}$ such that all strongly fair runs from γ satisfy formula φ; in other words, the set of configurations of $\mathcal{P}$ satisfying φ. We characterize these sets using the product system.

Theorem 14. *A configuration γ of $\mathcal{P}$ is in $[\![\exists\rho.\,\varphi]\!]$ if and only if (γ, ℓ_0) is in*

$$\mathcal{S}_{\mathcal{W}} := \mathsf{pre}^*\left(\bigcup_{(F,G)\in\mathcal{W}} \overline{\mathsf{pre}^*(\mathcal{C}_F)} \cap \overline{\mathsf{pre}^*(\overline{\mathsf{pre}^*(\mathcal{C}_G)})}\right)$$

Proof. Let $\gamma \in \Gamma$. By Lemma 13 and Remark 10, $\gamma \in [\![\exists\rho.\,\varphi]\!]$ if and only if there is a strongly fair accelerated run α from γ such that $\alpha \models \varphi$. Let G denote the graph whose vertices are the configurations of the product system reachable from (γ, ℓ_0) and where there is an edge from c to c' whenever $c \xrightarrow{t} c'$ for some $t \in \Delta$. We claim that there is a strongly fair accelerated run α from γ such that $\alpha \models \varphi$ if and only if there is a bottom SCC S of G reachable from (γ, ℓ_0) that is *winning*, *i.e.*, such that there is $(F, G) \in \mathcal{W}$ for which $S \cap \mathcal{C}_G \neq \emptyset$ but $S \cap \mathcal{C}_F = \emptyset$.

The arguments are the same as in the proof of Proposition 5, but with accelerated semantics in $\mathcal{P}$. If we have such an SCC S, it is easy to build a protocol-fair run ρ of $\mathcal{PS}$ that goes to S and visits all configurations in S infinitely often. We let $\alpha := \mathsf{pr}(\rho)$; α is strongly fair and, because S is winning, $\alpha \models \varphi$. Suppose now that we have a strongly fair accelerated run α such that $\alpha \models \varphi$. Let ρ be the run of $\mathcal{PS}$ such that $\mathsf{pr}(\rho) = \alpha$; ρ is protocol-fair. Let S be the SCC visited infinitely often in ρ; S is bottom and ρ visits infinitely often all configurations in S. Indeed, the same arguments as in the proof of Proposition 5 apply, except that we rely on strong fairness of the accelerated run, which makes no difference since strong fairness is defined the same for accelerated and non-accelerated runs.

It remains to prove that there is a winning bottom SCC S reachable from (γ, ℓ_0) if and only if $(\gamma, \ell_0) \in \mathcal{S}_{\mathcal{W}}$. Suppose first that there is such an SCC S; let $c \in S$ and let $(F, G) \in \mathcal{W}$ such that $S \cap \mathcal{C}_G \neq \emptyset$ and $S \cap \mathcal{C}_F = \emptyset$. We have $(\gamma, \ell_0) \in \mathsf{pre}^*(c)$. Since S is bottom and $S \cap \mathcal{C}_F = \emptyset$, we have $\mathsf{post}^*(c) \cap \mathcal{C}_F = \emptyset$ and so $c \in \overline{\mathsf{pre}^*(\mathcal{C}_F)}$. We also have $S = \mathsf{post}^*(S)$, and because $S \cap \mathcal{C}_G \neq \emptyset$, we have $\mathsf{post}^*(S) \subseteq \mathsf{pre}^*(\mathcal{C}_G)$; therefore $S \cap \mathsf{pre}^*(\overline{\mathsf{pre}^*(\mathcal{C}_G)}) = \emptyset$. This proves that $c \in \overline{\mathsf{pre}^*(\mathcal{C}_F)} \cap \overline{\mathsf{pre}^*(\overline{\mathsf{pre}^*(\mathcal{C}_G)})}$; therefore $(\gamma, \ell_0) \in \mathcal{S}_{\mathcal{W}}$. Suppose now that $(\gamma, \ell_0) \in \mathcal{S}_{\mathcal{W}}$. Let $(F, G) \in \mathcal{W}$, $c \in \mathsf{post}^*((\gamma, \ell_0))$ such that $c \in \overline{\mathsf{pre}^*(\mathcal{C}_F)} \cap \overline{\mathsf{pre}^*(\overline{\mathsf{pre}^*(\mathcal{C}_G)})}$. Let S be an SCC reachable from c. We claim that S is winning. Because $S \subseteq \mathsf{post}^*(c)$, we have $S \cap \mathcal{C}_F = \emptyset$. Also, if we had $S \cap \mathcal{C}_G \neq \emptyset$ then any configuration $c_S \in S$ would be in $\overline{\mathsf{pre}^*(\mathcal{C}_G)}$, so that c would be in $\mathsf{pre}^*(\overline{\mathsf{pre}^*(\mathcal{C}_G)})$, a contradiction. $\square$

4.4 *K*-blind Sets

Observe that $\mathcal{P} \models^{\forall} \varphi$ is false if and only if there exists an initial configuration of $\mathcal{P}$ in the set $[\![\exists \rho. \neg\varphi]\!]$. We show, using the characterization of Theorem 14, that set $[\![\exists \rho. \neg\varphi]\!]$ is "nice" in the following sense: if it is non-empty, then it contains a configuration with a bounded number of agents for a doubly exponential bound $B(\varphi, \mathcal{P})$. Checking $\mathcal{P} \models^{\forall} \varphi$ is then achieved by exploring the finite reachability graph of the product system for configurations with less than $B(\varphi, \mathcal{P})$ agents. Moreover, we will show that the set of configurations satisfying a monadic HyperLTL$\setminus$X formula can be decomposed into a Boolean combination of sets of the form $[\![\exists \rho. \varphi]\!]$, for φ an LTL$\setminus$X formula. This will allow us to check $\mathcal{P} \models^{\forall} \psi$ by repeated applications of the exploration procedure described above. We start by formalizing our "nice" sets.

Let $K \in \mathbb{N}$. A set $S \subseteq \Gamma$ of configurations of $\mathcal{P}$ is K*-blind* when, for all $\gamma \in \Gamma$ and $q \in Q$ such that $\gamma(q) \geqslant K$, $\gamma \in S$ if and only if $\gamma + \vec{q} \in S$. Similarly, a set $\mathcal{S} \subseteq \mathcal{C}$ of configurations of $\mathcal{PS}$ is K*-blind* when, for all $(\gamma, \ell) \in \mathcal{C}$ and $q \in Q$ such that $\gamma(q) \geqslant K$, $(\gamma, \ell) \in \mathcal{S}$ if and only if $(\gamma + \vec{q}, \ell) \in \mathcal{S}$.

Example 15. The set $\mathcal{I}$ is 1-blind, because $\gamma \in \mathcal{I}$ if and only if $\gamma(q)$ is non-zero when $q \in I$ and zero otherwise. For the same reason, the set $\mathcal{C}_0 \subseteq \mathcal{C}$ defined above is 1-blind. Also, for all $L \subseteq \mathcal{L}$, the set $\mathcal{C}_L$ is 0-blind.

Lemma 16. *Let $\mathcal{S}_1$ a K_1-blind set and $\mathcal{S}_2$ a K_2-blind set of $\mathcal{PS}$. Then $\mathcal{S}_1 \star \mathcal{S}_2$ is a $\max(K_1, K_2)$-blind set for $\star \in \{\cup, \cap\}$. Additionally, $\overline{\mathcal{S}_1}$ is a K_1-blind set.*

The previous result states that K-blind sets are closed under Boolean operations. Next, we find that K-blind sets are closed under reachability if we enlarge K.

Theorem 17. *Let $\mathcal{S}$ be a K'-blind set of $\mathcal{PS}$. Then $\mathsf{post}^*(\mathcal{S})$ and $\mathsf{pre}^*(\mathcal{S})$ are K-blind sets for $K := |Q|^2 \max(K', 2B)$ where $B = |\mathcal{L}|^{3^{|Q|^2+2} \cdot 2(\log(|Q|^2+2)+1)|Q|^2}$.*

This theorem crucially relies on the immediate observation assumption, its proof is technical and presented in Section 5. Note that K is doubly-exponential in $|Q|$ but polynomial in $|\mathcal{L}|$ and in K', so that this bound is doubly-exponential in $|\varphi|$ if we let $\mathcal{A} = \mathcal{A}_\varphi$ using Theorem 4. Let us apply this result to $[\![\exists\rho.\, \varphi]\!]$:

Lemma 18. *Set $[\![\exists\rho.\, \varphi]\!]$ is K-blind with K doubly-exponential in $|\mathcal{P}|$ and $|\varphi|$.*

Proof. By Theorem 14 we find that $[\![\exists\rho.\, \varphi]\!] \times \{\ell_0\} = \mathcal{S}_{\mathcal{W}}$. The sets $\mathcal{C}_F$ and $\mathcal{C}_G$ are 0-blind for each pair $(F, G) \in \mathcal{W}$. The result follows by iterative applications of Theorem 17 and Lemma 16. □

4.5 LTL and HyperLTL Verification

We now apply the results from the previous sections to the verification of LTL\X and verification of monadic HyperLTL\X for IOPP; we prove that both problems are decidable and in 2-EXPSPACE. For LTL\X, Lemma 18 shows that we only need to check emptiness of a K-blind set for K bounded doubly-exponentially.

Theorem 19. *Verification of LTL\X for IOPP is in 2-EXPSPACE, and the same is true for its existential variant.*

Proof. By Savitch's Theorem, we can present a non-deterministic procedure. Let φ be an LTL\X formula, and $\mathcal{P}$ an IOPP. We construct $\mathcal{A}_\varphi$ using Theorem 4; for this, we pay a doubly-exponential cost in $|\varphi|$, which is the most costly part of the procedure. We work in the product system $\mathcal{PS} := (\mathcal{P}, \mathcal{A}_\varphi)$.

Observe that $\mathcal{P} \models^\forall \varphi$ if and only if $\mathcal{I} \cap [\![\exists\rho.\, \neg\varphi]\!] = \emptyset$, so that it suffices to consider the existential variant. We therefore want to decide whether $[\![\exists\rho.\, \varphi]\!] \cap \mathcal{I} \neq \emptyset$. The set $\mathcal{I}$ is 1-blind; by Lemma 18 and Lemma 16, $\mathcal{I} \cap [\![\exists\rho.\, \varphi]\!]$ is K-blind for K doubly-exponential in the size of $\mathcal{P}$ and in the size of φ.

Hence, $\mathcal{I} \cap [\![\exists\rho.\, \varphi]\!] \neq \emptyset$ if and only if it contains γ_0 such that $\gamma_0(q) \leqslant K$ for all $q \in Q$. We guess such a configuration γ_0. We can write γ_0 in binary, and thus in exponential space. Checking if $\gamma_0 \in \mathcal{I}$ is immediate. By Theorem 14, we can check if $\gamma_0 \in [\![\exists\rho.\, \varphi]\!]$ by checking whether, in the product system $\mathcal{PS} = (\mathcal{P}, \mathcal{A}_\varphi)$, $(\gamma_0, \ell_0) \in \mathcal{S}_{\mathcal{W}} = \bigcup_{(F,G) \in \mathcal{W}} \mathsf{pre}^* \left(\overline{\mathsf{pre}^*(\mathcal{C}_F)} \cap \overline{\mathsf{pre}^*(\overline{\mathsf{pre}^*(\mathcal{C}_G)})}\right)$.

We guess a Rabin pair $(F, G) \in \mathcal{W}$. We only need to consider configurations in $\mathcal{C}_{\gamma_0} := \{(\gamma, \ell) \in \mathcal{C} \mid |\gamma| = |\gamma_0|\}$. Given a set $\mathcal{S} \subseteq \mathcal{C}$ whose membership can be checked in 2-EXPSPACE for configurations in $\mathcal{C}_{\gamma_0}$, checking whether a configuration $c \in \mathcal{C}_{\gamma_0}$ is in $\mathsf{pre}^*(\mathcal{S})$ can also be done in 2-EXPSPACE: guess a run starting at c, step by step. After each step, check if the current configuration c' is in $\mathcal{S}$. We only remember the previous configuration and the current one; checking the step can be done in 2-EXPSPACE because we have constructed $\mathcal{A}_\varphi$ and because, in the protocol, a step corresponds to simple arithmetic operations. For each $H \in \{F, G\}$, checking whether a configuration $c \in \mathcal{C}_{\gamma_0}$ is in $\mathcal{C}_H$ is easy. Therefore, checking whether $c \in \mathcal{C}_{\gamma_0}$ is in $\mathsf{pre}^*(\mathcal{C}_H)$ can be done in 2-EXPSPACE. By iterating this technique and treating Boolean operations in a natural manner, we check whether $(\gamma_0, \ell_0) \in \mathsf{pre}^*(\overline{\mathsf{pre}^*(\mathcal{C}_F) \cap \mathsf{pre}^*(\overline{\mathsf{pre}^*(\mathcal{C}_G)})})$. □

Let ψ be a HyperLTL formula over Δ, we write $[\![\psi]\!] := \{\gamma \in \Gamma \mid \gamma \models \psi\}$. We show that $[\![\psi]\!]$ can be written as a Boolean combination of sets of the form $[\![\exists \rho.\, \varphi]\!]$ with φ an LTL formula. Set $[\![\psi]\!]$ is then K-blind, for K doubly-exponential, as a Boolean combination of K-blind sets.

Lemma 20. *Let $\psi = Q_1 \rho_1. \ldots Q_k \rho_k. \varphi$ be a monadic HyperLTL$\setminus\mathsf{X}$ formula. Set $[\![\psi]\!]$ is K-blind for K doubly-exponential in $|\mathcal{P}|$ and $|\varphi|$.*

Proof. We show K-blindness where K is the bound obtained when applying Lemma 18 on $\mathcal{P}$ and on a formula of size linear in $|\varphi|$. Hence, the bound does not depend on the number of quantifiers of ψ. We proceed by induction on the number of quantifiers $k \geqslant 1$. The base case $k = 1$ is proved by Lemma 18. Let $k \geqslant 2$; suppose that the result holds for any monadic HyperLTL formula with $k-1$ quantifiers. Let $\psi = Q_1 \rho_1. Q_2 \rho_2. \ldots Q_k \rho_k. \varphi$ with φ described as a Boolean combination of φ_1 to φ_n, each referring to a single run variable. Note that $[\![\psi]\!] = \Gamma \setminus [\![\mathsf{neg}(\psi)]\!]$, where $\mathsf{neg}(\psi)$ is the formula obtained from ψ by transforming $\forall$ quantifiers into $\exists$ and vice versa, and by replacing the inner formula φ by $\neg\varphi$. Therefore, we may assume that $Q_1 = \exists$.

Suppose w.l.o.g. that φ_1 to φ_m are the formulas that refer to ρ_1. For every valuation $\nu : [1, m] \to \{\mathit{true}, \mathit{false}\}$, let $\mathsf{Ev}_\nu := \bigwedge_{i=1}^{m} \varphi_i(\rho) \Leftrightarrow \nu(i)$; note that $\mathsf{Ev}_\nu(\rho)$ only has run variable ρ. Let $\varphi[\nu]$ denote the formula φ simplified assuming that, for all $i \in [1, m]$, φ_i has truth value $\nu(i)$. Note that ρ_1 does not appear in $\varphi[\nu]$. Let $\psi_\nu := Q_2 \rho_2 \ldots Q_k \rho_k.\, \varphi[\nu]$. Let $\gamma \in \Gamma$; $\gamma \in [\![\exists \rho_1.\, \mathsf{Ev}_\nu]\!]$ is equivalent to the existence of $\rho_1 \in \mathsf{FRuns}(\gamma)$ such that, for all $i \in [1, m]$, $\rho_1 \models \varphi_i$ iff $\nu(i)$ is true. In words, $\gamma \in [\![\exists \rho_1.\, \mathsf{Ev}_\nu]\!]$ whenever there is $\rho_1 \in \mathsf{FRuns}(\gamma)$ that yields valuation ν. Also, ψ_ν corresponds to ψ simplified under the assumption that run variable ρ_1 yields valuation ν; run variable ρ_1 does not appear in ψ_ν and ψ_ν does not need quantifier Q_1. We deduce that $[\![\psi]\!] = \bigcup_{\nu:[1,m]\to\{\mathit{true},\mathit{false}\}} [\![\exists \rho_1.\, \mathsf{Ev}_\nu]\!] \cap [\![\psi_\nu]\!]$.

For every ν, ψ_ν only has $k-1$ quantifiers; by induction hypothesis, $[\![\psi_\nu]\!]$ is K-blind. This also holds for $[\![\exists \rho_1.\, \mathsf{Ev}_\nu]\!]$ because Ev_ν has size at most linear in $|\varphi|$. Thanks to Lemma 16, we obtain that $[\![\psi]\!]$ is K-blind. □

We can now extend Theorem 19 to monadic HyperLTL$\setminus\mathsf{X}$.

Theorem 21. *Verification of monadic HyperLTL\X for immediate observation population protocols is in 2-*EXPSPACE.

Proof. Again, we present a non-deterministic procedure. Let ψ be a HyperLTL\X formula; as in the proof of Theorem 19, we may consider the existential case only, where one asks whether $[\![\psi]\!] \cap \mathcal{I} \neq \emptyset$. By Lemma 20 and Lemma 16, $[\![\psi]\!] \cap \mathcal{I}$ is K-blind for some doubly-exponential K, so that $[\![\psi]\!] \cap \mathcal{I} \neq \emptyset$ if and only if there is $\gamma \in [\![\psi]\!] \cap \mathcal{I} \cap \Gamma_{\leqslant K}$ where $\Gamma_{\leqslant K} := \{\gamma \mid \forall q, \gamma(q) \leqslant K\}$. We guess such a $\gamma \in \Gamma_{\leqslant K}$. We can write γ in binary, and thus in exponential space. It is easy to check that $\gamma \in \mathcal{I}$. Let $\psi = Q_1\rho_1 \ldots Q_k\rho_k.\varphi$ with φ described as a Boolean combination of φ_1 to φ_n, each referring to a single run variable. For each $j \in [1, k]$, let ℓ_j be the number of φ_i that refer to run variable ρ_j. From the proof of Lemma 20, we can compute a *simple expression* for $[\![\psi]\!]$ in the form of a Boolean combination of *elementary sets* of the form $[\![\exists\rho.\,\varphi']\!]$. Moreover, with a straightforward induction, this simple expression is composed of at most $O(2^{\ell_1+\cdots+\ell_k})$ elementary sets, because the union over the possible valuations has 2^{ℓ_j} disjuncts during induction step j; also, each elementary set formula has size linear in $|\varphi|$. We compute, in exponential time, this simple expression. We check if $\gamma \in [\![\psi]\!]$ by evaluating membership of γ in each elementary set with Theorem 19 using doubly-exponential space, and then evaluating the simple expression. □

5 A Structural Bound in Product Systems

This section is devoted to proving Theorem 17. We rely on the theory of well-quasi-orders (see, *e.g.*, [16]). A *quasi-order* is a set equipped with a transitive and symmetric relation. In a quasi-order $(E, \preceq)$, a set $S \subseteq E$ is *upward-closed* (resp. *downward-closed*) when, for all $s \in S$, for all $t \in E$, if $s \preceq t$ then $t \in S$ (resp. if $t \preceq s$ then $s \in S$); also, $\uparrow S := \{t \in E \mid \exists s \in S, s \preceq t\}$ is its *upward-closure* and $\downarrow S := \{t \in E \mid \exists s \in S, t \preceq s\}$ its *downward-closure*. A *well-quasi-order* is a quasi-order $(E, \preceq)$ such that, for every infinite sequence $(x_i)_{i\in\mathbb{N}}$ of elements of E, there is $i < j$ such that $x_i \preceq x_j$. In a well-quasi-order $(E, \preceq)$, any upward-closed set S has a finite set of minimal elements $\mathsf{basis}(S)$, and $S = \uparrow\mathsf{basis}(S)$.

5.1 Transfer Flows

We fix a product system $\mathcal{PS} = (\mathcal{P}, \mathcal{A})$ with $\mathcal{P} =: (Q, \Delta, I)$ and $\mathcal{A} =: (\mathcal{L}, T, \ell_0, \mathcal{W})$. We prove Theorem 17 using *transfer flows*, an abstraction representing the possibilities offered by sequences of transitions. Let $\mathbb{N}_\# := \mathbb{N} \cup \{\#\}$; we extend $(\mathbb{N}, \leqslant)$ to $(\mathbb{N}_\#, \leqslant)$ where $\#$ is incomparable with integers: for all $x \in \mathbb{N}_\#$, $x \sim \#$ iff $x = \#$ for $\sim \in \{\leqslant, \geqslant\}$. We extend addition by $\# + x = x$ for all $x \in \mathbb{N}_\#$.

Definition 22. *A* transfer flow *is a triplet* $\mathsf{tf} = (f, \ell, \ell')$ *where* $f : Q^2 \to \mathbb{N}_\#$ *and* $\ell, \ell' \in \mathcal{L}$. *We denote by* $\mathcal{F}$ *the set of all transfer flows.*

Intuitively, (f, ℓ, ℓ') represents possible finite runs of $\mathcal{PS}$, with f the transfer of agents in $\mathcal{P}$ and ℓ, ℓ' the start and end states in $\mathcal{A}$. Having $f(q_1, q_2) = \#$

represents the impossibility to send agents from q_1 to q_2, while $f(q_1, q_2) = n$ represents the need to send at least n agents from q_1 to q_2; in this case, any number in $[n, +\infty[$ can be sent. The values ℓ, ℓ' are called the *control part* of tf, while the function f is called the *agent part* of tf. Given a transfer flow $\mathsf{tf} = (f, \ell, \ell') \in \mathcal{F}$, we define its *weight* by $\mathsf{weight}(\mathsf{tf}) := \sum_{q,q'} f(q, q')$.

We define a partial order $\preceq$ on $\mathcal{F}$ as follows. For $\mathsf{tf}_1 = (f_1, \ell_1, \ell_1')$ and $\mathsf{tf}_2 = (f_2, \ell_2, \ell_2')$, we let $\mathsf{tf}_1 \preceq \mathsf{tf}_2$ when $\ell_1 = \ell_1'$, $\ell_2 = \ell_2'$ and, for all q, q', $f_1(q, q') \leqslant f_2(q, q')$. In particular, this requires that, for all q, q', $f_1(q, q') = \#$ if and only if $f_2(q, q') = \#$. It is easy to see that $(\mathcal{F}, \preceq)$ is a well-quasi-order. We highlight the following rule of thumb: *smaller transfer flows are more powerful.* Indeed, when $\mathsf{tf}_1 \preceq \mathsf{tf}_2$, for q, q' such that $f_1(q, q'), f_2(q, q') \neq \#$, $f_1(q, q') \leqslant f_2(q, q')$: tf_1 allows to send from q to q' any number of agents in $[f_1(q, q'), +\infty[$ while tf_2 allows to send from q to q' any number of agents in $[f_2(q, q'), +\infty[\subseteq [f_1(q, q'), +\infty[$.

Definition 23. *Given $c_1 = (\gamma_1, \ell_1), c_2 = (\gamma_2, \ell_2) \in \mathcal{C}$ and $\mathsf{tf} = (f, \ell, \ell') \in \mathcal{F}$, we let $c_1 \xhookrightarrow{\mathsf{tf}} c_2$ when $\ell_1 = \ell$, $\ell_2 = \ell'$ and there is a* step witness *$g : Q^2 \to \mathbb{N}_\#$ such that $f(q, q') \leqslant g(q, q')$ for all $q, q' \in Q$, $\gamma_1(q) = \sum_{q'} g(q, q')$ for all $q \in Q$ and $\gamma_2(q) = \sum_{q'} g(q', q)$ for all $q \in Q$.*

Note that if $c_1 \xhookrightarrow{\mathsf{tf}} c_2$, then $c_1 \xhookrightarrow{\mathsf{tf}'} c_2$ for all $\mathsf{tf}' \preceq \mathsf{tf}$: again, smaller transfer flows are more powerful. Intuitively, g corresponds to a transfer of agents in $\mathcal{PS}$ concretizing $c_1 \xhookrightarrow{\mathsf{tf}} c_2$. We now build transfer flows corresponding to transitions of $\mathcal{PS}$. For each $t = (q_1, q_2) \to (q_1, q_3) \in \Delta$, we define the set $F[t] \subseteq \mathcal{F}$ that contains all transfer flows (f, ℓ, ℓ') such that $T(\ell, t) = \ell'$ holds where T is the transition function of the Rabin automaton and:

- if $q_1 \neq q_2$ or $q_1 \neq q_3$ then $f(q_1, q_1) \geqslant 1, f(q_2, q_3) \geqslant 1$;
- if $q_1 = q_2 = q_3$ then $f(q_1, q_1) \geqslant 2$;
- for all $q \neq q_1$ such that $(q, q) \neq (q_2, q_3)$, $f(q, q) \geqslant 0$;
- for all $q \neq q'$ such that $(q, q') \neq (q_2, q_3)$, $f(q, q') = \#$.

That is, at least one agent is in q_1, some agents are sent from q_2 to q_3 and the control part is changed according to t. The set $F[t]$ is upward-closed with respect to $\preceq$: the number of agents going from q_2 to q_3 can be arbitrarily large, which corresponds to an accelerated step of $\mathcal{P}$ using transition t.

Lemma 24. *For all $c, c' \in \mathcal{C}$, $t \in \Delta$, $c \xrightarrow{t} c'$ iff there is $\mathsf{tf} \in F[t]$ s.t. $c \xhookrightarrow{\mathsf{tf}} c'$.*

We define the product set $\mathsf{tf}_1 \otimes \mathsf{tf}_2 \subseteq \mathcal{F}$ of two transfer flows. This set is meant to encode the possibilities given by using tf_1 followed by tf_2. Let $\mathsf{tf}_1 = (f_1, \ell_1, \ell_1'), \mathsf{tf}_2 = (f_2, \ell_2, \ell_2') \in \mathcal{F}$. If $\ell_1' \neq \ell_2$, then we set $\mathsf{tf}_1 \otimes \mathsf{tf}_2 = \emptyset$. Assume now $\ell_1' = \ell_2$. The set $\mathsf{tf1} \otimes \mathsf{tf}_2$ contains all transfer flows of the form (h, ℓ_1, ℓ_2') for which there is a *product witness* $H : Q^3 \to \mathbb{N}_\#$ such that:

(prod.i) for all (q_1, q_3), $\sum_{q_2} H(q_1, q_2, q_3) = h(q_1, q_3)$;
(prod.ii) for all (q_1, q_2), $\sum_{q_3} H(q_1, q_2, q_3) \geqslant f_1(q_1, q_2)$;
(prod.iii) for all (q_2, q_3), $\sum_{q_1} H(q_1, q_2, q_3) \geqslant f_2(q_2, q_3)$.

In particular, for all q_1, q_2, $f_1(q_1, q_2) = \#$ if and only if, for all q_3, $H(q_1, q_2, q_3) = \#$. Similarly, $f_2(q_2, q_3) = \#$ if and only if, for all q_1, $H(q_1, q_2, q_3) = \#$. We extend $\otimes$ to sets of transfer flows: for $F, F' \subseteq \mathcal{F}$, $F \otimes F' := \bigcup_{\mathsf{tf} \in F, \mathsf{tf}' \in F'} \mathsf{tf} \otimes \mathsf{tf}'$.

Lemma 25. *Let* $\mathsf{tf}_1, \mathsf{tf}_2, \mathsf{tf}_3 \in \mathcal{F}$. *We have the following properties:*

(25.i) *the set* $\mathsf{tf}_1 \otimes \mathsf{tf}_2$ *is upward-closed with respect to* $\preceq$;
(25.ii) *for all* $\mathsf{tf}_1' \preceq \mathsf{tf}_1$ *and* $\mathsf{tf}_2' \preceq \mathsf{tf}_2$, $\mathsf{tf}_1 \otimes \mathsf{tf}_2 \subseteq \mathsf{tf}_1' \otimes \mathsf{tf}_2'$;
(25.iii) $\otimes$ *is associative:* $(\mathsf{tf}_1 \otimes \mathsf{tf}_2) \otimes \mathsf{tf}_3 = \mathsf{tf}_1 \otimes (\mathsf{tf}_2 \otimes \mathsf{tf}_3)$;
(25.iv) *for every* $\mathsf{tf} \in \mathsf{basis}(\mathsf{tf}_1 \otimes \mathsf{tf}_2)$, $\mathsf{weight}(\mathsf{tf}) \leqslant \mathsf{weight}(\mathsf{tf}_1) + \mathsf{weight}(\mathsf{tf}_2)$.

Example 26. Consider Fig. 1. Let $\mathsf{tf}_1{=}(f_1, \ell_1, \ell_2)$ and $\mathsf{tf}_2{=}(f_2, \ell_2, \ell_3)$, with $f_1(q_1, q_2){=}2$, $f_2(q_2, q_3){=}3$, $f_1(q, q){=}f_2(q, q){=}0$ for all q, $f_2(q_2, q_1){=}0$ and all other values equal to $\#$. Let $\mathsf{tf}{=}(f, \ell_1, \ell_3)$, with $f(q_1, q_1){=}1$, $f(q_1, q_3){=}1$, $f(q_2, q_3){=}2$, $f(q_2, q_2){=}f(q_3, q_3) = f(q_1, q_2){=}f(q_2, q_1){=}0$ and $f(q, q'){=}\#$ for all other (q, q'). We have $\mathsf{tf} \in \mathsf{tf}_1 \otimes \mathsf{tf}_2$. Indeed, we have a product witness H defined by $H(q_1, q_2, q_1){=}1$, $H(q_1, q_2, q_3){=}1$, $H(q_2, q_2, q_3){=}2$, $H(q_1, q_2, q_2){=}H(q_2, q_2, q_1)$ ${=}H(q_2, q_2, q_2){=}H(q_3, q_3, q_3){=}H(q_1, q_1, q_1){=}0$ and all other values equal to $\#$. In fact, tf is minimal for $\preceq$ in $\mathsf{tf}_1 \otimes \mathsf{tf}_2$.

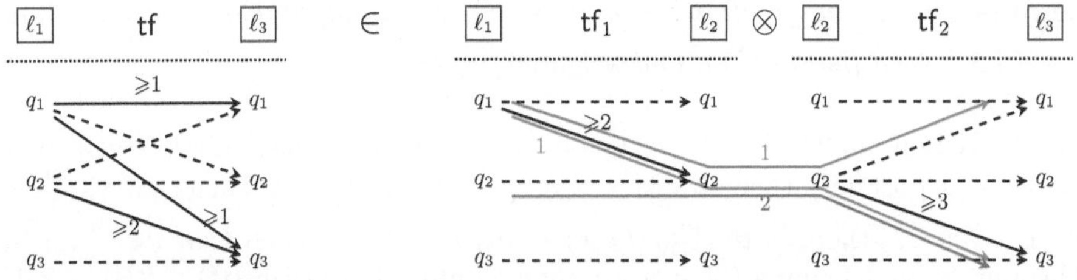

Fig. 1. Dashed arrows correspond to value 0, no arrow corresponds to $\#$. The product witness H is represented with colored arrows. We do not depict H when its value is 0.

Given a sequence $t_1 \dots t_k$ of transitions, we let $F[t_1 \dots t_k] := F[t_1] \otimes F[t_2] \otimes \dots \otimes F[t_k]$. For the empty sequence ϵ, we define $F[\epsilon]$ as the set of (f, ℓ, ℓ') where $\ell = \ell'$, $f(q, q) \in \mathbb{N}$ for all q and $f(q, q') = \#$ for all $q \neq q'$. For all upward-closed sets $F \subseteq \mathcal{F}$, we have $F \otimes F[\epsilon] = F[\epsilon] \otimes F = F$. Observe that, for every $t_1 \dots t_k$, all transfer flows $(f, \ell, \ell') \in F[t_1 \dots t_k]$ are such that $f(q, q) \in \mathbb{N}$ for all q.

Lemma 27. *For all* $k \geqslant 0$, *for all* $t_1, t_2 \dots, t_k \in \Delta$, *and for all* $c, c' \in \mathcal{C}$, $c \xrightarrow{t_1 \dots t_k} c'$ *if and only if there exists* $\mathsf{tf} \in F[t_1 \dots t_k]$ *such that* $c \overset{\mathsf{tf}}{\hookrightarrow} c'$.

Proof (sketch). The proof is by induction on k. The difficult case is $k = 2$: there exists $\mathsf{tf} \in \mathsf{tf}_1 \otimes \mathsf{tf}_2$ s.t. $c_1 \overset{\mathsf{tf}}{\hookrightarrow} c_3$ iff there exists $c_2 \in \mathcal{C}$ s.t. $c_1 \overset{\mathsf{tf}_1}{\hookrightarrow} c_2 \overset{\mathsf{tf}_2}{\hookrightarrow} c_3$. First, if there is $\mathsf{tf} \in \mathsf{tf}_1 \otimes \mathsf{tf}_2$ for which $c_1 \overset{\mathsf{tf}}{\hookrightarrow} c_3$ then we let H be a witness for $\mathsf{tf} \in \mathsf{tf}_1 \otimes \mathsf{tf}_2$, and we build the multiset μ_2 of c_2 by $\mu_2 : q_2 \in Q \mapsto \sum_{q_1, q_3} H(q_1, q_2, q_3)$.

Conversely, given $c_2 \in \mathcal{C}$ such that $c_1 \xhookrightarrow{\mathsf{tf}_1} c_2 \xhookrightarrow{\mathsf{tf}_2} c_3$, let h_1 be a step witness for $c_1 \xhookrightarrow{\mathsf{tf}_1} c_2$ and h_2 for $c_2 \xhookrightarrow{\mathsf{tf}_1} c_3$; we build a product witness $H : Q^3 \to \mathbb{N}_\#$ such that $\sum_{q_3} H(q_1, q_2, q_3) = h_1(q_1, q_2)$ and $\sum_{q_1} H(q_1, q_2, q_3) = h_2(q_2, q_3)$, which is possible because $\sum_{q_1} h_1(q_1, q_2) = \sum_{q_3} h_2(q_2, q_3) = \mu_2(q_2)$. □

Given $T \subseteq \Delta^*$, we let $F[T] := \bigcup_{w \in T} F[w]$. For all $k \geqslant 0$, we denote by $\Delta^{\leqslant k} \subseteq \Delta^*$ the set of sequences of length at most k. Let $m = |Q|$ and $M = |\mathcal{L}|$. We prove Theorem 17 using the following theorem, which we prove in Section 5.3.

Theorem 28 (Structural theorem). *Let* $\mathsf{B} := (M+1)^{3^{m^2+2} \cdot 2(\log(m^2+2)+1)m^2}$. *We have* $F[\Delta^{\leqslant B}] = F[\Delta^*]$ *and elements of* $\mathsf{basis}(F[\Delta^*])$ *have norm at most* $2B$.

5.2 Proof of Theorem 17

Again, we write $m = |Q|$ and $M = |\mathcal{L}|$. Let $K' \geqslant 0$, $K := m^2 \max(K', 2B)$ and $\mathcal{S}$ a K'-blind set. We prove that $\mathsf{post}^*(\mathcal{S})$ is K-blind; the proof for $\mathsf{pre}^*(\mathcal{S})$ is similar. We start with the following observation.

Lemma 29. *A configuration c is in* $\mathsf{post}^*(\mathcal{S})$ *if and only if there are* $c_\mathcal{S} \in \mathcal{S}$ *and* $\mathsf{tf} \in F[\Delta^*]$ *such that* $c_\mathcal{S} \xhookrightarrow{\mathsf{tf}} c$ *and* $\mathsf{weight}(\mathsf{tf}) \leqslant 2B$.

Proof. By Lemma 27, if we have such $c_\mathcal{S}$ and tf then $c \in \mathsf{post}^*(\mathcal{S})$. Conversely, if $c = (\gamma, \ell) \in \mathsf{post}^*(\mathcal{S})$, there are $c_\mathcal{S} = (\gamma_\mathcal{S}, \ell_\mathcal{S}) \in \mathcal{S}$ and $w \in \Delta^*$ such that $\gamma_\mathcal{S} \xrightarrow{w} \gamma$. By Lemma 27, there is $\mathsf{tf} = (f, \ell_\mathcal{S}, \ell) \in F[w] \subseteq F[\Delta^*]$ such that $c_\mathcal{S} \xhookrightarrow{\mathsf{tf}} c$; by Theorem 28 and Lemma (25.iv), one may assume that $\mathsf{weight}(\mathsf{tf}) \leqslant 2B$. □

Let $c = (\gamma, \ell) \in \mathcal{C}$ and $q \in Q$ be such that $\gamma(q) \geqslant K$; we show that $(\gamma, \ell) \in \mathsf{post}^*(\mathcal{S})$ if and only if $(\gamma + \vec{q}, \ell) \in \mathsf{post}^*(\mathcal{S})$. First, suppose that $c = (\gamma, \ell) \in \mathsf{post}^*(\mathcal{S})$. Let $\mathsf{tf}, c_\mathcal{S} = (\gamma_\mathcal{S}, \ell_\mathcal{S})$ obtained thanks to Lemma 29. Let $g : Q^2 \to \mathbb{N}_\#$ be a step witness for $c_\mathcal{S} \xhookrightarrow{\mathsf{tf}} c$. We have $\sum_{r \in Q} g(r, q) = \gamma(q) \geqslant K$. By the pigeonhole principle, there is r such that $g(r, q) \geqslant \frac{K}{m^2} \geqslant K'$ therefore $\gamma_\mathcal{S}(r) \geqslant K'$. Let g' be such that $g'(q_1, q_2) = g(q_1, q_2)$ for all $(q_1, q_2) \neq (r, q)$ and $g'(r, q) = g(r, q) + 1$; g' is a witness that $(\gamma_\mathcal{S} + \vec{r}, \ell_\mathcal{S}) \xhookrightarrow{\mathsf{tf}} (\gamma + \vec{q}, \ell)$. Thanks to Lemma 27, this proves that $(\gamma_\mathcal{S} + \vec{r}, \ell_\mathcal{S}) \xrightarrow{*} (\gamma + \vec{q}, \ell)$. Because $\mathcal{S}$ is K'-blind, we conclude that $(\gamma + \vec{q}, \ell) \in \mathsf{post}^*(\mathcal{S})$. Conversely, suppose that $(c + \vec{q}, \ell) \in \mathsf{post}^*(\mathcal{S})$. With the same reasoning as above, we obtain $c_\mathcal{S} = (\gamma_\mathcal{S}, \ell_\mathcal{S}) \in \mathcal{S}, \mathsf{tf} = (f, \ell_\mathcal{S}, \ell), g, r$ such that g is a witness that $c_\mathcal{S} \xhookrightarrow{\mathsf{tf}} c$. By the pigeonhole principle, there is r such that $g(r, q) \geqslant K' + 1$ and $g(r, q) \geqslant 2B + 1 > f(r, q)$. Because $\mathcal{S}$ is K'-blind and $\gamma_\mathcal{S}(r) \geqslant g(r, q) \geqslant K' + 1$, we have $(\gamma_\mathcal{S} - \vec{r}, \ell_\mathcal{S}) \in \mathcal{S}$. Let $g'(q_1, q_2) = g(q_1, q_2)$ for all $(q_1, q_2) \neq (r, q)$ and $g'(r, q) = g(r, q) - 1$. Because $g' \geqslant f$, g' is a step witness that $(\gamma_\mathcal{S} - \vec{r}, \ell_\mathcal{S}) \xhookrightarrow{\mathsf{tf}} (\gamma, \ell)$. Thanks to Lemma 27, this proves that $(\gamma, \ell) \in \mathsf{post}^*(\mathcal{S})$.

5.3 Proving the Structural Theorem with Descending Chains

To prove Theorem 28, we use a result bounding the length of descending chains in $\mathbb{N}^d$ from [34,42]. We recall the result and some definitions. Let $d \geqslant 1$. Given $\vec{v}$ of $\mathbb{N}^d$ and $i \in [1,d]$, we denote by $\vec{v}(i)$ its i-th component. Let $\leqslant_\times$ be the order over $\mathbb{N}^d$ such that $\vec{u} \leqslant_\times \vec{v}$ if and only if, for all $i \in [1,d]$, $\vec{u}(i) \leqslant \vec{v}(i)$. The obtained $(\mathbb{N}^d, \leqslant_\times)$ is a well-quasi-order (Dickson's lemma [17]). A *descending chain* is a sequence $D_0 \supsetneq D_1 \supsetneq D_2 \ldots$ of sets $D_k \subseteq \mathbb{N}^d$ that are downward-closed for $\leqslant_\times$. Because $(\mathbb{N}^d, \leqslant_\times)$ is a well-quasi-order, all descending chains have finite length, *i.e.*, are of the form $D_0, \ldots, D_\ell$ with $\ell \in \mathbb{N}$. To bound the length of descending chains [34,42] we need the sequence to be *controlled* and *ω-monotone*.

We extend $\mathbb{N}$ to $\mathbb{N}_\omega := \mathbb{N} \cup \{\omega\}$ with $n < \omega$ for all $n \in \mathbb{N}$. Given $\vec{v} \in \mathbb{N}_\omega^d$, its *norm* $||\vec{v}||$ is the largest $\vec{v}(i)$ that is not ω. An *ideal* I is the downward-closure in $\mathbb{N}^d$ of a vector $\vec{v} \in \mathbb{N}_\omega^d$, *i.e.*, $I = \downarrow\{\vec{v}\} \cap \mathbb{N}^d$; its *norm* $||I||$ is $||\vec{v}||$. A downward-closed set $D \subseteq \mathbb{N}^d$ is canonically represented as a finite union of ideals; its *norm* $||D||$ is the maximum of the norms of its ideals. Given $N > 0$ and a descending chain (D_k), we call (D_k) *N-controlled* when, for all k, $||D_k|| \leqslant (k+1)N$. In a descending chain (D_k), an ideal I is proper at step k if I is in the canonical representation of D_k but $I \not\subseteq D_{k+1}$. The sequence (D_k) is *ω-monotone* if, when an ideal I_{k+1} represented by some vector $\vec{v}_{k+1}$ is proper at step $k+1$, there is I_k that is proper at step k and that is represented by $\vec{v}_k$ such that, for all $i \in [1,d]$, if $\vec{v}_{k+1}(i) = \omega$ then $\vec{v}_k(i) = \omega$.

Theorem 30 ([42]). *Let $d, n > 0$. Every descending chain (D_k) of $\mathbb{N}^d$ that is n-controlled and ω-monotone has length at most $n^{3^d(\log(d)+1)}$.*

We now use this bound to prove Theorem 28. Recall that we write $m = |Q|$ and $M = |\mathcal{L}|$. Let $d := m^2 + 2$ and $N := M^2 \cdot 2^{m^2} = |\mathcal{L}^2 \times 2^{Q^2}|$. We fix two arbitrary bijective mappings $\theta : \mathcal{L}^2 \times 2^{Q^2} \to [1, N]$ and $\mathsf{index} : Q^2 \to [1, m^2]$. We map transfer flows to sets of elements of $\mathbb{N}^d$ with $\chi : \mathcal{F} \to 2^{\mathbb{N}^d}$. Let $\mathsf{tf} = (f, \ell, \ell') \in \mathcal{F}$ and $S := \{(q, q') \mid f(q, q') = \#\}$. A vector $\vec{v} \in \mathbb{N}^d$ is in $\chi(\mathsf{tf})$ when:

- for all (q, q') such that $f(q, q') \neq \#$, $\vec{v}(\mathsf{index}(q, q')) = f(q, q')$;
- $\vec{v}(m^2 + 1) = \theta(\ell, \ell', S)$;
- $\vec{v}(m^2 + 2) = N + 1 - \theta(\ell, \ell', S)$.

Note that there is no restriction to $\vec{v}(i)$ when the corresponding pair $(q, q') = \mathsf{index}^{-1}(i)$ is such that $f(q, q') = \#$. Also, if $\vec{v} \in \chi(\mathsf{tf})$ and $\vec{u} \in \chi(\mathsf{tf}')$ are such that $\vec{v} \leqslant_\times \vec{u}$, then $\vec{u}(m^2+1) = \vec{v}(m^2+1)$ and $\vec{u}(m^2+2) = \vec{v}(m^2+2)$, so that tf and tf' have the same states of $\mathcal{L}$ and the same $\#$ components. For $\mathsf{tf} \neq \mathsf{tf}'$, we have $\chi(\mathsf{tf}), \chi(\mathsf{tf}') \neq \emptyset$ but $\chi(\mathsf{tf}) \cap \chi(\mathsf{tf}') = \emptyset$, a property that we call *strong injectivity* of χ. The vectors of $\mathbb{N}^d \cap \chi(\mathcal{F})$ are exactly those whose last two components are strictly positive and sum to $N+1$. We build a decreasing chain (D_k) such that $D_k \cap \chi(\mathcal{F}) = \chi(\mathcal{F} \setminus F[\Delta^{\leqslant k}])$.

Let V_0 denote the set of vectors $\vec{v}$ such that either $(\vec{v}(m^2+1), \vec{v}(m^2+2)) = (N+1, 0)$ or $(\vec{v}(m^2+1), \vec{v}(m^2+2)) = (0, N+1)$. For technical reasons (related to ω-monotonicity), we will enforce that $D_k \cap V_0 = \emptyset$ for every k. Note that $V_0 \cap \chi(\mathcal{F}) = \emptyset$: vectors in V_0 have no relevance in terms of transfer flows. For all

$k \geqslant 0$, let $U_k := \uparrow\chi(F[\Delta^{\leqslant k}]) \cup V_0$, and let $D_k = \mathbb{N}^d \setminus U_k$; (D_k) is a decreasing chain because all D_k are downward-closed and $F[\Delta^{\leqslant k}] \subseteq F[\Delta^{\leqslant k+1}]$ for all k.

Lemma 31. *For all k, $U_k \cap \chi(\mathcal{F}) = \chi(F[\Delta^{\leqslant k}])$ and $D_k \cap \chi(\mathcal{F}) = \chi(\mathcal{F} \setminus F[\Delta^{\leqslant k}])$.*

Note that if $D_{k+1} = D_k$ then, by Lemma 31, $\chi(F[\Delta^{\leqslant k+1}]) = \chi(F[\Delta^{\leqslant k}])$ and, by injectivity of χ, $F[\Delta^{\leqslant k+1}] = F[\Delta^{\leqslant k}]$. This means that if $D_{k+1} = D_k$ then $F[\Delta^{\leqslant k}]$ is stable under product by $F[t]$ for all t, hence that $F[\Delta^*] = F[\Delta^{\leqslant k}]$. Let L be the smallest $k \in \mathbb{N}$ such that $D_k \neq D_{k-1}$; it exists because $(\mathbb{N}^d, \leqslant_\times)$ is a well-quasi-order. To prove Theorem 28, we want $L \leqslant N^{3^d(\log(d)+1)}$. To apply Theorem 30, we need to prove that (D_k) is $(N+1)$-controlled and ω-monotone.

Transfer flows in $\mathsf{basis}(F[\Delta])$ have weight bounded by 2. Let $\mathsf{tf} \in \mathsf{basis}(F[\Delta^{\leqslant k}])$, there are $\ell \leqslant k$ and $\mathsf{t}_1, \ldots, \mathsf{t}_\ell \in \mathsf{basis}(F[\Delta])$ such that $\mathsf{tf} \in \mathsf{t}_1 \otimes \ldots \otimes \mathsf{t}_\ell$. A straightforward induction using (25.ii) proves that $\mathsf{weight}(\mathsf{tf}) \leqslant 2\ell \leqslant 2k$. This proves that minimal elements of $F[\Delta^{\leqslant k}]$ have weight bounded by $2k$. In turn, this bounds the norm of minimal elements of U_k by $\max(N+1, 2k)$. Because $D_k = \mathbb{N}^d \setminus U_k$, this last bound applies to the norm of D_k.

Lemma 32. *(D_k) is $(N+1)$-controlled and ω-monotone.*

We apply Theorem 30 on (D_k) to prove that (D_k) and (U_k) stabilize at index at most $(N+1)^{3^d(\log(d)+1)} \leqslant (M+1)^{3^{m^2+2} \cdot 2(\log(m^2+2)+1)m^2} = B$, so that $F[\Delta^*] = F[\Delta^{\leqslant B}]$. By above, transfer flows in $\mathsf{basis}(F[\Delta^{\leqslant k}])$ have weight bounded by k, therefore transfer flows of $\mathsf{basis}(F[\Delta^*]) = \mathsf{basis}(F[\Delta^{\leqslant B}])$ have weight at most $2B$. This concludes the proof of Theorem 28.

6 Conclusion

When compared to the NEXPTIME result for LTL\X verification of shared-memory systems with pushdown machines [29], our 2-EXPSPACE LTL result may seem weak. However, their techniques are quite specific, while ours are generic, enabling us to go from LTL to monadic HyperLTL with little extra work. Additionally, we believe transfer flows, K-blind sets and the results thereof apply to other problems and systems, such as reconfigurable broadcast networks [15] or asynchronous shared-memory systems [22], which enjoy a similar monotonicity property to IOPP.

Most problems considered in this paper are undecidable; this was to be expected for infinite-state systems. However, our decidability result (Theorem 21) sheds light on a decidable fragment, suggesting that further research on verification of hyperproperties for infinite-state systems should be pursued.

Acknowledgement. We thank the reviewers for their helpful comments and suggestions to improve readability. This publication is part of the grant PID2022-138072OB-I00, funded by MCIN, FEDER, UE. This work has been partially supported by the PRODIGY Project (TED2021-132464B-I00) funded by MCIN and the European Union NextGeneration and by the ESF, as well as by a research grant from Nomadic Labs and the Tezos Foundation.

References

1. Alistarh, D., Gelashvili, R.: Recent Algorithmic Advances in Population Protocols. SIGACT News **49**(3), 63–73 (2018). https://doi.org/10.1145/3289137.3289150
2. Angluin, D., Aspnes, J., Diamadi, Z., Fischer, M.J., Peralta, R.: Computation in Networks of Passively Mobile Finite-state Sensors. Distributed Comput. **18**(4), 235–253 (2006). https://doi.org/10.1007/s00446-005-0138-3
3. Angluin, D., Aspnes, J., Eisenstat, D., Ruppert, E.: The Computational Power of Population Protocols. Distributed Comput. **20**(4), 279–304 (2007). https://doi.org/10.1007/s00446-007-0040-2
4. Baier, C., Katoen, J.: Principles of model checking. MIT Press (2008)
5. Baker, H.G.: Rabin's proof of the undecidability of the reachability set inclusion problem of vector addition systems. Massachusetts Institute of Technology, Project MAC (1973)
6. van Bergerem, S., Guttenberg, R., Kiefer, S., Mascle, C., Waldburger, N., Weil-Kennedy, C.: Verification of Population Protocols with Unordered Data. In: 51st International Colloquium on Automata, Languages, and Programming, ICALP 2024. LIPIcs, vol. 297, pp. 156:1–156:20. Schloss Dagstuhl - Leibniz-Zentrum für Informatik (2024). https://doi.org/10.4230/LIPICS.ICALP.2024.156
7. Beutner, R., Finkbeiner, B.: Software Verification of Hyperproperties Beyond k-Safety. In: Proc. of the 34th Int'l Conf. on Computer Aided Verification (CAV'22), Part I. LNCS, vol. 13371, pp. 341–362. Springer (2022). https://doi.org/10.1007/978-3-031-13185-1_17
8. Blondin, M., Ladouceur, F.: Population Protocols with Unordered Data. In: 50th International Colloquium on Automata, Languages, and Programming, ICALP 2023, July 10-14, 2023, Paderborn, Germany. LIPIcs, vol. 261, pp. 115:1–115:20. Schloss Dagstuhl - Leibniz-Zentrum für Informatik (2023). https://doi.org/10.4230/LIPICS.ICALP.2023.115
9. Bonakdarpour, B., Prabhakar, P., Sánchez, C.: Model Checking Timed Hyperproperties in Discrete-Time Systems. In: Proc. of the 12th NASA Formal Methods Symposium (NFM'20). LNCS, vol. 12229, pp. 311–328. Springer (2020)
10. Bonakdarpour, B., Sánchez, C., Schneider, G.: Monitoring Hyperproperties by Combining Static Analysis and Runtime Verification. In: Proc. of the 8th Int'l Symposium on Leveraging Applications of Formal Methods, Verification and Validation (ISoLA'18), Part II. LNCS, vol. 11245, pp. 8–27. Springer (2018)
11. Clarkson, M.R., Finkbeiner, B., Koleini, M., Micinski, K.K., Rabe, M.N., Sánchez, C.: Temporal Logics for Hyperproperties. In: Proc. of the 3rd Conference on Principles of Security and Trust (POST 2014). LNCS, vol. 8414, pp. 265–284. Springer (2014). https://doi.org/10.1007/978-3-642-54792-8_15
12. Clarkson, M.R., Schneider, F.B.: Hyperproperties. Journal of Computer Security **18**(6), 1157–1210 (2010). https://doi.org/10.3233/JCS-2009-0393
13. Coenen, N., Finkbeiner, B., Hahn, C., Hofmann, J.: The Hierarchy of Hyperlogics. In: Proc. 34th LICS. pp. 1–13. IEEE (2019). https://doi.org/10.1109/LICS.2019.8785713
14. Czerwinski, W., Orlikowski, L.: Reachability in Vector Addition Systems is Ackermann-complete. In: 2021 IEEE 62nd Annual Symposium on Foundations of Computer Science (FOCS). IEEE (2022). https://doi.org/10.1109/focs52979.2021.00120
15. Delzanno, G., Sangnier, A., Traverso, R., Zavattaro, G.: On the Complexity of Parameterized Reachability in Reconfigurable Broadcast Net-

works. In: IARCS Annual Conference on Foundations of Software Technology and Theoretical Computer Science, FSTTCS 2012. LIPIcs, vol. 18, pp. 289–300. Schloss Dagstuhl - Leibniz-Zentrum für Informatik (2012). https://doi.org/10.4230/LIPICS.FSTTCS.2012.289
16. Demri, S., Finkel, A., Goubault-Larrecq, J., Schmitz, S., Schnoebelen, P.: Well-Quasi-Orders for Algorithms. Lecture Notes, MPRI Course 2.9.1 – 2017/2018 (2017), https://wikimpri.dptinfo.ens-cachan.fr/lib/exe/fetch.php?media=cours:upload:poly-2-9-1v02oct2017.pdf
17. Dickson, L.E.: Finiteness of the Odd Perfect and Primitive Abundant Numbers with n Distinct Prime Factors. American Journal of Mathematics **35**(4), 413–422 (1913), http://www.jstor.org/stable/2370405
18. Elsässer, R., Radzik, T.: Recent Results in Population Protocols for Exact Majority and Leader Election. Bulletin of the EATCS **126** (2018)
19. Esparza, J.: Population Protocols: Beyond Runtime Analysis. In: Reachability Problems - 15th International Conference, RP 2021, Liverpool, UK, October 25-27, 2021, Proceedings. Lecture Notes in Computer Science, vol. 13035, pp. 28–51. Springer (2021). https://doi.org/10.1007/978-3-030-89716-1_3
20. Esparza, J., Ganty, P., Leroux, J., Majumdar, R.: Model Checking Population Protocols. In: 36th IARCS Annual Conference on Foundations of Software Technology and Theoretical Computer Science, FSTTCS 2016, December 13-15, 2016, Chennai, India. LIPIcs, vol. 65, pp. 27:1–27:14. Schloss Dagstuhl - Leibniz-Zentrum für Informatik (2016). https://doi.org/10.4230/LIPICS.FSTTCS.2016.27
21. Esparza, J., Ganty, P., Leroux, J., Majumdar, R.: Verification of Population Protocols. Acta Informatica **54**(2), 191–215 (2017). https://doi.org/10.1007/S00236-016-0272-3
22. Esparza, J., Ganty, P., Majumdar, R.: Parameterized Verification of Asynchronous Shared-Memory Systems. In: Computer Aided Verification - 25th International Conference, CAV 2013. Lecture Notes in Computer Science, vol. 8044, pp. 124–140. Springer (2013). https://doi.org/10.1007/978-3-642-39799-8_8
23. Esparza, J., Kretínský, J., Sickert, S.: One Theorem to Rule Them All: A Unified Translation of LTL into ω-Automata. CoRR **abs/1805.00748** (2018), http://arxiv.org/abs/1805.00748
24. Esparza, J., Raskin, M.A., Weil-Kennedy, C.: Parameterized Analysis of Immediate Observation Petri Nets. In: Application and Theory of Petri Nets and Concurrency - 40th International Conference, PETRI NETS 2019, Aachen, Germany, June 23-28, 2019, Proceedings. Lecture Notes in Computer Science, vol. 11522, pp. 365–385. Springer (2019). https://doi.org/10.1007/978-3-030-21571-2_20
25. Etessami, K.: A note on a question of Peled and Wilke regarding stutter-invariant LTL. Inf. Process. Lett. **75**(6), 261–263 (2000). https://doi.org/10.1016/S0020-0190(00)00113-7
26. Farzan, A., Vandikas, A.: Automated Hypersafety Verification. In: Proc. of CAV 2019. LNCS, vol. 11561, pp. 200–218 (2019). https://doi.org/10.1007/978-3-030-25540-4 11
27. Finkbeiner, B., Rabe, M.N., Sánchez, C.: A Temporal Logic for Hyperproperties. CoRR **abs/1306.6657** (2013), http://arxiv.org/abs/1306.6657
28. Fischer, M.J., Ladner, R.E.: Propositional Dynamic Logic of Regular Programs. J. Comput. Syst. Sci. **18**(2), 194–211 (1979). https://doi.org/10.1016/0022-0000(79)90046-1
29. Fortin, M., Muscholl, A., Walukiewicz, I.: Model-Checking Linear-Time Properties of Parametrized Asynchronous Shared-Memory Pushdown Systems. In: Computer

Aided Verification - 29th International Conference, CAV 2017, Heidelberg, Germany, July 24-28, 2017, Proceedings, Part II. Lecture Notes in Computer Science, vol. 10427, pp. 155–175. Springer (2017). https://doi.org/10.1007/978-3-319-63390-9_9
30. Goguen, J.A., Meseguer, J.: Security Policies and Security Models. In: IEEE Symposium on Security and Privacy. pp. 11–20. IEEE Computer Society (1982). https://doi.org/10.1109/SP.1982.10014
31. Gutsfeld, J.O., Müller-Olm, M., Ohrem, C.: Propositional Dynamic Logic for Hyperproperties. In: Proc. 31st CONCUR. pp. 50:1–50:22. LIPIcs 171, Schloss Dagstuhl - Leibniz-Zentrum für Informatik (2020). https://doi.org/10.4230/LIPIcs.CONCUR.2020.50
32. Hack, M.: The Equality Problem for Vector Addition Systems is Undecidable. Theor. Comput. Sci. **2**(1), 77–95 (1976). https://doi.org/10.1016/0304-3975(76)90008-6
33. Jancar, P., Valusek, J.: Structural Liveness of Immediate Observation Petri Nets. Fundam. Informaticae **188**(3), 179–215 (2022). https://doi.org/10.3233/FI-222146
34. Lazic, R., Schmitz, S.: The ideal view on Rackoff's coverability technique. Inf. Comput. **277**, 104582 (2021). https://doi.org/10.1016/J.IC.2020.104582
35. Leroux, J.: The Reachability Problem for Petri Nets is Not Primitive Recursive. In: 2021 IEEE 62nd Annual Symposium on Foundations of Computer Science (FOCS). IEEE (2022). https://doi.org/10.1109/focs52979.2021.00121
36. McLean, J.D.: A General Theory of Composition for a Class of "Possibilistic" Properties. IEEE Trans. Software Eng. **22**(1), 53–67 (1996). https://doi.org/10.1109/32.481534
37. Minsky, M.L.: Computation: Finite and Infinite Machines. Prentice-Hall, Inc. (1967)
38. Peled, D.A., Wilke, T.: Stutter-Invariant Temporal Properties are Expressible Without the Next-Time Operator. Inf. Process. Lett. **63**(5), 243–246 (1997). https://doi.org/10.1016/S0020-0190(97)00133-6
39. Pnueli, A.: The Temporal Logic of Programs. In: Proc. of the 18th IEEE Symp. on Foundations of Computer Science (FOCS'77). pp. 46–67. IEEE CS Press (1977)
40. Rabe, M.N.: A temporal logic approach to information-flow control. Ph.D. thesis, Saarland University (2016)
41. Sabelfeld, A., Sands, D.: A per model of secure information flow in sequential programs. Higher Order Symbolic Computation **14**(1), 59–91 (2001)
42. Schmitz, S., Schütze, L.: On the Length of Strongly Monotone Descending Chains over ℕ^d. In: 51st International Colloquium on Automata, Languages, and Programming, ICALP 2024, July 8-12, 2024, Tallinn, Estonia. LIPIcs, vol. 297, pp. 153:1–153:19. Schloss Dagstuhl - Leibniz-Zentrum für Informatik (2024). https://doi.org/10.4230/LIPICS.ICALP.2024.153
43. Shemer, R., Gurfinkel, A., Shoham, S., Vizel, Y.: Property Directed Self Composition. In: Proc. of CAV'19. LNCS, vol. 11560, pp. 161–179. Springer (2019). https://doi.org/10.1007/978-3-030-25540-4 9
44. Sistla, A.P., Vardi, M.Y., Wolper, P.: The Complementation Problem for Büchi Automata with Applications to Temporal Logic. Theoretical Computer Science **49**, 217–237 (1987). https://doi.org/10.1016/0304-3975(87)90008-9
45. Sousa, M., Dillig, I.: Cartesian Hoare logic for verifying k-safety properties. In: Proc. of ACM SIGPLAN Conference on Programming Language Design and Implementation (PLDI'16). ACM (2016). https://doi.org/10.1145/2908080.2908092

46. Unno, H., Terauchi, T., Koskinen, E.: Constraint-based Relational Verification. In: Proc. of CAV 2021. LNCS, vol. 12759, pp. 742—-766. Springer (2021). https://doi.org/10.1007/978-3-030-81685-8 35
47. Wang, Y., Zarei, M., Bonakdarpour, B., Pajic, M.: Statistical Verification of Hyperproperties for Cyber-Physical Systems. ACM Transactions on Embedded Computing systems **18**(5s), 92:1–92:23 (2019)
48. Zdancewic, S., Myers, A.C.: Observational Determinism for Concurrent Program Security. In: Proc. 16th IEEE CSFW-16. pp. 29–43. IEEE Computer Society (2003). https://doi.org/10.1109/CSFW.2003.1212703

BiGKAT: An Algebraic Framework forRelational Verification of ProbabilisticPrograms

Leandro Gomes[1](✉), Patrick Baillot[1], and Marco Gaboardi[2]

[1] Université de Lille, CNRS, Inria, Centrale Lille, UMR 9189 CRIStAL, F-59000 Lille, France
{leandrogomes.moreiragomes,patrick.baillot}@univ-lille.fr
[2] Boston University, Boston, USA
gaboardi@bu.edu

Abstract. This work is devoted to formal reasoning on relational properties of probabilistic imperative programs. Relational properties are properties which relate the execution of two programs (possibly the same one) on two initial memories. We aim at extending the algebraic approach of Kleene Algebras with Tests (KAT) to relational properties of probabilistic programs. For that we consider the approach of Guarded Kleene Algebras with Tests (GKAT), which can be used for representing probabilistic programs, and define a relational version of it, called Bi-guarded Kleene Algebras with Tests (BiGKAT) together with a semantics. We show that the setting of BiGKAT is expressive enough to encode a finitary version of probabilistic Relational Hoare Logic (pRHL) (without the While rule), a program logic that has been introduced in the literature for the verification of relational properties of probabilistic programs. We also discuss the additional expressivity brought by BiGKAT.

Keywords: Kleene algebra with tests · Relational reasoning · Probabilistic programs · Hoare logic

1 Introduction

Formal verification of program properties has triggered a variety of methods, among which the algebraic approach of Kleene Algebra with Tests (KAT) stands out as an elegant, simple and automatizable framework [18,16]. It is closely related to modeling with finite automata and has stimulated the development of techniques from coalgebra for reasoning about program behavior, for instance based on bisimulation checking [12]. It has also been implemented in a library for the Coq proof-assistant [19]. Among the properties one might want to check on programs, some important ones are expressed by relating the execution of two programs on two initial states, or of the same program on two initial states. They are called *relational properties* or *2-properties*. One can think for instance of simulation properties, refinements, or extensional equivalence. Another example is non-interference: assume the variables are divided into public ones and private ones, a program satisfies non-interference if the final value of public variables after an execution only depends on the initial value of public variables.

P. A. Abdulla and D. Kesner (Eds.): FoSSaCS 2025, LNCS 15691, pp. 243–264, 2025.
https://doi.org/10.1007/978-3-031-90897-2_12

Actually in a large number of situations the software systems one wants to verify are not deterministic but admit a probabilistic behaviour. Think for instance of randomized algorithms, cryptography, network programming or differential privacy. In those scenarios many crucial properties are also relational ones. For instance in cryptography one can express the fact that a randomized encryption scheme is safe as a probabilistic non-interference property: a public variable is assigned a ciphered value, obtained from a private variable, and we want to ensure that one cannot distinguish between two ciphered values computed from the same private initial state. Similarly in differential privacy (see e.g. [5]), in order to protect private data one might want to verify that two executions of a given program on two databases that differ only by one individual give indistinguishable result.

In order to express and prove relational properties on imperative programs some specific methods have been introduced. First in the deterministic case let us mention Relational Hoare Logic [10], that extends the classic Floyd-Hoare logic approach to reason on pairs of programs. This approach has been upgraded to the setting of probabilistic relational Hoare Logic (pRHL) by Barthe and coauthors [6]. It has then been extensively applied to the verification of cryptographic schemes, in particular through the development of the Certicrypt [9] and Easycrypt [3] tools.

However one would still benefit from additional techniques for the automation and the understandability of such reasoning methods. In particular one difficulty with (probabilistic) relational Hoare Logic is to find a suitable alignment of the two programs in order to be able, in a second step, to find the intermediate properties needed for the proof (see [1]). Algebraic methods coming from Kleene algebra with tests are promising in these respects. In particular they facilitate the reasoning on simple program transformations. Such direction is already addressed with the introduction of BiKAT [1], allowing to apply the KAT approach to reasoning on pairs of programs.

Our goal is thus to introduce a KAT approach to reason on relational properties of probabilistic programs. Unfortunately standard KAT techniques cannot be applied directly to probabilistic programs, since there is no known probabilistic interpretation for KAT. To handle this question, recent progress was made by the introduction of Guarded Kleene Algebra with tests (GKAT) [22], in which non-deterministic union and iteration are replaced by guarded union and iteration, while being sufficiently expressive to model imperative programming languages. The main motivation for the introduction of GKAT was initially to design a more efficient version of KAT where the complexity of the decision procedure is reduced, but it was also shown that GKAT admits a probabilistic model that can be used to interpret probabilistic programs.

Our strategy is thus to adapt the relational extension BiKAT to the setting of GKAT, in order to apply this relational approach to pairs of probabilistic programs. In practice we will consider programs of an imperative language extended with some probabilistic primitives. We call the corresponding calculus BiGKAT and in this paper define for it a syntax, a denotational semantics and a theory, which we prove to be sound. We would like to demonstrate the expressivity of our framework by showing how probabilistic relational Hoare Logic reasoning can be encoded in it, in an analogous way as (standard) Hoare logic can be encoded in KAT [17] and Relational Hoare Logic in BiKAT [1] (see 1). In the present work we make a first step in this direction, by showing that a finitary version of pRHL (without the While rule) can be encoded.

Property nature	Unary	Relational deterministic	Relational probabilistic
Hoare logic	HL	RHL	(finitary) pRHL
Algebra	KAT	BiKAT	**BiGKAT**
Examples of properties	Partial correctness	Validation of program transformations	Probabilistic non-interference

Table 1. Program logics and algebras

Outline. We first recall the definition of Guarded Kleene algebra with tests GKAT (Sect. 2), then introduce BiGKAT (Sect. 3) and describe its syntax, semantics, theory, as well as en encoding of a subset of pRHL. We present en example in Sect. 4 before discussing in Sect. 5 the differences between BiGKAT and pRHL and more generally related work in Sect. 6. We finish in Sect. 7 with conclusions and perspectives of future work.

Because of space constraints some proofs are omitted in this paper but can be found in the Appendix of the technical report [15].

2 Guarded Kleene algebra with tests

This section recalls the language and the semantics of *Guarded Kleene Algebra with Tests (GKAT)* [22], an abstraction of imperative programs where conditionals ($c_1 +_b c_2$) and loops ($c^{(b)}$) are expressions guarded by Boolean predicates b. As explained before, the structure is a restriction of KAT in which we are not allowed to freely use operators $+$ and * to build expressions, i.e. GKAT does not allow nondeterminism. Although less expressive than KAT, GKAT offers two advantages: decidability in (almost) linear time (compared to PSPACE complexity of decidability in KAT), and better foundation for probabilistic applications. Although the first one was the main motivation to introduce the structure [22], we are more interested in the second advantage for the purpose of this paper.

2.1 Syntax

The syntax of GKAT is defined with a set of *actions* Σ and a finite set of primitive tests $\mathtt{T}$, which are disjoint. We denote actions by a and primitive tests by p. The sets of Boolean expressions $\mathtt{BExp}$ (also called tests) and GKAT expressions $\mathtt{Exp}$ (also called programs) are then defined by the following grammars:

$b, b_1, b_2 \in \texttt{BExp} ::=$

|	0	**false**
|	1	**true**
|	$p \in \texttt{T}$	p
|	$b_1 \cdot b_2$	b_1 **and** b_2
|	$b_1 + b_2$	b_1 **or** b_2
|	$\bar{b}$	**not** b

$c, c_1, c_2 \in \texttt{Exp} ::=$

|	$a \in \Sigma$	**do** a
|	$b \in \texttt{BExp}$	**assert** b
|	$c_1 \cdot c_2$	$c_1; c_2$
|	$c_1 +_b c_2$	**if** b **then** c_1 **else** c_2
|	$c^{(b)}$	**while** b **do** c

where, for any $b, b_1, b_2 \in \texttt{BExp}$, operators $\cdot$, $+$ and $^-$ denote conjunction, disjunction and negation, respectively, and, for any $c, c_1, c_2 \in \texttt{Exp}$, the operator $\cdot$ denotes sequential composition. The notations on the r.h.s. are given to help intuition and will sometimes be used when writing programs. Command **skip** will be a shorthand for **assert** 1, which is encoded by the Boolean expression 1. The precedence of the operators is the usual one, i.e. the operator $\cdot$ has higher precedence than operator $+_b$, and $()^{(b)}$ has higher precedence than $\cdot$.[3] To simplify the writing, we often omit the operator $\cdot$ by writing c_1c_2 for the expression $c_1 \cdot c_2$, for any $c_1, c_2 \in \texttt{Exp}$.

We are interested in using GKAT for representing probabilistic programs. For that, let us first fix a few definitions. Given a set S, $\mathcal{D}(S)$ is the set of *probability sub-distributions*[4] over S with discrete support, i.e. the set of functions $\mu : S \to [0,1]$ such that $Supp(\mu) = \{x \in S \mid \mu(x) > 0\}$ is discrete and μ sums up to at most 1, i.e. $\sum_{s \in S} \mu(s) \leq 1$. In particular, the *Dirac* distribution $\delta_s \in \mathcal{D}(S)$ is the following map: $w \to \begin{cases} 1, \text{ if } w = s \\ 0, \text{ otherwise.} \end{cases}$

Example 1 (Imperative programming language). Take a set $\texttt{Var}$ of variables and a set $\texttt{Distr}$ of sub-distributions over $\mathbb{R}$ with discrete support. Consider a simple imperative programming language defined by the following grammar:

$$\begin{aligned}
&\textit{terms } t \in \texttt{Terms} ::= x \in \texttt{Var} \mid r \in \mathbb{R} \mid t_1 + t_2 \mid t_1 - t_2 \mid t_1 \times t_2 \\
&\textit{distributions } d \in \texttt{Distr} \\
&\textit{tests } b \in \texttt{Tests} ::= \textbf{false} \mid \textbf{true} \mid t_1 < t_2 \mid t_1 = t_2 \mid \textbf{not } b \mid b_1 \textbf{ and } b_2 \mid b_1 \textbf{ or } b_2 \\
&\textit{commands } c \in \texttt{Comm} ::= \textbf{skip} \mid x \leftarrow t \mid x \overset{\$}{\leftarrow} d \mid c_1; c_2 \mid \textbf{if } b \textbf{ then } c_1 \textbf{ else } c_2 \mid \textbf{while } b \textbf{ do } c
\end{aligned}$$

This language can be modeled in GKAT by taking as sets of actions and primitive tests respectively $\Sigma = \{x \leftarrow t,\ x \overset{\$}{\leftarrow} d \mid x \in \texttt{Var}, t \in \texttt{Terms}, d \in \texttt{Distr}\}$ and $\texttt{T} = \{t_1 < t_2,\ t_1 = t_2 \mid t_1, t_2 \in \texttt{Terms}\}$[5] The first action evaluates term t and

[3] For example the GKAT expression $c_1^{(b_1)} \cdot c_2 +_{b_2} c_3$ reads as $((c_1^{(b_1)}) \cdot c_2) +_{b_2} c_3$.

[4] Some examples of distributions are the tossing of a fair coin, with probability 0.5 for 0 and 1, and the (discrete version of the) Laplacian distribution $\mathcal{L}_p(a)$ centered in a with parameter p. The density function of $\mathcal{L}_p(a)$ is given by $\frac{1}{2p}\exp(\frac{|x-a|}{p})$.

[5] Note that technically speaking according to the definition of GKAT the set $\texttt{T}$ should be chosen finite, which is not the case here, but as observed in [22] Sect. 2.3 Example 2.5 we can use a finite subset $\texttt{T}'$ of $\texttt{T}$ for reasoning on pairwise equivalence of programs which terminate.

assigns the result to x; the second one samples from d and assigns the result to x.

Observe that while programs c may be probabilistic, due to the use of samplings, the tests b as for them are deterministic, i.e. they do not use any probabilistic primitives. In particular the conditional branching in programs is only done on deterministic tests.

2.2 Semantics

We now present the semantic interpretation of GKAT that we will be using, the *Probabilistic model* [22] [6]. We first review some basic concepts needed for the semantics. Given a statement ϕ over a set S, the *Iverson bracket* $[\phi]$ is the function on S taking value 1 on $s \in S$ if $\phi(s)$ is true and 0 if it is false. Typical models of probabilistic imperative programming languages interpret programs as *Markov kernels* on a set S, i.e. maps from S to probability distributions. The semantic model defined below interprets programs as *sub-Markov* kernels F, G... i.e. Markov kernels on sub-distributions.

Two particular examples of sub-Markov kernel on S are id_S and 0_S respectively defined by, for any s, $s' \in S$: $id_S(s) = \delta_s$ and $0_S(s)(s') = 0$.

The composition of two sub-Markov kernels F, G is $F; G$ defined by $(F; G)(s)(s') = \sum_{s'' \in S} F(s)(s'') \times G(s'')(s')$. This composition is associative, admits id_S as identity and 0_S as absorbing element: for any F, $F; id_S = id_S; F = F$ and $F; 0_S = 0_S; F = 0_S$.

Definition 1 (Probabilistic interpretation). *Let* $i = (S, eval, sat)$ *be a triple:*

- S *is a set of states,*
- *for each action* $a \in \Sigma$, $eval(a) : S \to \mathcal{D}(S)$ *is a sub-Markov kernel,*
- *for each primitive test* $p \in T$, $sat(p) \subseteq S$ *is a set of states.*

We define $sat^\dagger : \texttt{BExp} \to 2^S$ as the lifting of $sat : \texttt{T} \to 2^S$ to arbitrary Boolean expressions over BExp. The *probabilistic interpretation* of an expression c with respect to i is the sub-Markov kernel $\mathcal{P}_i[\![c]\!] : S \to \mathcal{D}(S)$ defined as follows:

$$\mathcal{P}_i[\![a]\!] := eval(a) \qquad \mathcal{P}_i[\![c_1 +_b c_2]\!](s) := \begin{cases} \mathcal{P}_i[\![c_1]\!](s), \text{ if } s \in sat^\dagger(b) \\ \mathcal{P}_i[\![c_2]\!](s), \text{ if } s \in sat^\dagger(\bar{b}) \end{cases}$$

$$\mathcal{P}_i[\![b]\!](s) := \begin{cases} \delta_s, \text{ if } s \in sat^\dagger(b) \\ 0, \text{ otherwise} \end{cases} \qquad \mathcal{P}_i[\![c^{(b)}]\!](s)(s') := \lim_{n \to \infty} \mathcal{P}_i[\![(c +_b 1)^n \cdot \bar{b}]\!](s)(s')$$

$$\mathcal{P}_i[\![c_1 \cdot c_2]\!] := \mathcal{P}_i[\![c_1]\!]; \mathcal{P}_i[\![c_2]\!]$$

Intuitively $\mathcal{P}_i[\![c]\!](s)(s')$ is the probability that the execution of c on initial state s terminates on state s', and $\sum_{s' \in S} \mathcal{P}_i[\![c]\!](s)(s')$ is the probability that the execution of c on initial state s terminates (we then also say that it is a successful

[6] Note that more interpretations of GKAT are presented in [22], namely a relational model and a trace model.

execution). Observe thus that we really need to consider sub-distributions and not only distributions. Note that the definition implies that $\mathcal{P}_i[\![1]\!] = id_S$ and $\mathcal{P}_i[\![0]\!] = 0_S$. In the sequel, when the interpretation i is fixed we will sometimes write $[\![c]\!]$ instead of $\mathcal{P}_i[\![c]\!]$.

One can observe that:

$$\mathcal{P}_i[\![c \cdot b]\!](s_1)(s_2) = \begin{cases} \mathcal{P}_i[\![c]\!](s_1)(s_2), \text{ if } s_2 \in sat^\dagger(b) \\ 0, \text{ otherwise} \end{cases} \tag{1}$$

$$\mathcal{P}_i[\![b \cdot c]\!](s_1)(s_2) = \begin{cases} \mathcal{P}_i[\![c]\!](s_1)(s_2), \text{ if } s_1 \in sat^\dagger(b) \\ 0, \text{ otherwise} \end{cases} \tag{2}$$

In the following we will consider programs over a finite set of variables $\mathtt{Var}$ and the set of states will be the set of *memories*, that we denote by m, i.e. functions in $\mathtt{Var} \to D$ where D is the domain of values (we can take for instance $D = \mathbb{Q}$, the rational numbers). If $x \in \mathtt{Var}$ and m is a memory, then $m[x \leftarrow t]$ is the memory identical to m except that it maps x to the evaluation of t in memory m. The interpretation of actions $a \in \mathbf{Exp}$ as sub-Markov kernels is then given by $eval(x \leftarrow t)(m) := \delta_{m[x \leftarrow t]}$ and $eval(x \xleftarrow{\$} d)(m) := \sum_{t \in Supp(d)} d(t) \cdot \delta_{m[x \leftarrow t]}$.

Example 2. Let us consider the example of the uniform distribution over the two-elements boolean set $Bool = \{tt, ff\}$ (or unbiased coin), that we call *dbool*: $dbool(tt) = dbool(ff) = 1/2$.

Then we have $eval(x \xleftarrow{\$} dbool)(m) = 1/2 \cdot \delta_{m[x \leftarrow tt]} + 1/2 \cdot \delta_{m[x \leftarrow ff]}$.

2.3 Axioms

The theory of GKAT introduced in [22] is given by the axioms from Fig. 1. Note

$$c +_b c = c \tag{3}$$
$$c_1 +_b c_2 = c_2 +_{\neg b} c_1 \tag{4}$$
$$(c_1 +_{b_1} c_2) +_{b_2} c_3 = c_1 +_{b_1 \cdot b_2} (c_2 +_{b_2} c_3) \tag{5}$$
$$c_1 +_b c_2 = b \cdot c_1 +_b c_2 \tag{6}$$
$$c_1 \cdot c_3 +_b c_2 \cdot c_3 = (c_1 +_b c_2) \cdot c_3 \tag{7}$$
$$(c_1 \cdot c_2) \cdot c_3 = c_1 \cdot (c_2 \cdot c_3) \tag{8}$$
$$0 \cdot c = 0 \tag{9}$$
$$c \cdot 0 = 0 \tag{10}$$
$$1 \cdot c = c \tag{11}$$
$$c \cdot 1 = c \tag{12}$$
$$c^{(b)} = c \cdot c^{(b)} +_b 1 \tag{13}$$
$$(c +_{b_2} 1)^{(b_1)} = (b_2 \cdot c)^{(b_1)} \tag{14}$$
$$\frac{c_3 = c_1 \cdot c_3 +_b c_2}{c_3 = c_1^{(b)} \cdot c_2} \text{ if } E(c_1) = 0 \tag{15}$$

Fig. 1. Axiomatisation of Guarded Kleene algebra with tests

in particular for the fixpoint axiom (15). Intuitively, it says that if expression c_3 chooses (using guard b) between executing c_1 and looping again, and executing c_2, then c_3 is a b-guarded loop followed by c_2. However, the rule is not sound in general (see [22] for more details). In order to overcome such limitation, the side condition $E(c_1) = 0$ is introduced, ensuring that command c_1 is productive, i.e. that it performs some action. To this end, the function E is inductively defined as follows: $E(b) := b$, $E(a) := 0$, $E(c_1 +_b c_2) := b \cdot E(c_1) + \neg b \cdot E(c_2)$, $E(c_1 \cdot c_2) := E(c_1) \cdot E(c_2)$, $E(c^{(b)}) := \neg b$. We can see $E(c)$ as the weakest test that guarantees that command c terminates successfully but does not perform any action.

Moreover, note particularly the following observation: in KAT the encoding $c_1; (b; c_2 + \neg b; c_3) = c_1; b; c_2 + c_1; \neg b; c_3$ is not an **if-then-else** statement; it is rather a nondeterministic choice between executing c_1, then testing b and executing c_2, and executing c_1, then testing $\neg b$ and executing c_3. That is why left distributivity does not hold in GKAT for any $c \in \mathrm{Exp}$; it only holds for the particular case of $b \in \mathrm{BExp}$, i.e. if b is a test.

In [15] we list additional derivable equations in GKAT, also given in [22].

We already mentioned that GKAT does not allow to construct an arbitrary program by using freely the nondeterministic choice operator $+$, allowing only guarded choice $+_b$, for any $b \in \mathrm{BExp}$. However, the $+$ operator is included in the grammar of BExp, representing the Boolean disjunction. Since BExp$\subseteq$Exp, the grammar allows to write expressions as $b_1 +_b b_2$, for any $b \in \mathrm{BExp}$.

By Boolean reasoning, we can observe that $b \cdot b + \neg b \cdot \neg b = 1$. Such property will be useful later to prove the soundness of R-Case rule (39).

3 Bi-guarded Kleene algebra with tests

We will define an algebra called BiGKAT, based on GKAT and which will allow us to reason on relations between two probabilistic programs, that we refer to as left and right programs. Just as GKAT it will be defined by a grammar of tests and a grammar of expressions. They can be thought of as tests over a product state space $S \times S$ and probabilistic programs over the same product state space. We will give them a semantics of sub-Markov kernels over $S \times S$.

3.1 Syntax

The syntax of BiGKAT is defined using that of GKAT and is additionnally parameterized by a set of actions $\mathbb{\Sigma}$ and a finite set of primitive tests $\mathbb{P}$ which are disjoint. Elements of $\mathbb{\Sigma}$ are denoted as A and those of $\mathbb{P}$ are denoted as P. A typical choice will be to consider in $\mathbb{P}$ some tests relating variables of the two programs on the two state spaces such as for instance $[x = x']$, and to consider in $\mathbb{\Sigma}$ some couplings between random assignments. The language of *Bi-guarded Kleene algebra with tests (BiGKAT)* consists of expressions in $\ddot{\mathrm{E}}\mathrm{xp}$ constructed from GKAT expressions c, c' in Exp, as follows:

$$B, B_1, B_2 \in \ddot{\mathrm{B}}\mathrm{Exp} ::= \ddot{0} \mid \ddot{1} \mid P \mid \langle b| \mid |b'\rangle \mid \bar{\neg} B \mid B_1 \mathbin{;} B_2 \mid B_1 \oplus B_2$$

$$C, C_1, C_2 \in \ddot{\mathrm{E}}\mathrm{xp} ::= A \mid \langle c| \mid |c\rangle \mid \{c \underset{C}{\mid} c'\} \mid B \mid C_1 \mathbin{;} C_2 \mid C_1 \oplus_B C_2 \mid C^{(B)}$$

We will sometimes omit $\fatsemi$ and write C_1C_2 for $C_1 \fatsemi C_2$, and $\overline{B}$ for $\neg B$.

We define the notation $\langle \_|\_ \rangle$ as $\langle c|c' \rangle \stackrel{\text{def}}{=} \langle c| \fatsemi |c' \rangle$.

Note that in the expression $\{c \underset{C}{|} c'\}$, c and c' belong to Exp (so GKAT) while the bottom C belongs to $\ddot{\text{E}}$xp (BiGKAT). The intuition behind this expression is that C is a joint kernel over $S \times S$, whose left and right projections are respectively c and c', and that $\{c \underset{C}{|} c'\}$ denotes C itself. We will make this intuition more precise in the following section by defining the semantic interpretation by sub-Markov kernels.

3.2 Semantics of BiGKAT

As for interpreting GKAT we were using sub-Markov kernels over a given set S, now for BiGKAT for interpreting pairs of programs we will consider sub-Markov kernels over the product S^2. We will denote sub-Markov kernels on S by lower-case letters c, c', c_1 ...and those on S^2 by upper-case letters C, C', C_1, D .... In order to recover the interpretation of the left and right programs we will need the notion of projections or marginals. For that, given a sub-distribution μ on S^2 we denote its left and right marginals respectively as: $\Pi_1(\mu)(s) = \sum_{s' \in S} \mu(s, s')$, $\Pi_2(\mu)(s') = \sum_{s \in S} \mu(s, s')$. Moreover if E is a subset of S^2, denote $\Pi_1(E) = \{s / \exists s' \in S, (s, s') \in E\}$ and $\Pi_2(E) = \{s' / \exists s \in S, (s, s') \in E\}$.

Example 3. Let us consider the set $Bool = \{tt, ff\}$ and the distributions on the product $Bool^2$ defined on Fig. 2 (the subscripts p, s, a are respectively for product, symmetric and antisymmetric). The array notation here means that the value of $\mu_p(x, x')$ is given by the coefficient on the line (resp. column) given by x (resp. x').

μ_p :

$x \backslash x'$	tt	ff
tt	1/4	1/4
ff	1/4	1/4

μ_s :

$x \backslash x'$	tt	ff
tt	1/2	0
ff	0	1/2

μ_a :

$x \backslash x'$	tt	ff
tt	0	1/2
ff	1/2	0

Fig. 2. Three distributions on $Bool^2$

By computing the left and right marginals, one obtains the following equalities, where $dbool$ is the uniform distribution on $Bool$ (see Example 2): $\Pi_1(\mu_p) = \Pi_1(\mu_s) = \Pi_1(\mu_a) = dbool$, $\quad \Pi_2(\mu_p) = \Pi_2(\mu_s) = \Pi_2(\mu_a) = dbool$.

A simple way of constructing sub-Markov kernels on S^2 is by using products. The product of two sub-Markov kernels c and c' on S is defined as $c \times c'$: $(s_1, s_1') \to ((s_2, s_2') \to c(s_1)(s_2) \times c'(s_1')(s_2'))$.

Let us show an example of sub-Markov kernel on S^2 in the case where S is a state of memories, as in Sect. 1. Assume μ is a distribution over a product space D^2 where D is a domain for a variable x (for instance $D = Bool$, as in Example 3). We define the sub-Markov kernel C_μ in a way similar to $eval(x \stackrel{\$}{\leftarrow} d)$ in Sect. 1 but over S^2: $C_\mu(m, m') = \sum_{(t,t') \in Supp(\mu)} \mu(t, t') \cdot \delta_{(m[x \leftarrow t], m'[x' \leftarrow t'])}$. In other words: $C_\mu(m,m')(m_1,m_1') = 0$ if there exists $y \neq x$ such that $m_1(y) \neq m(y)$ or $y' \neq x'$ such that $m_1'(y') \neq m'(y')$, and otherwise $C_\mu(m,m')(m_1,m_1') = \mu(t,t')$ where $t = m_1(x)$, $t' = m_1'(x')$. So Example 3 for instance allows to define the sub-Markov kernels C_{μ_p}, C_{μ_s}, C_{μ_a}.

Let us now consider the marginals of the sub-distributions $C_\mu(m,m')$:

$$\begin{aligned}\Pi_1(C_\mu(m,m'))(m_1) &= \sum_{m_1' \in S} \sum_{(t,t') \in Supp(\mu)} \mu(t,t') \cdot \delta_{(m[x \leftarrow t], m'[x' \leftarrow t'])}(m_1, m_1') \\ &= \begin{cases} 0 & \text{, if } \forall (t,t') \in Supp(\mu), m_1 \neq m[x \leftarrow t] \\ \sum_{t'/(t,t') \in Supp(\mu)} \mu(t,t') = \Pi_1(\mu)(t) & \text{, if } m_1 = m[x \leftarrow t] \end{cases}\end{aligned}$$

$$\begin{aligned}\Pi_1(C_\mu(m,m')) &= \sum_{t \in \Pi_1(Supp(\mu))} \Pi_1(\mu)(t) \cdot \delta_{m[x \leftarrow t]} \\ \Pi_2(C_\mu(m,m')) &= \sum_{t' \in \Pi_2(Supp(\mu))} \Pi_2(\mu)(t') \cdot \delta_{m'[x' \leftarrow t']}\end{aligned}$$

Example 4. In the case where μ is anyone of the three distributions μ_p, μ_s, μ_a of Example 3 one thus obtains, following the definition in Example 2:

$$\Pi_1(C_\mu(m,m')) = eval(x \stackrel{\$}{\leftarrow} dbool)(m), \;\; \Pi_2(C_\mu(m,m')) = eval(x' \stackrel{\$}{\leftarrow} dbool)(m')$$

Definition 2. *We say that two sub-Markov kernels C_1, C_2 on S^2 are equivalent, denoted by $C_1 \equiv C_2$, if $\forall_{(s,s') \in S^2}, i = 1, 2, \Pi_i(C_1(s,s')) = \Pi_i(C_2(s,s'))$.*

For instance we can deduce from Example 4 that: $C_{\mu_p} \equiv C_{\mu_s} \equiv C_{\mu_a}$.

Definition 3. *We say that a sub-Markov kernel C on S^2 is separable if there exists sub-Markov kernels $\mathcal{L}^C(.)$, $\mathcal{R}^C(.)$ on S such that, for all (s,s'): $\Pi_1(C(s,s')) = \mathcal{L}^C(s)$ and $\Pi_2(C(s,s')) = \mathcal{R}^C(s')$.*

Example 4 shows that C_{μ_p}, C_{μ_s} and C_{μ_a} are separable. Now, given a probabilistic interpretation $\mathcal{P}_i[\![.]\!]$ of GKAT we want to define a probabilistic interpretation $\overline{\mathcal{P}_i}[\![.]\!]$ of BiGKAT.

Definition 4 (Probabilistic interpretation of BiGKAT). *A probabilistic interpretation of BiGKAT is defined by a probabilistic interpretation $i = (State, eval, sat)$ of GKAT and two functions Eval and Sat such that:*

- *for each action $A \in \ddot{\Sigma}$, $Eval(A) : S^2 \to \mathcal{D}(S^2)$ is a sub-Markov kernel,*
- *for each primitive test $P \in \ddot{\mathbb{P}}$, $Sat(P) \subseteq S^2$ is a set of states.*

We want to define the interpretation $\overline{\mathcal{P}_i}[\![B]\!]$ and $\overline{\mathcal{P}_i}[\![C]\!]$ of expressions in B̈Exp and Ëxp. This interpretation will actually only be partially defined, that is to say in some cases $\overline{\mathcal{P}_i}[\![C]\!]$ is undefined.

For B in B̈Exp, $\overline{\mathcal{P}_i}[\![B]\!]$ is defined in the same way as $\mathcal{P}_i[\![b]\!]$ for b in BExp, using Sat.

For $A \in \Sigma$, $\overline{\mathcal{P}_i}[\![A]\!]$ is defined as $\overline{\mathcal{P}_i}[\![A]\!] = \text{Eval}(A)$. The interpretations of constructs $\overline{\mathcal{P}_i}[\![C_1 \fatsemi C_2]\!]$, $\overline{\mathcal{P}_i}[\![C_1 \oplus_B C_2]\!]$, $\overline{\mathcal{P}_i}[\![C^{(B)}]\!]$ are defined in the same way as the interpretations of the similar constructs of GKAT expressions with $\mathcal{P}_i[\![.]\!]$, except that the state is now S^2 instead of S.

There thus only remains to define $\overline{\mathcal{P}_i}[\![\langle c|]\!]$, $\overline{\mathcal{P}_i}[\![|c\rangle]\!]$ and $\overline{\mathcal{P}_i}[\![\{c \underset{C}{|} c'\}]\!]$.

$\overline{\mathcal{P}_i}[\![\langle c|]\!]$ is defined as $\overline{\mathcal{P}_i}[\![\langle c|]\!] = \mathcal{P}_i[\![c]\!] \times id_S$.
$\overline{\mathcal{P}_i}[\![|c\rangle]\!]$ is defined as $\overline{\mathcal{P}_i}[\![|c\rangle]\!] = id_S \times \mathcal{P}_i[\![c]\!]$.
$\overline{\mathcal{P}_i}[\![\{c \underset{C}{|} c'\}]\!]$ is defined only if $\overline{\mathcal{P}_i}[\![C]\!]$ is defined and if we have for all s, $s' \in S$:

$$\Pi_1(\overline{\mathcal{P}_i}[\![C]\!](s,s')) = \mathcal{P}_i[\![c]\!](s), \quad \Pi_2(\overline{\mathcal{P}_i}[\![C]\!](s,s')) = \mathcal{P}_i[\![c']\!](s'), \tag{16}$$

and in this case we define $\overline{\mathcal{P}_i}[\![\{c \underset{C}{|} c'\}]\!] = \overline{\mathcal{P}_i}[\![C]\!]$.

As a consequence we have:

Lemma 1. *For all c, c' in Exp we have* $\overline{\mathcal{P}_i}[\![\langle c|c'\rangle]\!] = \mathcal{P}_i[\![c]\!] \times \mathcal{P}_i[\![c']\!]$.

Now, remember the interpretation of composition by pre- and postconditions in GKAT, $\mathcal{P}_i[\![c \cdot b]\!]$ and $\mathcal{P}_i[\![b \cdot c]\!]$, given in (1) and (2). It also holds in the same way for BiGKAT, as the interpretations are the same as sub-Markov kernels over S^2. As a consequence we have, for the sub-distributions of Example 3:

Example 5. $C_{\mu_s} = C_{\mu_s}\ \overline{\mathcal{P}_i}[\![[x = x']]\!]$, $\quad C_{\mu_a} = C_{\mu_a}\ \overline{\mathcal{P}_i}[\![[x = \overline{x'}]]\!]$ (where $\overline{x'}$ is the negation of x).

3.3 Theory of BiGKAT

Now we give a list of axioms on BiGKAT expressions, using predicates $=$ and $\equiv$, that will be interpreted by equality and equivalence on sub-Markov kernels.

- The functions $\langle \_| : \text{Exp} \to \ddot{\text{Exp}}$, $|\_\rangle : \text{Exp} \to \ddot{\text{Exp}}$ satisfy, for any $b_1, b_2, b \in B$, $c_1, c_2, c \in C$, the following properties:

$$\langle 0| = |0\rangle = \ddot{0} \tag{17}$$
$$\langle 1| = |1\rangle = \ddot{1} \tag{18}$$
$$\langle b_1 + b_2| = \langle b_1| \oplus \langle b_2| \tag{19}$$
$$\langle \neg b| = \ddot{\neg}\langle b| \tag{20}$$
$$\langle c_1 \cdot c_2| = \langle c_1| \fatsemi \langle c_2| \tag{21}$$
$$\langle c_1 +_b c_2| = \langle c_1| \oplus_{\langle b|} \langle c_2| \tag{22}$$
$$\langle c^{(b)}| = \langle c|^{\langle b|} \tag{23}$$

  Similarly for $|\_\rangle$. We say that $\langle \_|$ and $|\_\rangle$ are *homomorphisms.* The operators have the same precedence as in GKAT.
- The following property on $\langle . |$ and $| . \rangle$:

$$\forall_{c_1, c_2 \in \text{Exp}},\ \langle c_1| \fatsemi |c_2\rangle = |c_2\rangle \fatsemi \langle c_1| \tag{24}$$

  The operators have the same precedence as in GKAT. For readability we use interchangeably the same notation for operators in GKAT and BiGKAT, i.e. operators $\cdot$, $\neg$ and $+_e$, for any $e \in B$, and constants 1 and 0 in GKAT

stand for $\fatsemi$, $\bar{\neg}$, $\oplus_{\langle e|}$ ($\oplus_{|e\rangle}$), $\ddot{1}$ and $\ddot{0}$, respectively. Often we go even further and omit the operator $\cdot$ and we write $\langle c_1|\langle c_2|$ ($|c_1\rangle|c_2\rangle$) for $\langle c_1|\cdot\langle c_2|$ ($|c_1\rangle\cdot|c_2\rangle$).
Note that property (24) states a commutativity property between programs that run in parallel, but we do not have in general $\langle c_1|\fatsemi\langle c_2| = \langle c_2|\fatsemi\langle c_1|$ for c_1, c_2 in $\ddot{\text{E}}\text{xp}$ (and similarly for operator $|\_\rangle$).
- The following properties on $\{.\mid.\}$:

$$\{c \underset{C}{\mid} c'\} = C \tag{25}$$

$$\{c_1 \underset{C_1}{\mid} c_1'\} \fatsemi \{c_2 \underset{C_2}{\mid} c_2'\} = \{c_1 \cdot c_2 \underset{C_1 \fatsemi C_2}{\mid} c_1' \cdot c_2'\} \tag{26}$$

- GKAT axioms for $=$ (see Fig. 1) on GKAT expressions,
- GKAT axioms for $=$ (see Fig. 1) written in the language of BiGKAT for BiGKAT expressions:
for instance equation (3) becomes $C \oplus_B C = C$ and (8) becomes $(C_1 \fatsemi C_2) \fatsemi C_3 = C_1 \fatsemi (C_2 \fatsemi C_3)$.
- Axioms for $\equiv$:

$$C_1 = C_2 \Rightarrow C_1 \equiv C_2 \tag{27}$$

$$C_1 \equiv C_2 \Rightarrow C_3 \fatsemi C_1 \equiv C_3 \fatsemi C_2 \tag{28}$$

$$C_1 \equiv C_2 \Rightarrow C_1 \fatsemi \{c \underset{C_3}{\mid} c'\} \equiv C_2 \fatsemi \{c \underset{C_3}{\mid} c'\} \tag{29}$$

We call this list of axioms the *theory of BiGKAT*. We say that a formula $C_1 = C_2$ (resp. $C_1 \equiv C_2$) is well-defined for an interpretation $\overline{\mathcal{P}_i}[\![.]\!]$ if $\overline{\mathcal{P}_i}[\![C_1]\!]$ and $\overline{\mathcal{P}_i}[\![C_2]\!]$ are both defined. An implicative formula $F_1 \Rightarrow F_2$ is well-defined for $\overline{\mathcal{P}_i}[\![.]\!]$ if both F_1 and F_2 are well-defined.

Now we say that an interpretation $\overline{\mathcal{P}_i}[\![.]\!]$ satisfies a formula F if F is well-defined for $\overline{\mathcal{P}_i}[\![.]\!]$ and if moreover:

1. if F is of the fom $C_1 = C_2$ (resp. $C_1 \equiv C_2$) then $\overline{\mathcal{P}_i}[\![C_1]\!] = \overline{\mathcal{P}_i}[\![C_2]\!]$ (resp. $\overline{\mathcal{P}_i}[\![C_1]\!] \equiv \overline{\mathcal{P}_i}[\![C_2]\!]$)
2. if F is of the form $C_1 = C_2 \Rightarrow C_3 \equiv C_4$ and if $\overline{\mathcal{P}_i}[\![C_1]\!] = \overline{\mathcal{P}_i}[\![C_2]\!]$, then $\overline{\mathcal{P}_i}[\![C_3]\!] \equiv \overline{\mathcal{P}_i}[\![C_4]\!]$ (and similarly for other implicative formulas built with $=$ and $\equiv$).

An example of formula which can in some cases not be well-defined is equation (25), because for $\overline{\mathcal{P}_i}[\![\{c \underset{C}{\mid} c'\}]\!]$ to be defined its projections need to satisfy the equations (16).

Proposition 1. *The interpretation $\overline{\mathcal{P}_i}[\![.]\!]$ of BiGKAT expressions by sub-Markov kernels on the product space S^2 defined in Sect. 3.2 satisfies all well-defined axioms of the theory of BiGKAT.*

Proof. The properties (17) - (23) are obtained directly from the semantics of BiGKAT (Definition 4). The proofs of properties (24), (26), (27) - (29) are in [15].

For GKAT axioms (Fig. 1) formulated for BiGKAT, note that since the probabilistic interpretation of BiGKAT expressions $\overline{\mathcal{P}_i}[\![.]\!]$ is the same as the one of GKAT (on product space S^2), we naturally have that such interpretation satisfies the axioms of Fig. 1 written in the language of BiGKAT. □

We give in [15] some properties as consequences of the theory of BiGKAT.

We now derive proofs in BiGKAT in the following way. We assume given an interpretation $\overline{\mathcal{P}_i}[\![.]\!]$ and a finite subset of elements A of Σ together with some equations: $A = \{c \underset{A}{|} c'\}$ and $B \,⨾\, A = B \,⨾\, A \,⨾\, B'$ which are well-defined and satisfied by $\overline{\mathcal{P}_i}[\![.]\!]$. Then we can use these equations and any formula of the theory of BiGKAT which is well-defined and satisfied by $\overline{\mathcal{P}_i}[\![.]\!]$ to derive new formulas. Proposition 1 then ensures that any formula derived in this way is satisfied by $\overline{\mathcal{P}_i}[\![.]\!]$.

Note that additionnally, if we allow ourselves to use in the proof any formula $C = C'$ which is satisfied by $\overline{\mathcal{P}_i}[\![.]\!]$ (semantic hypothesis), we still preserve the fact that the formulas derived are satisfied by $\overline{\mathcal{P}_i}[\![.]\!]$.

Example 6. Consider again the set space of memories. Consider two elements A_s, A_a of Σ with equations:

$$A_s = \{x \overset{\$}{\leftarrow} dbool \underset{A_s}{|} x' \overset{\$}{\leftarrow} dbool\} \qquad A_s = A_s \,⨾\, [x = x'] \tag{30}$$

$$A_a = \{x \overset{\$}{\leftarrow} dbool \underset{A_a}{|} x' \overset{\$}{\leftarrow} dbool\} \qquad A_a = A_a \,⨾\, [x = \overline{x'}] \tag{31}$$

Then if we choose $\overline{\mathcal{P}_i}[\![A_s]\!] = C_{\mu_s}$ and $\overline{\mathcal{P}_i}[\![A_a]\!] = C_{\mu_a}$, we know by Example 4 and Example 5 that $\overline{\mathcal{P}_i}[\![.]\!]$ satisfies the formulas above. Let us give a small example of proof:

$$\begin{aligned} A_s \oplus_{\langle y=tt|} (A_a \,⨾\, \langle x \leftarrow \overline{x}|) &= (A_s \,⨾\, [x = x']) \oplus_{\langle y=tt|} (A_a \,⨾\, [x = \overline{x'}] \,⨾\, \langle x \leftarrow \overline{x}|) (30), (31) \\ &= (A_s \,⨾\, [x = x']) \oplus_{\langle y=tt|} (A_a \,⨾\, \langle x \leftarrow \overline{x}| \,⨾\, [x = x']) \\ &= (A_s \oplus_{\langle y=tt|} (A_a \,⨾\, \langle x \leftarrow \overline{x}|)) \,⨾\, [x = x'] \text{ (7) for BiGKAT} \end{aligned}$$

3.4 Encoding of a subsystem of pRHL into BiGKAT

We want to encode (a part of) probabilistic relational Hoare logic (pRHL) in BiGKAT. Recall that Hoare logic can be encoded in KAT [17] by encoding a Hoare triple $\{\phi\}\; c\; \{\psi\}$ as a KAT equation $\phi \cdot c = \phi \cdot c \cdot \psi$. The intuitive meaning of the equation is that testing ψ after executing $\phi \cdot c$ is always redundant. This approach has been extended to the relational setting with RHL and BiKAT in [1]. To define such an encoding for pRHL and BiGKAT we need first to recall the definition and semantics of pRHL judgements and proofs [6]. We consider the language of probabilistic programs of Example 1, a state space S of memories and a probabilistic interpretation $\mathcal{P}_i[\![.]\!]$.

Let us first define the *range* of a subdistribution μ on S^2: $range(\mu) = \{(m, m') \in S \;/\; \mu(m, m') > 0\}$.

A pRHL *judgement* is a tuple of the form $c \sim c' : \phi \Rightarrow \psi$ where c, c' are programs and ϕ, ψ are predicates on S^2 (relations on states). For simplicity we will also denote as ϕ the subset of elements in S^2 satisfying ϕ.

One says that the pRHL judgement is *valid in the interpretation* $\mathcal{P}_i[\![.]\!]$, denoted as $\models_i c \sim c' : \phi \Rightarrow \psi$ if:

for any $(m, m') \in \phi$, there exists a subdistribution μ on S^2 such that: $\Pi_1(\mu) = \mathcal{P}_i[\![c]\!](m)$, $\Pi_2(\mu) = \mathcal{P}_i[\![c']\!](m')$, and $range(\mu) \subseteq \psi$. One then says that programs c and c' are equivalent with respect to precondition ϕ and postcondition ψ. If the interpretation i is fixed we write $\models$ instead of $\models_i$.

Following the above interpretation, we encode the pRHL judgment in BiGKAT as follows:

$$\exists_{C \in \ddot{\text{E}}\text{xp}} \cdot \phi \,;\, \{c \underset{C}{|} c'\} = \phi \,;\, \{c \underset{C}{|} c'\} \,;\, \psi \tag{32}$$

where $\phi, \psi \in \ddot{\text{B}}\text{Exp}$ and $c, c' \in \text{Exp}$.

- Note that we do not use the encoding $\phi \,;\, \{c \underset{C}{|} c'\} \leq \phi \,;\, \{c \underset{C}{|} c'\} \,;\, \psi$ since in GKAT and BiGKAT there is no natural notion of order $\leq$ as in KAT [18,16];
- We do not use either the encoding $\phi \,;\, \{c \underset{C}{|} c'\} \,;\, \neg\psi = 0$. In KAT, $\phi \cdot c = \phi \cdot c \cdot \psi$ is equivalent to $\phi \cdot c \cdot \neg\psi = 0$, but this cannot be proved in the same way in GKAT and the equivalence might not hold. We only have the implication $(\phi \cdot c = \phi \cdot c \cdot \psi) \Rightarrow (\phi \cdot c \cdot \neg\psi = 0)$, and we choose as encoding the stronger property. This encoding aligns with the intuitive interpretation of the validity of a Hoare triple, i.e. from a state satisfying the pre-condition ϕ, each execution of c, c', if it halts, it leads to a state satisfying the post-condition ψ.

Observe that the semantic interpretation of (32) is the same as $\models_i c \sim c' : \phi \Rightarrow \psi$.

We display on Fig. 3 the rules of pRHL defined in [6][7], except the rule for *While*, that we replace by an iteration rule (*R-Iter rule*) which we will explain below. This is a subsystem of pRHL, but we keep here the name pRHL for convenience.

We use different notation for pre and post conditions (ϕ, ψ) and for guards ($\langle b|$, $|b'\rangle$). Note in particular the side condition $\phi \Rightarrow b \ddot{=} b'$ in rule *R-Cond*, where the right-hand side $b \ddot{=} b'$ is equivalent to $\langle b|b'\rangle + \langle\neg b|\neg b'\rangle$, so the following holds

$$\phi\langle b|\neg b'\rangle = 0 \quad \phi\langle\neg b|b'\rangle = 0$$

These equalities assure that the predicates b and b' are evaluated to the same value on both left and right programs [1]. In particular, for the *R-Cond* rule it means that the same branch is executed for right-hand side and left-hand side programs. Observe that similarly as for Hoare logic, some rules of pRHL, namely axiom rules *R-Assign*, *R-Assign left* and *R-Rand assign* (see [15]) do not depend on pRHL judgements as premises but rather on an interpretation of actions and predicates, and a semantic condition (for *R-Rand assign*). Thus

[7] There are also one-sided versions of some of these rules, which we list in [15].

– *R-Assign rule:*

$$\frac{}{x \leftarrow v \sim x' \leftarrow v' : \phi[v/x, v'/x'] \Rightarrow \phi}$$

– *R-Seq rule:*

$$\frac{c_1 \sim c_1' : \phi \Rightarrow \psi \quad c_2 \sim c_2' : \psi \Rightarrow \xi}{c_1 \cdot c_2 \sim c_1' \cdot c_2' : \phi \Rightarrow \xi}$$

– *R-Rand assign rule:*

$$\frac{h \lhd (d, d') \quad \phi = \forall v \in Supp(d).\psi[v/x, h(v)/x']}{x \overset{\$}{\leftarrow} d \sim x' \overset{\$}{\leftarrow} d' : \phi \Rightarrow \psi}$$

– *R-Cond rule:*

$$\frac{\phi \Rightarrow b \doteq b' \quad c_1 \sim c_1' : \phi \wedge \langle b| \wedge |b'\rangle \Rightarrow \psi \quad c_2 \sim c_2' : \phi \wedge \langle \neg b| \wedge |\neg b'\rangle \Rightarrow \psi}{\textbf{if } b \textbf{ then } c_1 \textbf{ else } c_2 \sim \textbf{if } b' \textbf{ then } c_1' \textbf{ else } c_2' : \phi \Rightarrow \psi}$$

– *R-Iter rule:*

$$\frac{n \in \mathbb{N} \quad c \sim c' : \phi \Rightarrow \phi}{c^n \sim (c')^n : \phi \Rightarrow \phi}$$

– *R-Sub rule:*

$$\frac{\phi' \Rightarrow \phi \quad c \sim c' : \phi \Rightarrow \psi \quad \psi \Rightarrow \psi'}{c \sim c' : \phi' \Rightarrow \psi'}$$

– *R-Case rule:*

$$\frac{c \sim c' : \phi \wedge \phi' \Rightarrow \psi \quad c \sim c' : \phi \wedge \neg\phi' \Rightarrow \psi}{c \sim c' : \phi \Rightarrow \psi}$$

Fig. 3. Probabilistic Relational Hoare Logic rules (pRHL)

we do not expect to derive their encoding as an equation valid in the theory of BiGKAT. Instead, when we deal with examples we will consider a particular interpretation and thus reason on equalities of expressions in the model. The first rule derives a valid Hoare triple with the substitution of variables x, x' by expressions v, v', respectively; the second one derives a valid triple with samplings over distributions d, d'. The *R-Iter* rules means that if the execution of program c does not change ϕ, then the composition of c n times does not change ϕ. In *R-Iter* on Fig. 3, $c^n = c \cdot c^{n-1}$, where $c^1 = c$.

Let us explain the R-Rand assign rule: $h \lhd (d, d')$ means that h is a *coupling* between distributions d, d', that is to say a bijective function from $Supp(d)$ to $Supp(d')$ such that: for every $v \in Supp(d)$, $P_{x \sim d}[x = v] = P_{x \sim d'}[x = h(v)]$. For instance μ_s and μ_a in Ex.6 respectively correspond to couplings $h = id$ and negation on *Bool*.

Now, to show that the rules of Fig. 3 are sound in BiGKAT, we interpret them as in Fig. 4[8], by using the encoding of pRHL judgements as BiGKAT equations defined previously.

Our goal is now to prove that the rules above are valid in BiGKAT. In this approach, showing that a pRHL rule is sound in BiGKAT will consist in proving

[8] Note that the encoding of the one-sided rules are listed in [15]

– *R-Assign rule*:

$$\phi[v/x, v'/x']\{x \leftarrow v \mid_{C} x' \leftarrow v'\} = \phi[v/x, v'/x'] \cdot \{x \leftarrow v \mid_{C} x' \leftarrow v'\} \cdot \phi \tag{33}$$

– *R-Rand assign rule*:

$$\begin{aligned} & h \lhd (d, d') \wedge \phi = \forall v \in Supp(d).\psi[v/x, h(v)/x'] \\ & \Rightarrow \phi \cdot \{x \xleftarrow{\$} d \mid_{C} x' \xleftarrow{\$} d'\} = \phi \cdot \{x \xleftarrow{\$} d \mid_{C} x' \xleftarrow{\$} d'\} \cdot \psi \end{aligned} \tag{34}$$

– *R-Seq rule*:

$$\begin{aligned} & \phi \cdot \{c_1 \mid_{C_1} c_1'\} = \phi \cdot \{c_1 \mid_{C_1} c_1'\} \cdot \psi \ \wedge \ \psi \cdot \{c_2 \mid_{C_2} c_2'\} = \psi \cdot \{c_2 \mid_{C_2} c_2'\} \cdot \xi \\ & \Rightarrow \ \phi \cdot \{c_1 \cdot c_2 \mid_{C_1 C_2} c_1' \cdot c_2'\} = \phi \cdot \{c_1 \cdot c_2 \mid_{C_1 C_2} c_1' \cdot c_2'\} \cdot \xi \end{aligned} \tag{35}$$

– *R-Cond rule*:

$$\begin{aligned} & \phi \lesssim b \doteq b' \ \wedge \ \phi \cdot \langle b | b' \rangle \cdot \{c_1 \mid_{C_1} c_1'\} = \phi \cdot \langle b | b' \rangle \cdot \{c_1 \mid_{C_1} c_1'\} \cdot \psi \ \wedge \\ & \phi \cdot \ \langle \neg b | \neg b' \rangle \cdot \{c_2 \mid_{C_2} c_2'\} = \phi \cdot \ \langle \neg b | \neg b' \rangle \cdot \{c_2 \mid_{C_2} c_2'\} \cdot \psi \\ & \Rightarrow \ \phi \cdot \{c_1 +_b c_2 \mid_{C_1 \oplus_{\langle b|b' \rangle} C_2} c_1' +_{b'} c_2'\} = \phi \cdot \{c_1 +_b c_2 \mid_{C_1 \oplus_{\langle b|b' \rangle} C_2} c_1' +_{b'} c_2'\} \cdot \psi \end{aligned} \tag{36}$$

– *R-Iter rule*:

$$n \in \mathbb{N} \wedge \phi \cdot \{c \mid_{C} c'\} = \phi \cdot \{c \mid_{C} c'\} \cdot \phi \Rightarrow \phi \cdot \{c^n \mid_{C^n} (c')^n\} = \phi \cdot \{c^n \mid_{C^n} (c')^n\} \cdot \phi \tag{37}$$

– *R-Sub rule*:

$$\phi' \leq \phi \ \wedge \ \phi \cdot \{c \mid_{C} c'\} = \phi \cdot \{c \mid_{C} c'\} \cdot \psi \ \wedge \ \psi \leq \psi' \ \Rightarrow \ \phi' \cdot \{c \mid_{C} c'\} = \phi' \cdot \{c \mid_{C} c'\} \cdot \psi' \tag{38}$$

– *R-Case rule*:

$$\begin{aligned} & \phi \cdot \phi' \cdot \{c \mid_{C} c'\} = \phi \cdot \phi' \cdot \{c \mid_{C} c'\} \cdot \psi \ \wedge \phi \cdot \neg\phi' \cdot \{c \mid_{C} c'\} = \phi \cdot \neg\phi' \cdot \{c \mid_{C} c'\} \cdot \psi \\ & \Rightarrow \phi \cdot \{c \mid_{C} c'\} = \phi \cdot \{c \mid_{C} c'\}\psi \end{aligned} \tag{39}$$

Fig. 4. Encoding of pRHL rules in BiGKAT

that the conjunction of the BiGKAT equations encoding the premises of the pRHL rule implies the equation encoding the conclusion of the rule.

Finally we obtain the main result on the soundness of pRHL rules in BiGKAT.

Theorem 1 (Soundness of pRHL in BiGKAT). *The encoding of pRHL rules (Fig. 3) R-Seq, R-Cond, R-Sub, R-Case and R-Iter displayed in Fig. 4 can be derived by proofs in BiGKAT.*

4 Example

In this section we use the framework presented before to reason about invariance features of probabilistic programs.

Example 7. Consider a program c^9 encoded as the GKAT term

$$c = (b \xleftarrow{\$} dbool \cdot ((y \leftarrow y\ xor\ tt) +_{[b=tt]} 1) +_{[x=tt]} (b \leftarrow ff)) \cdot (y \leftarrow y\ xor\ b)$$

Consider also a second copy denoted as c'. We prove the invariance of variables y, y', relational predicate $[y = y']$, over executions of c, c', which corresponds to the following pRHL judgment $\vdash c \sim c' : [y = y'] \Rightarrow [y = y']$. In order to simplify the writing we denote $d_1 = b \xleftarrow{\$} dbool; ((y \leftarrow y\ xor\ tt) +_{[b=tt]} 1), d_2 = b \leftarrow ff$ and $c_2 = (y \leftarrow y\ xor\ b)$, so that $c = (d_1 +_{[x=tt]} d_2) \cdot c_2$. We then use some equational reasoning to obtain in GKAT $(d_1 +_{[x=tt]} d_2) \cdot c_2 =_{(7)} (d_1 \cdot c_2) +_{[x=tt]} (d_2 \cdot c_2)$.

In order to define C the BiGKAT expression showing the analog of the pRHL judgement above, we will distinguish 4 subcases depending on the evaluation of $\langle [x = tt] | [x' = tt] \rangle$: (1)$x = tt, x' = tt$, (2)$x \neq tt, x' = tt$, (3)$x = tt, x' \neq tt$ and (4)$x \neq tt, x' \neq tt$.

For that we will define 4 expressions C_{ij} $(i = 0, 1,\ j = 0, 1)$ and C as:

$$C = (C_{11} \oplus_{|[x'=tt]\rangle} C_{10}) \oplus_{\langle [x=tt]|} (C_{01} \oplus_{|[x'=tt]\rangle} C_{00})$$

Assume temporarily that this is done, then we have:

$$\begin{aligned}
[y = y']C &= [y = y'](C_{11} \oplus_{|[x'=tt]\rangle} C_{10}) \oplus_{\langle [x=tt]|} (C_{01} \oplus_{|[x'=tt]\rangle} C_{00}) \\
&= [y = y'](\langle [x = tt] | [x' = tt] \rangle\ C_{11} \oplus_{|[x'=tt]\rangle} C_{10}) \oplus_{\langle [x=tt]|} (C_{01} \oplus_{|[x'=tt]\rangle} C_{00}) \\
&= ([y = y'] \langle [x = tt] | [x' = tt] \rangle\ C_{11} \oplus_{|[x'=tt]\rangle} [y = y'] C_{10}) \\
&\quad \oplus_{\langle [x=tt]|} ([y = y'] C_{01} \oplus_{|[x'=tt]\rangle} [y = y'] C_{00})
\end{aligned}$$

where the last equality is obtained by one of the GKAT derivable equations listed in [15]. So if we can prove that

$$[y = y'] \langle [x = tt] | [x' = tt] \rangle\ C_{11} = [y = y'] \langle [x = tt] | [x' = tt] \rangle\ C_{11} [y = y'],$$

and similarly for the other C_{ij}, then we will be able to deduce that $[y = y']C = [y = y']C[y = y']$, by using repeatedly axiom (7) for BiGKAT.

We present here the proof of the property for C_{11} (subcase (1)), the most delicate one, and leave the proofs for the other C_{ij} to [15].

subcase (1):

Using assumptions $x = tt$, $x' = tt$ for the left and right programs we obtain $[x = tt]((d_1 \cdot c_2) +_{[x=tt]} (d_2 \cdot c_2)) = [x = tt](d_1 \cdot c_2)$ and $[x' = tt]((d_1' \cdot c_2') +_{[x'=tt]} (d_2' \cdot c_2')) = [x' = tt](d_1' \cdot c_2')$. As d_1 and d_1' contain a sampling we choose to use a coupling in order to obtain a postcondition. We use the constant A_s defined in Example 6 with its interpretation $A_s = \{b \xleftarrow{\$} dbool \mid_{A_s} b' \xleftarrow{\$} dbool\}$, $A_s = A_s \fatsemi [b = b']$.

9 Its code written in the programming language of Example 1 is in [15].

Denote $e_1 = (y \leftarrow y\ xor\ tt) +_{[b=tt]} 1$, $e_1' = (y' \leftarrow y'\ xor\ tt) +_{[b'=tt]} 1$ and let $C_{11} = A_s \fatsemi \langle e_1 \cdot c_2 | e_1' \cdot c_2' \rangle = A_s \fatsemi \langle e_1 | e_1' \rangle \fatsemi \langle c_2 | c_2' \rangle = A_s \fatsemi [b = b'] \fatsemi \langle e_1 | e_1' \rangle \fatsemi \langle c_2 | c_2' \rangle$. By using R-Cond rule (36) one can obtain: $[b = b'] \fatsemi \langle e_1 | e_1' \rangle = [b = b'] \fatsemi \langle e_1 | e_1' \rangle \fatsemi [b = b']$. We thus get $C_{11} = A_s \fatsemi [b = b'] \fatsemi \langle e_1 | e_1' \rangle \fatsemi [b = b'] \fatsemi \langle c_2 | c_2' \rangle$. Moreover we have by semantic hypothesis: $[y = y'] \fatsemi A_s = [y = y'] \fatsemi A_s \fatsemi [y = y']$ and $[y = y'] \fatsemi \langle e_1 | e_1' \rangle = [y = y'] \fatsemi \langle e_1 | e_1' \rangle \fatsemi [y = y']$, so we obtain $[y = y']C_{11} = [y = y'] \fatsemi A_s \fatsemi [b = b'] \fatsemi [y = y'] \fatsemi \langle e_1 | e_1' \rangle \fatsemi [b = b'] \fatsemi [y = y'] \fatsemi \langle c_2 | c_2' \rangle$. Now, recall that $c_2 = (y \leftarrow y\ xor\ b)$. We thus have $[b = b'] \fatsemi [y = y'] \fatsemi \langle c_2 | c_2' \rangle = [b = b'] \fatsemi [y = y'] \fatsemi \langle c_2 | c_2' \rangle \fatsemi [y = y']$. So from the two last equalities we deduce: $[y = y']C_{11} = [y = y'] \fatsemi A_s \fatsemi [b = b'] \fatsemi [y = y'] \fatsemi \langle e_1 | e_1' \rangle \fatsemi [b = b'] \fatsemi [y = y'] \fatsemi \langle c_2 | c_2' \rangle \fatsemi [y = y']$. So finally: $[y = y'] \fatsemi C_{11} = [y = y'] \fatsemi C_{11} \fatsemi [y = y']$. This was the property expected for C_{11}.

5 Discussion: comparison between BiGKAT and pRHL

We have seen that BiGKAT is at least as expressive as pRHL without *While*, since rules of the latter can be encoded in the former. Here we want to illustrate that BiGKAT is in some aspects more expressive than pRHL.

First, BiGKAT allows to derive some rules on pRHL judgements that are valid in the probabilistic model but that cannot be derived within the system pRHL, as defined in [6]. Consider the candidate rule below:

$$\frac{c \sim (c_1' +_b c_2')c_3' : \phi \Rightarrow \psi}{c \sim c_1'c_3' +_b c_2'c_3' : \phi \Rightarrow \psi}$$

It can be derived in BiGKAT, however it cannot be derived as a sequence of pRHL rules (Fig. 3), simply because if we read any pRHL rule bottom-up the programs in the premises are subterms of the programs in the conclusion. More generally we can derive in BiGKAT:

$$\frac{c_1 \sim c_1' : \phi \Rightarrow \psi \qquad c_1 = c_2 \qquad c_1' = c_2'}{c_2 \sim c_2' : \phi \Rightarrow \psi}$$

where premises $c_1 = c_2$ are given by any GKAT axioms. These rules extend in some sense pRHL with rewriting of programs according to GKAT axioms. This might be useful for some usages of pRHL (see for instance [2]).

A second point is that as BiGKAT, contrarily to pRHL, explicitly indicates for couplings the "witnesses" Markov kernels on S^2, it allows to express rules that cannot be written in pRHL. For instance, the following rule, displayed in pRHL format, is (trivially) derivable in BiGKAT:

$$\frac{\phi \fatsemi \{c \underset{C}{|} c'\} = \phi \fatsemi \{c \underset{C}{|} c'\} \fatsemi \psi_1 \qquad \phi \fatsemi \{c \underset{C}{|} c'\} = \phi \fatsemi \{c \underset{C}{|} c'\} \fatsemi \psi_2}{\phi \fatsemi \{c \underset{C}{|} c'\} = \phi \fatsemi \{c \underset{C}{|} c'\} \fatsemi (\psi_1 \wedge \psi_2)} \tag{40}$$

However the corresponding candidate pRHL rule is unsound:

$$\frac{c \sim c' : \phi \Rightarrow \psi_1 \qquad c \sim c' : \phi \Rightarrow \psi_2}{c \sim c' : \phi \Rightarrow \psi_1 \wedge \psi_2}$$

As a counter-example consider Example 6 and take $c = x \xleftarrow{\$} dbool$, $c' = x \xleftarrow{\$} dbool$, $\phi = 1$, $\psi_1 = [x = x']$, $\psi_2 = [x = \overline{x'}]$. Then $\psi_1 \wedge \psi_2$ is false, hence the conclusion is not valid. But the BiGKAT rule (40) could not be applied because the two witnesses A_s and A_a are not the same C. On this point we plan to study the possible links between BiGKAT and [7].

Finally, a third point is that BiGKAT allows to express judgements with assertions which cannot be expressed by a single pRHL judgement. Consider for example an equality $\phi \,;\, \langle c_1|c_1'\rangle \,;\, [x = x'] \,;\, \langle c_2|c_2'\rangle = \phi \,;\, \langle c_1|c_1'\rangle \,;\, [x = x'] \,;\, \langle c_2|c_2'\rangle \,;\, \psi$. It says that, assuming precondition ϕ for the pair of programs c_1c_2 and $c_1'c_2'$, if moreover after execution of $\langle c_1|c_1'\rangle$ the assertion $[x = x']$ holds, then postcondition ψ is satisfied. This cannot be expressed by a single pRHL judgement.

6 Related work

A seminal approach to relational analysis of programs is due to [10], with the introduction of Relational Hoare logic (RHL). The system takes as the central ingredient the interpretation of program properties as relations over memories, allowing, for example, to prove correctness of program transformations for compiler optimisations. An algebraic approach to this system was taken in [1], by introducing BiKAT, a relational extension of KAT to model relational reasoning on programs, being able to express *all exists* properties, i.e. 'for any run of one program there exists a run of the other such that ...'. It was shown in this paper that both the syntax and the deductive system of RHL, including *all-exists* corresponding versions [13,11], can be interpreted in BiKAT.

Probabilistic relational Hoare logic (pRHL), a probabilistic variant of RHL, was introduced by Barthe and coauthors in [6], motivated by the certification of cryptographic proofs. One goal of our approach is to subsume pRHL into algebraic reasoning, taking inspiration from BiKAT. Rather than using KAT, we build our approach on GKAT [22]. The main advantage of this structure for the purpose of this paper is the easier representation of probabilistic programming languages due to the absence of nondeterminism. The work of [22] also introduced the probabilistic model of GKAT based on sub-Markov kernels. The main model of the structure presented in this paper is naturally a relational version of such model. GKAT was also investigated further in [21], which in particular provides a semantics for which the equational theory is complete. The work [14] addressed the application of GKAT to unary (non-relational) properties of probabilistic programs, by investigating the relationships with the probabilistic Hoare logic aHL of [4].

In this work we have considered one probabilistic interpretation of GKAT to model the class of programs we wanted to address. By considering other more complex structures we would be able to increase the set of possible programs to analyse, an therefore capture a greater variety of examples. We could take, for instance, ProbGKAT [20], as our base structure, allowing to represents programs with probabilistic branching.

Another approach to relational reasoning of programs is approximate probabilistic relational Hoare logic (apRHL) [8], for the verification of differential privacy. The term 'approximate' refers here to the parameters associated to reasoning on judgments, which are related to the distance between the probabilistic distributions generated by the probabilistic programs.

7 Conclusion and perspectives

In this work we have introduced BiGKAT, a variant of KAT allowing to reason on relational properties of probabilistic programs, based on GKAT, provided a semantics for it based on sub-Markov kernels and a theory allowing to derive proofs. We have illustrated the expressivity of BiGKAT by proving how a subsystem of probabilistic relational Hoare logic [6] (with the *while* rule replaced by an iteration rule) can be soundly encoded in it.

In future work we want to extend this encoding to the full pRHL, including the *while* rule. Another interesting path, aligned with what we expressed in the previous section, would be to build a relational version of ProbGKAT, with the goal of extending the set of possible examples to programs with probabilistic branching. Moreover, while usually avoided, and always difficult, one possible direction for future work could be to consider a language with both nondeterminism and probabilities [23], capturing also more application scenarios. That could be indeed another direction to pursue in the future.

For the purpose of this work, we were focused on reasoning about properties of probabilistic non-interference over pairs of programs, i.e. *2-properties*. However, we believe that the generalization to a n-relational framework could be possible to handle properties such as n-safety.

Another natural path would be an algebraic approach to subsume approximate probabilistic relational Hoare logic (apRHL) [8], in order to provide an equational way of reasoning on differential privacy. This direction could use the approach of [14] which gives a way to define an approximate version of GKAT.

Acknowledgements Work partially funded by ANR Project HOPR (ANR-24-CE48-5521-01) from the French National Research Agency, by the project R-TALENT-21-001-BAILLOT from I-Site Université Lille Nord-Europe and by National Funds through FCT - Fundação para a Ciência e a Tecnologia, I.P. (Portuguese Foundation for Science and Technology) within the project IBEX, with reference 10.54499/PTDC/CCI-COM/4280/2021.

References

1. Timos Antonopoulos, Eric Koskinen, Ton Chanh Le, Ramana Nagasamudram, David A. Naumann, and Minh Ngo. An algebra of alignment for relational verification. *Proc. ACM Program. Lang.*, 7(POPL):573–603, 2023. doi:10.1145/3571213.

2. Martin Avanzini, Gilles Barthe, Benjamin Grégoire, Georg Moser, and Gabriele Vanoni. Hopping proofs of expectation-based properties: Applications to skiplists and security proofs. *Proc. ACM Program. Lang.*, 8(OOPSLA1):784–809, 2024. doi:10.1145/3649839.
3. Gilles Barthe, François Dupressoir, Benjamin Grégoire, César Kunz, Benedikt Schmidt, and Pierre-Yves Strub. Easycrypt: A tutorial. In Alessandro Aldini, Javier López, and Fabio Martinelli, editors, *Foundations of Security Analysis and Design VII - FOSAD 2012/2013 Tutorial Lectures*, volume 8604 of *Lecture Notes in Computer Science*, pages 146–166. Springer, 2013. doi:10.1007/978-3-319-10082-1_6.
4. Gilles Barthe, Marco Gaboardi, Benjamin Grégoire, Justin Hsu, and Pierre-Yves Strub. A program logic for union bounds. In Ioannis Chatzigiannakis, Michael Mitzenmacher, Yuval Rabani, and Davide Sangiorgi, editors, *43rd International Colloquium on Automata, Languages, and Programming, ICALP 2016, July 11-15, 2016, Rome, Italy*, volume 55 of *LIPIcs*, pages 107:1–107:15. Schloss Dagstuhl - Leibniz-Zentrum für Informatik, 2016. doi:10.4230/LIPIcs.ICALP.2016.107.
5. Gilles Barthe, Marco Gaboardi, Justin Hsu, and Benjamin C. Pierce. Programming language techniques for differential privacy. *ACM SIGLOG News*, 3(1):34–53, 2016. doi:10.1145/2893582.2893591.
6. Gilles Barthe, Benjamin Grégoire, and Santiago Zanella Béguelin. Formal certification of code-based cryptographic proofs. In Zhong Shao and Benjamin C. Pierce, editors, *Proceedings of the 36th ACM SIGPLAN-SIGACT Symposium on Principles of Programming Languages, POPL 2009, Savannah, GA, USA, January 21-23, 2009*, pages 90–101. ACM, 2009. doi:10.1145/1480881.1480894.
7. Gilles Barthe, Benjamin Grégoire, Justin Hsu, and Pierre-Yves Strub. Coupling proofs are probabilistic product programs. In Giuseppe Castagna and Andrew D. Gordon, editors, *Proceedings of the 44th ACM SIGPLAN Symposium on Principles of Programming Languages, POPL 2017, Paris, France, January 18-20, 2017*, pages 161–174. ACM, 2017. doi:10.1145/3009837.3009896.
8. Gilles Barthe, Boris Köpf, Federico Olmedo, and Santiago Zanella Béguelin. Probabilistic relational reasoning for differential privacy. *ACM Trans. Program. Lang. Syst.*, 35(3):9:1–9:49, 2013. doi:10.1145/2492061.
9. Santiago Zanella Béguelin, Gilles Barthe, Benjamin Grégoire, and Federico Olmedo. Formally certifying the security of digital signature schemes. In *30th IEEE Symposium on Security and Privacy (S&P 2009), 17-20 May 2009, Oakland, California, USA*, pages 237–250. IEEE Computer Society, 2009. doi:10.1109/SP.2009.17.
10. Nick Benton. Simple relational correctness proofs for static analyses and program transformations. In Neil D. Jones and Xavier Leroy, editors, *Proceedings of the 31st ACM SIGPLAN-SIGACT Symposium on Principles of Programming Languages, POPL 2004, Venice, Italy, January 14-16, 2004*, pages 14–25. ACM, 2004. doi:10.1145/964001.964003.
11. Raven Beutner. Automated software verification of hyperliveness. In Bernd Finkbeiner and Laura Kovács, editors, *Tools and Algorithms for the Construction and Analysis of Systems*, pages 196–216, Cham, 2024. Springer Nature Switzerland.
12. Filippo Bonchi and Damien Pous. Checking NFA equivalence with bisimulations up to congruence. In Roberto Giacobazzi and Radhia Cousot, editors, *The 40th Annual ACM SIGPLAN-SIGACT Symposium on Principles of Programming Languages, POPL '13, Rome, Italy - January 23 - 25, 2013*, pages 457–468. ACM, 2013. doi:10.1145/2429069.2429124.

13. Robert Dickerson, Qianchuan Ye, Michael K. Zhang, and Benjamin Delaware. RHLE: Modular deductive verification of relational forall exists properties. In Ilya Sergey, editor, *Programming Languages and Systems - 20th Asian Symposium, APLAS 2022, Auckland, New Zealand, December 5, 2022, Proceedings*, volume 13658 of *Lecture Notes in Computer Science*, pages 67–87. Springer, 2022. doi:10.1007/978-3-031-21037-2_4.
14. Leandro Gomes, Patrick Baillot, and Marco Gaboardi. A Kleene algebra with tests for union bound reasoning about probabilistic programs. In *33rd EACSL Annual Conference on Computer Science Logic, CSL 2025*, volume 326. Schloss Dagstuhl - Leibniz-Zentrum für Informatik, 2025. to appear. URL: https://hal.science/hal-04196675.
15. Leandro Gomes, Patrick Baillot, and Marco Gaboardi. BiGKAT: an algebraic framework for relational verification of probabilistic programs. Technical report, 2025. URL: https://hal.science/hal-04017128.
16. D. Kozen. Kleene algebra with tests. *ACM Trans. on Prog. Lang. and Systems*, 19(3):427–443, 1997. doi:10.1145/256167.256195.
17. D. Kozen. On Hoare logic and Kleene algebra with tests. *ACM Trans. on Comp. Logic*, 1(212):1–14, 2000. URL: http://dl.acm.org/citation.cfm?id=343378, doi:10.1109/LICS.1999.782610.
18. Dexter Kozen. Kleene algebra with tests and commutativity conditions. In Tiziana Margaria and Bernhard Steffen, editors, *Tools and Algorithms for Construction and Analysis of Systems, Second International Workshop, TACAS '96, Passau, Germany, March 27-29, 1996, Proceedings*, volume 1055 of *Lecture Notes in Computer Science*, pages 14–33. Springer, 1996. doi:10.1007/3-540-61042-1_35.
19. Damien Pous. Kleene algebra with tests and coq tools for while programs. In Sandrine Blazy, Christine Paulin-Mohring, and David Pichardie, editors, *Interactive Theorem Proving - 4th International Conference, ITP 2013, Rennes, France, July 22-26, 2013. Proceedings*, volume 7998 of *Lecture Notes in Computer Science*, pages 180–196. Springer, 2013. doi:10.1007/978-3-642-39634-2_15.
20. Wojciech Rozowski, Tobias Kappé, Dexter Kozen, Todd Schmid, and Alexandra Silva. Probabilistic guarded KAT modulo bisimilarity: Completeness and complexity. In Kousha Etessami, Uriel Feige, and Gabriele Puppis, editors, *50th International Colloquium on Automata, Languages, and Programming, ICALP 2023, July 10-14, 2023, Paderborn, Germany*, volume 261 of *LIPIcs*, pages 136:1–136:20. Schloss Dagstuhl - Leibniz-Zentrum für Informatik, 2023. doi:10.4230/LIPIcs.ICALP.2023.136.
21. Todd Schmid, Tobias Kappé, Dexter Kozen, and Alexandra Silva. Guarded Kleene algebra with tests: Coequations, coinduction, and completeness. In Nikhil Bansal, Emanuela Merelli, and James Worrell, editors, *48th International Colloquium on Automata, Languages, and Programming, ICALP 2021, July 12-16, 2021, Glasgow, Scotland (Virtual Conference)*, volume 198 of *LIPIcs*, pages 142:1–142:14. Schloss Dagstuhl - Leibniz-Zentrum für Informatik, 2021. doi:10.4230/LIPIcs.ICALP.2021.142.
22. Steffen Smolka, Nate Foster, Justin Hsu, Tobias Kappé, Dexter Kozen, and Alexandra Silva. Guarded Kleene algebra with tests: verification of uninterpreted programs in nearly linear time. *Proc. ACM Program. Lang.*, 4(POPL):61:1–61:28, 2020. doi:10.1145/3371129.
23. Daniele Varacca and Glynn Winskel. Distributing probability over nondeterminism. *Math. Struct. Comput. Sci.*, 16(1):87–113, 2006. doi:10.1017/S0960129505005074.

A General Completeness Theorem for Skip-Free Star Algebras

Tobias Kappé[1] and Todd Schmid[2(✉)]

[1] LIACS, Leiden University, Leiden, The Netherlands
t.w.j.kappe@liacs.leidenuniv.nl
[2] Bucknell University, Lewisburg, USA
t.schmid@bucknell.edu

Abstract. We consider process algebras with branching parametrized by an equational theory T, and show that it is possible to axiomatize bisimilarity under certain conditions on T. Our proof abstracts an earlier argument due to Grabmayer and Fokkink (LICS'20), and yields new completeness theorems for skip-free process algebras with probabilistic (guarded) branching, while also covering existing completeness results.

1 Introduction

Regular expressions generated by a set of action symbols *Act* are classically interpreted as regular languages, i.e., sets of words over *Act* obtained by the usual union, concatenation, and Kleene star operations [11]. Inspired by the introduction of process algebra as a formalization of communicating and concurrent processes, Milner gave an interpretation of regular expressions in terms of nondeterministic machine *behaviours*, i.e., up to bisimilarity [15]. The labelled transition system obtained from a regular expression is constructed by interpreting the union operation as nondeterministic choice instead of language union and the Kleene star operation as a (finite) loop instead of repeated language concatenation. Although introduced many years apart, Milner's interpretation of regular expressions is now known to be equivalent to Antimirov's derivative construction [2].

Milner transformed Salomaa's axioms for language equivalence of regular expressions [24] into sound axioms for his behavioural semantics, and left completeness as an open problem. This problem was only recently solved by Grabmayer [5], who was able to reduce the problem to results in a prior collaborative work with Fokkink [6]. In op. cit., the authors prove that Milner's axioms are complete when restricted to a subset of regular expressions that they call *1-free star expressions*, which are regular expressions in which 1 (the unit for concatenation) does not appear, and the Kleene star is replaced by a binary star operation $r_1 * r_2$.

Around the same time, Smolka et al. [29] introduced *guarded Kleene algebra with tests* (or *GKAT*) for reasoning about simple imperative programs. Essentially, GKAT is a restriction of *Kleene algebra with tests* (or *KAT*) [12] to programs

P. A. Abdulla and D. Kesner (Eds.): FoSSaCS 2025, LNCS 15691, pp. 265–286, 2025.
https://doi.org/10.1007/978-3-031-90897-2_13

constructed out of if-then-else and while.[3] In [29], the authors proposed an infinitary axiomatization of equivalence between GKAT programs. They left open whether a finitary version of these axioms, similar to Milner's, is also complete.

More recently, a step towards proving completeness of GKAT was taken in [9]. In that paper, it was shown that the finitary axiomatization of GKAT proposed in [29] is complete when restricted to GKAT programs that are 1-free in a sense similar to 1-free star expressions. The program that acts like 1 in GKAT is skip, so the authors of [9] refer to these programs as *skip-free* GKAT programs. To achieve this result, it was necessary to analyze some tricky steps in the original completeness proof from [6], and adapt them to the setting of skip-free GKAT.

In this paper, we study a unifying generalization of 1-free star expressions and skip-free GKAT programs that we call *skip-free unified-star expressions*. This generalization is based on the observation in [28,25] that 1-free star expressions and skip-free GKAT programs can be seen as instances of process algebras parametrized by *equational theories*, which capture computational effects such as nondeterminism and probability via their corresponding monads [18,19].

Our main contribution is a complete axiomatization for each process algebra in this class, provided the equational theory satisfies two conditions that we refer to as *admitting a support* and *malleability*. These properties essentially allow us to employ the strategy used by Grabmayer and Fokkink in [6], but in a much more abstract setting. The equational theories that correspond to 1-free star expressions and skip-free GKAT admit a support and are malleable, and so our abstract completeness theorem generalizes the results from [6,9]. We furthermore obtain several new completeness theorems for bisimilarity semantics of process algebras considered in the literature, including one for the skip-free fragment of probabilistic regular expressions studied in [22] and another for a skip-free variation of the probabilistic version of GKAT studied in [21].

We assume the reader is familiar with the basics of category theory, i.e., functors, universal properties, and natural transformations (see, for example, [20]). Omitted proofs can be found in the full version of the paper [8].

2 1-free Star Expressions

To set the stage, we begin with a brief description of Grabmayer and Fokkink's completeness theorem and its proof technique [6]. Fix a set of *action symbols* *Act*. The set of *1-free star expressions* *StExp* is generated by the grammar below.

$$r, r_1, r_2 ::= 0 \mid a \in Act \mid r_1 + r_2 \mid r_1 \cdot r_2 \mid r_1 * r_2 \qquad (1)$$

The operators $+$ and $\cdot$ are to be interpreted as nondeterministic choice and sequential composition, respectively. The expression $r_1 * r_2$ is intended to mean "run r_1 some number of times, and then run r_2" (cf. the regular expression $r_1^* r_2$).

[3] The syntax and significance of this fragment of KAT was already noticed by Kozen and Tseng in [13]. The innovations of [29] are the axiomatization and complexity results, as well as a number of different interpretations of GKAT.

$$\frac{}{a \xrightarrow{a} \checkmark} \quad \frac{r_1 \xrightarrow{a} \xi}{r_1 + r_2 \xrightarrow{a} \xi} \quad \frac{r_2 \xrightarrow{a} \xi}{r_1 + r_2 \xrightarrow{a} \xi} \quad \frac{r_1 \xrightarrow{a} s}{r_1 r_2 \xrightarrow{a} s r_2} \quad \frac{r_1 \xrightarrow{a} \checkmark}{r_1 r_2 \xrightarrow{a} r_2}$$

$$\frac{r_1 \xrightarrow{a} s}{r_1 * r_2 \xrightarrow{a} s(r_1 * r_2)} \quad \frac{r_1 \xrightarrow{a} \checkmark}{r_1 * r_2 \xrightarrow{a} r_1 * r_2} \quad \frac{r_2 \xrightarrow{a} s}{r_1 * r_2 \xrightarrow{a} s}$$

Fig. 1. Rules defining the transition structure of the syntactic chart $(StExp, \delta)$. In the above, $a \in Act$, $r_1, r_2, s \in StExp$, and $\xi \in \checkmark + StExp$.

In the absence of parentheses, $*$ takes precedence over $\cdot$, which is evaluated before $+$. So, $a + b \cdot c * d$ should be read as $a + (b \cdot (c * d))$. We typically write $r_1 r_2$ instead of $r_1 \cdot r_2$. The process semantics for 1-free star expressions can be phrased in terms of a variant of labelled transition systems called *charts*.

Definition 2.1. *A* chart *is a pair* (X, δ) *consisting of a set of* states X *and a* transition function $\delta \colon X \to \mathcal{P}(Act \times (\checkmark + X))$. *Here,* $\mathcal{P}$ *is the finite powerset functor,* $\checkmark$ *denotes the set* $\{\checkmark\}$, *and* $+$ *is disjoint union. Given* $x, y \in X$ *and* $a \in Act$, *we write* $x \xrightarrow{a}_\delta y$ *if* $(a, y) \in \delta(x)$, *and* $x \xrightarrow{a}_\delta \checkmark$ *if* $(a, \checkmark) \in \delta(x)$.

Immediately we see that δ can be either specified as a function or as a set of *transition relations* $\xrightarrow{a}_\delta \subseteq X \times (\checkmark + X)$, one for each $a \in Act$. We usually drop subscripts if the transition function can be inferred from context. We write $\to$ for the union of the transition relations and say that x *transitions* to y if $x \to y$.

Given (X, δ) and $U \subseteq X$, define $\langle U \rangle_\delta = \{y \mid \exists x \in U, x \to_\delta^* y\}$. The *subchart of* (X, δ) *generated by* U is $(\langle U \rangle_\delta, \delta_U)$, where $\delta_U : \langle U \rangle_\delta \to \mathcal{P}(Act \times (\checkmark + \langle U \rangle_\delta))$ is simply δ restricted to $\langle U \rangle_\delta$; we also write $\langle U \rangle_\delta$ for this chart. When $x \in X$, we may write $\langle x \rangle_\delta$ for $\langle \{x\} \rangle_\delta$, and speak of *the subchart of* (X, δ) *generated by* x.

The set of 1-free star expressions itself carries a chart structure, whose transitions are derived inductively from the inference rules in Fig. 1, in a way that is reminiscent of Antimirov's automaton construction [2]. We use $(StExp, \delta)$ to denote this chart structure, and refer to it as the *syntactic chart*.

Every 1-free star expression r generates a finite subchart $\langle r \rangle$ of $(StExp, \delta)$: the *syntactic chart* of r. The equivalence for 1-free star expressions that Grabmayer, Fokkink, and Milner sought to axiomatize is *bisimilarity*, defined as follows.

Definition 2.2. *A* bisimulation *between charts* (X, δ_X) *and* (Y, δ_Y) *is a relation* $R \subseteq X \times Y$ *such that for any* $(x, y) \in R$ *and* $a \in Act$, *(1)* $x \xrightarrow{a} \checkmark$ *if and only if* $y \xrightarrow{a} \checkmark$, *(2) if* $x \xrightarrow{a} x'$, *then there is a* $(x', y') \in R$ *such that* $y \xrightarrow{a} y'$, *and (3) if* $y \xrightarrow{a} y'$, *then there is a* $(x', y') \in R$ *such that* $x \xrightarrow{a} x'$. *We write* $x \leftrightarroweq y$ *and say that* x *and* y *are* bisimilar *if* $(x, y) \in R$ *for some bisimulation* R.

Note that the relation $\leftrightarroweq$ is the largest bisimulation between two charts. Furthermore, $\leftrightarroweq$ is an equivalence relation when restricted to a single chart. Another notion that we will rely on is that of a *chart homomorphism*.

Definition 2.3. *A* homomorphism *of charts* $h \colon (X, \delta_X) \to (Y, \delta_Y)$ *is a function* $h \colon X \to Y$ *such that the graph of* h *is a bisimulation.*

$$x + y = y + x \qquad (x+y)z = xz + yz \qquad x * y = x(x*y) + y$$
$$(x+y)+z = x+(y+z) \qquad (xy)z = x(yz) \qquad \frac{x = yx + z}{x = y * z}$$
$$x + x = x = x + 0 \qquad 0x = 0$$

Fig. 2. The axioms proposed by Grabmayer and Fokkink in [6]. The theory SL^* consists of the axioms above and equational logic (not pictured above).

It is straightforward to show that charts and chart homomorphisms form a category, i.e., identity maps $\mathrm{id}\colon (X,\delta) \to (X,\delta)$ are homomorphisms and any composition of homomorphisms is a homomorphism. Furthermore, bisimilarity can be characterized using chart homomorphisms, via the following lemma.

Lemma 2.4. *Given states $x \in X$ and $y \in Y$ of charts (X,δ_X) and (Y,δ_Y), $x \mathrel{\underline{\leftrightarrow}} y$ if and only if there is a third chart (Z,δ_Z) and chart homomorphisms $h\colon (X,\delta_X) \to (Z,\delta_Z)$ and $k\colon (X,\delta_X) \to (Z,\delta_Z)$ such that $h(x) = k(y)$.*

This characterization is useful in several ways; for one, it connects the existing notion of bisimilarity of charts to the more abstract definition in the next section. Another consequence of Lemma 2.4 is that for charts (U,δ_U) and (X,δ) with $U \subseteq X$, (U,δ_U) is a subchart of (X,δ) generated by U if and only if the inclusion map $U \hookrightarrow X$ is a chart homomorphism. Furthermore, for any $r, s \in \mathit{StExp}$, $r \mathrel{\underline{\leftrightarrow}} s$ as states of (StExp,δ) iff they are bisimilar as states of $\langle r \rangle$ and $\langle s \rangle$ respectively.

Axiomatization. Milner showed that bisimilarity is a congruence with respect to the regular expression operations [15]. This led him to give axioms for bisimilarity using a variation on Salomaa's axioms for language equivalence of regular expressions [24]. Grabmayer and Fokkink adapted these axioms for 1-free regular expressions in [6], which we recall in Fig. 2.[4] For reasons that will become clear in Section 3, we write SL^* to denote the theory in the figure.

Definition 2.5. *Given $r_1, r_2 \in \mathit{StExp}$, we write $\mathsf{SL}^* \vdash r_1 = r_2$ if there is a derivation of the equation $r_1 = r_2$ from the axioms of SL^*.*

The following theorem was the main result in [6].

Theorem 2.6 (Soundness and completeness of SL^*). *Let $r_1, r_2 \in \mathit{StExp}$. Then $\mathsf{SL}^* \vdash r_1 = r_2$ if and only if $r_1 \mathrel{\underline{\leftrightarrow}} r_2$ as states in (StExp,δ).*

The forward direction is called *soundness*, and it can be proven by induction on derivations. The reverse direction is *completeness*, and requires a more involved proof with several steps. To arrive at the completeness result, some mathematical machinery has to be developed, and two particularly treacherous proofs have to be worked out. In the remainder of this section, we give an overview of the necessary techniques as preparation for our abstract approach in the sequel.

[4] Grabmayer and Fokkink included the additional equation $(x * y)z = x * (yz)$ in their axiomatization, but this is derivable from the other axioms.

The Completeness Proof. The first step on our journey is to cast a chart as a system of equations and to study its solutions. Formally, given a chart (X, δ), we treat each state $x \in X$ as an unknown, and add the formal equation $x = a_1x_1 + \cdots + a_nx_n + b_1 + \cdots + b_m$, with the right-hand side determined by the transition function at x, $\delta(x) = \{(a_1, x_1), \ldots, (a_n, x_n), (b_1, \checkmark), \ldots, (b_m, \checkmark)\}$.

A solution to the system of equations for (X, δ) maps states to expressions in a way that satisfies the equations, up to equivalence. To formalize this, we need some notation: for a finite set of 1-free star expressions $S = \{r_1, \ldots, r_n\}$, we write $\sum_{r \in S} r$ to denote $r_1 + \cdots + r_n$. This is well-defined up to equivalence, because the axioms in Fig. 2 include commutativity and associativity. If $S = \{r \in \mathit{StExp} \mid P(r)\}$ for some finite predicate P, we write $\sum_{P(r)} r$ instead of $\sum_{r \in S} r$.

Definition 2.7. *A* solution *to a chart* (X, δ) *is a map* $\phi \colon X \to \mathit{StExp}$ *such that for any* $x \in X$, $\mathsf{SL}^* \vdash \phi(x) = \sum_{x \xrightarrow{a} y} a\phi(y) + \sum_{x \xrightarrow{b} \checkmark} b$. *Two solutions* ϕ, ψ *are* equivalent *if for any* $x \in X$, $\mathsf{SL}^* \vdash \phi(x) = \psi(x)$. *A chart* (X, δ) *is said to have a* unique *solution if it has exactly one solution up to equivalence.*

Example 2.8. Let $X = \{x, x'\}$ and suppose (X, δ_X) is the chart below.

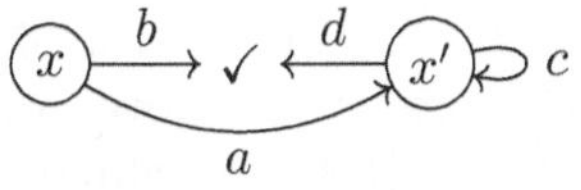

A solution to (X, δ_X) is a function $\phi : X \to \mathit{StExp}$ such that $\mathsf{SL}^* \vdash \phi(x) = a \cdot \phi(x') + b$ and $\mathsf{SL}^* \vdash \phi(x') = c \cdot \phi(x') + d$. The trick to solving this system is to apply the last axiom in Figure 2 to the second equation, and choose $\phi(x') = c * d$. If we also set $\phi(x) = b + a(c * d)$, then ϕ is a solution by construction.

The following property corresponds to [27, Lemma 2.2] and [27, Theorem 2.2], which characterize solutions to charts as chart homomorphisms into the quotient of *StExp* by provable equivalence. Given $r_1, r_2 \in \mathit{StExp}$, write $r_1 \equiv r_2$ to denote that $\mathsf{SL}^* \vdash r_1 = r_2$, and write $[r_1]_\equiv$ to denote the $\equiv$-equivalence class of r_1.

Proposition 2.9. *There is a unique* $[\delta]_\equiv \colon \mathit{StExp}/{\equiv} \to \mathcal{P}(\mathit{Act} \times (\checkmark + \mathit{StExp}/{\equiv}))$ *such that the quotient* $[-]_\equiv \colon \mathit{StExp} \to \mathit{StExp}/{\equiv}$ *is a chart homomorphism from* (StExp, δ) *to* $(\mathit{StExp}/{\equiv}, [\delta]_\equiv)$. *Moreover,* $\phi \colon X \to \mathit{StExp}$ *is a solution to a chart* (X, δ) *if and only if* $[-]_\equiv \circ \phi \colon (X, \delta) \to (\mathit{StExp}/{\equiv}, [\delta]_\equiv)$ *is a chart homomorphism.*

Proposition 2.9 has several consequences, including the following two observations, which appear as [6, Proposition 5.1] and [6, Proposition 2.9].

Proposition 2.10. *Let* $h \colon (X, \delta_X) \to (Y, \delta_Y)$ *be a chart homomorphism, and let* $\phi \colon Y \to \mathit{StExp}$ *be a solution to* (Y, δ_Y). *Then* $\phi \circ h$ *is a solution to* (X, δ_X). *Also, for any* $r \in \mathit{StExp}$, *the inclusion map* $\langle r \rangle \hookrightarrow \mathit{StExp}$ *is a solution to* $\langle r \rangle$.

Grabmayer and Fokkink's completeness proof strategy requires the construction of a distinguished class of charts $\mathcal{C}$ satisfying all of the following properties:

(Expressivity) For each $r \in \mathit{StExp}$, the chart $\langle r \rangle$ is in $\mathcal{C}$.
(Closure) $\mathcal{C}$ is *closed under homomorphic images.* That is, if there is a surjective chart homomorphism $(X, \delta_X) \to (Y, \delta_Y)$ and $(X, \delta_X) \in \mathcal{C}$, then $(Y, \delta_Y) \in \mathcal{C}$.
(Solvability) Every chart $(X, \delta) \in \mathcal{C}$ admits a unique solution.

The restriction of surjectivity in the second property is relatively mild: any chart homomorphism $h : (X, \delta_X) \to (Y, \delta_Y)$ can be restricted to a homomorphism onto its image via standard techniques [23]. If a class satisfying these properties exists, then the completeness of SL^* for bisimilarity can be argued as follows.

Proof (Theorem 2.6, completeness direction). Let $r_1, r_2 \in \mathit{StExp}$ and suppose $r_1 \leftrightarroweq r_2$. Then there is a chart (Z, δ) and chart homomorphisms $h_i \colon \langle r_i \rangle \to (Z, \delta)$ for $i \in \{1, 2\}$ such that $h_1(r_1) = h_2(r_2)$. Let $z = h_1(r_1) = h_2(r_2)$, and note that $\langle z \rangle = h_1(\langle r_1 \rangle) = h_2(\langle r_2 \rangle)$. In particular, $\langle z \rangle$ is the homomorphic image of $\langle r_1 \rangle$. By **(Expressivity)**, $\langle r_1 \rangle \in \mathcal{C}$. By **(Closure)**, $\langle z \rangle \in \mathcal{C}$, because it is the homomorphic image of $\langle r_1 \rangle$. By **(Solvability)**, $\langle r_1 \rangle$, $\langle r_2 \rangle$, and $\langle z \rangle$ all admit unique solutions. Let $\phi \colon \langle z \rangle \to \mathit{StExp}$ be the solution to $\langle z \rangle$. By Proposition 2.10, $\phi \circ h_i$ is a solution to $\langle r_i \rangle$ for $i \in \{1, 2\}$. Since the inclusions $\langle r_i \rangle \hookrightarrow \mathit{StExp}$ are also solutions, by uniqueness $\mathsf{SL}^* \vdash r_1 = \phi \circ h_1(r_1) = \phi \circ h_2(r_2) = r_2$, as desired.

Well-layered Charts and Solutions. For the rest of the section, we focus on a class $\mathcal{C}$ that satisfies the three properties above. Milner already observed that not every chart admits a solution [15]. This stands in sharp contrast with the situation for regular expressions considered *up to language equivalence*, where every automaton can be transformed into an equivalent regular expression by Kleene's theorem. To address this, Grabmayer and Fokkink proposed *LLEE charts*, which were later refined into *well-layered* charts in [27].

Definition 2.11 ([27, Definition 4.1]). *Let (X, δ) be a chart. An* entry/body labelling *of (X, δ) is a partition of $\to_\delta \subseteq X \times X$ into two relations:* loop entry *transitions, denoted $\to_{\mathsf{e}}$, and* body *transitions, denoted $\to_{\mathsf{b}}$.*

We write $x \curvearrowright y$ and say that x loops-around-to *y if there is a sequence of transitions $x \to_{\mathsf{e}} x_1 \to_{\mathsf{b}} \cdots \to_{\mathsf{b}} x_n = y$ such that $x \notin \{x_1, \ldots, x_n\}$.*

An entry/body labelling of (X, δ) is well-layered *if it satisfies the following additional properties. Write $(-)^+$ for transitive closure.*

1. *We do not have $x \to_{\mathsf{b}}^+ x$ for any $x \in X$.*
2. *For any $x, y \in X$, if $x \to_{\mathsf{e}} y$, then $y \to_{\mathsf{b}}^+ x$.*
3. *The directed graph $(X, \curvearrowright)$ is acyclic.*
4. *For any $x, y \in X$, if $x \curvearrowright y$, then we do not have $y \to \checkmark$.*

A chart is well-layered *if (1) it has a well-layered entry-body labelling, and (2) is* locally finite *in the sense that for any $x \in X$, $\langle x \rangle$ is finite.*

As we have already seen, (StExp, δ) is locally finite. The equivalence between LLEE charts and well-layered charts is explained in [27, Remark 4.1].

The loop entry transitions in a well-layered entry/body labelling are to be interpreted as the first transition that enters a program loop. This is formally captured by the entry/body labelling on $(StExp, \delta)$ given as follows: loop entry transitions are those that can be derived from the rules

$$\frac{r_1 \to \checkmark}{r_1 * r_2 \to_{\mathsf{e}} r_1 * r_2} \qquad \frac{r_1 \to s \quad s \to^{+} \checkmark}{r_1 * r_2 \to_{\mathsf{e}} s(r_1 * r_2)} \qquad \frac{r_1 \to_{\mathsf{e}} s}{r_1 r_2 \to_{\mathsf{e}} s r_2} \tag{2}$$

Body transitions are all other transitions. The following result is stated as [27, Lemma 4.1], but follows directly from [6, Proposition 3.7] and the equivalence between LLEE charts and well-layered charts.

Proposition 2.12. *The entry/body labelling of the chart $(StExp, \delta)$ given in* (2) *is well-layered. It follows that $\langle r \rangle$ is well-layered for any $r \in StExp$.*

The second statement in Proposition 2.12 follows from the first: it is a direct consequence of Definition 2.11 and the properties of subcharts that if $\to_{\mathsf{e}}, \to_{\mathsf{b}}$ is a well-layered entry/body labelling of (X, δ), then $\to_{\mathsf{e}} \cap U^2, \to_{\mathsf{b}} \cap U^2$ is also a well-layered entry/body labelling of the generated subchart $\langle U \rangle_\delta$.

If we choose the class of well-layered charts for $\mathcal{C}$, then Proposition 2.12 is **(Expressivity)**. **(Closure)** is stated below as Theorem 2.13, which is a mild strengthening of [6, Theorem 6.9] that appears as [27, Theorem 4.1].

Theorem 2.13 (Grabmayer-Fokkink). *The class of well-layered charts is* closed under homomorphic images.

Theorem 2.13 is the crux of the completeness proof in [6], and took an enormous amount of ingenuity to prove. We will rely on this result later, in Section 5, when we prove our general completeness theorem.

Finally, let us discuss **(Solvability)**, i.e., existence and uniqueness of solutions, which is the last of the pieces needed in the completeness proof for 1-free star expressions. Grabmayer and Fokkink offer the following inductively defined formula for computing the unique solution to a well-layered chart at each state.

Definition 2.14. *Given a well-layered entry/body labelling $\to_{\mathsf{e}}, \to_{\mathsf{b}}$ of a chart (X, δ), define the following two quantities: given $x \in X$, let $|x|_{en} = \max\{n \mid \exists y \in Y, x \curvearrowright^n y\}$ and $|x|_{bo} = \max\{n \mid \exists y \in X, x \to_{\mathsf{b}}{}^n y\}$.*

We define $\phi_\delta : X \to StExp$ inductively on $|x|_{bo}$ as follows:

$$\phi_\delta(x) = \Big(\sum_{x \xrightarrow{a}_e x} a + \sum_{\substack{x \xrightarrow{a}_e y \\ x \neq y}} a\ t_\delta(y, x) \Big) * \Big(\sum_{x \xrightarrow{a} \checkmark} a + \sum_{x \xrightarrow{a}_{\mathsf{b}} y} a \phi_\delta(y) \Big) \tag{3}$$

in which we define for each pair of states such that $x \curvearrowright y$ the term $t_\delta(y, x)$ below, by induction on the lexicographical ordering of $\mathbb{N} \times \mathbb{N}$:

$$t_\delta(y, x) = \Big(\sum_{y \xrightarrow{a}_{\mathsf{e}} y} a + \sum_{\substack{y \xrightarrow{a}_e z \\ y \neq z}} a\ t_\delta(z, y) \Big) * \Big(\sum_{y \xrightarrow{a}_{\mathsf{b}} x} a + \sum_{\substack{y \xrightarrow{a}_{\mathsf{b}} z \\ x \neq z}} a\ t_\delta(z, x) \Big) \tag{4}$$

The following corresponds to [6, Proposition 5.5] and [6, Proposition 5.8], and establishes **(Solvability)** for well-layered charts.

Proposition 2.15. *Let* (X, δ) *be a well-layered chart with entry/body labeling* $\to_{\mathsf{e}}, \to_{\mathsf{b}}$. *Then* ϕ_δ *as derived from* $\to_{\mathsf{e}}, \to_{\mathsf{b}}$ *in Definition 2.14 is the unique solution to* (X, δ). *In particular,* ϕ_δ *does not depend on* $\to_{\mathsf{e}}, \to_{\mathsf{b}}$ *up to equivalence.*

3 Equational Theories and M-systems

1-free star expressions denote processes that can be composed nondeterministically using $+$, and include the constant 0 for the process without outgoing transitions. The axioms involving $+$ and 0 are precisely those of a join-semilattice with bottom (see Instantiation 1 below). At the same time, the *free* semilattice with bottom generated by a set X is $\mathcal{P}(X)$, the finite powerset of X. This monad also pops up in the transition functions for charts, which are of the form $X \to \mathcal{P}(Act \times (\checkmark + X))$.

This is not a coincidence, and in this section we study the connection more formally. First, we recall *equational theories*, and touch on a well-known correspondence to *free algebra constructions*. This leads to an abstracted notion of a chart, parametrized by an equational theory, with its own notion of bisimilarity. The sections that follow develop these ideas to obtain a parametrized axiomatization.

Equational Theories. A *signature* is a set of *operation symbols* S paired with a function $\mathrm{ar}\colon S \to \mathbb{N}$. The value $\mathrm{ar}(\sigma)$ is called the *arity* of $\sigma \in S$. The set of *S-terms over X*, S^*X, is the smallest set of formal expressions containing X, and such that if $\sigma \in S$ with $\mathrm{ar}(\sigma) = n$ and $t_1, \dots, t_n \in S^*X$, then $\sigma(t_1, \dots, t_n) \in S^*X$.

Given $t \in S^*X$ and $\nu\colon X \to S^*Y$, we write $t(\nu)$ to denote the term in S^*Y obtained by replacing each x in t with $\nu(x)$. We will often write $t = t(x_1, \dots, x_n)$ for distinct $x_1, \dots, x_n$ to signal that the variables in t are among $x_1, \dots, x_n$, and more compactly $t(\vec{x})$ where $\vec{x} = (x_1, \dots, x_n)$. Given $t = t(\vec{x})$ and $t_1, \dots, t_n \in S^*Y$, we write $t(t_1, \dots, t_n)$ for the term $t(\nu)$ where $\nu\colon X \to S^*Y$ is such that $\nu(x_i) = t_i$.

Fix a set of *variables Var* and a signature S. Any relation on $S^*\,Var$ can be seen as a set of *(formal) equations* over S-terms. Given S-terms t, s over Var and a set of equations E, we write $\mathsf{E} \vdash t = s$ if $t = s$ can be derived from the equations in E and the laws of equational logic (reflexivity, symmetry, transitivity, substitution, and congruence). An *equational theory* for S is a set of equations that is closed under the inference rules of equational logic. A set of equations E is an *axiomatization* of T if T is the smallest equational theory containing E. In the future, we abuse terminology and simply refer to E as an equational theory when we are in fact referring to the equational theory it axiomatizes.

An *S-algebra* is a pair (X, ρ) consisting of a set X and for each $\sigma \in S$ an operation $\sigma^\rho\colon X^{\mathrm{ar}(\sigma)} \to X$. An S-algebra homomorphism $h\colon (X, \rho_X) \to (Y, \rho_Y)$ is a function $h\colon X \to Y$ such that $h(\sigma^{\rho_X}(x_1, \dots, x_n)) = \sigma^{\rho_Y}(h(x_1), \dots, h(x_n))$.

Given an S-algebra (X, ρ) and $t \in S^*X$, we can evaluate t to $t^\rho \in X$ by setting $x^\rho = x$ and $\sigma(t_1, \dots, t_n)^\rho = \sigma^\rho(t_1^\rho, \dots, t_n^\rho)$. An S-algebra (X, ρ) *satisfies* a set of equations E, written $(X, \rho) \models E$, if $(t_1, t_2) \in E$ implies $t_1^\rho = t_2^\rho$.

We can give an equivalent description of S-algebras using some categorical language. If we write **Set** for the category of sets and functions, we can form the signature functor $\Sigma_S = \bigsqcup_{\sigma \in S}\{\sigma\} \times \mathrm{Id}^{\mathrm{ar}(\sigma)}$ out of a signature S, where Id is the identity functor on **Set**. Then an S-algebra is the same data as a function $\rho \colon \Sigma_S X \to X$. We usually abuse notation and simply write S instead of Σ_S.

Definition 3.1. *Let* T *be an equational theory for the signature* S. *A* free-algebra construction *for* T *is a triple* (M, η, ρ) *consisting of an endofunctor* M *on* **Set** *and natural transformations* $\eta \colon \mathrm{Id} \Rightarrow M$ *and* $\rho \colon SM \Rightarrow M$ *such that* $(MX, \rho_X) \models \mathsf{T}$ *for all* X*, satisfying the following: given an* S*-algebra* (Y, μ) *that satisfies* T*, and given* $f \colon X \to Y$*, there is a unique* S*-algebra homomorphism* $f^{\#} \colon (MX, \rho_X) \to (Y, \mu)$ *such that* $f^{\#} \circ \eta = f$*, called the* Kleisli extension *of* f.

Any two free-algebra constructions for the same equational theory are isomorphic, in the sense that there exist a natural isomorphism between their functors that interacts well with the other structure. Thus, we often refer to a free-algebra construction for T as *the* free-algebra construction for T.

Remark 3.2. Free-algebra constructions for equational theories as we have described them have a close relationship with *finitary monads* on **Set**. In particular, the free-algebra constructions for an equational theory are in one-to-one correspondence with monads *presented by* the theory [4], and it is known that a monad is finitary if and only if it has an equational presentation [1].

We will use these instantiations of Definition 3.1 in the rest of the paper.

Instantiation 1. The *theory of semilattices (with bottom)* is the equational theory SL for the signature $S = \{+, 0\}$ axiomatized by the following equations:

$$x + 0 = x \qquad x + x = x \qquad x + y = y + x \qquad x + (y + z) = (x + y) + z$$

If we define $\eta_X(x) = \{x\}$ and $\rho_X \colon S\mathcal{P}X \to \mathcal{P}X$ by $V_1 +^{\rho_X} V_2 = V_1 \cup V_1$ and $0^{\rho_X} = \emptyset$, then $(\mathcal{P}, \{-\}, \rho)$ is a free-algebra construction for SL (keep in mind that we use $\mathcal{P}$ for the *finite* powerset). The Kleisli extension of $f \colon X \to Y$ with $(Y, +^{\mu}, 0^{\mu})$ a semilattice is $f^{\#} \colon \mathcal{P}X \to Y$ defined by $f^{\#}(\{x_1, x_2, \dots, x_n\}) = f(x_1) +^{\mu} f(x_2) +^{\mu} \cdots +^{\mu} f(x_n)$, with the empty sum being 0^{μ}. The theory of semilattices describes the branching type of 1-free star expressions.

Instantiation 2. Let T be a finite set of *primitive tests* and let BA be the free Boolean algebra on T, i.e., the set of Boolean expressions generated from T using $\bot$, $\wedge$ and $\neg$, modulo the axioms of Boolean algebra. The signature S of *guarded algebra* has a constant 0 and a binary operation $+_b$ for each $b \in BA$. Its theory GA is axiomatized by the following equations for all $b, c \in BA$:

$$x +_b x = x \qquad x +_b y = y +_{\neg b} x \qquad (x +_b y) +_c z = x +_{b \wedge c} (y +_c z)$$

Let At be the set of atomic elements of the Boolean algebra BA and note that this set is finite. Then the free algebra construction for GA is the triple $(\mathcal{R}, \eta, \rho)$ where

$\mathcal{R}$ is the *reader-with-exception functor* $\mathcal{R}X = (\bot + X)^{At}$, and for each $\alpha \in At$, $\eta_X(x)(\alpha) = x$, $(\theta_1 +_b^{\rho_X} \theta_2)(\alpha) = \textbf{if } \alpha \leq b \textbf{ then } \theta_1(\alpha) \textbf{ else } \theta_2(\alpha)$, and $0^{\rho_X}(\alpha) = \bot$. It is helpful to think of $x +_b y$ as notation for **if** b **then** x **else** y. The theory of guarded algebra captures the branching of skip-free GKAT programs [9].

Instantiation 3. The signature S of *(positive) convex algebra* consists of one constant symbol 0 and a binary operation $\oplus_p$ for each $p \in [0,1]$. Its theory CA is axiomatized by the following equations for all $p, q \in [0,1]$:

$$x \oplus_1 y = x \qquad x \oplus_p x = x \qquad x \oplus_p y = y \oplus_{(1-p)} x$$
$$(x \oplus_p y) \oplus_q z = x \oplus_{pq} (y \oplus_{(q(1-p)/(1-pq))}) \quad (\text{if } pq < 1)$$

Let $\mathcal{D}X$ denote the set of finitely supported probability distributions on X, and for each $x \in X$ define the *Dirac delta* distribution $\delta_x(y) = \textbf{if } x = y \textbf{ then } 1 \textbf{ else } 0$. Then the triple $(\mathcal{D}(\bot + (-)), \eta, \rho)$ is a free algebra construction for CA [32], where $\eta_X(x) = \delta_x$, $\theta_1 \oplus_p^{\rho_X} \theta_2 = p\theta_1 + (1-p)\theta_2$, and $0^{\rho_X} = \delta_\bot$. The theory of convex algebra captures the branching of probabilistic processes [31].

M-systems. Fix an equational theory T for the signature S and a free-algebra construction (M, η, ρ) for T. At the core of a free-algebra construction (M, η, ρ) for T is its endofunctor M on **Set**. By replacing the finite powerset functor $\mathcal{P}$ in the transition function type for charts $X \to \mathcal{P}(Act \times (\checkmark + X))$ with M, we obtain a transition function type for processes whose branching is captured by T.

Definition 3.3. *Let B_M be the endofunctor $M(Act \times (\checkmark + (-)))$ on* **Set**. *Then an M-*system *is a pair (X, β) consisting of a set of* states *X and a* transition function *$\beta \colon X \to B_M X$. A* homomorphism *of M-systems $h \colon (X, \beta_X) \to (Y, \beta_Y)$ is a function $h \colon X \to Y$ such that $B_M(h) \circ \beta_X = \beta_Y \circ h$.*

M-systems are precisely B_M-coalgebras, and homomorphisms of M-systems are precisely B_M-coalgebra homomorphisms. This unlocks many useful results from coalgebra [23]. In particular, M-systems and their homomorphisms form a category, which we will call $\mathrm{Coalg}(B_M)$. The theory of coalgebras also prescribes a general notion of bisimilarity, which instantiates to Definition 2.2 via Lemma 2.4.

Definition 3.4. *Let (X, β_X) and (Y, β_Y) be M-systems. We call $x \in X$ and $y \in Y$* bisimilar *(written $x \leftrightarroweq y$) if there is an M-system (Z, β_Z) and homomorphisms $h \colon (X, \beta_X) \to (Z, \beta_Z)$ and $k \colon (Y, \beta_Y) \to (Z, \beta_Z)$ such that $h(x) = k(y)$.*

A standard argument tells us that bisimilarity is an equivalence relation [7].

4 Skip-free Star Expressions

Grabmayer and Fokkink introduced 1-free star expressions as a specification language for processes with non-deterministic branching behaviour captured by charts. As we saw, charts coincide with $\mathcal{P}$-systems. Moreover, the algebraic signature of the theory of semilattices — for which $\mathcal{P}$ provides the free algebra

$$\gamma(a) = \eta(a, \checkmark) \qquad \gamma(\sigma(\vec{e})) = \sigma^\rho(\gamma(\vec{e})) \qquad \gamma(e_1e_2) = t^\rho((\vec{b}, e_2), (\vec{a}, \vec{f}e_2))$$
$$\gamma(e_1^{(s)}e_2) = s^\rho\left(t^\rho((\vec{b}, e_1^{(s)}e_2), (\vec{a}, \vec{f}(e_1^{(s)}e_2))), \gamma(e_2)\right)$$

Fig. 3. The definition of (Exp, γ) w.r.t. the free-algebra construction (M, η, ρ). Above, $e_i \in Exp$, $a \in Act$, and $s = s(u, v) \in S^*\,Var$. In the formulas for e_1e_2 and $e_1^{(s)}e_2$ above, $\gamma(e_1) = t^\rho = t^\rho((\vec{b}, \checkmark), (\vec{a}, \vec{f}))$. In the last formula, we have used red to indicate the portion of the formula that corresponds to the s in $e_1^{(s)}e_2$.

construction — suggests a syntax for behaviour expressed by well-layered charts, as well some axioms. In this section, we expand on these ideas, using equational theories and free algebra constructions to develop a syntax for M-systems, as well as a candidate set of axioms for equivalence expressed in this syntax.

For the remainder of this section, we fix an equational theory T for the signature S, which admits a free algebra construction (M, η, ρ).

Definition 4.1. *The* skip-free unified-star expressions Exp *are generated by*

$$Exp \ni e_i ::= \sigma(e_1, \ldots, e_n) \mid a \in Act \mid e_1 \cdot e_2 \mid e_1^{(s)}e_2$$

In the above, $\sigma \in S$ *with* $\mathrm{ar}(\sigma) = n$ *and* $s = s(u, v)$ *is an* S*-term in two variables.*

The small-step semantics of skip-free star expressions is captured by giving Exp the structure of an M-system. This automatically equips these expressions with a notion of equivalence, namely bisimilarity as states in this system. We call the M-system (Exp, γ), defined recursively in Fig. 3, the *syntactic M-system.*

Note that we use the following shorthands from now on: $(\vec{a}, \vec{e})$ denotes the list of tuples $(a_1, x_1), \ldots, (a_n, x_n)$; $\vec{be}$ denotes the list $b_1e_1, \ldots, b_ne_n$; $\vec{e}f$ denotes the list $e_1f, \ldots, e_nf$; and generally, where h is a function defined on the set $\{x_1, \ldots, x_n\}$, $h(\vec{x})$ denotes the list $h(x_1), \ldots, h(x_n)$.

Intuitively, the skip-free star expressions of the form $a \in Act$ and $e_1 \cdot e_2$ have the usual interpretation: a is the process that emits a and accepts, and $e_1 \cdot e_2$ is the sequential composition of e_1 and e_2. As before, we usually write e_1e_2 instead of $e_1 \cdot e_2$. The expression $\sigma(e_1, \ldots, e_n)$ denotes the process that branches into $e_1, \ldots, e_n$ and whose outgoing transitions carry the structure defined by σ.

Example 4.2. Let $\mathsf{T} = \mathsf{SL}$ be the theory of join-semilattices with bottom discussed in Instantiation 1. The free-algebra construction (M, η, ρ) for T is $(\mathcal{P}, \{-\}, \rho)$, and the behaviour of (Exp, γ) corresponds closely to that of $(StExp, \delta)$ (cf. Figure 1).

For instance, $\gamma(a) = \eta(a, \checkmark) = \{(a, \checkmark)\}$, and similarly $\gamma(b) = \{(b, \checkmark)\}$. This matches the transitions of $a, b \in StExp$ as prescribed by δ. Furthermore, $\gamma(a + b) = \gamma(a) +^\rho \gamma(b) = \gamma(a) \cup \gamma(b) = \{(a, \checkmark), (b, \checkmark)\}$, which matches $\delta(a + b)$.

If we look at $(a + b)c \in Exp$, then Figure 3 tells us that because $\gamma(a + b) = \eta(a, \checkmark) +^\rho \eta(b, \checkmark)$, we have $\gamma((a + b)c) = \eta(a, c) +^\rho \eta(b, c) = \{(a, c), (b, c)\}$; this also corresponds to the transitions exiting $(a + b)c \in StExp$ as specified by δ.

$$(\mathsf{T})\ \frac{\mathsf{T} \vdash t(\vec{x}) = s(\vec{x})}{t(\vec{e}) = s(\vec{e})} \qquad \begin{array}{ll} \text{(A)} & e(fg) = (ef)g \\ \text{(D)} & t(\vec{e})f = t(\vec{e}f) \\ \text{(U)} & e^{(s)}f = s(e(e^{(s)}f), f) \end{array} \qquad \text{(RSP)}\ \frac{g = s(eg, f)}{g = e^{(s)}f}$$

Fig. 4. The theory T^* that axiomatizes bisimilarity of star expressions. Above, $e_1, \ldots, e_n \in \mathit{Exp}$, $e, f, g \in \mathit{Exp}$, $t = t(\vec{v}) \in S^* \mathit{Var}$, and $s(u, v) \in S^* \mathit{Var}$ is a binary term with free variables u, v. Recall that the notation $\vec{e}f$ means $e_1 f, \ldots, e_n f$.

The process denoted $e_1^{(s)} e_2$, given by the *s-star* of e_1 and e_2, is a bit more complicated: it represents the process that loops on the branches of e_1 wherever the variable u appears in $s(u, v)$, and otherwise moves on to the branches of e_2 wherever the variable v appears. In the future, if $s(u, v) = \sigma(u, v)$ for some binary operation $\sigma \in S$, we will write $e_1^{(\sigma)} e_2$ in place of $e_1^{(\sigma(u,v))} e_2$.

Example 4.3. Let M, T and the free-algebra construction be as in the previous example. We can recover the behaviour of the binary star operator from 1-free regular expressions as a particular instance of skip-free unified-star expressions.

For example, if we look at $(a + b) * c$, then Figure 1 tells us that

$$\gamma(a * b) = \{(a, (a + b) * c), (b, (a + b) * c), (c, \checkmark)\}$$

At the same time, if we consider $a^{(+)}b$, then Figure 3 tells us that because $\gamma(a + b) = \eta(a, \checkmark) +^{\rho} \eta(b, \checkmark)$, we have that

$$\begin{aligned} \gamma((a + b)^{(+)}c) &= (\eta(a, (a + b)^{(+)}c) +^{\rho} \eta(b, (a + b)^{(+)}c)) +^{\rho} \eta(c, \checkmark) \\ &= \{(a, (a + b)^{(+)}c), (b, (a + b)^{(+)}c), (c, \checkmark)\} \end{aligned}$$

In fact, the above can be used to show that $(a + b) * c$ as a state in (StExp, δ) is bisimilar to $(a + b)^{(+)}c$ as a state in (Exp, γ).

Up to equivalence, there are three more choices for $s = s(u, v)$ in $e_1^{(s)} e_2$, namely the variables u and v, and the constant 0. The behaviours obtained by these expressions can still be modelled by 1-free regular expressions: $a^{(u)}b$ corresponds to $a * 0$, $a^{(v)}b$ corresponds to $0 * a$, and $a^{(0)}b$ is simply 0.

Given M-systems (U, β_U) and (X, β), (U, β_U) is a *subsystem* of (X, β) if $U \subseteq X$ and the inclusion map $(U, \beta_U) \hookrightarrow (X, \beta)$ is a homomorphism of M-systems. The syntactic M-system has a finite subsystem for each expression.

Proposition 4.4. *For each $e \in \mathit{Exp}$, there is a smallest finite subsystem $\langle e \rangle$ of (Exp, γ) containing e, called the subsystem* generated by *e.*

We now present a set of axioms that aims to capture bisimilarity in (Exp, γ).

Definition 4.5. *The* skip-free unified-star theory T^* *consists of the axioms in Fig. 4 and the laws of equational logic. We write $\mathsf{T}^* \vdash e_1 = e_2$ if $e_1 = e_2$ is derivable from T^* and say that e_1 and e_2 are* provably equivalent.

Intuitively, the first inference rule in Fig. 4 says that two terms that are equivalent up to T are equivalent as branching structures. The rules (A) and (D) express standard properties of sequential composition: associativity and right distributivity over branches. The rules (U) and (RSP) state that $e_1^{(s)}e_2$ is the unique process z that satisfies the recursion equation $z = s(e_1 z, e_2)$.

Theorem 4.6 (Soundness). *Let $e_1, e_2 \in Exp$. If $\mathsf{T}^* \vdash e_1 = e_2$, then $e_1 \leftrightarrow e_2$.*

Even working with an arbitrary equational theory, a number of interesting equivalences can be proven using the axioms of T^*. For example, $\mathsf{T}^* \vdash e_1^{(s)}(e_2e_3) = (e_1^{(s)}e_2)e_3$ is a consequence of (A), (D), (U), and (RSP). Also, if $\mathsf{T} \vdash s_1 = s_2$, then $\mathsf{T}^* \vdash e_1^{(s_1)}e_2 = e_1^{(s_2)}e_2$, which is a consequence of (T), (U), and (RSP).

In Examples 4.2 and 4.3, we have already seen that for SL, we recover a system that corresponds to 1-free star expressions. For GA and CA, we are in a similar situation: every s-star operation $e_1^{(s)}e_2$ is equivalent (up to the unified-star axioms) to a binary star, i.e., either $e_1^{(+_b)}e_2$ or $e_1^{(\oplus_p)}e_2$ respectively. These correspond precisely to the binary stars in the syntax for skip-free GKAT [9] and in the probabilistic regular expressions in [22]. Thus, our parametrized class of skip-free process algebras does indeed capture both skip-free GKAT and the algebra of skip-free probabilistic regular expressions (modulo bisimilarity).

5 Completeness

We are now ready to start proving the completeness of T^* w.r.t. bisimilarity in (Exp, γ), at least for a large class of equational theories T. We follow the same steps as Grabmayer and Fokkink, adapting and generalizing each step to other equational theories. Specifically, we will construct a class $\mathcal{C}$ of M-systems that satisfies analogues of **(Expressivity)**, **(Closure)** and **(Solvability)**.

Note that we are not able to prove completeness of T^* for arbitrary T. We will discuss which equational theories we have to restrict our attention to later in this section. We delay the restriction for now because the first few results presented below are true for arbitrary equational theories.

Solutions to M-systems. The completeness proof to follow casts M-systems as systems of equations of a certain form. For a given M-system (X, β), we associate each state $x \in X$ with an equation $x = t(\vec{b}, \vec{a}\vec{x})$ with unknowns $x_1, \ldots, x_n$, where $t \in S^*X$ and $\beta(x) = t^\rho((\vec{b}, \checkmark), (\vec{a}, \vec{x}))$.

Definition 5.1. *A* solution *to an M-system (X, β) is a map $\phi \colon X \to Exp$ such that for any $x \in X$ with $\beta(x) = t^\rho((\vec{b}, \checkmark), (\vec{a}, \vec{x}))$, $\mathsf{T}^* \vdash \phi(x) = t(\vec{b}, \vec{a}\phi(\vec{x}))$. Two solutions ϕ, ψ are* equivalent *if $\mathsf{T}^* \vdash \phi(x) = \psi(x)$ for all $x \in X$. The M-system (X, β)* admits a unique solution *if it has exactly one solution up to equivalence.*

Note that the specific choice of t in Definition 5.1 does not change the space of solutions to (X, β). The following result is analogous to Proposition 2.9.

Proposition 5.2. *There is a unique M-system structure $(Exp/{\equiv}, [\gamma]_{\equiv})$ such that the quotient map $[-]_{\equiv}\colon (Exp,\gamma) \to (Exp/{\equiv}, [\gamma]_{\equiv})$ is a homomorphism of M-systems. Moreover, for any M-system (X,β), a map $\phi\colon X \to Exp$ is a solution if and only if $[-]_{\equiv} \circ \phi\colon (X,\beta) \to (Exp/{\equiv}, [\gamma]_{\equiv})$ is a homomorphism of M-systems.*

As before, Proposition 5.2 has the following immediate consequences.

Proposition 5.3. *Let $h\colon (X,\beta_X) \to (Y,\beta_Y)$ be a homomorphism of M-systems, and let $\phi\colon Y \to Exp$ be a solution to (Y,β_Y). Then $\phi \circ h$ is a solution to (X,β_X). Furthermore, for any $e \in Exp$, the inclusion map $\langle e\rangle \hookrightarrow Exp$ is a solution to $\langle e\rangle$.*

We would now like to reuse Theorem 2.13, which says that well-layered charts are closed under homomorphic images, and Definition 2.14, which computes the canonical solution to a well-layered chart. To do this, we need to restrict T.

Support. We start by generalizing the notion of well-layeredness. Here, the idea is that we need a way to talk about how the states of an M-system are connected. This is encapsulated by the notion of *support*, defined below.

Definition 5.4. *A* support *for T is a natural transformation $\mathsf{supp}\colon M \Rightarrow \mathcal{P}$ s.t. (1) $\mathsf{supp}\circ\eta = \{-\}$ and (2) for $\sigma \in S$ and $t_1,\dots,t_n \in S^*X$, $\mathsf{supp}(\sigma^\rho(t_1^\rho,\dots,t_n^\rho)) \subseteq \mathsf{supp}(t_1^\rho) \cup \dots \cup \mathsf{supp}(t_n^\rho)$. T is* supported *if a support for T exists.*

Intuitively, an equational theory is supported if every term has a well-defined set of "essential variables" up to provable equivalence. For example, in the theory of semilattices SL, the set of essential variables of a term $x_1 + \dots + x_n$ is precisely $\{x_1,\dots,x_n\}$ — i.e., the support is the identity transformation on $\mathcal{P}$. For guarded algebra (GA), the support takes a function $\theta\colon At \to \bot + X$ to its image without $\bot$, $\mathsf{supp}(\theta) = \theta(At) \setminus \{\bot\}$. For convex algebra, the support of $\theta \in \mathcal{D}(\bot + X)$ is its support as a subprobability distribution, $\mathsf{supp}(\theta) = \{x \in X \mid \theta(x) > 0\}$.

Not every equational theory is supported. For a trivial nonexample, the theory axiomatized by $\mathsf{E} = \{(u,v)\}$ identifies all terms and has the constant functor $M = \{\star\}$ and $\eta_X(x) = \star$ in its free-algebra construction. The only natural transformation $\lambda\colon M \Rightarrow \mathcal{P}$ maps $\star$ to $\emptyset$, which does not satisfy $\lambda \circ \eta = \{-\}$.

Remark 5.5. The existence of a support does not depend on the choice of (M,η,ρ), since all free-algebra constructions for T are isomorphic. So, if one (M,η,ρ) has a support, then all do. However, more than one support may exist.

In an M-system corresponding to a supported equational theory, branching is essentially given by transitions, in the sense of charts, with extra structure. The *underlying chart* of an M-system is the chart obtained by forgetting this structure. This extends the notion of well-layered charts to M-systems.

Definition 5.6. *Let supp be a support for T. The* underlying chart *of an M-system (X,β) is the chart $\mathsf{supp}_*(X,\beta) = (X, \mathsf{supp}_X \circ \beta)$. An M-system is called* well-layered *if its underlying chart is well-layered.*

For the remainder of this section, we shall assume that T is supported. In an M-system (X, β), write $x \xrightarrow{a}_\beta \xi$ if $x \xrightarrow{a}_\beta \xi$ in its underlying chart, and $x \to_\beta y$ if $x \xrightarrow{a}_\beta y$ for some a. By Definition 5.6, a well-layered M-system (X, β) admits an entry/body labelling $\to_{\mathsf{e}}$, $\to_{\mathsf{b}}$ of $\mathsf{supp}_*(X, \beta)$ that satisfies Definition 2.11. We can then write $x \curvearrowright y$ for states $x, y \in X$ if $x \curvearrowright y$ in this entry/body labelling, and furthermore define $|x|_{en}$ and $|x|_{bo}$ as in Definition 2.14 for $x \in X$.

With this candidate class of M-systems, we can recover **(Expressivity)**.

Proposition 5.7. *Let* supp *be a support for* T. *Then* (Exp, γ) *is well-layered. Consequently, for any* $e \in Exp$, *the* M*-system* $\langle e \rangle$ *generated by* e *is well-layered.*

Recall that M-systems are just B_M-coalgebras. The transformation of M-systems, being a natural transformation, defines a functor $\mathsf{supp}_* \colon \mathrm{Coalg}(B_M) \to \mathrm{Coalg}(B_{\mathcal{P}})$ that has some nice properties [23]. The following is true for general F-coalgebras for an endofunctor on **Set**, and so we state it in full generality.

Lemma 5.8. *Let* F *and* G *be endofunctors on* **Set** *and let* $\lambda \colon F \Rightarrow G$ *be a natural transformation. Define* $\lambda_* \colon \mathrm{Coalg}(F) \to \mathrm{Coalg}(G)$ *to be the functor with* $\lambda_*(X, \beta) = (X, \lambda_X \circ \beta)$ *and* $\lambda_*(h) = h$ *for any coalgebra homomorphism* h. *Let* $\mathcal{C}$ *be a class of* G*-coalgebras that is closed under homomorphic images. Then* $\lambda^{-1}\mathcal{C} = \{(X, \beta) \mid \lambda_*(X, \beta) \in \mathcal{C}\}$ *is also closed under homomorphic images.*

In our situation, $F = B_M$, $G = B_{\mathcal{P}}$, and $\lambda = \mathsf{supp}_*$. As we have defined it, the class of well-layered M-systems is precisely the inverse image of supp_*. From Theorem 2.13 and Lemma 5.8, we immediately obtain a version of **(Closure)**:

Theorem 5.9. *Well-layered* M*-systems are closed under homomorphic images.*

Malleability. The point of well-layered systems is that we can solve them uniquely. To this end, we want to replay the strategy from Definition 2.14. The following notion helps to do that, by letting us isolate variables into a subterm.

Definition 5.10. *We say that* T *is* malleable *if for any set* X, *any partition* $U + V = X$, *and any term* $t \in S^*X$, *there are terms* $t_1 \in S^*U$, $t_2 \in S^*V$, *and a term* $s = s(u, v) \in S^*\,Var$ *such that* $\mathsf{T} \vdash t = s(t_1, t_2)$.

Example 5.11. In the case of GA, if $t = x +_b (y +_c z)$ and $U = \{x, y\}$ while $V = \{z\}$, then we can choose $s = u +_{b \vee c} v$, $t_1 = x +_b y$ and $t_2 = z$ to find that $\mathsf{T} \vdash t = s(t_1, t_2)$. More generally, due to the associativity and commutativity properties of SL, GA, and CA, all three of these equational theories are malleable.[5]

Remark 5.12. Not all equational theories are malleable. For a trivial example, consider the theory $\mathsf{T} = \emptyset$ for a signature with two binary operations $\star, \bullet$. Clearly, the term $x \star (y \bullet z)$ is not of the form $s(t_1(x, y), t_2(z))$ for any terms t_1, t_2, s. In Section 6, we will also see an example of a richer equational theory, which captures branching that mixes nondeterminism and probability, that is not malleable.

[5] These are not necessary conditions for malleability, though. In Section 6, we will see an example of a malleable equational theory that does not enjoy associativity.

For the rest of this section, we assume that T is malleable. We can now use this to recover the solution strategy for well-layered charts in Definition 2.14.

Definition 5.13. *Let $\to_{\mathsf{e}}, \to_{\mathsf{b}}$ be a well-layered entry/body labelling of (X, β). We define the* canonical solution *to (X, β)* given by *$\to_{\mathsf{e}}, \to_{\mathsf{b}}$ as follows. Let*

$$\beta(x) = s^{\rho}\Big(t_1^{\rho}((\vec{a}, x), (\vec{b}, \vec{x})),\ t_2^{\rho}((\vec{c}, \vec{y}), (\vec{d}, \checkmark))\Big)$$

where $\vec{x}$ is a vector such that $x \neq x_i$ and $x \to_{\mathsf{e}} x_i$ for each i, and $\vec{y}$ is a vector such that $x \to_{\mathsf{b}} y_j$ for each j. By induction on $|x|_{bo} \in \mathbb{N}$, we define

$$\phi_\beta(x) = \Big(t_1(\vec{a}, b_1\tau_\beta(x_1, x), \ldots, b_n\tau_\beta(x_n, x))\Big)^{(s)}\Big(t_2(c_1\phi_\beta(y_1), \ldots, c_m\phi(y_m), \vec{d})\Big)$$

where, by induction on $(|x|_{en}, |y|_{bo})$ in the lexicographical ordering of $\mathbb{N} \times \mathbb{N}$, for each pair of states such that $x \curvearrowright y$ we define $\tau_\beta(y, x)$ as follows. First, let

$$\beta(y) = s^{\rho}\Big(t_1^{\rho}((\vec{a}, y), (\vec{b}, \vec{x})),\ t_2^{\rho}((\vec{c}, x), (\vec{d}, \vec{y}))\Big)$$

where $\vec{x}$ is a vector such that $y \neq x_i$ and $y \to_{\mathsf{e}} x_i$ for each i, and $\vec{y}$ is a vector such that $x \neq y_k$ and $y \to_{\mathsf{b}} y_k$ for each k. Then

$$\tau_\beta(y, x) = \Big(t_1(\vec{a}, b_1\tau_\beta(x_1, y), \ldots, b_n\tau_\beta(x_n, y))\Big)^{(s)}\Big(t_2(\vec{c}, d_1\tau_\beta(y_1, x), \ldots, d_m\tau_\beta(y_m, x))\Big)$$

In the above, we assumed that $\beta(x)$ and $\beta(y)$ were in a specific form. This is where malleability comes in: we partitioned the support of $\beta(x)$ into pairs that correspond to self loops $x \to_{\mathsf{e}} x$ or loop entry transitions $x \to_{\mathsf{e}} x_i$, and those that come from body transitions $x \to_{\mathsf{b}} y_j$ or accepting transitions $x \to \checkmark$. Malleability assures us that we can write $\beta(x)$ as described, and similarly for $\beta(y)$.

Unravelling the definitions in the case of SL^*, one obtains the canonical solution formula in Definition 2.14, which appeared in [6]. In this case also, ϕ_β is the unique solution to a well-layered M-system, so we recover **(Solvability)**.

Proposition 5.14. *Let (X, β) be a well-layered M-system, with entry/body labeling $\to_{\mathsf{e}}, \to_{\mathsf{b}}$. Then the canonical solution ϕ_β given by $\to_{\mathsf{e}}, \to_{\mathsf{b}}$ is the unique solution to (X, β). In particular, up to T^*, ϕ_β does not depend on $\to_{\mathsf{e}}, \to_{\mathsf{b}}$.*

Completeness. Following the same steps in the first completeness proof we saw (of Theorem 2.6) with $\mathcal{C}$ the class of well-layered M-systems (and replacing the word "chart" with "M-system" everywhere), we can apply Propositions 2.12 and 5.14 and Theorem 5.9 to obtain the main result of the paper.

Theorem 5.15 (Completeness). *Let T be a supported malleable theory for the signature S. Given $e_1, e_2 \in \mathit{Exp}$, if $e_1 \leftrightarrow e_2$, then $\mathsf{T}^* \vdash e_1 = e_2$.*

6 Examples and Nonexamples

As we have hinted at, the theory of semilattices SL, guarded algebra GA, and convex algebra CA are supported and malleable, and therefore fit our framework. But, as we have already seen, some equational theories do not admit support, and others are not malleable. In this section, we discuss the scope of our story.

The following result gives a sufficient condition for malleability.

Proposition 6.1. *Let* T *be an equational theory for a signature* S *consisting of constants and binary operations.* T *is malleable if both of the following hold:*

1. Skew commutativity. *For any binary operation* $\sigma \in S$, *there is a binary operation* $\tau \in S$ *such that* $\mathsf{T} \vdash \sigma(x, y) = \tau(y, x)$.
2. Skew associativity. *For any binary operations* $\sigma_1, \sigma_2 \in S$, *there are binary operations* $\tau_1, \tau_2 \in S$ *such that* $\mathsf{T} \vdash \sigma_1(x, \sigma_2(y, z)) = \tau_1(\tau_2(x, y), z)$.

It is immediate from the axioms of SL, GA, and CA that all three satisfy these properties. However, some malleable theories are not skew-associative.

Instantiation 4. The theory of *guarded convex algebra* GC consists of one constant symbol 0, one binary operation $+_b$ for each Boolean expression $b \in BA$ (see Inst. 2), one binary operation $\oplus_p$ for each $p \in [0, 1]$, and is axiomatized by the equations of GA, CA, and the *distribution law* $x \oplus_p (y +_b z) = (x \oplus_p y) +_b (x \oplus_p z)$. Its free-algebra construction is given by $(\mathcal{D}(\bot + (-))^{At}, \eta, \rho)$ where $\eta_X(x)(\alpha) = \delta_x$; $(\chi_1 +_b^\rho \chi_2)(\alpha)(x) = \chi_1(\alpha)(x)$ if $\alpha \leq b$, and $(\chi_1 +_b^\rho \chi_2)(\alpha)(x) = \chi_2(\alpha)(x)$ otherwise; $(\chi_1 \oplus_p^\rho \chi_2)(\alpha)(x) = p\chi_1(\alpha)(x) + (1-p)\chi_2(\alpha)(x)$, and $0^\rho(\alpha) = \delta_\bot$.

Guarded convex algebra is a supported malleable theory that is not skew associative. Indeed, we can take $\mathsf{supp}\colon \mathcal{D}(\bot + (-))^{At} \Rightarrow \mathcal{P}$ to be $\mathsf{supp}_X(\chi) = \bigcup_{\alpha \in At}\{x \in X \mid \chi(\alpha)(x) > 0\}$. It is not difficult to show that this is a natural transformation that satisfies the requirements of a support. To see why GC is not skew-associative, consider the term $x +_{0.5} (y +_b z)$. A simple case analysis reveals that it is not equivalent to $\tau_1(\tau_2(x, y), z)$ for any binary operations τ_1 and τ_2.

We illustrate the proof that GC is malleable in the special case of $X = \{x, y, z\}$ and $At = \{\alpha_1, \alpha_2\}$. Consider $\chi = \theta_1 +_{\alpha_1} \theta_2$ for some probability distributions θ_1, θ_2 on X. The convex algebra $\mathcal{D}(\bot + X)$ can be visualized as the 3-simplex in $\mathbb{R}^4$ [30] with extremal points $\delta_x, \delta_y, \delta_z, \delta_\bot$. We only need one of its faces, the convex hull of $\delta_x, \delta_y, \delta_z$, depicted in (5). There, χ represents two points, one for each of α_1, α_2. To obtain terms $s(u, v)$, $t_1(x, y)$ and $t_2(z)$ such that $\chi = s^\rho(t_1^\rho, t_2^\rho)$, draw straight lines from δ_z through θ_1 to the segment between δ_x and δ_y.

The endpoints of the drawn lines represent distributions obtained from terms of the form $r_1(x, y), r_2(x, y)$, i.e., $\theta_1' = r_1^\rho(x, y)$ and $\theta_2' = r_2^\rho(x, y)$. Then $\chi = (\theta_1' \oplus_p \delta_z) +_{\alpha_1} (\theta_2' \oplus_q \delta_z)$ for some $p, q \in [0, 1]$. If we choose $s(u, v) = (u \oplus_p v) +_{\alpha_1} (u \oplus_q v)$, $t_1(x, y) = r_1(x, y) +_{\alpha_1} r_2(x, y)$ and $t_2 = z$, then $\chi = s^\rho(t_1^\rho, t_2^\rho)$.

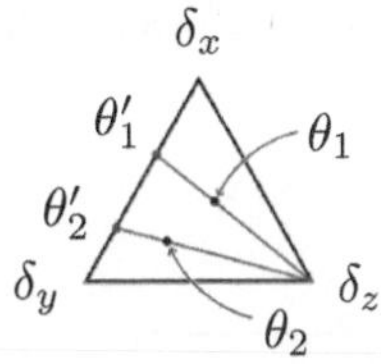

(5)

Guarded convex algebra is the equational theory GC underlying the recently introduced *probabilistic guarded Kleene algebra with tests* (or *ProbGKAT*) [21,25].

However, the skip-free universal-star fragment of GC is not the obvious skip-free fragment of ProbGKAT, because the latter allows only the binary stars $e_1^{(s)}e_2$ for $s(u,v) = u +_b v$ and $s = u \oplus_p v$. In contrast, skip-free unified-star expressions for GC allow *mixed loops*, like $e_1^{(s)}e_2$ with $s = u \oplus_p (u +_b v)$, which enters the loop body with probability p, and with probability $1-p$ does this iff b is true.

Instantiation 5. A *(unital) semiring* is a set $\mathbb{S}$ equipped with two constants 0 and 1 and two binary operations + and × (written as juxtaposition) such that $(\mathbb{S}, +, 0)$ is a commutative monoid, $(\mathbb{S}, \times, 1)$ is a monoid, and the distributive laws $p(q+r) = pq + pr$ and $(p+q)r = pr + qr$ hold. The theory $\mathbb{S}\mathsf{Mod}$ of *semimodules* over a semiring $\mathbb{S}$ has a signature consisting of a constant 0, a binary operation $\oplus$, and a unary operation $p \cdot (-)$ for each $p \in \mathbb{S}$. The axioms of $\mathbb{S}\mathsf{Mod}$ state that $\oplus$ is commutative, associative, and has 0 as a neutral element (the commutative monoid axioms), as well as $0 \cdot x = 0$, $1 \cdot x = x$, $p \cdot (q \cdot x) = (pq) \cdot x$, $p \cdot (x \oplus y) = (p \cdot x) \oplus (p \cdot y)$, and $(p+q) \cdot x = (p \cdot x) \oplus (q \cdot x)$, for any $p, q \in \mathbb{S}$.

Given a function $\theta\colon X \to \mathbb{S}$, define $\mathsf{supp}_X(\theta) = \{x \in X \mid \theta(x) \neq 0\}$. The free algebra construction for $\mathbb{S}\mathsf{Mod}$ is given by $(\mathcal{O}_\mathbb{S}, \eta, \rho)$, where $\mathcal{O}_\mathbb{S}X = \{\theta\colon X \to \mathbb{S} \mid \mathsf{supp}_X(\theta) \text{ is finite}\}$; $\eta(x)(y) = \textbf{if } x = y \textbf{ then } 1 \textbf{ else } 0$; and where $0^\rho(x) = 0$, $(\theta_1 \oplus^\rho \theta_2)(x) = \theta_1(x) + \theta_2(x)$, and $(p \cdot^\rho \theta)(x) = p\theta(x)$. An $\mathcal{O}_\mathbb{S}$-system is essentially a weighted transition system with weights that live in $\mathbb{S}$.

The theory $\mathbb{S}\mathsf{Mod}$ is supported malleable: $\mathsf{supp}_X(\theta)$ is finite for each $\theta \in \mathcal{O}_\mathbb{S}X$ by definition, so we obtain a natural transformation $\mathsf{supp}\colon \mathcal{O}_\mathbb{S} \Rightarrow \mathcal{P}$ that clearly satisfies the requirements of a support for $\mathbb{S}\mathsf{Mod}$. To see malleability, observe that up to $\mathbb{S}\mathsf{Mod}$, every term $t(\vec{x}, \vec{y})$ with disjoint $\vec{x}, \vec{y}$ is equivalent to one of the form $[(p_1 \cdot x_1) \oplus \cdots \oplus (p_n \cdot x_n)] \oplus [(q_1 \cdot y_1) \oplus \cdots \oplus (q_m \cdot y_m)]$ for some $p_i, q_j \in \mathbb{S}$.

An Unfortunate Nonexample. Several authors have taken an interest in mixing nondeterminism with probability [4,10,14,16,34]. A natural choice for the underlying equational theory in this case is the theory of *convex semilattices* CS [4], which consists of SL, CA and the distributive law $x \oplus_p (y + z) = (x \oplus_p y) + (x \oplus_p z)$. The free-algebra construction for CS is $(\mathcal{C}, \eta, \rho)$ where $\mathcal{C}X$ is the set of convex subsets of $\mathcal{D}(\bot + (-))$ that include $\delta_\bot$, $\eta_X(x) = \{p\delta_x + (1-p)\delta_\bot \mid p \in [0,1]\}$, $0^{\rho_X} = \{\delta_\bot\}$, $U \oplus_p^{\rho_X} V = \{p\theta_1 + (1-p)\theta_2 \mid \theta_1 \in U, \theta_2 \in V\}$, and $U +^{\rho_X} V = \mathrm{conv}(U \cup V)$ is the convex hull of $U \cup V$ [4] (see also [25, Example 4.1.14]).

The theory of convex semilattices admits the obvious support but, despite the similarity to GC, it is not malleable. Indeed, there are no terms $s(u,v)$, $t_1(x,y)$, and $t_2(z)$ such that $\mathsf{CS} \vdash s(t_1(x,y), t_2(z)) = x + (y \oplus_{\frac{1}{2}} z)$. To see why, recall that the space of probability distributions on $X = \{x, y, z\}$ can be identified with a face of the 3-simplex, depicted as a black triangle in (6).

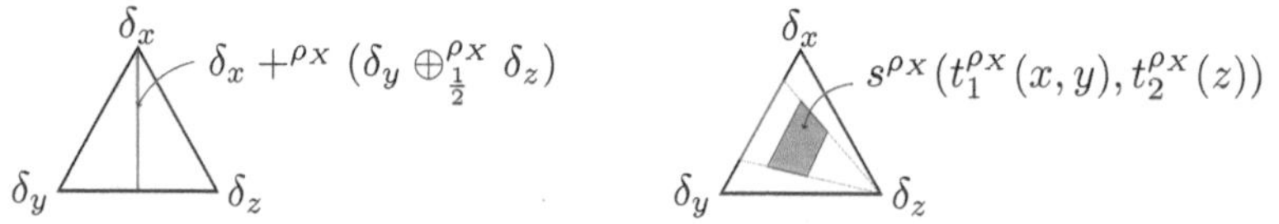

(6)

The line down the middle of the left triangle in (6) is the convex set of probability distributions that corresponds to the term $x + (y \oplus_{\frac{1}{2}} z)$ (cf. [17, Fig. 1]). The highlighted region in the center of the right triangle is the general shape that any convex set of the form $s(t_1(x,y),t_2(z))$ must take. No convex set of this form is equal to the line segment in (6). Thus, CS is not malleable.

On the other hand, the theory of convex semilattices still obtains a syntax of skip-free unified-star expressions, a bisimilarity semantics, and a sound set of axioms CS^* from the framework presented in the paper. So, we ask: is CS^* complete with respect to bisimilarity of skip-free unified-star expressions?

7 Discussion and Future Work

Given a supported and malleable equational theory T, we can derive a notion of bisimilarity and a complete axiomatization for "skip-free" processes with T-branching. This framework recovers existing completeness theorems for SL^* (the result by Grabmayer and Fokkink [6]) and GA^* (skip-free GKAT up to bisimilarity [9]), and yields new ones for CA^* (1-free probabilistic regular expressions up to bisimilarity [22]) and GC^* (a slightly generalized skip-free ProbGKAT [21]).

We would like our framework to abstract the completeness theorem of regular expressions up to bisimilarity [5]. This could settle the open completeness problem for full GKAT up to bisimilarity [26], and possibly its trace semantics [29].

The theories we consider all contain a constant 0, which stands for the *deadlocked* process, which does not allow any branching and satisfies $0e = 0$. It would be interesting to see what would be necessary to guarantee that $e0 = 0$. This would make 0 act like the "predictable failure" studied by Baeten and Bergstra in [3]. Earlier work in the setting of GKAT has shown that completeness for this extended system can be derived from the original [26,9].

Unified star-expressions give a middle ground between a Kleene-star and general recursion, by focusing on loops that derive from T. We would also like to investigate the hierarchy of expressiveness of star-expressions for n-ary operators. Perhaps such an extension would help to get a completeness theorem for CS^*.

A natural question to ask is which equational theories are supported malleable. For example, all of our examples are skew commutative (in the sense of Proposition 6.1), but it is currently not clear if this is a necessary condition. Also, conspicuously, the distributive law in GC allowed us to mix GA with CA to produce a malleable theory, while the distributive law in CS did not. We would like know if this is related to the existence of a distributive law of monads $\mathcal{DR} \Rightarrow \mathcal{RD}$ and the lack of a distributive law $\mathcal{DP} \Rightarrow \mathcal{PD}$ [34], or at least to the *composite theories* of [17] that guarantee the existence of a distributive law.

Finally, because bisimilarity coincides with provable equivalence for the equational theories satisfying our constraints, it is also a congruence. We wonder whether this implies the existence of a distributive law à la Turi and Plotkin [33].

Acknowledgements T. Kappé was partially supported by the European Union's Horizon 2020 research and innovation programme under grant no. 101027412 (VERLAN), and partially by the Dutch research council (NWO) under grant no. VI.Veni.232.286 (ChEOpS).

References

1. Adamek, J., Rosicky, J.: Locally Presentable and Accessible Categories. London Mathematical Society Lecture Note Series, Cambridge University Press (1994)
2. Antimirov, V.M.: Partial derivates of regular expressions and finite automata constructions. In: STACS. pp. 455–466 (1995). https://doi.org/10.1007/3-540-59042-0_96
3. Baeten, J.C.M., Bergstra, J.A.: Process algebra with a zero object. In: CONCUR. pp. 83–98 (1990). https://doi.org/10.1007/BFB0039053
4. Bonchi, F., Sokolova, A., Vignudelli, V.: Presenting convex sets of probability distributions by convex semilattices and unique bases ((co)algebraic pearls). In: CALCO. pp. 11:1–11:18 (2021). https://doi.org/10.4230/LIPICS.CALCO.2021.11
5. Grabmayer, C.: Milner's proof system for regular expressions modulo bisimilarity is complete: Crystallization: Near-collapsing process graph interpretations of regular expressions. In: LICS. pp. 34:1–34:13 (2022). https://doi.org/10.1145/3531130.3532430
6. Grabmayer, C., Fokkink, W.J.: A complete proof system for 1-free regular expressions modulo bisimilarity. In: LICS. pp. 465–478 (2020). https://doi.org/10.1145/3373718.3394744
7. Jacobs, B.: Introduction to Coalgebra: Towards Mathematics of States and Observation, Cambridge Tracts in Theoretical Computer Science, vol. 59. Cambridge University Press (2016). https://doi.org/10.1017/CBO9781316823187
8. Kappé, T., Schmid, T.: A general completeness theorem for skip-free star algebras (2025), https://arxiv.org/abs/2501.15325
9. Kappé, T., Schmid, T., Silva, A.: A complete inference system for skip-free guarded Kleene algebra with tests. In: ESOP. pp. 309–336 (2023). https://doi.org/10.1007/978-3-031-30044-8_12
10. Keimel, K., Plotkin, G.D.: Mixed powerdomains for probability and nondeterminism. Log. Methods Comput. Sci. **13**(1) (2017). https://doi.org/10.23638/LMCS-13(1:2)2017
11. Kleene, S.C.: Representation of events in nerve nets and finite automata. Automata studies **34**, 3–41 (1956)
12. Kozen, D.: Kleene algebra with tests. ACM Trans. Program. Lang. Syst. **19**(3), 427–443 (1997). https://doi.org/10.1145/256167.256195
13. Kozen, D., Tseng, W.D.: The Böhm-Jacopini theorem is false, propositionally. In: MPC. pp. 177–192 (2008). https://doi.org/10.1007/978-3-540-70594-9_11
14. Liell-Cock, J., Staton, S.: Compositional imprecise probability: A solution from graded monads and markov categories. In: POPL. pp. 1596–1626 (2025). https://doi.org/10.1145/3704890
15. Milner, R.: A complete inference system for a class of regular behaviours. J. Comput. Syst. Sci. **28**(3), 439–466 (1984). https://doi.org/10.1016/0022-0000(84)90023-0
16. Mislove, M.W., Ouaknine, J., Worrell, J.: Axioms for probability and nondeterminism. In: EXPRESS. pp. 7–28 (2003). https://doi.org/10.1016/J.ENTCS.2004.04.019
17. Piróg, M., Staton, S.: Backtracking with cut via a distributive law and left-zero monoids. J. Funct. Program. **27**, e17 (2017). https://doi.org/10.1017/S0956796817000077

18. Plotkin, G.D., Power, J.: Semantics for algebraic operations. In: MFCS. pp. 332–345 (2001). https://doi.org/10.1016/S1571-0661(04)80970-8
19. Plotkin, G.D., Power, J.: Notions of computation determine monads. In: FOSSACS. pp. 342–356 (2002). https://doi.org/10.1007/3-540-45931-6_24
20. Riehl, E.: Category theory in context. Aurora Dover Modern Math Originals, Dover Publications, Inc., Mineola, NY (2016)
21. Rozowski, W., Kappé, T., Kozen, D., Schmid, T., Silva, A.: Probabilistic guarded KAT modulo bisimilarity: Completeness and complexity. In: ICALP. vol. 261, pp. 136:1–136:20 (2023). https://doi.org/10.4230/LIPICS.ICALP.2023.136
22. Rozowski, W., Silva, A.: A completeness theorem for probabilistic regular expressions. In: LICS. pp. 66:1–66:14. ACM (2024). https://doi.org/10.1145/3661814.3662084
23. Rutten, J.J.M.M.: Universal coalgebra: a theory of systems. Theor. Comput. Sci. **249**(1), 3–80 (2000). https://doi.org/10.1016/S0304-3975(00)00056-6
24. Salomaa, A.: Two complete axiom systems for the algebra of regular events. J. ACM **13**(1), 158–169 (1966). https://doi.org/10.1145/321312.321326
25. Schmid, T.: Coalgebraic Completeness Theorems for Effectful Process Algebras. Ph.D. thesis, University College London (2024)
26. Schmid, T., Kappé, T., Kozen, D., Silva, A.: Guarded Kleene algebra with tests: Coequations, coinduction, and completeness. In: ICALP. pp. 142:1–142:14 (2021). https://doi.org/10.4230/LIPIcs.ICALP.2021.142
27. Schmid, T., Rot, J., Silva, A.: On star expressions and coalgebraic completeness theorems. In: CALCO. pp. 242–259 (2021). https://doi.org/10.4204/EPTCS.351.15
28. Schmid, T., Rozowski, W., Silva, A., Rot, J.: Processes parametrised by an algebraic theory. In: ICALP. pp. 132:1–132:20 (2022). https://doi.org/10.4230/LIPICS.ICALP.2022.132
29. Smolka, S., Foster, N., Hsu, J., Kappé, T., Kozen, D., Silva, A.: Guarded Kleene algebra with tests: verification of uninterpreted programs in nearly linear time. In: POPL. pp. 61:1–61:28 (2020). https://doi.org/10.1145/3371129
30. Sokolova, A., Woracek, H.: Congruences of convex algebras. Journal of Pure and Applied Algebra **219**(8), 3110–3148 (2015). https://doi.org/j.jpaa.2014.10.005
31. Stark, E.W., Smolka, S.A.: A complete axiom system for finite-state probabilistic processes. In: Proof, Language, and Interaction, Essays in Honour of Robin Milner. pp. 571–596. MIT Press (2000)
32. Świrszcz, T.: Monadic functors and categories of convex sets. Institute of Mathematics, Polish Academy of Sciences (1974)
33. Turi, D., Plotkin, G.D.: Towards a mathematical operational semantics. In: LICS. pp. 280–291 (1997). https://doi.org/10.1109/LICS.1997.614955
34. Varacca, D., Winskel, G.: Distributing probability over non-determinism. Mathematical Structures in Computer Science **16**(1), 87–113 (2006). https://doi.org/10.1017/S0960129505005074

Sharing and Linear Logic with Restricted Access

Pablo Barenbaum[1](✉) and Eduardo Bonelli[2]

[1] Universidad Nacional de Quilmes (CONICET) and Instituto de Ciencias de la Computación, Universidad de Buenos Aires, Buenos Aires, Argentina
pbarenbaum@dc.uba.ar

[2] Stevens Institute of Technology, Hoboken, USA

Abstract. The two Girard translations provide two different means of obtaining embeddings of Intuitionistic Logic into Linear Logic, corresponding to different lambda-calculus calling mechanisms. The translations, mapping $A \to B$ respectively to $!A \multimap B$ and $!(A \multimap B)$, have been shown to correspond respectively to call-by-name and call-by-value.

In this work, we split the *of-course* modality of linear logic into two modalities, written "!" and "•". Intuitively, the modality "!" specifies a subproof that can be duplicated and erased, but may not necessarily be "accessed", *i.e.* interacted with, while the combined modality "!•" specifies a subproof that can moreover be accessed. The resulting system, called MSCLL, enjoys cut-elimination and is conservative over MELL.

We study how restricting access to subproofs provides ways to control *sharing* in evaluation strategies. For this, we introduce a term-assignment for an intuitionistic fragment of MSCLL, called the $\lambda^{!\bullet}$-calculus, which we show to enjoy subject reduction, confluence, and strong normalization of the simply typed fragment. We propose three sound and complete translations that respectively simulate call-by-name, call-by-value, and a variant of call-by-name that *shares* the evaluation of its arguments (similarly as in call-by-need). The translations are extended to simulate the Bang-calculus, as well as weak reduction strategies.

Keywords: Linear Logic, Lambda Calculus, Sharing, Calling Mechanisms, Call-By-Value, Call-By-Name, Call-By-Need

1 Introduction

The propositions-as-types correspondence links *computation* and *logic*, relating types with propositions, programs with proofs, and program evaluation with proof normalization. The prime example is the simply typed λ-calculus, which corresponds to intuitionistic propositional logic. The correspondence has been extended to many other calculi and logics, including Linear Logic (LL) [16]. Linear Logic proposes a *resource conscious* approach to logic, in that only formulae prefixed with an *exponential modality* can be duplicated and erased. In LL, there are two exponential modalities: *of-course* ("!"), allowing duplication/erasure on the left, and *why-not* ("?"), allowing duplication/erasure on the right. These modalities recover the ability to duplicate and erase formulae in a controlled way, making LL a suitable language to model resource-sensitive phenomena such as concurrency, memory management, and computational complexity.

P. A. Abdulla and D. Kesner (Eds.): FoSSaCS 2025, LNCS 15691, pp. 287–307, 2025.
https://doi.org/10.1007/978-3-031-90897-2_14

Girard [16] discusses two possible ways of embedding intuitionistic logic into LL, mapping the intuitionistic implication $A \rightarrow B$ respectively to $!A \multimap B$ and to $!A \multimap !B$, where "$\multimap$" stands for the linear implication[3]. Maraist *et al.* [22] observed that these translations can be used to extend the propositions-as-types correspondence to provide a logical foundation for *evaluation mechanisms*.

The most well-known evaluation mechanism for the λ-calculus is perhaps *call-by-name* (CBN), in which arguments to functions are re-evaluated upon each use. The theory of CBN has been thoroughly developed, *e.g.* in Barendregt's book [11]. On the other hand, in the *call-by-value* (CBV) evaluation mechanism [26], arguments to functions are evaluated once and for all; then their value can be recalled upon each use. Call-by-value is less deeply studied in the literature than CBN, but it has been gaining attention since its theory is subtle and corresponds more closely to the evaluation mechanism behind most programming languages.

Embeddings Encode Evaluation Mechanisms. The first Girard translation $(\cdot)^{\mathsf{N}}$ operates on *formulae* by mapping $A \rightarrow B$ to $!A^{\mathsf{N}} \multimap B^{\mathsf{N}}$. It operates on *terms* in such a way that a λ-calculus application $t\,s$ is mapped to $t^{\mathsf{N}}\,!s^{\mathsf{N}}$ in a *linear* λ-calculus[4]. Here "!" is a *term constructor*, which corresponds to an instance of the !-*introduction* rule, also called *promotion*. The key point is that the argument s of an application is prefixed with "!". This enables the (arbitrary) term s^{N} to be freely copied or discarded in t^{N}, as dictated by CBN. In this sense, the first translation provides a logical foundation for the CBN evaluation mechanism.

The second Girard translation $(\cdot)^{\mathsf{V}}$ operates on formulae by mapping the intuitionistic implication $A \rightarrow B$ to $!A^{\mathsf{V}} \multimap !B^{\mathsf{V}}$. It operates on terms in such a way that a λ-calculus application $t\,s$ is mapped to $\mathtt{der}(t^{\mathsf{V}})\,s^{\mathsf{V}}$. Here, $\mathtt{der}(\cdot)$ stands for an appropriate operation in the target language that corresponds to the !-*elimination* rule, also called *dereliction*. The key point is that the argument is not prefixed with "!", which means that it must be evaluated before being consumed, as dictated by CBV. This translation prefixes *values*, such as $\lambda x.\,r$, with a promotion, resulting in $!(\lambda x.\,r^{\mathsf{V}})$. Consequently, only values will end up being copied or discarded.

Linear Logic with Restricted Access. In this work, we propose a logical system that arises from splitting each exponential modality into two new modalities. We refer to this new system as *Linear Logic with Restricted Access*, and formally as MSCLL, to reflect that we study the fragment with **M**ultiplicative, **S**haring, and a**C**cess connectives. The *exponential* of-course ("!") modality of LL is split into two modalities in MSCLL: a *sharing* modality "!" and an *access* modality "•", that "grants" access to a subproof. The *sharing of-course* of MSCLL turns out to be weaker than the *exponential of-course* of LL but, by abuse of notation, we denote it using the same symbol "!". Dually, the exponential why-not ("?") modality of LL is split into a sharing modality "?" and an access

[3] Sometimes the second embedding is defined as mapping $A \rightarrow B$ to $!(A \multimap B)$. This is just an apparent difference, which is canceled out by adjusting the translation of sequents accordingly.

[4] Girard's original translations target Linear Logic presented in sequent calculus style. We follow here Maraist *et al.* [22], in which the target language is presented as a linear λ-calculus, in natural deduction style.

modality "∘" in MSCLL. The resulting system MSCLL is a conservative extension of Multiplicative Exponential Linear Logic (MELL), in which the combined modality "$!\bullet$" plays the role of the exponential of-course modality of LL. The operational intuition is that the sharing modality "!" in MSCLL specifies an expression that may be duplicated and erased, but it may not necessarily be used or *accessed*. The combined modality "$!\bullet$" specifies an expression that may, additionally, be accessed, *i.e.* its contents can be made to interact with the surrounding computational context.

Embedding Call-by-Name and Call-by-Value. Following, we revisit Girard's translations, but this time targetting the $\lambda^{!\bullet}$-*calculus*, a linear λ-calculus based on MSCLL. Our CBN translation $(\cdot)^{\mathsf{N}}$ operates on formulae by mapping $A \to B$ to $!\bullet A^{\mathsf{N}} \multimap B^{\mathsf{N}}$. It operates on terms by mapping a λ-calculus application $t\,s$ to $t^{\mathsf{N}}\,!\bullet s^{\mathsf{N}}$, indicating that the argument may be freely copied, discarded, and accessed. This mimics the original translation of CBN to LL. Our CBV translation $(\cdot)^{\mathsf{V}}$ operates on formulae by mapping $A \to B$ to $!\bullet A^{\mathsf{V}} \multimap !\bullet B^{\mathsf{V}}$. It operates on terms by leaving the translation of the argument of an application intact, mapping $t\,s$ to $\mathtt{req}(\mathtt{der}(t^{\mathsf{V}}))\,s^{\mathsf{V}}$. As before, $\mathtt{der}(\cdot)$ is an appropriate operator corresponding to !-elimination, while $\mathtt{req}(\cdot)$ stands for $\bullet$-elimination. At the same time, the translation maps values such as $\lambda x.\,t$ (roughly[5]) to $!\bullet\lambda x.\,t^{\mathsf{V}}$. Thus an argument s^{V} cannot be copied, discarded, or accessed at all unless later, after further evaluation, it takes the form $!\bullet\mathsf{v}$, for some value v, in which case v can be copied. This too mimics the original translation from CBV to LL. The CBN and CBV translations are proved to be sound and complete, in the sense that two λ-terms are interconvertible in the source language if and only if they are mapped to interconvertible terms in the target language.

Call-by-Sharing. The two translations above suggest a "missing link" translation $(\cdot)^{\mathsf{S}}$ that maps $A \to B$ to $!\bullet A^{\mathsf{S}} \multimap \bullet B^{\mathsf{S}}$. This translation cannot be expressed directly in a linear λ-calculus, because the "$\bullet$" modality is used as a stand-alone operator. A λ-calculus application $t\,s$ is now mapped to $t^{\mathsf{S}}\,!s^{\mathsf{S}}$, meaning that the argument can be copied and discarded, but *not accessed* yet. A value such as $\lambda x.\,r$ is now (roughly) mapped to $\bullet\lambda x.\,r^{\mathsf{S}}$. Thus an argument s^{S} cannot be accessed unless, after further evaluation, it takes the form $\bullet\mathsf{v}$, for some value v, in which case v can be accessed. The fact that arguments cannot be accessed until they become values means that the evaluation mechanism can keep *references* to a single *shared* copy of the argument, until it becomes accessible. This translation suggests an evaluation mechanism that we dub *call-by-sharing* (CBS).

Call-by-sharing bears a strong resemblance to call-by-need (CBNd), an evaluation mechanism introduced by Wadsworth in 1971 [28]. Both in CBNd and in CBS, arguments that are not used may be discarded without being evaluated. A *reference* to a shared argument may be freely copied, but the argument itself can only be copied after it has been evaluated to a value. Nevertheless, there are some subtle differences between CBNd and CBS, and in particular CBS achieves less sharing than CBNd (see the discussion in Section 5). Unfortunately, there does not seem to be a way to embed CBNd into $\lambda^{!\bullet}$.

[5] Intuitionistic variables will be mapped to modal variables.

Bang Calculus. Another approach towards providing a common framework to explain CBV and CBN is the Bang-calculus [15]. It is an untyped lambda calculus that has explicit constructors in the syntax for promotion (!-introduction) and dereliction (!-elimination). It was motivated by the fact that Girard's original CBN and CBV translations of the intuitionistic logic into LL made use of logical exponentials (promotion and dereliction) that were not reflected in the syntax. The aim was thus to introduce an intermediate formalism between lambda calculus and proof nets, a graphical notation for LL proofs [16], that allows explicit use of "boxes" to mark values. Soundness and completeness of these translations with respect to reduction was proved by Guerrieri and Manzonetto [17] for slightly different notion of reduction for the Bang-calculus than that of [15]. The Bang-calculus can in fact be embedded into our $\lambda^{!\bullet}$-calculus and this embedding is both sound and complete.

Contributions. A summary of the contributions are as follows:

1. The introduction of a new logic called MSCLL ("Linear Logic with Restricted Access) that enjoys cut-elimination and is conservative over MELL. It provides a split of each exponential modality into a *sharing modality* and an *access modality*.
2. A term assignment for MSCLL, the $\lambda^{!\bullet}$*-calculus*, which operationally distinguishes between two kinds of expressions. *Sharable expressions* can be discarded, and *references* to shared expressions can be duplicated, but they cannot be accessed, and thus they can remain shared. *Sharable accessible expressions* can moreover be accessed, and they are copied whenever access to them is requested. This distinction allows to formulate the CBS evaluation mechanism.
3. Translations from CBV and CBN to $\lambda^{!\bullet}$ that are sound, complete and preserve normal forms.
4. The presentation of the *call-by-sharing* calculus (CBS), and a translation from CBS to $\lambda^{!\bullet}$ that is sound, complete and preserves normal forms.
5. A weak *evaluation mechanism* for $\lambda^{!\bullet}$ that can simulate weak evaluation strategies in CBN, CBV and CBS, with soundness and completeness results.

Structure of the paper. We review some background notions in Section 2. We present MSCLL in Section 3 and a term assignment for this logic, the $\lambda^{!\bullet}$-calculus, in Section 4. Definitions of the CBV, CBN, and CBS calculi and their translations to $\lambda^{!\bullet}$ are presented in Section 5 together with results of soundness, completeness, and preservation of normal forms. Section 6 presents a notion of weak evaluation for $\lambda^{!\bullet}$ which is shown to simulate weak CBV, CBN, and CBS evaluation. Finally, we conclude and discuss future work. Full proofs are available in the companion report [10]. A further section detailing a sound and complete embedding from the Bang-calculus into the $\lambda^{!\bullet}$-calculus has been omitted due to lack of space, and can also be found in the companion report.

2 Preliminary Notions

In this section we present some background notions and results that we use throughout the paper.

Recall that an *abstract rewriting system* (ARS) is a pair $\mathcal{X} = (X, \to_{\mathcal{X}})$ where X is a set and $\to_{\mathcal{X}} \subseteq X^2$ is a binary relation called *reduction*. We write $\to^*_{\mathcal{X}}$ for the reflexive–transitive closure of $\to_{\mathcal{X}}$, $\to^+_{\mathcal{X}}$ for the transitive closure, $\to^=_{\mathcal{X}}$ for the reflexive closure, $\leftrightarrow_{\mathcal{X}}$ for the symmetric closure, $\leftarrow_{\mathcal{X}}$ or $\to^{-1}_{\mathcal{X}}$ for the inverse relation, and $\to^n_{\mathcal{X}}$ for the composition of $\to_{\mathcal{X}}$ with itself n times. An ARS is confluent (CR) if ${}^*_{\mathcal{X}}\!\!\leftarrow \to^*_{\mathcal{X}} \subseteq \to^*_{\mathcal{X}} {}^*_{\mathcal{X}}\!\!\leftarrow$ and strongly normalizing (SN) if there are no infinite reductions $x_1 \to_{\mathcal{X}} x_2 \to_{\mathcal{X}} \ldots$.

Abstract Results on Translations Given ARSs $\mathcal{X} = (X, \to_{\mathcal{X}})$ and $\mathcal{Y} = (Y, \to_{\mathcal{Y}})$, a *translation* $T : \mathcal{X} \to \mathcal{Y}$ is a function $T : X \to Y$, also written T, by abuse of notation. A translation is *sound* if $x_1 \to^*_{\mathcal{X}} x_2$ implies $T(x_1) \to^*_{\mathcal{Y}} T(x_2)$ for all $x_1, x_2 \in X$, and *complete* if $T(x_1) \to^*_{\mathcal{Y}} T(x_2)$ implies $x_1 \to^*_{\mathcal{X}} x_2$ for all $x_1, x_2 \in X$. The following are easy results on a translation T:

Proposition 2.1 (Conditions for soundness). *If $x_1 \to_{\mathcal{X}} x_2$ implies $T(x_1) \to^*_{\mathcal{Y}} T(x_2)$ for every $x_1, x_2 \in X$, then T is sound.*

Theorem 2.1 (Conditions for completeness). *Let $Y' \subseteq Y$ and let $T^{-1} : Y' \to X$ be a function. Suppose that T^{-1} is the left-inverse of T,* i.e. *for all $x \in X$ we have that $T(x) \in Y'$ and $T^{-1}(T(x)) = x$. Suppose moreover that T^{-1} simulates reduction,* i.e. *for all $y_1 \in Y'$ and $y_2 \in Y$ such that $y_1 \to_{\mathcal{Y}} y_2$, we have that $y_2 \in Y'$ and $T^{-1}(y_1) \to^*_{\mathcal{X}} T^{-1}(y_2)$. Then T is complete.*

The Linear Substitution Calculus The LSC is a refinement of the λ-calculus with explicit substitutions, introduced by Accattoli and Kesner [1,3] as a variation over a calculus by Milner [24]. The set of LSC *terms* ($\mathcal{T}_{\mathsf{LSC}}$) is defined as follows, where $[x/s]$ is called an *explicit substitution* (ES):

$$t, s, \ldots ::= x \mid \lambda x.\, t \mid t\, s \mid t[x/s]$$

A *context* is a term with a single occurrence of a *hole* "$\Box$". A *substitution context* is a list of ESs. Formally:

General contexts	$\mathsf{C} ::= \Box \mid \lambda x.\, \mathsf{C} \mid \mathsf{C}\, t \mid t\, \mathsf{C} \mid \mathsf{C}[x/t] \mid t[x/\mathsf{C}]$
Substitution contexts	$\mathsf{L} ::= \Box \mid \mathsf{L}[x/t]$

Free and *bound* occurrences of variables are defined as expected, and $\mathsf{fv}(t)$ denotes the set of free variables of t. Terms are defined up to α-renaming of bound variables.

We write $\mathsf{C}\langle t\rangle$ for the *variable-capturing* substitution of $\Box$ in C by t. We write $\mathsf{C}\langle\!\langle t\rangle\!\rangle$ for the *capture-avoiding* substitution of $\Box$ in C by t. For example, if $\mathsf{C} = \lambda x.\, \Box$ then $\mathsf{C}\langle x\rangle = \lambda x.\, x$, while $\mathsf{C}\langle\!\langle x\rangle\!\rangle = \lambda y.\, x \neq \lambda x.\, x$. In the case of substitution contexts, we usually write $t\mathsf{L}$ rather than $\mathsf{L}\langle t\rangle$. The *domain* of L, written $\mathsf{dom}(\mathsf{L})$, is the set of variables bound by L. The LSC has three rewriting rules, closed by congruence under arbitrary contexts:

$$(\lambda x.\, t)\mathsf{L}\, s \to_{\mathsf{db}} t[x/s]\mathsf{L} \qquad \mathsf{C}\langle\!\langle x\rangle\!\rangle[x/t] \to_{\mathsf{ls}} \mathsf{C}\langle\!\langle t\rangle\!\rangle[x/t] \qquad t[x/s] \to_{\mathsf{gc}} t \text{ (if } x \notin \mathsf{fv}(t))$$

Distance beta (db) performs a β-step but creates an ES rather than doing meta-level substitution, *linear substitution* (ls) replaces a *single* occurrence of x to a term t when

x is bound to t by an ES, and *garbage collection* (gc) removes an unreachable ES. Reduction in LSC is the union $\to_{\mathsf{lsc}} := \to_{\mathsf{db}} \cup \to_{\mathsf{ls}} \cup \to_{\mathsf{gc}}$.

The rewriting rules of LSC are said to operate "at a distance" due to their peculiar use of contexts. This avoids reduction getting stuck in the presence of explicit substitutions. For example, if the left-hand side of db were declared to be $(\lambda x.\, t)\, s$, then an expression such as $(\lambda x.\, t)[y/r]\, s$ would not be a redex. This notation has a strong connection to proof nets [2].

We write $\Gamma \vdash_{\mathsf{lsc}} t : A$ is t has type A under the typing context Γ, with standard simple type assignment rules. Recall from [19, Theorem 6.11] that simply typed terms are SN:

Theorem 2.2. *If $\Gamma \vdash_{\mathsf{lsc}} t : A$, there are no infinite reduction sequences $t \to_{\mathsf{lsc}} t_1 \to_{\mathsf{lsc}} \ldots$.*

3 Linear Logic with Restricted Access

We start by defining a one-sided sequent calculus presentation for a linear logic with *multiplicative* connectives ($\otimes$, $⅋$, called *tensor* and *par*), *sharing* modalities ($!$, $?$, called *of-course* and *why-not*), and *access* modalities ($\bullet$, $\circ$, called *grant* and *demand*), which we dub MSCLL.

Formulae and Sequent Calculus Presentation. We assume given a denumerable set of atomic formulae $(\alpha, \beta, \ldots)$, each of which has a corresponding negative version $(\overline{\alpha}, \overline{\beta}, \ldots)$. The set of formulae is given by the grammar:

$$A, B, \ldots ::= \alpha \mid \overline{\alpha} \mid A \otimes B \mid A ⅋ B \mid\, !A \mid ?A \mid \bullet A \mid \circ A$$

Linear negation is the involutive operator $(\cdot)^{\perp}$ given by:

$$\begin{array}{cccc} \alpha^{\perp} := \overline{\alpha} & \overline{\alpha}^{\perp} := \alpha & (A \otimes B)^{\perp} := A^{\perp} ⅋ B^{\perp} & (A ⅋ B)^{\perp} := A^{\perp} \otimes B^{\perp} \\ (!A)^{\perp} := ?A^{\perp} & (?A)^{\perp} := !A^{\perp} & (\bullet A)^{\perp} := \circ A^{\perp} & (\circ A)^{\perp} := \bullet A^{\perp} \end{array}$$

Sequents are of the form $\vdash \Gamma$, where Γ is a finite *multiset* of formulae (note that we do not include an explicit exchange rule). If $\Gamma = A_1, \ldots, A_n$ and $\mathfrak{m}$ is one of the modalities, we write $\mathfrak{m}\Gamma$ to stand for $\mathfrak{m}A_1, \ldots, \mathfrak{m}A_n$, so for instance $\circ\Gamma = \circ A_1, \ldots, \circ A_n$. Derivable sequents are given inductively by the following rules:

$$\frac{}{\vdash A, A^{\perp}}\,\mathsf{ax} \qquad \frac{\vdash \Gamma, A \quad \vdash \Delta, A^{\perp}}{\vdash \Gamma, \Delta}\,\mathsf{cut} \qquad \frac{\vdash \Gamma, A \quad \vdash \Delta, B}{\vdash \Gamma, \Delta, A \otimes B}\,\otimes \qquad \frac{\vdash \Gamma, A, B}{\vdash \Gamma, A ⅋ B}\,⅋ \qquad \frac{\vdash ?\Gamma, A}{\vdash ?\Gamma, !A}\,!\mathsf{p}$$

$$\frac{\vdash \Gamma}{\vdash \Gamma, ?A}\,?\mathsf{w} \qquad \frac{\vdash \Gamma, ?A, ?A}{\vdash \Gamma, ?A}\,?\mathsf{c} \qquad \frac{\vdash \Gamma, \circ A}{\vdash \Gamma, ?\circ A}\,?\circ\mathsf{d} \qquad \frac{\vdash \Gamma, A}{\vdash \Gamma, \bullet A}\,\bullet \qquad \frac{\vdash \Gamma, A}{\vdash \Gamma, \circ A}\,\circ$$

Most rules are standard rules from multiplicative-exponential linear logic (MELL), including the standard weakening (?w), contraction (?c), and promotion (!p) rules. The atypical rule is dereliction (?∘d), which requires the conclusion to be $\Gamma, ?\circ A$ instead of the usual $\Gamma, ?A$. The $\bullet$ and $\circ$ rules are (trivial) introduction rules for $\bullet$ and $\circ$.

As discussed in the introduction, the intuition is that a proof of $!A$ lies inside a box which may be duplicated or erased, but it may not necessarily be possible to *access* the contents of the box, which means that having a proof does not necessarily enable one to interact with its contents. A proof of $!\bullet A$ lies inside a box which may both duplicated, erased, and accessed. Informally speaking, the combined modalities $!\bullet A$ and $?\circ A$ in MSCLL play the role of the usual $!A$ and $?A$ modalities in MELL.

Remark 3.1. If we define linear equivalence of formulae $A \multimapboth B$ as usual in linear logic, it is immediate to show that $\vdash A \multimapboth \circ A$ holds, but in general $\vdash ?A \multimapboth ?\circ A$ does not hold. Hence linear equivalence is not a congruence with respect to the sharing modalities.

Basic Properties. Consider the mapping $(\cdot)^\bullet$ from formulae in MELL to formulae in MSCLL which replaces each occurrence of "!" with "$!\bullet$", each occurrence of "?" with "$?\circ$", and leaves the other connectives unaltered. The following theorem is easy to prove by induction on the derivation of the sequents:

Theorem 3.1 (Conservativity). $\vdash \Gamma$ *holds in* MELL *if and only if* $\vdash \Gamma^\bullet$ *holds in* MSCLL

By the usual techniques, one can show that MSCLL enjoys cut-elimination:

Theorem 3.2 (Cut elimination). *If* $\vdash \Gamma$ *is provable in* MSCLL*, then there is a derivation of* $\vdash \Gamma$ *without instances of the* `cut` *rule.*

4 A Sharing Linear λ-Calculus

In this section, we present a *sharing* linear λ-calculus based on MSCLL, called the $\lambda^{!\bullet}$-calculus. The relationship between the $\lambda^{!\bullet}$-calculus and MSCLL is akin to that between linear λ-calculi and MELL. In particular, typing rules for the $\lambda^{!\bullet}$-calculus are presented in natural deduction (rather than sequent calculus) style, and, furthermore, the $\lambda^{!\bullet}$-calculus is intuitionistic (rather than classical).

Syntax and Typing System We assume given denumerable sets of *linear variables* $(a, b, \ldots)$ and *unrestricted variables* $(u, v, \ldots)$. The set of *types* $(A, B, \ldots)$ and the set $\mathcal{T}_\bullet$ of $\lambda^{!\bullet}$*-terms* $(t, s, \ldots)$, or just *terms*, are given by:

$$A ::= \alpha \mid A \multimap B \mid \bullet A \mid !A \qquad t ::= a \mid u \mid \lambda a.t \mid t\,s \mid \bullet t \mid \mathtt{req}(t) \mid !t \mid t[u/s]$$

A term may be a *linear* or an *unrestricted variable*, an *abstraction* $\lambda a.t$ (binding a linear variable), an *application* $t\,s$, an access *grant* $\bullet t$, an access *request* $\mathtt{req}(t)$, a *promotion* $!t$ or a *substitution* $t[u/s]$ (binding an unrestricted variable).

Free and bound occurrences of variables are defined as expected. We write $\mathsf{fv}(t)$ for the set of free variables of t. Terms are defined up to α-renaming of bound variables. By convention, we assume that $!t[x/s]$ stands for $!(t[x/s])$. Similarly, $\bullet t[x/s]$, and $\lambda a.\,t[x/s]$ stand, respectively, for $\bullet(t[x/s])$, and $\lambda a.\,(t[x/s])$.

Unrestricted typing environments $(\Delta, \Delta', \ldots)$ are partial functions mapping unrestricted variables to types, written $u_1 : A_1, \ldots, u_n : A_n$. *Linear typing environments* $(\Gamma, \Gamma', \ldots)$ map linear variables to types, written $a_1 : A_1, \ldots, a_n : A_n$. We assume that typing environments have finite domain.

Typing judgments are of the form $\Delta;\Gamma \vdash t : A$. Derivable judgments are defined inductively by the following rules. The types of unrestricted variables in Δ may be thought of as being implicitly prefixed by "!•", as attested by the rule sub.

$$\frac{}{\Delta; a : A \vdash a : A}\,\texttt{lvar} \qquad \frac{}{\Delta, u : A; \cdot \vdash u : \bullet A}\,\texttt{uvar} \qquad \frac{\Delta;\Gamma_1 \vdash t : A \multimap B \quad \Delta;\Gamma_2 \vdash s : A}{\Delta;\Gamma_1,\Gamma_2 \vdash t\,s : B}\,\texttt{app}$$

$$\frac{\Delta;\Gamma, a : A \vdash t : B}{\Delta;\Gamma \vdash \lambda a.t : A \multimap B}\,\texttt{abs} \qquad \frac{\Delta;\Gamma \vdash t : A}{\Delta;\Gamma \vdash \bullet t : \bullet A}\,\texttt{grant} \qquad \frac{\Delta;\Gamma \vdash t : \bullet A}{\Delta;\Gamma \vdash \mathtt{req}(t) : A}\,\texttt{request}$$

$$\frac{\Delta; \cdot \vdash t : A}{\Delta; \cdot \vdash !t : !A}\,\texttt{prom} \qquad \frac{\Delta, u : A;\Gamma_1 \vdash t : B \quad \Delta;\Gamma_2 \vdash s : !\bullet A}{\Delta;\Gamma_1,\Gamma_2 \vdash t[u/s] : B}\,\texttt{sub}$$

Example 4.1. $\lambda a.(!\bullet !u)[u/a]$ has type $!\bullet A \multimap !\bullet !\bullet A$, under the empty contexts.

Logical Soundness. Types of $\lambda^{!\bullet}$ encode *formulae* of MSCLL, while terms encode *proofs*. Indeed, consider the translation $(\cdot)^\star$ on types below, and the following result:

$$\alpha^\star := \alpha \qquad (A \multimap B)^\star := A^{\star\perp} \parr B^\star \qquad (\bullet A)^\star := \bullet A^\star \qquad (!A)^\star := !A^\star$$

Proposition 4.1 (Logical Soundness of $\lambda^{!\bullet}$). *If $\Delta;\Gamma \vdash t : A$ holds in $\lambda^{!\bullet}$, then $\vdash ?\circ(\Delta^{\star\perp}), \Gamma^{\star\perp}, A^\star$ holds in* MSCLL.

Note that, by design, $\lambda^{!\bullet}$ is not intended to be *complete* with respect to MSCLL. For example, the type $(!\alpha \multimap !\alpha \multimap \beta) \multimap !\alpha \multimap \beta$ is not inhabited in $\lambda^{!\bullet}$, whereas the corresponding formula $(!\alpha \otimes !\alpha \otimes \overline{\beta}) \parr ?\overline{\alpha} \parr \beta$ is provable in MSCLL.

Reduction Semantics Let us write $\mathsf{Ctxs}_\bullet$ for the the set of $\lambda^{!\bullet}$-*contexts* $(\mathrm{C}, \mathrm{C}', \ldots)$, or just *contexts*, which are $\lambda^{!\bullet}$-terms with a single occurrence of a *hole* "□", and $\mathsf{SCtxs}_\bullet$ for the the set of *substitution contexts* $(\mathrm{L}, \mathrm{L}', \ldots)$, which are lists of ESs:

$$\mathrm{C} ::= \square \mid \lambda a.\mathrm{C} \mid \mathrm{C}\,t \mid t\,\mathrm{C} \mid \bullet\mathrm{C} \mid \mathtt{req}(\mathrm{C}) \mid !\mathrm{C} \mid \mathrm{C}[u/t] \mid t[u/\mathrm{C}] \qquad \mathrm{L} ::= \square \mid \mathrm{L}[u/t]$$

We write $t\{a := s\}$ for the capture-avoiding substitution of the free occurrences of a in t by s. The domain of a substitution context ($\mathsf{dom}(\mathrm{L})$), and the plugging of a term into a context, both with capture ($\mathrm{C}\langle t\rangle$ and $t\mathrm{L}$) and avoiding capture ($\mathrm{C}\langle\!\langle t\rangle\!\rangle$), are defined similarly as for the LSC (see Section 2).

There are four **rewriting rules**, closed by compatibility under arbitrary contexts:

$$\begin{array}{rll}
(\lambda a.t)\mathrm{L}\,s & \to_{\bullet\mathsf{db}} & t\{a := s\}\mathrm{L} \\
\mathtt{req}((\bullet t)\mathrm{L}) & \to_{\bullet\mathsf{req}} & t\mathrm{L} \\
\mathrm{C}\langle\!\langle u\rangle\!\rangle[u/(!(\bullet t)\mathrm{L}_1)\mathrm{L}_2] & \to_{\bullet\mathsf{ls}} & \mathrm{C}\langle\!\langle(\bullet t)\mathrm{L}_1\rangle\!\rangle[u/!(\bullet t)\mathrm{L}_1]\mathrm{L}_2 \\
t[u/(!s)\mathrm{L}] & \to_{\bullet\mathsf{gc}} & t\mathrm{L} \quad (\text{if } u \notin \mathsf{fv}(t))
\end{array}$$

Reduction in $\lambda^{!\bullet}$ is defined as the union $\to_\bullet := \to_{\bullet\mathsf{db}} \cup \to_{\bullet\mathsf{req}} \cup \to_{\bullet\mathsf{ls}} \cup \to_{\bullet\mathsf{gc}}$.

The $\bullet\mathsf{db}$ rule is a distant β-rule. The calculus does not assume that terms are typable; but, in typable terms, there is exactly one occurrence of a in the body of $\lambda a.t$ by linearity. The $\bullet\mathsf{req}$ rule requests access to a term. The $\bullet\mathsf{ls}$ rule substitutes a single occurrence of u by $(\bullet t)\mathrm{L}_1$, provided that u is bound to a term of the form $(!(\bullet t)\mathrm{L}_1)\mathrm{L}_2$. Note that the subterm $(\bullet t)\mathrm{L}_1$ is *copied* by $\bullet$ls, while L_2 is moved outside so that its bindings remain shared. The $\bullet$gc rule may erase an unused substitution if u is bound to a term of the form $(!s)\mathrm{L}$. Allowing to erase *any* unused substitution, *e.g.* with the usual gc rule of LSC, would break subject reduction, as it could lead to weakening (erasure) of a linear variable.

The $\bullet$gc rule requires that it first be evaluated to a *sharable* term, of the form $!s'$, or more in general $(!s')\mathrm{L}$. A sharable term may not, a priori, be accessed. For example, in $(\mathtt{req}(u)\,\mathtt{req}(u))[u/!v]$ the two occurrences of u cannot be substituted by v, because the $\bullet$ls rule requires the argument to be a *sharable accessible* term, of the form $!\bullet s'$, or more in general $(!(\bullet s')\mathrm{L}_1)\mathrm{L}_2$.

A first routine result is:

Proposition 4.2 (Subject Reduction). *If $\Delta;\Gamma \vdash t : A$ and $t \to_\bullet s$, then $\Delta;\Gamma \vdash s : A$.*

Confluence. Confluence is tricky since standard techniques are not immediately applicable. A variant of Tait–Martin-Löf's method [11, § 3.2] would require to define parallel reduction, which is not immediate due to the fact that rewrite rules operate at a distance. The interpretation method, used for confluence of the structural lambda calculus [4], a close relative of LSC, does not apply since *full composition, i.e.* $t[x/s] \to^* t\{x := s\}$, does not hold for $\lambda^{!\bullet}$. An axiomatic rewriting approach based on residuals and orthogonality fails for $\lambda^{!\bullet}$ too. A simple example is the following, where we use labels α and β to mark redexes:

$$t[v/(!\bullet s)\mathrm{L}] \leftarrow^{\alpha}_{\bullet\mathsf{gc}} t[u^{\alpha}/(!v^{\beta})[v/(!\bullet s)\mathrm{L}]] \to^{\beta}_{\bullet\mathsf{ls}} t[u^{\alpha}/(!\bullet s)[v/(!\bullet s)]\mathrm{L}] \to^{\alpha}_{\bullet\mathsf{gc}} t[v/!\bullet s]\mathrm{L}$$

where, however, $t[v/(!\bullet s)\mathrm{L}] \neq t[v/!\bullet s]\mathrm{L}$. Nevertheless, confluence holds for $\lambda^{!\bullet}$ modulo the congruence generated by $t[u/s[v/r]] \equiv t[u/s][v/r]$, provided that $v \notin \mathsf{fv}(t)$. To prove this, we develop a theory of residuals for $\lambda^{!\bullet}$ modulo $\equiv$. We then resort to the axiomatic rewriting framework due to Melliès [23], verifying that $\lambda^{!\bullet}$ modulo $\equiv$ can be modeled as an *orthogonal axiomatic rewriting system*, which entails confluence (see [23, Theorem 2.4]):

Proposition 4.3 (Confluence). *$\lambda^{!\bullet}$ modulo $\equiv$ is confluent.*

Strong Normalization. Typable terms in $\lambda^{!\bullet}$ are strongly normalizing. To prove this, we reduce SN of typed $\lambda^{!\bullet}$ to SN of simply typed LSC.

Let us write $\to_{\bullet\mathrm{i}}$ for $\to_\bullet \setminus \to_{\bullet\mathsf{gc}}$ and $\to_{\mathsf{lsci}}$ for $\to_{\mathsf{lsc}} \setminus \to_{\bullet\mathsf{gc}}$. Garbage collection $\to_{\bullet\mathsf{gc}}$ can be *postponed*, so to show there are no infinite reduction sequences $t \to_\bullet t_1 \to_\bullet t_2 \ldots$ we may assume without loss of generality that the sequence consists of $\to_{\bullet\mathrm{i}}$ steps.

We start by defining a translation $[\![\cdot]\!]$ from $\lambda^{!\bullet}$ to LSC. Let $\mathbf{1}$ be any inhabited type in simply typed LSC, and $*$ a closed inhabitant of $\mathbf{1}$ in normal form. Types and terms are translated as follows (where z is assumed to be fresh in the "$\bullet$" case):

$$[\![\alpha]\!] := \alpha \qquad [\![A \multimap B]\!] := [\![A]\!] \to [\![B]\!] \qquad [\![\bullet A]\!] := \mathbf{1} \to [\![A]\!] \qquad [\![!A]\!] := [\![A]\!]$$

$$\begin{array}{llll} [\![a]\!] := a & [\![u]\!] := u & [\![\bullet t]\!] := \lambda z.\,[\![t]\!] & [\![\mathtt{req}(t)]\!] := [\![t]\!]\,* \\ [\![\lambda a.\,t]\!] := \lambda a.\,[\![t]\!] & [\![t\,s]\!] := [\![t]\!]\,[\![s]\!] & [\![!t]\!] := [\![t]\!] & [\![t[x/s]]\!] := [\![t]\!][x/[\![s]\!]] \end{array}$$

An unrestricted variable $u : A$ in the environment is translated as $u : \mathbf{1} \to [\![A]\!]$, while a linear variable $a : A$ is translated as $a : [\![A]\!]$. It is easy to show that the translation preserves typing, in the sense that $\Delta; \Gamma \vdash t : A$ implies $[\![\Delta]\!], [\![\Gamma]\!] \vdash_{\mathsf{lsc}} [\![t]\!] : [\![A]\!]$.

To conclude, it would suffice to show that LSC simulates $\lambda^{!\bullet}$ reduction; more precisely that $t \to_{\bullet} s$ implies $[\![t]\!] \to^{+}_{\mathsf{lsc}} [\![s]\!]$. Unfortunately, this is not the case; for instance, in the following example, the $\to_{\bullet\mathsf{ls}}$ rule of $\lambda^{!\bullet}$ *extrudes* the $[y/z]$ outside to share the binding, while the $\to_{\mathsf{ls}}$ rule of LSC makes a *copy* of $[y/z]$:

$$x[x/(!\bullet y)[y/z]] \to_{\bullet\mathsf{ls}} (\bullet y)[x/!\bullet y][y/z] \qquad x[x/t[y/z]] \to_{\mathsf{ls}} (t[y/z])[x/t[y/z]]$$

To address this, we define a binary relation $\Rrightarrow$ on LSC terms called *fusion*, that allows to "extrude and fuse" ESs, given by the reflexive, transitive, and contextual closure of the following rules, avoiding capture:

$$t[x/s] \Rrightarrow t \ \ (\text{if } x \notin \mathsf{fv}(t)) \qquad t[x/s][y/s] \Rrightarrow t\{x := y\}[y/s] \qquad \mathsf{C}\langle t[x/s]\rangle \Rrightarrow \mathsf{C}\langle t\rangle[x/s]$$

The two key technical properties are, first, that the $\to_{\mathsf{lsci}}$ simulates $\to_{\bullet\mathsf{i}}$ up to fusion, more precisely that $t \to_{\bullet\mathsf{i}} s$ implies $[\![t]\!] \to^{+}_{\mathsf{lsci}}\Rrightarrow [\![s]\!]$. Second, that fusion can be postponed, *i.e.* that $\Rrightarrow\to_{\bullet\mathsf{i}} \subseteq \to^{+}_{\bullet\mathsf{i}}\Rrightarrow$. Thus an infinite reduction sequence $t \to_{\bullet\mathsf{i}} t_1 \to_{\bullet\mathsf{i}} t_2 \ldots$ can be mapped to $[\![t]\!] \to^{+}_{\mathsf{lsci}}\Rrightarrow [\![t_1]\!] \to^{+}_{\mathsf{lsci}}\Rrightarrow [\![t_2]\!] \ldots$ and by postponing fusion we obtain an infinite reduction in LSC. If t is typable in $\lambda^{!\bullet}$ then $[\![t]\!]$ is typable in LSC, contradicting Thm. 2.2. To sum up:

Theorem 4.1 (Termination). *If t is typable in $\lambda^{!\bullet}$ then t is $\to_{\bullet}$-SN.*

To conclude this section, we remark that $\to_{\bullet}$-normal forms can be characterized inductively. We omit the characterization for lack of space, but see Section **??** for details.

5 Embedding CBN, CBV and CBS

In this section, we recall known CBN and CBV calculi, and we introduce a sharing variant of call-by-name we dub *call-by-sharing* (CBS). Second, we define translations that provide embeddings into $\lambda^{!\bullet}$, and study their properties.

Our first step is to give precise definitions of each of the calculi we will work with. These calculi operate on LSC terms, and some of them also use the notions of *strict values* ($\mathtt{v}, \mathtt{w}, \ldots$) and *lax values* ($\mathtt{v}^{+}, \mathtt{w}^{+}, \ldots$), defined as follows:

$$\text{Strict values} \quad \mathtt{v} ::= \lambda x.\, t \qquad \text{Lax values} \quad \mathtt{v}^{+} ::= x \mid \lambda x.\, t$$

We define the following rewriting rules on LSC terms, closed by compatibility under arbitrary contexts:

$$\begin{array}{rlrl}
(\lambda x.\, t)\mathsf{L}\, s & \to_{\mathsf{db}} t[x/s]\mathsf{L} & \mathsf{C}\langle\!\langle x\rangle\!\rangle[x/t] & \to_{\mathsf{ls}} \mathsf{C}\langle\!\langle t\rangle\!\rangle[x/t] \\
\mathsf{C}\langle\!\langle x\rangle\!\rangle[x/\mathtt{v}\mathsf{L}] & \to_{\mathsf{lsv}} \mathsf{C}\langle\!\langle \mathtt{v}\rangle\!\rangle[x/\mathtt{v}]\mathsf{L} & \mathsf{C}\langle\!\langle x\rangle\!\rangle[x/\mathtt{v}\mathsf{L}] & \to_{\mathsf{lsw}} \mathsf{C}\langle\!\langle \mathtt{v}\mathsf{L}\rangle\!\rangle[x/\mathtt{v}\mathsf{L}] \\
t[x/s] & \to_{\mathsf{gc}} t \quad (\text{if } x \notin \mathsf{fv}(t)) & t[x/\mathtt{v}^{+}\mathsf{L}] & \to_{\mathsf{gcv+}} t\mathsf{L} \quad (\text{if } x \notin \mathsf{fv}(t))
\end{array}$$

Definition 5.1 (Notions of reduction). *The relations corresponding to* CBN *($\to_{\mathsf{N}}$),* CBV *($\to_{\mathsf{V}}$), and* CBS *($\to_{\mathsf{S}}$) reduction are defined by:*

$$\to_{\mathsf{N}} := \to_{\mathsf{db}} \cup \to_{\mathsf{ls}} \cup \to_{\mathsf{gc}} \qquad \to_{\mathsf{V}} := \to_{\mathsf{db}} \cup \to_{\mathsf{lsv}} \cup \to_{\mathsf{gcv+}} \qquad \to_{\mathsf{S}} := \to_{\mathsf{db}} \cup \to_{\mathsf{lsw}} \cup \to_{\mathsf{gc}}$$

The call-by-name (CBN) calculus $\to_{\mathsf{N}}$ corresponds to usual reduction in LSC [3].

The call-by-value (CBV) calculus $\to_{\mathsf{V}}$ is a variant of Accattoli and Paolini's value-substitution calculus [5] (VSC), with two differences. First, the rule $\to_{\mathsf{lsv}}$ is *linear* in that it substitutes one occurrence of a variable x at a time, while the corresponding rule of VSC substitutes all occurrences of x at once. This difference allows us to present the calculi in a uniform way. Second, $\to_{\mathsf{lsv}}$ allows substituting variables for *strict values* (abstractions), while VSC allows substituting *lax values* (both abstractions and variables). This is necessary to be able to define a *complete* embedding (see also Rem. 5.2).

The call-by-sharing (CBS) calculus $\to_{\mathsf{S}}$ is a sharing variant of CBN, in which the argument may be discarded without being evaluated (using gc). At the same time, the evaluation of the argument is *shared*, in the sense that lsw only allows copying arguments when they have been evaluated to the form vL. The CBS calculus bears a strong resemblance to the call-by-need λ-calculus of Ariola *et al.* [6], which can be obtained by changing $\to_{\mathsf{lsw}}$ to $\to_{\mathsf{lsv}}$, *i.e.* call-by-need is $\to_{\mathsf{Nd}} := \to_{\mathsf{db}} \cup \to_{\mathsf{lsv}} \cup \to_{\mathsf{gc}}$ (see for instance [8]). Note that CBS achieves *less* sharing that CBNd, because the $\to_{\mathsf{lsw}}$ rule makes two copies of L, whereas $\to_{\mathsf{lsv}}$ keeps a single shared copy of L. Unfortunately, it does not seem possible to give a sound embedding of CBNd into $\lambda^{!\bullet}$.

Remark 5.1. The reduction relations above are *calculi*, *i.e.* orientations of equational theories, not *evaluation* mechanisms. We shall turn our attention to evaluation in Section 6.

Next, we describe the translations $(\cdot)^{\mathsf{N}}$, $(\cdot)^{\mathsf{V}}$, and $(\cdot)^{\mathsf{S}}$. Each translation maps a simple type into a $\lambda^{!\bullet}$-type, an LSC term into a $\lambda^{!\bullet}$-term, and an LSC typing judgment into an $\lambda^{!\bullet}$ typing judgment.

Embedding Call-by-Name The CBN translation $(\cdot)^{\mathsf{N}}$ is defined on types and terms by:

$$\alpha^{\mathsf{N}} := \alpha \qquad (A \to B)^{\mathsf{N}} := !\bullet A^{\mathsf{N}} \multimap B^{\mathsf{N}}$$

$$x^{\mathsf{N}} := \mathtt{req}(x) \qquad (\lambda x.\, t)^{\mathsf{N}} := \lambda a.\, t^{\mathsf{N}}[x/a] \qquad (t\, s)^{\mathsf{N}} := t^{\mathsf{N}}\, !\bullet s^{\mathsf{N}} \qquad t[x/s]^{\mathsf{N}} := t^{\mathsf{N}}[x/!\bullet s^{\mathsf{N}}]$$

In the abstraction case, a is assumed to be fresh, *i.e.* $a \notin \mathsf{fv}(t^{\mathsf{N}})$. The translation is extended to typing environments: $(x_1 : A_1, \ldots, x_n : A_n)^{\mathsf{N}} := x_1 : A_1^{\mathsf{N}}, \ldots, x_n : A_n^{\mathsf{N}}$, and judgments: $(\Gamma \vdash t : A)^{\mathsf{N}} := \Gamma^{\mathsf{N}}; \cdot \vdash t^{\mathsf{N}} : A^{\mathsf{N}}$.

Proposition 5.1 (CBN **typing).** *If $\Gamma \vdash t : A$ then $\Gamma^{\mathsf{N}}; \cdot \vdash t^{\mathsf{N}} : A^{\mathsf{N}}$.*

Lemma 5.1 (CBN **simulation).** *If $t \to_{\mathsf{N}} s$ then $t^{\mathsf{N}} \to_{\bullet}^{*} s^{\mathsf{N}}$. Furthermore, the reduction uses either at least one, and at most two $\to_{\bullet}$ steps.*

Proof. By induction on the derivation of $t \to_{\mathsf{N}} s$. The interesting cases are when there is a db, ls, or gc step at the root. If $(\lambda x.\, t)\mathsf{L}\, s \to_{\mathsf{db}} t[x/s]\mathsf{L}$, then:

$$((\lambda x.\, t)\mathsf{L}\, s)^{\mathsf{N}} = (\lambda a.\, t^{\mathsf{N}}[x/a])\mathsf{L}^{\mathsf{N}}\, !\bullet s^{\mathsf{N}} \to_{\bullet\mathsf{db}} t^{\mathsf{N}}[x/!\bullet s^{\mathsf{N}}]\mathsf{L}^{\mathsf{N}} = (t[x/s]\mathsf{L})^{\mathsf{N}}$$

If $\mathsf{C}\langle\!\langle x\rangle\!\rangle[x/t] \to_{\mathsf{ls}} \mathsf{C}\langle\!\langle t\rangle\!\rangle[x/t]$, then:

$$\begin{aligned}\mathsf{C}\langle\!\langle x\rangle\!\rangle[x/t]^{\mathsf{N}} = \mathsf{C}^{\mathsf{N}}\langle\!\langle \mathtt{req}(x)\rangle\!\rangle[x/!\bullet t^{\mathsf{N}}] &\to_{\bullet\mathsf{ls}} \mathsf{C}^{\mathsf{N}}\langle\!\langle \mathtt{req}(\bullet t^{\mathsf{N}})\rangle\!\rangle[x/!\bullet t^{\mathsf{N}}] \\ &\to_{\bullet\mathsf{req}} \mathsf{C}^{\mathsf{N}}\langle\!\langle t^{\mathsf{N}}\rangle\!\rangle[x/!\bullet t^{\mathsf{N}}] = \mathsf{C}\langle\!\langle t\rangle\!\rangle[x/t]^{\mathsf{N}}\end{aligned}$$

If $t[x/s] \rightarrow_{\mathsf{gc}} t$ with $x \notin \mathsf{fv}(t)$, then $t[x/s]^{\mathsf{N}} = t^{\mathsf{N}}[x/!\bullet s^{\mathsf{N}}] \rightarrow_{\bullet\mathsf{gc}} t^{\mathsf{N}}$, where that $x \notin \mathsf{fv}(t^{\mathsf{N}})$ because $\mathsf{fv}(t^{\mathsf{N}}) = \mathsf{fv}(t)$. ∎

For completeness, we define an **inverse** CBN **translation**. We define a subset $\mathcal{T}_{\bullet}^{\mathsf{N}} \subseteq \mathcal{T}_{\bullet}$, containing the closure by $\rightarrow_{\bullet}$-reduction of the image of $(\cdot)^{\mathsf{N}}$:

$$\underline{t}, \underline{s}, \ldots ::= \mathtt{req}(u) \mid \mathtt{req}(\bullet\underline{t}) \mid \lambda a.\, \underline{t}[u/a] \mid \underline{t}\,!\bullet\underline{s} \mid \underline{t}[u/!\bullet\underline{s}]$$

where, in the production $\underline{t} ::= \lambda a.\, \underline{t}[u/a]$ we assume that a is fresh, that is, $a \notin \mathsf{fv}(\underline{t})$. The *inverse* CBN *translation* is a function $(\cdot)^{-\mathsf{N}} : \mathcal{T}_{\bullet}^{\mathsf{N}} \rightarrow \mathcal{T}_{\mathsf{LSC}}$ defined as follows, by induction on the derivation of a term with the grammar above[6].

$$\mathtt{req}(x)^{-\mathsf{N}} := x \qquad \mathtt{req}(\bullet\underline{t})^{-\mathsf{N}} := \underline{t}^{-\mathsf{N}} \qquad (\lambda a.\, \underline{t}[x/a])^{-\mathsf{N}} := \lambda x.\, \underline{t}^{-\mathsf{N}}$$
$$(\underline{t}\,!\bullet\underline{s})^{-\mathsf{N}} := \underline{t}^{-\mathsf{N}}\, \underline{s}^{-\mathsf{N}} \qquad \underline{t}[x/!\bullet\underline{s}]^{-\mathsf{N}} := \underline{t}^{-\mathsf{N}}[x/\underline{s}^{-\mathsf{N}}]$$

It is easy to check that $(\cdot)^{-\mathsf{N}}$ is the left-inverse of $(\cdot)^{\mathsf{N}}$, *i.e.* if $t \in \mathcal{T}_{\mathsf{LSC}}$ then $t^{\mathsf{N}} \in \mathcal{T}_{\bullet}^{\mathsf{N}}$ and $(t^{\mathsf{N}})^{-\mathsf{N}} = t$. Moreover:

Lemma 5.2 (Inverse CBN **simulation).** *Let $\underline{t} \in \mathcal{T}_{\bullet}^{\mathsf{N}}$ and $s \in \mathcal{T}_{\bullet}$ such that $\underline{t} \rightarrow_{\bullet} s$. Then $s \in \mathcal{T}_{\bullet}^{\mathsf{N}}$ and $\underline{t}^{-\mathsf{N}} \rightarrow_{\mathsf{N}}^{=} s^{-\mathsf{N}}$.*

Using the abstract soundness and completeness results (Prop. 2.1, Thm. 2.1) together with the lemmas above, we obtain:

Theorem 5.1 (Sound and complete CBN **embedding).** *Given terms $t, s \in \mathcal{T}_{\mathsf{LSC}}$, $t \rightarrow_{\mathsf{N}}^{*} s$ if and only if $t^{\mathsf{N}} \rightarrow_{\bullet}^{*} s^{\mathsf{N}}$. Moreover, t is in $\rightarrow_{\mathsf{N}}$-normal form iff t^{N} is in $\rightarrow_{\bullet}$-normal form.*

Embedding Call-by-Value The CBV translation $(\cdot)^{\mathsf{V}}$ is defined on types and terms by:

$$\alpha^{\mathsf{V}} := \alpha \qquad (A \rightarrow B)^{\mathsf{V}} := !\bullet A^{\mathsf{V}} \multimap !\bullet B^{\mathsf{V}}$$

$$x^{\mathsf{V}} := !x \qquad (\lambda x.\, t)^{\mathsf{V}} := !\bullet\lambda a.\, t^{\mathsf{V}}[x/a] \qquad (t\, s)^{\mathsf{V}} := \mathtt{req}(u)[u/t^{\mathsf{V}}]\, s^{\mathsf{V}} \qquad t[x/s]^{\mathsf{V}} := t^{\mathsf{V}}[x/s^{\mathsf{V}}]$$

where, as for CBN, a is assumed to be fresh in the abstraction case. The translation is extended to typing environments: $(x_1 : A_1, \ldots, x_n : A_n)^{\mathsf{V}} := x_1 : A_1^{\mathsf{V}}, \ldots, x_n : A_n^{\mathsf{V}}$, and judgments: $(\Gamma \vdash t : A)^{\mathsf{V}} := \Gamma^{\mathsf{V}}; \cdot \vdash t^{\mathsf{V}} : !\bullet A^{\mathsf{V}}$.

Proposition 5.2 (CBV **typing).** *If $\Gamma \vdash t : A$ then $\Gamma^{\mathsf{V}}; \cdot \vdash t^{\mathsf{V}} : !\bullet A^{\mathsf{V}}$.*

Lemma 5.3 (CBV **simulation).** *If $t \rightarrow_{\mathsf{V}} s$ then $t^{\mathsf{V}} \rightarrow_{\bullet}^{*} s^{\mathsf{V}}$. Furthermore, the reduction uses at least one, and at most four $\rightarrow_{\bullet}$ steps.*

We turn our attention to **completeness** for $(\cdot)^{\mathsf{V}}$. A first comment is that the CBV translation only turns out to be complete up to garbage collection. More precisely, soundness with respect to reduction holds, in the sense that $t \rightarrow_{\mathsf{V}}^{*} s$ implies $t^{\mathsf{V}} \rightarrow_{\bullet}^{*} s^{\mathsf{V}}$, but completeness only holds in the following weak form: $t^{\mathsf{V}} \rightarrow_{\bullet}^{*} s^{\mathsf{V}}$ implies $t \rhd_{\mathsf{V}}^{*} s$, where $\rhd_{\mathsf{V}} := (\rightarrow_{\mathsf{V}} \cup \rightarrow_{\mathsf{gcv+}}^{-1})$. Resorting to confluence, it is possible recover "plain" soundness and completeness of the translation, *i.e.* with respect to the equational theory and not to reduction; more precisely $t \leftrightarrow_{\mathsf{V}}^{*} s$ if and only if $t^{\mathsf{V}} \leftrightarrow_{\bullet}^{*} s^{\mathsf{V}}$. Besides:

[6] Observe that the derivation is unique since, as can be easily seen, the grammar is unambiguous.

Remark 5.2. The study of completeness motivates the fact that in CBV the ls rule can substitute only *strict* values (abstractions) while the $\mathsf{gcv+}$ rule can erase *lax* values (abstractions and variables). To allow substituting variables, the translation of a variable x should be a term of the form $!\bullet t$. A preliminary version of this work used a CBV translation $(\cdot)^{\mathsf{V+}}$ similar to $(\cdot)^{\mathsf{V}}$ but with $x^{\mathsf{V+}} := !\bullet\mathtt{req}(x)$. However, $(\cdot)^{\mathsf{V+}}$ is not complete. In fact, it can be checked $x[x/t]\,s \leftrightarrow^*_{\mathsf{V}} t\,s$ does not hold in general (because t may not be convertible to a value), while it can be seen that $(x[x/t]\,s)^{\mathsf{V+}} \leftrightarrow^*_\bullet (t\,s)^{\mathsf{V+}}$ always holds. On the other hand, if the $\mathsf{gcv+}$ rule were not allowed to erase variables, this would again lead to incompleteness, as $x[y/z] \leftrightarrow^*_{\mathsf{V}} x$ would not hold, but $(x[y/z])^{\mathsf{V}} \leftrightarrow^*_\bullet x^{\mathsf{V}}$ would hold, since $(x[y/z])^{\mathsf{V}} = !x[y/!z] \to_{\bullet\mathsf{gc}} !x = x^{\mathsf{V}}$.

Next, we define an **inverse CBV translation**. First, we define a subset $\mathcal{T}^{\mathsf{V}}_\bullet \subseteq \mathcal{T}_\bullet$, containing the closure by $\to_\bullet$-reduction of the image of $(\cdot)^{\mathsf{V}}$, as well as a subset $\mathsf{SCtxs}^{\mathsf{V}}_\bullet \subseteq \mathsf{SCtxs}_\bullet$:

$$\underline{t}, \underline{s}, \ldots ::= !x \mid !\bullet\lambda a.\,\underline{t}[x/a] \mid \mathtt{req}(u)[u/\underline{t}] \mid \mathtt{req}(\bullet\lambda a.\,\underline{t}[x/a]) \mid \lambda a.\,\underline{t}[x/a] \mid \underline{t}\,\underline{s} \mid \underline{t}[u/\underline{s}]$$
$$\underline{\mathsf{L}} ::= \Box \mid \underline{\mathsf{L}}[u/\underline{t}]$$

where in the occurrences of $\lambda a.\,\underline{t}[x/a]$ we assume that a is fresh. The *inverse* CBV *translation* is a function $(\cdot)^{-\mathsf{V}} : \mathcal{T}^{\mathsf{V}}_\bullet \to \mathcal{T}_{\mathsf{LSC}}$ defined as follows, by induction on the derivation of a term with the (unambiguous) grammar above:

$$\begin{array}{rlrlrl} (!x)^{-\mathsf{V}} &:= x & (!\bullet\lambda a.\,\underline{t}[x/a])^{-\mathsf{V}} &:= \lambda x.\,\underline{t}^{-\mathsf{V}} & \underline{t}\,\underline{s}^{-\mathsf{V}} &:= \underline{t}^{-\mathsf{V}}\,\underline{s}^{-\mathsf{V}} \\ (\mathtt{req}(\bullet\lambda a.\,\underline{t}[x/a]))^{-\mathsf{V}} &:= \lambda x.\,\underline{t}^{-\mathsf{V}} & (\lambda a.\,\underline{t}[x/a])^{-\mathsf{V}} &:= \lambda x.\,\underline{t}^{-\mathsf{V}} & & \\ (\mathtt{req}(u)[u/\underline{t}])^{-\mathsf{V}} &:= \underline{t}^{-\mathsf{V}} & \underline{t}[u/\underline{s}]^{-\mathsf{V}} &:= \underline{t}^{-\mathsf{V}}[u/\underline{s}^{-\mathsf{V}}] & & \end{array}$$

It is easy to check that $(\cdot)^{-\mathsf{V}}$ is the left-inverse of $(\cdot)^{\mathsf{V}}$.

Lemma 5.4 (Inverse CBV simulation, up to gcv+). *Let $\underline{t} \in \mathcal{T}^{\mathsf{V}}_\bullet$ and $s \in \mathcal{T}_\bullet$ such that $\underline{t} \to_\bullet s$. Then $s \in \mathcal{T}^{\mathsf{V}}_\bullet$ and $\underline{t}^{-\mathsf{V}} \rhd^*_{\mathsf{V}} s^{-\mathsf{V}}$, where $\rhd_{\mathsf{V}} := (\to_{\mathsf{V}} \cup \to^{-1}_{\mathsf{gcv+}})$.*

Theorem 5.2 (Sound and complete CBV embedding). *Given terms $t, s \in \mathcal{T}_{\mathsf{LSC}}$:*

1. *$t \to^*_{\mathsf{V}} s$ implies $t^{\mathsf{V}} \to^*_\bullet s^{\mathsf{V}}$*
2. *$t^{\mathsf{V}} \to^*_\bullet s^{\mathsf{V}}$ implies $t \rhd^*_{\mathsf{V}} s$ where $\rhd_{\mathsf{V}} := (\to_{\mathsf{V}} \cup \to^{-1}_{\mathsf{gcv+}})$.*
3. *$t \leftrightarrow^*_{\mathsf{V}} s$ if and only if $t^{\mathsf{V}} \leftrightarrow^*_\bullet s^{\mathsf{V}}$*
4. *t is in $\to_{\mathsf{V}}$-normal form iff t^{V} is in $\to_\bullet$-normal form.*

Remark 5.3. Arrial [7] suggests an additional CBV translation mapping $A \to B$ to $!(A \multimap !B)$. In our setting, this means mapping $A \to B$ to $!\bullet(A \multimap !\bullet B)$. This translation is also sound and complete but still requires $\rhd_{\mathsf{V}}$ to obtain completeness.

Embedding Call-by-Sharing The CBS translation $(\cdot)^{\mathsf{S}}$ is defined on types and terms by:

$$\alpha^{\mathsf{S}} := \alpha \qquad (A \to B)^{\mathsf{S}} := !\bullet A^{\mathsf{S}} \multimap \bullet B^{\mathsf{S}}$$

$$x^{\mathsf{S}} := x \qquad (\lambda x.\,t)^{\mathsf{S}} := \bullet\lambda a.\,t^{\mathsf{S}}[x/a] \qquad (t\,s)^{\mathsf{S}} := \mathtt{req}(t^{\mathsf{S}})\,!s^{\mathsf{S}} \qquad t[x/s]^{\mathsf{S}} := t^{\mathsf{S}}[x/!s^{\mathsf{S}}]$$

where, as before, a is assumed to be fresh in the abstraction case. The translation is extended to typing environments: $(x_1 : A_1, \ldots, x_n : A_n)^{\mathsf{S}} := x_1 : A_1^{\mathsf{S}}, \ldots, x_n : A_n^{\mathsf{S}}$, and judgments: $(\Gamma \vdash t : A)^{\mathsf{S}} := \Gamma^{\mathsf{S}}; \cdot \vdash t^{\mathsf{S}} : \bullet A^{\mathsf{S}}$.

Proposition 5.3 **(CBS typing).** *If $\Gamma \vdash t : A$ then $\Gamma^{\mathsf{S}}; \cdot \vdash t^{\mathsf{S}} : \bullet A^{\mathsf{S}}$.*

Lemma 5.5 **(CBS simulation).** *If $t \to_{\mathsf{S}} s$ then $t^{\mathsf{S}} \to_{\bullet}^{*} s^{\mathsf{S}}$.*

Proof. By induction on the derivation of $t \to_{\mathsf{S}} s$. The interesting cases are when there is a db, lsv, or gc step at the root.
If $(\lambda x. t)\mathtt{L}\, s \to_{\mathsf{db}} t[x/s]\mathtt{L}$, then:

$$((\lambda x. t)\mathtt{L}\, s)^{\mathsf{S}} = \mathtt{req}((\bullet\lambda a. t^{\mathsf{S}}[x/a])\mathtt{L}^{\mathsf{S}})\,!s^{\mathsf{S}} \to_{\bullet\mathsf{req}} (\lambda a. t^{\mathsf{S}}[x/a])\mathtt{L}^{\mathsf{S}}\,!s^{\mathsf{S}}$$
$$\to_{\bullet\mathsf{db}} t^{\mathsf{S}}[x/!s^{\mathsf{S}}]\mathtt{L}^{\mathsf{S}} = (t[x/s]\mathtt{L})^{\mathsf{S}}$$

If $\mathsf{C}\langle\langle x\rangle\rangle[x/\mathsf{vL}] \to_{\mathsf{lsw}} \mathsf{C}\langle\langle \mathsf{vL}\rangle\rangle[x/\mathsf{vL}]$, note that $\mathsf{v} = \lambda y. t$ so $\mathsf{v}^{\mathsf{S}} = \bullet\lambda a. t^{\mathsf{S}}[y/a]$. Then:

$$(\mathsf{C}\langle\langle x\rangle\rangle[x/\mathsf{vL}])^{\mathsf{S}} = \mathsf{C}^{\mathsf{S}}\langle\langle x\rangle\rangle[x/!(\bullet(\lambda a. t^{\mathsf{S}}[y/a])\mathtt{L}^{\mathsf{S}})]$$
$$\to_{\bullet\mathsf{ls}} \mathsf{C}^{\mathsf{S}}\langle\langle(\bullet(\lambda a. t^{\mathsf{S}}[y/a]))\mathtt{L}^{\mathsf{S}}\rangle\rangle[x/!(\bullet(\lambda a. t^{\mathsf{S}}[y/a]))\mathtt{L}^{\mathsf{S}}] = (\mathsf{C}\langle\langle \mathsf{vL}\rangle\rangle[x/\mathsf{vL}])^{\mathsf{S}}$$

If $t[x/s] \to_{\mathsf{gc}} t$, where $x \notin \mathsf{fv}(t)$, then $(t[x/s])^{\mathsf{S}} = t^{\mathsf{S}}[x/!s^{\mathsf{S}}] \to_{\bullet\mathsf{gc}} t^{\mathsf{S}}$, where $x \notin \mathsf{fv}(t^{\mathsf{S}})$ because $\mathsf{fv}(t^{\mathsf{S}}) = \mathsf{fv}(t)$. ■

For completeness, we define an **inverse CBS translation**. First, we define a subset $\mathcal{T}_{\bullet}^{\mathsf{S}} \subseteq \mathcal{T}_{\bullet}$, containing the closure by $\to_{\bullet}$-reduction of the image of $(\cdot)^{\mathsf{S}}$, as well as a subset $\mathsf{SCtxs}_{\bullet}^{\mathsf{S}} \subseteq \mathsf{SCtxs}_{\bullet}$, as follows:

$$\underline{t}, \underline{s}, \ldots ::= x \mid \bullet\lambda a. \underline{t}[x/a] \mid \mathtt{req}(\underline{t}) \mid \underline{t}[u/!\underline{s}] \mid \lambda a. \underline{t}[x/a] \mid \underline{t}\,!\underline{s}$$
$$\underline{\mathtt{L}} ::= \Box \mid \underline{\mathtt{L}}[u/!\underline{t}]$$

where, in the productions involving a subterm of the form $\lambda a. \underline{t}[x/a]$, we assume that a is fresh, that is, $a \notin \mathsf{fv}(\underline{t})$.

The *inverse* CBS *translation* is a function $(\cdot)^{-\mathsf{S}} : \mathcal{T}_{\bullet}^{\mathsf{S}} \to \mathcal{T}_{\mathsf{LSC}}$ defined as follows, by induction on the derivation of a term with the (unambiguous) grammar above:

$$x^{-\mathsf{S}} := x \qquad (\bullet\lambda a. \underline{t}[x/a])^{-\mathsf{S}} := \lambda x. \underline{t}^{-\mathsf{S}} \qquad \mathtt{req}(\underline{t})^{-\mathsf{S}} := \underline{t}^{-\mathsf{S}}$$
$$(\lambda a. \underline{t}[x/a])^{-\mathsf{S}} := \lambda x. \underline{t}^{-\mathsf{S}} \qquad \underline{t}[u/!\underline{s}]^{-\mathsf{S}} := \underline{t}^{-\mathsf{S}}[u/\underline{s}^{-\mathsf{S}}] \qquad \underline{t}\,!\underline{s}^{-\mathsf{S}} := \underline{t}^{-\mathsf{S}}\,\underline{s}^{-\mathsf{S}}$$

It is easy to check that $(\cdot)^{-\mathsf{S}}$ is the left-inverse of $(\cdot)^{\mathsf{S}}$.

Lemma 5.6 **(Inverse CBS simulation).** *Let $\underline{t} \in \mathcal{T}_{\bullet}^{\mathsf{S}}$ and $s \in \mathcal{T}_{\bullet}$ such that $\underline{t} \to_{\bullet} s$. Then $s \in \mathcal{T}_{\bullet}^{\mathsf{S}}$ and $\underline{t}^{-\mathsf{S}} \to_{\mathsf{Nd}}^{=} s^{-\mathsf{S}}$.*

Theorem 5.3 **(Sound and complete CBS embedding).** *Let $t, s \in \mathcal{T}_{\mathsf{LSC}}$. Then $t \to_{\mathsf{S}}^{*} s$ if and only if $t^{\mathsf{S}} \to_{\bullet}^{*} s^{\mathsf{S}}$. Moreover, t is in $\to_{\mathsf{S}}$-normal form iff t^{S} is in $\to_{\bullet}$-normal form.*

As previously mentioned, it does not seem possible to embed Wadsworth's call-by-need (*i.e.* CBNd) in $\lambda^{!\bullet}$. One could imagine a variant of $\lambda^{!\bullet}$ that includes the following $\bullet\mathsf{ls}'$ rule rather than $\bullet\mathsf{ls}$:

$$\mathsf{C}\langle\langle u\rangle\rangle[u/(!(\bullet t)\mathtt{L}_1)\mathtt{L}_2] \mapsto_{\bullet\mathsf{ls}'} \mathsf{C}\langle\langle \bullet t\rangle\rangle[u/!(\bullet t)]\mathtt{L}_1\mathtt{L}_2$$

The resulting calculus allows embeddings from CBN, CBV, and CBNd. However, it is not well-behaved, as confluence fails. Let $\Omega := (\lambda a. a\, a)(\lambda a. a\, a)$. For example, $x[y/!u][u/!(\bullet v)[v/\Omega]] \to_{\bullet\mathsf{gc}} x[u/!(\bullet v)[v/\Omega]] \to_{\bullet\mathsf{gc}} x$. But also $x[y/!u][u/!(\bullet v)[v/\Omega]] \to_{\bullet\mathsf{ls}} x[y/!\bullet v][u/!(\bullet v)][v/\Omega] \to_{\bullet\mathsf{gc}} x[u/!(\bullet v)][v/\Omega] \to_{\bullet\mathsf{gc}} x[v/\Omega]$.

6 Simulating Weak Evaluation Strategies

In Section 5, we have shown that CBN, CBV, and CBS calculi can be embedded in the $\lambda^{!\bullet}$-calculus. Reduction in these calculi is intended to capture *equivalence*, rather than *evaluation*, of programs. That is, these calculi are orientations of CBN, CBV, and CBS equational theories rather than evaluation mechanisms.

Reduction in the calculi of Section 5 is closed by arbitrary contexts. *E.g.* in the CBV *calculus*, a step $(\lambda x. y)\, \lambda z. t \to_{\mathsf{V}} (\lambda x. y)\, \lambda z. t'$ is allowed if $t \to_{\mathsf{V}} t'$, while typically call-by-value *evaluation* would proceed to contract the outermost redex.

In this section, we first define *weak evaluation* relations $\rightsquigarrow^{\mathsf{N}}$, $\rightsquigarrow^{\mathsf{V}}$, and $\rightsquigarrow^{\mathsf{S}}$ for CBN, CBV, and CBS respectively. Recall that evaluation is called *weak* if it does not proceed inside the bodies of λ-abstractions. Second, we define an evaluation relation $\rightsquigarrow$ for $\lambda^{!\bullet}$, which is also "weak" in that it does not reduce inside λ-abstractions, boxes ($\bullet$), nor promotions (!). Finally, we show that evaluation according to $\rightsquigarrow^{\mathsf{N}}$, $\rightsquigarrow^{\mathsf{V}}$, and $\rightsquigarrow^{\mathsf{S}}$ can be simulated by $\rightsquigarrow$ via the translations already introduced in Section 5.

Weak CBN Evaluation The one-step weak CBN evaluation judgment is of the form $t \rightsquigarrow^{\mathsf{N}}_{\rho} t'$, where $t, t' \in \mathcal{T}_{\mathsf{LSC}}$ and the set of CBN-*rulenames* $(\rho, \rho', \ldots)$ is given by $\rho ::= \mathsf{db} \mid \varsigma(x, t) \mid \mathsf{ls} \mid \mathsf{gc}$. Weak CBN evaluation is the union $\rightsquigarrow^{\mathsf{N}} := \rightsquigarrow^{\mathsf{N}}_{\mathsf{db}} \cup \rightsquigarrow^{\mathsf{N}}_{\mathsf{ls}} \cup \rightsquigarrow^{\mathsf{N}}_{\mathsf{gc}}$, excluding auxiliary $\varsigma(x, t)$ steps. It is defined by the following inductive rules:

$$\frac{}{(\lambda x. t)\mathsf{L}\, s \rightsquigarrow^{\mathsf{N}}_{\mathsf{db}} t[x/s]\mathsf{L}}\ \mathtt{E^N\text{-}db} \qquad \frac{}{x \rightsquigarrow^{\mathsf{N}}_{\varsigma(x,t)} t}\ \mathtt{E^N}\text{-}\varsigma \qquad \frac{t \rightsquigarrow^{\mathsf{N}}_{\varsigma(x,s)} t'}{t[x/s] \rightsquigarrow^{\mathsf{N}}_{\mathsf{ls}} t'[x/s]}\ \mathtt{E^N\text{-}ls}$$

$$\frac{x \notin \mathsf{fv}(t)}{t[x/s] \rightsquigarrow^{\mathsf{N}}_{\mathsf{gc}} t}\ \mathtt{E^N\text{-}gc} \qquad \frac{t \rightsquigarrow^{\mathsf{N}}_{\rho} t'}{t\, s \rightsquigarrow^{\mathsf{N}}_{\rho} t'\, s}\ \mathtt{E^N\text{-}app} \qquad \frac{t \rightsquigarrow^{\mathsf{N}}_{\rho} t' \quad x \notin \mathsf{fv}(\rho)}{t[x/s] \rightsquigarrow^{\mathsf{N}}_{\rho} t'[x/s]}\ \mathtt{E^N\text{-}subL}$$

The $\mathtt{E^N\text{-}db}$, $\mathtt{E^N\text{-}ls}$, and $\mathtt{E^N\text{-}gc}$ rules derive *root reduction* steps. The $\mathtt{E^N\text{-}app}$ and $\mathtt{E^N\text{-}subL}$ rules correspond to congruence closure below weak head evaluation contexts. The side condition in the $\mathtt{E^N\text{-}subL}$ rule is to avoid unwanted variable capture. The somewhat atypical $\mathtt{E^N}\text{-}\varsigma$ rule derives steps of the form $t \rightsquigarrow^{\mathsf{N}}_{\varsigma(x,s)} t'$, which substitute a single free occurrence of x (in evaluation position) by s. This rule works in synchrony with $\mathtt{E^N\text{-}ls}$ to allow ls steps: for example, $x\, x\, y \rightsquigarrow^{\mathsf{N}}_{\varsigma(x,t)} t\, x\, y$ and $(x\, x\, y)[x/t] \rightsquigarrow^{\mathsf{N}}_{\mathsf{ls}} (t\, x\, y)[x/t]$. This is inspired the formulation of strong call-by-need of Balabonski *et al.* [9].

It is straightforward to show that $\rightsquigarrow^{\mathsf{N}} \subseteq \to_{\mathsf{N}}$. Note also that $\rightsquigarrow^{\mathsf{N}}$ is non-deterministic, although confluent. The source of non-determinism is that gc steps can be performed in any order. For example $((\lambda x. x)\, y)[z/s]$ reduces both with a db and with a gc step.

Weak CBV Evaluation The one-step weak CBV evaluation judgment is of the form $t \rightsquigarrow^{\mathsf{V}}_{\rho} t'$, where $t, t' \in \mathcal{T}_{\mathsf{LSC}}$ and the set of CBV-*rulenames* $(\rho, \rho', \ldots)$ is given by $\rho ::= \mathsf{db} \mid \varsigma(x, \mathrm{v}) \mid \mathsf{lsv} \mid \mathsf{gcv+}$. Weak CBV evaluation is $\rightsquigarrow^{\mathsf{V}} := \rightsquigarrow^{\mathsf{V}}_{\mathsf{db}} \cup \rightsquigarrow^{\mathsf{V}}_{\mathsf{lsv}} \cup \rightsquigarrow^{\mathsf{V}}_{\mathsf{gcv+}}$, excluding auxiliary $\varsigma(x, \mathrm{v})$ steps. It is defined by the following inductive rules:

$$\frac{}{(\lambda x. t)\mathsf{L}\, s \rightsquigarrow^{\mathsf{V}}_{\mathsf{db}} t[x/s]\mathsf{L}}\ \mathtt{E^V\text{-}db} \qquad \frac{}{x \rightsquigarrow^{\mathsf{V}}_{\varsigma(x,\mathrm{v})} \mathrm{v}}\ \mathtt{E^V}\text{-}\varsigma$$

$$\frac{x \notin \mathsf{fv}(t)}{t[x/\mathsf{v}^{+}\mathsf{L}] \rightsquigarrow^{\mathsf{V}}_{\mathsf{gcv+}} t\mathsf{L}}\ \mathsf{E}^{\mathsf{V}}\texttt{-gcv+} \quad \frac{t \rightsquigarrow^{\mathsf{V}}_{\varsigma(x,\mathsf{v})} t'}{t[x/\mathsf{vL}] \rightsquigarrow^{\mathsf{V}}_{\mathsf{lsv}} t'[x/\mathsf{v}]\mathsf{L}}\ \mathsf{E}^{\mathsf{V}}\texttt{-lsv} \quad \frac{t \rightsquigarrow^{\mathsf{V}}_{\rho} t'}{t\,s \rightsquigarrow^{\mathsf{V}}_{\rho} t'\,s}\ \mathsf{E}^{\mathsf{V}}\texttt{-app}$$

$$\frac{t \rightsquigarrow^{\mathsf{V}}_{\rho} t' \quad x \notin \mathsf{fv}(\rho)}{t[x/s] \rightsquigarrow^{\mathsf{V}}_{\rho} t'[x/s]}\ \mathsf{E}^{\mathsf{V}}\texttt{-subL} \quad \frac{s \rightsquigarrow^{\mathsf{V}}_{\rho} s'}{t[x/s] \rightsquigarrow^{\mathsf{V}}_{\rho} t[x/s']}\ \mathsf{E}^{\mathsf{V}}\texttt{-subR}$$

Similar remarks as for CBN apply, in particular $\rightsquigarrow^{\mathsf{V}} \subseteq \to_{\mathsf{V}}$. Rules E^{V}-db, E^{V}-lsv, and E^{V}-gcv+ are root reduction rules, while E^{V}-app, E^{V}-subL, and E^{V}-subR correspond to congruence closure rules. The E^{V}-ς rule plays a similar role as the analogue rule in CBN, but only allows substituting variables for *strict values*. In this notion of CBV evaluation, arguments of applications are not evaluated. The restriction that the argument is a value is not imposed to contract a β-like redex, but rather to perform the substitution. These ideas can already be found in the λ_{CBV}-calculus of [18]. Note that in CBV evaluation ($\rightsquigarrow^{\mathsf{V}}$) there is a second source of non-determinism, namely that E^{V}-subL and E^{V}-subR overlap, so the body and the argument of an ES can be evaluated concurrently.

Weak CBS Evaluation The one-step weak CBS evaluation judgment is of the form $t \rightsquigarrow^{\mathsf{S}}_{\rho} t'$, where $t, t' \in \mathcal{T}_{\mathsf{LSC}}$ and the set of CBS-*rulenames* $(\rho, \rho', \ldots)$ is given by $\rho ::= \mathsf{db} \mid \varsigma(x, \mathsf{vL}) \mid \iota(x) \mid \mathsf{lsw} \mid \mathsf{gc}$. Weak CBS evaluation is the union $\rightsquigarrow^{\mathsf{S}} := \rightsquigarrow^{\mathsf{S}}_{\mathsf{db}} \cup \rightsquigarrow^{\mathsf{S}}_{\mathsf{lsw}} \cup \rightsquigarrow^{\mathsf{S}}_{\mathsf{gc}}$, excluding auxiliary $\varsigma(x, \mathsf{vL})$ and $\iota(x)$ steps. It is defined by the following inductive rules:

$$\frac{}{(\lambda x.\, t)\mathsf{L}\, s \rightsquigarrow^{\mathsf{S}}_{\mathsf{db}} t[x/s]\mathsf{L}}\ \mathsf{E}^{\mathsf{S}}\texttt{-db} \quad \frac{}{x \rightsquigarrow^{\mathsf{S}}_{\varsigma(x,\mathsf{vL})} \mathsf{vL}}\ \mathsf{E}^{\mathsf{S}}\texttt{-}\varsigma \quad \frac{}{t[u/x] \rightsquigarrow^{\mathsf{S}}_{\varsigma(x,\mathsf{vL})} t[u/\mathsf{vL}]}\ \mathsf{E}^{\mathsf{S}}\texttt{-}\varsigma_2$$

$$\frac{}{x \rightsquigarrow^{\mathsf{S}}_{\iota(x)} x}\ \mathsf{E}^{\mathsf{S}}\texttt{-}\iota \quad \frac{t \rightsquigarrow^{\mathsf{S}}_{\varsigma(x,\mathsf{vL})} t'}{t[x/\mathsf{vL}] \rightsquigarrow^{\mathsf{S}}_{\mathsf{lsw}} t'[x/\mathsf{vL}]}\ \mathsf{E}^{\mathsf{S}}\texttt{-lsw} \quad \frac{x \notin \mathsf{fv}(t)}{t[x/s] \rightsquigarrow^{\mathsf{S}}_{\mathsf{gc}} t}\ \mathsf{E}^{\mathsf{S}}\texttt{-gc}$$

$$\frac{t \rightsquigarrow^{\mathsf{S}}_{\rho} t' \quad x \notin \mathsf{fv}(\rho)}{t[x/s] \rightsquigarrow^{\mathsf{S}}_{\rho} t'[x/s]}\ \mathsf{E}^{\mathsf{S}}\texttt{-subL} \quad \frac{t \rightsquigarrow^{\mathsf{S}}_{\rho} t'}{t\,s \rightsquigarrow^{\mathsf{S}}_{\rho} t'\,s}\ \mathsf{E}^{\mathsf{S}}\texttt{-app} \quad \frac{t \rightsquigarrow^{\mathsf{S}}_{\iota(x)} t \quad s \rightsquigarrow^{\mathsf{S}}_{\rho} s'}{t[x/s] \rightsquigarrow^{\mathsf{S}}_{\rho} t[x/s']}\ \mathsf{E}^{\mathsf{S}}\texttt{-subR}$$

Similar remarks as for CBN apply, in particular $\rightsquigarrow^{\mathsf{S}} \subseteq \to_{\mathsf{S}}$. Rules E^{S}-db, E^{S}-lsw, and E^{S}-gc are root reduction rules, while E^{S}-app, E^{S}-subL, and E^{S}-subR are congruence closure rules. The rule E^{S}-ς plays a similar role as the analogue rules in CBN and CBV, but only allows substituting variables for terms of the form vL (known as *answers* in the literature). The rule E^{S}-ς_2 is a variant of E^{S}-ς that acts on the argument of an ES; this rule is not strictly necessary for evaluation, but it is crucial for the embedding into $\lambda^{!\bullet}$ to be complete. The E^{S}-ι rule is used in synchrony with the congruence rules to derive steps that are always of the form $t \rightsquigarrow^{\mathsf{S}}_{\iota(x)} t$ indicating that x occurs in t in an evaluation position. This is used to check whether x it a *needed* variable. For example, $x\,y \rightsquigarrow^{\mathsf{S}}_{\iota(x)} x\,y$ and $z\,z \rightsquigarrow^{\mathsf{S}}_{\varsigma(z,\mathsf{v})} \mathsf{v}\,z$, so the fact that x is needed on the left triggers the evaluation of the argument: $(x\,y)[x/z\,z] \rightsquigarrow^{\mathsf{S}}_{\varsigma(z,\mathsf{v})} (x\,y)[x/\mathsf{v}\,z]$. From this, one obtains the substitution step $(x\,y)[x/z\,z][z/\mathsf{v}] \rightsquigarrow^{\mathsf{S}}_{\mathsf{lsw}} (x\,y)[x/\mathsf{v}\,z][z/\mathsf{v}]$.

Weak $\lambda^{!\bullet}$-Calculus Evaluation The one-step weak $\lambda^{!\bullet}$ evaluation judgment is of the form $t \rightsquigarrow_{\rho} t'$, where $t, t' \in \mathcal{T}_{\bullet}$, and the set of *rulenames* $(\rho, \rho', \ldots)$ is given by $\rho ::= \bullet\mathsf{db} \mid \varsigma(u, (\bullet t)\mathsf{L}) \mid \iota(u) \mid \bullet\mathsf{ls} \mid \bullet\mathsf{gc} \mid \bullet\mathsf{req}$. Weak $\lambda^{!\bullet}$ evaluation is the union $\rightsquigarrow := \rightsquigarrow_{\bullet\mathsf{db}} \cup \rightsquigarrow_{\bullet\mathsf{ls}} \cup \rightsquigarrow_{\bullet\mathsf{gc}}$, excluding auxiliary $\varsigma(x, \mathsf{vL})$ and $\iota(x)$ steps. It is defined by the following inductive rules:

$$\frac{}{(\lambda a.\, t)\mathsf{L}\; s \rightsquigarrow_{\bullet\mathsf{db}} t\{a := s\}\mathsf{L}}\ \texttt{E}^{\bullet}\texttt{-db} \qquad \frac{}{u \rightsquigarrow_{\varsigma(u,(\bullet t)\mathsf{L})} (\bullet t)\mathsf{L}}\ \texttt{E}^{\bullet}\texttt{-}\varsigma \qquad \frac{}{!u \rightsquigarrow_{\varsigma(u,(\bullet t)\mathsf{L})} !(\bullet t)\mathsf{L}}\ \texttt{E}^{\bullet}\texttt{-!}\varsigma$$

$$\frac{}{u \rightsquigarrow_{\iota(u)} u}\ \texttt{E}^{\bullet}\texttt{-}\iota \qquad \frac{t \rightsquigarrow_{\varsigma(u,(\bullet s)\mathsf{L}_1)} t'}{t[u/(!(\bullet s)\mathsf{L}_1)\mathsf{L}_2] \rightsquigarrow_{\bullet\mathsf{ls}} t'[u/!(\bullet s)\mathsf{L}_1]\mathsf{L}_2}\ \texttt{E}^{\bullet}\texttt{-ls} \qquad \frac{t \rightsquigarrow_{\rho} t'}{t\, s \rightsquigarrow_{\rho} t'\, s}\ \texttt{E}^{\bullet}\texttt{-app}$$

$$\frac{u \notin \mathsf{fv}(t)}{t[u/(!s)\mathsf{L}] \rightsquigarrow_{\bullet\mathsf{gc}} t\mathsf{L}}\ \texttt{E}^{\bullet}\texttt{-gc} \qquad \frac{}{\mathtt{req}((\bullet t)\mathsf{L}) \rightsquigarrow_{\bullet\mathsf{req}} t\mathsf{L}}\ \texttt{E}^{\bullet}\texttt{-req}\bullet \qquad \frac{t \rightsquigarrow_{\rho} t'}{\mathtt{req}(t) \rightsquigarrow_{\rho} \mathtt{req}(t')}\ \texttt{E}^{\bullet}\texttt{-req}$$

$$\frac{t \rightsquigarrow_{\rho} t' \quad u \notin \mathsf{fv}(\rho)}{t[u/s] \rightsquigarrow_{\rho} t'[u/s]}\ \texttt{E}^{\bullet}\texttt{-esL} \qquad \frac{s \rightsquigarrow_{\rho} s'}{t[u/s] \rightsquigarrow_{\rho} t[u/s']}\ \texttt{E}^{\bullet}\texttt{-esR} \qquad \frac{t \rightsquigarrow_{\iota(u)} t \quad s \rightsquigarrow_{\rho} s'}{t[u/!s] \rightsquigarrow_{\rho} t[u/!s']}\ \texttt{E}^{\bullet}\texttt{-es!}$$

Weak $\lambda^{!\bullet}$ evaluation is a sub-ARS of $\lambda^{!\bullet}$, in the sense that $\rightsquigarrow \subseteq \rightarrow_{\bullet}$. Rules $\texttt{E}^{\bullet}\texttt{-db}$, $\texttt{E}^{\bullet}\texttt{-ls}$, $\texttt{E}^{\bullet}\texttt{-gc}$, and $\texttt{E}^{\bullet}\texttt{-req}\bullet$ are root reduction rules, while $\texttt{E}^{\bullet}\texttt{-app}$, $\texttt{E}^{\bullet}\texttt{-req}$, $\texttt{E}^{\bullet}\texttt{-esL}$, and $\texttt{E}^{\bullet}\texttt{-esR}$ are congruence rules. Rule $\texttt{E}^{\bullet}\texttt{-}\varsigma$ plays a similar role as the analogue rules in CBN, CBV, and CBNd, but only allows substituting a variable by a term of the form $(\bullet t)\mathsf{L}$. Rule $\texttt{E}^{\bullet}\texttt{-}\iota$ plays a similar role as the analogue rule in CBNd, used to check whether a variable is in evaluation position. Note that there are no congruence rules below λ-abstraction, box ($\bullet$), nor promotion (!). Evaluation can proceed below promotion in two particular cases. First, the $\texttt{E}^{\bullet}\texttt{-!}\varsigma$ rule allows to perform substitution immediately below a promotion; for instance $(!u)[u/!\bullet v] \rightsquigarrow_{\bullet\mathsf{ls}} (!\bullet v)[u/!\bullet v]$. Second, the $\texttt{E}^{\bullet}\texttt{-es!}$ rule allows evaluation below a promotion when a term is the argument of a "needed" substitution; for instance $(uv)[u/!((\lambda a.\, a)(\bullet w))] \rightsquigarrow_{\bullet\mathsf{db}} (uv)[u/!\bullet w] \rightsquigarrow_{\bullet\mathsf{ls}} ((\bullet w)v)[u/!\bullet w]$.

The $\lambda^{!\bullet}$-calculus is designed with the goal in mind of providing a *unifying framework* for call-by-name, call-by-value, and call-by-sharing. As a matter of fact, weak $\lambda^{!\bullet}$ evaluation simulates weak CBN, CBV, and CBS evaluation via the $(\cdot)^{\mathsf{N}}$, $(\cdot)^{\mathsf{V}}$, and $(\cdot)^{\mathsf{S}}$ translations introduced in Section 5:

Theorem 6.1 (Simulation and inverse simulation of evaluation).

	Soundness	Completeness
CBN	*If* $t \rightsquigarrow^{\mathsf{N}} s$ *then* $t^{\mathsf{N}} \rightsquigarrow^{*} s^{\mathsf{N}}$.	*If* $t^{\mathsf{N}} \rightsquigarrow s$ *then* $s \in \mathcal{T}_{\bullet}^{\mathsf{N}}$ *and* $t\, (\rightsquigarrow^{\mathsf{N}})^{=}\, s^{-\mathsf{N}}$.
CBV	*If* $t \rightsquigarrow^{\mathsf{V}} s$ *then* $t^{\mathsf{V}} \rightsquigarrow^{*} s^{\mathsf{V}}$.	*If* $t^{\mathsf{V}} \rightsquigarrow s$ *then* $s \in \mathcal{T}_{\bullet}^{\mathsf{V}}$ *and* $t\, (\blacktriangleright^{\mathsf{V}})^{=}\, s^{-\mathsf{V}}$.
CBS	*If* $t \rightsquigarrow^{\mathsf{S}} s$ *then* $t^{\mathsf{S}} \rightsquigarrow^{*} s^{\mathsf{S}}$.	*If* $t^{\mathsf{S}} \rightsquigarrow s$ *then* $s \in \mathcal{T}_{\bullet}^{\mathsf{S}}$ *and* $t\, (\rightsquigarrow^{\mathsf{S}})^{=}\, s^{-\mathsf{S}}$.

In the CBV *case, completeness only holds for an extended relation* $\blacktriangleright^{\mathsf{V}}$*, defined as* $\rightsquigarrow^{\mathsf{V}}$ *but adding an inverse garbage collection rule that derives* $t \blacktriangleright^{\mathsf{V}} t[x/\mathsf{v}^{+}]$ *if* $x \notin \mathsf{fv}(t)$.

7 Related Work and Conclusions

Related Work. The seminal work [22] is the first work to have related Girard's embeddings of intuitionistic logic into LL with evaluation mechanisms[7]. Call-by-push-value

[7] Although the paper mentions some other authors that had already hinted at this.

(CBPV) [20,21] is a calculus that distinguishes *values* from *computations* and allows to subsume both the CBV and CBN evaluation mechanisms. Ehrhard [14] studied the connection between CBPV [20,21] and LL, producing a calculus which was later modified to become the Bang-calculus [15]. CBV and CBN translations to the Bang-calculus were studied in [15]. Soundness and completeness of these translations with respect to reduction was proved by Guerrieri and Manzonetto [17] for a slightly different notion of reduction for the Bang-calculus than that of [15]. The CBV translation does not preserve normal forms; an amended translation that does was studied in [12,13]. Intuitionistic truth in terms of classical provability underlies Gödel's embedding of intuitionistic logic into (classical) S4. In [27], the authors consider a program similar to that of CBPV but where that target language is a modal lambda calculus. Promotion and dereliction are recast as boxing and unboxing and CBV and CBN are described in terms of a so called *call-by-box* evaluation mechanism [27].

Conclusions. This work introduces MSCLL, a Sharing Linear Logic. It arises from splitting each exponential modality (!/?) into a sharing modality (!/?) and a cloning modality (•/∘). MSCLL is conservative over MELL and enjoys cut-elimination. The usual embeddings of intuitionistic logic into LL can be restated in the setting of $\lambda^{!\bullet}$, a Sharing Linear λ-calculus derived from MSCLL. The decomposition of the of-course modality allows us to define an embedding of intuitionistic logic into $\lambda^{!\bullet}$, corresponding to a *call-by-need* λ-calculus CBS. The following table summarizes the, sound and complete, embeddings studied in Section 5:

	$A \to B$	x	$\lambda x. t$	$t\,s$	$t[x/s]$
CBN, $(\cdot)^{\mathsf{N}}$	$!\bullet A^{\mathsf{N}} \to B^{\mathsf{N}}$	$\mathtt{req}(x)$	$\lambda a. t^{\mathsf{N}}[x/a]$	$t^{\mathsf{N}}\,!\bullet s^{\mathsf{N}}$	$t^{\mathsf{N}}[x/!\bullet s^{\mathsf{N}}]$
CBV, $(\cdot)^{\mathsf{V}}$	$!\bullet A^{\mathsf{V}} \to !\bullet B^{\mathsf{V}}$	$!x$	$!\bullet\lambda a. t^{\mathsf{V}}[x/a]$	$\mathtt{req}(u)[u/t^{\mathsf{V}}]\,s^{\mathsf{V}}$	$t^{\mathsf{V}}[x/s^{\mathsf{V}}]$
CBS, $(\cdot)^{\mathsf{S}}$	$!\bullet A^{\mathsf{S}} \to \bullet B^{\mathsf{S}}$	x	$\bullet\lambda a. t^{\mathsf{S}}[x/a]$	$\mathtt{req}(t^{\mathsf{S}})\,!s^{\mathsf{S}}$	$t^{\mathsf{S}}[x/!s^{\mathsf{S}}]$

A weak evaluation mechanism can be defined for $\lambda^{!\bullet}$ that simulates weak evaluation in the original calculi in a sound and complete way. Moreover, MSCLL also admits a sound and complete embedding of the Bang-calculus (see the companion report [10]).

There are several avenues worth pursuing. First, developing an appropriate notion of proof nets and semantics for MSCLL, which perhaps would help clarify the somewhat intriguing interaction between the sharing and access modalities. Second, studying operational properties of the $\lambda^{!\bullet}$-calculus such as standardization (as developed for LSC [3]) and solvability. Additionally, one can consider extending weak evaluation in $\lambda^{!\bullet}$ to *strong* evaluation, to simulate strong CBN/CBV/CBS evaluation. Also, our use of multiple exponentials is reminiscent of subexponentials [25], where instead of one pair of of-course and why-not modalities one introduces a family of them, each of which cannot be proven equivalent to any other. Further work is required to determine if there is a rigorous connection with subexponentials.

It should be noted that our original motivation to study MSCLL was to try to provide a unified logical account of CBN, CBV, and CBNd. In [22], an attempt was made at embedding CBNd in a linear λ-calculus, but the target language had to be changed to become *affine*, allowing weakening of arbitrary propositions.

Acknowledgments The first author was partially supported by project grants PUNQ 2219/22 and PICT-2021-I-INVI-00602.

References

1. Accattoli, B.: An abstract factorization theorem for explicit substitutions. In: Tiwari, A. (ed.) 23rd International Conference on Rewriting Techniques and Applications (RTA'12), RTA 2012, May 28 - June 2, 2012, Nagoya, Japan. LIPIcs, vol. 15, pp. 6–21. Schloss Dagstuhl - Leibniz-Zentrum für Informatik (2012). https://doi.org/10.4230/LIPICS.RTA.2012.6, https://doi.org/10.4230/LIPIcs.RTA.2012.6
2. Accattoli, B.: Proof nets and the linear substitution calculus. In: Fischer, B., Uustalu, T. (eds.) Theoretical Aspects of Computing - ICTAC 2018 - 15th International Colloquium, Stellenbosch, South Africa, October 16-19, 2018, Proceedings. Lecture Notes in Computer Science, vol. 11187, pp. 37–61. Springer (2018). https://doi.org/10.1007/978-3-030-02508-3_3, https://doi.org/10.1007/978-3-030-02508-3_3
3. Accattoli, B., Bonelli, E., Kesner, D., Lombardi, C.: A nonstandard standardization theorem. In: Jagannathan, S., Sewell, P. (eds.) The 41st Annual ACM SIGPLAN-SIGACT Symposium on Principles of Programming Languages, POPL '14, San Diego, CA, USA, January 20-21, 2014. pp. 659–670. ACM (2014). https://doi.org/10.1145/2535838.2535886, https://doi.org/10.1145/2535838.2535886
4. Accattoli, B., Kesner, D.: The structural *lambda*-calculus. In: Dawar, A., Veith, H. (eds.) Computer Science Logic, 24th International Workshop, CSL 2010, 19th Annual Conference of the EACSL, Brno, Czech Republic, August 23-27, 2010. Proceedings. Lecture Notes in Computer Science, vol. 6247, pp. 381–395. Springer (2010). https://doi.org/10.1007/978-3-642-15205-4_30, https://doi.org/10.1007/978-3-642-15205-4_30
5. Accattoli, B., Paolini, L.: Call-by-value solvability, revisited. In: Schrijvers, T., Thiemann, P. (eds.) Functional and Logic Programming - 11th International Symposium, FLOPS 2012, Kobe, Japan, May 23-25, 2012. Proceedings. Lecture Notes in Computer Science, vol. 7294, pp. 4–16. Springer (2012). https://doi.org/10.1007/978-3-642-29822-6_4, https://doi.org/10.1007/978-3-642-29822-6_4
6. Ariola, Z.M., Felleisen, M., Maraist, J., Odersky, M., Wadler, P.: A call-by-need lambda calculus. In: Conference Record of POPL'95: 22nd ACM SIGPLAN-SIGACT Symposium on Principles of Programming Languages, San Francisco, California, USA, January 23-25, 1995. pp. 233–246 (1995)
7. Arrial, V.: A deeper study of $\lambda!$-calculus simulations. In: Fuhs, C. (ed.) Proceedings of the 11th International Workshop on Higher-Order Rewriting (HOR 2023) (July 2023)
8. Balabonski, T., Barenbaum, P., Bonelli, E., Kesner, D.: Foundations of strong call by need. Proc. ACM Program. Lang. **1**(ICFP), 20:1–20:29 (2017). https://doi.org/10.1145/3110264, https://doi.org/10.1145/3110264
9. Balabonski, T., Lanco, A., Melquiond, G.: A strong call-by-need calculus. Log. Methods Comput. Sci. **19**(1) (2023). https://doi.org/10.46298/LMCS-19(1:21)2023, https://doi.org/10.46298/lmcs-19(1:21)2023
10. Barenbaum, P., Bonelli, E.: Sharing and linear logic with restricted access (extended version). CoRR **abs/2501.16576** (2025). https://doi.org/10.48550/ARXIV.2501.16576, https://doi.org/10.48550/arXiv.2501.16576
11. Barendregt, H.P.: The lambda calculus - its syntax and semantics, Studies in logic and the foundations of mathematics, vol. 103. North-Holland (1985)
12. Bucciarelli, A., Kesner, D., Ríos, A., Viso, A.: The bang calculus revisited. In: Nakano, K., Sagonas, K. (eds.) Functional and Logic Programming - 15th International Symposium, FLOPS 2020, Akita, Japan, September 14-16, 2020, Proceedings. Lecture Notes

in Computer Science, vol. 12073, pp. 13–32. Springer (2020). https://doi.org/10.1007/978-3-030-59025-3_2, https://doi.org/10.1007/978-3-030-59025-3_2
13. Bucciarelli, A., Kesner, D., Ríos, A., Viso, A.: The bang calculus revisited. Inf. Comput. **293**, 105047 (2023). https://doi.org/10.1016/J.IC.2023.105047, https://doi.org/10.1016/j.ic.2023.105047
14. Ehrhard, T.: Call-by-push-value from a linear logic point of view. In: Thiemann, P. (ed.) Programming Languages and Systems - 25th European Symposium on Programming, ESOP 2016, Held as Part of the European Joint Conferences on Theory and Practice of Software, ETAPS 2016, Eindhoven, The Netherlands, April 2-8, 2016, Proceedings. Lecture Notes in Computer Science, vol. 9632, pp. 202–228. Springer (2016). https://doi.org/10.1007/978-3-662-49498-1_9, https://doi.org/10.1007/978-3-662-49498-1_9
15. Ehrhard, T., Guerrieri, G.: The bang calculus: an untyped lambda-calculus generalizing call-by-name and call-by-value. In: Cheney, J., Vidal, G. (eds.) Proceedings of the 18th International Symposium on Principles and Practice of Declarative Programming, Edinburgh, United Kingdom, September 5-7, 2016. pp. 174–187. ACM (2016). https://doi.org/10.1145/2967973.2968608, https://doi.org/10.1145/2967973.2968608
16. Girard, J.: Linear logic. Theor. Comput. Sci. **50**, 1–102 (1987). https://doi.org/10.1016/0304-3975(87)90045-4, https://doi.org/10.1016/0304-3975(87)90045-4
17. Guerrieri, G., Manzonetto, G.: The bang calculus and the two girard's translations. In: Ehrhard, T., Fernández, M., de Paiva, V., de Falco, L.T. (eds.) Proceedings Joint International Workshop on Linearity & Trends in Linear Logic and Applications, Linearity-TLLA@FLoC 2018, Oxford, UK, 7-8 July 2018. EPTCS, vol. 292, pp. 15–30 (2018). https://doi.org/10.4204/EPTCS.292.2, https://doi.org/10.4204/EPTCS.292.2
18. Herbelin, H., Zimmermann, S.: An operational account of call-by-value minimal and classical lambda-calculus in "natural deduction" form. In: Curien, P. (ed.) Typed Lambda Calculi and Applications, 9th International Conference, TLCA 2009, Brasilia, Brazil, July 1-3, 2009. Proceedings. Lecture Notes in Computer Science, vol. 5608, pp. 142–156. Springer (2009). https://doi.org/10.1007/978-3-642-02273-9_12, https://doi.org/10.1007/978-3-642-02273-9_12
19. Kesner, D., Conchúir, S.Ó.: Milner's lambda-calculus with partial substitutions. CoRR **abs/2312.13270** (2023). https://doi.org/10.48550/ARXIV.2312.13270, https://doi.org/10.48550/arXiv.2312.13270
20. Levy, P.B.: Call-By-Push-Value: A Functional/Imperative Synthesis, Semantics Structures in Computation, vol. 2. Springer (2004)
21. Levy, P.B.: Call-by-push-value: Decomposing call-by-value and call-by-name. High. Order Symb. Comput. **19**(4), 377–414 (2006). https://doi.org/10.1007/S10990-006-0480-6, https://doi.org/10.1007/s10990-006-0480-6
22. Maraist, J., Odersky, M., Turner, D.N., Wadler, P.: Call-by-name, call-by-value, call-by-need and the linear lambda calculus. Theor. Comput. Sci. **228**(1-2), 175–210 (1999). https://doi.org/10.1016/S0304-3975(98)00358-2, https://doi.org/10.1016/S0304-3975(98)00358-2
23. Melliès, P.A.: Description abstraite des systémes de réécriture. Ph.D. thesis, Université Paris VII (1996)
24. Milner, R.: Local bigraphs and confluence: Two conjectures: (extended abstract). In: Amadio, R.M., Phillips, I. (eds.) Proceedings of the 13th International Workshop on Expressiveness in Concurrency, EXPRESS 2006, Bonn, Germany, August 26, 2006. Electronic Notes in Theoretical Computer Science, vol. 175, pp. 65–73. Elsevier (2006). https://doi.org/10.1016/J.ENTCS.2006.07.035, https://doi.org/10.1016/j.entcs.2006.07.035
25. Nigam, V., Miller, D.: Algorithmic specifications in linear logic with subexponentials. In: Porto, A., López-Fraguas, F.J. (eds.) Proceedings of the 11th International ACM SIGPLAN Conference on Principles and Practice of Declarative Programming, September 7-9, 2009,

Coimbra, Portugal. pp. 129–140. ACM (2009). https://doi.org/10.1145/1599410.1599427, https://doi.org/10.1145/1599410.1599427
26. Plotkin, G.D.: Call-by-name, call-by-value and the lambda-calculus. Theor. Comput. Sci. **1**(2), 125–159 (1975). https://doi.org/10.1016/0304-3975(75)90017-1, https://doi.org/10.1016/0304-3975(75)90017-1
27. Santo, J.E., Pinto, L., Uustalu, T.: Modal embeddings and calling paradigms. In: Geuvers, H. (ed.) 4th International Conference on Formal Structures for Computation and Deduction, FSCD 2019, June 24-30, 2019, Dortmund, Germany. LIPIcs, vol. 131, pp. 18:1–18:20. Schloss Dagstuhl - Leibniz-Zentrum für Informatik (2019). https://doi.org/10.4230/LIPICS.FSCD.2019.18, https://doi.org/10.4230/LIPIcs.FSCD.2019.18
28. Wadsworth, C.P.: Semantics and pragmatics of the lambda-calculus. Ph.D. thesis, Oxford University (1971)

A Diagrammatic Algebra for Program Logics

Filippo Bonchi[1], Alessandro Di Giorgio[3,4](✉), and Elena Di Lavore[1,2]

[1] University of Pisa, Pisa, Italy
[2] University of Oxford, Oxford, UK
[3] Tallinn University of Technology, Tallinn, Estonia
aless@taltech.ee
[4] University College London, London, UK

Abstract. Tape diagrams provide a convenient graphical notation for arrows of rig categories, i.e., categories equipped with two monoidal products, $\oplus$ and $\otimes$. In this work, we introduce Kleene-Cartesian rig categories, namely rig categories where $\otimes$ provides a Cartesian bicategory, while $\oplus$ a Kleene bicategory. We show that the associated tape diagrams can conveniently deal with Hoare logic.

Keywords: Rig categories · Cartesian Bicategories · Program logics.

1 Introduction

The calculus of relations, originally introduced by De Morgan and Peirce in the late 19th century, is an ancestor of first order logic that has been revitalised by Tarski in 1941. With the dawn of program logics, the calculus of relations – extended with transitive closure – was early recognised [53] to play a key role.

Around the same time, Lawvere was pioneering categorical logic by introducing the concept of *functorial semantics* [47]. Given an algebraic theory T (in the sense of universal algebra, i.e., a signature Σ and a set of equations E), one can freely generate a Cartesian category $\mathcal{L}_T$. Models, in the standard algebraic sense, correspond one-to-one with Cartesian functors F from $\mathcal{L}_T$ to **Set**, the category of sets and functions. More generally, models of the theory in any Cartesian category **C** are represented by Cartesian functors $F : \mathcal{L}_T \to \mathbf{C}$. However this approach fails if one tries to apply it to relational theories by choosing **C** as **Rel**, the category of sets and relations, because the Cartesian product of sets is not the categorical product in **Rel**.

A refinement of Lawvere's method for relational structures has been recently proposed in [13,15,26]. Starting from a *monoidal signature*, one can freely generate a *Cartesian bicategory* [40] and define models as morphisms into $(\mathbf{Rel}, \otimes, 1)$, the monoidal category of relations, where the monoidal product $\otimes$ is the Cartesian product of sets. This framework captures regular theories, i.e., those involving the $\{\exists, \wedge, \top\}$-fragment of first order logic. More recently [10], this approach was extended to full first order logic by deriving negation from the interaction of Cartesian and linear bicategories [21].

In this paper, we extend the Cartesian bicategory framework in a different direction: program logics. In the last decades, there has been an explosion of program logics and many researchers felt the need for more systematic approaches. Our proposal is based on relational and categorical algebra. We propose tape diagrams as an "assembly language" for interpreting various program logics (Remark 3). While the inference rules

P. A. Abdulla and D. Kesner (Eds.): FoSSaCS 2025, LNCS 15691, pp. 308–330, 2025.
https://doi.org/10.1007/978-3-031-90897-2_15

for each of these logics are usually defined by the ingenuity of the researchers, in our approach such rules follow from the laws of Kleene-Cartesian rig categories. These laws arise from the interaction of canonical categorical structures on the category of sets and relations. Crucially, the same approach has lead to identify various categorical structures corresponding to various well-known logics (Figure 1).

	Logic	Categorical structure
[47]	Equational logic	Cartesian category
[15]	Regular logic	Cartesian bicategory
[11]	Coherent logic	Finite-biproduct Cartesian bicategory
[10]	First-order logic	First-order bicategory
This work	Program logic	Kleene-Cartesian rig category

Fig. 1: Categorical structures correspond to logics.

An idea, originating at least from Bainbridge [4], is to model data flow using the Cartesian product of relations, $(\mathbf{Rel}, \otimes, 1)$, and control flow using a different monoidal structure on relations: $(\mathbf{Rel}, \oplus, 0)$. In this second structure, the monoidal product $\oplus$ is the disjoint union of sets, which acts both as a coproduct and a product, hence, a *biproduct*. Both monoidal categories are *traced* [39]: the trace in $(\mathbf{Rel}, \otimes, 1)$ represents feedback, while in $(\mathbf{Rel}, \oplus, 0)$ –the focus of our work– it provides *iteration* [54].

Our first step is to extract from $(\mathbf{Rel}, \oplus, 0)$ the categorical structures essential for modelling control flow, which we term *Kleene bicategories*. Essentially, a Kleene bicategory is a poset-enriched traced monoidal category where the monoidal product $\oplus$ is a biproduct, and the induced natural comonoid [27] is *right adjoint* to the natural monoid. The trace must satisfy a posetal variant of the so-called *uniformity* condition [20,33]. The term "Kleene" is justified because every Kleene bicategory forms a (typed) Kleene algebra in Kozen's sense [41,43] (Corollary 1), while any Kleene algebra canonically gives rise, through the biproduct completion [48], to a Kleene bicategory.

To model control and data flow within a unified categorical structure, we employ *rig categories* [46], categories equipped with two monoidal products, $\oplus$ and $\otimes$, where $\otimes$ distributes over $\oplus$. We define *Kleene-Cartesian rig* (kc rig) *categories*, where $\oplus$ and $\otimes$ exhibit the structures of Kleene and Cartesian bicategories, respectively. To construct the freely generated kc rig category (Theorem 2), we extend *tape diagrams* [11], a diagrammatic notation recently introduced for rig categories. Intuitively, tape diagrams are *string diagrams* [38] in which other string diagrams are nested: the inner diagrams model data flow, and the outer ones model control flow. On one hand, this offers an intuitive unified picture of Bainbridge's idea; on the other it allows for visualising the laws of kc rig categories (Figures 2, 3 and 4) in a way that enlights several monoidal algebras occurring in different types of systems [56,22,3,28,18,36,16,12,14,51,29].

We then introduce *Kleene-Cartesian theories* and their models that, like in Lawvere's approach, coincide with functors (Proposition 3). We illustrate an example of a Kleene-Cartesian theory which is not first order: Peano's axiomatisation of natural numbers. We demonstrate how imperative programs and their logics [35,44,24,50,2] – even more sophisticated ones, like [7], where the interaction of data and control flow play a key role – can be encoded within Kleene-Cartesian tape diagrams. In particular, we show that the rules of Hoare logic follow from the laws of kc rig categories (Proposition 4). Finally, the framework is expressive enough to capture the positive fragment of the calculus of relations with transitive closure, which is the departure of our journey.

The full version of the paper is found in [9] which contains the missing proofs.

2 The calculus of relations

We commence by recalling the positive fragment of the calculus of relations with reflexive and transitive closure (CR). Its syntax is given by the grammar on the left, where R is taken from a given set Σ of generating symbols. Beyond the usual relational composition ;, union $\cup$, intersection $\cap$ and their units id, $\bot$ and $\top$, the calculus features two unary operations: the opposite $(\cdot)^\dagger$ and the reflexive and transitive closure $(\cdot)^*$. Composition and identities are defined, for sets X, Y, Z, and relations $R \subseteq X \times Y$, $S \subseteq Y \times Z$, as

$$
\begin{aligned}
E ::= R \mid id \mid E;E \mid \quad & (1)\\
E^\dagger \mid \top \mid E \cap E \mid \quad & (2)\\
E^* \mid \bot \mid E \cup E \quad & (3)
\end{aligned}
$$

$$R;S \stackrel{\text{def}}{=} \{(x,z) \mid \exists y. \in Y. (x,y) \in R \wedge (y,z) \in S\} \text{ and } id_X \stackrel{\text{def}}{=} \{(x,x) \mid x \in X\},$$

the opposite as $R^\dagger \stackrel{\text{def}}{=} \{(y,x) \mid (x,y) \in R\}$, while for $R \subseteq X \times X$, its reflexive and transitive closure is $R^* \stackrel{\text{def}}{=} \bigcup_{n \in \mathbb{N}} R^n$ where $R^0 \stackrel{\text{def}}{=} id_X$ and $R^{n+1} \stackrel{\text{def}}{=} R; R^n$.

Its semantics, illustrated below, is defined wrt a *relational interpretation* $\mathcal{I}$, that is, a set X together with a binary relation $\rho(R) \subseteq X \times X$ for each $R \in \Sigma$.

$$
\begin{array}{lll}
\langle R \rangle_{\mathcal{I}} \stackrel{\text{def}}{=} \rho(R) & \langle id \rangle_{\mathcal{I}} \stackrel{\text{def}}{=} id_X & \langle E_1; E_2 \rangle_{\mathcal{I}} \stackrel{\text{def}}{=} \langle E_1 \rangle_{\mathcal{I}}; \langle E_2 \rangle_{\mathcal{I}} \\
\langle E^\dagger \rangle_{\mathcal{I}} \stackrel{\text{def}}{=} \langle E \rangle_{\mathcal{I}}^\dagger & \langle \bot \rangle_{\mathcal{I}} \stackrel{\text{def}}{=} \{\} & \langle E_1 \cup E_2 \rangle_{\mathcal{I}} \stackrel{\text{def}}{=} \langle E_1 \rangle_{\mathcal{I}} \cup \langle E_2 \rangle_{\mathcal{I}} \\
\langle E^* \rangle_{\mathcal{I}} \stackrel{\text{def}}{=} \langle E \rangle_{\mathcal{I}}^* & \langle \top \rangle_{\mathcal{I}} \stackrel{\text{def}}{=} X \times X & \langle E_1 \cap E_2 \rangle_{\mathcal{I}} \stackrel{\text{def}}{=} \langle E_1 \rangle_{\mathcal{I}} \cap \langle E_2 \rangle_{\mathcal{I}}
\end{array}
$$

Two expressions E_1, E_2 are said to be *equivalent*, written $E_1 \equiv_{\mathsf{CR}} E_2$, iff $\langle E_1 \rangle_{\mathcal{I}} = \langle E_2 \rangle_{\mathcal{I}}$ for all interpretations $\mathcal{I}$. For instance, $(R^*)^\dagger \equiv_{\mathsf{CR}} (R^\dagger)^*$. Inclusion, denoted by $\leq_{\mathsf{CR}}$, is defined analogously by replacing $=$ with $\subseteq$. Axiomatisations and decidability of $\equiv_{\mathsf{CR}}$ have been studied focusing on several different fragments: see [52] and the references therein. Particularly interesting are the *allegorical fragment*, consisting of (1) and (2), and the *Kleene fragment* consisting of (1) and (3).

Our starting observation is that these two fragments arise from two different traced monoidal structures on **Rel**, the category of sets and relations: $(\mathbf{Rel}, \otimes, 1)$ and $(\mathbf{Rel}, \oplus, 0)$. In the former, the monoidal product $\otimes$ is given by the cartesian product of sets and, for relations $R: X_1 \to Y_1$, $S: X_2 \to Y_2$, $R \otimes S: X_1 \otimes X_2 \to Y_1 \otimes Y_2$ is defined as

$$R \otimes S \stackrel{\text{def}}{=} \{((x_1,x_2),(y_1,y_2)) \mid (x_1,y_1) \in R \text{ and } (x_2,y_2) \in S\} \text{ with unit } 1 \stackrel{\text{def}}{=} \{\bullet\}.$$

In $(\mathbf{Rel}, \oplus, 0)$, $0 \stackrel{\text{def}}{=} \{\}$, $\oplus$ on sets is their disjoint union and $R \oplus S : X_1 \oplus X_2 \to Y_1 \oplus Y_2$ is

$$R \oplus S \stackrel{\text{def}}{=} \{((x_1,1),(y_1,1)) \mid (x_1,y_1) \in R\} \cup \{((x_2,2),(y_2,2)) \mid (x_2,y_2) \in S\}.$$

Here, we tag with 1 and 2 the elements of the disjoint union of two arbitrary sets.

For all sets X, the unique function $!_X : X \to 1$ and the pairing $\langle id_X, id_X \rangle \stackrel{\text{def}}{=} \blacktriangleleft_X : X \to X \otimes X$ form a comonoid in $(\mathbf{Rel}, \otimes, 1)$. Similarly the unique function $ⵗ_X : 0 \to X$ and the copairing $[id_X, id_X] \stackrel{\text{def}}{=} \triangleright_X \colon X \oplus X \to X$ form a monoid in $(\mathbf{Rel}, \oplus, 0)$. By taking their opposite relations, we obtain in total the two (co)monoid structures illustrated below.

$$\begin{array}{llll} \blacktriangleleft_X \stackrel{\text{def}}{=} \{(x,\ (x,x)) \mid x \in X\} & !_X \stackrel{\text{def}}{=} \{(x,\bullet) \mid x \in X\} \subseteq X \times 1 & \blacktriangleright_X \stackrel{\text{def}}{=} \blacktriangleleft_X^{\dagger} & \text{¡}_X \stackrel{\text{def}}{=} !_X^{\dagger} \\ \triangleright_X \stackrel{\text{def}}{=} \{((x,1),\ x) \mid x \in X\} \cup \{((x,2),\ x) \mid x \in X\} & \text{⫯}_X \stackrel{\text{def}}{=} \{\} & \triangleleft_X \stackrel{\text{def}}{=} \triangleright_X^{\dagger} & \text{⫰}_X \stackrel{\text{def}}{=} \text{⫯}_X^{\dagger} \end{array} \quad (4)$$

The black (co)monoids give to $(\mathbf{Rel}, \otimes, 1)$ the structure of a *Cartesian bicategory* [19], while the white ones give to $(\mathbf{Rel}, \oplus, 0)$ the structure of, what we named, a *Kleene bicategory*. These are illustrated in the next two sections.

3 Cartesian Bicategories

All bicategories considered in this paper are *poset enriched symmetric monoidal categories*: every homset carries a partial order $\leq$, and composition ; and monoidal product $\odot$ are monotone. A *poset enriched symmetric monoidal functor* is a symmetric monoidal functor that preserves the order $\leq$. The notion of *adjoint arrows*, which will play a key role, amounts to the following: for $f \colon X \to Y$ and $g \colon Y \to X$, f is *left adjoint* to g, or g is *right adjoint* to f, written $f \dashv g$, if $id_X \leq f ; g$ and $g ; f \leq id_Y$. We extend such terminology to pairs of arrows: (a, b) is left adjoint to (c, d) iff $a \dashv c$ and $b \dashv b$.

All monoidal categories and functors considered throughout this paper are tacitly assumed to be strict [48], i.e. $(X \odot Y) \odot Z = X \odot (Y \odot Z)$ and $I \odot X = X = X \odot I$ for all objects X, Y, Z. This is harmless: strictification [48] allows to transform any monoidal category into a strict one, enabling the sound use of string diagrams. In this and in the next section we will use the string diagrammatic notation for traced monoidal categories from [55]. The unfamiliar reader may check e.g. [9, Sec. 2]. In particular multiplication, unit, comultiplication and counit of the various (co)monoids, always tacitly assumed to be (co)commutative, will be drawn hereafter respectively, as

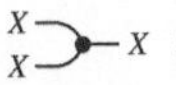 $: X \odot X \to X$  $: I \to X$  $: X \to X \odot X$  $: X \to I.$

Definition 1. A Cartesian bicategory *is a poset enriched symmetric monoidal category* $(\mathbf{C}, \otimes, 1)$ *and, for every object* X *in* $\mathbf{C}$, *a monoid* $(\blacktriangleright_X, \text{¡}_X)$ *and a comonoid* $(\blacktriangleleft_X, !_X)$ *satisfying the usual coherence conditions (see e.g. [55, Table 4.7]) such that*
1. $(\blacktriangleleft_X, !_X)$ *is left adjoint to* $(\blacktriangleright_X, \text{¡}_X)$*;*
2. arrows $f \colon X \to Y$ *are lax comonoid morphisms:* $f ; \blacktriangleleft_Y \leq \blacktriangleleft_X ; (f \otimes f)$ *and* $f ; !_Y \leq !_X$*;*
3. $(\blacktriangleleft_X, !_X)$ *and* $(\blacktriangleright_X, \text{¡}_X)$ *form special Frobenius algebras (see e.g. [45]);*

A morphism of Cartesian bicategories *is a poset enriched symmetric monoidal functor preserving monoids and comonoids.*

The archetypal example of a Cartesian bicategory is $(\mathbf{Rel}, \otimes, 1)$ with $\blacktriangleleft_X$, $!_X$, $\blacktriangleright_X$, ¡_X defined as in (4). Simple computations confirm that all the laws of Definition 1 are satisfied. The operations of CR in (2) can be defined in any Cartesian bicategory, as

$f \sqcap g \stackrel{\text{def}}{=}$  $\top \stackrel{\text{def}}{=}$  $f^{\dagger} \stackrel{\text{def}}{=}$ 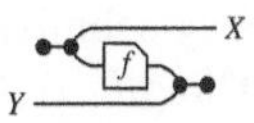 (5)

$(f \sqcap g) \sqcap h = f \sqcap (g \sqcap h)$	$f \sqcap g = g \sqcap f$	$f \sqcap \top = f$	$f \sqcap f = f$
$(f \sqcap g); h \leq (f; h \sqcap g; h)$	$h; (f \sqcap g) \leq (h; f \sqcap h; g)$	$f; \top \leq \top \geq \top; f$	
$(f; g)^\dagger = g^\dagger; f^\dagger$	$(f \otimes g)^\dagger = f^\dagger \otimes g^\dagger$	$(id_X)^\dagger = id_X$	$(f^\dagger)^\dagger = f$

Table 1: Derived laws in Cartesian bicategories.

for all objects X, Y and arrows $f, g \colon X \to Y$. The reader can easily check that, in **Rel**, these correspond to the intersection $f \sqcap g$, the top relation $X \times Y$ and the opposite relation $f^\dagger$, respectively. From now on, we will depict a morphism $f \colon X \to Y$ as $x -\!\boxed{f}\!- y$, and use $y -\!\boxed{f}\!- x$ as syntactic sugar for $f^\dagger$.

Proposition 1. *In any Cartesian bicategory, the laws in Table 1 hold.*

An arrow $f \colon X \to Y$ is said to be *single valued* iff satisfies (SV) below, *total* iff satisfies (TOT), *injective* iff satisfies (INJ) and *surjective* iff satisfies (SUR).

$$x -\!\boxed{}\!\!<^Y_Y \; \leq \; x -\!\boxed{}\!\!\mathsf{C}^Y_Y \qquad \text{(SV)} \qquad\qquad y -\!\boxed{f}\!\boxed{f}\!- y \; \leq \; y \text{———} y \qquad (6)$$

$$x \text{———}\!\bullet \; \leq \; x -\!\boxed{}\!\!\bullet \qquad \text{(TOT)} \qquad\qquad x \text{———} x \; \leq \; x -\!\boxed{}\!\boxed{}\!- x \qquad (7)$$

$$^X_X\!\boxed{}\!\!\mathrel{-} y \; \leq \; {}^X_X\!\!\succ\!\boxed{}\!- y \qquad \text{(INJ)} \qquad\qquad x -\!\boxed{}\!\boxed{}\!- x \; \leq \; x \text{———} x \qquad (8)$$

$$\bullet\!\text{———} y \; \leq \; \bullet\!-\!\boxed{}\!- y \qquad \text{(SUR)} \qquad\qquad y \qquad y \; \leq \; y -\!\boxed{}\!\boxed{}\!- y \qquad (9)$$

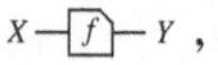

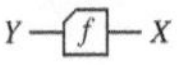

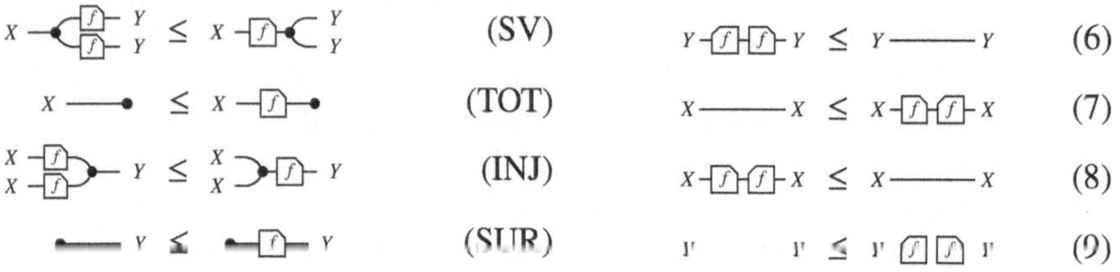

Lemma 1. *In a Cartesian bicategory, an arrow $f \colon X \to Y$ is single valued iff (6), it is total iff (7), it is injective iff (8) and it is surjective iff (9).*

Any Cartesian bicategory is self-dual compact closed and thus *traced*. Later on, to deal with *tests* in imperative programs we will use *coreflexives*, namely arrows $f \colon X \to X$ such that $f \leq id_X$. In Cartesian bicategories, they enjoy several useful properties:

Lemma 2. *In a Cartesian bicategory, the following hold:*

1. *f is a coreflexive iff f is transitive ($f; f \leq f$), symmetric ($f^\dagger \leq f$) and single valued.*
2. *Coreflexives $X \to X$ are in bijective correspondece with arrows $1 \to X$.*
3. *For all coreflexives $f, g \colon X \to X$, $f; g = f \sqcap g$; Moreover, for $f' \colon 1 \to X$ corresponding to f and $i \colon 1 \to X$, $i \sqcap f' = i; f$.*

4   Kleene Bicategories

Now, we introduce Kleene bicategories and show that their laws capture the complete axiomatisation of Kleene algebras from [41]. Recall that a category **C** is *enriched over join-semilattices* if every homset carries a join $\sqcup$ semilattice with bottom $\perp$ and composition ; distributes over it (see (10) and (11) in Table 2). A *(typed) Kleene algebra* [41,43] is a category enriched over join-semilattices equipped with a *Kleene star operator*, namely a family of operations $(\cdot)^* \colon \mathbf{C}[X, X] \to \mathbf{C}[X, X]$ such that for all $f \colon X \to X$, $r \colon X \to Y$ and $l \colon Y \to X$ the four laws in (12) hold.

$(f \sqcup g) \sqcup h = f \sqcup (g \sqcup h)$	$f \sqcup g = g \sqcup f$	$f \sqcup \bot = f$	$f \sqcup f = f$	(10)
$(f \sqcup g); h = (f; h \sqcup g; h)$	$h; (f \sqcup g) = (h; f \sqcup h; g)$	$f; \bot = \bot = \bot; f$		(11)
$id_X \sqcup f ; f^* \leq f^*$ $\quad id_X \sqcup f^* ; f \leq f^*$	$f ; r \leq r \implies f^* ; r \leq r$	$l ; f \leq l \implies l ; f^* \leq l$		(12)

Table 2: Axioms of (Typed) Kleene Algebras

Definition 2. *A* finite biproduct *(shortly, fb)* category with idempotent convolution *is a poset enriched symmetric monoidal category* $(\mathbf{C}, \oplus, 0)$ *and, for every object X in* $\mathbf{C}$*, a monoid* $(\triangleright_X, ¡_X)$ *and a comonoid* $(\triangleleft_X, !_X)$ *satisfying the usual coherence conditions s.t.:*
1. $(\triangleleft_X, !_X)$ *is right adjoint to* $(\triangleright_X, ¡_X)$*;*
2. arrows $f \colon X \to Y$ *are both monoid and comonoid morphisms:* $f; \triangleleft_Y = \triangleleft_X; (f \oplus f)$*,* $f; !_Y = !_X$*,* $\triangleright_Y; f = (f \oplus f); \triangleright_X$ *and* $¡_X; f = ¡_Y$*.*

Remark 1. A finite biproduct category is defined as above but without the poset enrichment and the adjointness condition: 0 is both final and initial object and $\oplus$ is both a categorical product and coproduct, i.e., a biproduct [27]. While any finite biproduct category is enriched over commutative monoids, the additional conditions in Definition 2 guarantee that this is enriched over join semilattices where $\sqcup$ and $\bot$ are defined for all objects X, Y and arrows $f, g \colon X \to Y$ as on the left below.

$f \sqcup g \stackrel{\text{def}}{=}$  $\quad \bot \stackrel{\text{def}}{=}$  $\quad f^* \stackrel{\text{def}}{=}$ 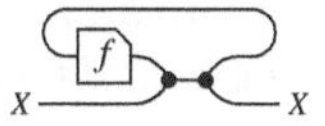 (13)

The reader can easily check that $(\mathbf{Rel}, \oplus, 0)$ with $\triangleleft_X$, $!_X$, $\triangleright_X$, $¡_X$ defined as in (4) is a finite biproduct category with idempotent convolution and that $\sqcup$ and $\bot$ give union and empty relation. Unfortunately, finite biproduct categories with idempotent convolution do not have enough structure to deal with $(\cdot)^*$: differently from Cartesian bicategories they are not necessarily traced. Such structure has to be explicitely added:

Definition 3. *A* Kleene bicategory $\mathbf{C}$ *is both a finite biproduct category with idempotent convolution and a poset enriched traced monoidal category such that*

1. for all objects X, the trace tr_X *satisfies the axiom* $\mathsf{tr}_X(\triangleright_X; \triangleleft_X) \leq id_X$*;*
2. the trace is posetal uniform: for all $f \colon S \oplus X \to S \oplus Y$ *and* $g \colon T \oplus X \to T \oplus Y$*,*
(AU1) if $\exists r \colon S \to T$ *such that* $f; (r \oplus id_Y) \leq (r \oplus id_X); g$*, then* $\mathsf{tr}_S f \leq \mathsf{tr}_S g$*;*
(AU2) if $\exists r \colon T \to S$ *such that* $(r \oplus id_X); f \leq g; (r \oplus id_Y)$*, then* $\mathsf{tr}_S f \leq \mathsf{tr}_S g$*.*

A morphism of Kleene bicategory *is a poset enriched symmetric monoidal functor preserving monoids, comonoids and traces.*

The laws in (AU1) and (AU2) are the posetal extension of the uniformity condition for traces (see e.g. [33]). To the best of our knowledge, they have never been studied. Instead, the axioms in 1 already appeared in the literature (see e.g. [51]). Like in any finite biproduct category with trace (see e.g. [20]), in a Kleene bicategory one can define for each endomorphism $f \colon X \to X$, a morphism $f^* \colon X \to X$ as in (13). The distinguishing property of Kleene bicategories is that $(\cdot)^*$ is a Kleene star operator: see (12). Viceversa, any Kleene star operation gives rise to a trace satisfying the laws of Kleene bicategories.

Theorem 1. *Let* **C** *be a fb category with idempotent convolution.* **C** *is a Kleene bicategory iff* **C** *has a Kleene-star operator.*

Corollary 1. *All Kleene bicategories are typed Kleene algebras.*

The opposite does not hold: not all Kleene algebras are monoidal categories. Nevertheless, from a Kleene algebra, one can canonically build a Kleene bicategory by means of the *matrix construction*, aka *biproduct completion* [23,48]. See [9, Sec. 6.3]

5 Rig Categories

We have seen that **Rel** carries two monoidal categories $(\mathbf{Rel}, \otimes, 1)$ and $(\mathbf{Rel}, \oplus, 0)$. The appropriate setting for studying their interaction is given by rig categories [46,37].

Definition 4. A rig category *is a category* **C** *with two symmetric monoidal structures* $(\mathbf{C}, \otimes, 1)$ *and* $(\mathbf{C}, \oplus, 0)$ *and natural isomorphisms*

$$\delta^l_{X,Y,Z} \colon X \otimes (Y \oplus Z) \to (X \otimes Y) \oplus (X \otimes Z) \qquad \lambda^\bullet_X \colon 0 \otimes X \to 0$$

$$\delta^r_{X,Y,Z} \colon (X \oplus Y) \otimes Z \to (X \otimes Z) \oplus (Y \otimes Z) \qquad \rho^\bullet_X \colon X \otimes 0 \to 0$$

satisfying certain coherence axioms [46]. A rig category is said to be right strict *when* $\lambda^\bullet, \rho^\bullet$ *and* δ^r *are all identity natural isomorphisms. A* right strict rig functor *is a strict symmetric monoidal functor for both* $\otimes$ *and* $\oplus$ *preserving* δ^l*.*

A sesquistrict rig category *is a functor* $H \colon \mathbf{S} \to \mathbf{C}$*, where* **S** *is a discrete category and* **C** *is a right strict rig category, such that for all* $A \in \mathbf{S}$

$$\delta^l_{H(A),X,Y} \colon H(A) \otimes (X \oplus Y) \to (H(A) \otimes X) \oplus (H(A) \otimes Y)$$

is an identity morphism. Given $H \colon \mathbf{S} \to \mathbf{C}$ *and* $H' \colon \mathbf{S}' \to \mathbf{C}'$ *two sesquistrict rig categories, a* sesquistrict rig functor *from* H *to* H' *is a pair* $(\alpha \colon \mathbf{S} \to \mathbf{S}', \beta \colon \mathbf{C} \to \mathbf{C}')$*, with* α *a functor and* β *a strict rig functor, such that* $\alpha; H' = H; \beta$*.*

Intuitively, a sesquistrict rig category is right strict and, partially (just for the selected class of objects **S**), left strict. For more details, we refer the reader to [11, Sec. 4].

From any rig category **C**, one can construct its (right) strictification $\overline{\mathbf{C}}$ [37] and then embed $ob(\mathbf{C})$, the discrete category of the objects of **C**, into $\overline{\mathbf{C}}$. The embedding $ob(\mathbf{C}) \to \overline{\mathbf{C}}$ forms a sesquistrict category and it is equivalent (as a rig category) to the original **C** [11, Corollary 4.5]. Through the paper, when dealing with a rig category **C**, we will often implicitly refer to the equivalent sesquistrict $ob(\mathbf{C}) \to \overline{\mathbf{C}}$.

Hereafter, we write $A, B, \dots$ for the elements of a fixed set $\mathcal{S}$ and $U, V, \dots$ for the words in $\mathcal{S}^\star$. Given two words $U, V \in \mathcal{S}^\star$, we write $U \otimes V$, shortly UV, for their concatenation and 1 for the empty word. We use $P, Q, \dots$ for words of words in $(\mathcal{S}^\star)^\star$ and we write $P \oplus Q$ for their concatenation and 0 for the empty word of words. Beyond $\oplus$, one can define $\otimes$ on $(\mathcal{S}^\star)^\star$: for all $P = U_0 \oplus \dots \oplus U_n$ and $Q = V_0 \oplus \dots \oplus V_m$

$$P \otimes Q \stackrel{\text{def}}{=} U_0V_0 \oplus \dots \oplus U_0V_n \oplus \dots \oplus U_nV_0 \oplus \dots \oplus U_nV_m \tag{14}$$

For instance, $(A \oplus B) \otimes (C \oplus D)$ is $(A \otimes C) \oplus (A \otimes D) \oplus (B \otimes C) \oplus (B \otimes D)$. One can readily see that elements in $(\mathcal{S}^\star)^\star$ are in bijective correspondence with non-commutative *polynomials* with variables in $\mathcal{S}$. Consequently, we will call *monomials* the words in $\mathcal{S}^\star$.

Given a set of sorts $\mathcal{S}$, a *monoidal signature* is a tuple $(\mathcal{S}, \Sigma, ar, coar)$ where ar and $coar$ assign to each symbol $s \in \Sigma$ an arity and a coarity in $\mathcal{S}^\star$. A *rig signature* is the same but with arity and coarity in $(\mathcal{S}^\star)^\star$. An *interpretation* $\mathcal{I}$ of a rig signature $(\mathcal{S}, \Sigma, ar, coar)$ in a sesquistrict rig category $H \colon \mathbf{M} \to \mathbf{D}$ is a pair of functions $(\alpha_{\mathcal{S}} \colon \mathcal{S} \to Ob(\mathbf{M}), \alpha_\Sigma \colon \Sigma \to Ar(\mathbf{D}))$ such that, for all $s \in \Sigma$, $\alpha_\Sigma(s)$ is an arrow having as domain and codomain $(\alpha_{\mathcal{S}}; H)^\sharp(ar(s))$ and $(\alpha_{\mathcal{S}}; H)^\sharp(coar(s))$. Here, $(\alpha_{\mathcal{S}}; H)^\sharp$ stands for inductive extension of $\alpha_{\mathcal{S}}; H \colon \mathcal{S} \to Ob(\mathbf{D})$ to $(\mathcal{S}^\star)^\star$.

Definition 5. *Let $(\mathcal{S}, \Sigma, ar, coar)$ (simply Σ for short) be a rig signature. A sesquistrict rig category $H \colon \mathbf{M} \to \mathbf{D}$ is said to be* freely generated *by Σ if there is an interpretation $(\alpha_{\mathcal{S}}, \alpha_\Sigma)$ of Σ in H such that for every sesquistrict rig category $H' \colon \mathbf{M}' \to \mathbf{D}'$ and every interpretation $(\alpha'_{\mathcal{S}} \colon \mathcal{S} \to Ob(\mathbf{M}'), \alpha'_\Sigma \colon \Sigma \to Ar(\mathbf{D}'))$ there exists a unique sesquistrict rig functor $(\alpha \colon \mathbf{M} \to \mathbf{M}', \beta \colon \mathbf{D} \to \mathbf{D}')$ such that $\alpha_{\mathcal{S}}; \alpha = \alpha'_{\mathcal{S}}$ and $\alpha_\Sigma; \beta = \alpha'_\Sigma$.*

6 Kleene-Cartesian Bicategories

We have seen that Cartesian bicategories provide enough structure for the allegorical fragment of CR, while Kleene bicategories for its Kleene fragment. For the whole CR, we make the Cartesian and Kleene bicategory structures interact as rig categories.

Definition 6. *A* Kleene-Cartesian rig category *(shortly kc rig) is a poset enriched rig category* **C** *such that*

1. $(\mathbf{C}, \oplus, 0)$ *is a Kleene bicategory;*
2. $(\mathbf{C}, \otimes, 1)$ *is a Cartesian bicategory;*
3. *the (co)monoids satisfy the following coherence conditions:*

$$\begin{array}{ll} \blacktriangleleft_{X \oplus Y} = (\blacktriangleleft_X \oplus \blacktriangleleft_Y); (id_{XX} \oplus \text{⫯}_{XY} \oplus \text{⫯}_{YX} \oplus id_{YY}); (\delta^{-l}_{X,X,Y} \oplus \delta^{-l}_{Y,X,Y}) & !_{X \oplus Y} = (!_X \oplus !_Y); \rhd_1 \\ \blacktriangleright_{X \oplus Y} = (\blacktriangleright_X \oplus \blacktriangleright_Y); (id_{XX} \oplus \text{⫰}_{XY} \oplus \text{⫰}_{YX} \oplus id_{YY}); (\delta^{l}_{X,X,Y} \oplus \delta^{l}_{Y,X,Y}) & ¡_{X \oplus Y} = \lhd_1; (¡_X \oplus ¡_Y) \end{array} \quad (15)$$

A morphism of kc rig-categories *is a poset enriched rig functor that is a morphism of both Kleene and Cartesian bicategories.*

The axioms in (15) rule the interaction of black and white (co)monoids. Recall that in Cartesian and Kleene bicategories, homsets carry respectively meet and join semilattices. In any kc rig category, they carry distributive lattices; also $(\cdot)^\dagger$ distributes over $\oplus$, while $(\cdot)^*$ only distributes laxly over $\otimes$. As expected $(\cdot^*)^\dagger = (\cdot^\dagger)^*$.

Proposition 2. *The laws in Table 3 hold in any kc rig category.*

Corollary 2. *Any kc rig category is a typed Kleene algebra with converse [17,8].*

We have already seen that $(\mathbf{Rel}, \oplus, 0)$ is a Kleene bicategory and $(\mathbf{Rel}, \otimes, 1)$ is a Cartesian bicategory. To conclude that **Rel** is a kc rig category, it is enough to check the coherence conditions of the two (co)monoids defined in (4).

$f \sqcap (g \sqcup h) = (f \sqcap g) \sqcup (f \sqcap h)$	$f \sqcap \bot = \bot$	$f \sqcup (g \sqcap h) = (f \sqcup g) \sqcap (f \sqcup h)$		$f \sqcup \top = \top$
$(f \otimes g)^* \leq f^* \otimes g^*$	$(f \oplus g)^\dagger = f^\dagger \oplus g^\dagger$	$(f \sqcap g)^* \leq f^* \sqcap g^*$	$(f \sqcup g)^\dagger = f^\dagger \sqcup g^\dagger$	$(f^\dagger)^* = (f^*)^\dagger$

Table 3: Derived laws in kc rig categories.

6.1 Kleene-Cartesian Tape Diagrams

We identify the kc rig category freely generated by a rig signature $(\mathcal{S}, \Sigma)$. This is described as *Kleene-Cartesian tape diagrams*, which provide a syntax for kc rig categories. These are strictly more expressive than the calculus of relations: thanks to the monoidal product $\otimes$, they can deal with n-ary function symbols, e.g. "add" in Equation (22), data flow and relational Hoare logic (Remark 3).

Thanks to Theorem 4.9 in [11], we can restrict, without loss of generality, to the simpler case where Σ is just a monoidal signature. Our work extends [11, Section 7] that identifies a freely generated *fb-cb rig category*, shortly a kc rig category without traces ($\oplus$ is just a fb category with idempotent convolution).

As explained in Section 5, we can consider the sesquistrict rig categories having as sets of objects $(\mathcal{S}^\star)^\star$. For arrows, consider the following two-layer grammar

$$\begin{array}{l} c ::= id_A \mid id_1 \mid \underline{s} \mid \sigma_{A,B} \mid c;c \mid c \otimes c \mid !_A \mid \blacktriangleleft_A \mid ¡_A \mid \blacktriangleright_A \\ \mathsf{t} ::= id_U \mid id_0 \mid \overline{\underline{c}} \mid \sigma^{\oplus}_{U,V} \mid \mathsf{t};\mathsf{t} \mid \mathsf{t} \oplus \mathsf{t} \mid \delta_U \mid \triangleleft_U \mid ?_U \mid \triangleright_U \mid \mathsf{tr}_U \mathsf{t} \end{array} \tag{16}$$

where $s \in \Sigma$, $A, B \in \mathcal{S}$ and $U, V \in \mathcal{S}^\star$. The terms of the first row, called *circuits*, intuitively represent arrows of a Cartesian bicategory. The terms of the second row, called *tapes*, represent arrows of a Kleene bicategory. As expected, we only consider those terms to which is possible to associate source and target objects: the reader may check the simple type system in [9, Table 1]. Constants and operations in (16) can be extended to arbitrary polynomials in $(\mathcal{S}^\star)^\star$. For instance, $\mathsf{tr}_P \mathsf{t}$ is inductively defined as $\mathsf{tr}_0(\mathsf{t}) \stackrel{\text{def}}{=} \mathsf{t}$ and $\mathsf{tr}_{U \oplus P}(\mathsf{t}) \stackrel{\text{def}}{=} \mathsf{tr}_P \mathsf{tr}_U(\mathsf{t})$. For the other definitions see [11].

Particularly interesting is the fact that one can define $\otimes$ on tapes: for $\mathsf{t}_1 : P \to Q$, $\mathsf{t}_2 : R \to S$, $\mathsf{t}_1 \otimes \mathsf{t}_2 \stackrel{\text{def}}{=} \mathsf{L}_P(\mathsf{t}_2); \mathsf{R}_S(\mathsf{t}_1)$ where $\mathsf{L}_P(\cdot)$, $\mathsf{R}_S(\cdot)$ are the left and right whiskerings that can be defined by extending [11, Def 5.7] with an extra case for the trace: $\mathsf{L}_U(\mathsf{tr}_V \mathsf{t}) \stackrel{\text{def}}{=} \mathsf{tr}_{UV} \mathsf{L}_U(\mathsf{t})$ and $\mathsf{R}_U(\mathsf{tr}_V \mathsf{t}) \stackrel{\text{def}}{=} \mathsf{tr}_{VU} \mathsf{R}_U(\mathsf{t})$. Left distributors $\delta^l_{P,Q,R} : P \otimes (Q \oplus R) \to (P \otimes Q) \oplus (P \otimes R)$ and $\otimes$-symmetries $\sigma^{\otimes}_{P,Q} : P \otimes Q \to Q \otimes P$ are defined as in [11, Table 4]. The (co)monoids of $\otimes$ as in [11, (20),(21)].

Next, we impose the laws of kc rig categories on tapes. However, this should be done carefully, in order to properly tackle the two uniformity laws (AU1) and (AU2) which are implications and not (in)equalities. Let $\mathbb{I}$ be a a set of pairs $(\mathsf{t}_1, \mathsf{t}_2)$ of tapes with the same domain and codomain. We define $\leq_{\mathbb{I}}$ to be the set generated by the following inference system (where $\mathfrak{t} \leq_{\mathbb{I}} \mathfrak{s}$ is a shorthand for $(\mathfrak{t}, \mathfrak{s}) \in \leq_{\mathbb{I}}$).

$$\frac{\mathfrak{t}_1 \,\mathbb{I}\, \mathfrak{t}_2}{\mathfrak{t}_1 \leq_{\mathbb{I}} \mathfrak{t}_2}(\mathbb{I}) \qquad \frac{-}{\mathfrak{t} \leq_{\mathbb{I}} \mathfrak{t}}(r) \qquad \frac{\mathfrak{t}_1 \leq_{\mathbb{I}} \mathfrak{t}_2 \quad \mathfrak{t}_2 \leq_{\mathbb{I}} \mathfrak{t}_3}{\mathfrak{t}_1 \leq_{\mathbb{I}} \mathfrak{t}_3}(t)$$

$$\frac{\mathfrak{t}_1 \leq_{\mathbb{I}} \mathfrak{t}_2 \quad \mathfrak{s}_1 \leq_{\mathbb{I}} \mathfrak{s}_2}{\mathfrak{t}_1;\mathfrak{s}_1 \leq_{\mathbb{I}} \mathfrak{t}_2;\mathfrak{s}_2}(;) \qquad \frac{\mathfrak{t}_1 \leq_{\mathbb{I}} \mathfrak{t}_2 \quad \mathfrak{s}_1 \leq_{\mathbb{I}} \mathfrak{s}_2}{\mathfrak{t}_1 \oplus \mathfrak{s}_1 \leq_{\mathbb{I}} \mathfrak{t}_2 \oplus \mathfrak{s}_2}(\oplus) \qquad \frac{\mathfrak{t}_1 \leq_{\mathbb{I}} \mathfrak{t}_2 \quad \mathfrak{s}_1 \leq_{\mathbb{I}} \mathfrak{s}_2}{\mathfrak{t}_1 \otimes \mathfrak{s}_1 \leq_{\mathbb{I}} \mathfrak{t}_2 \otimes \mathfrak{s}_2}(\otimes) \tag{17}$$

$$\frac{\mathfrak{s}_2 \leq_{\mathbb{I}} \mathfrak{s}_1 \quad \mathfrak{t}_1;(\mathfrak{s}_1 \oplus id) \leq_{\mathbb{I}} (\mathfrak{s}_2 \oplus id);\mathfrak{t}_2}{\mathsf{tr}_{S_1}\mathfrak{t}_1 \leq_{\mathbb{I}} \mathsf{tr}_{S_2}\mathfrak{t}_2}(u_r) \qquad \frac{\mathfrak{s}_2 \leq_{\mathbb{I}} \mathfrak{s}_1 \quad (\mathfrak{s}_1 \oplus id);\mathfrak{t}_1 \leq_{\mathbb{I}} \mathfrak{t}_2;(\mathfrak{s}_2 \oplus id)}{\mathsf{tr}_{S_1}\mathfrak{t}_1 \leq_{\mathbb{I}} \mathsf{tr}_{S_2}\mathfrak{t}_2}(u_l)$$

The first six laws ensure that $\leq_{\mathbb{I}}$ is a precongruence (w.r.t. ;, $\oplus$ and $\otimes$) containing $\mathbb{I}$. The last two rules force the uniformity laws: observe that, while in (AU1) and (AU2) the same arrow r occurs in both the left and the right- hand-side of the premises, here r is replaced by two different but related tapes $\mathfrak{s}_1$ and $\mathfrak{s}_2$. This technicality is needed to guarantee uniformity in the category resulting from the following construction.

We take $\mathbb{KC}$ to be the set of all pairs of tapes containing the axioms in Table 7[5] and define $\leq_{\mathbb{KC}}$ according to (17). We fix $\sim_{\mathbb{KC}} \stackrel{\text{def}}{=} \leq_{\mathbb{KC}} \cap \geq_{\mathbb{KC}}$. With these definitions we can construct the category of Kleene-Cartesian tapes $\mathbf{KCT}_\Sigma$: objects are polynomials in $(\mathcal{S}^\star)^\star$ with $\oplus$ and $\otimes$ defined as in (14); arrows are $\sim_{\mathbb{KC}}$-equivalence classes of tapes; every homset $\mathbf{KCT}_\Sigma[P, Q]$ is ordered by $\leq_{\mathbb{KC}}$. The construction of $\mathbf{KCT}_\Sigma$ gives rise to a sesquistrict kc rig category. More importantly, $\mathbf{KCT}_\Sigma$ is the freely generated one.

Theorem 2. *$\mathbf{KCT}_\Sigma$ is the free sesquistrict kc rig category generated by $(\mathcal{S}, \Sigma)$.*

In [11], it is shown that a key feature of tapes is that they can be drawn nicely in 2 dimensions despite representing arrows of rig categories. Indeed, both circuits and tapes can be drawn as string diagrams. Note however that *inside* tapes, there are string diagrams. Thus, the grammar in (16) can be graphically rendered as follows.

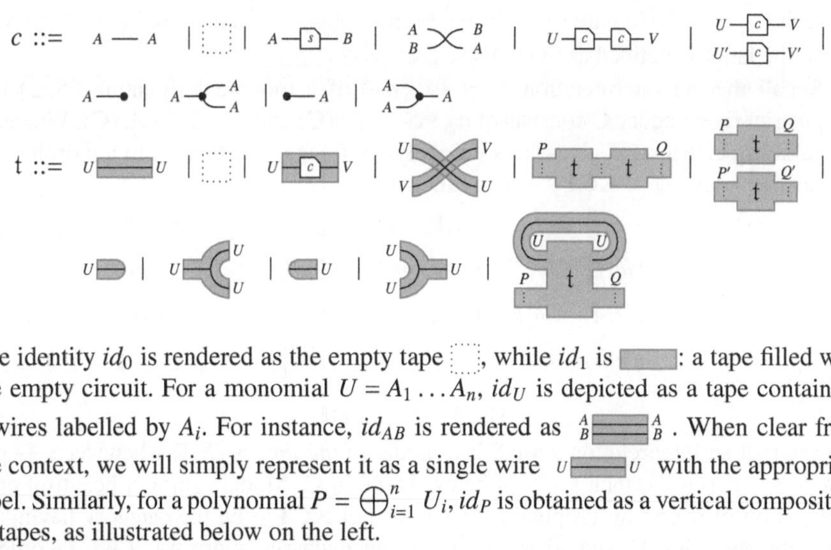

The identity id_0 is rendered as the empty tape , while id_1 is 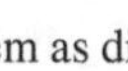: a tape filled with the empty circuit. For a monomial $U = A_1 \dots A_n$, id_U is depicted as a tape containing n wires labelled by A_i. For instance, id_{AB} is rendered as . When clear from the context, we will simply represent it as a single wire with the appropriate label. Similarly, for a polynomial $P = \bigoplus_{i=1}^n U_i$, id_P is obtained as a vertical composition of tapes, as illustrated below on the left.

[5] These rules are at the end of the paper, since we will soon draw them as diagrams.

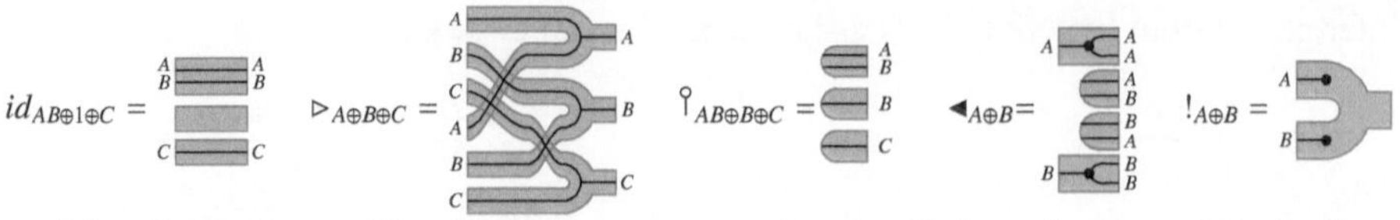

The diagonal $\lhd_U : U \to U \oplus U$ is represented as a splitting of tapes, while the bang $\delta_U : U \to 0$ is a tape closed on its right boundary. Codiagonals and cobangs are represented in the same way but mirrored along the y-axis. Exploiting the usual coherence conditions e.g. in [55, Table 4.7], we can construct (co)diagonals and (co)bangs for arbitrary polynomials. For example, $\rhd_{A \oplus B \oplus C}$ and $\text{ꝗ}_{AB \oplus B \oplus C}$ are depicted as the second and third diagrams above.

The copier $\blacktriangleleft_A : A \to A \otimes A$ is represented as a splitting of wires, while the discharger $!_A : A \to 1$ is a wire closed on the right. From the coherence laws in (15), one can build (co)copiers and (co)dischargers for arbitrary polynomials. For instance, $\blacktriangleleft_{A \oplus B} : A \oplus B \to (A \oplus B) \otimes (A \oplus B) = AA \oplus AB \oplus BA \oplus BB$ and $!_{A \oplus B} : A \oplus B \to 1$ are drawn as the last two diagrams above. For an arbitrary tape diagram $\mathsf{t} : P \to Q$ we write 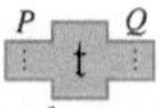.

The graphical representation embodies several axioms such as those of monoidal categories and several axioms for traces. Those axioms which are not implicit in the graphical representation are illustrated in Figures 2 and 3. Figure 4 illustrates the uniformity laws in the form of tape diagrams.

6.2 Kleene-Cartesian Theories and Functorial Semantics

A *Kleene-Cartesian theory*, shortly kc theory, is a pair $(\Sigma, \mathbb{I})$ where Σ is a monoidal signature and $\mathbb{I}$ is a set of pairs $(\mathsf{t}_1, \mathsf{t}_2)$ of tapes with same domain and codomain. Hereafter, we think of each pair $(\mathsf{t}_1, \mathsf{t}_2)$ as an inequation $\mathsf{t}_1 \leq \mathsf{t}_2$, but the results that we develop in this section trivially hold also for equations: it is enough to add in $\mathbb{I}$ a pair $(\mathsf{t}_2, \mathsf{t}_1)$ for each $(\mathsf{t}_1, \mathsf{t}_2) \in \mathbb{I}$. Hereafter we always keep implicit $\mathbb{KC}$ and we write $\leq_{\mathbb{I}}$ for $\leq_{\mathbb{KC} \cup \mathbb{I}}$ where the latter is defined as in (17). We fix $\sim_{\mathbb{I}} \stackrel{\text{def}}{=} \leq_{\mathbb{I}} \cap \geq_{\mathbb{I}}$.

Recall that an interpretation $\mathcal{I} = (\alpha_{\mathcal{S}}, \alpha_{\Sigma})$ of a monoidal signature $(\mathcal{S}, \Sigma)$ in a sesquistrict rig category $\mathbf{C}$ consists of $\alpha_{\mathcal{S}} : \mathcal{S} \to Ob(\mathbf{C})$ and $\alpha_{\Sigma} : \Sigma \to Ar(\mathbf{C})$. Whenever $\mathbf{C}$ is a kc rig category, $\mathcal{I}$ gives rises uniquely, by freeness of $\mathbf{KCT}_{\Sigma}$, to a morphism of kc rig categories $[\![\cdot]\!]_{\mathcal{I}} : \mathbf{KCT}_{\Sigma} \to \mathbf{C}$ defined as:

$$
\begin{array}{lllll}
[\![s]\!]_{\mathcal{I}} = \alpha_{\Sigma}(s) & [\![id_A]\!]_{\mathcal{I}} = id_{\alpha_{\mathcal{S}}(A)} & [\![\blacktriangleleft_A]\!]_{\mathcal{I}} = \blacktriangleleft_{\alpha_{\mathcal{S}}(A)} & [\![!_A]\!]_{\mathcal{I}} = !_{\alpha_{\mathcal{S}}(A)} & [\![c;d]\!]_{\mathcal{I}} = [\![c]\!]_{\mathcal{I}} ; [\![d]\!]_{\mathcal{I}} \\
\left[\!\left[\sigma^{\otimes}_{A,B}\right]\!\right]_{\mathcal{I}} = \sigma^{\otimes}_{\alpha_{\mathcal{S}}(A),\alpha_{\mathcal{S}}(B)} & [\![id_1]\!]_{\mathcal{I}} = id_1 & [\![\blacktriangleright_A]\!]_{\mathcal{I}} = \blacktriangleright_{\alpha_{\mathcal{S}}(A)} & [\![\text{i}_A]\!]_{\mathcal{I}} = \text{i}_{\alpha_{\mathcal{S}}(A)} & [\![c \otimes d]\!]_{\mathcal{I}} = [\![c]\!]_{\mathcal{I}} \otimes [\![d]\!]_{\mathcal{I}} \\
[\![\boxed{c}]\!]_{\mathcal{I}} = [\![c]\!]_{\mathcal{I}} & [\![id_U]\!]_{\mathcal{I}} = id_{\alpha^{\sharp}_{\mathcal{S}}(U)} & [\![\lhd_U]\!]_{\mathcal{I}} = \lhd_{\alpha^{\sharp}_{\mathcal{S}}(U)} & \left[\!\left[\delta_U\right]\!\right]_{\mathcal{I}} = \delta_{\alpha^{\sharp}_{\mathcal{S}}(U)} & [\![\mathfrak{s};\mathsf{t}]\!]_{\mathcal{I}} = [\![\mathfrak{s}]\!]_{\mathcal{I}} ; [\![\mathsf{t}]\!]_{\mathcal{I}} \\
\left[\!\left[\sigma^{\oplus}_{U,V}\right]\!\right]_{\mathcal{I}} = \sigma^{\oplus}_{\alpha^{\sharp}_{\mathcal{S}}(U),\alpha^{\sharp}_{\mathcal{S}}(V)} & [\![id_0]\!]_{\mathcal{I}} = id_0 & [\![\rhd_U]\!]_{\mathcal{I}} = \rhd_{\alpha^{\sharp}_{\mathcal{S}}(U)} & \left[\!\left[\text{ꝗ}_U\right]\!\right]_{\mathcal{I}} = \text{ꝗ}_{\alpha^{\sharp}_{\mathcal{S}}(U)} & [\![\mathfrak{s} \oplus \mathsf{t}]\!]_{\mathcal{I}} = [\![\mathfrak{s}]\!]_{\mathcal{I}} \oplus [\![\mathsf{t}]\!]_{\mathcal{I}}
\end{array}
$$

$$
[\![\mathsf{tr}_U \mathsf{t}]\!]_{\mathcal{I}} = \mathsf{tr}_{\alpha^{\sharp}_{\mathcal{S}}(U)} [\![\mathsf{t}]\!]_{\mathcal{I}}
$$

We say that an intepretation $\mathcal{I}$ of Σ is *a model of the theory* $(\Sigma, \mathbb{I})$ whenever $[\![\cdot]\!]_{\mathcal{I}}$ preserves $\leq_{\mathbb{I}}$: if $\mathsf{t}_1 \leq_{\mathbb{I}} \mathsf{t}_2$, then $[\![\mathsf{t}_1]\!]_{\mathcal{I}}$ is below $[\![\mathsf{t}_2]\!]_{\mathcal{I}}$ in $\mathbf{C}$. Models enjoy a beautiful characterisation provided by Proposition 3 below. Let $\mathbf{KCT}_{\Sigma,\mathbb{I}}$ be the category having the same objects as $\mathbf{KCT}_{\Sigma}$ and arrows $\sim_{\mathbb{I}}$-equivalence classes of arrows of $\mathbf{KCT}_{\Sigma}$ ordered by $\leq_{\mathbb{I}}$. Since $\mathbf{KCT}_{\Sigma}$ is a kc rig category, then also $\mathbf{KCT}_{\Sigma,\mathbb{I}}$ is so.

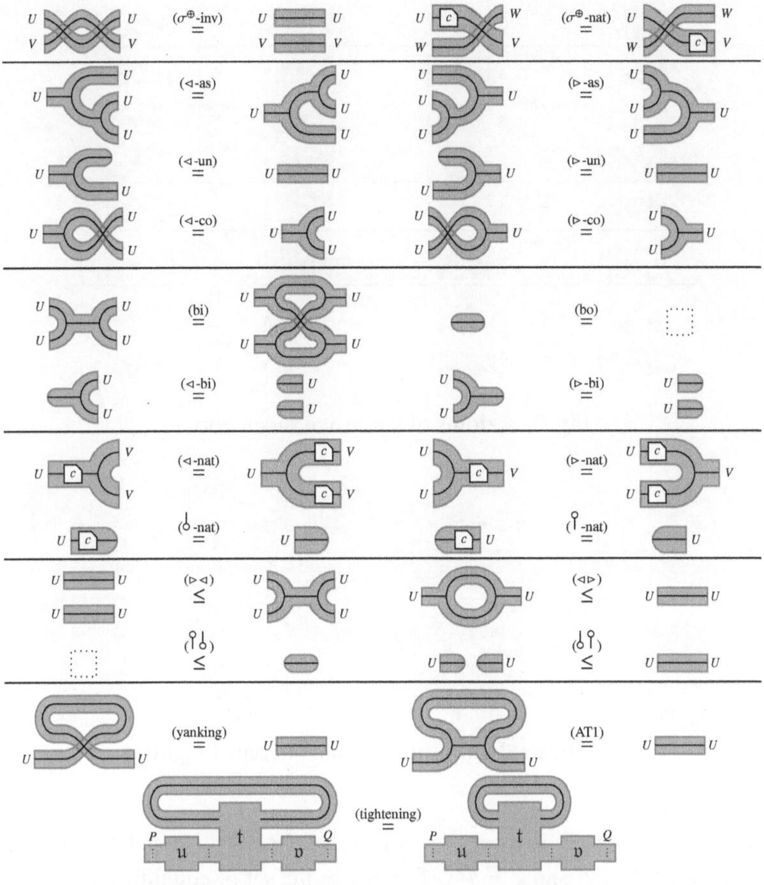

Fig. 2: Tape axioms for Kleene bicategory

Proposition 3. *Let $(\Sigma, \mathbb{I})$ be a kc tape theory and* **C** *a sesquistrict kc rig category. Models of $(\Sigma, \mathbb{I})$ are in bijective correspondence with morphisms of sesquistrict kc rig categories from* $\mathbf{KCT}_{\Sigma,\mathbb{I}}$ *to* **C**.

Example 1 (Functions). Let $\mathcal{S}$ be a set of sorts and $\Sigma \stackrel{\text{def}}{=} \{f : U \to A\}$ for some $A \in \mathcal{S}$ and $U \in S^{\star}$. Let $\mathbb{I}$ be the set of the two equalities in (18). An interpretation $\mathcal{I}$ of $(\mathcal{S}, \Sigma)$ in **Rel**, consists of a set $\alpha_{\mathcal{S}}(A_i)$ for each $A_i \in \mathcal{S}$ and a relation $\alpha_{\Sigma}(f) \subseteq \alpha_{\mathcal{S}}^{\sharp}(U) \times \alpha_{\mathcal{S}}(A)$. $\mathcal{I}$ is a model of $(\Sigma, \mathbb{I})$ iff $\alpha_{\Sigma}(f)$ is single valued (SV) and total (TOT), i.e., a function.

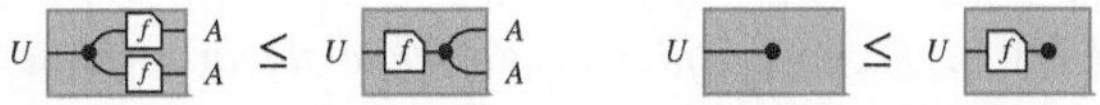 (18)

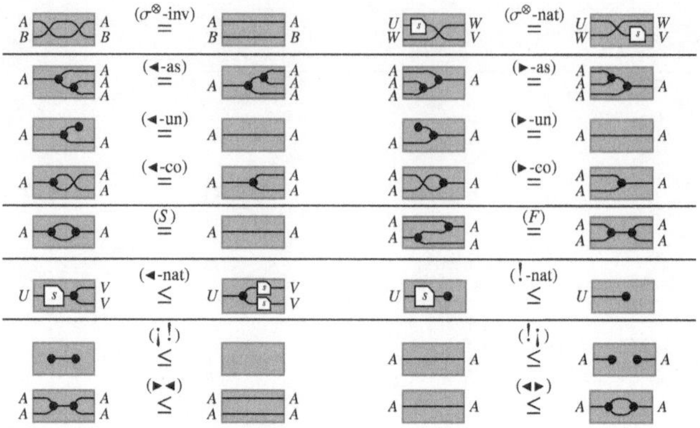

Fig. 3: Axioms of Cartesian bicategories

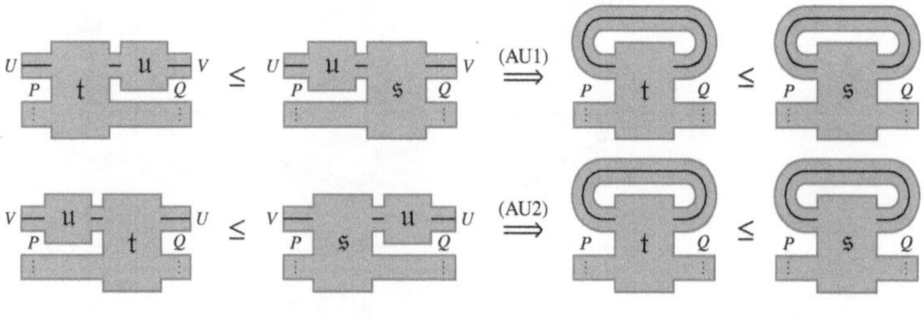

Fig. 4: Posetal uniformity axioms in tape diagrams.

Example 2 (KAT). Let $\mathcal{P}$ be a set of predicate symbols $R\colon U \to 1$ for $U \in \mathcal{S}^\star$. Take $\bar{\mathcal{P}} \stackrel{\text{def}}{=} \{\bar{R}\colon U \to 1 \mid R \in \mathcal{P}\}$ and $\Sigma \stackrel{\text{def}}{=} \mathcal{P} \cup \bar{\mathcal{P}}$. Let $\mathbb{I}$ be the set of equalities below.

(19)

An interpretation $\mathcal{I}$ of $(\mathcal{S},\Sigma)$ in **Rel** is a set $\alpha_{\mathcal{S}}(A_i)$ for each $A_i \in \mathcal{S}$ together with predicates $R \subseteq X \times 1$ and $\bar{R} \subseteq X \times 1$ for $X \stackrel{\text{def}}{=} \alpha^\sharp_{\mathcal{S}}(U)$. $\mathcal{I}$ is a model iff, for all R, $\bar{R}$ is its set-theoretic complement: by (13) and (5), the equalities above asserts that $R \cup \bar{R} = X$ and $R \cap \bar{R} = \{\}$. By Proposition 2, $\mathbf{KCT}_{\Sigma,\mathbb{I}}[U, 1]$ carries a distributive lattice and even a Boolean algebra when defining $\neg P$, for all $P\colon U \to 1$ as follows.

$$\neg R \stackrel{\text{def}}{=} \bar{R} \quad \neg\top \stackrel{\text{def}}{=} \bot \quad \neg(P \sqcup Q) \stackrel{\text{def}}{=} \neg P \sqcap \neg Q \quad \neg\bar{R} \stackrel{\text{def}}{=} R \quad \neg\bot \stackrel{\text{def}}{=} \top \quad \neg(P \sqcap Q) \stackrel{\text{def}}{=} \neg P \sqcup \neg Q \tag{20}$$

Consider the set C that contains, for each $P\colon U \to 1$, the associated coreflexive:

$c(P) \stackrel{\text{def}}{=}$ 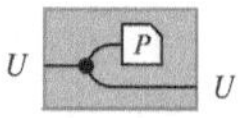 (drawn as ) (21)

Again coreflexives form a Boolean algebra $(C, \sqcup, ;, \neg, \bot, id)$ where composition ; acts as $\sqcap$: see Lemma 2.3. Moreover, by Corollary 1, $\mathbf{KCT}_{\Sigma,\mathbb{I}}[U, U]$ is a Kleene algebra. Thus $(\mathbf{KCT}_{\Sigma,\mathbb{I}}[U, U], C, \sqcup, ;, (\cdot)^*, \bot, id, \neg)$ is a *Kleene algebra with tests* [42].

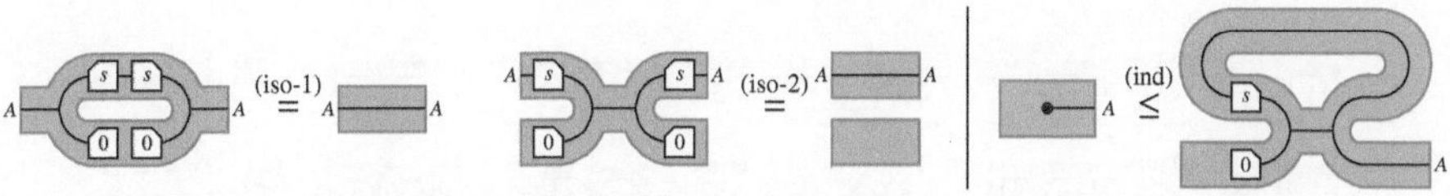

Fig. 5: The Kleene-Cartesian theory of Peano.

Remark 2. Recall CR from Section 2. The set Σ of generating symbols corresponds to a monoidal signature where $\mathcal{S} = \{A\}$ and each symbol has both arity and coarity A. A relational interpretation of CR is exactly an interpretation $\mathcal{I}$ of the monoidal signature $(\mathcal{S}, \Sigma)$ into **Rel**. One can define an inductive encoding $\mathcal{E}(\cdot)$ from CR to $\mathbf{KCT}_\Sigma$ by using (5) and (13). A simple inductive argument confirms that, for all interpretations $\mathcal{I}$ and expressions $E \in$ CR, $\langle E \rangle_{\mathcal{I}} = [\![\mathcal{E}(E)]\!]_{\mathcal{I}}$. Thus, if $\mathcal{E}(E_1) \leq_{\mathbb{KC}} \mathcal{E}(E_2)$, then $E_1 \leq_{\mathsf{CR}} E_2$.

7 The Kleene-Cartesian Theory of Peano

In this section we illustrate a further example of kc theory: Peano's axiomatisation of natural numbers. Recall that such axiomatisation is not a first order theory [32].

We commence by fixing the signature: $\mathcal{S} \stackrel{\text{def}}{=} \{A\}$ and $\Sigma \stackrel{\text{def}}{=} \{$ $\boxed{0}\!-\!A$, $A\!-\!\boxed{s}\!-\!A$ $\}$. An interpretation of Σ in **Rel** consists of a set X (i.e., $\alpha_{\mathcal{S}}(A)$), a relation $0 \subseteq 1 \times X$ (i.e., $\alpha_\Sigma($ $\boxed{0}\!-\!A$ $)$) and a relation $s \subseteq X \times X$ (i.e., $\alpha_\Sigma($ $A\!-\!\boxed{s}\!-\!A$ $)$). The set of axioms $\mathbb{P}$ consists of those in Figure 5.

From a universal-algebraic perspective, natural numbers are the smallest set X such that X is isomorphic to $X \oplus 1$. The two leftmost axioms in Figure 5 force $[s, 0] \stackrel{\text{def}}{=} (s \oplus 0); \triangleright_X$ to be an isomorphism of type $X \oplus 1 \to X$: (iso-1) states that $[s, 0]^\dagger ; [s, 0] = id_X$; (iso-2) that $[s, 0] \,;\, [s, 0]^\dagger = id_{X \oplus 1}$; the rightmost the fact that it is the smallest: i.e. $X \subseteq 0 \,;\, s^* \colon 1 \to X$. One can similarly obtain any algebraic data type.

We illustrate that $(\Sigma, \mathbb{P})$ is equivalent to Peano's axiomatisation of natural numbers. Possibly, the most interesting axiom is the principle of induction: (ind-princ) in Figure 6. This follows easily from posetal uniformity and (ind).

Theorem 3 (Principle of Induction). *Let* $(\Pi, \mathbb{Q})$ *be a kc theory, such that* $\Sigma \subseteq \Pi$ *and* $\mathbb{P} \subseteq \mathbb{Q}$. *For all* $P \colon 1 \to A$ *in* $\mathbf{KCT}_{\Pi,\mathbb{Q}}$, (ind-princ) *in Figure 6 holds*

Proof. Observe that the following holds:

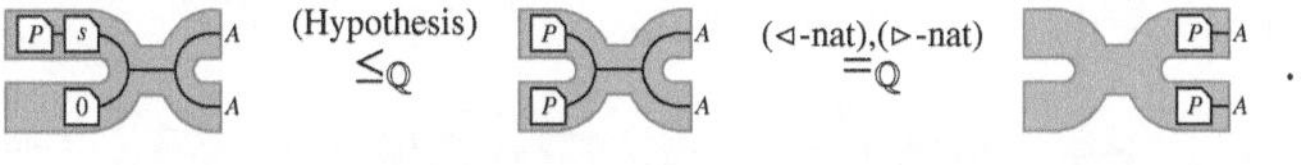

Thus, by (AU2) the inclusion below holds and the derivation concludes the proof.

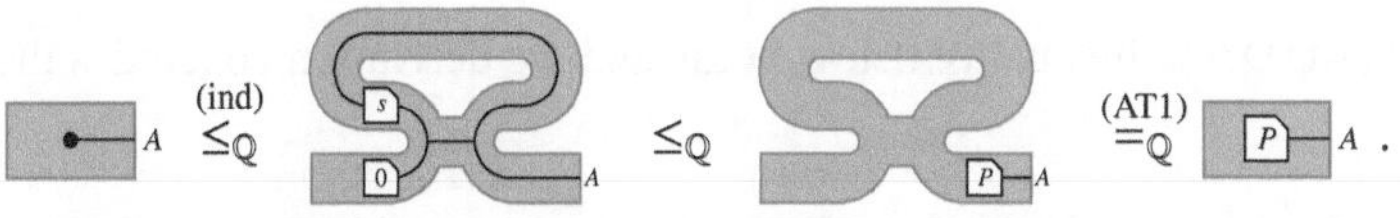

The other Peano's axioms state that 0 is a natural number, s is an injective function and that 0 is *not* the successor of any natural number. These are illustrated by means of

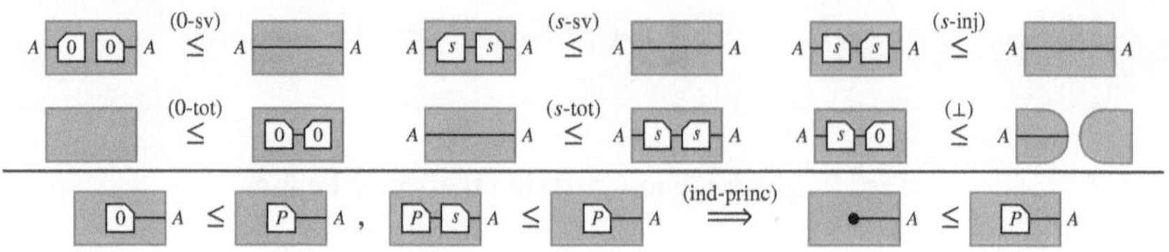

Fig. 6: Peano's theory of the natural numbers.

tapes in Figure 6, where we use the characterisation of total, single valued and injective relations provided by Lemma 1. Observe that $(\bot)$ states that $\{x \in X \mid (x, 0) \in s\} \subseteq \emptyset$.

Lemma 3. *The laws in Figures 5 and 6 are equivalent.*

In [25], Dedekind showed that any two models of Peano's axioms are isomorphic, and thus any model of $(\Sigma, \mathbb{P})$ is isomorphic to the one on natural numbers. To give to the reader a taste of how one can program with tapes, we now illustrate how to start to encode arithmetic within $(\Sigma, \mathbb{P})$. The tape for addition is illustrated on the right. Such tape can be thought as a simple imperative program:

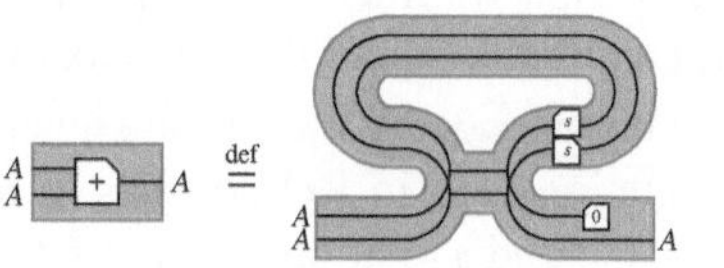 (22)

```
add(x,y) = while (x>0) { x:=x-1; y:=y+1 }; return y
```

The variable **x** corresponds to the top wire in (22), while y to the bottom one. At any iteration, the program checks whether **x** is 0, in which case it returns y, or the successor of some number, in which case **x** takes such number, while y takes its own successor.

In [9, Lemma 10.3] we show that (22) satisfies the usual inductive definition of addition, namely:

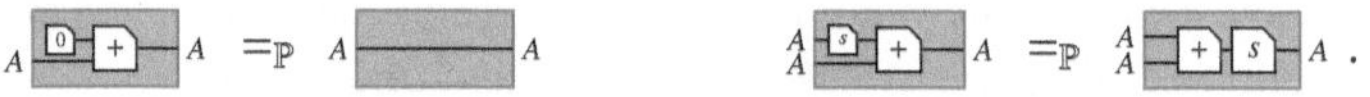

While, it is straightforward that 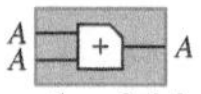 terminates on all possible inputs, it is interesting to see how this can be proved within the kc theory $(\Sigma, \mathbb{P})$.

Lemma 4. *The tape* 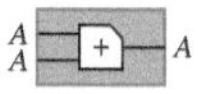 *is total, i.e.* 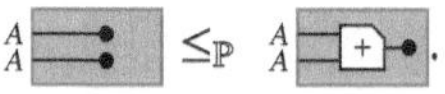.

Proof. First observe that the following holds:

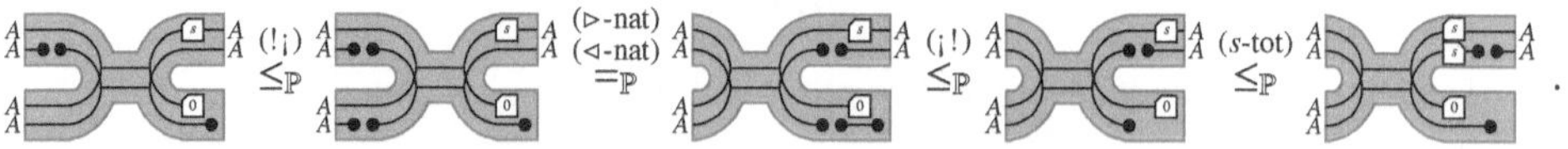

Then, by (AU1), the inequality below holds and the derivation concludes the proof.

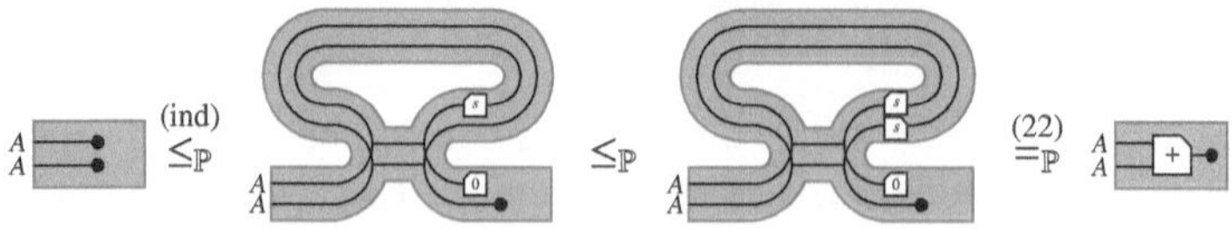

$$\frac{-}{\Gamma, x\colon A, \Delta \vdash x\colon A}\ (\textsc{var}) \qquad \frac{\Gamma \vdash e_i\colon A_i \quad f\colon A_1 \otimes \cdots \otimes A_n \to A}{\Gamma \vdash f(e_1,\ldots,e_n)\colon A}\ (\textsc{op})$$

$$\frac{}{\Gamma \vdash \top\colon 1}\ (\textsc{top}) \qquad \frac{\Gamma \vdash P\colon 1 \quad \Gamma \vdash Q\colon 1}{\Gamma \vdash (P \wedge Q)\colon 1}\ (\textsc{and}) \qquad \frac{\Gamma \vdash e_i\colon A_i \quad R\colon A_1 \otimes \ldots \otimes A_n \to 1}{\Gamma \vdash \bar{R}(e_1,\ldots e_n)\colon 1}\ (\bar{\mathrm{R}})$$

$$\frac{}{\Gamma \vdash \bot\colon 1}\ (\textsc{bot}) \qquad \frac{\Gamma \vdash P\colon 1 \quad \Gamma \vdash Q\colon 1}{\Gamma \vdash (P \vee Q)\colon 1}\ (\textsc{or}) \qquad \frac{\Gamma \vdash e_i\colon A_i \quad R\colon A_1 \otimes \ldots \otimes A_n \to 1}{\Gamma \vdash R(e_1,\ldots e_n)\colon 1}\ (\mathrm{R})$$

$$\frac{}{\Gamma \vdash \mathsf{abort}}\ (\textsc{abort}) \qquad \frac{}{\Gamma \vdash \mathsf{skip}}\ (\textsc{skip}) \qquad \frac{\Gamma = \Gamma', x\colon A, \Delta' \qquad \Gamma \vdash e\colon A}{\Gamma \vdash x := e}\ (\textsc{assn})$$

$$\frac{\Gamma \vdash C \quad \Gamma \vdash D}{\Gamma \vdash C; D}\ (;) \qquad \frac{\Gamma \vdash P\colon 1 \quad \Gamma \vdash C}{\Gamma \vdash \mathsf{while}\, P\, \mathsf{do}\, C}\ (\textsc{while}) \qquad \frac{\Gamma \vdash P\colon 1 \quad \Gamma \vdash C \quad \Gamma \vdash D}{\Gamma \vdash \mathsf{if}\, P\, \mathsf{then}\, C\, \mathsf{else}\, D}\ (\textsc{if})$$

Table 4: Type system for expressions, predicates and commands.

8 Diagrammatic Hoare Logic

We now illustrate how tape diagrams can provide a comfortable setting to reason about imperative programs. For the sake of generality, we avoid to fix basic types and operations and, rather, we work parametrically with respect to a triple $(\mathcal{S}, \mathcal{F}, \mathcal{P})$: $\mathcal{S}$ is a set of sorts, representing basic types; $\mathcal{F}$ is a set of function symbols, equipped with an arity in $\mathcal{S}^*$ and a coarity in $\mathcal{S}$; $\mathcal{P}$ is a set of predicate symbols equipped just with an arity in $\mathcal{S}^*$. The coarity of predicates is fixed to be 1.

We take the signature $\Sigma \stackrel{\text{def}}{=} \mathcal{F} \cup \mathcal{P} \cup \bar{\mathcal{P}}$ where $\bar{\mathcal{P}}$ is as in Example 2. The set $\mathbb{I}$ contains, for all $f\colon U \to A$ in $\mathcal{F}$, the axioms in (18) and, for each $R\colon U \to 1$ in $\mathcal{P}$, those in (19). One may add to $\mathbb{I}$ other axioms, e.g., those of $\mathbb{P}$ in Section 7.

We consider terms generated by the following grammar

$$\begin{aligned} e &::= x \mid f(e_1,\ldots,e_n) \\ P &::= R(e_1,\ldots,e_n) \mid \bar{R}(e_1,\ldots,e_n) \mid \top \mid \bot \mid P \vee P \mid P \wedge P \\ C &::= \mathsf{abort} \mid \mathsf{skip} \mid \mathsf{if}\, P\, \mathsf{then}\, C\, \mathsf{else}\, D \mid \mathsf{while}\, P\, \mathsf{do}\, C \mid C; D \mid x := e \end{aligned}$$

where $f \in \mathcal{F}$, $R \in \mathcal{P}$ and x is taken from a fixed set of variables. As usual, e are expressions, P predicates and C commands. Negation of predicates can be expressed as in (20). In order to encode terms into diagrams, we need to make copying and discarding of variables explicit; we thus define a simple type systems with judgement of the form

$$\Gamma \vdash e\colon A \qquad \Gamma \vdash P\colon 1 \qquad \Gamma \vdash C$$

where A is a sort in $\mathcal{S}$ and Γ is a *typing context*, i.e., an ordered sequence $x_1 : A_1, \ldots x_n : A_n$, where all the x_i are distinct variables and $A_i \in \mathcal{S}$. The type system is in Table 4.

The encoding of terms into diagrams, presented in Table 5, is defined inductively on the typing rules. For instance, in the encoding of the assignment $\Gamma \vdash x := e$, the context is $\Gamma = \Gamma', x\colon A, \Delta'$, according to the typing rule (Assn).

The encoding $\mathcal{E}(\cdot)$ maps well typed expressions $\Gamma \vdash e\colon A$, predicates $\Gamma \vdash P\colon 1$ and commands $\Gamma \vdash C$ into, respectively, diagrams of the following types

$$\mathcal{E}(\Gamma) \to A \qquad \mathcal{E}(\Gamma) \to 1 \qquad \mathcal{E}(\Gamma) \to \mathcal{E}(\Gamma)$$

where for $\Gamma = x_1 : A_1, \ldots, x_n : A_n$, we fix $\mathcal{E}(\Gamma) \stackrel{\text{def}}{=} A_1 \otimes \ldots \otimes A_n$. In the encoding of expressions and predicates, we use $\blacktriangleleft^n_{\mathcal{E}(\Gamma)}$ to copy n times the content of variables in

$$
\begin{aligned}
\mathcal{E}(\Gamma', x\colon A, \Delta' \vdash x\colon A) &\stackrel{\text{def}}{=} !_{\mathcal{E}(\Gamma)} \otimes id_A \otimes !_{\mathcal{E}(\Delta)} \\
\mathcal{E}(\Gamma \vdash f(e_1, \dots, e_n)\colon A) &\stackrel{\text{def}}{=} \blacktriangleleft^n_{\mathcal{E}(\Gamma)}; (\mathcal{E}(\Gamma \vdash e_1) \otimes \dots \otimes \mathcal{E}(\Gamma \vdash e_n)); f \\
\hline
\mathcal{E}(\Gamma \vdash R(e_1, \dots e_n)\colon 1) &\stackrel{\text{def}}{=} \blacktriangleleft^n_{\mathcal{E}(\Gamma)}; (\mathcal{E}(\Gamma \vdash e_1\colon 1) \otimes \dots \otimes \mathcal{E}(\Gamma \vdash e_n\colon 1)); R \\
\mathcal{E}(\Gamma \vdash \bar{R}(e_1, \dots e_n)\colon 1) &\stackrel{\text{def}}{=} \blacktriangleleft^n_{\mathcal{E}(\Gamma)}; (\mathcal{E}(\Gamma \vdash e_1\colon 1) \otimes \dots \otimes \mathcal{E}(\Gamma \vdash e_n\colon 1)); \bar{R} \\
\mathcal{E}(\Gamma \vdash \top\colon 1) &\stackrel{\text{def}}{=} !_{\mathcal{E}(\Gamma)} \qquad \mathcal{E}(\Gamma \vdash \bot\colon 1) \stackrel{\text{def}}{=} ¡_{\mathcal{E}(\Gamma)} ; ¡_1 \\
\mathcal{E}(\Gamma \vdash P \vee Q\colon 1) &\stackrel{\text{def}}{=} \lhd_{\mathcal{E}(\Gamma)}; (\mathcal{E}(\Gamma \vdash P\colon 1) \oplus \mathcal{E}(\Gamma \vdash P\colon 1)); \rhd_1 \\
\mathcal{E}(\Gamma \vdash P \wedge Q\colon 1) &\stackrel{\text{def}}{=} \blacktriangleleft_{\mathcal{E}(\Gamma)}; (\mathcal{E}(\Gamma \vdash P\colon 1) \otimes \mathcal{E}(\Gamma \vdash P\colon 1)) \\
\hline
\mathcal{E}(\Gamma \vdash \mathsf{abort}) &\stackrel{\text{def}}{=} \bot \qquad \mathcal{E}(\Gamma \vdash \mathsf{skip}) \stackrel{\text{def}}{=} id_{\mathcal{E}(\Gamma)} \\
\mathcal{E}(\Gamma \vdash C; D) &\stackrel{\text{def}}{=} \mathcal{E}(\Gamma \vdash C); \mathcal{E}(\Gamma \vdash D) \\
\mathcal{E}(\Gamma \vdash \mathsf{if}\, P\, \mathsf{then}\, C\, \mathsf{else}\, D) &\stackrel{\text{def}}{=} (c(\mathcal{E}(\Gamma \vdash P\colon 1); \mathcal{E}(\Gamma \vdash C)) \sqcup (c(\mathcal{E}(\Gamma \vdash \neg P)); \mathcal{E}(\Gamma \vdash D\colon 1)) \\
\mathcal{E}(\Gamma \vdash \mathsf{while}\, P\, \mathsf{do}\, C) &\stackrel{\text{def}}{=} (c(\mathcal{E}(\Gamma \vdash P\colon 1)); \mathcal{E}(\Gamma \vdash C))^*; c(\mathcal{E}(\Gamma \vdash \neg P\colon 1)) \\
\mathcal{E}(\Gamma \vdash x := e) &\stackrel{\text{def}}{=} (\blacktriangleleft_{\mathcal{E}(\Gamma')} \otimes id_A \otimes \blacktriangleleft_{\mathcal{E}(\Delta')}); (id_{\mathcal{E}(\Gamma')} \otimes \mathcal{E}(\Gamma \vdash e\colon A) \otimes id_{\mathcal{E}(\Delta')})
\end{aligned}
$$

Table 5: Encoding of expressions, predicates and commands into diagrams.

$$
\frac{}{\{P\}\mathsf{skip}\{P\}}\ (\mathsf{skip}) \qquad \frac{}{\{P[e/x]\}x := e\{P\}}\ (\mathsf{assn}) \qquad \frac{P_1 \subseteq P_2 \quad \{P_2\}C\{Q_2\} \quad Q_2 \subseteq Q_1}{\{P_1\}C\{Q_1\}}\ (\subseteq)
$$

$$
\frac{\{P\}C\{Q\} \quad \{Q\}D\{R\}}{\{P\}C; D\{R\}}\ (\mathsf{seq}) \qquad \frac{\{P \wedge B\}C\{Q\} \quad \{P \wedge \neg B\}D\{Q\}}{\{P\}\,\mathsf{if}\, B\, \mathsf{then}\, C\, \mathsf{else}\, D\{Q\}}\ (\mathsf{if}) \qquad \frac{\{P \wedge B\}C\{P\}}{\{P\}\,\mathsf{while}\, B\, \mathsf{do}\, C\{P \wedge \neg B\}}\ (\mathsf{while})
$$

Table 6: Rules of Hoare logic

I. Formally, $\blacktriangleleft^n_U \colon U \to U^n$ is defined as $\blacktriangleleft^0_U \stackrel{\text{def}}{=} !_U$ and $\blacktriangleleft^{n+1}_U \stackrel{\text{def}}{=} \blacktriangleleft_U; (\blacktriangleleft^n_U \otimes id_U)$. Guards in commands are encoded using $c(P)$, defined in (21). The encoding of commands is pretty standard, see e.g. [42], with the only exception of assignment that is depicted as the diagram on the right where, for the sake of readability, we label wires directly with the typing contexts rather than their encodings.

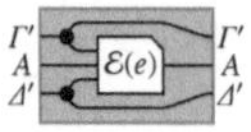

Note that, in this case, $\mathcal{E}(\cdot)$ exploits the structure of Cartesian bicategories to properly take into account data flow. For instance, for $\Gamma = x : A, y : A, z : A$, the encoding maps $\Gamma \vdash x := z; y := z$ and $\Gamma \vdash y := z; x := z$ into the same tape.

$\mathcal{E}(\Gamma \vdash x := z; y := z) =$ 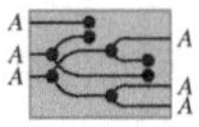 $\stackrel{(\blacktriangleleft\text{-un})}{=_{\mathbb{I}}}$ 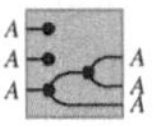 $\stackrel{(\blacktriangleleft\text{-un})}{=_{\mathbb{I}}}$ 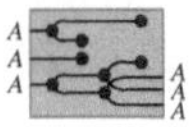 $= \mathcal{E}(\Gamma \vdash y := z; x := z).$

The following lemma illustrates the interaction of substitutions with the encoding.

Lemma 5. *Let $\Gamma' = \Gamma, x\colon A, \Delta$. If $\Gamma' \vdash P\colon 1$ and $\Gamma' \vdash t\colon A$, then*

$\mathcal{E}(\Gamma' \vdash P[t/x]\colon 1) =$ 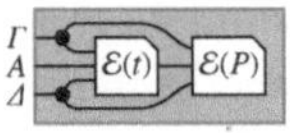

Hoare logic [35] is one of the most influential languages to reason about imperative programs. Its rules –in the version appearing in [57]– are in Figure 6. In partial correctness, a triple $\{P\}C\{Q\}$ asserts that if, starting from a state satisfying the precondition P, the execution of C terminates, then the resulting state satisfies the postcondition Q.

The following result shows that

Proposition 4. *If $\{P\}C\{Q\}$ is derivable as in Table 6, then $\mathcal{E}(P)^\dagger ; \mathcal{E}(C) \leq_{\mathbb{I}} \mathcal{E}(Q)^\dagger$.*

Proof. By induction on the rules in Table 6. We illustrate only two cases: the others can be found in the proof of [9, Proposition 11.5]. For (assn), observe that by Lemma 5, $\mathcal{E}(P[e/x])^\dagger \,;\mathcal{E}(x := e)$ is the leftmost diagram below. Thus:

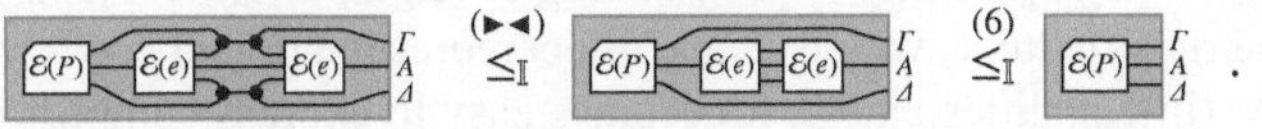

For (while), observe that the following holds:

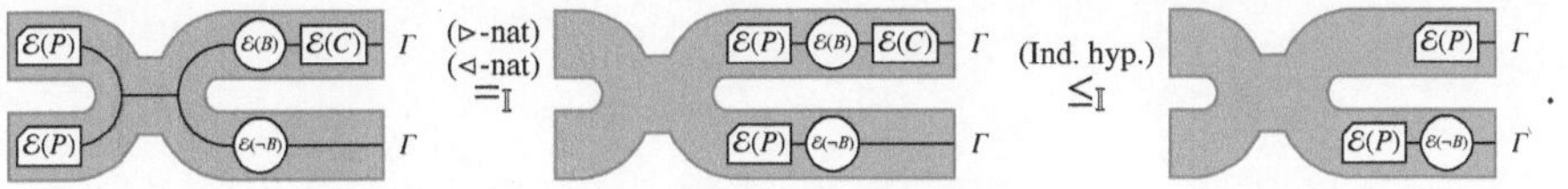

Then, by (AU2), the inequality below holds and the derivation concludes the proof.

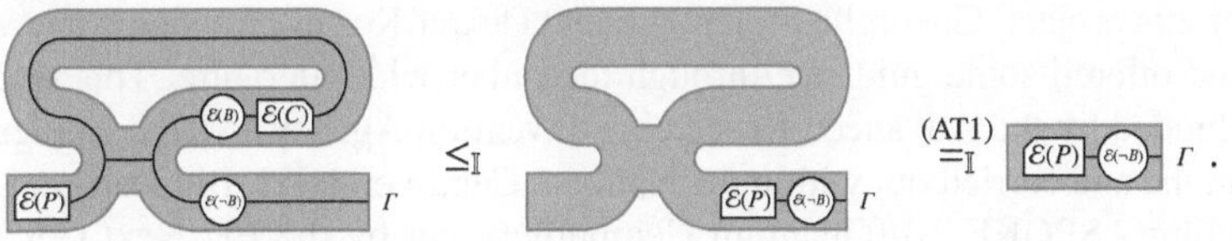

Remark 3 (Other Program Logics). The above result proves a syntactic correspondence amongst the deduction systems in Table 6 and $\leq_{\mathbb{I}}$. To prove such result we did not need to explicitly give the semantics of command or prove that our encoding preserves the semantics. However, assuming that, for a fixed interpretation $\mathcal{I}$, $[\![\mathcal{E}(\Gamma \vdash C)]\!]_{\mathcal{I}}$ is the intended relational semantics of a command C, one can immediately see that the correspondence at the semantic level –illustrated below on the top-left corner– holds.

$$\{P\}C\{Q\} \text{ iff } \left[\!\!\left[\mathcal{E}(P)^\dagger \,; \mathcal{E}(C)\right]\!\!\right]_{\mathcal{I}} \subseteq \left[\!\!\left[\mathcal{E}(Q)^\dagger\right]\!\!\right]_{\mathcal{I}} \qquad \langle\!\langle P \rangle\!\rangle C \langle\!\langle Q \rangle\!\rangle \text{ iff } [\![\mathcal{E}(P)]\!]_{\mathcal{I}} \subseteq [\![\mathcal{E}(C)\,; \mathcal{E}(Q)]\!]_{\mathcal{I}}$$

$$[P]C[Q] \text{ iff } \left[\!\!\left[\mathcal{E}(P)^\dagger \,; \mathcal{E}(C)\right]\!\!\right]_{\mathcal{I}} \supseteq \left[\!\!\left[\mathcal{E}(Q)^\dagger\right]\!\!\right]_{\mathcal{I}} \qquad (P)C(Q) \text{ iff } [\![\mathcal{E}(P)]\!]_{\mathcal{I}} \supseteq [\![\mathcal{E}(C)\,; \mathcal{E}(Q)]\!]_{\mathcal{I}}$$

The other correspondences concern other logics: $[P]C[Q]$ are triples of incorrectness logic [50], $\langle\!\langle P \rangle\!\rangle C \langle\!\langle Q \rangle\!\rangle$ of sufficient incorrectness [2] and $(P)C(Q)$ of necessary [24]. Interestingly, quadruples of *relational Hoare logics* [7] can be characterised as

$$\left[\!\!\left[\mathcal{E}(\Gamma_1, \Gamma_2 \vdash P \colon 1)^\dagger \,; \mathcal{E}(\Gamma_1 \vdash C_1) \otimes \mathcal{E}(\Gamma_2 \vdash C_2)\right]\!\!\right]_{\mathcal{I}} \subseteq \left[\!\!\left[\mathcal{E}(\Gamma_1, \Gamma_2 \vdash Q \colon 1)^\dagger\right]\!\!\right] \qquad (23)$$

9 Concluding remarks

We introduced Kleene bicategories and proved that they form Kleene algebras in Kozen's sense (Corollary 1). By examining their interaction with Cartesian bicategories, we developed Kleene-Cartesian rig categories and characterised the free one as tape diagrams (Theorem 2). Furthermore, we showed that tape diagrams can express Kleene algebra with tests [42], Peano's natural numbers, and various program logics. We proved that the rules of Hoare logic follow from the structure of kc rig categories (Proposition 4).

Regarding Kleene bicategories, while we are not the first to explore uniform traces over biproduct categories (see, e.g., [20]), the use of *posetal* uniformity and the correspondence with Kozen's axioms are, to the best of our knowledge, novel contributions.

Computer scientists usually find control flow and data flow graphs intuitive. Tapes diagram combine the two in a single diagrammatic representation equipped with compositional semantics. Another key distinction of tape diagrams, compared to other relational or categorical approaches to program logics (e.g., [44,1,31,49,34,5]), lies in the $\otimes$ monoidal and rig structures, which enable direct representation of *product programs* [6]. Through tape (in)equalities and $\otimes$, it becomes easy to express complex properties like *non-interference* [30]. Exploring how such properties can be proved by the laws of kc rig categories remains a promising direction for future research.

Acknowledgements. The authors would like to acknowledge Alessio Santamaria, Chad Nester and the students of the ACT school 2022 for several useful discussions at early stage of this project. Gheorghe Stefanescu and Dexter Kozen provided some wise feedback and offered some guidance through the rather wide literature. This research was partly funded by the Advanced Research + Invention Agency (ARIA) Safeguarded AI Programme and carried out within the National Centre on HPC, Big Data and Quantum Computing - SPOKE 10 (Quantum Computing) and by the EU Next-GenerationEU - National Recovery and Resilience Plan (NRRP) – MISSION 4 COMPONENT 2, INVESTMENT N. 1.4 – CUP N. I53C22000690001 and by the EPSRC grant No. EP/V002376/1. Bonchi is supported by the Ministero dell'Università e della Ricerca of Italy grant PRIN 2022 PNRR No. P2022HXNSC - RAP (Resource Awareness in Programming). Di Giorgio is supported by the EU grant No. 101087529.

$$(f;g);h = f;(g;h) \qquad id_X;f = f = f;id_Y \qquad (f_1 \odot f_2);(g_1 \odot g_2) = (f_1;g_1) \odot (f_2;g_2)$$
$$id_I \odot f = f = f \odot id_I \qquad (f \odot g) \odot h = f \odot (g \odot h)$$
$$\sigma^{\odot}_{A,B};\sigma^{\odot}_{B,A} = id_{A\odot B} \qquad (s \odot id_Z);\sigma^{\odot}_{Y,Z} = \sigma^{\odot}_{X,Z};(id_Z \odot s)$$

$$\mathsf{tr}_U((id \oplus u)\,;\mathsf{t}\,;(id \oplus v)) = u\,;\mathsf{tr}_U\mathsf{t}\,;v \qquad \mathsf{tr}_U(\mathsf{t} \oplus \mathfrak{s}) = \mathsf{tr}_U\mathsf{t} \oplus \mathfrak{s}$$
$$\mathsf{tr}_V\mathsf{tr}_U\mathsf{t} = \mathsf{tr}_{U\oplus V}\mathsf{t} \qquad \mathsf{tr}_0\mathsf{t} = \mathsf{t}$$
$$\mathsf{tr}_V(\mathsf{t}\,;(u \oplus id)) = \mathsf{tr}_U((u \oplus id)\,;\mathsf{t}) \qquad \mathsf{tr}_U\sigma^{\oplus}_{U,U} = id_U$$

$$\overline{\underline{\blacktriangleleft_A;(id_A \otimes \blacktriangleleft_A)}} = \overline{\underline{\blacktriangleleft_A;(\blacktriangleleft_A \otimes id_A)}} \qquad \overline{\underline{\blacktriangleleft_A;(!_A \otimes id_A)}} = \overline{\underline{id_A}} \qquad \overline{\underline{\blacktriangleleft_A;\sigma^{\otimes}_{A,A}}} = \overline{\underline{\blacktriangleleft_A}}$$
$$\overline{\underline{(id_A \otimes \blacktriangleright_A);\blacktriangleright_A}} = \overline{\underline{(\blacktriangleright_A \otimes id_A);\blacktriangleright_A}} \qquad \overline{\underline{(¡_A \otimes id_A);\blacktriangleright_A}} = \overline{\underline{id_A}} \qquad \overline{\underline{\sigma^{\otimes}_{A,A};\blacktriangleright_A}} = \overline{\underline{\blacktriangleright_A}}$$
$$\overline{\underline{(id_A \otimes \blacktriangleleft_A);(\blacktriangleright_A \otimes id_A)}} = \overline{\underline{\blacktriangleright_A;\blacktriangleleft_A}} \qquad \overline{\underline{id_A}} = \overline{\underline{\blacktriangleleft_A;\blacktriangleright_A}} \qquad \overline{\underline{s;\blacktriangleleft_V}} \leq \overline{\underline{\blacktriangleleft_U;(s \otimes s)}} \qquad \overline{\underline{s;!_V}} \leq \overline{\underline{!_U}}$$
$$\overline{\underline{id_A}} \leq \overline{\underline{\blacktriangleleft_A;\blacktriangleright_A}} \qquad \overline{\underline{\blacktriangleright_A;\blacktriangleleft_A}} \leq \overline{\underline{id_{A\otimes A}}} \qquad \overline{\underline{id_A}} \leq \overline{\underline{!_A;¡_A}} \qquad \overline{\underline{¡_A;!_A}} \leq \overline{\underline{id_1}}$$

$$\lhd_U;(id_U \oplus \lhd_U) = \lhd_U;(\lhd_U \oplus id_U) \qquad \lhd_U;(⫰_U \oplus id_U) = id_U \qquad \lhd_U;\sigma^{\oplus}_{U,U} = \lhd_U$$
$$(id_U \oplus \rhd_U);\rhd_U = (\rhd_U \oplus id_U);\rhd_U \qquad (⫯_U \oplus id_U);\rhd_U = id_U \qquad \sigma^{\oplus}_{U,U};\rhd_U = \rhd_U$$
$$\rhd_U;\lhd_U = \lhd_{U\oplus U};(\rhd_U \oplus \rhd_U) \qquad ⫯_U;⫰_U = id_0 \qquad ⫯_U;\lhd_U = ⫯_{U\oplus U} \qquad \lhd_U;⫰_U = ⫰_{U\oplus U}$$
$$\overline{\underline{c}};⫰_V = ⫰_U \qquad \overline{\underline{c}};\lhd_V = \lhd_U;(\overline{\underline{c}} \oplus \overline{\underline{c}}) \qquad ⫯_U;\overline{\underline{c}} = ⫯_V \qquad \rhd_U;\overline{\underline{c}} = (\overline{\underline{c}} \oplus \overline{\underline{c}});\rhd_V$$
$$id_U \geq \lhd_U;\rhd_U \qquad \rhd_U;\lhd_U \geq id_{U\oplus U} \qquad id_U \geq ⫰_U;⫯_U \qquad ⫯_U;⫰_U \geq id_0$$

$$\mathsf{tr}_U(\rhd_U;\lhd_U) = id_U \qquad \overline{\underline{id_U}} = id_U \qquad \overline{\underline{c;d}} = \overline{\underline{c}}\,;\overline{\underline{d}}$$

Table 7: Axioms for Kleene-Cartesian Tapes. For each axiom $l = r$ in top-left corner, the set $\mathbb{KC}$ contains the pairs (l, r) and (r, l) where $\odot$ and I are replaced by $\oplus$ and 0 and the pairs $(\overline{\underline{l}}, \overline{\underline{r}})$ and $(\overline{\underline{r}}, \overline{\underline{l}})$ where $\odot$ and I are replaced by $\otimes$ and 1. In the rest, for each $l \leq r$ $\mathbb{KC}$ contains a pair (l, r) and, additionally, the pair (r, l) in case of an axiom $l = r$.

References

1. Aguirre, A., Katsumata, S.y.: Weakest preconditions in fibrations. Electronic Notes in Theoretical Computer Science **352**, 5–27 (2020)
2. Ascari, F., Bruni, R., Gori, R., Logozzo, F.: Sufficient incorrectness logic: SIL and separation SIL (2023)
3. Backens, M.: The zx-calculus is complete for stabilizer quantum mechanics. New Journal of Physics **16**(9), 093021 (2014)
4. Bainbridge, E.S.: Feedback and generalized logic. Information and Control **31**(1), 75–96 (1976)
5. Barrett, C., Castle, D., Heijltjes, W.: The relational machine calculus. In: Sobocinski, P., Lago, U.D., Esparza, J. (eds.) Proceedings of the 39th Annual ACM/IEEE Symposium on Logic in Computer Science, LICS 2024, Tallinn, Estonia, July 8-11, 2024. pp. 9:1–9:15. ACM (2024). https://doi.org/10.1145/3661814.3662091
6. Barthe, G., Crespo, J.M., Kunz, C.: Relational verification using product programs. In: International Symposium on Formal Methods. pp. 200–214. Springer (2011)
7. Benton, N.: Simple relational correctness proofs for static analyses and program transformations. ACM SIGPLAN Notices **39**(1), 14–25 (2004)
8. Bloom, S.L., Ésik, Z., Stefanescu, G.: Notes on equational theories of relations. Algebra Universalis **33**(1), 98–126 (1995)
9. Bonchi, F., Di Giorgio, A., Di Lavore, E.: A diagrammatic algebra for program logics (2024), https://arxiv.org/abs/2410.03561
10. Bonchi, F., Di Giorgio, A., Haydon, N., Sobocinski, P.: Diagrammatic algebra of first order logic. In: Sobocinski, P., Lago, U.D., Esparza, J. (eds.) Proceedings of the 39th Annual ACM/IEEE Symposium on Logic in Computer Science, LICS 2024, Tallinn, Estonia, July 8-11, 2024. pp. 16:1–16:15. ACM (2024). https://doi.org/10.1145/3661814.3662078
11. Bonchi, F., Di Giorgio, A., Santamaria, A.: Deconstructing the calculus of relations with tape diagrams. Proceedings of the ACM on Programming Languages **7**(POPL), 1864–1894 (2023)
12. Bonchi, F., Holland, J., Piedeleu, R., Sobociński, P., Zanasi, F.: Diagrammatic algebra: From linear to concurrent systems. Proceedings of the ACM on Programming Languages **3**(POPL), 25:1–25:28 (Jan 2019). https://doi.org/10.1145/3290338
13. Bonchi, F., Pavlovic, D., Sobocinski, P.: Functorial semantics for relational theories. arXiv preprint arXiv:1711.08699 (2017)
14. Bonchi, F., Piedeleu, R., Sobociński, P., Zanasi, F.: Graphical affine algebra. In: Proceedings of the 34th Annual ACM/IEEE Symposium on Logic in Computer Science (LICS). pp. 1–12 (2019)
15. Bonchi, F., Seeber, J., Sobocinski, P.: Graphical Conjunctive Queries. In: Ghica, D., Jung, A. (eds.) 27th EACSL Annual Conference on Computer Science Logic (CSL 2018). Leibniz International Proceedings in Informatics (LIPIcs), vol. 119, pp. 13:1–13:23. Schloss Dagstuhl–Leibniz-Zentrum fuer Informatik, Dagstuhl, Germany (2018). https://doi.org/10.4230/LIPIcs.CSL.2018.13
16. Bonchi, F., Sobocinski, P., Zanasi, F.: Full abstraction for signal flow graphs. ACM SIGPLAN Notices **50**(1), 515–526 (2015)
17. Brunet, P., Pous, D.: Kleene algebra with converse. In: Relational and Algebraic Methods in Computer Science: 14th International Conference, RAMiCS 2014, Marienstatt, Germany, April 28–May 1, 2014. Proceedings 14. pp. 101–118. Springer (2014)
18. Bruni, R., Melgratti, H., Montanari, U.: Connector algebras, Petri nets, and BIP. In: International Andrei Ershov Memorial Conference on Perspectives of System Informatics. pp. 19–38. Springer (2011)

19. Carboni, A., Walters, R.F.C.: Cartesian bicategories I. Journal of Pure and Applied Algebra **49**, 11–32 (1987)
20. Căzănescu, V.E., Ştefănescu, G.: Feedback, iteration, and repetition. In: Mathematical aspects of natural and formal languages, pp. 43–61. World Scientific (1994)
21. Cockett, J.R.B., Koslowski, J., Seely, R.A.: Introduction to linear bicategories. Mathematical Structures in Computer Science **10**(2), 165–203 (2000)
22. Coecke, B., Duncan, R.: Interacting quantum observables: categorical algebra and diagrammatics. New Journal of Physics **13**(4), 043016 (2011)
23. Coecke, B., Selby, J., Tull, S.: Two Roads to Classicality **266**, 104–118 (Feb 2018). https://doi.org/10.4204/EPTCS.266.7
24. Cousot, P., Cousot, R., Fähndrich, M., Logozzo, F.: Automatic inference of necessary preconditions. In: Giacobazzi, R., Berdine, J., Mastroeni, I. (eds.) Verification, Model Checking, and Abstract Interpretation, 14th International Conference, VMCAI 2013, Rome, Italy, January 20-22, 2013. Proceedings. Lecture Notes in Computer Science, vol. 7737, pp. 128–148. Springer (2013). https://doi.org/10.1007/978-3-642-35873-9_10
25. Dedekind, R.: The nature and meaning of numbers (1888)
26. Fong, B., Spivak, D.: String diagrams for regular logic (extended abstract). In: Baez, J., Coecke, B. (eds.) Applied Category Theory 2019. Electronic Proceedings in Theoretical Computer Science, vol. 323, p. 196–229. Open Publishing Association (Sep 2020). https://doi.org/10.4204/eptcs.323.14
27. Fox, T.: Coalgebras and cartesian categories. Communications in Algebra **4**(7), 665–667 (1976). https://doi.org/10.1080/00927877608822127
28. Fritz, T.: A presentation of the category of stochastic matrices. CoRR **abs/0902.2554** (2009)
29. Ghica, D.R., Jung, A.: Categorical semantics of digital circuits. In: 2016 Formal Methods in Computer-Aided Design (FMCAD). pp. 41–48 (2016). https://doi.org/10.1109/FMCAD.2016.7886659
30. Goguen, J.A., Meseguer, J.: Unwinding and inference control. In: 1984 IEEE Symposium on Security and Privacy. pp. 75–75. IEEE (1984)
31. Goncharov, S., Schröder, L.: A relatively complete generic hoare logic for order-enriched effects. In: 2013 28th Annual ACM/IEEE Symposium on Logic in Computer Science. pp. 273–282. IEEE (2013)
32. Harsanyi, J.C.: Mathematics, the empirical facts, and logical necessity. Erkenntnis **19**(1), 167–192 (1983)
33. Hasegawa, M.: The uniformity principle on traced monoidal categories. Electronic Notes in Theoretical Computer Science **69**, 137–155 (2003)
34. Hasuo, I.: Generic weakest precondition semantics from monads enriched with order. Theoretical Computer Science **604**, 2–29 (2015)
35. Hoare, C.A.R.: An axiomatic basis for computer programming. Communications of the ACM **12**(10), 576–580 (1969)
36. John C. Baez, Brandon Coya, F.R.: Props in network theory. CoRR **abs/1707.08321** (2017), http://arxiv.org/abs/1707.08321
37. Johnson, N., Yau, D.: Bimonoidal categories, e_n-monoidal categories, and algebraic k-theory (2022), https://nilesjohnson.net/En-monoidal.html
38. Joyal, A., Street, R.: The geometry of tensor calculus, I. Advances in Mathematics **88**(1), 55–112 (Jul 1991). https://doi.org/10.1016/0001-8708(91)90003-P
39. Joyal, A., Street, R., Verity, D.: Traced monoidal categories. Math Procs Cambridge Philosophical Society **119**(3), 447–468 (4 1996)
40. Katis, P., Sabadini, N., Walters, R.F.: Bicategories of processes. Journal of Pure and Applied Algebra **115**(2), 141–178 (1997)
41. Kozen, D.: A completeness theorem for kleene algebras and the algebra of regular events. Information and Computation **110**, 366–390 (1994)

42. Kozen, D.: Kleene algebra with tests. ACM Transactions on Programming Languages and Systems (TOPLAS) **19**(3), 427–443 (1997)
43. Kozen, D.: Typed Kleene algebra. Tech. Rep. TR98-1669, Computer Science Department, Cornell University (March 1998)
44. Kozen, D.: On Hoare logic and Kleene algebra with tests. Trans. Computational Logic **1**(1), 60–76 (July 2000)
45. Lack, S.: Composing PROPs. Theory and Application of Categories **13**(9), 147–163 (2004)
46. Laplaza, M.L.: Coherence for distributivity. In: Kelly, G.M., Laplaza, M., Lewis, G., Mac Lane, S. (eds.) Coherence in Categories. pp. 29–65. Lecture Notes in Mathematics, Springer, Berlin, Heidelberg (1972). https://doi.org/10.1007/BFb0059555
47. Lawvere, F.W.: Functorial Semantics of Algebraic Theories. Ph.D. thesis, Columbia University, New York, NY, USA (1963)
48. Mac Lane, S.: Categories for the Working Mathematician, Graduate Texts in Mathematics, vol. 5. Springer-Verlag, New York, second edn. (1978)
49. Martin, U., Mathiesen, E.A., Oliva, P.: Hoare logic in the abstract. In: International Workshop on Computer Science Logic. pp. 501–515. Springer (2006)
50. O'Hearn, P.W.: Incorrectness logic. Proceedings of the ACM on Programming Languages **4**(POPL), 1–32 (2019)
51. Piedeleu, R., Zanasi, F.: A Finite Axiomatisation of Finite-State Automata Using String Diagrams. Logical Methods in Computer Science **Volume 19, Issue 1** (Feb 2023). https://doi.org/10.46298/lmcs-19(1:13)2023
52. Pous, D.: On the positive calculus of relations with transitive closure. In: Niedermeier, R., Vallée, B. (eds.) 35th Symposium on Theoretical Aspects of Computer Science, STACS 2018, February 28 to March 3, 2018, Caen, France. LIPIcs, vol. 96, pp. 3:1–3:16. Schloss Dagstuhl - Leibniz-Zentrum für Informatik (2018). https://doi.org/10.4230/LIPIcs.STACS.2018.3
53. Pratt, V.R.: Semantical considerations on floyd-hoare logic. In: 17th Annual Symposium on Foundations of Computer Science (sfcs 1976). pp. 109–121. IEEE (1976)
54. Selinger, P.: A note on Bainbridge's power set construction. Ann Arbor **1001**, 48109–1109 (1998)
55. Selinger, P.: A survey of graphical languages for monoidal categories. In: New structures for physics, pp. 289–355. Springer (2010)
56. Stein, D., Staton, S.: Probabilistic programming with exact conditions. Journal of the ACM (2023)
57. Winskel, G.: The formal semantics of programming languages: an introduction. MIT press (1993)

Fair Quantitative Games[*,**]

Ashwani Anand, Satya Prakash Nayak, Ritam Raha, Irmak Sağlam[(✉)], and Anne-Kathrin Schmuck

Max Planck Institute for Software Systems, Kaiserslautern, Germany
{ashwani,sanayak,rraha,isaglam,akschmuck}@mpi-sws.org

Abstract. We examine two-player games over finite weighted graphs with quantitative (mean-payoff or energy) objective, where one of the players additionally needs to satisfy a fairness objective. The specific fairness we consider is called *strong transition fairness*, given by a subset of edges of one of the players, which asks the player to take fair edges infinitely often if their source nodes are visited infinitely often. We show that when fairness is imposed on player 1, these games fall within the class of previously studied ω-regular mean-payoff and energy games. On the other hand, when the fairness is on player 2, to the best of our knowledge, these games have not been previously studied. We provide gadget-based algorithms for fair mean-payoff games where fairness is imposed on either player, and for fair energy games where the fairness is imposed on player 1. For all variants of fair mean-payoff and fair energy (under unknown initial credit) games, we give pseudo-polynomial algorithms to compute the winning regions of both players. Additionally, we analyze the strategy complexities required for these games. Our work is the first to extend the study of strong transition fairness, as well as gadget-based approaches, to the quantitative setting. We thereby demonstrate that the simplicity of strong transition fairness, as well as the applicability of gadget-based techniques, can be leveraged beyond the ω-regular domain.

1 Introduction

Games on graphs serve as a formal and effective framework for automatically synthesizing correct-by-design software in cyber-physical systems (CPS). In this setting, one player represents a controller aiming to ensure a high-level logical specification in response to the actions of an adversarial player representing the external environment. Two-player graph games abstract the strategic reactive behavior of autonomous systems, such as robots [37, 46, 52] or cars [40, 41], and can be employed to synthesize *correct-by-design* strategies through the algorithmic process of reactive synthesis. Such strategies function as logical controllers to ensure the required behavioral guarantees, ensuring the system meets its safety and performance goals over its strategic interactions with its environment.

In the CPS context, these game-based abstractions have been augmented with additional information, such as assumptions about the players' strategic

* The extended version of this paper can be found at [8].

** All authors are partially supported by the DFG project 89792660 as part of TRR 248 – CPEC, and by the Emmy Noether Grant SCHM 3541/1-1.

P. A. Abdulla and D. Kesner (Eds.): FoSSaCS 2025, LNCS 15691, pp. 331–354, 2025.
https://doi.org/10.1007/978-3-031-90897-2_16

behavior induced by the underlying physical processes [5, 34], and quantitative objectives that allow to optimize strategies w.r.t. given performance metrics [9]. These enhancements enable reactive synthesis algorithms to compute control strategies that are more aligned with the specific requirements of CPS applications. However, these extensions typically come with increased computational costs, hindering their practical applicability. In this paper, we show that a particular subclass of such games – namely, *energy* and *mean-payoff games* under *strong-transition-fairness constraints* – distinctly posses favorable computational properties. We refer to these games as *fair energy* and *fair mean-payoff games.*

1.1 Background

Strong Transition Fairness. Strong transition fairness [10, 34, 44] is defined over specific edges, referred to as *fair edges*, and restricts strategies to use a fair edge infinitely often if its source vertex is visited infinitely often. Strong transition fairness constraints are less expressive than general fairness, typically expressed via a qualitative objective called the *Streett objective* [48]. Yet, they naturally arise in areas such as resource management [19], abstractions of continuous-time physical processes for planning [6, 26, 31] and controller synthesis [38, 42, 50]. At the same time, games with strong transition fairness conditions offer more favorable computational properties compared to those with Streett fairness, making them a compelling subject of study.

Despite their strong motivation from CPS applications and their favorable computational properties, strong transition fairness has so far been considered only in the context of qualitative objectives, such as parity [35, 45] or Rabin [11, 49], where it is shown to be computationally inexpensive, preserving the overall complexity of the game and inducing negligible computational cost. This paper shows that these property carry over to quantitative games.

Energy & Mean-Payoff Games. Quantitative objectives, such as energy or mean-payoff, allow capturing strategic limitations induced by constrained resources, such as the power usage of embedded components, the buffer size of a networking element [28], or string selections with limited storage [53].

Mean-payoff games, introduced in [33], assign integer weights to edges, representing the payoffs received by player 1 (the controller) and paid by player 2 (the environment) when an edge is taken. These games are played over infinitely many rounds, with player 1 aiming to maximize the long run average value of the edges traversed, while player 2 aims to minimize this value.

Energy games, introduced more recently in [18], also assign integer weights to edges, this time representing energy gains or losses. In this setting, the controller's goal is to construct an infinite path where the total energy in every prefix remains non-negative, while the environment aims to prevent this.

It is known that determining the winner in both energy and mean-payoff games is in $\mathsf{NP} \cap \mathsf{coNP}$ [54]. The state-of-the-art algorithms for mean-payoff games have runtime $\mathcal{O}(n^2mW)$ (and a runtime of $\mathcal{O}(nmW)$ for the threshold problem), where n is the number of nodes, m is the number of edges and W is the maximum absolute weight in the game arena [29]. For energy games, the best algorithms are either deterministic with the run time $\mathcal{O}(\min(mnW, mn2^{n/2}\log W))$ [16, 32] or randomized with the run time $2^{\mathcal{O}(\sqrt{n\log n})}$ [12].

Combining Quantitative & Qualitative Objectives. Conceptually, *fair energy* and *fair mean-payoff games* considered in this paper combine quantitative (i.e., energy or mean-payoff) with qualitative (i.e. strong transition fairness) obligations to constrain the moves of the players in the resulting game.

Combinations of quantitative and qualitative objectives have been studied for several variants of mean-payoff and energy objectives combined with general ω-regular goals [13, 14, 18, 21–24, 30, 36]. In particular, energy parity games were introduced in [21], demonstrating these games' inclusion in NP ∩ coNP and polynomial equivalence to mean-payoff parity games, previously studied in [24]. Additionally, the interplay between quantitative objectives (in both optimization and threshold variants) and Boolean constraints has been investigated in a broad range of settings [1–4, 14, 15, 17, 22, 23, 27, 51], highlighting the richness and diversity of approaches in this area.

It is known that strong transition fairness conditions can be translated into a classical Streett condition, i.e., a subclass of (qualitative) ω-regular objectives, by transforming each fair edge $e = (u, v)$ into a Streett pair $(\{u\}, \{e\})$ in the original game arena. If only player 1 (the controller) is constrained by the fairness, this transformation results in an energy (resp. mean-payoff) Streett game. Following [7], Energy Streett games[1] can be solved in one of the following ways:

- ▷ Encode the Streett energy objective as a μ-calculus formula and solve the formula using symbolic fixed-point computations: The computation takes $\mathcal{O}(n\cdot [b+1]^{\lfloor \frac{d}{2}+1 \rfloor})$ symbolic steps, where $b = (d+1)\cdot\big(((n^2+n)\frac{n}{2}! - 1)W\big)$ is an upper bound on the credit a Streett energy game requires, where d is the number of Streett pairs, which in our case corresponds to the number of fair edges. This complexity can be simplified to $\mathcal{O}\left((\frac{n}{2}!\cdot W)^{\frac{m}{2}}\right)$ by omitting the polynomial factors where m is the number of edges; which is super-exponential.
- ▷ Translate the energy Streett game into a Streett game by encoding the domain of energy levels in the state space. This explodes the state space by multiplying the number of states with the above-mentioned upper-bound b. Using state-of-the-art algorithms for Streett games [43], this gives an algorithm with worst-case super-exponential runtime of $\mathcal{O}((\frac{n}{2}!W)^{m+1}m!)$.

If, however, player 2 (the environment) is constrained by strong-transition fairness, the above reduction to energy (resp. mean-payoff) Streett games does not apply as the energy or mean-payoff objectives are not symmetric for both players. In this case, the given energy (resp. mean payoff) game is enhanced with a *fairness assumption* weakening the opponent. While such games have been investigated for qualitative objectives [35], this paper is the first to consider quantitative games under such fairness assumptions.

[1] This also gives a solution for mean-payoff Streett games via their polynomial reduction to energy games [21].

1.2 Contributions

Based on the above discussion, we distinguish between *1-fair* and *2-fair* energy (resp. mean-payoff) games, emphasizing which player *is* additionally constrained by the strong transition fairness condition. We provide a unified framework to solve both classes of games via a novel, gadget-based approach. Our gadgets enable the translation of fair quantitative games into regular quantitative games on a linearly larger state space. This allows solving these games in pseudo-polynomial time, significantly improving the computational complexity for *1-fair* quantitative games, compared to the naive approaches outlined above. The origins of these gadget-based techniques can be traced back to the works in [25] and [20], where similar gadgets were employed to convert stochastic parity and Rabin games with almost-sure winning conditions into regular parity and Rabin games. In [35], a gadget-based technique was developed that reduces fair parity games to (linearly larger) regular parity games, demonstrating a quasi-polynomial run time for these games. Our work extends this approach to fair games with quantitative objectives. Concretely, our contributions are as follows.

Fair Mean-Payoff Games:

▷ *Complexity.* Using gadgets, we reduce (1- and 2-) fair mean-payoff games to regular mean-payoff games on a game arena with linearly larger number of nodes and edges, and a maximum absolute weight of $\mathcal{O}(n^2W)$, where n is the number of nodes and W is the maximum absolute weight in the fair game arena. Following the state-of-the-art algorithm of [29], we obtain a complexity of $\mathcal{O}(n^4mW)$ for solving the fair mean-payoff games for optimal value problem, and a complexity of $\mathcal{O}(n^3mW)$ for threshold problem.

▷ *Determinacy.* The above reduction establishes that fair mean-payoff games are determined irrespective of which player has fairness constraints.

▷ *Strategies.* We show that, in 1-fair mean-payoff games, player 1 in general needs infinite memory strategies to win, although finite memory suffices for achieving suboptimal values. Memoryless strategies suffice for player 2. In 2-fair mean-payoff games, player 2 has finite memory winning strategies, whereas memoryless strategies suffice for player 1.

Fair Energy Games:

▷ *Determinacy.* In contrast to fair mean-payoff games, we show that fair energy games (with unknown initial credit) are not determined in general. In particular, 2-fair energy games are not determined, whereas 1-fair energy games are. We argue that the lack of determinacy prevents us from constructing similar gadgets to reduce 2-fair energy games to regular energy games.

▷ *Complexity.* We introduce a gadget that reduces *1-fair energy games* to regular energy games on an arena with linearly larger number of nodes and edges, and a maximum absolute weight of $\mathcal{O}(n^3W)$. Using the state-of-the-art algorithm of [32], we obtain a complexity of $\mathcal{O}(n^4mW)$ for solving these games. For *2-fair energy games*, we provide simple algorithms to compute the winning regions separately for both players. We show that in 2-fair energy games, player 1's winning region is the same as in the corresponding energy game where fair edges are treated as regular edges. Meanwhile, player 2's winning region is the same as in the corresponding 2-fair mean-payoff games with threshold 0. These reductions yield an $\mathcal{O}(n^3mW)$ algorithm for solving 2-fair energy games.

▷ *Strategies.* We show that in fair energy games, the player restricted by transition fairness has finite memory winning strategies whereas memoryless strategies suffice for the respective other player.

In summary, our results show that strong transition fairness can be seamlessly incorporated into quantitative games without significant computational overhead. In fact, just as in the qualitative setting, this simple form of fairness comes *virtually for free* in the quantitative setting. Our key conceptual insight is that both the simplicity of strong transition fairness, and the effectiveness of gadget-based approaches extend beyond the ω-regular domain.

2 Preliminaries

In this section, we introduce the basic notations used throughout the paper. We write $\mathbb{Q}$ and $\mathbb{N}$ to denote the set of rational numbers and the set of natural numbers including 0, respectively.

2.1 Weighted Game Arena

A two-player weighted game arena is a tuple $G = (Q, E, w)$, where $Q = Q_1 \uplus Q_2$ is a finite set of nodes, $E \subseteq Q \times Q$ is a set of edges, and $w : E \to [-W, W]$ is a weight function that assigns an integer weight to each edge in G, where $W \in \mathbb{N}$ is the maximum absolute weight appearing in G. The nodes are partitioned into two sets, Q_1 and Q_2, where Q_i is the set of nodes controlled by Player i for $i \in \{1, 2\}$ and $Q_1 \cap Q_2 = \emptyset$. For a node q, we write qE to denote the set $\{e \in E \mid e = (q, q') \text{ for } q' \in Q\}$ of all outgoing edges from q and $E(q)$ to denote all successors of q. W.l.o.g., we assume that all nodes have out-degree at least one, i.e., $E(q) \neq \emptyset$ for all $q \in Q$. In rest of the paper, we use circles and squares in a figure to denote nodes controlled by player 1 and player 2, respectively.

Plays. A *play* $\tau = q_0 q_1 \ldots \in Q^\omega$ on G is an infinite sequence of nodes starting from q_0 such that, $\forall i \geq 0, (q_i, q_{i+1}) \in E$. We use notations $\tau[0; i] = q_0 \ldots q_i$, and $\tau[i; j] = q_i \ldots q_j$ to denote the finite *prefix* and *infix* of the play τ, respectively. A node q is said to *appear infinitely often* in τ, i.e., $q \in \mathtt{Inf}(\tau)$ if $\forall i, \exists j \geq i$, such that $q_j = q$. We naturally extend the notion of appearing infinitely often in a play for the edges. We let $\mathtt{plays}(G)$ denote the set of all plays on G, and let $\mathtt{plays}(G, q)$ denote the set of all plays starting from node q.

Strategies. A *strategy* σ for player $i \in \{1, 2\}$ (or, a player i-strategy) is a function $\sigma : Q^* \cdot Q_i \mapsto E$ where for all $H \cdot q \in Q^* \cdot Q_i$, $\sigma(H \cdot q) \in qE$. Intuitively, from every node $q \in Q_i$, a strategy for player i assigns an outgoing edge of that node based on a history $H \in Q^*$.

For a player i strategy σ, a play $\tau = q_0 q_1 \ldots$ is called a σ-play if it conforms with σ, i.e., for all $j \in \mathbb{N}$ with $q_j \in Q_i$, it holds that $\sigma(q_0 \ldots q_j) = (q_j, q_{j+1})$. Given a strategy σ we denote the restriction of sets $\mathtt{plays}(G)$ and $\mathtt{plays}(G, q)$

to σ-plays with $\texttt{plays}_\sigma(G)$ and $\texttt{plays}_\sigma(G, q)$. Similarly, $\texttt{play}_{\sigma,\pi}(G, q)$ denotes the unique play from q conforming with player 1 and player 2 strategies σ and π.

Let M be a set called *memory*. A player i strategy σ with memory M can be represented as a tuple (M, m_0, α, β), where $m_0 \in M$ is the initial memory value, $\alpha : M \times Q \to M$ is the update function, and $\beta : M \times Q_i \to Q$ is the function prescribing next state. Intuitively, if the current node is a player i node q and m is the current memory value, the strategy σ selects $q' = \beta(m, q)$ as the next node and updates the memory to $\alpha(m, q)$. If M is finite, then we call σ a *finite memory strategy*; otherwise it is an *infinite memory strategy.* Formally, given a history $H \cdot q \in Q^* \cdot Q_i$, $\sigma(H \cdot q) = \beta(\hat{\alpha}(m_0, H), q)$, where $\hat{\alpha}$ extends α to sequences of nodes canonically. A strategy is called *memoryless* or *positional* if $|M| = 1$. For such memoryless strategy σ, it holds that $\sigma(H_1 \cdot q) = \sigma(H_2 \cdot q)$ for every history $H_1, H_2 \in Q^*$. We denote it as $\sigma(q)$ for convenience. For any positional strategy σ of player i, we define G_σ to be the subgame arena (Q, E', w) of G, where $E' = \{(q, \sigma(q)) \mid q \in Q_i\} \cup \{qE \mid q \in Q_{3-i}\}$. Note that in this subgame arena, each player i node has exactly one successor according to the strategy σ.

Weighted Games and Objectives. A *weighted game* is a tuple (G, φ), where G is a weighted game arena and $\varphi \subseteq Q^\omega$ is an *objective* for player 1. A play τ is *winning* for player 1 if $\tau \in \varphi$, else it is winning for player 2. A player i strategy σ is *winning* from some node q, if all σ-plays starting from q are winning for player i. We denote the winning regions of player i by $\texttt{Win}_i(G, \varphi)$, or shortly $\texttt{Win}_i$, when (G, φ) is clear from the context. Formally, the winning regions are defined as follows where Σ_i is the set of all player i strategies:

$$q \in \texttt{Win}_1(G, \varphi) \iff \exists \sigma \in \Sigma_1. \forall \pi \in \Sigma_2,\ \texttt{play}_{\sigma,\pi}(G, q) \in \varphi \tag{1}$$

$$q \in \texttt{Win}_2(G, \varphi) \iff \exists \pi \in \Sigma_2. \forall \sigma \in \Sigma_1,\ \texttt{play}_{\sigma,\pi}(G, q) \notin \varphi \tag{2}$$

A game (G, φ) is called *determined*, if $Q = \texttt{Win}_1 \cup \texttt{Win}_2$.

Given a finite infix $H = q_0 \dots q_k$ of a play in G, we denote the (total) weight and the average weight of H by $w(H) = \sum_{i=0}^{k-1} w(q_i, q_{i+1})$ and $\texttt{avg}(H) = \frac{w(H)}{k}$, respectively. Furthermore, we denote the (limit) average weight of a play τ by $\texttt{avg}(\tau) = \liminf_{i\to\infty} \texttt{avg}(\tau[0; i])$.

Mean-Payoff Games. A *mean-payoff game* is a weighted game with a mean-payoff objective for a given threshold value $v \in \mathbb{Q}$, which is defined as $\texttt{MP}_v = \{\tau \in Q^\omega \mid \texttt{avg}(\tau) \geq v\}$. Intuitively, a play is winning for player 1 if the average weight of the play is above a certain threshold value v. Note that, mean-payoff games are *prefix independent* as any finite prefix of a play τ does not affect the limit average weight of τ.

In this work, without loss of generality, we focus on mean-payoff objective $\texttt{MP}_0$ with threshold value $v = 0$. It is easy to see that for any value v, the weighted mean-payoff game $(G, \texttt{MP}_v)$ with $G = (Q, E, w)$ can be reduced to a mean-payoff game $(G', \texttt{MP}_0)$ by subtracting v from the weights of all edges in G, i.e., $G' = (Q, E, w')$ where $w'(q, q') = w(q, q') - v$ for all $(q, q') \in E$.

The *optimal value* of a mean-payoff game is the maximum value v for which player 1 has a winning strategy with threshold v. A winning strategy of player 1 that achieves this optimal value is called an *optimal value strategy*. The *optimal value problem* is to compute the optimal values in a mean-payoff game.

Energy Games. An *energy objective* w.r.t. a given initial credit $c \in \mathbb{N}$ is defined as $\mathtt{En}_c = \{\tau \in Q^\omega \mid c + w(\tau[0;i]) \geq 0,\ \forall i \in \mathbb{N}\}$. Intuitively, a play belongs $\mathtt{En}_c$ if the total weight ('energy level' starting from c) of the play remains non-negative along the play. An *energy game* is a weighted game with an energy objective with unknown initial credit, denoted by $\mathtt{En}$, where player 1 wins from some node q if there exists an initial credit $c \in \mathbb{N}$ such that she can ensure the objective $\mathtt{En}_c$ from q. Formally, the winning regions are defined as follows, for $\varphi_c = \mathtt{En}_c$:

$$q \in \mathtt{Win}_1 \iff \exists c \in \mathbb{N}.\, \exists \sigma \in \Sigma_1.\, \forall \pi \in \Sigma_2,\ \mathtt{play}_{\sigma,\pi}(G,q) \in \varphi_c \tag{3}$$

$$q \in \mathtt{Win}_2 \iff \exists \pi \in \Sigma_2.\, \forall c \in \mathbb{N}.\, \forall \sigma \in \Sigma_1,\ \mathtt{play}_{\sigma,\pi}(G,q) \notin \varphi_c \tag{4}$$

Note that, in contrast to mean-payoff games, the energy objective is not a *prefix-independent* objective as the total weight of each prefix has to be non-negative along any play. However, it is known that such energy games are log-space equivalent to mean-payoff games [16]. Furthermore, it is known that both energy games and mean-payoff games are *positionally determined* [18, 33] i.e., $Q = \mathtt{Win}_1 \cup \mathtt{Win}_2$ and both players have positional winning strategies.

2.2 Fair Game Arena

A *fair game arena* is a tuple (G, E_f), where G is a weighted game arena as defined above and $E_f \subseteq E$ is a given set of 'fair edges'. We call the non-fair edges $E \setminus E_f$ 'regular', and a node q 'fair' if it is the source node of a fair edge, i.e., $qE \cap E_f \neq \emptyset$. Let Q_f be the set of all fair nodes in G. We investigate the scenario where all fair nodes are owned by the same player. If $Q_f \subseteq Q_1$, we call the game arena *1-fair* and if $Q_f \subseteq Q_2$, we call it *2-fair*. For brevity, we sometimes denote the fair game arena (G, E_f) by G and keep the set of fair edges E_f implicit. Given a fair node q, we write qE_f to denote the set $\{e \in E_f \mid e = (q, q') \text{ for } q' \in Q\}$ of all fair outgoing edges from q and $E_f(q) = \{q' \mid (q, q') \in E_f\}$ to denote the set of successors of q via fair edges.

A play τ is *fair* if for all nodes $q \in \mathtt{Inf}(\tau)$ and for all edges $e \in qE$, if $e \in E_f$, then $e \in \mathtt{Inf}(\tau)$. We let $\mathtt{plays}^f(G)$ denote the set of all fair plays on (G, E_f). We say a strategy σ for player i is fair if $\mathtt{plays}_\sigma(G) \subseteq \mathtt{plays}^f(G)$.

Fair Mean-Payoff/Energy Games. An *i-fair mean-payoff game* with objective $\varphi = \mathtt{MP}_v$ is a tuple (G, E_f, φ^f), where (G, E_f) is an i-fair game arena, and φ^f is the objective with the respective fairness condition defined as follows:

$$\varphi^f = \begin{cases} \varphi \cap \mathtt{plays}^f(G) & \text{if } (G, E_f) \text{ is 1-fair} \\ \varphi \cup \mathtt{plays}(G) \setminus \mathtt{plays}^f(G) & \text{if } (G, E_f) \text{ is 2-fair} \end{cases} \tag{5}$$

Intuitively, a play is winning for player i in i-fair game if it satisfies both the fairness condition and the player's corresponding objective, i.e., φ for player 1 and $\neg\varphi$ for player 2.

Similarly, an *i-fair energy game* with energy objective $\mathtt{En}$ is a tuple $(G, E_f, \mathtt{En}^f)$ where the winning regions are defined as in (3)-(4) for $\varphi_c = \mathtt{En}_c^f$.

3 Example: Mean-Payoff vs Fair Mean-Payoff Games

In this section, we depict the intricacy of fair mean-payoff games in comparison to regular mean-payoff games via the following example:

Consider the game arena depicted in Fig. 1 with both the nodes belonging to player 1. With regular mean-payoff objective with threshold 0, player 1 has a simple optimal positional strategy: at q, play the self-loop (q, q) and from p play the edge (p, q). This strategy is winning for player 1 because the limit average weight of any play conforming to this strategy is 1. Note that the strategy is also an optimal value strategy as the optimal value of the game is 1.

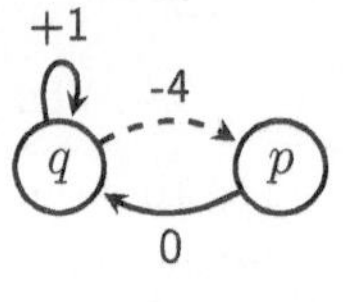

Fig. 1

Now let us consider the game arena to be 1-fair with a fair edge (q, p). Now the fair mean-payoff objective $\mathtt{MP}_0^f$ for player 1 is to ensure that the limit average weight of the play is ≥ 0 along with the fairness condition that says if q appears infinitely often in a play, so does the edge (q, p).

First note that there is no positional strategy that can ensure both fairness and mean-payoff objective and hence winning for player 1. Now, consider the following finite memory strategy for player 1: the strategy takes the self-loop (q, q) for k times before taking (q, p) once, and this sequence of $k + 1$ choices is repeated forever. The resulting play uses the fair edge (q, p) infinitely often and hence it is fair. For $k \geq 4$, this strategy is winning as the limit average of the plays conforming to this strategy is at least 0. In fact, we can show that for all $\epsilon > 0$, it is possible to choose a large enough k, such that player 1 ensures a limit average weight of at least $1 - \epsilon$.

Finally, consider the following infinite-memory strategy for player 1: the strategy is played in rounds; in round $i \geq 0$, the strategy plays (q, q) for i times, then plays (q, p) once and progresses to round $i+1$. This fair strategy ensures player 1 achieves a limit average weight of value 1, and it is therefore an optimal value strategy. However, it is not hard to see that there is no finite-memory optimal value strategy for player 1 in this game. In the following sections, we formally present how to solve fair mean-payoff games and discuss the memory requirements for each player to achieve these winning strategies.

4 Solving Fair Mean-Payoff Games

In this section, we present algorithms for solving fair mean-payoff games. The algorithms transform a fair mean-payoff game into a 'regular' mean-payoff game using *gadgets*. We introduce two gadgets depending on which player owns the fair edges as depicted in Fig. 2. Given an i-fair mean-payoff game G, we replace all fair nodes with the corresponding i-fair gadgets and obtain an equivalent mean-payoff game G' such that $(G, E_f, \mathtt{MP}_0^f)$ is winning for player i if the mean-payoff game $(G', \mathtt{MP}_0)$ is winning for player i. In particular, we prove the following.

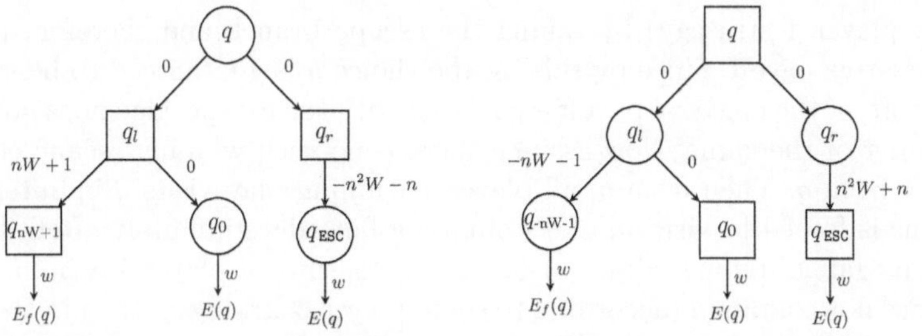

(a) 1-fair mean-payoff gadget (b) 2-fair mean-payoff gadget

Fig. 2: Gadgets for converting fair mean-payoff games to mean-payoff games. Edges are labeled with their weights. The edges labeled with w represent: an edge from a gadget node to a node $q' \in E(q)$ or $E_f(q)$ carry the weight $w(q, q')$.

Theorem 1. *Let* $(G, E_f, \texttt{MP}_0^f)$ *be a fair mean-payoff game with threshold* 0, *where* $G = (Q, E, w)$, $w : E \to [-W, W]$, *and* $|Q| = n$. *Then there exists a mean-payoff game* $G' = (Q', E', w')$ *where* $Q' \supseteq Q$, $|Q'| \leq 6n$ *and* $w' : E' \to [-W', W']$ *with* $W' = n^2W + n$ *such that* $\texttt{Win}_i(G) = \texttt{Win}_i(G') \cap Q$ *for* $i \in \{1, 2\}$.

Standard algorithms for mean-payoff games [29] with Thm. 1 gives an $\mathcal{O}(n^3mW)$ time algorithm (with $|E| = m$) for solving fair mean-payoff games with threshold and an $\mathcal{O}(n^4mW)$ time algorithm for the optimal value problem. As regular mean-payoff games are determined, we also get the following from Thm. 1:

Theorem 2. *Fair mean-payoff games are determined.*

The rest of this section discusses the proof of Thm. 1 for 1-fair mean-payoff games in Sec. 4.1 and for 2-fair mean-payoff games in Sec. 4.2. The detailed proof of Thm. 1 for 2-fair mean-payoff games is provided in [8, App. B]. Finally, Sec. 4.3 discusses the memory requirements of strategies in fair mean-payoff games resulting from the gadget-based reductions.

4.1 Proof of Thm. 1 for 1-fair Mean-Payoff Games

Gadget Construction & Intuition. Let $(G, E_f, \texttt{MP}_0^f)$ be a 1-fair mean-payoff game. We construct the *gadget game* G' by replacing every fair node in $q \in Q_f$ in G, with its 3-step gadget presented in Fig. 2a. That is, the incoming edges of q is redirected to node q at the root, and the outgoing edges on the third step lead to $E_f(q)$ and $E(q)$, the fair successors and successors of q, respectively. The formal construction of G' can be found in [8, App. A].

For notational convenience, for each $q \in Q_f$, we denote the leftmost branch $q \to q_l \to q_{\text{nW+1}}$ of the gadget as *fair branch* $\texttt{br}^q_{\texttt{fair}}$, the middle branch $q \to q_l \to q_0$ as the *simulation branch* $\texttt{br}^q_{\texttt{sim}}$ and the rightmost branch $q \to q_r \to q_{\texttt{ESC}}$ as the *escape branch* $\texttt{br}^q_{\texttt{esc}}$. Intuitively, from a fair node q, player 1 can escape to a part of the game that doesn't contain q and is player 1 winning (according to the current player 1 strategy) by taking the escape branch and therefore paying a high negative payoff. Since by this escape choice q is guaranteed to be seen only finitely often, the negative payoff associated with the escape edge does not change the winner of the game. Now assume there is no such winning escape choice for player 1 from q. Then from q, all player 1 winning play visits q infinitely often, and thus is forced to visit all of q's outgoing fair edges infinitely often. If taking one of its fair outgoing edges pushes the game into a player 2 winning region that does not contain q (according to some player 2 strategy), then in the gadget game player 2 can pay a high positive payoff and choose the fair branch with the correct successor to escape to a player 2 winning region. So, in a way "*fair branch is the escape branch for player 2*". However, player 2 can take this escape branch to win from q, *only if* player 1 cannot take her escape branch to win. The difference in the amplitude of weights between the escape branches of the two players (i.e. $\mathtt{br}^q_{\mathtt{esc}}$ and $\mathtt{br}^q_{\mathtt{fair}}$) stems from this hierarchy between the escape choices of the players. If neither player has an escape choice, then the middle branch $\mathtt{br}^q_{\mathtt{sim}}$ faithfully simulates the fair mean-payoff game without adding any additional weights to the play via the gadget edges to alter the winner.

Using this intuition, we now prove both directions of Thm. 1 separately for 1-fair mean-payoff games.

1. Proof of ($\mathtt{Win_1}(\mathbf{G'}) \cap \mathbf{Q} \subseteq \mathtt{Win_1}(\mathbf{G})$): Let σ' be a positional player 1 strategy that wins q in G' for the regular mean-payoff objective. We construct a player 1 strategy σ that wins q in G.

Consider the subgame arena $G'_{\sigma'}$ of G'. All cycles in $G'_{\sigma'}$ are non-negative; else player 2 can force to eventually only visit a negative cycle and therefore construct a play in $G'_{\sigma'}$ with negative mean-payoff value.

Using σ', we construct σ in G as follows:

a. If $q' \notin Q_f$, then set σ to be positional at q' with $\sigma(q') = \sigma'(q')$.
b. If $\sigma'(q') = q_r$ for some fair node $q' \in Q_f$, then set σ to be positional at q' with $\sigma(q') = \sigma'(q'_{\mathtt{ESC}})$. Intuitively this says if for a fair node q', σ' asks to choose the escape branch of the gadget, then σ follows σ' to choose that branch and moves to the corresponding 'escape node' in G.
c. If $\sigma'(q') = q_l$ for some fair node $q' \in Q_f$, then we set σ to be an infinite memory strategy at q' that is played in rounds: consider an order on the fair successors of q'. In round $i \geq 0$, σ chooses the 'preferred successor' $\sigma'(q'_0)$ exactly i times and then chooses the ith (modulo $|E_f(q')|$) fair successor.

Now we prove that all $\tau \in \mathtt{plays}_\sigma(G)$ are winning for player 1 in G. In particular, we will show that τ is fair and $\mathtt{avg}(\tau) \geq 0$.

For a play $\tau = q^0 q^1 \ldots \in \mathtt{plays}_\sigma(G)$, we construct its *extension* τ' in G' inductively as follows: we start with q^0 and $i = 0$ and with increasing i,

- If $q^i \notin Q_f$, append τ' with q^{i+1};
- If $\sigma(H \cdot q^i)$ is defined using Item b. i.e., $\sigma'(q^i) = q^i_r$ then append τ' with $\mathtt{br}^{q^i}_{\mathtt{esc}} \to q^{i+1}$. In particular, we append τ' with $q^i_r \cdot q^i_{\mathtt{ESC}} \cdot q^{i+1}$.
- If $\sigma(H \cdot q^i)$ is defined using Item c., then

- if $q^{i+1} = \sigma'(q_0^i)$, append τ' with $\mathtt{br}_{\mathtt{sim}}^{q^i} \to q^{i+1}$ i.e., with $q_l^i \cdot q_0^i \cdot q^{i+1}$;
- else, append τ' with $\mathtt{br}_{\mathtt{fair}}^{q^i} \to q^{i+1}$, i.e. $q_l^i \cdot q_{\mathrm{nW}+1}^i \cdot q^{i+1}$.

Clearly, τ' is a play in $G'_{\sigma'}$; hence it contains only non-negative cycles. Further, the restriction of τ' to the nodes Q of G, denoted by $\tau'|_Q$, is exactly τ.

Let a 'tail' of τ denote an infinite suffix $\hat{\tau}$ of τ that contains only the nodes and edges visited infinitely often in τ; and let $\hat{\tau}'$ be the extension of τ in G'. We denote the subgraph of $G'_{\sigma'}$ restricted to the nodes and edges in $\hat{\tau}'$ by $G'_{\hat{\tau}'}$.

We first show that τ is fair. It is sufficient to show that $\hat{\tau}$ does not contain fair nodes p for which $\sigma(p)$ is defined via Item b., because for all the other fair nodes in $\hat{\tau}$, τ visits all their fair successors infinitely often (Item c.). Let p be a node for which $\sigma(p)$ is defined via Item b.. Then $\sigma'(p) = p_r$. As σ' is positional, only the escape branch of p exists in $G'_{\sigma'}$, carrying the weight $-n^2W - n$. Since p is visited infinitely often, there is a (simple) cycle in $G'_{\sigma'}$ that passes through p, and thus sees the weight $-n^2W - n$. In order to maintain a non-negative weight, the cycle needs to contain n-many fair branches (carrying weight $nW+1$), which is not possible since there are at most $n-1$ many fair branches in $G'_{\sigma'}$. With this we conclude that τ is fair.

Now we prove that $\mathtt{avg}(\tau) \geq 0$. First let us make an observation about the cycles in $G'_{\sigma'}$ that use only simulation branches.

Observation (sim-cycles) *We call cycles* $\mathbf{c}'$ *in* $G'_{\sigma'}$ *that do not contain any fair or escape branches* sim-cycles. *That is,* $\mathbf{c}'$ *contains only simulation branches. Being part of* $G'_{\sigma'}$, $\mathbf{c}$ *has non-negative weight. Since all the gadget branches in* $\mathbf{c}'$ *carry* 0 *weight, the restriction* $\mathbf{c} = \mathbf{c}'|_Q$ *is a non-negative weight cycle in* G.

We saw that $\hat{\tau}$ does not contain nodes defined using Item b.. I.e., for all fair nodes p in $\hat{\tau}$, σ is defined via Item c., and $G'_{\hat{\tau}'}$ contains the branches $\mathtt{br}_{\mathtt{fair}}^p$ and/or $\mathtt{br}_{\mathtt{sim}}^p$. Thus, all cycles in $\hat{\tau}'$ are either sim-cycles, or visit a fair branch.

If $\hat{\tau}$ does not contain any fair nodes, $\hat{\tau} = \hat{\tau}'$, and $\mathtt{avg}(\tau) \geq 0$ easily follows. Now we assume $\hat{\tau}$ contains at least one fair node, and use the following easy-to-verify remark to show $\mathtt{avg}(\tau) \geq 0$:

Remark 1. Given a play ρ and $\epsilon > 0$, if there exists a $k \in \mathbb{N}$ and a tail $\hat{\rho}$ of ρ such that every k-length infix $\hat{\rho}_k$ of $\hat{\rho}$ satisfies $\mathtt{avg}(\hat{\rho}_k) \geq -\epsilon$, then $\mathtt{avg}(\rho) \geq -\epsilon$.

Let $\epsilon > 0$ be arbitrary. Let $k \in \mathbb{N}$ such that $\frac{n^2W}{k} < \epsilon$. Since $\hat{\tau}$ has at least one fair node, Item c. is triggered infinitely often to construct the strategy σ. Therefore, the round (in the sense of Item c.) will grow unboundedly bigger. Thus, for any k, we can get a tail $\hat{\tau}$ of τ that is in round k. In $\hat{\tau}$, for every fair node, their edge to their preferred successor is taken at least k-many times between any two occurrences of their other (fair) outgoing edges. Then in $\hat{\tau}'$, the simulation branches of fair nodes are visited at least k-many times in between two occurrences of fair branches. Therefore, in $\hat{\tau}'$ each infix $\hat{\tau}'_k$ of length k visits the fair branch of each node at most once. Consequently, there are at most n-many simple cycles in $\hat{\tau}'_k$ that visits a fair branch, and all its other simple cycles are sim-cycles. Now, total weight of n-many simple cycles that are not sim-cycles are $\geq -n^2W$, and since all the sim-cycles have non-negative weight, $\mathtt{avg}(\tau_k) \geq \frac{-n^2W}{k} \geq -\epsilon$. Using the fact from the previous paragraph, we get that $\mathtt{avg}(\tau) \geq -\epsilon$. This implies that for any $\epsilon > 0$, $\mathtt{avg}(\tau) \geq -\epsilon$. Hence, $\mathtt{avg}(\tau) \geq 0$.

This concludes the proof of the first direction.

2. Proof of $(\mathtt{Win_2}(\mathbf{G'}) \cap \mathbf{Q} \subseteq \mathtt{Win_2}(\mathbf{G}))$**:** Analogous to the previous part, we consider a positional winning strategy π' of player 2 in G' and construct, this time, a *positional* winning strategy π for player 2 in G. Note that, none of the fair nodes in G belong to player 2 and hence the construction of π is quite straight forward: for all $q \in Q_2$, we set $\pi(q) = \pi'(q)$. We define $G'_{\pi'}$ as the subgame restricted to π', as before.

For a play $\tau = q^0q^1 \ldots \in \mathtt{plays}_\pi(G)$, we define its *extension* τ' in G' inductively as follows: we start from with q^0 and $i = 0$, and with increasing i,

- If $q^i \notin Q_f$, append τ' with q^{i+1}; Otherwise,
- If $\mathtt{br}^{q^i}_{\mathtt{sim}}$ exists in $G'_{\pi'}$, append τ' with $q^i_l \cdot q^i_0 \cdot q^{i+1}$;
- If $\mathtt{br}^{q^i}_{\mathtt{sim}}$ does not exist in $G'_{\pi'}$:
  - If $q^{i+1} = \pi'(q^i_{\mathrm{nW}+1})$, append τ' with the fair branch i.e. $q^i_l \cdot q^i_{\mathrm{nW}+1} \cdot q^{i+1}$,
  - Otherwise, append it with the escape branch, i.e. $q^i_r \cdot q^i_{\mathsf{ESC}} \cdot q^{i+1}$.

We define the tail of τ as $\hat{\tau}$ and its extension $\hat{\tau}'$, analogous to the previous part. We denote the subgraph of $G'_{\pi'}$ restricted to the nodes and edges in $\hat{\tau}'$ by $G'_{\hat{\tau}'}$.

Let $\tau \in \mathtt{plays}_\pi(G)$ be a play in G that conforms with π. If τ is not fair, it is automatically won by player 2. Therefore, assume that τ is fair. We show that all the simple cycles in $\hat{\tau}$ has negative weight. For this, we use the fact that its extension τ' is a play in $G'_{\pi'}$ and hence all the cycles in $G'_{\hat{\tau}'}$ have negative weight. Since the length of simple cycles are bounded by n, this gives us $\mathtt{avg}(\tau) < 0$. The main argument is the absence of fair branches in $G'_{\hat{\tau}'}$, introduced in Claim 1.

Claim 1 *If τ is fair, there exist no fair branches ($\mathtt{br}_{fair}$) in $G'_{\hat{\tau}'}$.*

Before proving the claim, we discuss how it implies that all simple cycles in $\hat{\tau}$ are negative, resulting $\mathtt{avg}(\tau) < 0$. From the definition of extension, absence of fair branches in $G'_{\hat{\tau}'}$ implies that either (i) for all $q \in Q_f$ only $\mathtt{br}^q_{\mathtt{sim}}$ exists in $G'_{\hat{\tau}'}$, or (ii) for some $q \in Q_f$, only $\mathtt{br}^q_{\mathtt{esc}}$ exists in $\hat{\tau}'$. By construction of τ', case (ii) above implies the existence of a fair node $q \in Q_f$ that appears infinitely often in τ but does not take all its fair outgoing edges, in particular $q \to \pi'(q_{\mathrm{nW}+1}) \in E_f(q)$ infinitely often. This makes τ unfair, contradicting our assumption. In case of (i), since all simple cycles $\mathbf{c}'$ in $\hat{\tau}'$ are negative and none of the gadget nodes contribute any weight due to only $\mathtt{br}_{\mathtt{sim}}$ existing in $\hat{\tau}'$, we conclude that the cycles $\mathbf{c} = \mathbf{c}'|_Q$ in $\hat{\tau}$ have negative weight. This implies that $\mathtt{avg}(\tau) < 0$.

By the above-mentioned argument, proving Claim 1 concludes the second direction of the proof.

Proof of Claim 1. As discussed above, whenever $\mathtt{br}^q_{\mathtt{esc}}$ is in $G'_{\hat{\tau}'}$ for a node q, $\mathtt{br}^q_{\mathtt{fair}}$ is also in it. Now we remove all the escape branches from $G'_{\hat{\tau}'}$ and obtain the subgraph S. Note that S has no dead-ends, since for all q for which $\mathtt{br}^q_{\mathtt{esc}}$ got removed from $G'_{\hat{\tau}'}$, $\mathtt{br}^q_{\mathtt{fair}}$ was also in $G'_{\hat{\tau}'}$.

For all nodes q in S, either $\mathtt{br}^q_{\mathtt{fair}}$ or $\mathtt{br}^q_{\mathtt{sim}}$ is in S. As a subgraph of $G'_{\pi'}$, all cycles in S have negative weight. It is easy to see that a fair branch cannot lie on a simple cycle in S, because fair branches carry weight $nW + 1$ and the other edges come either from $\mathtt{br}_{\mathtt{sim}}$ carrying weight 0, or from G carrying weight $\geq -W$. Hence, a simple cycle containing a fair branch would have positive weight.

Now assume fair branches exist in $\hat{\tau}'$, but not on cycles of S. Then all cycles in S consist only of simulation branches. However, nodes with simulation branches have the same outgoing edges in $G'_{\hat{\tau}'}$ and in S. Since every node in S is visited infinitely often in τ, τ will eventually enter cycles of S and never leave them, since these nodes have the same outgoing edges in $G'_{\hat{\tau}'}$. This implies fair branches are not infinitely often visited in τ, and concludes the proof of Claim 1. □

4.2 Proof of Thm. 1 for 2-fair Mean-Payoff Games

Observing page constrains, we provide the proof of Thm. 1 for 2-fair mean-payoff games in [8, App. B] and only discuss its differences to Sec. 4.1 here.

Interestingly, the player 2 gadget for 2-fair games (Fig. 2b) is exactly the dual of the player 1 gadget for 1-fair games (Fig. 2a). This is surprising, as the winning objectives of player 1 and 2 are not exactly dual in these games. While player 1 in 1-fair games is expected to win the vertices with optimal value 0 (that is, the best value a player 1 strategy can achieve from these nodes is 0, and getting ϵ-close to the best possible value isn't sufficient to win); for player 2 to win a vertex in a 2-fair game, it is sufficient for him to get ϵ-close to the best possible value he can achieve from this vertex. As we will see in the next chapter, these differences in the objectives' behavior w.r.t. optimal values reflect drastically in the required strategy sizes. Namely, in 2-fair games finite strategies are sufficient whereas in 1-fair games infinite strategies are required to win.

It is therefore surprising that this imbalance in objectives does not demand any changes in the gadget's structure for proving Thm. 1 in the case of 2-fair mean-payoff games. The main difference w.r.t. the proof form Sec. 4.1 lies in the construction of winning strategies for the fair player. In particular, the proof for 1-fair games (in Sec. 4.1) is slightly more complicated due to the required infinite strategy construction for player 2. On the other hand, the proof of 2-fair games (in [8, App. B]) reveals that a finite memory is sufficient for a winning player 2 strategy in 2-fair games.

4.3 Strategy Complexity for Fair Mean-Payoff Games

We list an overview of results on strategy requirements for fair mean-payoff games. Most of these results follow from the proofs of Thm. 1. We discussed in Sec. 3 that player 1 may need infinite memory to achieve an optimal value in a 1-fair mean-payoff game. Lem. 1 shows that player 1 can reach ϵ-close to the optimal value with finite memory strategies.

Lemma 1. *Given a 1-fair mean-payoff game $(G, E_f, \mathtt{MP}^f)$, let the optimal value from some node q_0 is v. Then, for all $\epsilon > 0$, there exists a finite-memory strategy of player 1 that is winning from q_0 in fair mean-payoff game $(G, E_f, \mathtt{MP}(v - \epsilon))$.*

Proof. W.l.o.g, we show this for $v = 0$. In the proof of Thm. 1, we construct an infinite memory strategy σ for player 1 where at i^{th} round, player 1 plays the preferred successor i times and then plays a fair successor. Any σ play has the following property: for any ϵ, there exists a tail of the play such that the average weight of every k-length infix of the tail is at least $-\epsilon$. We fix this k and modify the strategy construction in Item c. to play the preferred successor k times and then play a fair successor in every round. By Rem. 1, this finite memory strategy ensures the mean-payoff value of any play is at least $-\epsilon$. □

The above lemma entails that in a 1-fair mean-payoff game with threshold value 0, player 1 has a finite memory strategy from a node $q \in \mathtt{Win}_1$ if the optimal value is strictly larger than 0. On the other hand, in 2-fair mean-payoff games getting ϵ-close to the best value achievable from a vertex is always sufficient to ensure winning for player 2. Intuitively, for this reason, player 2 has finite memory winning strategies in 2-fair mean-payoff games.

Lemma 2. *Given a 2-fair mean-payoff game* $(G, E_f, \mathtt{MP}^f)$, *for all* $q \in \mathtt{Win}_2$, *there exists a finite-memory strategy of player* 2 *that is winning from* q.

Finally, the proofs of Thm. 1 from Sec. 4.1 (resp. [8, App. B]) reveal the existence of memoryless winning player 2 (resp. player 1) strategies in 1-fair (resp. 2-fair) mean-payoff games.

Lemma 3. *For all fair mean-payoff games, memoryless strategies are sufficient for the player who does not own the fair nodes.*

5 Solving Fair Energy Games

In this section, we aim to extend our gadget-based approach to solve fair energy games. Intuitively, gadget-based approaches transfer the determinacy of regular (qualitative or quantitative) objectives to their fair variants, as demonstrated in [35] for fair parity games and by the proofs of Thm. 1 for fair mean-payoff games. Conceptually, these techniques build on finding winning strategies for either player in the fair game – based on their winning strategies in the gadget game – with similar infinite behavior. Therefore, before attempting a gadget-based approach for fair games, we need to ensure that these games are determined.

In this section, we show that 2-fair energy games are *not determined*, whereas 1-fair energy games are. While due to the above-mentioned reason we cannot construct gadgets for 2-fair energy games, we give simple algorithms to compute both $\mathtt{Win}_1$ and $\mathtt{Win}_2$. For 1-fair energy games, we observe that the gadget in Fig. 2b falls short, and we present new gadgets.

5.1 Discussion on Determinacy of Energy Games

Whenever φ_c is a Borel set, the following holds due to Borel determinacy [39]:

$$\forall \sigma \in \Sigma_1.\, \exists \pi \in \Sigma_2,\ \mathtt{play}_{\sigma,\pi}(G, q) \notin \varphi_c \Leftrightarrow \exists \pi \in \Sigma_2.\, \forall \sigma \in \Sigma_1,\ \mathtt{play}_{\sigma,\pi}(G, q) \notin \varphi_c$$

As $\mathtt{En}_c$ and $\mathtt{En}_c^f$ are Borel sets, the equation holds, showing that energy and fair energy games are determined under a fixed credit c. Furthermore, this equality combined with the negation of Eq. (1) defining $\mathtt{Win}_1$, yields the following formulation of $Q \setminus \mathtt{Win}_1$:

$$q \notin \mathtt{Win}_1 \iff \forall c \in \mathbb{N}. \exists \pi \in \Sigma_2. \forall \sigma \in \Sigma_1, \mathtt{play}_{\sigma,\pi}(G, q) \notin \varphi_c \tag{6}$$

Again, this formulation holds for both energy and fair energy games. Clearly it follows that a (fair) energy game is determined if and only if Eq. (4) is equivalent to Eq. (6). In fact, if we can restrict the quantification over c in Eq. (6) to a finite set, then we can swap $\forall c \in \mathbb{N}$ and $\exists \pi \in \Sigma_2$, which yields the desired equivalence, showing the game is determined. This is indeed the case in energy games: Whenever player 1 wins from a node, it wins with initial credit nW. Dually, whenever player 2 wins w.r.t. initial credit nW, it wins against every initial credit. That is, the quantification over the c can be restricted to $[0, nW]$.

Using this trick, we show in Sec. 5.3 that 1-fair energy games are determined. However, we demonstrate in the next section that this does not hold for 2-fair energy games and that these games are not determined.

5.2 2-fair Energy Games

We start by showing that 2-fair energy games are not determined.

Theorem 3. *2-fair energy games are not determined.*

Proof. Recall that energy games are determined if and only if Eq. (4) is equivalent to Eq. (6). We provide a counterexample to this equivalence by the game graph in Fig. 3. Any fair strategy π for player 2 takes the edge (q, q') after finitely many steps, say after c steps. Then, the unique strategy σ for player 1, which takes the self-loop on q', wins the play $\mathtt{play}_\sigma(G, q)$ with respect to the initial credit c. Thus, player 2 has no winning strategy. However, player 1 has no winning strategy either since for any initial credit c, there exists a player 2 strategy that ensures the energy of every play drops below 0. Namely, any strategy that takes the self-loop of q more than c times achieves this. □

This occurs as player 2 can delay taking his fair edges enough to violate any given c while still satisfying fairness condition in the suffix. Hence, if player 2 can force a negative cycle, then he can use it to violate any initial credit. In contrast, player 1's objective remains the same as in the regular energy game: she wins from a node iff she can prevent player 2 from forcing a negative cycle from that node. This leads to the following result, which is proven in [8, App. C].

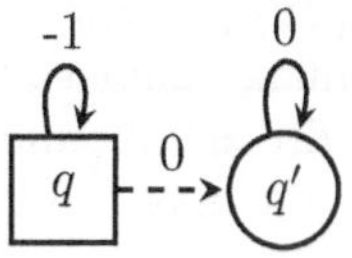

Fig. 3

Theorem 4. *Given a 2-fair energy game $(G, E_f, \mathtt{En})$, the set of winning nodes for player 1 is the same as the set of winning nodes for player 1 in the (regular) energy game $(G, \mathtt{En})$.*

Lemma 4. *In 2-fair energy games, player 1 has memoryless winning strategies.*

We note that Lem. 4 is a corollary of Thm. 4. As 2-fair energy games are not determined, $\mathtt{Win}_2 \neq Q \setminus \mathtt{Win}_1$. We present a simple reduction to compute $\mathtt{Win}_2$ and resulting memory requirements of strategies as a corollary of its proof.

Lemma 5. *Given a 2-fair energy game* $(G, E_f, \mathtt{En})$, *the set of winning nodes for player* 2 *is the same as the set of winning nodes for player* 2 *in the 2-fair mean-payoff game* $(G, E_f, \mathtt{MP}_0^f)$.

Proof. Take a player 2 strategy π that wins from q in the 2-fair mean-payoff game. Then, for any $\tau \in \mathtt{plays}_\pi(G, q)$, $\mathtt{avg}(\tau) = -\epsilon$ for some positive ϵ. That is, there exists an infinite subsequence I of $\mathbb{N}$ such that for all $i \in I$, $w(\tau[0; i]) < -i \cdot \epsilon$. Thus, the energy of τ drops unboundedly. For the opposite direction, due to the determinacy of mean-payoff games and Lem. 3, it suffices to show that for each positional player 1 strategy σ, there exists a player 2 strategy π such that $\tau = \mathtt{play}_{\sigma,\pi}(G, q)$ has negative limit average weight. Let σ be a positional player 1 strategy and π a player 2 strategy that wins q in the 2-fair energy game. Consider $\tau = \mathbf{p} \cdot \hat{\tau}$, where $\hat{\tau}$ is a tail of τ consisting of infinitely often visited nodes and edges of τ. Since the energy of $\hat{\tau}$ drops unboundedly, there exists a negative cycle $\mathbf{c}$ in $\hat{\tau}$ that visits all nodes and (fair) edges of $\hat{\tau}$. Let π_σ be a player 2 strategy that mimics π until $\mathbf{c}$ is seen, and then repeats $\mathbf{c}$. Then $\rho = \mathtt{play}_{\sigma,\pi_\sigma}(G, q)$ has a tail $\mathbf{c}^\omega$, and since $w(\mathbf{c}) < 0$, we have $\mathtt{avg}(\rho) < 0$. □

Lemma 6. *Player 2 has finite memory winning strategies in 2-fair energy games.*

5.3 1-fair Energy Games

In this section, we prove that 1-fair energy games are determined and using that introduce a gadget to reduce 1-fair energy games to regular energy games.
The determinacy of 1-fair energy games does not follow from prior work on ω-regular energy games [7] as the definition of $\mathtt{Win}_2$ there uses Eq. (6) instead of Eq. (4). We establish determinacy by showing that, as in energy games, one can restrict the quantification over $c \in \mathbb{N}$ to a finite set (namely to $c \in [0, nW]$) in Eq. (6). That is, if there exists a player 2 strategy π that satisfies $\forall \sigma \in \Sigma_1, \mathtt{play}_{\sigma,\pi}(G, q) \notin \mathtt{En}_{nW}^f$, then there exists a winning player 2 strategy $\pi^{\mathtt{ext}}$ (winning against all c) as the energy level of plays conforming to this strategy $\pi^{\mathtt{ext}}$ drops unboundedly. It then follows from the discussion in Sec. 5.1 that 1-fair energy games are determined. The proof of Thm. 5 can be found in [8, App. D].

Theorem 5. *1-fair energy games are determined.*

Next we will show that the gadget in Fig. 4 turns 1-fair energy games into regular energy games. The difference of this gadget from Fig. 2a mostly lies in the different treatment of the zero-weight-cycles. We first discuss briefly why such cycles are important, and then introduce this gadget.

The importance of 0-cycles in 1-fair energy games. It is known that the winning regions of a player in energy games and mean-payoff games (with objective $\mathtt{MP}_0$) coincide [16]. This is not true for the 1-fair variants of these games, as player 1 may need infinite memory to win in 1-fair mean-payoff games (see Sec. 3 and 4.1), but not in 1-fair energy games. However, the equivalence of winning regions still holds in fair games without 0-cycles. Intuitively, if there are no 0-cycles, for a winning player 1 (player 2) strategy σ (π) in the fair energy game, the energy of each play τ that conforms with σ (π) grows (drops) unboundedly, which makes τ also winning in the fair mean-payoff game. For instance, in a 1-fair energy game with a player 1 node with a single fair edge with 0 weight is winning for player 1; but if we add another fair edge to this node with weight -1, it becomes losing for player 1. However, when we consider the graph under the mean-payoff objective, both nodes remain in $\mathtt{Win}_1$.

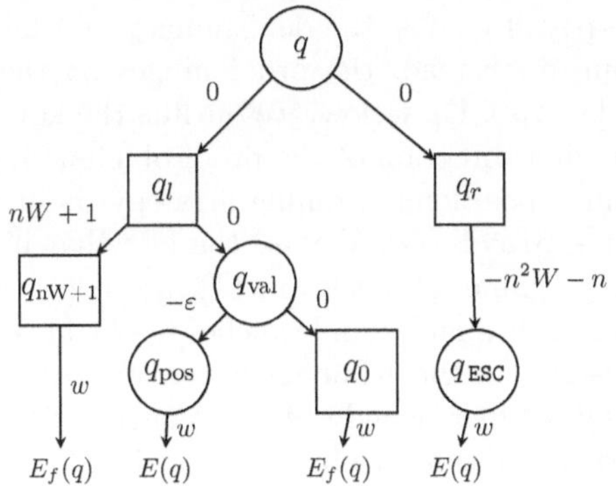

Fig. 4: 1-fair energy gadget where $\epsilon = 1/(n+1)$

Therefore, a 1-fair energy gadget should distinguish between two kinds of nodes, A and B, both of which achieve mean-payoff value 0 but node A (in $\mathtt{Win}_1$) achieves value 0 due to a strategy that visits only 0-cycles while node B (in $\mathtt{Win}_2$) achieves it due to an infinite memory strategy that sees a 0-cycle more and more often. The middle branch of the gadget in Fig. 4 serves this purpose. The gadget matches Fig. 2a on the leftmost ($\mathtt{br}_{\mathtt{fair}}$) and rightmost ($\mathtt{br}_{\mathtt{esc}}$) branches, which still serve as the escape branches for player 1 and player 2, as in Sec. 4.1. The middle branch, called the *value branch* $\mathtt{br}^q_{\mathtt{val}}$ distinguishes nodes of type A and B. Intuitively, if both players do not have escape strategies from q, then the value branch is taken. If player 1 can visit a positive cycle from q, then she wins from q since intuitively she can take this cycle enough times before visiting its fair edges. In this case, she chooses the left successor $q_{\mathtt{pos}}$ of $q_{\mathtt{val}}$, which is called the *positive value branch* $\mathtt{br}^q_{\mathtt{pos}}$ and visits the weight $-\epsilon$. The game is won by player 1 despite the weight $-\epsilon$, due to the positive cycle. On the other hand, if player 1 is winning from q but not by visiting a positive cycle, then she is winning as all the live edges lead to 0-cycles (Node A). In this case, player 1 takes the right successor q_0 of $q_{\mathtt{val}}$, which is called the *zero value branch* $\mathtt{br}^q_0$, and wins. If neither of these cases holds for a node q (Node B), player 2 wins from both successors of the value branch.

Theorem 6. *Let* $(G, E_f, \mathtt{MP}_0^f)$ *be a 1-fair energy game where* $G = (Q, E, w)$, $w : E \to [-W, W]$, *and* $|Q| = n$. *Then there exists an energy game* $G' = (Q', E', w')$ *where* $Q' \supseteq Q$, $|Q'| \leq 8n$ *and* $w' : E' \to [-W', W']$ *with* $W' = (n^2W + n)(n+1)$ *such that* $\mathtt{Win}_i(G) = \mathtt{Win}_i(G') \cap Q$ *for* $i \in \{1, 2\}$.

We obtain the game G' in Thm. 6 by replacing each fair node in the 1-fair energy game G via the gadget in Fig. 4, as in Sec. 4.1. We note that the size of the game given in Thm. 6 is w.r.t. the equivalent energy game on integer weights, obtained by multiplying every weight in the game with $1/\epsilon = n+1$. The state-of-the-art algorithm for energy games [16] and Thm. 6 imply that 1-fair energy games can be solved in time $O(n^4mw)$. Furthermore, the proof reveals that player 1 has finite memory strategies (a local memory of $log(n^3W+n)$ each node) and player 2 has positional strategies.

Unlike fair mean-payoff games, the determinacy of 1-fair energy games *does not* follow from Thm. 6. Instead, the proof hinges on the determinacy result itself, as detailed in [8, App. D]. Below, we outline the key ideas.

Recall that in the first direction of the proof of Thm. 1, we derive a player 1 strategy σ in G from a positional winning strategy σ' in G', reasoning about the tails $\hat{\tau}$ of plays $\tau \in \mathtt{plays}_\sigma(G)$. We use the fact that in a mean-payoff (and similarly, in an energy) game, all cycles in $G'_{\sigma'}$ are non-negative. This gives us that the cycles $\mathbf{c}'$ using only simulation branches in $G'_{\sigma'}$ have projections $\mathbf{c} = \mathbf{c}'|_G$ with non-negative weight. In the 1-fair mean-payoff case, with infinite memory we let σ take these cycles increasingly often to ensure that the projection play $\tau = \tau'|_Q$ has non-negative mean-payoff.

Now, in 1-fair energy games, the cycles $\mathbf{c}'$ in $G'_{\sigma'}$ that visit a positive-value branch have *positive* weight projections $\mathbf{c} = \mathbf{c}'|_Q$, due to the $-\epsilon$ weights. We refer to the choice of the positive-value branch, $\sigma'(q_{\mathbf{pos}})$ as the 'preferred successor' of q that induces positive weight cycles in G. Thus, player 1 can adopt a strategy σ in G that repeatedly selects the preferred successors at fair nodes to accumulate sufficient positive weight before traversing the fair edges to satisfy fairness. As a result, the weight of every tail $\hat{\tau}$ of all plays $\tau \in \mathtt{plays}_\sigma(G)$ is monotonically increasing, and bounded below by some $-d$. However, unlike mean-payoff games, energy games require reasoning about prefixes of plays. Specifically, we need to ensure that the weight of every prefix of each play conforming to σ is bounded below by $-c$ for some $c \in \mathbb{N}$. This is challenging, as the prefix lengths are unbounded, making the determinacy result crucial.

As these games are determined, player 2 wins from a node q if and only if it has a winning strategy π that wins against all initial credits c and all player 1 strategies (Eq. (4)). Therefore, the weight of all plays in $\mathtt{plays}_\pi(G)$ drops unboundedly. Dually, a player 1 strategy λ is winning if all plays in $\mathtt{plays}_\lambda(G)$ are fair with weights bounded below (not necessarily by the same c). Then for all $\tau = \mathbf{u} \cdot \hat{\tau}$, the length of the finite prefix $\mathbf{u}$, together with the lower bound $-d$ of the tail $\hat{\tau}$ naturally yields a lower bound of $-d - W \cdot |\mathbf{u}|$ for the weight of τ.

The proof of the second direction is similar to the case of 1-fair mean-payoff, where we construct a positional winning player 2 strategy π in G based on a positional winning strategy π' in G'. This shows that memoryless strategies suffice for player 2. We give an alternative, simpler proof of this in [8, App. D].

Theorem 7. *In 1-fair energy games, player 1 has finite memory and player 2 has memoryless winning strategies.*

6 Conclusion

In this work, we study the complexity of solving mean-payoff and energy games under strong transition fairness assumptions. We show that when combined with quantitative goals, fairness comes for free in the sense that the complexity of solving these games do not become computationally expensive. We discussed the determinacy and strategy complexity of these games and provided gadget-based techniques to solve them.

A possible future direction is to study the complexity of solving the doubly fair counterpart of these games, where both players are restricted by fairness constraints on their moves, similar to the extension of fair ω-regular games in [35].

In the ω-regular setting, a slight extension of strong transition fairness called *group fairness* is sufficient to make the problem NP-hard, even when combined with the reachability objective [47]. Group fairness is defined via tuples $(G_1, E_1), \ldots, (G_k, E_k)$ where for each $i \in [1, k]$, G_i is a subset of player 2 nodes, and E_i is a subset of edges such that the source node of each edge $e \in E_i$ belongs to G_i. A play τ is called *group fair* if, for each $v \in V$ and $i \in [1, k]$, whenever $v \in \mathtt{Inf}(\tau) \cap G_i$, there exists an $e \in E_i$ that is taken infinitely often in τ.

In a group fair game with an ω-regular objective φ and tuples $(G_1, E_1), \ldots, (G_k, E_k)$ defining the group fairness, player 1 wins a play τ if τ satisfies φ or *is not group fair.* Being a subclass of Streett fairness (or, strong fairness), group fairness already unravels the full complexity of Streett fairness[2]. Since reachability can be expressed as a mean-payoff/energy objective, this NP-hardness result carries over to 2-group-fair[3] mean-payoff games, making these games NP-complete. 2-group-fair energy games again suffer from a lack of determinacy, since they subsume 2-fair energy games. Similarly, as the dual of reachability, the safety objective can be expressed as a mean-payoff/energy objective. This allows us to derive a coNP lower bound for 1-group-fair[4] mean-payoff and 1-group-fair energy games, making these games coNP-complete. However, the exploration of more general notions of fairness – such as other subclasses of Streett fairness – that could provide better complexity bounds or lead to elegant solution algorithms for qualitative or quantitative games remains a significant open area of research.

---

[2] Note that in a game where fairness is inflicted on player 2 (as in the group fairness definition), the fairness condition is a subclass of Rabin objective rather than Streett. Thus, the derived complexity lower bound is NP rather than coNP.

[3] By this, we indicate a game with group fairness where all G_i are player 2 nodes, as in the original definition of group fairness.

[4] By this, we indicate a game with group fairness where all G_i are player 1 nodes. player 1 wins a play τ of a 1-group-fair game with objective φ if τ satisfies φ and *is group fair.*

References

1. de Alfaro, L., Faella, M., Henzinger, T.A., Majumdar, R., Stoelinga, M.: Model checking discounted temporal properties. Theor. Comput. Sci. **345**(1), 139–170 (2005). https://doi.org/10.1016/J.TCS.2005.07.033
2. Almagor, S., Boker, U., Kupferman, O.: Formalizing and reasoning about quality. In: Fomin, F.V., Freivalds, R., Kwiatkowska, M.Z., Peleg, D. (eds.) Automata, Languages, and Programming - 40th International Colloquium, ICALP 2013, Riga, Latvia, July 8-12, 2013, Proceedings, Part II. Lecture Notes in Computer Science, vol. 7966, pp. 15–27. Springer (2013). https://doi.org/10.1007/978-3-642-39212-2_3
3. Almagor, S., Kuperberg, D., Kupferman, O.: The sensing cost of monitoring and synthesis. In: Harsha, P., Ramalingam, G. (eds.) 35th IARCS Annual Conference on Foundation of Software Technology and Theoretical Computer Science, FSTTCS 2015, December 16-18, 2015, Bangalore, India. LIPIcs, vol. 45, pp. 380–393. Schloss Dagstuhl - Leibniz-Zentrum für Informatik (2015). https://doi.org/10.4230/LIPICS.FSTTCS.2015.380
4. Almagor, S., Kupferman, O., Velner, Y.: Minimizing expected cost under hard boolean constraints, with applications to quantitative synthesis. In: Desharnais, J., Jagadeesan, R. (eds.) 27th International Conference on Concurrency Theory, CONCUR 2016, August 23-26, 2016, Québec City, Canada. LIPIcs, vol. 59, pp. 9:1–9:15. Schloss Dagstuhl - Leibniz-Zentrum für Informatik (2016). https://doi.org/10.4230/LIPICS.CONCUR.2016.9
5. Alur, R.: Principles of Cyber-Physical Systems. The MIT Press (2015)
6. Aminof, B., Giacomo, G.D., Rubin, S.: Stochastic fairness and language-theoretic fairness in planning in nondeterministic domains. In: Beck, J.C., Buffet, O., Hoffmann, J., Karpas, E., Sohrabi, S. (eds.) Proceedings of the Thirtieth International Conference on Automated Planning and Scheduling, Nancy, France, October 26-30, 2020. pp. 20–28. AAAI Press (2020)
7. Amram, G., Maoz, S., Pistiner, O., Ringert, J.O.: Energy mu-calculus: Symbolic fixed-point algorithms for omega-regular energy games. CoRR **abs/2005.00641** (2020)
8. Anand, A., Nayak, S.P., Raha, R., Sağlam, I., Schmuck, A.K.: Fair quantitative games (2025), https://arxiv.org/abs/2501.17255
9. Asl, H.J., Uchibe, E.: Estimating cost function of expert players in differential games: A model-based method and its data-driven extension. Expert Systems with Applications **255**, 124687 (2024). https://doi.org/https://doi.org/10.1016/j.eswa.2024.124687
10. Baier, C., Katoen, J.P.: Principles of Model Checking (Representation and Mind Series). The MIT Press (2008)
11. Banerjee, T., Majumdar, R., Mallik, K., Schmuck, A., Soudjani, S.: Fast symbolic algorithms for omega-regular games under strong transition fairness. TheoretiCS **2** (2023)
12. Björklund, H., Sandberg, S., Vorobyov, S.G.: A combinatorial strongly subexponential strategy improvement algorithm for mean payoff games. In: Fiala, J., Koubek, V., Kratochvíl, J. (eds.) Mathematical Foundations of Computer Science 2004, 29th International Symposium, MFCS 2004, Prague, Czech Republic, August 22-27, 2004, Proceedings. Lecture Notes in Computer Science, vol. 3153, pp. 673–685. Springer (2004). https://doi.org/10.1007/978-3-540-28629-5_52

13. Bloem, R., Chatterjee, K., Henzinger, T.A., Jobstmann, B.: Better quality in synthesis through quantitative objectives. In: Bouajjani, A., Maler, O. (eds.) Computer Aided Verification, 21st International Conference, CAV 2009, Grenoble, France, June 26 - July 2, 2009. Proceedings. Lecture Notes in Computer Science, vol. 5643, pp. 140–156. Springer (2009). https://doi.org/10.1007/978-3-642-02658-4_14
14. Bohy, A., Bruyère, V., Raskin, J.: Symblicit algorithms for optimal strategy synthesis in monotonic markov decision processes. In: Chatterjee, K., Ehlers, R., Jha, S. (eds.) Proceedings 3rd Workshop on Synthesis, SYNT 2014, Vienna, Austria, July 23-24, 2014. EPTCS, vol. 157, pp. 51–67 (2014). https://doi.org/10.4204/EPTCS.157.8
15. Bouyer, P., Markey, N., Matteplackel, R.M.: Averaging in LTL. In: Baldan, P., Gorla, D. (eds.) CONCUR 2014 - Concurrency Theory - 25th International Conference, CONCUR 2014, Rome, Italy, September 2-5, 2014. Proceedings. Lecture Notes in Computer Science, vol. 8704, pp. 266–280. Springer (2014). https://doi.org/10.1007/978-3-662-44584-6_19
16. Brim, L., Chaloupka, J., Doyen, L., Gentilini, R., Raskin, J.: Faster algorithms for mean-payoff games. Formal Methods Syst. Des. **38**(2), 97–118 (2011). https://doi.org/10.1007/S10703-010-0105-X
17. Bruyère, V., Filiot, E., Randour, M., Raskin, J.: Meet your expectations with guarantees: Beyond worst-case synthesis in quantitative games. Inf. Comput. **254**, 259–295 (2017). https://doi.org/10.1016/J.IC.2016.10.011
18. Chakrabarti, A., de Alfaro, L., Henzinger, T.A., Stoelinga, M.: Resource interfaces. In: Alur, R., Lee, I. (eds.) Embedded Software, Third International Conference, EMSOFT 2003, Philadelphia, PA, USA, October 13-15, 2003, Proceedings. Lecture Notes in Computer Science, vol. 2855, pp. 117–133. Springer (2003). https://doi.org/10.1007/978-3-540-45212-6_9
19. Chatterjee, K., de Alfaro, L., Faella, M., Majumdar, R., Raman, V.: Code aware resource management. Formal Methods Syst. Des. **42**(2), 146–174 (2013)
20. Chatterjee, K., de Alfaro, L., Henzinger, T.A.: The complexity of stochastic rabin and streett games'. In: Automata, Languages and Programming, 32nd International Colloquium, ICALP 2005, Lisbon, Portugal, July 11-15, 2005, Proceedings. Lecture Notes in Computer Science, vol. 3580, pp. 878–890. Springer (2005). https://doi.org/10.1007/11523468_71
21. Chatterjee, K., Doyen, L.: Energy parity games. CoRR **abs/1001.5183** (2010)
22. Chatterjee, K., Doyen, L.: Energy and mean-payoff parity markov decision processes. In: Murlak, F., Sankowski, P. (eds.) Mathematical Foundations of Computer Science 2011 - 36th International Symposium, MFCS 2011, Warsaw, Poland, August 22-26, 2011. Proceedings. Lecture Notes in Computer Science, vol. 6907, pp. 206–218. Springer (2011). https://doi.org/10.1007/978-3-642-22993-0_21
23. Chatterjee, K., Doyen, L.: Games and markov decision processes with mean-payoff parity and energy parity objectives. In: Kotásek, Z., Bouda, J., Cerná, I., Sekanina, L., Vojnar, T., Antos, D. (eds.) Mathematical and Engineering Methods in Computer Science - 7th International Doctoral Workshop, MEMICS 2011, Lednice, Czech Republic, October 14-16, 2011, Revised Selected Papers. Lecture Notes in Computer Science, vol. 7119, pp. 37–46. Springer (2011). https://doi.org/10.1007/978-3-642-25929-6_3
24. Chatterjee, K., Henzinger, T.A., Jurdzinski, M.: Mean-payoff parity games. In: 20th IEEE Symposium on Logic in Computer Science (LICS 2005), 26-29 June 2005, Chicago, IL, USA, Proceedings. pp. 178–187. IEEE Computer Society (2005). https://doi.org/10.1109/LICS.2005.26

25. Chatterjee, K., Jurdzinski, M., Henzinger, T.A.: Simple stochastic parity games. In: CSL. Lecture Notes in Computer Science, vol. 2803, pp. 100–113. Springer (2003)
26. Cimatti, A., Pistore, M., Roveri, M., Traverso, P.: Weak, strong, and strong cyclic planning via symbolic model checking. Artif. Intell. **147**(1-2), 35–84 (2003)
27. Clemente, L., Raskin, J.: Multidimensional beyond worst-case and almost-sure problems for mean-payoff objectives. In: 30th Annual ACM/IEEE Symposium on Logic in Computer Science, LICS 2015, Kyoto, Japan, July 6-10, 2015. pp. 257–268. IEEE Computer Society (2015). https://doi.org/10.1109/LICS.2015.33
28. Cole, R., Hariharan, R., Paterson, M., Zwick, U.: Tighter lower bounds on the exact complexity of string matching. SIAM Journal on Computing **24**(1), 30–45 (1995). https://doi.org/10.1137/S0097539793245829
29. Comin, C., Rizzi, R.: Improved pseudo-polynomial bound for the value problem and optimal strategy synthesis in mean payoff games. Algorithmica **77**(4), 995–1021 (Apr 2017). https://doi.org/10.1007/s00453-016-0123-1
30. Daviaud, L., Jurdzinski, M., Lazic, R.: A pseudo-quasi-polynomial algorithm for mean-payoff parity games. In: Dawar, A., Grädel, E. (eds.) Proceedings of the 33rd Annual ACM/IEEE Symposium on Logic in Computer Science, LICS 2018, Oxford, UK, July 09-12, 2018. pp. 325–334. ACM (2018). https://doi.org/10.1145/3209108.3209162
31. D'Ippolito, N., Rodríguez, N., Sardiña, S.: Fully observable non-deterministic planning as assumption-based reactive synthesis. J. Artif. Intell. Res. **61**, 593–621 (2018)
32. Dorfman, D., Kaplan, H., Zwick, U.: A Faster Deterministic Exponential Time Algorithm for Energy Games and Mean Payoff Games. In: 46th International Colloquium on Automata, Languages, and Programming (ICALP 2019). Leibniz International Proceedings in Informatics (LIPIcs), vol. 132, pp. 114:1–114:14. Schloss Dagstuhl–Leibniz-Zentrum fuer Informatik (2019). https://doi.org/10.4230/LIPIcs.ICALP.2019.114
33. Ehrenfeucht, A., Mycielski, J.: Positional strategies for mean payoff games. Int. J. Game Theory **8**(2), 109–113 (jun 1979). https://doi.org/10.1007/BF01768705
34. Francez, N.: Fairness. Springer, Berlin (1986)
35. Hausmann, D., Piterman, N., Saglam, I., Schmuck, A.: Fair ømega-regular games. In: FoSSaCS (1). Lecture Notes in Computer Science, vol. 14574, pp. 13–33. Springer (2024)
36. Hélouët, L., Markey, N., Raha, R.: Reachability games with relaxed energy constraints. In: Leroux, J., Raskin, J. (eds.) Proceedings Tenth International Symposium on Games, Automata, Logics, and Formal Verification, GandALF 2019, Bordeaux, France, 2-3rd September 2019. EPTCS, vol. 305, pp. 17–33 (2019). https://doi.org/10.4204/EPTCS.305.2
37. Kress-Gazit, H., Lahijanian, M., Raman, V.: Synthesis for robots: Guarantees and feedback for robot behavior. Annual Review of Control, Robotics, and Autonomous Systems **1**(1), 211–236 (2018)
38. Majumdar, R., Mallik, K., Schmuck, A., Soudjani, S.: Symbolic control for stochastic systems via parity games. CoRR **abs/2101.00834** (2021)
39. Martin, D.A.: Borel determinacy. Annals of Mathematics **102**(2), 363–371 (1975)
40. Mehdipour, N., Althoff, M., Tebbens, R.D., Belta, C.: Formal methods to comply with rules of the road in autonomous driving: State of the art and grand challenges. Automatica **152**, 110692 (2023). https://doi.org/https://doi.org/10.1016/j.automatica.2022.110692, https://www.sciencedirect.com/science/article/pii/S0005109822005568

41. Nilsson, P., Hussien, O., Balkan, A., Chen, Y., Ames, A.D., Grizzle, J.W., Ozay, N., Peng, H., Tabuada, P.: Correct-by-construction adaptive cruise control: Two approaches. IEEE Transactions on Control Systems Technology **24**(4), 1294–1307 (2016). https://doi.org/10.1109/TCST.2015.2501351
42. Nilsson, P., Ozay, N., Liu, J.: Augmented finite transition systems as abstractions for control synthesis. Discret. Event Dyn. Syst. **27**(2), 301–340 (2017)
43. Piterman, N., Pnueli, A.: Faster solutions of rabin and streett games. In: 21th IEEE Symposium on Logic in Computer Science (LICS 2006), 12-15 August 2006, Seattle, WA, USA, Proceedings. pp. 275–284. IEEE Computer Society (2006). https://doi.org/10.1109/LICS.2006.23
44. Queille, J.P., Sifakis, J.: Fairness and related properties in transition systems – a temporal logic to deal with fairness. Acta Inf. **19**(3), 195–220 (1983). https://doi.org/10.1007/BF00265555
45. Saglam, I., Schmuck, A.: Solving odd-fair parity games. In: FSTTCS. LIPIcs, vol. 284, pp. 34:1–34:24. Schloss Dagstuhl - Leibniz-Zentrum für Informatik (2023)
46. Scher, G., Kress-Gazit, H.: Warehouse automation in a day: From model to implementation with provable guarantees. In: 2020 IEEE 16th International Conference on Automation Science and Engineering (CASE). pp. 280–287 (2020). https://doi.org/10.1109/CASE48305.2020.9217012
47. Schmuck, A.K., Thejaswini, K.S., Sağlam, I., Nayak, S.P.: Solving two-player games under progress assumptions. In: Dimitrova, R., Lahav, O., Wolff, S. (eds.) Verification, Model Checking, and Abstract Interpretation. pp. 208–231. Springer Nature Switzerland (2024)
48. Streett, R.S.: Propositional dynamic logic of looping and converse. In: Proceedings of the Thirteenth Annual ACM Symposium on Theory of Computing. p. 375–383. STOC '81, Association for Computing Machinery (1981). https://doi.org/10.1145/800076.802492
49. Thejaswini, K.S.: Attractor decompositions for solving parity and Rabin games. Ph.D. thesis, Warwick (2023), https://wrap.warwick.ac.uk/id/eprint/187144/
50. Thistle, J.G., Malhamé, R.: Control of ω-automata under state fairness assumptions. Systems & control letters **33**(4), 265–274 (1998)
51. Velner, Y., Chatterjee, K., Doyen, L., Henzinger, T.A., Rabinovich, A.M., Raskin, J.: The complexity of multi-mean-payoff and multi-energy games. Inf. Comput. **241**, 177–196 (2015). https://doi.org/10.1016/J.IC.2015.03.001
52. Wong, K.W., Ehlers, R., Kress-Gazit, H.: Resilient, provably-correct, and high-level robot behaviors. IEEE Transactions on Robotics **34**(4), 936–952 (2018). https://doi.org/10.1109/TRO.2018.2830353
53. Zwick, U., Paterson, M.: The complexity of mean payoff games on graphs. Theor. Comput. Sci. **158**(1&2), 343–359 (1996). https://doi.org/10.1016/0304-3975(95)00188-3
54. Zwick, U., Paterson, M.: The complexity of mean payoff games on graphs. Theoretical Computer Science **158**(1), 343–359 (1996). https://doi.org/https://doi.org/10.1016/0304-3975(95)00188-3

Structural Liveness of Conservative Petri Nets

Petr Jančar[1(✉)], Jérôme Leroux[2], and Jiří Valůšek[1]

[1] Department of Computer Science, Faculty of Science, Palacký University, Olomouc, Czechia
petr.jancar@upol.cz

[2] Univ. Bordeaux, CNRS, Bordeaux INP, LaBRI, UMR 5800, F-33400Talence, France

Abstract. We show that the EXPSPACE-hardness result for structural liveness of Petri nets [Jančar and Purser, 2019] holds even for a simple subclass of conservative nets. As the main result we then show that for structurally live conservative nets the values of the least live markings are at most double exponential in the size of the nets, which entails the EXPSPACE-completeness of structural liveness for conservative Petri nets; the complexity of the general case remains unclear. As a proof ingredient with a potential of wider applicability, we present an extension of the known results bounding the smallest integer solutions of boolean combinations of linear (in)equations and divisibility constraints.

Keywords: Petri net · conservative net · structural liveness.

1 Introduction

Petri nets are a well-known model of a class of distributed systems; we can refer, e.g., to the monographs [25] or [3] for an introduction. The *reachability* problem, asking if a given target configuration is reachable from a given initial configuration, is a basic problem of system analysis; in the case of Petri nets this problem is famous for its computational complexity: its Ackermann-completeness has been only recently established (see [18,6] for the lower bound, and [19] for the upper bound). The *boundedness* problem, asking if the reachability set for a given initial configuration is finite, and the *liveness* problem, asking if no action can become dead, are among other standard analysis problems. While boundedness is known to be EXPSPACE-complete [5,24], liveness is tightly related to reachability [10], and is thus now known to be Ackermann-complete.

There are also natural structural versions of boundedness and liveness. The *structural boundedness* problem asks, given a Petri net, if the net is bounded for each possible initial configuration; the *structural liveness* problem asks, given a Petri net, if there is an initial configuration for which the net is live.

While structural boundedness is easily shown to be in PTIME (and is thus substantially easier than boundedness), structural liveness is only known to be EXPSPACE-hard and decidable [14]. In fact, the decidability result can be strengthened by the recent result on the home-space problem [12,13] that easily implies an Ackermannian upper bound for structural liveness as well, but the huge complexity gap still calls for a clarification.

P. A. Abdulla and D. Kesner (Eds.): FoSSaCS 2025, LNCS 15691, pp. 355–376, 2025.
https://doi.org/10.1007/978-3-031-90897-2_17

Our contribution. As a step towards clarifying the complexity of *structural liveness* for general Petri nets, we show the *EXPSPACE-completeness* in the case of *conservative nets*, which do not change a weighted sum of the tokens during their executions. We recall that the problem if a given net is conservative is also in PTIME, similarly as structural boundedness (see, e.g., [22]). A crucial notion in our proof is structural reversibility, called just *reversibility* in this paper; a net is reversible if there is a sequence of actions that contains each action at least once and whose effect is zero (it does not change the configuration when executed). Reversibility is also in PTIME, and it can be easily shown to be a necessary condition for structural liveness in the case of structurally bounded nets. Moreover, it is trivial that each conservative net is structurally bounded, and a straightforward application of Farkas' lemma shows that each reversible structurally bounded net is conservative. Hence our EXPSPACE-completeness result can be equivalently presented as the result for *structurally bounded nets.* A first natural step of our future research plan is to deal with a few subtle points that would allow us to extend the result to the whole class of reversible nets.

The lower bound, the EXPSPACE-hardness, is achieved by adapting the construction of [14] that shows a reduction from the EXPSPACE-complete word problem for commutative semigroups which can be also phrased as a coverability problem for reversible Petri nets [5,21,20]. We recall that coverability is a weaker form of reachability: it asks whether there is a reachable configuration that is component-wise at least as large as the target. Our adaptation, described in detail in the arxiv version of this paper, shows that we get the EXPSPACE-hardness of structural liveness even in the case of nets where each transition has precisely two input places and two output places (i.e., for the nets that naturally correspond to population protocols [1]).

Our main result is the EXPSPACE upper bound. The crucial step proves that for every structurally live conservative net there is a live configuration with an at most 2-exp (double exponential) number of tokens, which in principle matches the lower bound. We achieve this by showing that for any structurally live conservative net there is a quantifier-free Presburger formula which has a "small" (i.e. 2-exp) solution and for which all solutions are live configurations. More precisely, a solution of this formula presents a collection of at most exponentially many configurations that are mutually reachable and are chosen so that they witness that they are all live.

For showing the existence of the mentioned Presburger formula we suggest a way to present a witness for which reachability is safely replaced with conceptually much simpler *virtual reachability*, which allows us to have also negative numbers of tokens in (virtual) configurations. For expressing virtual reachability of reversible nets we use *linear systems*, i.e. boolean combinations of linear (in)equations and divisibility constraints. As a proof ingredient with a potential of wider applicability, we present exponential bounds on the least solutions of linear systems, which is an extension of the known results like those in [23] (which are also referred to in the survey [9]).

We try to perform our analysis of 2-exp functions at a level that allows us to derive the results with sufficient rigour but without technical details that we find unnecessary. To this aim we also introduce the notion of *RB-functions* ("Rackoff-bounded" functions, inspired by [24]) that constitute a special case of 2-exp functions with two variables. The class of RB-functions is closed under various operations including iteration, which gives us a lucid method to build the needed new RB-functions from already established ones.

Related research. This paper can be viewed as a continuation of the research line initiated by the paper [4] which explicitly indicated that even the decidability question for structural liveness of Petri nets had been still open. As already mentioned, now the decidability and the EXPSPACE-hardness are known [14], and the decidability can be strengthened by [12,13] that implies an Ackermannian upper bound. Another result of this research line shows the PSPACE-completeness of structural liveness for IO-nets (Immediate Observation Petri Nets) that were introduced in [8], inspired by a subclass of population protocols in [2]. The paper [8] does not consider structural liveness explicitly, but a PSPACE upper bound follows from its results on liveness immediately; an explicit self-contained proof of the PSPACE-completeness is given in [15].

There is a long list of papers that studied liveness for various subclasses of Petri nets, often exploring related structural properties (we can refer to the monographs [25,3,7] for examples). We can also name [11] as an example of a paper in which structural liveness is among the explicitly studied problems for a subclass of Petri nets. We will also use the known results on liveness for conservative nets, for which we can refer, e.g., to [22].

Organization of the paper. Section 2 gives basic notions and notation, introduces linear systems and RB-functions, and the main results. Section 3 proves the above mentioned 2-exp upper bound, also using the results that are proved separately in Sections 4 and 5. Section 6 discusses the EXPSPACE lower bound. *A longer version of this paper can be found at* https://arxiv.org/abs/2503.11590.

2 Basic Definitions, and Results

By $\mathbb{N}$, $\mathbb{N}_+$, and $\mathbb{Z}$ we denote the sets of nonnegative integers, positive integers, and integers, respectively. For $i, j \in \mathbb{Z}$ we put $[i, j] = \{i, i+1, \ldots, j\}$. The unary operation $|.|$ denotes the absolute value for numbers, the cardinality for sets, and the length for sequences. For a vector $x \in \mathbb{Z}^d$ ($d \in \mathbb{N}$), by $x(i)$ we denote the value of its component $i \in [1, d]$, hence $x = (x(1), x(2), \ldots, x(d))$. We use the *component-wise (partial) order* $\leq$ on $\mathbb{Z}^d$.

It will be clear from the context when a vector is understood as a column vector; e.g., if B is an $m \times n$ matrix, then in $Bx = b$ the vectors x, b are viewed as column vectors $n \times 1$ and $m \times 1$, respectively. By $\mathbf{0}$ we denote the zero vector whose type is clear from the context. Sometimes we consider vectors as elements of $\mathbb{Z}^J$ where J is a finite subset of $\mathbb{N}_+$; e.g., $\mathbb{N}^d$ is viewed as equal to $\mathbb{N}^{[1,d]}$. For $x \in \mathbb{Z}^d$ and $J \subseteq [1, d]$, by $x_{|J}$ we denote the (restricted) vector from $\mathbb{Z}^J$ satisfying $x_{|J}(i) = x(i)$ for each $i \in J$; for $X \subseteq \mathbb{Z}^d$ we put $X_{|J} = \{x_{|J} \mid x \in X\}$.

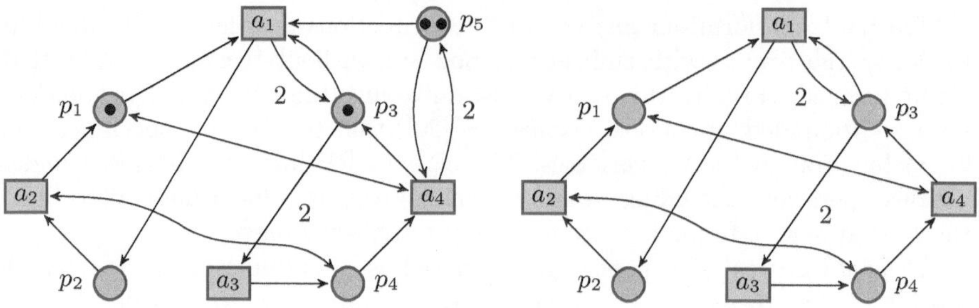

Fig. 1. Conservative net A_1 with a live configuration (left); restriction A_2 of A_1 (right).

We use "$\cdot$" for the standard multiplication, but $(x \cdot y)$ for $x, y \in \mathbb{Z}^d$ denotes the dot product $\sum_{i=1}^{d} x(i) \cdot y(i)$. The *rank* of an $m \times n$ matrix B is denoted by $\mathrm{rank}(B)$; hence $\mathrm{rank}(B) \leq \min\{m, n\}$.

We use the *norms* of $x \in \mathbb{Z}^J$ and finite sets $X \subseteq \mathbb{Z}^J$ (where $\mathbb{Z}^J$ can be $\mathbb{Z}^d$):

$$||x|| = \max_{i \in J} |x(i)|, \quad ||x||_1 = \sum_{i \in J} |x(i)|, \quad ||X|| = \max_{x \in X} ||x||, \quad ||X||_1 = \max_{x \in X} ||x||_1;$$

hence $0 \leq ||x|| \leq ||x||_1 \leq |J| \cdot ||x||$, and $0 \leq ||X|| \leq ||X||_1 \leq |J| \cdot ||X||$. We define $||D||$ and $||D||_1$ also *for matrices over* $\mathbb{Z}$, viewing them as the sets of their rows: for an $m \times n$ matrix B, $||B|| = \max_{i,j} |B_{ij}|$ and $||B||_1 = \max_{i \in [1,m]} \sum_{j=1}^{n} |B_{ij}|$.

For $X \subseteq \mathbb{Z}^d$, by X^* we denote the *submonoid of* $(\mathbb{Z}^d, +)$ *generated by* X, i.e. the set of finite sums of elements of X.

Petri Nets. A *Petri net A of dimension $d \in \mathbb{N}$*, a *d-dim net A* for short, is a finite set of pairs $a = (a_-, a_+) \in \mathbb{N}^d \times \mathbb{N}^d$ which are called *actions* (or *transitions*); we put $||A|| = \max_{a \in A} ||a||$ where $||a|| = ||\{a_+, a_-\}||$.

A *configuration* (or a *marking*) of A is a vector $x \in \mathbb{N}^d$, attaching the values $x(i)$ (the number of *tokens*) to the *components* (or *places*) $i \in [1, d]$. Each action $a = (a_-, a_+)$ has the associated *displacement*, namely $\Delta(a) = (a_+ - a_-) \in \mathbb{Z}^d$. A d-dim *net* A is *conservative* if there is $w \in (\mathbb{N}_+)^d$ such that $(\Delta(a) \cdot w) = 0$ for all $a \in A$; if $w(i) = 1$ for all $i \in [1, d]$, then A is 1-*conservative.*

Example. Figure 1(left) shows a conservative 5-dim net A_1 with 4 actions (and with $w = (1, 1, 1, 2, 1)$); e.g., $a_1 = ((a_1)_-, (a_1)_+) = ((1, 0, 1, 0, 1), (0, 1, 2, 0, 0))$, $\Delta(a_1) = (-1, 1, 1, 0, -1)$, $\Delta(a_3) = (0, 0, -2, 1, 0)$, ...

If $x = c + a_-$ and $y = c + a_+$ for some $c \in \mathbb{N}^d$, then we have $x \xrightarrow{a} y$; this defines the *relation* $\xrightarrow{a}$ on $\mathbb{N}^d$. (For A_1 in Figure 1 we have $(0, 2, 0, 1, 0) \xrightarrow{a_2} (1, 1, 0, 1, 0)$ but *not* $(0, 2, 0, 0, 0) \xrightarrow{a_2} (1, 1, 0, 0, 0)$.) For an action sequence $\sigma = a_1 a_2 \ldots a_k$, the relation $\xrightarrow{\sigma} \subseteq \mathbb{N}^d \times \mathbb{N}^d$ is the composition $\xrightarrow{a_1} \circ \xrightarrow{a_2} \cdots \circ \xrightarrow{a_k}$, with the *displacement* $\Delta(\sigma) = \sum_{j=1}^{k} \Delta(a_j)$. Hence $x \xrightarrow{\sigma} y$ implies $y = x + \Delta(\sigma)$ but not necessarily vice versa. To $x \xrightarrow{\sigma} y$ we also refer as to an *execution of A*. The *reachability relation of A* is $\xrightarrow{*} \subseteq \mathbb{N}^d \times \mathbb{N}^d$ where $x \xrightarrow{*} y$ if $x \xrightarrow{\sigma} y$ for some σ.

For a d-dim net A, the *virtual reachability relation of A* is the relation $\overset{*}{\dashrightarrow}$ on $\mathbb{Z}^d$ for which $x \overset{*}{\dashrightarrow} y$ if $y-x = \Delta(\sigma)$ for some action sequence σ; in this case we speak on a *virtual execution* $x \overset{\sigma}{\dashrightarrow} y$ *of* A. We define $A_\delta = \{\Delta(a) \mid a \in A\}$, and note that $x \overset{*}{\dashrightarrow} y$ iff $(y-x) \in (A_\delta)^*$; we further write A_δ^* instead of $(A_\delta)^*$. A net A is *structurally reversible*, called just *reversible* in this paper, if the monoid A_δ^* is a subgroup of $(\mathbb{Z}^d, +)$, i.e., if for every $a \in A$ we have $-\Delta(a) \in A_\delta^*$; in this case the virtual reachability is symmetric: $x \overset{*}{\dashrightarrow} y$ iff $y \overset{*}{\dashrightarrow} x$.

For a d-dim net A and $I \subseteq [1,d]$, the $|I|$-dim net $A_{|I}$ arises by the restriction of A to the components in I. We also refer to executions $x \xrightarrow{\sigma} y$ (or virtual executions $x \overset{\sigma}{\dashrightarrow} y$) of $A_{|I}$, implicitly assuming that $x, y \in \mathbb{N}^I$ (or $x, y \in \mathbb{Z}^I$) and that the actions in σ are restricted to I ($a \in A$ is in $A_{|I}$ viewed as $((a_-)_{|I}, (a_+)_{|I})$). We note that reversibility of A implies reversibility of $A_{|I}$; this implication does not hold for conservativeness. (In Figure 1, $A_2 = (A_1)_{|[1,4]}$ is not conservative since $\Delta(a_1 a_2) = (0,0,1,0)$.)

A *configuration* $x \in \mathbb{N}^d$ of a net A is *live* if for all $a \in A$ and x' such that $x \xrightarrow{*} x'$ there is y such that $x' \xrightarrow{*} y$ and $y \geq a_-$ (hence $y \xrightarrow{a} y + \Delta(a)$). A *net A* is *structurally live* if it has a live configuration. (A_1 in Figure 1 is an example.)

Linear Systems. Let $\mathbf{x}$ be a variable ranging over $\mathbb{Z}^d$, a *d-dim variable* for short. A constraint of the form $(\alpha \cdot \mathbf{x}) \sim c$ where $\alpha \in \mathbb{Z}^d$, $c \in \mathbb{Z}$, and $\sim\, \in \{=, \geq\}$ is an *equality constraint* if $\sim$ is $=$, and an *inequality constraint* if $\sim$ is $\geq$; it is a *homogeneous constraint* if $c = 0$. For $m \in \mathbb{N}_+$, a constraint $(\alpha \cdot \mathbf{x}) \equiv c \pmod{m}$, where $\alpha \in [0, m-1]^d$ and $c \in [0, m-1]$, is called a *divisibility constraint.* A *d-dim linear system S* is a propositional formula in which atomic propositions are equality, inequality, and divisibility constraints, for a fixed d-dim variable. The *set* $[[S]]$ of *solutions* of S consists of the vectors $x \in \mathbb{Z}^d$ satisfying S; if $[[S]] \neq \emptyset$, then S is *satisfiable*. By $||S||$ we mean the least $s \in \mathbb{N}$ such that $\max\{||\alpha||, |c|\} \leq s$ for all equality and inequality constraints $(\alpha \cdot \mathbf{x}) \sim c$ in S; moreover, $\mathrm{lcm}(S)$ is the *least common multiple* of all m occurring in $(\bmod\, m)$ in divisibility constraints in S, stipulating $\mathrm{lcm}(S) = 1$ if there are no such constraints.

Given a set $M \subseteq \mathbb{N}^d$, by $\min_{\leq}(M)$ we denote the set of minimal vectors in M w.r.t. the (component-wise) order $\leq$. Since $\leq$ is a well quasi order on $\mathbb{N}^d$, the set $\min_{\leq}(M)$ is finite. In [23], Pottier provided several bounds for minimal nonnegative solutions of a conjunction of homogeneous equality constraints. We now recall a bound that is particularly useful for us.

Lemma 1 ([23]). *Let $M = \{x \in \mathbb{N}^n \mid Bx = \mathbf{0}\}$ where B is an $m \times n$ matrix over $\mathbb{Z}$. Then $X = \min_{\leq}(M \setminus \{\mathbf{0}\})$ is a finite set such that $M = X^*$. Moreover the following bound holds, where $r = \mathrm{rank}(B)$:*

$$||X||_1 \leq (1 + ||B||_1)^r.$$

We note that the "rank" bound r in Lemma 1 satisfies $r \leq \min\{m, n\}$. There is also a bound on the solutions of a conjunction of inequality constraints, $Bx \geq b$, in [23], but with the exponent m; this is not convenient for us when m, the number of constraints, is much larger than $n = d$, the dimension related to our problem. Therefore in Section 4 we provide a proof of the following theorem:

Theorem 2. *Any satisfiable d-dim linear system S has a solution $x \in \mathbb{Z}^d$ such that $\|x\|_1 \leq \text{lcm}(S) \cdot (d+1) \cdot \left(2 + d + d^2 \cdot ||S||\right)^{2d+1}$.*

Remark. Given a structurally live conservative d-dim net A, we will later (in the proof of Theorem 5) use a d'-dim linear system S where $d' = d + d \cdot 2^d$; it is crucial for us that Theorem 2 then yields a solution that is at most 2-exp in d. By [16, Theorem 3.12] we could derive that for S, even in the case of no divisibility constraints, the size of the minimal automaton encoding $[[S]]$ in binary is bounded by $(2 + 2 \cdot ||S||)^{|S|}$ where $|S|$ is the number of constraints in S; in our case $|S| \geq 2^d$. The bounds on the values of minimal solutions of S that can be derived from the shortest accepting paths of that automaton are thus not smaller than $2^{(2+2\cdot||S||)^{|S|}}$, which is 3-exp in d.

Linear Systems for Subgroups of $\mathbb{Z}^d$. In the sequel, by a *group* we mean a *subgroup* of $(\mathbb{Z}^d, +)$, which is also called a *lattice* in this context. The group *spanned* by a set $X \subseteq \mathbb{Z}^d$ is the monoid $(X \cup -X)^*$, i.e., the set of finite sums of elements of $X \cup -X$. In Section 5 we prove the following theorem that provides a way to encode any group by a linear system, with a bound on its size.

Theorem 3. *Let L be the group spanned by a finite set $X \subseteq \mathbb{Z}^d$. There exists a linear system S such that $[[S]] = L$ and the following bounds hold:*

$$||S|| \leq d \cdot d! \cdot \|X\|^d \quad \textit{and} \quad \text{lcm}(S) \leq d! \cdot \|X\|^d.$$

We thus get a corollary characterizing virtual reachability of reversible nets:

Corollary 4. *For every reversible d-dim net A there is a 2d-dim linear system S_A such that $[[S_A]] = \{(x, y) \in \mathbb{Z}^d \times \mathbb{Z}^d \mid x \overset{*}{\dashrightarrow} y\}$ and we have*

$$||S_A|| \leq d \cdot d! \cdot \|A_\delta\|^d \quad \textit{and} \quad \text{lcm}(S_A) \leq d! \cdot \|A_\delta\|^d.$$

Proof. We recall that $x \overset{*}{\dashrightarrow} y$ iff $y - x \in A_\delta^*$, and that A_δ^* is a subgroup of $(\mathbb{Z}^d, +)$ when A is reversible. Let S be the linear system, with a variable $\mathbf{x}$ ranging over $\mathbb{Z}^d$, guaranteed by Theorem 3 for $L = A_\delta^*$; we note that L is spanned by A_δ. From S we create S_A with variables $(\mathbf{x}, \mathbf{y})$ ranging over $\mathbb{Z}^d \times \mathbb{Z}^d$ by replacing every occurrence of $\mathbf{x}$ by $\mathbf{y} - \mathbf{x}$. It follows that $[[S_A]]$ is the set of pairs $(x, y) \in \mathbb{Z}^d \times \mathbb{Z}^d$ such that $x \overset{*}{\dashrightarrow} y$, and that $||S_A|| = ||S||$ and $\text{lcm}(S_A) = \text{lcm}(S)$. □

RB-functions. We call a function $f : \mathbb{N}^2 \to \mathbb{N}$ an *RB-function* (from "Rackoff bounded") if there are $c \in \mathbb{N}$ and a polynomial $p : \mathbb{N} \to \mathbb{N}$ so that

$$f(m, d) \leq (c + m)^{2^{p(d)}} \tag{1}$$

for all $m, d \in \mathbb{N}$. Using this notion, we state our main theorem:

Theorem 5. *There is an RB-function f with the following property: for every structurally live conservative Petri net A of dimension d there is a live configuration $x \in \mathbb{N}^d$ with $||x|| \leq f(||A||, d)$.*

Corollary 6. *Structural liveness for conservative nets is EXPSPACE-complete.*

Proof. Given a structurally live conservative d-dim net A, we can guess a live configuration $x \in \mathbb{N}^d$ guaranteed by Theorem 5, whose binary presentation fits in exponential space. Since verifying livenes of x can be done in polynomial space w.r.t. the binary encoding of A and x (see, e.g., Lemma 5 in [22]), we have the upper bound. The lower bound follows by adapting the reduction from [14], as discussed in Section 6 and demonstrated in the arxiv version of this paper. □

3 Upper Bound (Proof of Theorem 5)

Convention on RB-functions. It is straightforward to derive that the set of RB-functions (defined by (1)) is closed under the standard operations of sum, product, and composition, when the composition is defined by $f \circ g\,(m,d) = f(g(m,d),d)$. Moreover, if $f(m,d)$ is an RB-function, then $f'(m,d) = f^{(d)}(m,d)$ is an RB-function as well, when $f^{(i)}$ denotes $f \circ f \cdots \circ f$ where f occurs i times. We use these facts implicitly when deriving the existence of RB-functions. For convenience, we also assume that each *RB-function* f that we consider satisfies $m \leq f(m,d)$, and is *nondecreasing*, i.e. $(m \leq m' \wedge d \leq d') \Rightarrow f(m,d) \leq f(m',d')$.

Convention on using "small" and "large". In the following proof, if we consider a d-dim net A and say that a *value* $k \in \mathbb{N}$ is *small*, then we mean that $k \leq f(||A||, d)$ for an RB-function f (independent of A) whose existence is clear from the context or will be clarified later, while its concrete form might be left implicit. When we say that a *value* $k \in \mathbb{N}$ is *(sufficiently) large*, then we analogously mean that $k \geq f(||A||, d)$ for a suitable RB-function f. We use the same convention for Petri nets with states (PNSs) that are introduced later; in such cases the size $||A||$ in $f(||A||, d)$ is replaced with $||G||$ for the respective PNS G.

Overview of the proof. In Section 3.1 we show that structurally live conservative nets are reversible, and that for any $x \overset{*}{\dashrightarrow} y$ in a reversible net there is a virtual execution from x to y that consists of small segments that stepwise approach the target y from the start x. Section 3.2 introduces Petri nets with states (PNSs) that handle the case when some components in live configurations are necessarily small, in which case their values can be viewed as "control states". In Section 3.3 we show how to extract a suitable PNS G with a small set of control states when given a structurally live conservative net A. Section 3.4 then shows how to reduce virtual reachability in the extracted PNS G to virtual reachability in a small net A^G_{sc} whose actions correspond to simple cycles in the control unit of G. Sections 3.5 and 3.6 give a characterization of large nonlive configurations of G in terms of virtual reachability in A^G_{sc}. Finally, Section 3.7 uses this characterization to prove Theorem 5 by defining a related linear system and applying

Corollary 4 and Theorem 2. We remark that we use the name *lemma* for a few important ingredients of the main proof; the proofs of lemmas use facts captured by *propositions.*

3.1 Virtual reachability in reversible nets

By a *bottom SCC* (strongly connected component) *of* a *net* A we mean a nonempty set X of configurations where for each $x \in X$ we have $\{y \mid x \xrightarrow{*} y\} = X$.

Proposition 7 (Finite bottom SCCs, and reversibility).
Given a conservative net A*, the reachability set* $R(x) = \{y \mid x \xrightarrow{*} y\}$ *of any configuration* x *of* A *is finite and subsumes a bottom SCC* $X \subseteq R(x)$*. If* A *is, moreover, structurally live, then* A *is reversible (due to a live bottom SCC).*

We recall that the virtual reachability of any reversible net A is symmetric ($x \overset{*}{\dashrightarrow} y$ implies $y \overset{*}{\dashrightarrow} x$), since A_δ^* is a group in this case; we can thus segment virtual executions $x \overset{*}{\dashrightarrow} y$ into small parts that are "directed" from the start x to the target y, as we now explain.

For any $d \in \mathbb{N}$, we define the function $\text{SIGN} : \mathbb{Z}^d \to \{-1, 0, 1\}^d$ so that for all $x \in \mathbb{Z}^d$ and $i \in [1, d]$ we have $\text{SIGN}(x)(i) = -1$ if $x(i) < 0$, $\text{SIGN}(x)(i) = 0$ if $x(i) = 0$, and $\text{SIGN}(x)(i) = 1$ if $x(i) > 0$. On $\mathbb{Z}^d$ we define the partial order $\preceq$:

$$x \preceq y \text{ if } \text{SIGN}(x) = \text{SIGN}(y) \text{ and } |x(i)| \leq |y(i)| \text{ for all } i \in [1, d].$$

For $X \subseteq \mathbb{Z}^d$, $\min_\preceq(X)$ denotes the set of minimal elements of X w.r.t. $\preceq$; this set is finite, and for each $x \in X$ there is $y \in \min_\preceq(X)$ such that $y \preceq x$, since $\preceq$ is a wqo (well quasi order).

Proposition 8 ("Directed" decomposition of group elements).
If $L \subseteq \mathbb{Z}^d$ *is a group and* $y \in L$*, then* $y = z + (y - z)$ *where* $z \in \min_\preceq(L) \subseteq L$*,* $z \preceq y$*, and* $(y - z) \in L$*; here for each* $i \in [1, d]$ *we have: if* $y(i) = 0$*, then* $0 = z(i) = y(i)$*; if* $y(i) > 0$*, then* $0 < z(i) \leq y(i)$*; and if* $y(i) < 0$*, then* $0 > z(i) \geq y(i)$*, which also entails that* $||y - z|| < ||y||$ *when* $y \neq \mathbf{0}$*.*

Corollary 9 (Segmenting virtual executions for reversible nets).
Given a reversible net A *(hence* A_δ^* *is a group), if* $x \overset{*}{\dashrightarrow} y$ *then there are* $z_1, z_2, \ldots, z_k$ *in* $\min_\preceq(A_\delta^*)$ *such that*

$$x \overset{*}{\dashrightarrow} (x + z_1) \overset{*}{\dashrightarrow} (x + z_1 + z_2) \cdots \overset{*}{\dashrightarrow} (x + z_1 + z_2 \cdots + z_k) = y \qquad (2)$$

and for each $j \in [1, k]$ *we have* $z_j \preceq y - (x + z_1 + z_2 \cdots + z_{j-1})$*.*

The next lemma shows that $||\min_\preceq(A_\delta^*)||$ is small (hence the steps in (2) are small). In its precise form (shown in the arxiv version), the lemma might find applications in more general contexts; it does not assume that A_δ^* is a group, so it is applicable to all Petri nets. We note that the precise bound is exponential, not double exponential, but the claim here is sufficient at our level of analysis.

Lemma 10 (For any finite $A \subseteq \mathbb{Z}^d$, $||\min_\preceq(A^*)||$ is small).
There is an RB-function f *with the following property:*
for any finite set $A \subseteq \mathbb{Z}^d$ *we have* $||\min_\preceq(A^*)|| \leq f(||A||, d)$*.*

3.2 Petri nets with states (PNSs)

It will turn out that it is convenient for our proof (of Theorem 5) when there is a live bottom SCC X containing a configuration x with all components $x(i)$ being sufficiently large (since (non)reachability between configurations x, y with all components being large coincides with their virtual (non)reachability, as is expressed more precisely later).

But liveness is not monotonic (if x is live and $x \leq y$, then y is not necessarily live, even in the case of conservative nets), and we cannot assume that structurally live nets have live configurations with all components being large; we must also handle the cases when some components, constituting a set $I \subseteq [1,d]$, are small in all configurations in the respective live bottom SCC X. In this case it will be convenient to view the restriction $X_{|I}$ as a set of (control) states, and to present any configuration $x \in X$ as the pair (p, x') where $p = x_{|I}$ and $x' = x_{|J}$ for $J = [1,d] \smallsetminus I$. (For the structurally live conservative net A_1 in Figure 1 it is not hard to check that each live configuration has (very) small values in the components (places) p_1, p_2, p_3, p_4, while the value in p_5 can be arbitrarily large.)

This leads us to (a special type of) the notion of Petri nets with states.

Petri Nets with States (PNSs). Given a bottom SCC X of a conservative d-dim net A, for any $I \subseteq [1,d]$ we get a *Petri net with states (PNS)* $G_{(X,I)}$ as described below; for them we will also use a result from [17] where a more general definition of PNSs is given.

We view $G_{(X,I)}$ as a tuple (an "enhanced graph") (Q, A, E) where $Q = X_{|I}$ is the set of *states*, A is the underlying Petri net, and E is the set of *edges* $(p, a, q) \in Q \times A \times Q$ such that $p \xrightarrow{a} q$ in the restricted net $A_{|I}$.

For $G = G_{(X,I)} = (Q, A, E)$ we say that G is of *dimension* d (inherited from A), or that G is a *d-dim PNS*. We define the *norm of* G as $||G|| = \max\{||Q||, ||A||\}$ (which might be much smaller than $||X||$). By the set $\textsc{Conf}(G)$ of *configurations* of the PNS G we mean the set $\{x \in \mathbb{N}^d \mid x_{|I} \in Q\}$ that is equivalently presented as $Q \times \mathbb{N}^J$, where $J = [1,d] \smallsetminus I$.

Remark. The dimension of $G = G_{(X,I)} = (Q, A, E)$ in the above notation could be naturally defined as $|J|$ (which corresponds to the notion of dimension in the case of vector addition systems with states) but we define it as d to stress that the underlying net A is always the primary object for us.

For each action $a \in A$, the relation $\xrightarrow{a}_G$ is the restriction of $\xrightarrow{a} \subseteq \mathbb{N}^d \times \mathbb{N}^d$ (related to A) to the set $\textsc{Conf}(G) \times \textsc{Conf}(G)$. The notation $(p, x) \xrightarrow{a} (q, y)$ refers to $\xrightarrow{a}_G$ implicitly. The notation $(p, x) \xrightarrow{\sigma} (q, y)$, for action sequences σ, and $(p, x) \xrightarrow{*} (q, y)$ refers to *executions of* G, which are implicitly based on the relations $\xrightarrow{a}_G$ (and thus constitute a subset of the set of executions of A).

For the graph (Q, A, E) (with labelled edges), we use the standard notions of *paths*, *cycles*, *simple cycles*, and their *displacements*: the displacement of a path $(p_0, a_1, p_1)(p_1, a_2, p_2) \cdots (p_{k-1}, a_k, p_k)$ is $\Delta(a_1 \cdots a_k) \in \mathbb{Z}^d$ (where $\Delta(a_1 \cdots a_k)_{|I} = p_k - p_0$). We note that our definition guarantees that the graph $G_{(X,I)}$ is strongly connected, and that there is a cycle that visits all states in Q and all edges in E and has the displacement $\mathbf{0}$ (since $G_{(X,I)}$ has arisen from a bottom SCC X).

We remark that the PNS $G_{(X,\emptyset)}$ has a single state and corresponds to the original net A, with the set of configurations (isomorphic with) $\mathbb{N}^d$. On the other hand, $\textsc{Conf}(G_{(X,[1,d])})$ is finite, since it is (isomorphic with) X.

Reversibility of PNSs, and Restricted PNSs. By the above definition, each PNS $G = G_{(X,I)} = (Q, A, E)$ is *reversible* in the sense that for every edge $(p, a, q) \in E$ there is a path from q to p labelled with σ such that $\Delta(a) + \Delta(\sigma) = \mathbf{0}$.

For technical reasons we will also consider PNSs arising from the PNSs $G_{(X,I)}$ as follows: For a d-dim PNS $G = G_{(X,I)} = (Q, A, E)$ and $J' \subseteq J = [1, d] \smallsetminus I$, by $G_{|J'}$ we denote the *restricted PNS* arising from G by removing (ignoring) the components from $J \smallsetminus J'$; the set of configurations of $G_{|J'}$ is thus $Q \times \mathbb{N}^{J'}$, and the executions of $G_{|J'}$ are executions of $A_{|I \cup J'}$. We note that the reversibility of G entails that the PNS $G_{|J'}$ is also reversible.

Proper PNSs. Since we have defined $G_{(X,I)} = (Q, A, E)$ for any $I \subseteq [1, d]$, we cannot exclude that for $A_{|I}$ we have $p \xrightarrow{a} p'$ where $p \in Q$ and $p' \notin Q$; in other words, Q might not be a bottom SCC related to $A_{|I}$.

We say that a *PNS* $G = G_{(X,I)} = (Q, A, E)$ is *proper* if Q is a bottom SCC related to $A_{|I}$. We observe that in this case each execution of A from any x such that $x_{|I} \in Q$ is, in fact, also an execution of G.

We now show that $G_{(X,I)} = (Q, A, E)$ is proper if there is $x \in X$ for which $x_{|J}$, where $J = [1, d] \smallsetminus I$, is sufficiently large w.r.t. $||G||$ in all components. For $C \in \mathbb{N}$, by the *area* $\uparrow(C, \ldots, C)$ we mean the set $\{x \in \mathbb{N}^J \mid x(i) \geq C$ for all $i \in J\}$ when the respective set J is clear from the context.

Proposition 11 (Large "counters" in X induce that G is proper). *There is an RB-function f_{PROP} with the following property: for every d-dim PNS $G = G_{(X,I)} = (Q, A, E)$, if there is $x \in X$ with $x_{|J} \in \uparrow(C, \ldots, C)$ for $J = [1, d] \smallsetminus I$ and $C = f_{\text{PROP}}(||G||, d)$, then G is a proper PNS.*

Proof. (Idea.) Suppose $p \in Q = X_{|I}$, $p \xrightarrow{a} p'$ in $A_{|I}$, and $(q, x) \in X$ where $x \in \uparrow(C, \ldots, C)$ for a large value C. Since there is a short path (i.e., a path of a small length) $q \xrightarrow{\sigma} p$, we have $(q, x) \xrightarrow{\sigma} (p, x + \Delta(\sigma)) \xrightarrow{a} (p', x + \Delta(\sigma a)) \in X$, which implies $p' \in X_{|I}$. □

3.3 Extracting a proper PNS $G_{(X,I)}$ from a bottom SCC X

Lemma 12 (Small $G_{(X,I)}$ with large counters, by a given RB-function). *For any RB-function f there is an RB-function $\bar{f}$ with the following property: for any conservative d-dim net A and any bottom SCC X related to A there is a set $I \subseteq [1, d]$ such that*

1. *for the d-dim PNS $G = G_{(X,I)} = (Q, A, E)$ we have $||Q|| \leq \bar{f}(||A||, d)$;*
2. *there is $(q, x) \in X$ with $x \in \uparrow(C, \ldots, C)$ where $C = f(||G||, d)$.*

We can imagine that for each RB-function f we fix some $\bar{f}$ guaranteed by the lemma; we thus get the notion of *an f-extracted PNS related to (A, X)* by which we mean the PNS $G_{(X,I)}$ for a set $I \subseteq [1, d]$ guaranteed by this lemma for f and $\bar{f}$. Referring to Proposition 11, we note that if $f \geq f_{\text{PROP}}$ then each f-extracted PNS is proper.

To prove the lemma, we use a result on extractors from [17]. By a *d-dim extractor* λ we mean a tuple (or a sequence) $(\lambda_1, \lambda_2, \ldots, \lambda_d) \in (\mathbb{N}_+)^d$ where $\lambda_1 \leq \lambda_2 \leq \cdots \leq \lambda_d$. For technical convenience we also refer to λ_0 and λ_{d+1}, stipulating $\lambda_0 = 1$ and $\lambda_{d+1} = \lambda_d$. For $m \in \mathbb{N}$ we say that λ is *m-adapted* if $\lambda_{i+1} \geq \lambda_i + m \cdot (\lambda_i)^i$, for all $i = 0, 1, \ldots, d-1$.

Proposition 13 (a weaker version of Lemma 20 in Section 6 of [17]). *Let X be an SCC related to a d-dim net A (a bottom SCC of a conservative net is a particular case), and let $(\lambda_1, \lambda_2, \ldots, \lambda_d)$ be a d-dim extractor that is $||A||$-adapted. There exists a set $I \subseteq [1, d]$ satisfying the following conditions, where $J = [1, d] \setminus I$:*

a) $||X_{|I}|| < \lambda_{|I|}$ (hence for all $x \in X$ and $i \in I$ we have $x(i) < \lambda_{|I|}$);
b) there is $x \in X$ such that for all $j \in J$ we have $x(j) \geq \lambda_{|I|+1} - ||A|| \cdot |I| \cdot (\lambda_{|I|})^{|I|}$.

We note that if $I = \emptyset$ then by b) there is $x \in X$ with $x(i) \geq \lambda_1$ for all $i \in [1, d]$ and a) can be viewed as vacuous; or formally we put $X_{|\emptyset} = \{()\}$ and $||\{()\}|| = 0$, which entails $||X_{|\emptyset}|| < 1 = \lambda_0$. If $I = [1, d]$, then for all $x \in X$ and $i \in [1, d]$ we have $x(i) < \lambda_d$, and b) is vacuous.

Proof of Lemma 12. (Idea.) Let f be an RB-function. We define the three-argument function $\lambda(m, d, i)$ for $m, d \in \mathbb{N}$ and $i \in [1, d]$, written rather as $\lambda_i^{<m,d>}$:

- we put $\lambda_1^{<m,d>} = f(m, d)$,
- for $i = 1, 2, \ldots, d-1$ we put $\lambda_{i+1}^{<m,d>} = f(\lambda_i^{<m,d>}, d) + m \cdot i \cdot (\lambda_i^{<m,d>})^i$.

Given a conservative d-dim net A and a bottom SCC X related to A, we consider $I \subseteq [1, d]$, and the respective $G_{(X,I)}$, that is guaranteed by Proposition 13 for the extractor $(\lambda_1^{<||A||,d>}, \lambda_2^{<||A||,d>}, \ldots, \lambda_d^{<||A||,d>})$ which is clearly $||A||$-adapted. The function $\bar{f}(m, d) = \lambda_d^{<m,d>}$ is an RB-function such that the conditions 1 and 2 in the lemma are satisfied. □

3.4 Virtual reachability reduced from PNSs to nets

Given a d-dim PNS $G = G_{(X,I)}$, on the set $\text{Conf}(G)$ we also define the *virtual reachability relation*: for an action sequence σ we have $(p, x) \overset{\sigma}{\dashrightarrow} (q, y)$ if there is a path from p to q labelled with σ and for $J = [1, d] \setminus I$ we have $\Delta(\sigma)_{|J} = y - x$ (while $\Delta(\sigma)_{|I} = q - p$ by the above definitions). By $(p, x) \overset{*}{\dashrightarrow} (q, y)$ we denote that $(p, x) \overset{\sigma}{\dashrightarrow} (q, y)$ for some σ. This notation also applies to the PNSs $G_{|J'}$.

Remark. Even if $G = G_{(X,I)} = (Q, A, E)$ is proper, and thus all executions of A from any x with $x_{|I} \in Q$ are also executions of G, we might have that $x \overset{*}{\dashrightarrow} y$

holds for A but does not hold for G even if $x_{|I}, y_{|I} \in Q$: a virtual execution of A can "sink below zero" in any component while for G we require that the restriction of a virtual execution to I is a standard execution of $A_{|I}$.

Proposition 14 provides some conditions under which $(p, x) \overset{*}{\dashrightarrow} (q, y)$ implies $(p, x) \xrightarrow{*} (q, y)$. It follows from a result in [17] that was shown for more general reversible PNSs than our PNSs $G = G_{(X,I)}$ and their restrictions $G_{|J'}$.

Proposition 14 (Virtual and standard reachability, Lemma 5 in [17]). *There are RB-functions f_0 and f_{VR} with the following property: for any (reversible) d-dim PNS G, if $(p, x) \overset{*}{\dashrightarrow} (q, y)$ and both x and y are in $\uparrow(C, \ldots, C)$ for $C = f_0(||G||, d)$, then we also have $(p, x) \xrightarrow{*} (q, y)$ and, moreover, there is an execution $(p, x) \xrightarrow{\sigma} (q, y)$ where $|\sigma| \leq f_{\mathrm{VR}}(||G||, d) \cdot ||y - x||$.*

In fact, the lemma in [17] is more precise, and the respective RB-functions f_0, f_{VR} are exponential, not double exponential. Similarly as in Lemma 10 (to which Proposition 14 is closely related), we are rather generous at our level of analysis. We further use the notation

$$C^G_{\mathrm{V=R}} = f_0(||G||, d);$$

the virtual reachability relation $\overset{*}{\dashrightarrow}$ thus coincides with the reachability relation $\xrightarrow{*}$ in the area $\uparrow(C^G_{\mathrm{V=R}}, \ldots, C^G_{\mathrm{V=R}})$, i.e. for the pairs of configurations in $Q \times \uparrow(C^G_{\mathrm{V=R}}, \ldots, C^G_{\mathrm{V=R}})$ where Q is the state set of G. Moreover, if the maximal component-difference $||y - x||$ of two vectors x, y in this area is small, and $(p, x) \overset{*}{\dashrightarrow} (q, y)$, then there is also a short execution $(p, x) \xrightarrow{\sigma} (q, y)$.

A crux of virtual reachability in a PNS is captured by *cyclic virtual executions* $(p, x) \overset{*}{\dashrightarrow} (p, y)$; we will now relate them to a corresponding small net.

Given a PNS $G = G_{(X,I)} = (Q, A, E)$, where $J = [1, d] \setminus I$, we fix

$$\text{the } |J|\text{-dim net } A^G_{\mathrm{SC}} \tag{3}$$

in which the actions are defined so that their displacements constitute the set

$$(A^G_{\mathrm{SC}})_\delta = \{z \in \mathbb{Z}^d \mid \text{there is a simple cycle in } (Q, A, E) \text{ with the displacement } z\};$$

for each $z \in (A^G_{\mathrm{SC}})_\delta$ there is an action (z_-, z_+) in A^G_{SC} where z_+ and z_- arise from z and $-z$, respectively, by replacing all negative components with 0.

Proposition 15 (Virtual reachability for G and the small net A^G_{SC}). *For any PNS $G = (Q, A, E)$ we have: $(p, x) \overset{*}{\dashrightarrow} (p, y)$ iff $x \overset{*}{\dashrightarrow} y$ in A^G_{SC}. Moreover, $||A^G_{\mathrm{SC}}|| \leq ||A|| \cdot |Q|$ (hence $||A^G_{\mathrm{SC}}|| \leq f(||G||, d)$ for an RB-function f).*

3.5 Down reachability of dead configurations

To get another ingredient for the proof of Theorem 5, we look how nonlive configurations (p, x) with all components of x being large can reach configurations (q, y) in which some actions are dead.

We say that a *configuration* $x \in \mathbb{N}^d$ of a d-dim net A is *dead* if some action $a \in A$ is dead at x, i.e., a is disabled at each configuration in $R(x) = \{y \mid x \xrightarrow{*} y\}$. (In other contexts configurations are called dead if all actions are disabled but we use this weaker notion.) Hence a configuration x is live iff it cannot reach any dead configuration. Rackoff's result for coverability in [24] gives us:

Proposition 16 (Deadness determined by small components, by [24]). *There is an RB-function f_{DEAD} with the following property:*
For any d-dim net A and any configuration $y \in \mathbb{N}^d$, the f_{DEAD}-small components of y, namely the vector $y_{|\mathrm{S}_y}$ for $\mathrm{S}_y = \{i \in [1,d] \mid y(i) < f_{\text{DEAD}}(||A||, d)\}$, determine whether y is dead; moreover, if $y_{|\mathrm{S}_y}$ determines that y is dead, then each y' satisfying $(y')_{|\mathrm{S}_y} = y_{|\mathrm{S}_y}$ is dead.

Now we aim to prove Proposition 18; roughly speaking, it shows a situation when a start vector that is larger than but close to a target vector (in particular to a dead configuration) is also close w.r.t. the reachability distance. This result will be then used in the proof of Lemma 19. But we first recall another useful result from [17], for general Petri nets (that are not necessarily reversible).

Proposition 17 (Mutual reachability in nets, Theorem 2 in [17]). *There is an RB-function f_{MR} with the following property:*
for any d-dim net A, if $x \xrightarrow{*} y$ and $y \xrightarrow{*} x$ then there are executions $x \xrightarrow{\sigma_1} y$ and $y \xrightarrow{\sigma_2} x$ such that $|\sigma_1\sigma_2| \leq f_{\text{MR}}(||A||, d) \cdot ||x - y||$.

Proposition 18 (Down reachability in PNSs).
There is an RB-function f_{DR} with the following property:
for any proper d-dim PNS $G = G_{(X,I)} = (Q, A, E)$ and any $J' \subseteq [1,d] \setminus I$, if $(q,x) \xrightarrow{*} (q,y)$ for $G_{|J'}$ (hence $x, y \in \mathbb{N}^{J'}$), $x \in \uparrow(C^G_{\text{V=R}}, \dots, C^G_{\text{V=R}})$, and $x \geq y$ (y is "down" w.r.t. x), then there is an execution $(q,x) \xrightarrow{\sigma} (q,y)$ of $G_{|J'}$ where $|\sigma| \leq f_{\text{DR}}(||G||, d) \cdot ||x - y||$.

Proof. (Idea.) Let G, J', q, x, y satisfy the above assumptions, and let A' arise from $A_{|I \cup J'}$ by adding actions $(\mathbf{0}, \mathbf{e}_i)$ for all $i \in J'$ where $x(i) > y(i)$; $\mathbf{e}_i(i) = 1$ and $\mathbf{e}_i(j) = 0$ for $j \neq i$. For A' we thus have both $(q,x) \xrightarrow{*} (q,y)$ and $(q,y) \xrightarrow{*} (q,x)$, and $||A'|| \leq ||A||$; hence there is an execution $(q,x) \xrightarrow{\rho} (q,y)$ of A' where $|\rho| \leq f_{\text{MR}}(||A||, d) \cdot ||x - y||$ (by Proposition 17). We can assume ρ to be in the form $\rho = \rho_1\rho_2$ where ρ_1 contains precisely the added increasing actions $(\mathbf{0}, \mathbf{e}_i)$; thus $(q,x) \xrightarrow{\rho} (q,y)$ can be written as $(q,x) \xrightarrow{\rho_1} (q, x + \Delta(\rho_1)) \xrightarrow{\rho_2} (q,y)$ where $(q, x + \Delta(\rho_1)) \xrightarrow{\rho_2} (q,y)$ is an execution of $A_{|I \cup J'}$, and thus also an execution of $G_{|J'}$. By reversibility of $G_{|J'}$ and the assumption $(q,x) \xrightarrow{*} (q,y)$ we get $(q,x) \dashrightarrow^{*} (q, x + \Delta(\rho_1)) \xrightarrow{\rho_2} (q,y)$ and use Proposition 14 to finish the proof. □

3.6 (Virtual) reachability of quasi-dead configurations

We consider a PNS $G = G_{(X,I)} = (Q, A, E)$. When saying that a configuration (q,y) is dead, we mean that it is dead for A (at least one action $a \in A$ is dead). Given an RB-function f, we say that a *configuration* (p, x) is f-*quasi-dead* if $x \in \uparrow(C^G_{\text{V=R}}, \dots, C^G_{\text{V=R}})$ and there is $(p, x) \xrightarrow{\rho} (q, y)$ where (q, y) is dead and $|\rho| \le f(||G||, d)$.

Lemma 19 (Large nonlive reach quasi-dead configurations). *There are RB-functions f_1, f_2 for which the following claim is true. Every proper d-dim PNS $G = G_{(X,I)} = (Q, A, E)$ satisfies the following implication: if (p, x) is nonlive and $x \in \uparrow(C, \dots, C)$ for $C = f_2(||G||, d)$, then there is some f_1-quasi-dead configuration (p, x') such that $(p, x) \xrightarrow{*} (p, x')$.*

Proof. (Idea.) Let $G = G_{(X,I)} = (Q, A, E)$ be a proper d-dim PNS, where $J = [1, d] \smallsetminus I$, and let (p, x_0) be a nonlive configuration with $x_0 \in \uparrow(C, \dots, C)$ for a large number C. To prove the lemma, it suffices to show that $(p, x_0) \overset{*}{\dashrightarrow} (p, x_1)$, i.e. $x_0 \overset{*}{\dashrightarrow} x_1$ for the $|J|$-dim net A^G_{SC} (recall (3)), where $x_1 \in \uparrow(C^G_{\text{V=R}}, \dots, C^G_{\text{V=R}})$ and (p, x_1) can reach a dead configuration by a short execution.

We fix a dead configuration (q, y_0) such that $(p, x_0) \xrightarrow{*} (q, y_0)$, and a shortest path $p \xrightarrow{\pi} q$ in the graph (Q, A, E). The facts that $||\Delta(\pi)||$ is small, C is large and G is reversible entail that

$$(p, x_0) \xrightarrow{\pi} (q, y_1) \xrightarrow{*} (q, y_0),$$

where $y_1 = x_0 + \Delta(\pi)_{|J}$, and thus all components of y_1 are large. We assume that $y_0(i) < f_{\text{DEAD}}(||A||, d)$ for some $i \in J$, since otherwise also (q, y_1) is dead (recall Proposition 16) and we are done ((p, x_0) reaches a dead configuration, namely (q, y_1), by a short execution). Hence (we can assume that) $||y_1 - y_0||$ is large, and any virtual execution of A^G_{SC} demonstrating $y_0 \overset{*}{\dashrightarrow} y_1$ (which holds by reversibility of G and A^G_{SC}) is long. We fix a segmented virtual execution

$$y_0 \overset{*}{\dashrightarrow} (y_0 + z_1) \overset{*}{\dashrightarrow} (y_0 + z_1 + z_2) \cdots \overset{*}{\dashrightarrow} (y_0 + z_1 + \cdots + z_m) = y_1 \tag{4}$$

of A^G_{SC} as shown by (2) in Corollary 9. Proposition 15 and Lemma 10 show that $||A^G_{\text{SC}}||$ and $||\min_{\preceq}((A^G_{\text{SC}})^*_{\delta})||$ are small, and we note that for some small j the value $y_2 = y_0 + \sum_{i=1}^{j} z_i$ (reached by a short prefix of (4)) satisfies that both y_2 and $y_2 - \Delta(\pi)$ are in $\uparrow(C^G_{\text{V=R}}, \dots, C^G_{\text{V=R}})$ and, moreover, $(p, y_2 - \Delta(\pi)) \xrightarrow{\pi} (q, y_2)$. For $x_1 = y_2 - \Delta(\pi)$ we thus get $(p, x_0) \xrightarrow{*} (q, y_0) \overset{*}{\dashrightarrow} (q, y_2) \overset{*}{\dashrightarrow} (p, x_1)$, which entails $(p, x_0) \xrightarrow{*} (p, x_1)$ (since both x_0, x_1 are in $\uparrow(C^G_{\text{V=R}}, \dots, C^G_{\text{V=R}})$) and $(p, x_1) \xrightarrow{\pi} (q, y_2) \xrightarrow{*} (q, y_0)$ (since $(q, y_2) \overset{*}{\dashrightarrow} (q, y_1) \xrightarrow{*} (q, y_0)$ and both y_2 and y_1 are in $\uparrow(C^G_{\text{V=R}}, \dots, C^G_{\text{V=R}})$).

To finish the proof, it suffices to show that there is a short σ such that $(q, y_2) \xrightarrow{\sigma} (q, y')$ where (q, y') is a dead configuration (which does not immediately follow from $(q, y_2) \xrightarrow{*} (q, y_0)$). We recall that $||y_2 - y_0||$ is small (since j in the definition of y_2 is small), and y_2 is an intermediate vector for the pair y_0, y_1. We put $J_{\text{DOWN}} = \{i \in J \mid y_1(i) \ge y_0(i)\}$; hence for all $i \in J_{\text{DOWN}}$ we have $y_2(i) \ge y_0(i)$, and by applying Proposition 18 to $G_{|J_{\text{DOWN}}}$ we get a short σ such that $(q, (y_2)_{|J_{\text{DOWN}}}) \xrightarrow{\sigma} (q, (y_0)_{|J_{\text{DOWN}}})$. Since for $i \in J \smallsetminus J_{\text{DOWN}}$ we have $y_0(i) \ge y_2(i) \ge y_1(i)$ and these values are large, we get $(q, y_2) \xrightarrow{\sigma} (q, y')$ where $(y')_{|J_{\text{DOWN}}} = (y_0)_{|J_{\text{DOWN}}}$, which entails that (q, y') is dead (by Proposition 16). □

3.7 Proof of Theorem 5

We fix some RB-functions $f_1, f_2, \bar{f}$ with the following property:
For any conservative d-dim net A and any bottom SCC $X \subseteq \mathbb{N}^d$ related to A, any f_2-extracted PNS $G = G_{(X,I)} = (Q, A, E)$ satisfies the following conditions, for $J = [1,d] \setminus I$ and $C = f_2(||G||, d)$:

a) G is a proper PNS (i.e., Q is a bottom SCC for $A_{|I}$), and $||Q|| \leq \bar{f}(||A||, d)$;
b) there is $x \in X$ such that ($x_{|I} \in Q$ and) $x_{|J} \in \uparrow(C, \dots, C)$;
c) $C \geq C^G_{\text{V=R}}$;
d) for all $p \in Q$ and $x \in \uparrow(C, \dots, C)$, if (p, x) is nonlive (as a configuration of A) then there is $x' \in \uparrow(C^G_{\text{V=R}}, \dots, C^G_{\text{V=R}})$ such that $(p, x) \overset{*}{\dashrightarrow} (p, x')$ and (p, x') is f_1-quasi-dead.

The existence of such RB-functions $f_1, f_2, \bar{f}$ follows from Lemmas 19 and 12.

Based on $f_1, f_2, \bar{f}$, we aim to show that there is an RB-function f such that for each structurally live conservative d-dim net A there is a live configuration $x \in \mathbb{N}^d$ satisfying $||x|| \leq f(||A||, d)$. The existence of such a function f will follow by the further discussion.

We fix a structurally live conservative d-dim net A and a live bottom SCC $X \subseteq \mathbb{N}^d$, and consider an f_2-extracted PNS $G = G_{(X,I)} = (Q, A, E)$ (for which the above conditions a)-d) hold). If $J = \emptyset$ (i.e., $I = [1, d]$), then $Q = X$ and each $p \in Q$ is a live configuration of A satisfying $||p|| \leq \bar{f}(||A||, d)$ (and thus any RB-function $f \geq \bar{f}$ satisfies our aim in this case). Hence we further assume $J \neq \emptyset$, and we fix

$$\text{a live configuration } (p_0, x_0) \in Q \times \left(\mathbb{N}^J \cap \uparrow(C, \dots, C)\right) \text{ for } C = f_2(||G||, d).$$

(by b) we can choose (p_0, x_0) in the live SCC X). We say that $x \in \mathbb{N}^J$ is *good* if $x \in \uparrow(C, \dots, C)$ and (p_0, x) is live, and that $x' \in \mathbb{N}^J$ is *bad* if (p_0, x') is an f_1-quasi-dead configuration (which entails that $x' \in \uparrow(C^G_{\text{V=R}}, \dots, C^G_{\text{V=R}})$). Hence

$$x \in \mathbb{N}^J \text{ is good iff } x \in \uparrow(C, \dots, C) \text{ and there is no bad } x' \text{ such that } x \overset{*}{\dashrightarrow} x' \tag{5}$$

where we refer to the virtual reachability of the $|J|$-dim net A^G_{SC} defined around (3); the claim (5) follows from the condition d).

We will transform the characterization (5) of good vectors by negative virtual reachability constraints to a characterization by $2^{|J|}$ positive virtual reachability constraints that will allow us to use Corollary 4 and Theorem 2 for deriving the existence of small good vectors. To this aim we first define a $|J|$-dimensional extractor $\lambda = (\lambda_1, \lambda_2, \dots, \lambda_{|J|})$, now stipulating $\lambda_0 = 0$ and $\lambda_{|J|+1} = \lambda_{|J|}$:

$$\lambda_i = \max\{f_{\text{DEAD}}(||A||, d), C^G_{\text{V=R}}\} + ||A|| \cdot \left(f_1(||G||, d) + f_{\text{VR}}(||G||, d) \cdot \lambda_{i-1}\right)$$

for $i = 1, 2, \dots, |J|$ (hence $\lambda_1 = \max\{f_{\text{DEAD}}(||A||, d), C^G_{\text{V=R}}\} + ||A|| \cdot f_1(||G||, d)$). By f_{DEAD} and f_{VR} we refer to the RB-functions from Propositions 16 and 14, respectively. It is clear that $\lambda_{|J|}$ is small (i.e., $\lambda_{|J|} \leq f'(||A||, d)$ for an RB-function f' independent of A).

Claim 1. For each $x \in \mathbb{N}^J$ there is a unique maximal set $J_x^\lambda \subseteq J$ such that $x(i) < \lambda_{|J_x^\lambda|}$ for all $i \in J_x^\lambda$, and $x(i) \geq \lambda_{|J_x^\lambda|+1}$ for all $i \in J \smallsetminus J_x^\lambda$.

Claim 2. For all $x, x' \in \mathbb{N}^J \cap \uparrow(C^G_{\text{V=R}}, \ldots, C^G_{\text{V=R}})$ such that $J_x^\lambda = J_{x'}^\lambda = J'$ and $x_{|J'} \overset{*}{\dashrightarrow} (x')_{|J'}$ we have: if x' is bad, then (p_0, x) is a nonlive configuration.

For each $J' \subseteq J$, in the set $\{x \in \mathbb{N}^J \mid x \in \uparrow(C^G_{\text{V=R}}, \ldots, C^G_{\text{V=R}})$ and $x_0 \overset{*}{\dashrightarrow} x\}$ we fix a vector $x_{J'}$ with the least value $\text{DIF}(x_{J'}, J')$ ("difference from the J'-class") where for vectors $x \in \mathbb{N}^J$ we put $\text{DIF}(x, J') = \max(\text{ABOVE}(x, J') \cup \text{BELOW}(x, J'))$ for the sets $\text{ABOVE}(x, J') = \{x(i) - (\lambda_{|J'|} - 1) \mid i \in J', x(i) \geq \lambda_{|J'|}\}$ and $\text{BELOW}(x, J') = \{\lambda_{|J'|+1} - x(i) \mid i \in J \smallsetminus J', x(i) < \lambda_{|J'|+1}\}$ (where $\max(\emptyset) = 0$). Hence $\text{DIF}(x, J') \in \mathbb{N}$, and $\text{DIF}(x, J') = 0$ iff $J_x^\lambda = J'$; therefore $J_{x_{J'}}^\lambda \neq J'$ iff $\text{DIF}(x_{J'}, J') > 0$. We note that $(p_0, x_{J'})$ is a live configuration, for each $J' \subseteq J$ (since $(p_0, x_0) \xrightarrow{*} (p_0, x_{J'})$ and (p_0, x_0) is live).

For $B \in \mathbb{N}$ we define the equivalence $\equiv_{\leq B}$ on $\mathbb{N}^J$: $x \equiv_{\leq B} y$ if for each $i \in J$ we have either $x(i) > B$ and $y(i) > B$, or $x(i) = y(i)$. We define the following value B_0, which is small (also due to Lemma 10):

$$B_0 = \lambda_{|J|} + ||\min_{\preceq}(((A^G_{\text{SC}})_\delta)^*)||.$$

Claim 3. If $x \equiv_{\leq B_0} y$ then $J_x^\lambda = J_y^\lambda$. (This follows trivially from definitions.)
Claim 4. For any $\bar{x} \in \mathbb{N}^J \cap \uparrow(C, \ldots, C)$, if for each $J' \subseteq J$ there is $\bar{x}_{J'}$ such that $x \overset{*}{\dashrightarrow} x_{J'}$ and $x_{J'} \equiv_{\leq B_0} x_{J'}$, then x is good.

Now we observe that the conditions put on $\bar{x}$ in Claim 4 can be expressed by a linear system $\bar{S} = S_0 \wedge \bigwedge_{J' \subseteq J} S_{J'}$: All linear systems S_0 and $S_{J'}$, $J' \subseteq J$, share the same integer variables $\mathbf{x}_1, \mathbf{x}_2, \ldots, \mathbf{x}_{|J|}$, and each $S_{J'}$ has, moreover, its own variables $\mathbf{y}_1^{J'}, \mathbf{y}_2^{J'}, \ldots, \mathbf{y}_{|J|}^{J'}$; the system S_0 expresses $(\mathbf{x}_1, \mathbf{x}_2, \ldots, \mathbf{x}_{|J|}) \in \uparrow(C, \ldots, C)$, i.e., $S_0 = \bigwedge_{i \in J}(\mathbf{x}_i \geq C)$. Each $S_{J'}$ is based on a linear system $S_{A^G_{\text{SC}}}$ that is guaranteed by Corollary 4 and expresses $(\mathbf{x}_1, \mathbf{x}_2, \ldots, \mathbf{x}_{|J|}) \overset{*}{\dashrightarrow} (\mathbf{y}_1^{J'}, \mathbf{y}_2^{J'}, \ldots, \mathbf{y}_{|J|}^{J'})$ (for A^G_{SC}), to which it adds the constraints guaranteeing $(\mathbf{y}_1^{J'}, \mathbf{y}_2^{J'}, \ldots, \mathbf{y}_{|J|}^{J'}) \equiv_{\leq B_0} x_{J'}$ (hence the constraint $\mathbf{y}_i^{J'} = x_{J'}(i)$ for all $i \in J$ for which $(x_{J'})(i) \leq B_0$ and the constraint $\mathbf{y}_i^{J'} > B_0$ for all $i \in J$ for which $(x_{J'})(i) > B_0$).

The solutions of the system $\bar{S}$ are the tuples $(x, (y_{J'})_{J' \subseteq J}) \in \mathbb{N}^{|J|+|J| \cdot 2^{|J|}}$ satisfying $x \in \uparrow(C, \ldots, C)$ and, for all $J' \subseteq J$, $x \overset{*}{\dashrightarrow} y_{J'}$ and $y_{J'} \equiv_{\leq B_0} x_{J'}$. Since $(x_0, (x_{J'})_{J' \subseteq J}) \in [[\bar{S}]]$, the system $\bar{S}$ is satisfiable. The dimension of the linear system $\bar{S}$ is at most $d + d \cdot 2^d$, and it is a routine to verify that Corollary 4 and Theorem 2 guarantee a solution $(x, (y_{J'})_{J' \subseteq J})$ with the norm bounded by $f(||A||, d)$ for some RB-function f (independent of A).

The proof of Theorem 5 is thus finished. □

4 Small Solutions (Proof of Theorem 2)

We first extend Lemma 1 to equality constraints that are not homogeneous:

Lemma 20. *Let $N = \{y \in \mathbb{N}^d \mid Cy = c\}$ where C is a $k \times d$ matrix over $\mathbb{Z}$, and $c \in \mathbb{Z}^k$. Then $Y = \min_\leq(N)$ is a finite set such that $N = Y + \{x \in \mathbb{N}^d \mid Cx = 0\}$. Moreover the following bound holds where $r = \text{rank}(C)$:*

$$\|Y\|_1 \leq \|c\|_1(2 + r\|C\|)^r.$$

Proof. (Idea.) We first reduce the proof to the special case $k = \text{rank}(C)$ and $c \in \mathbb{N}^k$. Then we apply Lemma 1 to $Bx = \mathbf{0}$ where $x \in \mathbb{N}^{d+k}$, $B = [C| - I]$, and I is the $k \times k$ identity matrix. In fact $B(y, z) = \mathbf{0}$ where $(y, z) \in \mathbb{N}^d \times \mathbb{N}^k$ is equivalent to $Cy = z$. □

The following lemma handles a conjunction of inequality constraints.

Lemma 21. *Let $Z = \{z \in \mathbb{N}^d \mid Cz \geq c\}$ where C is a $k \times d$ matrix over $\mathbb{Z}$, and $c \in \mathbb{Z}^k$. The following bound holds where $r = \text{rank}(C)$:*

$$\|\min_\leq(Z)\|_1 \leq (2 + d\max\{||C||, ||c||\})^{2r+1}.$$

Proof. If Z is empty, the lemma is trivial. Otherwise, let $z \in \min_\leq(Z)$. We denote by C_j for $j \in [1, k]$ the jth row of C. We introduce $s = \max\{||C||, ||c||\}$ and $h = (s - 1) + s(1 + ds)^r$ and $J = \{j \in [1, k] \mid C_j z \leq h\}$. We also denote by C' the $|J| \times d$ matrix obtained from C by keeping rows in J. We introduce $N = \{y \in \mathbb{N}^d \mid C'y = C'z\}$, and $M = \{x \in \mathbb{N}^d \mid C'x = \mathbf{0}\}$. Since $z \in N$, there exists $y \in \min_\leq(N)$ such that $y \leq z$. Lemma 20 shows that $\|y\|_1 \leq dh(2 + rs)^r$ since $\text{rank}(C') \leq \text{rank}(C)$. Let $x = z - y$ and notice that $x \in M$.

Assume by contradiction that $x \neq \mathbf{0}$. In this case, there exists $x' \in \min_\leq(M \setminus \{\mathbf{0}\})$ such that $x' \leq x$. Lemma 1 shows that $\|x'\|_1 \leq (1 + ds)^r$. Let $z' = z - x'$. Since $z' = y + (x - x')$, it follows that $z' \in \mathbb{N}^d$. Let us prove that $Cz' \geq c$. Let $j \in [1, k]$ and let us prove that $C_j z' \geq c(j)$. If $j \in J$ then $C_j z' = C_j z \geq c(j)$. If $j \notin J$, we have $C_j z' = C_j z - C_j x' \geq h + 1 - s\|x'\|_1 \geq h + 1 - s(1 + ds)^r = s \geq c(j)$. It follows that $z' \in Z$ which contradicts the minimality of z. If follows that $x = \mathbf{0}$.

From $z = y$, we derive $\|z\|_1 \leq dh(2 + ds)^r$. From $h = (s - 1) + s(1 + ds)^r$, we get $\|z\|_1 \leq (2 + ds)^{2r+1}$. □

We strenghten Theorem 2 for linear systems without divisibility constraints:

Lemma 22. *Any satisfiable d-dim linear system S with no divisibility constraints has a solution $x \in \mathbb{Z}^d$ such that $\|x\|_1 \leq (2 + d + d \cdot ||S||)^{2d+1}$.*

Proof. (Idea.) By putting S in disjunctive normal form, we reduce the problem to a conjunction of inequality constraints. The proof then follows from Lemma 21. □

Proof of Theorem 2. We consider a satisfiable linear system S. We introduce $\ell = \text{lcm}(S)$. Euclidean divisor by ℓ shows that $\mathbb{Z}^d$ is the disjoint union of sets $a + \ell\mathbb{Z}^d$ where a ranges in the finite set of vectors $[0, \ell - 1]^d$. In particular $[[S]]$ is the disjoint union of the sets $[[S]] \cap (a + \ell\mathbb{Z}^d)$ where $a \in [0, \ell - 1]^d$. Since $[[S]]$ is non-empty, there exists a vector $a \in [0, \ell - 1]^d$ such that the set $[[S]] \cap (a + \ell\mathbb{Z}^d)$ is non empty. This last set can be denoted by a linear system S_a obtained from S by replacing:

- equality constraints $(\alpha \cdot \mathbf{x}) = c$ by $(\alpha \cdot \mathbf{x}) = \frac{c-(\alpha \cdot a)}{\ell}$ if ℓ divides $c - (\alpha \cdot a)$, and by false otherwise.
- inequality constraints $(\alpha \cdot \mathbf{x}) \geq c$ by $(\alpha \cdot \mathbf{x}) \geq \left\lceil \frac{c-(\alpha \cdot a)}{\ell} \right\rceil$.
- divisibility constraints $(\alpha \cdot \mathbf{x}) \equiv c \pmod{m}$ by the boolean value $(\alpha \cdot a) \equiv c \pmod{m}$.

Now, just observe that $[[S]] \cap (a + \ell\mathbb{Z}^d) = a + \ell[[S_a]]$ and in particular S_a is satisfiable. Notice that if $(\alpha \cdot \mathbf{x}) \sim c$ is a constraint of S where $\sim$ is the equality or the inequality, then $|(\alpha \cdot a)| \leq d(\ell - 1)||S||$ and $|c| \leq ||S||$. We deduce that $|c-(\alpha \cdot a)| \leq d\ell||S||$. We have proved the inequality $||S_a|| \leq d||S||$. As S_a does not contain any divisibility constraint, Lemma 22 shows that there exists a solution y of S_a such that $\|y\|_1 \leq (2+d+d||S_a||)^{2d+1}$. Observe that $x = a+\ell y$ is a solution of S. Moreover $\|x\|_1 \leq d(\ell-1)+\ell(2+d+d^2||S||)^{2d+1} \leq (d+1)\ell(2+d+d^2||S||)^{2d+1}$. □

5 Linear Systems for Groups

In this section, we introduce a linear system denoting a group spanned by a finite set of vectors. More precisely, we prove Theorem 3, all other results of this section are only used for proving that theorem.

We define the rank of a set $X \subseteq \mathbb{Z}^d$ as the maximal $r \in [0, d]$ such that there exists a finite sequence $x_1, \ldots, x_k \in X$ such that r is the rank of the $d \times k$ matrix $[x_1|\cdots|x_k]$. This rank is denoted as $\text{rank}(X)$. We recall some classical results from the book [26, Chapter 4]. In that book, groups L are assumed to be *full-rank*, i.e. $\text{rank}(L) = d$. In order to overcome this limitation, we introduce the notion of *maximal ranking set*. A set $K \subseteq \{1, \ldots, d\}$ is called a *maximal ranking set* for a set $X \subseteq \mathbb{Q}^d$ if $\text{rank}(X_{|K}) = |K| = \text{rank}(X)$, where $X_{|K} = \{x_{|K} \mid x \in X\}$. Classical linear algebra results shows that any set admits a maximal ranking set.

Example. The maximal ranking sets of $\{(1, 1)\}$ are $\{1\}$ and $\{2\}$, for $\{(1, 0)\}$ and $\{(1, 0), (0, 1)\}$ the unique maximal ranking sets are $\{1\}$ and $\{1, 2\}$ respectively.

A *Hermite decomposition* of (K, L) where K is a maximal ranking set of a group L is a sequence $(v_i)_{i\in K}$ of vectors in $\mathbb{N}^d$ that span the group L and such that for every $i, j \in K$, we have $v_j(i) < v_i(i)$ if $j < i$, $v_j(i) > 0$ if $j = i$, and $v_j(i) = 0$ if $j > i$. Recall that there exists a unique Hermite decomposition of (K, L). We denote by $\det_K(L)$ the product $\prod_{i\in K} v_i(i)$. This product corresponds to the determinant of the triangular matrix $(v_i(j))_{i,j\in K}$.

We first provide the following two lemmas that are obtained thanks to classical matrix operations that provide a way to reduce proofs to the special case of full-rank groups L.

Lemma 23. *Let L be a group spanned by a sequence of vectors in $[-m, m]^d$ for some $m \in \mathbb{N}$, let K be a maximal ranking set for L, and let r be the rank of L. The Hermite decomposition $(v_i)_{i\in K}$ of (K, L) satisfies $\det_K(L) \leq (r!)m^r$ and $||v_i|| \leq d(r!)m^r$ for every $i \in K$.*

Given a vector $x \in \mathbb{Z}^d$ and a positive integer $\ell \geq 1$, let us introduce that $[x]_\ell$ the unique vector in $[0, \ell - 1]^d$ such that $x \in [x]_\ell + \ell\mathbb{Z}^d$.

Lemma 24. *Let L be a group spanned by a sequence of vectors in $[-m, m]^d$ for some $m \in \mathbb{N}$, let K be a maximal ranking set for L, let $r = |K|$, and let $\ell = \det_K(L)$. Let $B = [L_{|K}]_\ell$. There exists a linear system S given as a conjunction of $d - r$ homogeneous equality constraints such that $\|S\| \leq r(r!)m^r$ and such that L is the set of solutions x of S satisfying $[x_{|K}]_\ell \in B$.*

We are now ready for proving Theorem 3. Let L be a group spanned by a finite set $X \subseteq [-m, m]^d$ for some $m \in \mathbb{N}$. We introduce a maximal ranking set K for L, let $r = |K|$, and $\ell = \det_K(L)$. Lemma 23 shows that $\ell \leq (d!)m^d$. Let $B = [L_{|K}]_\ell$. Lemma 24 shows that there exists a linear system S_0 given as a conjunction of $d-r$ homogeneous equality constraints such that $\|S_0\| \leq d(d!)m^d$ and such that L is the set of solutions x of S_0 satisfying $[x_{|K}]_\ell \in B$. It follows that the following linear system S satisfies $L = [[S]]$.

$$[\bigvee_{b \in B}(\bigwedge_{i=1}^d x(i) \in b(i) + \ell\mathbb{Z})] \wedge S_0$$

Notice that $\|S\| = \|S_0\|$ and $\mathrm{lcm}(S) = \ell$.

6 EXPSPACE-hardness (a counterpart of Theorem 5)

We recall the *EXPSPACE-complete uniform word problem for commutative semigroups* [21]: an instance consists of a finite alphabet Σ, a finite set of equations $u \equiv v$ for $u, v \in \Sigma^*$ and $u_0, v_0 \in \Sigma^*$, and we ask whether $u_0 \equiv v_0$. The crux of the high complexity is the fact that the commutative semigroup defined by $a \equiv b^{2^{2^n}}$ (which can be written in space $O(2^n)$ when using binary notation for exponents) can be embedded in a commutative semigroup of size $O(n)$ (even when using unary notation for exponents). In fact, by [21] even the (weaker) *coverability* version in which $u_0, v_0 \in \Sigma$ and we ask whether $u_0 \equiv v_0 w$ for some $w \in \Sigma^*$ is EXPSPACE-complete.

The mentioned problems can be naturally presented by reversible Petri nets, and [14] shows how a given instance of the *coverability problem for reversible Petri nets* can be transformed into a net A that is structurally live iff the instance is positive. In the arxiv version we show that we can transform such a net A so that it becomes 1-conservative, and even such that in each vector in the actions $(a_-, a_+) \in A$ two components have the value one while the other are zero. This reduction, together with the above mentioned fact on the equation $a \equiv b^{2^{2^n}}$, witnesses that *structural liveness of conservative nets is EXPSPACE-hard* and also that the *2-exp upper bound in Theorem 5 cannot be essentially improved.*

Acknowledgments. Jiří Valůšek was supported by Grant No. IGA_PrF_2024_024 of IGA of Palacký University Olomouc. We also thank the reviewers for their helpful comments.

References

1. Angluin, D., Aspnes, J., Diamadi, Z., Fischer, M.J., Peralta, R.: Computation in networks of passively mobile finite-state sensors. Distributed Comput. **18**(4), 235–253 (2006). https://doi.org/10.1007/s00446-005-0138-3
2. Angluin, D., Aspnes, J., Eisenstat, D., Ruppert, E.: The computational power of population protocols. Distributed Comput. **20**(4), 279–304 (2007). https://doi.org/10.1007/s00446-007-0040-2
3. Best, E., Devillers, R.: Petri Net Primer. Birkhäuser Cham (2024). https://doi.org/10.1007/978-3-031-48278-6, 545 pp.
4. Best, E., Esparza, J.: Existence of home states in Petri nets is decidable. Inf. Process. Lett. **116**(6), 423–427 (2016). https://doi.org/10.1016/j.ipl.2016.01.011
5. Cardoza, E., Lipton, R.J., Meyer, A.R.: Exponential Space Complete Problems for Petri Nets and Commutative Semigroups: Preliminary Report. In: Chandra, A.K., Wotschke, D., Friedman, E.P., Harrison, M.A. (eds.) Proceedings of the 8th Annual ACM Symposium on Theory of Computing, May 3-5, 1976, Hershey, Pennsylvania, USA. pp. 50–54. ACM (1976). https://doi.org/10.1145/800113.803630
6. Czerwinski, W., Orlikowski, L.: Reachability in Vector Addition Systems is Ackermann-complete. In: 62nd IEEE Annual Symposium on Foundations of Computer Science, FOCS 2021, Denver, CO, USA, February 7-10, 2022. pp. 1229–1240. IEEE (2021). https://doi.org/10.1109/FOCS52979.2021.00120
7. Desel, J., Esparza, J.: Free Choice Petri Nets. Cambridge Tracts in Theoretical Computer Science, Cambridge University Press (1995). https://doi.org/10.1017/CBO9780511526558
8. Esparza, J., Raskin, M.A., Weil-Kennedy, C.: Parameterized Analysis of Immediate Observation Petri Nets. In: Donatelli, S., Haar, S. (eds.) Application and Theory of Petri Nets and Concurrency - 40th International Conference, PETRI NETS 2019, Aachen, Germany, June 23-28, 2019, Proceedings. Lecture Notes in Computer Science, vol. 11522, pp. 365–385. Springer (2019). https://doi.org/10.1007/978-3-030-21571-2_20
9. Haase, C.: A survival guide to Presburger arithmetic. ACM SIGLOG News **5**(3), 67–82 (2018). https://doi.org/10.1145/3242953.3242964
10. Hack, M.: The Recursive Equivalence of the Reachability Problem and the Liveness Problem for Petri Nets and Vector Addition Systems. In: 15th Annual Symposium on Switching and Automata Theory, New Orleans, Louisiana, USA, October 14-16, 1974. pp. 156–164. IEEE Computer Society (1974). https://doi.org/10.1109/SWAT.1974.28
11. Hujsa, T., Devillers, R.: On Deadlockability, Liveness and Reversibility in Subclasses of Weighted Petri Nets. Fundam. Informaticae **161**(4), 383–421 (2018). https://doi.org/10.3233/FI-2018-1708
12. Jančar, P., Leroux, J.: The Semilinear Home-Space Problem Is Ackermann-Complete for Petri Nets. In: Pérez, G.A., Raskin, J. (eds.) 34th International Conference on Concurrency Theory, CONCUR 2023, September 18-23, 2023, Antwerp, Belgium. LIPIcs, vol. 279, pp. 36:1–36:17. Schloss Dagstuhl - Leibniz-Zentrum für Informatik (2023). https://doi.org/10.4230/LIPICS.CONCUR.2023.36
13. Jančar, P., Leroux, J.: On the Home-Space Problem for Petri Nets and its Ackermannian Complexity. Log. Methods Comput. Sci. **20**(4) (2024). https://doi.org/10.46298/LMCS-20(4:23)2024
14. Jančar, P., Purser, D.: Structural liveness of Petri nets is ExpSpace-hard and decidable. Acta Informatica **56**(6), 537–552 (2019). https://doi.org/10.1007/s00236-019-00338-6, https://doi.org/10.1007/s00236-019-00338-6

15. Jančar, P., Valůšek, J.: Structural Liveness of Immediate Observation Petri Nets. Fundam. Informaticae **188**(3), 179–215 (2023). https://doi.org/10.3233/FI-222146
16. Klaedtke, F.: Bounds on the automata size for Presburger arithmetic. ACM Trans. Comput. Log. **9**(2), 11:1–11:34 (2008). https://doi.org/10.1145/1342991.1342995
17. Leroux, J.: Distance Between Mutually Reachable Petri Net Configurations. In: Chattopadhyay, A., Gastin, P. (eds.) 39th IARCS Annual Conference on Foundations of Software Technology and Theoretical Computer Science, FSTTCS 2019, December 11-13, 2019, Bombay, India. LIPIcs, vol. 150, pp. 47:1–47:14. Schloss Dagstuhl - Leibniz-Zentrum für Informatik (2019). https://doi.org/10.4230/LIPIcs.FSTTCS.2019.47
18. Leroux, J.: The Reachability Problem for Petri Nets is Not Primitive Recursive. In: 62nd IEEE Annual Symposium on Foundations of Computer Science, FOCS 2021, Denver, CO, USA, February 7-10, 2022. pp. 1241–1252. IEEE (2021). https://doi.org/10.1109/FOCS52979.2021.00121
19. Leroux, J., Schmitz, S.: Reachability in Vector Addition Systems is Primitive-Recursive in Fixed Dimension. In: 34th Annual ACM/IEEE Symposium on Logic in Computer Science, LICS 2019, Vancouver, BC, Canada, June 24-27, 2019. pp. 1–13. IEEE (2019). https://doi.org/10.1109/LICS.2019.8785796
20. Mayr, E.W.: Some Complexity Results for Polynomial Ideals. J. Complex. **13**(3), 303–325 (1997). https://doi.org/10.1006/jcom.1997.0447
21. Mayr, E.W., Meyer, A.R.: The complexity of the word problems for commutative semigroups and polynomial ideals. Advances in mathematics **46**(3), 305–329 (1982). https://doi.org/10.1016/0001-8708(82)90048-2
22. Mayr, E.W., Weihmann, J.: A Framework for Classical Petri Net Problems: Conservative Petri Nets as an Application. In: Ciardo, G., Kindler, E. (eds.) Application and Theory of Petri Nets and Concurrency - 35th International Conference, PETRI NETS 2014, Tunis, Tunisia, June 23-27, 2014. Proceedings. Lecture Notes in Computer Science, vol. 8489, pp. 314–333. Springer (2014). https://doi.org/10.1007/978-3-319-07734-5_17
23. Pottier, L.: Minimal Solutions of Linear Diophantine Systems: Bounds and Algorithms. In: Book, R.V. (ed.) Rewriting Techniques and Applications, 4th International Conference, RTA-91, Como, Italy, April 10-12, 1991, Proceedings. Lecture Notes in Computer Science, vol. 488, pp. 162–173. Springer (1991). https://doi.org/10.1007/3-540-53904-2_94
24. Rackoff, C.: The Covering and Boundedness Problems for Vector Addition Systems. Theor. Comput. Sci. **6**, 223–231 (1978). https://doi.org/10.1016/0304-3975(78)90036-1
25. Reisig, W.: Understanding Petri Nets. Springer-Verlag (2013). https://doi.org/10.1007/978-3-642-33278-4, 230 pp.
26. Schrijver, A.: Theory of Linear and Integer Programming. John Wiley & Sons, Inc., USA (1986)

Two-Sorted Algebraic Decompositions of Brookes's Shared-State Denotational Semantics

Yotam Dvir[1(✉)], Ohad Kammar[2], Ori Lahav[1], and Gordon Plotkin[2]

[1] Tel Aviv University, Tel Aviv-Yafo, Israel
yotamdvir@mail.tau.ac.il, orilahav@tau.ac.il
[2] University of Edinburgh, Edinburgh, UK
ohad.kammar@ed.ac.uk, gdp@inf.ed.ac.uk

Abstract. We define a two sorted equational theory of algebraic effects that models concurrent shared state with preemptive interleaving, recovering Brookes's seminal 1996 trace-based model precisely. The decomposition allows us to analyse Brookes's model algebraically in terms of separate but interacting components. The multiple sorts partition terms into layers. We use two sorts: a "hold" sort for layers that disallow interleaving of environment memory accesses, analogous to holding a global lock on the memory; and a "cede" sort for the opposite. The algebraic signature comprises of independent interlocking components: two new operators that switch between these sorts, delimiting the atomic layers, thought of as acquiring and releasing the global lock; non-deterministic choice; and state-accessing operators. The axioms similarly divide cleanly: the delimiters behave as a closure pair; all operators are strict, and distribute over non-empty non-deterministic choice; and non-deterministic global state obeys Plotkin and Power's presentation of global state. Our representation theorem expresses the free algebras over a two-sorted family of variables as sets of traces with suitable closure conditions. When the held sort has no variables, we recover Brookes's trace semantics. We define several other single- and two-sorted theories to elucidate the connection to Brookes's model via translation embeddings and equivalences.

Keywords: shared state · concurrency · denotational semantics · monads · algebraic effects · equational theory · multi-sorted algebra · trace semantics · representability · join semilattices · closure pairs · mnemoids · global state

1 Introduction

We decompose Brookes's pioneering denotational model of concurrent shared state under preemptive interleaving [6] using algebraic effects [30]. This model possesses several desirable features in the area of denotational models for programming languages with concurrent features. (I) It is based on traces, an elementary sequential gadget. (II) It is fully compositional, as in traditional denotational semantics for shared-state [e.g. 15, 17]. Each syntactic programming construct, including parallel composition, has a corresponding semantic operation combining the meanings of its constituents. Such full compositionality

P. A. Abdulla and D. Kesner (Eds.): FoSSaCS 2025, LNCS 15691, pp. 377–398, 2025.
https://doi.org/10.1007/978-3-031-90897-2_18

contrasts with some recent models in this area that require additional 'semantic post-processing': some form of quotient, pruning of auxiliary mathematical constructs, reasoning up-to behavioural equivalence; or capture only sequential blocks, reasoning about the parallel composition on a separate layer [e.g. 7, 8, 19, 21]. (III) Subsequent variations and extensions [4, 38, 39], as well as adaptations to relaxed memory models [12, 13, 21], attest to its versatility, making it a cornerstone in the denotational semantics for concurrent languages with side-effects. (IV) It achieves a high level of abstraction, evident in the many compiler transformations that the model supports, including the most common memory access introductions and eliminations, and the laws of parallel programming. Moreover, Brookes showed the model to be fully abstract in a language extended with the `await` construct, which blocks execution until all memory locations contain a given tuple of values, and then atomically updates them to contain another tuple of values. This construct is not a natural programming construct, but is clearly suggested by Brookes's semantics.

Plotkin and Power's modern theory of *algebraic effects* [30] refines Moggi's monadic approach [26] with algebraic theories. The algebraic approach informs the monadic structure by identifying semantic counterparts to syntactic constructs and axiomatising their semantics equationally. The monadic structure emerges through the well-established connection between algebraic theories and monads [23] via *representation theorems*. For example: global state emerges by axiomatising memory lookup and update [30] and a representation theorem involving the state monad; non-determinism emerges by axiomatising semi-lattices and a representation theorem involving the powerdomains [15, 28]; and so on. The algebraic perspective may offer insights into the making of the denotational semantics. It can suggest methods for combining different effects and modularly augment a semantics with a given computational effect [17].

The connection between algebraic effects and concurrency has long been emphasised. For example, the ability to use algebraic effects, without any axioms, and their *effect handlers* [3, 32, 33] to allow users to define their own schedulers was the original motivation for their implementation in the OCaml programming language [9, 10, 35]. Nonetheless, exhibiting abstract models such as Brookes's algebraically via equational axiomatisation of syntactic constructs has proved challenging. Our own previous algebraic model [11] invalidates a key transformation, reflecting a fundamental limitation.

To overcome this limitation, we use multi-sorted algebraic theories, a direction that was raised in personal discussions since the earliest work on algebraic effects [30]. A multi-sorted algebraic term decomposes into layers. Our two sorts represent two modes of interaction between a program fragment and its concurrent environment. A "hold" sort ($\bullet$) provides a reasoning layer in which the environment may not interfere, whereas in the "cede" sort ($\circ$) it may. We provide two operators that switch between these sorts. Our core idea is to axiomatise these operators as a *closure pair*, an established order-theoretic special Galois-connection, the dual to the domain-theoretic embedding-projection pairs [1]. Additionally, we axiomatise strict distributivity of the closure pair over nondeterminism. The remaining axioms, all in the hold sort, are strikingly independent from these axioms. In our shared-state theory $\mathbb{S}$, the remaining axioms are precisely those of non-deterministic global state.

We prove, twice over, that $\mathbb{S}$ recovers Brookes's model in the cede sort. First, using sets of traces akin to Brookes's, we define a representation of $\mathbb{S}$. The representation recovers Brookes's model via the adjunction that forgets the hold sort. Second, we define three algebraic theories for Brookes's `await` and its sequential variant, relating them to global-state, shared-state, and each other via embeddings and equivalences. The theory for concurrent `await` is straightforwardly represented by Brookes's model, and embeds in the cede sort of $\mathbb{S}$.

Caveats In our development, we opt for mathematical simplicity whenever possible. For example, we use countable-join semilattices instead of finite-join semilattices to represent non-determinism. This choice streamlines the development leading up to the representation theorem, allowing us to use countable sets instead of finitely generated ones. We also do not treat recursion to avoid the complexity that a domain-theoretic account incurs. The resulting model—identical to Brookes's—coincides with the elided domain-theoretic model over discrete predomains. This model also supports iteration (i.e. `while`-loops) without change thanks to countable-joins. It also supports first-order recursion without change by equipping it with a domain-theoretic structure. These compromises let us focus on the core concepts, and provide a relatively elementary exposition and a clear presentation of the underlying idea, motivating future inquiry.

2 Overview

Equational theories study terms constructed from algebraic operators (§3.1). In Plotkin and Power's algebraic theory of effects, the operators represent fundamental program effects, and their arguments represent continuations. Equational axioms reflect fundamental relationships between the operators. The equations that hold in the theory, reflecting the semantics as a whole, are those that follow from its axiomatic presentation by equational logic (§3.2).

In the global-state theory, a theory for sequential stateful computation, the operators U and L represent updating and looking up bits in memory. For example, consider the global-state term $\mathsf{U}_{\mathtt{y},0}\, \mathsf{L}_{\mathtt{y}}(3, \mathsf{U}_{\mathtt{x},1}\, \mathsf{U}_{\mathtt{y},1}\, 7)$. After updating y to 0, the computation looks y up: if it finds 0, it returns 3; if it finds 1, it updates x and y to 1 in succession, then returns 7. Between the update and the lookup, the value at y cannot change. Therefore, the computation finds the value 0, and takes the left-hand continuation. The global-state axiom (UL) $\mathsf{U}_{\ell,b}\, \mathsf{L}_{\ell}(x_0, x_1) = \mathsf{U}_{\ell,b}\, x_b$ reflects this fact. By (UL), $\mathsf{U}_{\mathtt{y},0}\, \mathsf{L}_{\mathtt{y}}(3, \mathsf{U}_{\mathtt{x},1}\, \mathsf{U}_{\mathtt{y},1}\, 7) = \mathsf{U}_{\mathtt{y},0}\, 3$ holds in global state.

A representing monadic model for an equational theory (§3.3) interprets the algebraic operators as corresponding operations over the model's domain; such that each term, up-to equality in the theory, is represented uniquely in the domain. Interpretations respect the theory; that is, applying an operation to representations of terms results in the representation of the corresponding operator applied to said terms: $[\![O]\!]_{\mathrm{op}}([\![t_1]\!]_{\mathrm{term}}, \dots, [\![t_\alpha]\!]_{\mathrm{term}}) = [\![O(t_1, \dots, t_\alpha)]\!]_{\mathrm{term}}$.

For example, global-state terms are represented by memory-manipulating functions in the state-monad model. This model interprets update by precomposing a state update $[\![\mathsf{U}_{\ell,b}]\!]_{\mathrm{op}} f = f \circ [\ell \mapsto b]$; and interprets lookup by passing the input memory state σ along to the σ_ℓ-continuation $[\![\mathsf{L}_\ell]\!]_{\mathrm{op}}(f_0, f_1) = \lambda\sigma.\, f_{\sigma_\ell}\sigma$. In this way, global state recovers the (historically precedent) state monad.

The state monad does not account for concurrent interference. The monad underlying Brookes's denotational semantics does, by using sequences of transitions to denote potential behaviours (§6.1). Each transition $\langle\sigma, \rho\rangle$ in sequence means that the computation, by relying on exclusive access to the memory at state σ, can guarantee to provide the state ρ and yield to the environment.

Following the tradition of algebraic effects, we wish to recover Brookes's model using an equational theory for shared state. This is rather straightforward using a single-sorted theory B (§6.3) in which transitions appear as operators. However, this theory is dissatisfying, for two reasons. (I) Transitions do not correspond to familiar programming constructs, but to Brookes artificial `await` construct. (II) Conceptualising shared state as global state with concurrent interference, we expect the global-state effects to be present in a theory of shared state, and the equations between them to hold when interference is prohibited.

A more appropriate approach adds an operator Y to global state for yielding control to the environment. Otherwise, the computation has exclusive access to memory. This direction lead to a Brookes-like model [11]. However, unlike Brookes's model, it does not validate the Irrelevant Read Introduction (IRI) program transformation, which introduces a read instruction that possibly yields to the environment and discards the value it read. The IRI transformation is useful as a stepping stone to other practical transformations, such as common-subexpression elimination from conditionals, and consequently loops. Invalidating IRI seems to be a fundamental limitation of the yield-operator approach, as we show in the extended manuscript [14, §A]. In retrospect, we can pinpoint the issue to Y both releasing exclusive access to memory, and acquiring it back. Our key insight is that the mode of computation that cedes access to memory needs to be explicit, decomposing Y into a pair of mode-switching operators.

The remaining structure falls into place straightforwardly and naturally, in our proposed two-sorted theory for shared state $\mathbb{S}$ (§4). Each sort represents a computation mode: *hold* ($\bullet$) represents the computation's exclusive access to memory; *cede* ($\circ$) represents positions in which the environment may interfere. Each operator has a sort and expects each continuation to have a specific sort. In $\mathbb{S}$, update $\mathsf{U} : \bullet\langle\bullet\rangle$ and lookup $\mathsf{L} : \bullet\langle\bullet, \bullet\rangle$ are $\bullet$-sorted and expect $\bullet$-sorted continuations, allowing us to reason about interference-free stateful interactions.

The theory $\mathbb{S}$ also supports non-deterministic choice in both sorts. For example, the term $(\mathsf{U}_{\mathtt{x},1}\, 2 \vee \mathsf{U}_{\mathtt{x},1}\, 5)$ either updates `x` to `1` and returns 2, or updates `x` to `1` and returns 5. We axiomatise $\mathbb{S}$ such that each operator distributes over non-deterministic choice, e.g. $(\mathsf{U}_{\mathtt{x},1}\, 2 \vee \mathsf{U}_{\mathtt{x},1}\, 5) = \mathsf{U}_{\mathtt{x},1}(2 \vee 5)$. We order terms by potential behaviours $(l \geq r) := (l = l \vee r)$, a partial-order for $\mathbb{S}$-equality (§3.2).

The mode-switching operators of $\mathbb{S}$ are $\lhd : \circ\langle\bullet\rangle$ and $\rhd : \bullet\langle\circ\rangle$. That is, $\lhd$ is $\circ$-sorted and expects a $\bullet$-sorted continuation, and vice versa for $\rhd$. We think of them as delimiting atomic blocks, or acquiring and releasing an abstract global lock. We axiomatise them in $\mathbb{S}$ by strict distributivity over countable joins, i.e. (ND-$\lhd$) $\bigvee_{i<\alpha} \lhd x_i = \lhd \bigvee_{i<\alpha} x_i$ and (ND-$\rhd$) $\bigvee_{i<\alpha} \rhd x_i = \rhd \bigvee_{i<\alpha} x_i$; and:

Empty ($\lhd \rhd y = y$)**.** An empty atomic block has no observable effect.
Fuse ($\rhd \lhd x \geq x$)**.** Fusing atomic blocks eliminates potential interference.

These axiomatise $\lhd$ and $\rhd$ as an *(insertion)-closure pair* [e.g. 1].

Each $\mathbb{S}$-term denotes a set of sort-delimited traces (§5.1), which generalise Brookes's traces. For example, $t := \mathsf{U}_{\mathtt{y},0} \rhd \lhd \mathsf{L}_{\mathtt{y}}(3, \mathsf{U}_{\mathtt{x},1}\, \mathsf{U}_{\mathtt{y},1} \rhd 7)$ denotes a set that includes $\bullet\langle\left(\begin{smallmatrix}\mathtt{x}\mapsto 1\\ \mathtt{y}\mapsto 1\end{smallmatrix}\right), \left(\begin{smallmatrix}\mathtt{x}\mapsto 1\\ \mathtt{y}\mapsto 0\end{smallmatrix}\right)\rangle\langle\left(\begin{smallmatrix}\mathtt{x}\mapsto 1\\ \mathtt{y}\mapsto 1\end{smallmatrix}\right), \left(\begin{smallmatrix}\mathtt{x}\mapsto 0\\ \mathtt{y}\mapsto 0\end{smallmatrix}\right)\rangle\circ 7$. Indeed, we can read off a corresponding computation from t, initially holding the lock in the state $\left(\begin{smallmatrix}\mathtt{x}\mapsto 1\\ \mathtt{y}\mapsto 1\end{smallmatrix}\right)$ of both bits $\mathtt{1}$: the computation updates $\mathtt{y}$ to $\mathtt{0}$, then yields to the environment before looking $\mathtt{y}$ up; finding $\mathtt{1}$, it updates $\mathtt{x}$ and $\mathtt{y}$ to $\mathtt{1}$, then releases the lock and returns 7. Brookes's original traces correspond to those delimited by $\circ$ on both ends.

With these traces we define our two-sorted generalisation of Brookes's model, and prove that it represents $\mathbb{S}$ (§5.2). The $\circ$-sorted $\circ$-valued fragment (§6.2), which represents the "block-closed" terms, is Brookes's original model (§6.1).

We also provide an algebraic perspective on the representation, by $\circ$-embedding (§6.4) the transitions-theory B into $\mathbb{S}$ (§6.5). This embedding maps $\langle\sigma, \rho\rangle$ to $\lhd \{\!|\sigma, \rho|\!\} \rhd$, where $\{\!|\sigma, \rho|\!\} : \bullet\langle\bullet\rangle$ is defined by global-state operators (§5.2).

3 Preliminaries

We present a standard treatment of countably-infinitary multi-sorted equational theories and their free models [e.g. 2, 37], straightforwardly generalising the single-sorted case by assigning sorts to functions and their arguments. The reader may choose to skim/skip this section, consulting it as necessary.

3.1 Terms

We define the logical language of multi-sorted equational logic. The basic vocabulary of multi-sorted algebra is parameterised by a set **sort** whose elements $\square, \diamond$ we call *sorts*. We will mostly focus on the *single-sorted* case (**sort** $= \{\star\}$) and the *two-sorted* case (**sort** $= \{\bullet, \circ\}$). A **sort**-*scheme* $\vec{\square} \in$ Scheme **sort** is a countable sequence of sorts from **sort**, i.e. a finite sequence $\vec{\square} = \langle \square_0, \ldots, \square_{n-1}\rangle$ of length n, or countably infinite sequence $\vec{\square} = \langle \square_0, \square_1, \ldots\rangle$ of length ω, where $\square_i \in$ **sort** for all i. For example: the empty scheme $\mathbf{0} := \langle\rangle$ of length 0; and the constant schemes $\alpha \cdot \square := \langle\square\rangle_{i<\alpha}$ of length α. We write $\square$ for the scheme $1 \cdot \square$.

A **sort**-*sorted signature* $\Sigma = \langle \mathbf{op}_\Sigma, \mathbf{ar}_\Sigma\rangle$ consists of a set of *operators* $\mathbf{op}_\Sigma$ and an *arity* assignment $\mathbf{ar}_\Sigma : \mathbf{op}_\Sigma \to \mathbf{sort} \times \text{Scheme}\,\mathbf{sort}$. For $O \in \mathbf{op}_\Sigma$ with $\mathbf{ar}_\Sigma\, O = \langle \square, \langle\diamond_i\rangle_i\rangle$, we write $(O : \square\, \langle\diamond_i\rangle_{i<\alpha}) \in \Sigma$. The operator O will allow us to construct a $\square$-sort term with a tuple of terms, with the ith subterm having sort $\diamond_i$. For single-sorted arities (**sort** $= \{\star\}$), we write $O : \alpha$ for $O : \star\,(\alpha \cdot \star)$. A *signature* is a set $\mathbf{sort}_\Sigma$ and a $\mathbf{sort}_\Sigma$-sorted signature we also denote by Σ.

We will use the following signature to model non-deterministic choice:

Example 1. The *join semilattice* single-sorted signature $\mathtt{J}$ consists of two operators: *join* $\vee : 2$, i.e. $\vee : \star \langle \star, \star \rangle$; and *bottom* $\bot : 0$, i.e. $\bot : \star \langle \rangle$. □

To simplify the formulation of our representation theorem later, we generalise the signature to countable non-deterministic choice operators:

Example 2. The *countable-join semilattice* single-sorted signature $\mathtt{V}$ consists of an α-ary *choice* operator $\bigvee_{\alpha} : \alpha$ for every $\alpha \leq \omega$. In particular, the signature $\mathtt{J}$ is included with $\alpha = 2$ (join) and $\alpha = 0$ (bottom). □

The final example demonstrates the treatment for multiple sorts:

Example 3. The *finite dimensional transformations* signature $\mathtt{M}$ consists of a sort for each pair of natural numbers $\mathbf{sort}_{\mathtt{M}} := \{\mathbf{Hom}\,(m, n) \mid m, n \in \mathbb{N}\}$, an identity operator $\mathrm{Id}_n : \mathbf{Hom}\,(n, n) \langle \rangle$ for each $n \in \mathbb{N}$, and, for each triple $m, n, k \in \mathbb{N}$, a composition operator $(\circ_{m,n,k}) : \mathbf{Hom}\,(m, k) \langle \mathbf{Hom}\,(n, k), \mathbf{Hom}\,(m, n) \rangle$. □

A signature generates a language of algebraic terms as follows. A **sort**-*family* $\boldsymbol{X} \in \mathbf{Set}^{\mathbf{sort}}$ is an assignment of a set $\boldsymbol{X}_{\square}$, to each sort $\square \in \mathbf{sort}$. We identify $\mathbf{Set}^{\{\star\}} \cong \mathbf{Set}$, and use a set-like notation to specify families, e.g. $\boldsymbol{X} := \{x : \bullet, y, z : \circ\}$ is the two-sorted family $\boldsymbol{X}_{\bullet} := \{x\}$ and $\boldsymbol{X}_{\circ} := \{y, z\}$. We can turn every **sort**-family $\boldsymbol{X}$ into the set $\oint \boldsymbol{X} := \coprod_{\square \in \mathbf{sort}} \boldsymbol{X}_{\square}$ equipped with the injections $\mathrm{in}_{\square} : \boldsymbol{X}_{\square} \to \oint \boldsymbol{X}$. This construction is a special case of the Grothendieck construction, and lets us track the distinction between sets and families.

For a signature Σ and $\mathbf{sort}_{\Sigma}$-family $\boldsymbol{X} \in \mathbf{Set}^{\mathbf{sort}_{\Sigma}}$, define the $\mathbf{sort}_{\Sigma}$-family of Σ-*terms over* $\boldsymbol{X}$: $\mathrm{Term}^{\Sigma}\boldsymbol{X} \in \mathbf{Set}^{\mathbf{sort}_{\Sigma}}$, $\mathrm{Term}^{\Sigma}_{\square}\boldsymbol{X} := \{t \mid \boldsymbol{X} \vdash_{\Sigma} t : \square\}$ inductively:

$$\frac{(x : \square) \in \boldsymbol{X}}{\boldsymbol{X} \vdash_{\Sigma} x : \square} \qquad \frac{(O : \square \langle \diamond_i \rangle_{i<\alpha}) \in \Sigma \qquad \forall i.\, \boldsymbol{X} \vdash_{\Sigma} t_i : \diamond_i}{\boldsymbol{X} \vdash_{\Sigma} O \langle t_i \rangle_{i<\alpha} : \square}$$

Here, the elements $x \in \boldsymbol{X}_{\square}$, written $(x : \square) \in \boldsymbol{X}$, represent variables of sort $\square$. We may drop the set-brackets left of a trunstile, e.g. write $x : \bullet, y, z : \circ \vdash_{\Sigma} y : \circ$; and omit the sorts, especially in the single-sorted case, e.g. write $x, y \vdash_{\mathtt{J}} x \vee \bot$. For $t \in \mathrm{Term}^{\Sigma}_{\square}\boldsymbol{X}$, we write $\boldsymbol{X} \vdash_{\Sigma} \psi := t : \square$ to define ψ as t, e.g. $x, y \vdash_{\mathtt{J}} \psi := x \vee \bot$.

A **sort**-*sorted map* $f : \boldsymbol{X} \to \boldsymbol{Y}$ is a **sort**-indexed tuple of functions between the corresponding sets: $f_{\square} : \boldsymbol{X}_{\square} \to \boldsymbol{Y}_{\square}$, for every $\square \in \mathbf{sort}$. Our development utilises sorted maps extensively. A *(simultaneous) substitution* $\boldsymbol{X} \vdash_{\Sigma} \theta : \boldsymbol{Y}$ is a sorted function $\theta : \boldsymbol{Y} \to \mathrm{Term}^{\Sigma}\boldsymbol{X}$, specifying which $\square$-term $\boldsymbol{X} \vdash_{\Sigma} \theta_{\square} y : \square$ to substitute for each variable $y \in \boldsymbol{Y}_{\square}$. Each such substitution determines a sorted map $[\theta] : \mathrm{Term}\boldsymbol{Y} \to \mathrm{Term}\boldsymbol{X}$ inductively, which we write in post-fix notation:

$$(\boldsymbol{Y} \vdash_{\Sigma} y : \square)\,[\theta] := (\boldsymbol{X} \vdash_{\Sigma} \theta_{\square} y : \square) \qquad (\boldsymbol{Y} \vdash_{\Sigma} O \langle t_i \rangle_i)\,[\theta] := (\boldsymbol{X} \vdash_{\Sigma} O \langle t_i\,[\theta] \rangle_i)$$

3.2 Equational logic

A $\square$-*sorted* Σ-*equation in context* $\boldsymbol{X}$ is a pair $\langle l, r \rangle \in \mathrm{Term}^{\Sigma}_{\square}\boldsymbol{X}$ of $\square$-sorted Σ-terms over $\boldsymbol{X}$. We write this situation as $\boldsymbol{X} \vdash_{\Sigma} l = r : \square$, or just $l = r$, and call l the left-hand side (LHS) and r the right-hand side (RHS) of the equation. A *presentation* $\mathfrak{p}$ consists of a signature $\Sigma_{\mathfrak{p}}$ and *axioms*: a set $\mathrm{Ax}_{\mathfrak{p}}$ of Σ-equations.

$$\frac{\boldsymbol{X} \vdash_{\Sigma_\mathfrak{p}} t : \square}{\boldsymbol{X} \vdash_\mathfrak{p} t = t : \square} \qquad \frac{\boldsymbol{X} \vdash_\mathfrak{p} t_2 = t_1 : \square}{\boldsymbol{X} \vdash_\mathfrak{p} t_1 = t_2 : \square} \qquad \frac{\boldsymbol{X} \vdash_\mathfrak{p} t_1 = t_2 : \square \quad \boldsymbol{X} \vdash_\mathfrak{p} t_2 = t_3 : \square}{\boldsymbol{X} \vdash_\mathfrak{p} t_1 = t_3 : \square}$$

$$\frac{(\boldsymbol{X} \vdash_{\Sigma_\mathfrak{p}} t_1 = t_2 : \square) \in \mathrm{Ax}_\mathfrak{p}}{\boldsymbol{X} \vdash_\mathfrak{p} t_1 = t_2 : \square} \qquad \frac{\boldsymbol{Y} \vdash_\mathfrak{p} t_1 = t_2 : \square \quad \boldsymbol{X} \vdash_{\Sigma_\mathfrak{p}} \theta : \boldsymbol{Y}}{\boldsymbol{X} \vdash_\mathfrak{p} t_1[\theta] = t_2[\theta] : \square}$$

$$\frac{\boldsymbol{Y} \vdash_{\Sigma_\mathfrak{p}} t : \square \quad \boldsymbol{X} \vdash_{\Sigma_\mathfrak{p}} \theta, \theta' : \boldsymbol{Y} \quad \forall (y : \diamond) \in \boldsymbol{Y}. \boldsymbol{X} \vdash_\mathfrak{p} \theta_\diamond y = \theta'_\diamond y : \diamond}{\boldsymbol{X} \vdash_\mathfrak{p} t[\theta] = t[\theta'] : \square}$$

Fig. 1. Multi-sorted equational logic with countable arities

Example 4. The *join semilattice* presentation J consists of the signature $\Sigma_\mathsf{J} := \mathtt{J}$ of example 1, and the axioms Ax_J below:

(Associativity) $x \vee (y \vee z) = (x \vee y) \vee z$ (Idempotency) $x \vee x = x$
(Commutativity) $x \vee y = y \vee x$ (Neutrality) $x \vee \bot = x$ □

Example 5. The *countable-join semilattice* presentation V consists of the signature $\Sigma_\mathsf{V} := \mathtt{V}$ of example 2, and the axioms Ax_V:

(ND-return) $\bigvee_{i<1} x_i = x_0$
(ND-squash) $\bigvee_{i<\alpha} \bigvee_{j<\beta_i} x_{i,j} = \bigvee_{k<\gamma} x_{fk}$ where $f : \gamma \twoheadrightarrow \coprod_{i<\alpha} \beta_i$ □

Example 6. The *finite dimensional transformations* presentation M consists of the signature $\Sigma_\mathsf{M} := \mathtt{M}$ of example 3 and the axioms Ax_M below, suppressing the sort indices (each axiom scheme includes every possible instantiation):

(L-Id) $\mathrm{Id} \circ f = f$ (R-Id) $f \circ \mathrm{Id} = f$ (Assoc) $f \circ (g \circ h) = (f \circ g) \circ h$ □

Figure 1 presents the deductive system called *equational logic.* We say that a presentation $\mathfrak{p}$ *proves* an equation, writing $\boldsymbol{X} \vdash_\mathfrak{p} t_1 = t_2 : \square$, when it is derivable from $\mathrm{Ax}_\mathfrak{p}$ using these standard equational reasoning rules, namely: reflexivity, symmetry, transitivity, use of an axiom, substitution, and congruence. This logic is monotone: assuming more axioms allows us to prove more equations. The *algebraic theory* of a presentation $\mathfrak{p}$ is the smallest derivation-closed set of equations containing the axioms. We denote the theory of $\mathfrak{p}$ by $\mathfrak{p}$ as well.

Example 7. We can prove $\{x, y : \star\} \vdash_\mathsf{J} (x \vee \bot) \vee y = x \vee y : \star$ using an instance of Neutrality and reflexivity with the following instance of congruence:

$$\{z, y : \star\} \vdash_\mathsf{J} t := z \vee y : \star \qquad \theta_\star := \begin{pmatrix} z \mapsto x \vee \bot \\ y \mapsto y \end{pmatrix} \qquad \theta'_\star := \begin{pmatrix} z \mapsto x \\ y \mapsto y \end{pmatrix}$$ □

When a presentation $\mathfrak{p}$ proves the semi-lattice axioms in one of its sorts $\square$, then the encoding $(\boldsymbol{X} \vdash_{\Sigma_\mathfrak{p}} l \leq r : \square) := (\boldsymbol{X} \vdash_{\Sigma_\mathfrak{p}} l \vee r = r : \square)$ of inequations as equations in this sort is a preorder that is a partial order w.r.t. $\mathfrak{p}$-equality, i.e. $(\boldsymbol{X} \vdash_\mathfrak{p} s \leq t : \square) \wedge (\boldsymbol{X} \vdash_\mathfrak{p} t \leq s : \square) \implies (\boldsymbol{X} \vdash_\mathfrak{p} s = t : \square)$. We encode $(\geq)$ similarly. Due to the monotonicity property of equational logic, once we have included an axiomatisation of semi-lattices through a subset of the axioms, we may proceed to postulate inequations.

We also use a generalisation of distributivity axioms [18], reproducing familiar arithmetic distributivity equations such as $x \cdot \max\{y_1, y_2\} = \max\{x \cdot y_1, x \cdot y_2\}$, the distributivity of $(\cdot)$ over max in the right-hand-side position. The extended manuscript [14, §B] details the straightforward, but technical generalisation. The main message is as follows. In a given presentation $\mathfrak{p}$, if all operators distribute over binary joins in every position, the congruence rule is valid for inequations:

$$\frac{\boldsymbol{Y} \vdash_{\Sigma_{\mathfrak{p}}} t : \square \qquad \boldsymbol{X} \vdash_{\Sigma_{\mathfrak{p}}} \theta, \theta' : \boldsymbol{Y} \qquad \forall (y : \diamond) \in \boldsymbol{Y}. \boldsymbol{X} \vdash_{\mathfrak{p}} \theta_{\diamond} y \leq \theta'_{\diamond} y : \diamond}{\boldsymbol{X} \vdash_{\mathfrak{p}} t\,[\theta] \leq t\,[\theta'] : \square}$$

If a presentation $\mathfrak{p}$ supports semi-lattices in every sort and they distribute over binary joins in every positions, then we say that $\mathfrak{p}$ *supports inequational reasoning*. The theory of $\mathfrak{p}$ then admits Bloom's logic for ordered algebraic theories [5]. We let future work determine the most appropriate variety of inequational logic [29].

Going forward, all of our presentations support inequational reasoning in this sense, and all operators distribute over arbitrary non-empty joins, not just the binary ones. Moreover, they are all strict: $O(\bot, \dots, \bot) = \bot$ for every operator $(O : \square \langle \diamond_i \rangle_{i<\alpha}) \in \Sigma_{\mathfrak{p}}$. Such theories 'absorb' side-effects when their continuations diverge, an inherent 'partial correctness' property of Brookes's model.

3.3 Algebras and models

After presenting the proof theory—equational logic—let's turn to the model theory of universal algebra. A Σ-*algebra* $\mathbf{A}$ consists of a $\mathbf{sort}_\Sigma$-family $\underline{\mathbf{A}} \in \mathbf{Set}^{\mathbf{sort}_\Sigma}$, the *carrier*, and an assignment $\mathbf{A}\,[\![-]\!]_{\mathrm{op}}$, for each operator $(O : \square \langle \diamond_i \rangle_{i<\alpha}) \in \Sigma$, of an *operation* over this carrier: $\mathbf{A}\,[\![O]\!]_{\mathrm{op}} : (\prod_{i<\alpha} \underline{\mathbf{A}}_{\diamond_i}) \to \underline{\mathbf{A}}_{\square}$.

Example 8. For any set X, define the V-algebra $\mathbf{V}X$ by taking the carrier to be the set of countable (finite or infinite) X-subsets $\underline{\mathbf{V}X} := \mathcal{P}^{\aleph_0}(X)$, and interpret choice as union $\mathbf{V}X[\![\bigvee_\alpha]\!]_{\mathrm{op}} \langle D_i \rangle_{i<\alpha} := \bigcup_{i<\alpha} D_i$. □

Example 9. Define the M-algebra $\mathbf{M}$ by taking the carrier to be the set of real-valued matrices of the corresponding dimensions, $\underline{\mathbf{M}}_{\mathbf{Hom}(m,n)} := \mathbb{M}^{\mathbb{R}}_{m\times n}$, interpret the identity $\mathbf{M}[\![\mathrm{Id}_n]\!]_{\mathrm{op}} := I_n \in \mathbb{M}^{\mathbb{R}}_{n\times n}$ as the identity matrix, and composition $\mathbf{M}[\![(\circ)]\!]_{\mathrm{op}} := (\cdot)$ as matrix multiplication.

Let $\mathbf{A}$ be an M-algebra. Define the *opposite* algebra $\mathbf{A}^{\mathrm{op}}$ by exchanging dimensions. So $\underline{\mathbf{A}^{\mathrm{op}}_{\mathbf{Hom}(m,n)}} := \underline{\mathbf{A}}_{\mathbf{Hom}(n,m)}$, the same identity $\mathbf{A}^{\mathrm{op}}[\![\mathrm{Id}_n]\!]_{\mathrm{op}} := \mathbf{A}[\![\mathrm{Id}_n]\!]_{\mathrm{op}}$, and reversing composition $\mathbf{A}^{\mathrm{op}}[\![(\circ)]\!]_{\mathrm{op}}(A, B) := \mathbf{A}[\![(\circ)]\!]_{\mathrm{op}}(B, A)$. □

Example 10 (term algebra). The Σ-terms with variables from $\boldsymbol{X}$ carry a canonical algebra structure $\mathbf{F}^\Sigma \boldsymbol{X}$, given by $\underline{\mathbf{F}^\Sigma \boldsymbol{X}} := \mathrm{Term}^\Sigma \boldsymbol{X}$, with each O-term constructor as the corresponding O-operation: $(\mathbf{F}^\Sigma \boldsymbol{X})\,[\![O]\!]_{\mathrm{op}} \langle t_i \rangle_i := O \langle t_i \rangle_i$. □

A Σ-*algebra homomorphism* $\varphi : \mathbf{A} \to \mathbf{B}$ is a sorted-function $\varphi : \underline{\mathbf{A}} \to \underline{\mathbf{B}}$ that preserves the operations: $\varphi_\square(\mathbf{A}\,[\![O]\!]_{\mathrm{op}} (a_1, \dots, a_\alpha)) = \mathbf{B}\,[\![O]\!]_{\mathrm{op}} (\varphi_{\diamond_1} a_1, \dots, \varphi_{\diamond_\alpha} a_\alpha)$.

Example 11. Transposing real-valued matrices $(-)^\top : \mathbb{M}^{\mathbb{R}}_{m\times n} \to \mathbb{M}^{\mathbb{R}}_{n\times m}$ is a homomorphism $(-)^\top : \mathbf{M} \to \mathbf{M}^{\mathsf{op}}$, by the well-known identity $(A\cdot B)^\top = B^\top \cdot A^\top$. □

A Σ-algebra allows us to interpret every Σ-term, by assigning values to its variables. Formally, let $\mathbf{A}$ be a Σ-algebra. An $\boldsymbol{X}$*-environment in* $\mathbf{A}$ is a sorted function $e : \boldsymbol{X} \to \underline{\mathbf{A}}$. Given such an environment, interpret terms by induction:

$$\mathbf{A}[\![\boldsymbol{X} \vdash_\Sigma x : \square]\!]_{\mathsf{term}}\, e := e_\square x \qquad \mathbf{A}[\![O\,\langle t_i\rangle_i]\!]_{\mathsf{term}}\, e := \mathbf{A}[\![O]\!]_{\mathsf{op}} \left\langle \mathbf{A}[\![t_i]\!]_{\mathsf{term}}\, e\right\rangle_i$$

Example 12 (substitution). An $\boldsymbol{X}$-environment in $\mathbf{F}^\Sigma\boldsymbol{X}$ amounts to a substitution, and interpreting terms in $\mathbf{F}^\Sigma\boldsymbol{X}$ amounts to substitution. □

Example 13 (evaluation homomorphism). Evaluation using an $\boldsymbol{X}$-environment $e : \boldsymbol{X} \to \underline{\mathbf{A}}$ in a Σ-algebra $\mathbf{A}$ is a homomorphism $\mathbf{A}[\![-]\!]_{\mathsf{term}}e : \mathbf{F}^\Sigma\boldsymbol{X} \to \mathbf{A}$. □

A Σ-algebra $\mathbf{A}$ *validates* the equation $\boldsymbol{X} \vdash_\Sigma l = r : \square$ when evaluation in all environments equates its sides: $\mathbf{A}[\![l]\!]_{\mathsf{term}}e = \mathbf{A}[\![r]\!]_{\mathsf{term}}e$ for all $e : \boldsymbol{X} \to \underline{\mathbf{A}}$. We then write $\mathbf{A} \vdash \boldsymbol{X} \vdash_\Sigma l = r : \square$. A $\mathfrak{p}$*-model* is an algebra validating all of $\mathrm{Ax}_{\mathfrak{p}}$. The soundness theorem of equational logic states that every $\mathfrak{p}$-model validates all the equations in the algebraic theory of $\mathfrak{p}$.

Example 14. Referring to previous examples, the algebras $\mathbf{V}X$ are V-models, the algebras $\mathbf{M}$ and $\mathbf{M}^{\mathsf{op}}$ are M-models, and algebras of terms are $\emptyset$-models. □

Example 15. Consider the Σ_{J}-algebra $\mathbf{A}$ for which the carrier is the set of natural numbers $\underline{\mathbf{A}} := \mathbb{N}$, join interprets as addition $\mathbf{A}[\![\vee]\!]_{\mathsf{op}}(m, n) := m + n$, and bottom as zero $\mathbf{A}[\![\bot]\!]_{\mathsf{op}} := 0$. This is *not* a J-model, since, taking $e : \{x : \star\} \to \underline{\mathbf{A}}$ with $ex = 1$, we get $\mathbf{A}[\![x \vee x]\!]_{\mathsf{term}}e \neq \mathbf{A}[\![x]\!]_{\mathsf{term}}e$; and so $\mathbf{A} \not\vdash x : \star \vdash_{\mathsf{J}} x \vee x = x : \star$. □

We end this section with representations of free models. These are $\mathfrak{p}$-models whose elements represent the $\Sigma_{\mathfrak{p}}$-terms up-to provable equality in $\mathfrak{p}$.

A $\mathfrak{p}$*-model* $\langle \mathbf{A}, e\rangle$ *over a family* $\boldsymbol{X}$ consists of a $\mathfrak{p}$-model $\mathbf{A}$ and an $\boldsymbol{X}$-environment in it $e : \boldsymbol{X} \to \underline{\mathbf{A}}$. A *free* $\mathfrak{p}$-model $\langle \mathbf{A}, \mathrm{return}\rangle$ over a family $\boldsymbol{X}$ is then a $\mathfrak{p}$-model over $\boldsymbol{X}$ such that every environment in every $\mathfrak{p}$-model $e : \boldsymbol{X} \to \underline{\mathbf{B}}$ extends uniquely along return to a $\mathfrak{p}$-homomorphism $e^\# : \mathbf{A} \to \mathbf{B}$, i.e., for all $x \in \boldsymbol{X}_\square$, we have: $e^\#_\square(\mathrm{return}_\square\, a) = ea$. We then say that the algebra $\mathbf{A}$ *represents* $\boldsymbol{X}$-environments via the assignment $e \mapsto e^\#$, the corresponding *representation.*

The algebraic theory of effects [30] emphasises the role free models play in denotational semantics for programming languages with effects. In particular, given a free $\mathfrak{p}$-model over $\boldsymbol{X}$ for every family $\boldsymbol{X}$, one standardly obtains a monad suitable for the denotational semantics of a language with computational effects conforming to the operators in $\mathfrak{p}$.

Example 16. For any set X, the V-algebra $\mathbf{V}X$ given by the countable powerset in example 8 represents X-environments; together with $\mathrm{return}\, x := \{x\}$ it forms a free V-model over X. The representation assigns $e : X \to \underline{\mathbf{B}}$ to $e^\# : \mathbf{V}X \to \mathbf{B}$, defined $e^\# D := \mathbf{B}[\![\bigvee_{|D|}]\!]_{\mathsf{op}}\langle ex\rangle_{x\in D}$; how it enumerates D doesn't matter since $\mathbf{B}$ is a V-model. The data $\langle X \mapsto \underline{\mathbf{V}X}, \mathrm{return}, (-)^\#\rangle$ is a monad. □

4 Shared state

To define the equational theory of shared state, we first recall the standard, single sorted *(non-deterministic) global state* theory G [17, 25, 30]. The variant we present here has countable non-determinism, and the global state operators manipulate a common memory store $\mathbb{S} := \mathbb{L} \to \mathbb{B}$ with a finite set of locations $\mathbb{L} \neq \emptyset$ each storing a bit $\mathbb{B} := \{0, 1\}$. A larger finite set of storable-values would not be conceptually different. Infinite sets of storable-values or locations work similarly with more involved representation theorems. In concrete examples, we let $\mathbb{L} = \{\mathsf{x}, \mathsf{y}\}$ and use non-bracketed vectors for stores, e.g. $\substack{1\\0}$ denotes $\binom{\mathsf{x}\mapsto 1}{\mathsf{y}\mapsto 0}$.

The signature Σ_{G} consists of the countable-join semilattice operators (example 2), as well as two kinds of memory-access operators: *lookup* operators $\mathsf{L}_\ell : 2$, to look a location $\ell \in \mathbb{L}$ up and branch according to the value found; and *update* operators $\mathsf{U}_{\ell,b} : 1$, to update a location $\ell \in \mathbb{L}$ to the value $b \in \mathbb{B}$. The global state axioms Ax_{G} consists of the countable-join semilattice axioms (example 5), as well as the following:

Non-deterministic global state (omitting semilattice axioms)

(UL) $\mathsf{U}_{\ell,b}\,\mathsf{L}_\ell(x_0, x_1) = \mathsf{U}_{\ell,b}\,x_b$		(LU) $\mathsf{L}_\ell(\mathsf{U}_{\ell,0}\,x, \mathsf{U}_{\ell,1}\,x) = x$	
(UU) $\mathsf{U}_{\ell,b'}\,\mathsf{U}_{\ell,b}\,x = \mathsf{U}_{\ell,b}\,x$		(ND-U) $\bigvee_{i<\alpha} \mathsf{U}_{\ell,b}\,x_i = \mathsf{U}_{\ell,b} \bigvee_{i<\alpha} x_i$	
(UUc) $\mathsf{U}_{\ell,b}\,\mathsf{U}_{\ell',b'}\,x = \mathsf{U}_{\ell',b'}\,\mathsf{U}_{\ell,b}\,x$	where $\ell \neq \ell'$		

The induced algebraic theory G includes axioms of less succinct presentations of the same theory [25]. For example, lookup also distributes over binary join, so the theory admits inequational reasoning; consecutively looking the same location up can be merged, e.g. $x_0, x_1, y \vdash_{\mathsf{G}} \mathsf{L}_\ell(\mathsf{L}_\ell(x_0, x_1), y) = \mathsf{L}_\ell(x_0, y)$; and other combinations of looking-up and updating different locations commute, e.g. for any $\ell \neq \ell'$ we have $x_0, x_1 \vdash_{\mathsf{G}} \mathsf{L}_\ell(\mathsf{U}_{\ell',b}\,x_0, \mathsf{U}_{\ell',b}\,x_1) = \mathsf{U}_{\ell',b}\,\mathsf{L}_\ell(x_0, x_1)$.

Our two-sorted presentation $\mathbb{S}$ of *shared state* extends global state. Its sorts are $\mathbf{sort}_{\Sigma_{\mathbb{S}}} = \{\bullet, \circ\}$. The *hold* sort ($\bullet$) represents an uninterrupted sequence of memory accesses, whereas the *cede* sort ($\circ$) allows control to pass to the environment. The operators and the arities of the signature $\Sigma_{\mathbb{S}}$ consist of a copy of Σ_{G} at $\bullet$, a copy of Σ_{V} at $\circ$, and new operators $\lhd : \circ\langle\bullet\rangle$ and $\rhd : \bullet\langle\circ\rangle$.

The intuitive reading for algebraic effects is from the outside in. With this intuition, one interpretation of the operators $\lhd$ and $\rhd$ is to acquire and release a global lock. The hold sort ($\bullet$) represents the lock being held by one of the threads in the program. The cede sort ($\circ$) represents points in the execution in which one of the threads in the concurrent environment may acquire the lock. The sorts ensure exclusive access to the lock, and therefore to the store. In an alternative interpretation, these operators delimit atomic blocks; their sorts prevent nesting.

The shared state axioms $\mathrm{Ax}_{\mathbb{S}}$ include a copy of the (non-deterministic) global state axioms Ax_{G} at $\bullet$ and a copy of the countable-join semilattice axioms Ax_{V} at $\circ$. In particular, $\mathbb{S}$ proves the semi-lattice axioms in both sorts. It further includes standard strict distributivity axioms for the new unary operators:

Strict distributivity of $\lhd$ and $\rhd$

(ND-$\lhd$) $\bigvee_{i<\alpha} \lhd x_i = \lhd \bigvee_{i<\alpha} x_i$ (ND-$\rhd$) $\bigvee_{i<\alpha} \rhd x_i = \rhd \bigvee_{i<\alpha} x_i$

With these axioms, $\mathbb{S}$ supports inequational reasoning, which represents the semantic refinement relation used to validate program transformations [e.g. 11].

Finally, $\mathrm{Ax}_{\mathbb{S}}$ axiomatises $\lhd$ and $\rhd$ as an *(insertion)-closure pair* [e.g. 1]:

Closure pair (Empty) $\lhd \rhd y = y$ (Fuse) $\rhd \lhd x \geq x$

They are compatible with the global-lock interpretation:

Empty ($\lhd \rhd y = y$). Acquiring and immediately releasing the lock has no effect on the sequence of effects that can occur as a result of arbitrary interleavings.

Fuse ($\rhd \lhd x \geq x$). Releasing and immediately acquiring the lock only allows more behaviours. The environment may or may not interleave there.

To summerise, $\mathrm{Ax}_{\mathbb{S}} := \mathrm{Ax}^{\bullet}_{\mathsf{G}} \cup \mathrm{Ax}^{\circ}_{\mathsf{V}} \cup \{\text{ND-}\rhd, \text{ND-}\lhd\} \cup \{\text{Empty}, \text{Fuse}\}$.

Example 17. The $\Sigma_{\mathbb{S}}$-equations appearing below are named after corresponding transformations that may or may not be valid, depending on the setting (e.g. is there concurrency, and under what assumptions), all $\circ$-sorted over $\{x : \circ\}$:

$$\lhd \mathsf{L}_\ell(\rhd x, \rhd x) = x \qquad \text{(Irrelevant Read Intro \& Elim)}$$
$$\lhd \mathsf{U}_{\ell,b_1} \rhd \lhd \mathsf{U}_{\ell,b_2} \rhd x \geq \lhd \mathsf{U}_{\ell,b_2} \rhd x \qquad \text{(Write Elim)}$$
$$\lhd \mathsf{U}_{\ell,b_1} \rhd \lhd \mathsf{U}_{\ell,b_2} \rhd x \leq \lhd \mathsf{U}_{\ell,b_2} \rhd x \qquad \text{(Write Intro)}$$

Intuitively, Irrelevant Read Intro & Elim should be valid in our setting, as looking a value up is not observable by the environment, and the computation itself disregards the value. Write Elim should be valid too, because it is possible that the environment does not look ℓ up at the interference point between the updates on the LHS, covering the behaviour denoted by the RHS. On the other hand, Write Intro should be invalid in our setting because only on the LHS can a concurrently running thread look ℓ up and find b_1. Formally, we will show $\mathbb{S}$ does not prove Write Intro in example 25. Here we show $\mathbb{S}$ proves the other two:

$$\lhd \mathsf{L}_\ell(\rhd x, \rhd x) \overset{\mathrm{LU}}{=} \lhd \mathsf{L}_\ell\left(\mathsf{U}_{\ell,0}\, \mathsf{L}_\ell(\rhd x, \rhd x), \mathsf{U}_{\ell,1}\, \mathsf{L}_\ell(\rhd x, \rhd x)\right)$$
$$\overset{\mathrm{UL}}{=} \lhd \mathsf{L}_\ell\left(\mathsf{U}_{\ell,0} \rhd x, \mathsf{U}_{\ell,1} \rhd x\right) \overset{\mathrm{LU}}{=} \lhd \rhd x \overset{\mathrm{Empty}}{=} x$$
$$\lhd \mathsf{U}_{\ell,b_1} \rhd \lhd \mathsf{U}_{\ell,b_2} \rhd x \overset{\mathrm{Fuse}}{\geq} \lhd \mathsf{U}_{\ell,b_1}\, \mathsf{U}_{\ell,b_2} \rhd x \overset{\mathrm{UU}}{=} \lhd \mathsf{U}_{\ell,b_2} \rhd x$$

□

5 Representation

We now establish the representation theorem describing a free $\mathbb{S}$-model over any $\boldsymbol{X} \in \mathbf{Set}^{\{\bullet,\circ\}}$. Following Brookes [6], we use sets of traces to denote behaviours.

5.1 Sorted traces

A *sorted trace* starts with a sort ($\bullet$ or $\circ$) followed by a non-empty sequence of state transitions, and ending in a sorted value. The initial sort in the trace and the initial store in each transition represent assumptions the trace relies on from its concurrent and sequential environment. The final sort and value and the final store in each transition represent guarantees the trace makes to its environment.

Formally, a *(state) transition* is a pair $\langle \sigma, \rho \rangle \in \mathbb{S} \times \mathbb{S}$. Let $\xi^? \in (\mathbb{S} \times \mathbb{S})^*$ range over possibly empty sequences of transitions, and $\xi \in (\mathbb{S} \times \mathbb{S})^+$ range over non-empty ones. For any set X, define the set of *X-valued Brookes traces* $\mathsf{T}X := (\mathbb{S} \times \mathbb{S})^+ \times X$, also used in Brookes's model (§6). For any family $\boldsymbol{X} \in \mathbf{Set}^{\{\bullet,\circ\}}$ define the $\{\bullet,\circ\}$-sorted family $\mathbf{T}\boldsymbol{X}$ of *traces* $(\mathbf{T}\boldsymbol{X})_{\square} := \mathsf{T} \oint \boldsymbol{X}$. Then, for any sorted family $\boldsymbol{X} \in \mathbf{Set}^{\{\bullet,\circ\}}$, we define the set of *sorted traces over $\boldsymbol{X}$* by:

$$\mathbb{T}\boldsymbol{X} := \oint \mathbf{T}\boldsymbol{X} = \{\bullet,\circ\} \times (\mathbb{S} \times \mathbb{S})^+ \times \coprod_{\diamond \in \{\bullet,\circ\}} \boldsymbol{X}_{\diamond}$$

A *$\square$-sorted $\diamond$-valued trace* is one of the form $\square\xi\diamond x := \langle \square, \xi, \mathrm{in}_{\diamond}\, x \rangle$ in the set $\mathbb{T}\boldsymbol{X}$.

Example 18. $\bullet\langle \genfrac{}{}{0pt}{}{1}{1}, \genfrac{}{}{0pt}{}{1}{0} \rangle \langle \genfrac{}{}{0pt}{}{1}{1}, \genfrac{}{}{0pt}{}{0}{0} \rangle \circ 7 \in \mathbb{T}\boldsymbol{X}$, with $\boldsymbol{X}_{\circ} = \mathbb{N}$, is $\bullet$-sorted and $\circ$-valued. □

Intuitively, the trace $\square\xi\diamond x$ models a potential behaviour, or protocol, that a shared-state program phrase under preemptive interleaving concurrency can exhibit, or adhere to, given as a rely/guarantee sequence.

Example 19. The behaviour denoted by $\bullet\langle \genfrac{}{}{0pt}{}{1}{1}, \genfrac{}{}{0pt}{}{1}{0} \rangle \langle \genfrac{}{}{0pt}{}{1}{1}, \genfrac{}{}{0pt}{}{0}{0} \rangle \circ 7$ relies on the preceding environment for $\genfrac{}{}{0pt}{}{1}{1}$ and for the sequential environment to hold access to the store; then guarantees $\genfrac{}{}{0pt}{}{1}{0}$; then relies on $\genfrac{}{}{0pt}{}{1}{1}$; and finally guarantees $\genfrac{}{}{0pt}{}{0}{0}$, and returns 7 to the succeeding sequential environment, ceding exclusive store access. □

One can make these trace-semantic concepts more formal, for example, when formulating an adequacy proof w.r.t. an operational semantics. We will not define these concepts formally since we will not need the additional level of rigour, for example, because we appeal to the well-established adequacy of Brookes's model.

We implicitly understand the exclusive access to the store is ceded ($\circ$) between transitions. For example, for the trace $\bullet\langle \genfrac{}{}{0pt}{}{1}{1}, \genfrac{}{}{0pt}{}{1}{0} \rangle \langle \genfrac{}{}{0pt}{}{1}{1}, \genfrac{}{}{0pt}{}{0}{0} \rangle \circ 7$, we could write $\bullet\langle \genfrac{}{}{0pt}{}{1}{1}, \genfrac{}{}{0pt}{}{1}{0} \rangle \circ \langle \genfrac{}{}{0pt}{}{1}{1}, \genfrac{}{}{0pt}{}{0}{0} \rangle \circ 7$ for emphasis. A hypothetical $\bullet\langle \genfrac{}{}{0pt}{}{1}{1}, \genfrac{}{}{0pt}{}{1}{0} \rangle \bullet \langle \genfrac{}{}{0pt}{}{1}{1}, \genfrac{}{}{0pt}{}{0}{0} \rangle \circ 7$ would denote an impossible behaviour, making intermediate sorts redundant.

One of Brookes's innovations is that sets of traces should be closed under what we now call *(trace) deductions*. Specifically, Brookes identified two such deductions, given as binary relations called $\mathsf{stutter}$ ($\xrightarrow{\mathtt{st}}$) and mumble ($\xrightarrow{\mathtt{mu}}$), defined in such a way that if the program phrase can adhere to the source protocol (left of arrow), then it can adhere to the target protocol (right of arrow).

We define these deductions in our two-sorted setting. For convenience, we write $\square\xi_1^? \circ \xi_2^? \diamond x$ for the trace $\square\xi_1^?\xi_2^?\diamond x$ in which, intuitively, the lock is ceded ($\circ$) at the marked spot. Formally, we require that both (a) if $\xi_1^?$ is empty, then $\square = \circ$; and (b) if $\xi_2^?$ is empty, then $\diamond = \circ$. In particular, the requirement holds when both $\xi_1^?$ and $\xi_2^?$ are non-empty, where we implicitly assume the ceded sort between them; and in the case of a $\circ$-sorted $\circ$-valued trace, i.e. $\square = \circ = \diamond$.

Example 20. We have the following valid/invalid notations for $\bullet\langle {\substack{1\\1}}, {\substack{1\\0}}\rangle\langle {\substack{1\\1}}, {\substack{0\\0}}\rangle\circ 7$:

$$\text{valid: } \bullet\langle {\substack{1\\1}}, {\substack{1\\0}}\rangle\circ\langle {\substack{1\\1}}, {\substack{0\\0}}\rangle\circ 7 \quad \bullet\langle {\substack{1\\1}}, {\substack{1\\0}}\rangle\langle {\substack{1\\1}}, {\substack{0\\0}}\rangle\circ\circ 7 \qquad \text{invalid: } \bullet\circ\langle {\substack{1\\1}}, {\substack{1\\0}}\rangle\langle {\substack{1\\1}}, {\substack{0\\0}}\rangle\circ 7$$ □

We define the following *sorted stutter and mumble deductions*:

$$\square\xi_1^? \circ \xi_2^? \diamond x \xrightarrow{\texttt{st}} \square\xi_1^?\langle\sigma,\sigma\rangle\xi_2^?\diamond x \qquad \square\xi_1^?\langle\sigma,\rho\rangle\langle\rho,\theta\rangle\xi_2^?\diamond x \xrightarrow{\texttt{mu}} \square\xi_1^?\langle\sigma,\theta\rangle\xi_2^?\diamond x$$

The condition on stutter's source rules out deductions which implicitly cede access to the store to the concurrent environment at the ends of the trace. We will compare these deductions to Brookes's in §6.

Example 21. These deductions are valid, highlighting the change to the trace:

$$\bullet\langle {\substack{1\\1}}, {\substack{1\\0}}\rangle\langle {\substack{1\\1}}, {\substack{0\\0}}\rangle\circ 7 \xrightarrow{\texttt{st}} \bullet\langle {\substack{1\\1}}, {\substack{1\\0}}\rangle\langle {\substack{1\\1}}, {\substack{0\\0}}\rangle\langle {\substack{0\\1}}, {\substack{0\\1}}\rangle\circ 7 \qquad \bullet\langle {\substack{1\\1}}, {\substack{1\\0}}\rangle\langle {\substack{1\\0}}, {\substack{0\\0}}\rangle\circ 7 \xrightarrow{\texttt{mu}} \bullet\langle {\substack{1\\1}}, {\substack{0\\0}}\rangle\circ 7$$

However, thanks to the condition on stutter's source, this deduction is invalid:

$$\bullet\langle {\substack{1\\1}}, {\substack{1\\0}}\rangle\langle {\substack{1\\1}}, {\substack{0\\0}}\rangle\circ 7 \not\xrightarrow{\texttt{st}} \bullet\langle {\substack{0\\1}}, {\substack{0\\1}}\rangle\langle {\substack{1\\1}}, {\substack{1\\0}}\rangle\langle {\substack{1\\1}}, {\substack{0\\0}}\rangle\circ 7$$

The source protocol relies on the preceding sequential environment for ${\substack{1\\1}}$. We prohibit relaxing the protocol to rely on the concurrent environment for it. □

The stutter and mumble deductions follow the rely/guarantee intuition:

Stuttering ($\square\xi_1^? \circ \xi_2^? \diamond x \xrightarrow{\texttt{st}} \square\xi_1^?\langle\sigma,\sigma\rangle\xi_2^?\diamond x$) means a thread-pool also obeys the protocol that guarantees a state σ by relying on its environment for σ.

Mumbling ($\square\xi_1^?\langle\sigma,\rho\rangle\langle\rho,\theta\rangle\xi_2^?\diamond x \xrightarrow{\texttt{mu}} \square\xi_1^?\langle\sigma,\theta\rangle\xi_2^?\diamond x$) means a thread-pool that guarantees the store ρ it later relies on also obeys the protocol in which we exclude the environment's access to the store ρ at that point.

Sets of traces represent a non-deterministic choice between the behaviours that a program phrase may exhibit. For such a set K, define its *closure* under trace deduction $K^\dagger$ as the least set K' such that: $K \subseteq K'$; and if $\tau_1 \in K'$ and $\tau_1 \xrightarrow{\texttt{x}} \tau_2$ for $\texttt{x} \in \{\texttt{st}, \texttt{mu}\}$, then $\tau_2 \in K'$. According to the rely/guarantee intuition above, a program phrase that is compatible with a set of traces is also compatible with its closure. We therefore represent program phrases as *closed* sets, i.e. sets K such that $K = K^\dagger$. The closure $K^\dagger$ of a countable K is countably infinite—by stuttering indefinitely—unless K is a finite set of single-transition $\bullet$-sorted $\bullet$-valued traces, in which case K is already closed.

For a set of traces U and sort $\square \in \{\bullet, \circ\}$, define a $\{\bullet, \circ\}$-sorted family $\mathcal{P}^{\aleph_0}(U)$ by taking its $\square$ component to be the set $\mathcal{P}^{\aleph_0}_{\square}(U)$ of countable subsets of U whose elements are all $\square$-sorted. Similarly, define $\mathcal{P}^{\dagger}_{\square}(U) \subseteq \mathcal{P}^{\aleph_0}_{\square}(U)$ to be the set of *closed* countable subsets of U whose elements are all $\square$-sorted.

The *prefixing* function adds the given transition to each $\bullet$-sorted trace:

$$(\sigma,\rho) : \mathcal{P}^{\aleph_0}_{\bullet}(\mathbb{T}\boldsymbol{X}) \to \mathcal{P}^{\aleph_0}_{\bullet}(\mathbb{T}\boldsymbol{X}) \quad (\sigma,\rho)\,K := \{\bullet\langle\sigma,\theta\rangle\xi^?\diamond x \mid \bullet\langle\rho,\theta\rangle\xi^?\diamond x \in K\}$$

It lifts to closed sets, i.e. $K \in \mathcal{P}^{\dagger}_{\bullet}(\mathbb{T}\boldsymbol{X})$ implies that $(\sigma,\rho)\,K \in \mathcal{P}^{\dagger}_{\bullet}(\mathbb{T}\boldsymbol{X})$.

5.2 Representation theorem

For $\boldsymbol{X} \in \mathbf{Set}^{\{\bullet,\circ\}}$, define the $\Sigma_{\mathbb{S}}$-algebra of *$\boldsymbol{X}$-valued closed trace-sets* $\mathbf{R}\boldsymbol{X}$ as:

$$\begin{array}{ll} \underline{\mathbf{R}\boldsymbol{X}}_{\square} := \mathcal{P}^{\dagger}_{\square}(\mathbb{T}\boldsymbol{X}) & [\![\mathsf{U}_{\ell,b}]\!]_{\mathrm{op}}K := \bigcup_{\sigma\in\mathbb{S}}(\sigma,\sigma[\ell\mapsto b])\,K \\ [\![\bigvee_{i<\alpha}]\!]_{\mathrm{op}}K_i := \bigcup_{i<\alpha}K_i & [\![\mathsf{L}_\ell]\!]_{\mathrm{op}}(K_0,K_1) := \bigcup_{\sigma\in\mathbb{S}}(\sigma,\sigma)\,K_{\sigma_\ell} \\ [\![\lhd]\!]_{\mathrm{op}}K := \{\circ\xi\lozenge x \mid \bullet\xi\lozenge x\in K\}^{\dagger} & [\![\rhd]\!]_{\mathrm{op}}K := \{\bullet\langle\sigma,\sigma\rangle\xi\lozenge x \mid \sigma\in\mathbb{S}, \circ\xi\lozenge x\in K\}^{\dagger} \end{array}$$

Additionally, define $\mathrm{return} : \boldsymbol{X}\to\underline{\mathbf{R}\boldsymbol{X}}$ by $\mathrm{return}_{\square}\,x := \{\square\langle\sigma,\sigma\rangle\square x \mid \sigma\in\mathbb{S}\}^{\dagger}$.

The rest of this section establishes that the algebra $\langle\mathbf{R}\boldsymbol{X},\mathrm{return}\rangle$ over $\boldsymbol{X}$ is a free $\mathbb{S}$-model over $\boldsymbol{X}$. A key ingredient is *reification*: for any $\{\bullet,\circ\}$-sorted family $\boldsymbol{X}$, we define a sorted-function $\mathrm{reify} : \mathcal{P}^{\aleph_0}(\mathbb{T}\boldsymbol{X})\to\mathrm{Term}^{\Sigma_{\mathbb{S}}}\boldsymbol{X}$, choosing a representative term $t_2 := \mathrm{reify}\,[\![\boldsymbol{X}\vdash t_1]\!]_{\mathrm{term}}$ such that $\boldsymbol{X}\vdash_{\mathbb{S}} t_1 = t_2$. This use of countable choice is inessential, the mere existence of the defining term t_2 suffices.

First define for any $\ell\in\mathbb{L}$ and $b\in\mathbb{B}$ the *cell assertion* term $x:\bullet\vdash_{\Sigma_{\mathbb{S}}}\mathsf{A}_{\ell,b}\,x:\bullet$ that looks ℓ up and only continues if it holds b:

$$x:\bullet\vdash_{\Sigma_{\mathbb{S}}}\mathsf{A}_{\ell,0}\,x := \mathsf{L}_\ell(x,\bot):\bullet \qquad x:\bullet\vdash_{\Sigma_{\mathbb{S}}}\mathsf{A}_{\ell,1}\,x := \mathsf{L}_\ell(\bot,x):\bullet$$

Next, for any $\sigma,\rho\in\mathbb{S}$ we define the *open transition* $x:\bullet\vdash_{\Sigma_{\mathbb{S}}}\{\!\!\{\sigma,\rho\}\!\!\}\,x:\bullet$, as a term that asserts the state is σ, then updates the state to ρ, and returns x:

$$x:\bullet\vdash_{\Sigma_{\mathbb{S}}}\{\!\!\{\sigma,\rho\}\!\!\}\,x := \mathsf{A}_{\mathtt{l}_1,\sigma_{\mathtt{l}_1}}\dots\mathsf{A}_{\mathtt{l}_n,\sigma_{\mathtt{l}_n}}\,\mathsf{U}_{\mathtt{l}_1,\rho_{\mathtt{l}_1}}\dots\mathsf{U}_{\mathtt{l}_n,\rho_{\mathtt{l}_n}}\,x:\bullet \quad (\mathbb{L}=\{\mathtt{l}_1,\dots,\mathtt{l}_n\})$$

Now we can represent traces as terms. Define the $\Sigma_{\mathbb{S}}$-term reifying a trace $x:\lozenge\vdash_{\Sigma_{\mathbb{S}}}\underline{\square\xi\lozenge x}:\square$ by sequencing open transition as they are in ξ, separated by $\rhd\lhd$; and delimited by $\lhd$ on the left if $\square=\circ$ and by $\rhd$ on the right if $\lozenge=\circ$.

Example 22. $x:\circ\vdash_{\Sigma_{\mathbb{S}}}\underline{\bullet\langle\sigma,\rho\rangle\langle\sigma',\rho'\rangle\circ x} := \{\!\!\{\sigma,\rho\}\!\!\}\rhd\lhd\{\!\!\{\sigma',\rho'\}\!\!\}\rhd x:\bullet$ □

Trace deductions are sound w.r.t. this encoding, in the following sense:

Proposition 23. *Assume that τ_1 and τ_2 are $\square$-sorted traces over $\{x:\lozenge\}$, such that $\tau_1\xrightarrow{\mathtt{x}}\tau_2$ for $\mathtt{x}\in\{\mathtt{st},\mathtt{mu}\}$. Then $x:\lozenge\vdash_{\Sigma_{\mathbb{S}}}\underline{\tau_1}\geq\underline{\tau_2}:\square$.*

Finally, we reify a trace set by reifying its traces in a chosen enumeration:

$$\mathrm{reify}:\mathcal{P}^{\aleph_0}(\mathbb{T}\boldsymbol{X})\to\mathrm{Term}^{\Sigma_{\mathbb{S}}}\boldsymbol{X} \qquad \mathrm{reify}_{\square}K := \Big(\boldsymbol{X}\vdash_{\Sigma_{\mathbb{S}}}\bigvee\nolimits_{\tau\in K}\underline{\tau}:\square\Big)$$

By proposition 23, closure preserves reification: $\boldsymbol{X}\vdash_{\mathbb{S}}\mathrm{reify}_{\square}K = \mathrm{reify}_{\square}K^{\dagger}:\square$.

Using reification, we state the representation theorem [proof in 14, §C].

Theorem 24 ($\mathbb{S}$-representation). *The pair $\langle\mathbf{R}\boldsymbol{X},\mathrm{return}\rangle$ is a free $\mathbb{S}$-model over $\boldsymbol{X}$. Its representation sends environments $e:\boldsymbol{X}\to\underline{\mathbf{A}}$ to $\mathbb{S}$-homomorphisms $e^{\#}:\mathbf{R}\boldsymbol{X}\to\mathbf{A}$ by $e^{\#}_{\square}K := \mathbf{R}\boldsymbol{X}[\![\mathrm{reify}_{\square}K]\!]_{\mathrm{term}}e$. Moreover, for $\mathbf{A}=\mathbf{R}\boldsymbol{Y}$ we have:*

$$e^{\#}_{\square}K = \left\{\square\xi_1\xi_2\lozenge y \,\middle|\, \begin{array}{l}\square\xi_1\circ x\in K,\\ \circ\xi_2\lozenge y\in e_{\lozenge}x\end{array}\right\}^{\dagger} \cup \left\{\square\xi_1\langle\sigma,\theta\rangle\xi_2\lozenge y \,\middle|\, \begin{array}{l}\square\xi_1\langle\sigma,\rho\rangle\bullet x\in K,\\ \bullet\langle\rho,\theta\rangle\xi_2\lozenge y\in e_{\lozenge}x\end{array}\right\}^{\dagger}.$$

Example 25. The model $\mathbf{R}\{x:\circ\}$ invalidates Write Intro:

$$\mathbf{R}\{x:\circ\}[\![\lhd\,\mathsf{U}_{\ell,b_1}\rhd\lhd\,\mathsf{U}_{\ell,b_2}\rhd x]\!]_{\mathrm{term}}\mathrm{return}\neq\mathbf{R}\{x:\circ\}[\![\lhd\,\mathsf{U}_{\ell,b_2}\rhd x]\!]_{\mathrm{term}}\mathrm{return}$$

Every trace in the right-hand set has at most one state-changing transition. The left-hand set has traces with two. Therefore, $\mathbb{S}$ does not prove Write Intro. □

6 Recovering Brookes's model

The theory $\mathbb{S}$ recovers Brookes's model (§6.1). We recover it twice, using different strategies that offer different perspectives. The first transforms the monad induced by the representation of §5.2 along a right adjoint $(-)_\circ : \mathbf{Set}^{\{\bullet,\circ\}} \to \mathbf{Set}$ sending each $\{\bullet,\circ\}$-family $\boldsymbol{X}$ to the set $\boldsymbol{X}_\circ := \{x \mid (x : \circ) \in \boldsymbol{X}\}$ (§6.2). In the second, we define a single-sorted theory of transitions B that recovers Brookes's model straightforwardly (§6.3). In this theory, the transition operators correspond to Brookes's `await` construct. After swiftly introducing embedding translations (§6.4), we show that B embeds into $\mathbb{S}$. The embedding factors through another, two-sorted, theory of transitions Tr (§6.5).

6.1 Brookes's model

We designed our notions of traces, deduction, etc. from §5.1 based on the following model of Brookes [6], in which traces cannot hold exclusive memory access at their ends. In this model, ceding access is implicit.

For any set $X \in \mathbf{Set}$, recall the set of Brookes traces $\mathsf{T}X := (\mathbb{S} \times \mathbb{S})^+ \times X$ from §5.1. Writing ξx for $\langle \xi, x \rangle$, Brookes's stutter and mumble deductions are:

$$\xi_1^? \xi_2^? x \xrightarrow{\mathtt{st}} \xi_1^? \langle \sigma, \sigma \rangle \xi_2^? x \qquad \xi_1^? \langle \sigma, \rho \rangle \langle \rho, \theta \rangle \xi_2^? x \xrightarrow{\mathtt{mu}} \xi_1^? \langle \sigma, \theta \rangle \xi_2^? x$$

We reuse the notation $(-)^\dagger$ for closure under these deductions.

The difference between Brookes's deductions and our multi-sorted deductions is the maintenance of the sort on each end of the trace. In particular, Brookes's stutter does not need to explicitly allow interleaving at the relevant position in the source, because the environment may always interleave on either end.

Brookes's semantic domain $BX := \mathcal{P}^\dagger(\mathsf{T}X)$ forms a monad. The monadic unit is return $: X \to BX$, return $x := \{\langle \sigma, \sigma \rangle x \mid \sigma \in \mathbb{S}\}^\dagger$. The Kleisli extension $e^\# : BX \to BY$ of every $e : X \to BY$ is $e^\# K := \{\xi_1 \xi_2 y \mid \xi_1 x \in K, \xi_2 y \in ex\}^\dagger$. It interprets memory accesses, dereferencing ($\ell!$) and mutation ($\ell := b$), as follows:

$$[\![\ell!]\!] : \mathbb{1} \xrightarrow{\{\langle \sigma,\sigma \rangle \sigma_\ell \mid \sigma \in \mathbb{S}\}^\dagger} B\mathbb{B} \qquad [\![\ell := b]\!] : \mathbb{1} \xrightarrow{\{\langle \sigma,\sigma[\ell \mapsto b] \rangle \langle\rangle \mid \sigma \in \mathbb{S}\}^\dagger} B\mathbb{1}$$

These *generic effects* [31] correspond to these monadic algebraic operations:

$$\begin{aligned} [\![\mathsf{R}_\ell]\!] &: (BX)^2 \to BX & [\![\mathsf{R}_\ell]\!](K_0, K_1) &:= \{\langle \sigma, \sigma \rangle \xi x \mid \sigma \in \mathbb{S}, \xi x \in K_{\sigma_\ell}\}^\dagger \\ [\![\mathsf{W}_{\ell,b}]\!] &: BX \to BX & [\![\mathsf{W}_{\ell,b}]\!] K &:= \{\langle \sigma, \sigma[\ell \mapsto b] \rangle \xi x \mid \sigma \in \mathbb{S}, \xi x \in K\}^\dagger \end{aligned}$$

6.2 Recovery via an adjunction

In Brookes's model, yielding to the concurrent environment is implicit, and always allowed. From our two-sorted point-of-view, we expect the traces in Brookes's model to represent $\circ$-sorted $\circ$-valued traces.

There is an abstract construction that recovers the monad and its operations in §6.2 from our $\{\bullet,\circ\}$-sorted model. The functor $(-)_\circ : \mathbf{Set}^{\{\bullet,\circ\}} \to \mathbf{Set}$

has a left-adjoint $(-)^{\circ} : \mathbf{Set} \to \mathbf{Set}^{\{\bullet,\circ\}}$. This functor sends each set X to the $\{\bullet,\circ\}$-family $X^{\circ} := \{x : \circ \mid x \in X\}$, using the set-like notation for families we introduced in §3.1. Monads transform along adjoints, and transforming the monad obtained standardly from the representation of §5.2 along the adjunction above results in Brookes's model. Explicitly, denoting $B_{\circ}X := \underline{\mathbf{R}X^{\circ}}_{\circ} = \mathcal{P}^{\dagger}_{\circ}(\mathbb{T}X^{\circ})$, the resulting monad over **Set** is $\langle B_{\circ}, \mathrm{return}_{\circ}, (-)^{\#}_{\circ}\rangle$. This monad is isomorphic to Brookes's $\langle B, \mathrm{return}, (-)^{\#}\rangle$ above by way of removing $\circ$ from both ends of every trace. Thus, the Brookes model amounts to the free $\mathbb{S}$-model from §5.2 transformed along the adjunction $(-)^{\circ} \dashv (-)_{\circ}$. The monad $\mathbf{R}$ supports the following generic effects. The adjunction transforms them, via its natural bijection on homsets, into Brookes's generic effects for memory access:

$$[\![\ell !]\!] : \mathbb{1}^{\circ} \xrightarrow{[\![\lhd \mathsf{L}_{\ell}(\rhd 0, \rhd 1)]\!]} \mathbf{R}\mathbb{B}^{\circ} \qquad [\![\ell := b]\!] : \mathbb{1}^{\circ} \xrightarrow{[\![\lhd \mathsf{U}_{\ell,b} \rhd ()]\!]} \mathbf{R}\mathbb{1}^{\circ}$$

6.3 The single-sorted theory of transitions

There is a more direct, single-sorted presentation B for Brookes's model. It uses transitions as operators rather than lookup and update operators. The signature Σ_{B} consists of countable-joins Σ_{V} and a unary transition operator $\langle\sigma, \rho\rangle$ for every $\sigma, \rho \in \mathbb{S}$. The axioms Ax_{B} consist of the countable-join semilattice axioms Ax_{V}, strict distributivity axioms (ND-B) $\langle\sigma, \rho\rangle \bigvee_{i<\alpha} x_i = \bigvee_{i<\alpha} \langle\sigma, \rho\rangle x_i$, and:

Trace closure

(M) $\langle\sigma, \rho\rangle\langle\rho, \theta\rangle x \geq \langle\sigma, \theta\rangle x$ (S) $x \geq \langle\sigma, \sigma\rangle x$ (H) $\bigvee_{\sigma\in\mathbb{S}} \langle\sigma, \sigma\rangle x \geq x$

$$\frac{\forall \sigma.\, \xi_1^{?}\langle\sigma, \sigma\rangle\xi_2^{?} x \in K}{\xi_1^{?}\xi_2^{?} x \in K}$$

Fig. 2. The hush rule

The first two axiom schemes are algebraic counterparts to mumble and stutter. These alone do not recover Brookes's model—the representation theorem for the theory without the (H) axioms includes potentially-empty traces. The axiom (H) fails in this model, but holds in Brookes's. In the representation theorem for B it is tempting to require, along with closure under Brookes's mumble and stutter trace deductions, closure under hush: presented in fig. 2 for a set of traces K. However, there is no need, due to the non-emptiness of the traces. Indeed, either $\xi_1^{?}$ or $\xi_2^{?}$ must be non-empty for the rule to apply. Take σ to match an adjacent transition, and apply the mumble closure rule to obtain the required consequence. This nuanced observation exposing the hush rule would be hard to notice without this algebraic analysis.

To conclude, we formulate the representation theorem for B. Let $X \in \mathbf{Set}$. Define the Σ_{B}-algebra $\mathbf{B}X$ with carrier $\underline{\mathbf{B}X} := \mathcal{P}^{\dagger}(\mathbb{T}X)$ and interpretations:

$$\mathbf{B}X[\![\textstyle\bigvee_{i<\alpha}]\!]_{\mathrm{op}} K_i := \bigcup_{i<\alpha} K_i \qquad \mathbf{B}X[\![\langle\sigma, \rho\rangle]\!]_{\mathrm{op}} K := \{\langle\sigma, \rho\rangle\tau \mid \tau \in K\}^{\dagger}$$

Additionally, define $\mathrm{return} : X \to \underline{\mathbf{B}X}$ by $\mathrm{return}\, x := \lambda x.\, \{\langle\sigma, \sigma\rangle x \mid \sigma \in \mathbb{S}\}^{\dagger}$.

To prove that this is a free B-model, we use reification as in §5.2, though here reification is more straightforward. A trace is reified as itself, and sets of traces use countable-joins as before: $\text{reify}\,K := \left(\boldsymbol{X} \vdash_{\Sigma_{\mathsf{B}}} \bigvee_{\tau \in K} \underline{\tau} : \star\right)$. The monad obtained from the next proposition is Brookes's model:

Proposition 26. *The pair* $\langle \mathbf{B}X, \text{return}\rangle$ *is a free* B*-model over* X*, for which the representation sends* $e : X \to \underline{\mathbf{A}}$ *to* $e^{\#} : \mathbf{B}X \to \mathbf{A}$ *by* $e^{\#}_{\square} K := \mathbf{B}X[\![\text{reify}_{\square}\, K]\!]_{\text{term}} e$.

6.4 Translations and equivalences

We will need the following notions for relating presentations. Consider a map between two sort sets $\epsilon : \mathbf{sort}_1 \to \mathbf{sort}_2$. It lifts to $\epsilon : \mathbf{Set}^{\mathbf{sort}_2} \to \mathbf{Set}^{\mathbf{sort}_1}$ by precomposition: $(\epsilon \boldsymbol{Y})_{\square} := \boldsymbol{Y}_{\epsilon\square}$. It forms the object part of a geometric morphism between (pre)sheaf toposes, i.e., it has left and right adjoints. The left adjoint $\epsilon^* : \mathbf{Set}^{\mathbf{sort}_1} \to \mathbf{Set}^{\mathbf{sort}_2}$ is in this case $(\epsilon^* \boldsymbol{X})_{\diamond} := \coprod_{\epsilon\square = \diamond} \boldsymbol{X}_{\square}$. When ϵ is injective, the left adjoint is given by the simpler formula $\epsilon^* \boldsymbol{X} := \{x : \epsilon\square \mid x \in \boldsymbol{X}_{\square}\}$.

Example 27. The geometric morphism for the map $\star \mapsto \circ : \{\star\} \rightarrowtail \{\bullet, \circ\}$ is the forgetful functor $(-)_{\circ} : \mathbf{Set}^{\{\bullet,\circ\}} \to \mathbf{Set}^{\{\star\}} \cong \mathbf{Set}$. As we saw in §6.2, its left adjoint is $(-)^{\circ} : \mathbf{Set}^{\{\star\}} \to \mathbf{Set}^{\{\bullet,\circ\}}$. □

Let Σ_1 and Σ_2 be signatures and $\epsilon : \mathbf{sort}_{\Sigma_1} \to \mathbf{sort}_{\Sigma_2}$ a map between their sort sets. A *translation of signatures* $\mathbf{E} : \Sigma_1 \rightarrowtail \Sigma_2$ *along* ϵ is an assignment, to each $(O : \square\langle\diamond_i\rangle_{i<\alpha}) \in \Sigma_1$, of a term $\mathbf{E}O \in \text{Term}^{\Sigma_2}_{\epsilon\square}\,\{x_i : \epsilon\diamond_i \mid i < \alpha\}$. Such a translation yields a functor $\mathbf{E}_{\text{tln}} : \mathbf{Alg}\Sigma_2 \to \mathbf{Alg}\Sigma_1$, mapping a Σ_2-algebra $\mathbf{B}$ to:

$$\underline{\mathbf{E}_{\text{tln}}\mathbf{B}} := \epsilon\underline{\mathbf{B}} \qquad \mathbf{E}_{\text{tln}}\mathbf{B}\,[\![O : \square\langle\diamond_i\rangle_{i<\alpha}]\!]_{\text{op}}\,\langle b_i\rangle := \mathbf{B}\,[\![\mathbf{E}O]\!]_{\text{term}}\,\langle x_i \mapsto b_i\rangle_{i<\alpha}$$

For a given family $\boldsymbol{Y} \in \mathbf{Set}^{\mathbf{sort}_{\Sigma_2}}$, such a translation therefore extends uniquely to a Σ_1-homomorphism $(\mathbf{E}_{\text{tln}})_{\boldsymbol{Y}} : F_{\Sigma_1}\epsilon\boldsymbol{Y} \to \mathbf{E}_{\text{tln}}F_{\Sigma_2}\boldsymbol{Y}$.

Example 28. We have a translation $\mathbf{E} : \Sigma_{\mathsf{G}} \rightarrowtail \Sigma_{\mathbb{S}}$ along $\star \mapsto \bullet : \{\star\} \rightarrowtail \{\bullet, \circ\}$ that translates the Σ_{G}-operators using their respective copies in the $\bullet$ sort:

$$\begin{aligned}
\mathbf{E}(\textstyle\bigvee_{\alpha} : \alpha) &:= (\{x_i : \bullet \mid i < \alpha\} \vdash_{\Sigma_{\mathbb{S}}} \textstyle\bigvee_{i<\alpha} x_i &&: \bullet)\\
\mathbf{E}(\mathsf{L}_{\ell} : \mathbf{2}) &:= (\{x_0, x_1 : \bullet\} &&\vdash_{\Sigma_{\mathbb{S}}} \mathsf{L}_{\ell}(x_0, x_1) : \bullet)\\
\mathbf{E}(\mathsf{U}_{\ell,b} : \mathbf{1}) &:= (\{x_0 : \bullet\} &&\vdash_{\Sigma_{\mathbb{S}}} \mathsf{U}_{\ell,b}\, x_0 \quad : \bullet)
\end{aligned}$$

□

A translation of *presentations* $\mathbf{E} : \mathfrak{p}_1 \rightarrowtail \mathfrak{p}_2$ along ϵ is a translation of their signatures along ϵ that, moreover, preserves the provability of axioms:

$$(\boldsymbol{X} \vdash_{\Sigma_{\mathfrak{p}_1}} t_1 = t_2 : \square) \in \text{Ax}_{\mathfrak{p}_1} \implies \epsilon^*\boldsymbol{X} \vdash_{\mathfrak{p}_2} \mathbf{E}_{\text{tln}}t_1 = \mathbf{E}_{\text{tln}}t_2 : \epsilon\square$$

Example 29. The translation of global state into shared state from example 28 is a translation of presentations $\mathbf{E} : \mathsf{G} \rightarrowtail \mathbb{S}$. □

Translations along composable sort maps compose via substitution, and a translation $\mathbf{E} : \mathfrak{p} \rightarrowtail \mathfrak{p}$ along $\mathsf{id}_{\Sigma_{\mathfrak{p}}}$ is an *identity* translation when, for all terms $t \in \text{Term}^{\Sigma_{\mathfrak{p}}}_{\square}\boldsymbol{X}$, we have $\boldsymbol{X} \vdash_{\mathfrak{p}} \mathbf{E}_{\text{tln}}t = t : \square$. A translation $\mathbf{E} : \mathfrak{p}_1 \rightarrowtail \mathfrak{p}_2$ along ϵ is an *equivalence* if ϵ is a bijection, and there exists an embedding $\mathbf{E}^{-1} : \mathfrak{p}_2 \rightarrowtail \mathfrak{p}_1$ along ϵ^{-1}, such that $\mathbf{E} \circ \mathbf{E}^{-1}$ and $\mathbf{E}^{-1} \circ \mathbf{E}$ are identity translations. We then write $\mathfrak{p}_1 \simeq \mathfrak{p}_2$ and say that the presentations are *equivalent*. Two multi-sorted theories are equivalent iff their associated free-model monads are isomorphic.

6.5 Translation through the two-sorted theory of transitions

We define a two-sorted presentation Tgs of the *open* transitions $\{\!|\sigma, \rho|\!\}$ as sequential operators. The signature Σ_{Tgs} consists of countable-joins Σ_{V} and a unary open transition operator $\llparenthesis\sigma, \rho\rrparenthesis$ for $\sigma, \rho \in \mathbb{S}$. The axioms $\mathrm{Ax}_{\mathsf{Tgs}}$ consist of the countable-join semilattice axioms Ax_{V}, strict distributivity axioms (ND-T) $\llparenthesis\sigma, \rho\rrparenthesis \bigvee_{i<\alpha} x_i = \bigvee_{i<\alpha} \llparenthesis\sigma, \rho\rrparenthesis\, x_i$, and:

Open transition axioms	
(HS) $x = \bigvee_{\sigma \in \mathbb{S}} \llparenthesis\sigma, \sigma\rrparenthesis\, x$	(Seq$^{=}$) $\llparenthesis\sigma, \rho\rrparenthesis \llparenthesis\rho, \theta\rrparenthesis\, x = \llparenthesis\sigma, \theta\rrparenthesis\, x$
	(Seq$^{\neq}$) $\llparenthesis\sigma, \rho\rrparenthesis \llparenthesis\mu, \theta\rrparenthesis\, x = \bot \qquad \rho \neq \mu$

Translate $\mathbf{E}_{\mathsf{G}} : \mathsf{Tgs} \rightarrowtail \mathsf{G}$ by interpreting transitions as the open transitions from §5.2: $\mathbf{E}_{\mathsf{G}}\, \llparenthesis\sigma, \rho\rrparenthesis := (x_0 \vdash_{\Sigma_{\mathsf{G}}} \{\!|\sigma, \rho|\!\}\, x_0)$. Conversely, translate $\mathbf{E}_{\mathsf{Tgs}} : \mathsf{G} \rightarrowtail \mathsf{Tgs}$ as follows, similar to the representation of update and lookup from §5.2:

$$\mathbf{E}_{\mathsf{Tgs}} \mathsf{U}_{\ell,b} := (x_0 \vdash_{\Sigma_{\mathsf{Tgs}}} \textstyle\bigvee_{\sigma \in \mathbb{S}} \llparenthesis\sigma, \sigma[\ell \mapsto b]\rrparenthesis\, x_0) \quad \mathbf{E}_{\mathsf{Tgs}} \mathsf{L}_{\ell} := (x_0, x_1 \vdash_{\Sigma_{\mathsf{Tgs}}} \textstyle\bigvee_{\sigma \in \mathbb{S}} \llparenthesis\sigma, \sigma\rrparenthesis\, x_{\sigma_\ell})$$

Using the equivalence $\mathsf{Tgs} \simeq \mathsf{G}$ that these translations witness we can translate $\mathsf{B} \rightarrowtail \mathbb{S}$ along $\star \mapsto \circ$. We define a two-sorted presentation Tr, mimicking the definition of $\mathbb{S}$ but replacing the operators and axioms of G with those of Tgs in the hold ($\bullet$) sort: $\mathrm{Ax}_{\mathsf{Tr}} := \boxed{\mathrm{Ax}^{\bullet}_{\mathsf{Tgs}}} \cup \mathrm{Ax}^{\circ}_{\mathsf{V}} \cup \{\text{ND-}\triangleright, \text{ND-}\triangleleft\} \cup \{\text{Empty}, \text{Fuse}\}$. Extending the translations $\mathbf{E}_{\mathsf{Tgs}}$ and $\mathbf{E}_{\mathsf{G}}$ to all of the operators gives an equivalence $\mathsf{Tr} \simeq \mathbb{S}$. So Tr induces the same monad as $\mathbb{S}$, recovering Brookes's model.

Define the translation $\mathbf{E}_{\mathsf{Tr}} : \mathsf{B} \rightarrowtail \mathsf{Tr}$ along $\star \mapsto \circ$ by sending transitions to their delimited open counterparts: $\mathbf{E}_{\mathsf{Tr}}\langle\sigma, \rho\rangle := (x_0 : \circ \vdash_{\Sigma_{\mathsf{Tr}}} \triangleleft \llparenthesis\sigma, \rho\rrparenthesis \triangleright x_0 : \circ)$. Using $\mathsf{Tr} \simeq \mathbb{S}$ we get $\mathsf{B} \rightarrowtail \mathbb{S}$ (fig. 3). Brookes's model, as a free B-model, is thus the $\circ$-sorted fragment of $\mathbb{S}$ over $\circ$-variables, formally.

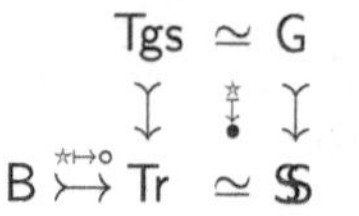

Fig. 3. Th. chart

7 Conclusion and further work

We presented an equational theory for shared state ($\mathbb{S}$). It separates reasoning into two layers. In the held layer ($\bullet$), we prohibit the concurrent environment from accessing memory, and we can reason about memory accesses by a pool of threads sequentially. In the ceded layer ($\circ$), the concurrent environment may interleave, and local memory access is forbidden. We also presented theories of transitions (B, Tgs, & Tr) and formally related them to (non-deterministic) global state (G) and shared state ($\mathbb{S}$). The single-sorted theory B recovers Brookes's model, but it does so by using Brookes's `await` construct, which we find unnatural; and it does not admit global state explicitly as a component of the theory. We believe that admitting global state will inform modelling other effects in the concurrent setting. Our theory $\mathbb{S}$ addresses these concerns. It admits the global state theory as-is, and axiomatises the mode-switching operators ($\triangleleft/\triangleright$) without explicit interaction with global state. This theory recovers Brookes's model exactly, in a principled manner: by transforming a monad and its operations along an adjunction; and, independently, through algebraic translations.

Our theory uses countable-join semilattices to recover Brookes's model. They can express iteration (i.e. `while`-loops). The same model admits first-order recursion, i.e. least-fixpoints of mutually-defined first-order functions, using the ω-complete partial order structure of the refinement order and the Scott-continuity of the semantics. We can support higher-order recursion by recourse to domain-theory, generalising algebraic theories using order-enriched theories. There are several standard variants, each with subtle logical trade-offs [29]. We can also restrict the semantics to terminating languages by restricting to finite joins, and using finitely-generated closed subsets for the representation.

We want to analyse Brookes's parallel composition operator algebraically. Brookes composed programs in parallel by interleaving traces from each thread. Initial results show we can define Brookes's parallel composition by simultaneous induction over terms. However, we would like to provide a more abstract account, by recourse to the universal property of free models. This abstraction may expose special properties of global state, or lead to a general parallel composition operation satisfying the expected laws of concurrent programming [16, 27, 34].

We would like to model more effects within this modular multi-sorted algebraic framework. These effects include: more advanced notions of state, such as dynamic allocation [20], higher-order memory cells [24, 36], and weak memory [12, 13]; control-flow effects such as exceptions and effect handlers [3]; and probabilistic programming with shared state [22].

If the multi-sorted approach does indeed generalise to more sophisticated effects, then it will be instructive to review its assumptions. For example, the strictness axioms impose a partial-correctness discipline: the semantics says nothing about the effect a diverging program has on its memory. Relaxing or removing strictness may give a model that allows us to reason about diverging programs.

Our two sorts limit access to the whole store. We would like to explore finer granularity. For example, a theory with per-location access limitation, with sorts for every finite subset $s \subseteq \mathbb{L}$ of locations, and operators $(\lhd_\ell : s\ \{\ell\}\, \langle s\cup\{\ell\}\rangle)$ and $(\rhd_\ell : s\cup\{\ell\}\, \langle s\ \{\ell\}\rangle)$. We expect the axiomatisation's design to require subtlety.

It may be interesting to to expose the sort discipline in the surface language through typing judgements, explicating regions that rule out data-races with the environment. It seems such judgements would rule out deadlocks structurally, and so may limit expressiveness. Whether this idea is useful remains to be seen.

In conclusion, our two-sorted decomposition of Brookes's seminal model provides new insights into its assumptions and components, and reveals new directions for modelling more advanced features involving concurrent shared state.

Acknowledgments. Supported by the Israel Science Foundation (grant number 814/22) and the European Research Council (ERC) under the European Union's Horizon 2020 research and innovation programme (grant agreement no. 851811); and by a Royal Society University Research Fellowship and Enhancement Award. For the purpose of Open Access the authors have applied a CC BY public copyright licence to any Author Accepted Manuscript version arising from this submission. We thank Danel Ahman, Andrej Bauer, Martín Escardó, Justus Matthiesen, Sam Staton, and Rob van Glabbeek for interesting and useful discussions and suggestions.

References

[1] Abramsky, S., Jung, A.: Domain Theory. In: Handbook of Logic in Computer Science, Oxford University Press (04 1995), ISBN 9780198537625, https://doi.org/10.1093/oso/9780198537625.003.0001

[2] Adámek, J., Rosický, J., Vitale, E.M.: Algebraic Theories: A Categorical Introduction to General Algebra. Cambridge Tracts in Mathematics, Cambridge University Press (2010)

[3] Bauer, A., Pretnar, M.: Programming with algebraic effects and handlers. J. Log. Algebraic Methods Program. **84**(1) (2015), https://doi.org/10.1016/J.JLAMP.2014.02.001

[4] Benton, N., Hofmann, M., Nigam, V.: Effect-dependent transformations for concurrent programs. In: PPDP, ACM (2016), https://doi.org/10.1145/2967973.2968602

[5] Bloom, S.L.: Varieties of ordered algebras. Journal of Computer and System Sciences **13**(2) (1976), ISSN 0022-0000, https://doi.org/10.1016/S0022-0000(76)80030-X

[6] Brookes, S.D.: Full abstraction for a shared-variable parallel language. Inf. Comput. **127**(2) (1996), https://doi.org/10.1006/inco.1996.0056

[7] Castellan, S., Clairambault, P., Winskel, G.: The parallel intensionally fully abstract games model of pcf. In: LICS (2015), https://doi.org/10.1109/LICS.2015.31

[8] Dodds, M., Batty, M., Gotsman, A.: Compositional verification of compiler optimisations on relaxed memory. In: ESOP, ETAPS, LNCS, vol. 10801, Springer (2018), https://doi.org/10.1007/978-3-319-89884-1_36

[9] Dolan, S., Eliopoulos, S., Hillerström, D., Madhavapeddy, A., Sivaramakrishnan, K.C., White, L.: Concurrent system programming with effect handlers. In: TFP, LNCS, vol. 10788, Springer (2017), https://doi.org/10.1007/978-3-319-89719-6_6

[10] Dolan, S., White, L., Sivaramakrishnan, K.C., Yallop, J., Madhavapeddy, A.: Effective concurrency with algebraic effects (2015), OCaml Workshop

[11] Dvir, Y., Kammar, O., Lahav, O.: An algebraic theory for shared-state concurrency. In: APLAS, LNCS, vol. 13658, Springer (2022), https://doi.org/10.1007/978-3-031-21037-2_1

[12] Dvir, Y., Kammar, O., Lahav, O.: A denotational approach to release/acquire concurrency. In: ESOP, ETAPS, LNCS, vol. 14577, Springer (2024), https://doi.org/10.1007/978-3-031-57267-8_5

[13] Dvir, Y., Kammar, O., Lahav, O.: A brookes-style denotational semantics for release/acquire concurrency. TOPLAS (Jan 2025), ISSN 0164-0925, https://doi.org/10.1145/3715096, just Accepted

[14] Dvir, Y., Kammar, O., Lahav, O., Plotkin, G.: Two-sorted algebraic decompositions of brookes's shared-state denotational semantics (2025), URL https://arxiv.org/abs/2501.15104

[15] Hennessy, M.C.B., Plotkin, G.D.: Full abstraction for a simple parallel programming language. In: Mathematical Foundations of Computer Science, Springer, Berlin, Heidelberg (1979), ISBN 978-3-540-35088-0

[16] Hoare, T.: Laws of programming: The algebraic unification of theories of concurrency. In: CONCUR 2014 – Concurrency Theory, Springer, Berlin, Heidelberg (2014), ISBN 978-3-662-44584-6

[17] Hyland, M., Plotkin, G.D., Power, J.: Combining effects: Sum and tensor. Theor. Comput. Sci. **357**(1-3) (2006), https://doi.org/10.1016/j.tcs.2006.03.013

[18] Hyland, M., Power, J.: Discrete Lawvere theories and computational effects. Theoretical Computer Science **366**(1) (2006), ISSN 0304-3975, https://doi.org/10.1016/j.tcs.2006.07.007, algebra and Coalgebra in Computer Science

[19] Jeffrey, A., Riely, J.: On Thin Air Reads: Towards an Event Structures Model of Relaxed Memory. LMCS **Volume 15, Issue 1** (Mar 2019), https://doi.org/10.23638/LMCS-15(1:33)2019

[20] Kammar, O., Levy, P.B., Moss, S.K., Staton, S.: A monad for full ground reference cells. In: LICS (2017), https://doi.org/10.1109/LICS.2017.8005109

[21] Kavanagh, R., Brookes, S.: A denotational semantics for SPARC TSO. In: MFPS, ENTCS, vol. 341, Elsevier (2018), https://doi.org/10.1016/j.entcs.2018.03.025

[22] Kozen, D.: Semantics of probabilistic programs. Journal of Computer and System Sciences **22**(3) (1981), ISSN 0022-0000, https://doi.org/10.1016/0022-0000(81)90036-2

[23] Lawvere, F.W.: Functorial Semantics of Algebraic Theories and Some Algebraic Problems in the context of Functorial Semantics of Algebraic Theories. Ph.D. thesis, Department of Mathematics (1963)

[24] Levy, P.B.: Possible world semantics for general storage in call-by-value. In: Computer Science Logic, Springer, Berlin, Heidelberg (2002), ISBN 978-3-540-45793-0

[25] Melliès, P.: Local states in string diagrams. In: RTA-TLCA, LNCS, vol. 8560, Springer (2014), https://doi.org/10.1007/978-3-319-08918-8_23

[26] Moggi, E.: Notions of computation and monads. Inf. Comput. **93**(1) (1991), https://doi.org/10.1016/0890-5401(91)90052-4

[27] Paquet, H., Saville, P.: Effectful semantics in bicategories: strong, commutative, and concurrent pseudomonads. LICS, Association for Computing Machinery, New York, NY, USA (2024), ISBN 9798400706608, https://doi.org/10.1145/3661814.3662130

[28] Plotkin, G.D.: A powerdomain for countable non-determinism. In: Automata, Languages and Programming, Springer, Berlin, Heidelberg (1982), ISBN 978-3-540-39308-5

[29] Plotkin, G.D.: Some Varieties of Equational Logic. Springer, Berlin, Heidelberg (2006), ISBN 978-3-540-35464-2, https://doi.org/10.1007/11780274_8

[30] Plotkin, G.D., Power, J.: Notions of computation determine monads. In: FOSSACS, ETAPS, LNCS, vol. 2303, Springer (2002), https://doi.org/10.1007/3-540-45931-6_24
[31] Plotkin, G.D., Power, J.: Algebraic operations and generic effects. Applied Categorical Structures **11**(3) (2003), ISSN 1572-9095, https://doi.org/10.1023/A:1023064908962
[32] Plotkin, G.D., Pretnar, M.: Handlers of algebraic effects. In: ESOP, ETAPS, LNCS, vol. 5502, Springer (2009), https://doi.org/10.1007/978-3-642-00590-9_7
[33] Plotkin, G.D., Pretnar, M.: Handling algebraic effects. Log. Methods Comput. Sci. **9**(4) (2013), https://doi.org/10.2168/LMCS-9(4:23)2013
[34] Rivas, E., Jaskelioff, M.: Monads with merging (Jun 2019), URL https://inria.hal.science/hal-02150199, working paper or preprint
[35] Sivaramakrishnan, K.C., Dolan, S., White, L., Kelly, T., Jaffer, S., Madhavapeddy, A.: Retrofitting effect handlers onto ocaml. In: PLDI, ACM (2021), https://doi.org/10.1145/3453483.3454039
[36] Sterling, J., Gratzer, D., Birkedal, L.: Denotational semantics of general store and polymorphism (2023), URL https://arxiv.org/abs/2210.02169
[37] Tarlecki, A.: Some nuances of many-sorted universal algebra: A review. Bull. EATCS **104** (2011), URL http://eatcs.org/beatcs/index.php/beatcs/article/view/121
[38] Turon, A.J., Wand, M.: A separation logic for refining concurrent objects. In: POPL, ACM (2011), https://doi.org/10.1145/1926385.1926415
[39] Xu, Q., de Roever, W.P., He, J.: The rely-guarantee method for verifying shared variable concurrent programs. Formal Aspects Comput. **9**(2) (1997), https://doi.org/10.1007/BF01211617

Model-Checking Real-Time Systems: Revisiting the Alternating Automaton Route

Patricia Bouyer[1], B. Srivathsan[2,3], and Vaishnavi Vishwanath[2(✉)]

[1] Université Paris-Saclay, CNRS, ENS Paris-Saclay, LMF, Gif-sur-Yvette, France
bouyer@lmf.cnrs.fr

[2] Chennai Mathematical Institute, Chennai, India
sri@cmi.ac.in, vaishnaviv@cmi.ac.in

[3] CNRS, ReLaX, IRL 2000, Siruseri, India

Abstract. Alternating timed automata (ATA) are an extension of timed automata, that are closed under complementation and hence amenable to logic-to-automata translations. Several timed logics, including Metric Temporal Logic (MTL), can be converted to equivalent 1-clock ATAs (1-ATAs). Satisfiability of an MTL formula reduces to checking emptiness of a 1-ATA. Furthermore, algorithms for 1-ATA emptiness can be adapted for model-checking timed automata models against 1-ATA specifications. However, existing emptiness algorithms for 1-ATA proceed by an extended region construction, and are not suitable for implementations.

In this work, we initiate the study of zone-based methods for 1-ATAs. The challenge here, as opposed to timed automata, is the fact that the zone graph may generate an unbounded number of variables. We first introduce a *deactivation operation* to the 1-ATA syntax that allows for an explicit deactivation of the clock in transitions. Using the deactivation operation, we improve the existing MTL-to-1-ATA conversion and present a fragment of MTL for which the equivalent 1-ATA generate a bounded number of variables. Secondly, we develop the idea of zones for 1-ATA and present an emptiness algorithm which explores a corresponding zone graph. For termination, a special entailment check between zones is necessary. Our main technical contributions are: (1) an algorithm for the entailment check using simple zone operations and (2) an NP-hardness for the entailment check in the general case. Finally, for 1-ATA which generate a bounded number of variables, we present a modified entailment check with quadratic complexity.

1 Introduction

Model-Checking Real-Time Systems. The task of verifying if a model A (typically a network of automata) satisfies a specification φ (a formula in a logic) can be cast as the emptiness of an automaton $A \times B_{\neg\varphi}$, where $B_{\neg\varphi}$ is an automaton recognizing the set of behaviours that violate the specification – in other words, an automaton corresponding to $\neg\varphi$. This is the well-known model-checking paradigm. In the context of real-time systems, *timed automata* [4] are

P. A. Abdulla and D. Kesner (Eds.): FoSSaCS 2025, LNCS 15691, pp. 399–421, 2025.
https://doi.org/10.1007/978-3-031-90897-2_19

the de-facto choice for the automaton model. For the logic part, metric temporal logic (MTL) is a natural extension of the widely used linear temporal logic (LTL), incorporating timing constraints in the modalities. However, timed logics cannot be easily converted to timed automata, since they are not closed under complementation.

A popular choice for logic-to-automata translations for timed logics, is the model of *1-clock Alternating Timed Automata (1-ATAs)* [33,29]. 1-ATAs are closed under complementation, and are incomparable in expressive power to (multi-clock) timed automata. The power of alternation enables resetting fresh copies of the single clock at different positions along the word. Therefore, while reading a timed word, a 1-ATA can keep track of the time elapsed from multiple positions in the word. This feature is quite convenient to handle future operators like U_I, an Until operator with an interval constraint I. Therefore, the model-checking problem reduces to checking the emptiness of $A \times B_{\neg\varphi}$ where A is a timed automaton and $B_{\neg\varphi}$ is a 1-ATA. This is a rather natural formulation of the model-checking problem in the real-time setting. Surprisingly, there is no tool that implements an algorithm to this problem, to the best of our knowledge. Decidability of this question follows from [31] and [29]. However it is based on extended regions and not suitable for practical implementations.

On the other hand, emptiness for timed automata has been extensively studied over the last three decades, and there are mature and well developed tools like UPPAAL [28], PAT [35], LTSMin [25], TChecker [22], Theta [36]. Algorithms for timed automata are based on *zones*. These are convenient data structures that symbolically capture reachable configurations of a timed automaton. Our goal is to adapt the best known methods from timed automata literature to solve the emptiness of $A \times B_{\neg\varphi}$. The crucial missing piece in this picture is a zone-based algorithm to handle 1-ATAs, like $B_{\neg\varphi}$. This is the subject of this paper.

The Challenge. Undoubtedly, the biggest challenge in developing a zone-based procedure for 1-ATA is the unbounded size of the configurations that it generates (illustrated in Fig. 4). A 1-ATA maintains several copies of its clock, each of them storing the time since a specific position. This not only adds technical difficulty in describing zone graphs, but it also makes some of the operations used in the computation algorithmically hard. In this work, we pinpoint the challenges and provide some efficient algorithmic solutions for some restricted cases of 1-ATA which arise from fragments of MTL.

Contributions. As a first contribution, we introduce a new *deactivation* operation for the clock, on transitions. When a clock is deactivated, the branch of the 1-ATA that passes through the transition maintains only the control state, and not the clock value (until the clock gets reactivated again through a reset). This explicit deactivation helps reducing the number of active clock copies maintained by the 1-ATA. To substantiate this idea, we improve the MTL-to-1-ATA construction of [33] using the deactivation operator and identify a fragment that induces 1-ATAs with *bounded width* – these are automata which contain a bounded number of active clock copies in any reachable configuration.

Next, we present a definition of a zone graph for the enhanced 1-ATAs with the deactivation operation, and provide an algorithm to compute the zone successors. For termination of the zone graph computation, we adapt the entailment relation between zones, studied in [1], to our setting. Our main technical contribution is a close study of this entailment check.

- We provide an algorithm for the entailment check using standard zone operations known from the timed automata literature.
- We prove that deciding non-entailment is NP-hard, by reducing the monotone 3-SAT problem to the non-entailment check.
- Then, we present a modified entailment check, with quadratic complexity, for 1-ATAs with bounded width.

Related Work. Recall the casting of the model-checking question as an emptiness of $A \times B_{\neg\varphi}$ where $B_{\neg\varphi}$ is a 1-ATA. Since timed automata tools are mature, some of the existing techniques convert $B_{\neg\varphi}$ into a timed automaton. MightyL [14] is a tool that converts Metric Interval Temporal Logic (MITL) into a network of timed automata, by observing special properties of the 1-ATA obtained from MITL [13]. Recently, a conversion of MITL into a network of generalized timed automata [2] has been developed [3]. Both these works deal with the MITL fragment. To the best of our knowledge, the only available logic-to-automaton procedure for the full MTL fragment is the MTL-to-1-ATA conversion of [33]. Hence, our work would be the first attempt at an implementable algorithm for the full MTL model-checking (over finite timed words).

The core zone based algorithm for timed automata reachability was proposed in [15]. Termination of the zone enumeration has been an intricate topic since then [9,7,23,21]. We refer the reader to the surveys [10,34] for a more detailed exposition on this topic. An important ingredient in the zone enumeration is a simulation operation between zones (similar in spirit to the entailment check as said above). For extended models of timed automata – like timed automata with diagonal constraints [9] and event-clock automata [6], there is a translation to classical timed automata. Therefore emptiness for these extended automata can be reduced to emptiness of a bigger timed automaton. However, previous works have observed that zone-based algorithms which work directly on the extended automata perform better than the zone-based algorithms that run on equivalent bigger timed automata: [21] shows it for timed automata with diagonal constraints; [2] introduces *generalized timed automata* which can model event-clock automata and provides a comparison of the direct method versus the conversion to timed automata. The main reason is that the conversion to timed automata adds many control states which cannot be handled well by the current zone-based algorithms. Algorithms which work directly on the extended automata encode the extra information using zones and manage the explosion using zone simulation techniques. For timed automata with diagonal constraints, the simulation check is NP-hard. Nevertheless an algorithm using basic zone operations has been proposed [21] and has been shown to work well on examples. This history of successful zone-based methods on extensions of timed automata encourages us to look for direct zone based methods for 1-ATAs as well.

Alternating Timed Automata appeared in [33,29]. With two clocks, emptiness is undecidable, simply because universality for 2-clock timed automata is undecidable [4]. Emptiness for 1-ATA is known to be decidable with a non-primitive recursive complexity – one reason why 1-ATAs have been ignored for direct manipulations. However, several timed logics have been compiled into 1-ATA [13,26] and multiple restrictions of 1-ATAs with better complexity, like very weak 1-ATA [14], 1-ATA with reset-free loops [26] have been independently considered in the literature.

In the untimed setting, emptiness of Alternating Finite Automata (AFA) can be performed by a *de-alternation* that computes an equivalent NFA on-the-fly. To deal with large state-spaces, techniques based on antichains have been extensively explored in the literature [38,17,18]. Essentially, the algorithms maintain an antichain of the state space, induced by various simulation preorders. Several heuristics based on preprocessing of the automaton [37] and making use of SAT solvers [24] have been investigated. A comparative study of various AFA emptiness algorithms can be found in [19].

In the timed setting, the closest to our work is [1], which presents a zone-based approach for answering universality of 1-clock timed automata. Our zone construction for 1-ATA builds on this approach. The entailment check between zones that we use has been proposed in [1]. However, there is no algorithm given for the entailment check. An approximation of the check is encoded as an SMT formula. In this work, we make a comprehensive study of this entailment check in terms of algorithms and complexity.

Organization. Section 2 defines the syntax and semantics of 1-ATA with the deactivation operation, and defines 1-ATAs with bounded width. Section 3 presents an improvement of the MTL-to-1-ATA construction and identifies a fragment of MTL which generates bounded width 1-ATAs through the new construction. Section 4 provides a comprehensive description of zone graphs for ATAs. Finally, in Section 5 we delve into an in-depth study of the entailment relation that is essential for termination of the zone graph computation.

All missing proofs can be found in the extended version of the paper [12].

2 1-ATA with a Deactivation Operator

In what follows, we use $\mathbb{N}, \mathbb{Z}$, and $\mathbb{R}_{\geq 0}$ to represent the set of natural numbers, integers and non-negative real numbers respectively. For $d \in \mathbb{R}_{\geq 0}$, we use $\lfloor d \rfloor$ to denote the integral part and $\text{fract}(d)$ to denote the fractional part of d. We will refer to the single clock of the alternating timed automaton as x. We define intervals as $I := [a,b] \mid [a,b) \mid (a,b] \mid (a,b) \mid [a,\infty) \mid (a,\infty)$ where $a, b \in \mathbb{N}$. We write $\mathcal{I}$ for the set of all such intervals. For a finite set S, we define $\Phi(S)$ as the set of formulas generated by the following grammar, where $s \in S$ and $I \in \mathcal{I}$:

$$\varphi = \textbf{true} \mid \textbf{false} \mid s \mid I \mid \varphi \wedge \varphi \mid \varphi \vee \varphi \mid x.\varphi \mid \overline{x}.\varphi$$

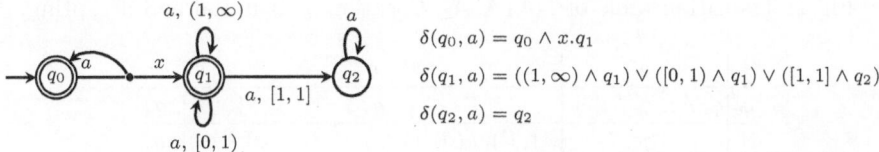

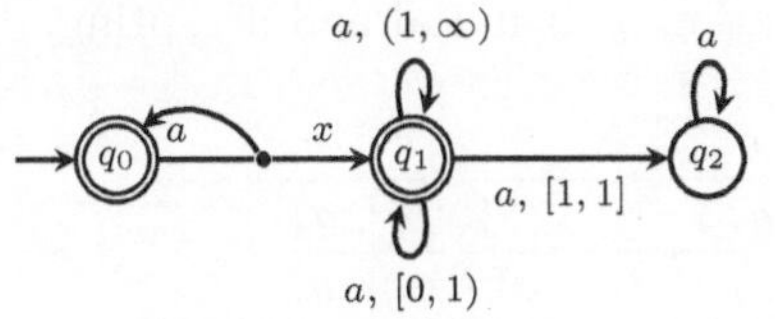

$$\delta(q_0, a) = q_0 \wedge x.q_1$$
$$\delta(q_1, a) = ((1, \infty) \wedge q_1) \vee ([0, 1) \wedge q_1) \vee ([1, 1] \wedge q_2)$$
$$\delta(q_2, a) = q_2$$

Fig. 1: 1-ATA $\mathcal{A}_1$. The transition function δ is given on the right. The transition $\delta(q_0, a) = q_0 \wedge x.q_1$ is depicted as two branches starting from q_0, one that leads back to q_0 and one that goes onto q_1. The edge to q_1 contains an x to say that the clock is set to 0 on this branch.

Here, I denotes the clock interval guard, $x.\varphi$ corresponds to resetting the clock (followed by applying φ), and $\overline{x}.\varphi$ corresponds to making the clock *inactive* and then applying φ.

A *timed word* over a finite alphabet Σ is a finite sequence of the form $(d_1, a_1) \dots (d_n, a_n)$, where $a_1, \dots, a_n \in \Sigma$ are events and $d_1, \dots, d_n \in \mathbb{R}_{\geq 0}$ are time delays. For example, d_2 is the time between a_1 and a_2, d_3 is between a_2 and a_3, and so on. Alternating timed automata were introduced independently in [32,33] and [29,30]. In this paper, we consider the model as defined in [32,33], with the added deactivation operator.

Definition 1 (1-ATA). *A one-clock alternating timed automaton (1-ATA) is given by $\mathcal{A} = (Q, \Sigma, q_0, F, \delta)$ where Q is a finite set of (control) locations, Σ is a finite alphabet, $q_0 \in Q$ is the initial location, $F \subseteq Q$ is a set of accepting locations, and $\delta : Q \times \Sigma \to \Phi(Q)$ is a partial function describing the transitions. We assume $\delta(q, a)$ is a formula in disjunctive normal form for $q \in Q$ and $a \in \Sigma$.*

Before we explain the formal semantics, we give two 1-ATA examples and an informal explanation of their mechanics. Fig. 1 shows 1-ATA $\mathcal{A}_1$ with $\Sigma = \{a\}$. It accepts all timed words such that no two as are at distance 1 apart. The initial location is q_0. On reading the first a, $\mathcal{A}_1$ opens two branches, one that comes back to q_0 and the other that resets clock x and goes to q_1. Similarly, for each of the next as that are read, the q_0 branch spawns into two. The purpose of the q_1 branch is to check if there is an a at distance 1 from the a that created the branch. If yes, the branch goes to a non-accepting state q_2. Fig. 2 illustrates 1-ATA $\mathcal{A}_2$ with $\Sigma = \{a, b, c\}$ that accepts timed words satisfying the property: for every a in the word, there is a c at a later position (with no timing constraints), and a b at a later position which is at 1 time unit away from a. Intuitively, at location q_a, whenever an a is read, two obligations are generated in the form of locations q_c and q_b. Since the only timing constraints are on q_b, we deactivate the clock while going to q_c. This is reflected by $\overline{x}.q_c$ in the transition $\delta(q_a, a)$. Location q_c waits for a c, and on seeing a c, the obligation is discharged (using $\delta(q_c, c) = \mathbf{true}$, the obligation q_c "disappears"). Location q_b waits for a b at time 1 from the a where the obligation appeared. The only accepting location is q_a. Therefore a configuration containing q_b or q_c will denote unfulfilled obligations, and hence will be non-accepting. We now present the formal semantics.

Fig. 2: Transition table of 1-ATA $\mathcal{A}_2$. Location q_a is initial and accepting.

	a	b	c
q_a	$q_a \wedge x.q_b \wedge \bar{x}.q_c$	q_a	q_a
q_b	q_b	$([1,1]) \vee ((0,1) \wedge q_b) \vee ((1,\infty) \wedge q_b)$	q_b
q_c	q_c	q_c	**true**

We fix a 1-ATA $\mathcal{A} = (Q, \Sigma, q_0, F, \delta)$ for the rest of this section. We define the set of valuations, or values of the clock x, as $\mathrm{Val} = \mathbb{R}_{\geq 0} \cup \{\bot\}$, where $\bot$ means the clock is inactive. A *state* of $\mathcal{A}$ is a pair (q, v), where $q \in Q$ and $v \in \mathrm{Val}$. The state (q, v) is said to be *active* if $v \in \mathbb{R}_{\geq 0}$ and inactive if $v = \bot$. We say $\mathit{val}((q, v)) = v$. We call $(q_0, 0)$ the initial state. We define $S = Q \times \mathrm{Val}$ as the set of all states of $\mathcal{A}$. Since transitions in a 1-ATA are formulas, potentially with conjunctions, each state (q, v) could result in a set of states after a transition. For example, if there is a transition $t := q_0 \wedge x.q_2$ from q, then (q, v) on t results in $\{(q_0, v), (q_2, 0)\}$. To formalize the idea, we first recursively define when a formula $\varphi \in \Phi(Q)$ is satisfied by a set of states $M \subseteq S$ on a value v, or $M \models_v \varphi$:

$M \not\models_v$ **false**	$M \models_v$ **true**
$M \models_v q$ if $(q, v) \in M$	$M \models_v I$ if $v = \bot$ or $v \in I$
$M \models_v \varphi_1 \wedge \varphi_2$ if $M \models_v \varphi_1$ and $M \models_v \varphi_2$	$M \models_v \varphi_1 \vee \varphi_2$ if $M \models_v \varphi_1$ or $M \models_v \varphi_2$
$M \models_v x.\varphi$ if $M \models_0 \varphi$	$M \models_v \bar{x}.\varphi$ if $M \models_\bot \varphi$

If $M \models_v \varphi$, we call M to be a *model* of φ on v. If $M \models_v \varphi$ and for all $M' \subsetneq M$, $M' \not\models_v \varphi$, then M is called a *minimal model* of φ on v.

For e.g., let $\varphi = q_0 \wedge x.q_1$, $M_1 = \{(q_0, 1), (q_1, 0), (q_2, 1)\}$, $M_2 = \{(q_0, 1), (q_1, 0)\}$, and $v = 1$ both $M_1, M_2 \models_v \varphi$, and M_2 is a minimal model on v for φ. Let $\varphi = [1, 2] \wedge x.q_1$, and $v = 0.2$. Since $v \notin [1, 2]$, there is no model for φ on v. Let $\varphi = x.([0, 1] \wedge q_1)$ and $v = 2.5$. Then $\{(q_1, 0)\}$ is a minimal model for φ on v.

A *configuration* γ of $\mathcal{A}$ is a finite set of states $\{(q_1, v_1), \ldots, (q_n, v_n)\} \subseteq S$. We call $\gamma_0 = \{(q_0, 0)\}$ the initial configuration. We say that a configuration γ is accepting if for *every* $(q, v) \in \gamma$, we have $q \in F$. The point is that each state with a non-accepting location is an *obligation* that is required to be discharged. Therefore, an accepting configuration is one containing no obligations. Observe that by definition of an accepting configuration, the *empty configuration* $\{\}$ consisting of no states is accepting. We define the location signature of a configuration $\gamma = \{(q_1, v_1), \ldots, (q_n, v_n)\}$ as the multiset $\text{loc-sign}(\gamma) = \{q_1, \ldots, q_n\}$.

Next we look at definition of the transitions. We define two kinds of transitions from γ:

Timed transitions. for $d \in \mathbb{R}_{\geq 0}$, we define $\gamma + d = \{(q, v + d) \mid (q, v) \in \gamma,\ v \in \mathbb{R}_{\geq 0}\} \cup \{(q, \bot) \mid (q, \bot) \in \gamma\}$. We add an edge $\gamma \xrightarrow{d} \gamma + d$ for all $d \in \mathbb{R}_{\geq 0}$.

Discrete transitions. let $\gamma = \{(q_1, v_1), \ldots, (q_n, v_n)\}$ and let $a \in \Sigma$; recall that $\delta(q_i, a)$ is a formula in disjunctive normal form for all $1 \leq i \leq n$; for each combination $C = (C_1, \ldots, C_n)$ such that C_i is one disjunct of $\delta(q_i, a)$, we add an edge $\gamma \xrightarrow{a,C} \gamma'$ if $\gamma' = \bigcup_{i=1}^{n} M_i$ where each M_i is a minimal model of C_i on v_i, for all $1 \leq i \leq n$.

For example, consider the 1-ATA $\mathcal{A}_1$ of Fig. 1. Let $\gamma = \{(q_0, 0.4), (q_1, 0.2)\}$. We have $\delta(q_0, a)$ to be a single clause and $\delta(q_1, a)$ to be a disjunction of three clauses. So, there are three possible combinations, out of which only $([0, 1) \wedge q_1)$ has an extension: $\gamma \xrightarrow{a,(q_0 \wedge x.q_1,\ [0,1)\wedge q_1)} \gamma'$ with $\gamma' = \{(q_0, 0.4), (q_1, 0), (q_1, 0.2)\}$.

We use $\gamma \xrightarrow{d,a,C} \gamma'$ as shorthand to mean there is some γ'' such that $\gamma \xrightarrow{d} \gamma'' \xrightarrow{a,C} \gamma'$. We define a run from γ as a sequence of transitions:

$$\gamma \xrightarrow{d_1} \gamma' \xrightarrow{a_1,C_1} \gamma_1 \xrightarrow{d_2} \gamma_1' \xrightarrow{a_2,C_2} \gamma_2 \ldots \gamma_{n-1} \xrightarrow{d_n} \gamma_{n-1}' \xrightarrow{a_n,C_{n-1}} \gamma_n$$

The run is accepting if γ_n is accepting. A timed word $(d_1, a_1) \ldots (d_n, a_n)$ is accepted by $\mathcal{A}$ if there is an accepting run of the above form from γ_0. We define the language of $\mathcal{A}$, or $L(\mathcal{A})$ to be the set of words accepted by $\mathcal{A}$.

Example 1. For the 1-ATA of Fig 1, $\{(q_0, 0)\}$ is the initial configuration. For timed word $w = (0.5, a)(0.7, a)$, an accepting run is

$$\{(q_0, 0)\} \xrightarrow{0.5} \{(q_0, 0.5)\} \xrightarrow{a,(q_0 \wedge x.q_1)} \{(q_0, 0.5), (q_1, 0)\} \xrightarrow{0.7} \{(q_0, 1.2), (q_1, 0.7)\}$$

$$\xrightarrow{a,((q_0 \wedge x.q_1),([0,1)\wedge q_1))} \{(q_0, 1.2), (q_1, 0), (q_1, 0.7)\}$$

Given a 1-ATA $\mathcal{A}$, the *emptiness problem* asks whether $L(\mathcal{A})$ is empty or not. To decide the emptiness problem, it is enough to find if there is some accepting run from γ_0 in the transition system above. This problem has been proven decidable with non-primitive recursive complexity in [29]. The decidability result relies on the construction of extended regions, on which a well-quasi order can be defined. In this paper, following standard techniques for timed automata [15,10], we design a symbolic zone-based approach to decide the emptiness problem for 1-ATA. Before we delve into it, we consider a special restriction of 1-ATAs.

1-ATA with Bounded Width. We define the *width* of a configuration γ to be the number of active states present in γ. For instance $\{(q_0, 1.2), (q_1, 0), (q_1, \bot)\}$ has width 2. From Example 1, we notice that the run can be extended to generate configurations of larger and larger width. Each transition out of q_0 generates two more states, and the existing states with q_1 remain. We say that a 1-ATA $\mathcal{A}$ has width k if every *reachable* configuration starting from the initial configuration of $\mathcal{A}$ has width $\leq k$. It can be checked that $\mathcal{A}_1, \mathcal{A}_2$ of Fig. 1 and 2 have unbounded width.

3 Improving the MTL-to-1-ATA Construction

In this section, we revisit the MTL-to-1-ATA construction of [32,33] and propose a modified construction using the deactivation operation. We identify a fragment of MTL which yields bounded width 1-ATAs with the proposed modification. The syntax of MTL is described using the grammar: $\varphi := a \mid \neg a \mid \varphi \wedge \varphi \mid \varphi \vee \varphi \mid \mathrm{X}_I\, \varphi \mid \varphi\, \mathrm{U}_I\, \varphi$, where $a \in \Sigma$ and I is an interval. Given a timed word $w = (d_1, a_1), \ldots, (d_n, a_n)$, a position $1 \leq i \leq n$ and an MTL formula φ, we define $(w,i) \models \varphi$ as follows:

- $(w,i) \models a$ if $a_i = a$, and $(w,i) \models \neg a$ if $a_i \neq a$
- $(w,i) \models \varphi_1 \wedge \varphi_2$ if $(w,i) \models \varphi_1$ and $(w,i) \models \varphi_2$; $(w,i) \models \varphi_1 \vee \varphi_2$ is similar.
- $(w,i) \models \mathrm{X}_I\, \varphi$ if $(w,i+1) \models \varphi$ and $d_{i+1} \in I$
- $(w,i) \models \varphi_1\, \mathrm{U}_I\, \varphi_2$ if there is some $i \leq k \leq n$ such that $(w,k) \models \varphi_2$, $\Sigma_{c=1}^{k} d_c - \Sigma_{c=1}^{i} d_c \in I$, and for all $i \leq j < k$, $(w,j) \models \varphi_1$

We say a timed word w satisfies formula φ, denoted as $w \models \varphi$, if $(w,1) \models \varphi$. Observe that the time stamp of the first letter is not relevant for the satisfaction of the formula, and therefore the semantics is translation invariant. We define $L(\varphi)$ as the set of timed words that satisfy φ.

MTL to 1-ATA Construction of [33]. Given a formula φ, [33] gives a method to construct a 1-ATA $\mathcal{A}_\varphi$ with the same language. Without loss of generality, we assume that φ is in negation normal form: all negations appear only at the atomic level. Locations of $\mathcal{A}_\varphi$ include all subformulas of φ, a location $(X_I\psi)^r$ for every subformula $X_I\psi$ of φ and and a special location φ_{init}. The initial location is φ_{init}. The transition relation δ is designed so that the following invariant is satisfied [33]: on a word $w = (d_1, a_1) \ldots (d_n, a_n)$ the presence of state $(\psi, 0)$ in the configuration occuring after reading a_j, ensures that $(w,j) \models \psi$. With this in mind, the transition relation δ is given as follows, for each $a \in \Sigma$:

- $\delta(\varphi_{init}, a) = x.\delta(\varphi, a)$
- $\delta(\psi_1 \vee \psi_2, a) = \delta(\psi_1, a) \vee \delta(\psi_2, a)$, and $\delta(\psi_1 \wedge \psi_2, a) = \delta(\psi_1, a) \wedge \delta(\psi_2, a)$
- $\delta(X_I\psi, a) = x.(X_I\psi)^r$, and $\delta((X_I\psi)^r, a) = I \wedge \delta(\psi, a)$,
- $\delta(\psi_1\, \mathrm{U}_I\, \psi_2, a) = (x.\delta(\psi_2, a) \wedge I) \vee (x.\delta(\psi_1, a) \wedge (\psi_1\, \mathrm{U}_I\, \psi_2))$
- $\delta(b, a) =$ **true**, if $b = a$, and **false** otherwise
- $\delta(\neg b, a) =$ **false**, if $b = a$ and **true** otherwise

Proposed Modification. The idea is that if ψ is a pure LTL formula, then in the state $(\psi, 0)$ generated to check ψ, the value of x is unnecessary. Hence, we deactivate the clock whenever we generate an obligation for a pure LTL formula. More precisely, we replace every occurrence of $x.\psi$ in δ above by:

- $\bar{x}.\psi$ if ψ is a pure LTL formula; and keep $x.\psi$ otherwise

Call the resulting 1-ATA $\mathcal{A}'_\varphi$. We can then say the following:

Lemma 1. *For an MTL formula φ and the constructed $\mathcal{A}'_\varphi$, $L(\varphi) = L(\mathcal{A}'_\varphi)$.*

Fig. 3 gives an example of the proposed construction for the formula $(\mathbf{F}a)\, \mathrm{U}_{[1,2]}\, c$ where $\mathbf{F}a$, a short form for $\mathbf{true}\, \mathrm{U}\, a$, is a pure LTL formula. In the figure, the transition system on the right depicts a run. In the original construction, each

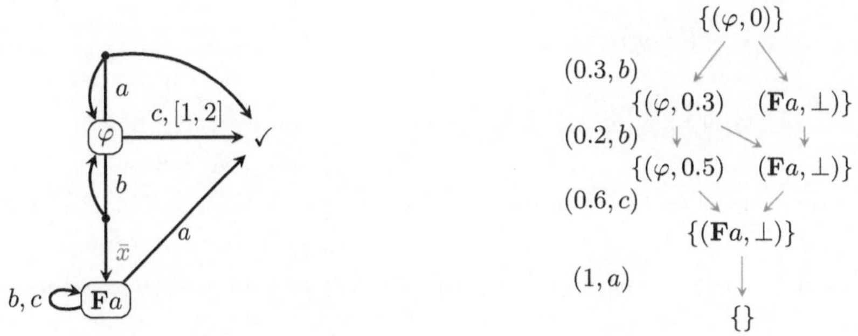

Fig. 3: Left: 1-ATA $\mathcal{A}'_\varphi$ corresponding to the MTL formula $\varphi := (\mathbf{F}a)\,\mathrm{U}_{[1,2]}\,c$. The construction of [33] gives the 1-ATA obtained by replacing the $\bar{x}$ above with the reset operation x. The initial location φ_{init} is not depicted for clarity. The ✓ is a placeholder for transitions going to **true**. Missing transitions are assumed to be **false**. Transitions to **true** (✓) deactivate the clock – $\bar{x}$ is not depicted for clarity. Right: The run of $\mathcal{A}'_\varphi$ on the word shown in blue, read from top to bottom. It leads to an accepting configuration {}.

b generates a new copy of location $\mathbf{F}a$ with an active clock, and makes the 1-ATA unbounded. On the other hand, with the deactivated clock, the resulting 1-ATA becomes bounded width, in fact, it has width 1 – the only active state checks for an occurrence of c within the interval $[1, 2]$ starting from the initial position. Motivated by this observation, we define a fragment of MTL for which our proposed construction results in 1-ATAs with bounded width. This fragment is inspired by the flat-MTL fragment defined in [11].

Definition 2 (One-sided MTL). *The syntax of one-sided MTL is defined as:*

$$\varphi := a \mid \neg a \mid \varphi \wedge \varphi \mid \varphi \vee \varphi \mid \mathrm{X}_I\,\varphi \mid \psi\,\mathrm{U}_I\,\varphi$$

where $a \in \Sigma$, $I \in \mathcal{I}$, and ψ is a purely LTL-formula.

The key idea is that in the U_I modality, the left branch is an LTL formula. For example, the formula $(\mathbf{F}a)\,\mathrm{U}_{[1,2]}\,c$, is a one-sided MTL formula. Notice that in the Flat-MTL fragment([11]), the Until formulas either have bounded intervals, or have pure LTL formulas on the left branch. Here, we only allow the latter kind of formulas. Also, this fragment is different from MITL([5]) as we do not forbid singular intervals. Here is the main result of this section:

Lemma 2. *For a one-sided MTL formula φ, there is a bound k_φ such that the 1-ATA $\mathcal{A}'_\varphi$ has width k_φ.*

4 A Zone Graph for 1-ATAs

We will now move on to solving the emptiness problem for 1-ATAs using zone based techniques. Algorithms based on zones are well known in the timed au-

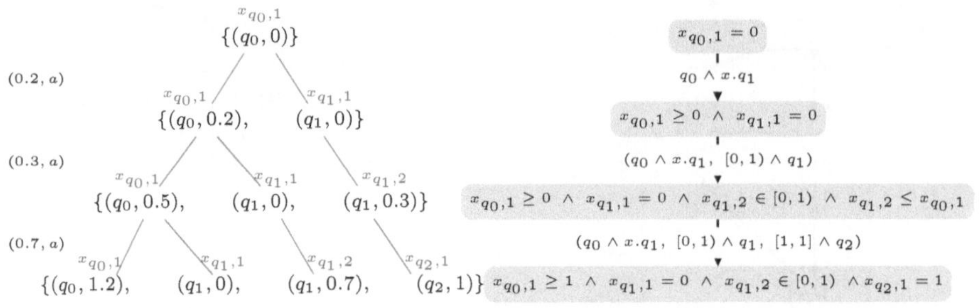

Fig. 4: A timed word $(0.2, a)(0.3, a)(0.7, a)$ shown in blue; the run of the 1-ATA $\mathcal{A}_1$ of Fig. 1 in the middle; part of the zone graph shown on the right. Variable names corresponding to states in a configuration are shown in red.

tomata literature. We begin this section with two examples that provide an overview of our zone graphs for 1-ATAs.

Example 2. Consider 1-ATA $\mathcal{A}_1$ of Fig. 1. Fig. 4 illustrates a run of $\mathcal{A}_1$ on a timed word $(0.2, a)(0.3, a)(0.7, a)$. The picture on the left (without the red annotations) gives the run. Observe that the size of the configurations keeps growing as we see more and more letters. We need a naming convention to represent each state of a configuration. In Fig. 4, the annotations in red give the variable names. For each location q of the 1-ATA, we have variables $x_{q,1}, x_{q,2}, \ldots$. At each configuration, we make use of a fresh variable as and when needed. For example, to compute the successor of configuration $\gamma := \{(q_0, 0.5), (q_1, 0), (q_1, 0.3)\}$, we pick one state at a time, pick one clause in an outgoing transition on a, and compute the minimal model. In the figure, we first pick $(q_0, 0.5)$ and compute minimal model w.r.t. $q_0 \wedge x.q_1$. This results in states, corresponding to variable names $x_{q_0,1}$ and $x_{q_1,1}$. Next, we pick $(q_1, 0)$ from γ, and the transition $[0, 1) \wedge q_1$. This gives the state $(q_1, 0.7)$ with variable associated being $x_{q_1,2}$. Finally, we pick $(q_1, 0.3)$ from γ, and the transition $[1, 1] \wedge q_2$. This gives the state $(q_2, 1)$ with variable associated being $x_{q_2,1}$. On the right of Fig. 4 is the zone graph built with this naming convention. Each zone collects the set of all configurations obtained by following the sequence of transitions given alongside the arrow. For example, the initial zone $x_{q_0,1} = 0$ says that the initial configuration is $(q_0, 0)$; the zone $x_{q_0,1} \geq 0 \wedge x_{q_1,1} = 0$ contains all configurations $\{(q_0, \theta), (q_1, 1)\}$ where $\theta \geq 0$.

Example 3. As a second example, consider the automaton $\mathcal{A}'_\varphi$ as depicted in Fig. 3, for $\varphi = (\mathbf{F}a)\,\mathrm{U}_{[1,2]}\,c$. This example shows how to deal with inactive states. Fig. 5 depictes the zone graph for $\mathcal{A}'_\varphi$. There are two remarks. Firstly, the zone graph maintains a zone over the active states, and a set of inactive variables. Successors are computed for all of them, but zone constraints are maintained only for the active states. Secondly, we explicitly keep a node for the "empty zone". This is important since it shows that all obligations have been discharged. In fact, for $\mathcal{A}'_\varphi$, the empty configuration is the only accepting configuration, and

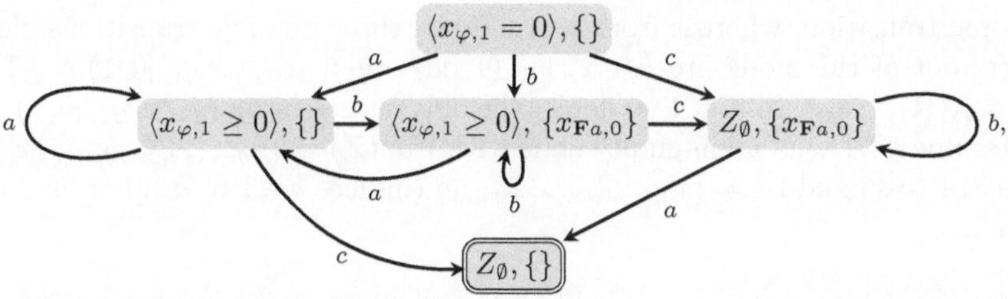

Fig. 5: Zone graph for the 1-ATA $\mathcal{A}'_\varphi$ illustrated in Fig. 3. Node with location φ_{init} is not depicted for clarity. $Z_\emptyset$ denotes an empty zone with $[\![Z_\emptyset]\!] = \emptyset$.

therefore the node corresponding to the empty configuration in the zone graph is the only accepting node (c.f. Fig. 5).

We now move on to a formal definition of the zone graph and the procedure to compute successors. Fix a 1-ATA $\mathcal{A} = (Q, \Sigma, q_0, F, \delta)$ for the rest of this section. We assume an infinite supply of variable names $x_{q,0}, x_{q,1}, \dots$ and $x'_{q,0}, x'_{q,1}, \dots$ for each location q of $\mathcal{A}$. For variables $x_{q,i}$ and $x'_{q,i}$, we define $\text{loc}(x_{q,i}) = q$, $\text{loc}(x'_{q,i}) = q$. For a set of variables X, we define its *location signature*, $\text{loc-sign}(X)$ to be the multi-set $\{\text{loc}(y) \mid y \in X\}$.

Zones and Nodes. A *zone* is a conjunction of constraints of the form $(y \sim k)$ or $(y - x \sim k)$, where $\sim \in \{<, \leq, >, \geq\}$, $k \in \mathbb{Z}$ and x, y are variables. We write $Var(Z)$ for the set of variables used in Z. We define $\text{loc-sign}(Z) := \text{loc-sign}(Var(Z))$. To deal with the inactive valuations, we define the nodes of a zone graph to be of the form (Z, IA) where Z is a zone and $\text{IA} \subseteq \{x_{q,0} \mid q \text{ is a location of } \mathcal{A}\}$ is a set of variables that are currently inactive. Notice that we always use variables with index 0 in IA, and hence there are finitely many choices for the IA component of nodes. We say $Var(Z, \text{IA}) = Var(Z) \cup \text{IA}$ and $\text{loc-sign}(Z, \text{IA}) = \text{loc-sign}(Z) \cup \text{loc-sign}(\text{IA})$. We will call the pair (Z, IA) a *node*.

We say that a configuration γ satisfies a node (Z, IA), denoted as $\gamma \models (Z, \text{IA})$, if there is a surjection $h : Var(Z, \text{IA}) \mapsto \gamma$ such that (i) $\text{loc}(y) = \text{loc}(h(y))$ for every $y \in Var(Z, \text{IA})$, (ii) for every $x_{q,0} \in \text{IA}$, $val(h(x_{q,0})) = \bot$ and (iii) replacing every variable $y \in Var(Z)$ with $val(h(y))$ satisfies all the zone constraints. We can now define $[\![(Z, \text{IA})]\!] = \{\gamma \mid \gamma \models (Z, \text{IA})\}$. This means we can look at zones as a representation of a set of configurations. For convenience, we will simply use (Z, IA) to refer to both the zone and the set $[\![(Z, \text{IA})]\!]$ of configurations. For instance, we will use $\gamma \in (Z, \text{IA})$ to mean $\gamma \in [\![(Z, \text{IA})]\!]$. Zones can be represented using Difference-Bound-Matrices (DBMs) [16]. We refer the reader to [8,10] for an exposition of algorithms for some of the standard operations on zones.

Computing Successors. Given a node (Z, IA) and a letter $a \in \Sigma$, we need to first pick an outgoing transition for each variable – more precisely, for the location corresponding to the variable. For instance, in Fig. 5, consider the node $\langle x_{\varphi,1} \geq 0\rangle, \{x_{\mathbf{F}a,0}\}$. On b, location φ has transition $\varphi \wedge \bar{x}.(\mathbf{F}a)$ and location $\mathbf{F}a$ has a loop back to it. Hence the successor is computed on the tuple $(\varphi \wedge \bar{x}.(\mathbf{F}a), \mathbf{F}a)$. In Fig. 4, consider node $\langle x_{q_0,1} \geq 0 \wedge x_{q_1,1} = 0\rangle$. From q_0 there is a unique outgoing transition, whereas from q_1, there are three possible transitions. So the targets out of this node are $(q_0 \wedge x.q_1, (1,\infty) \wedge q_1)$, $(q_0 \wedge x.q_1, [0,1) \wedge q_1)$ and $(q_0 \wedge x.q_1, [1,1] \wedge q_2)$. In Fig. 4 we depict the only successor on $(q_0 \wedge x.q_1, [0,1) \wedge q_1)$.

Assume (Z, IA) is a non-empty node with $Var(Z) = \{x_{q_1,i_1}, x_{q_2,i_2}, \ldots, x_{q_k,i_k}\}$ (indices 1 to k) and $\mathrm{IA} = \{x_{q_{k+1},0}, \ldots, x_{q_m,0}\}$ (indices $k+1$ to m). For an $a \in \Sigma$, we define:

$$\mathrm{target}((Z, \mathrm{IA}), a) = \{(C_1, \ldots, C_m) \mid C_j \text{ is a disjunct in } \delta(q_j, a) \text{ for } 1 \leq j \leq m\}$$

Pick $(C_1, \ldots, C_m)$. Our goal is to compute the successor node $(Z, \mathrm{IA}) \xrightarrow{a,(C_1,\ldots,C_m)} (Z', \mathrm{IA}')$. If some $C_j =$ **false** then according to our definition, there is no model for it, w.r.t. to any valuation. We discard such targets. Let us assume none of the C_j is **false**. We do the following sequence of operations

Time elapse. Compute node (Z_1, IA_1) where $[\![(Z_1, \mathrm{IA}_1)]\!] = \{\gamma + \delta \mid \delta \geq 0\}$ and $\mathrm{IA}_1 = \mathrm{IA}$. The zone Z_1 represents the closure of Z w.r.t. time successors. It can be computed using a standard DBM technique as in [8].

Guard intersection. For every active variable j ranging from 1 to k, and for every interval $I_j \in C_j$, add the constraints corresponding to $x_{q_j,i_j} \in I_j$. After adding all the constraints, we tighten the constraints using an operation known as *canonicalization* in the timed automata literature [8]. For instance if $x' = x, y' = y$ and $x = y$, we derive the constraint $x' = y'$. Let the resulting node be (Z_2, IA_2) with $\mathrm{IA}_2 = \mathrm{IA}_1$. We have $[\![(Z_2, \mathrm{IA}_2)]\!] := \{\gamma \in (Z_1, \mathrm{IA}_1) \mid \gamma$ satisfies all clock constraints in the transition $\}$.

Reset and move to new variables. Since we have dealt with intervals already, we can now assume that each C_j is a conjunction of atoms of the form q, $x.q$ and $\bar{x}.q$, or $C_j =$ **true**. If $C_j =$ **true** for all $1 \leq j \leq m$, the successor of (Z, IA) is a special empty node $(Z_\emptyset, \{\})$ where $Z_\emptyset$ denotes a zone with $[\![Z_\emptyset]\!] = \emptyset$ (see Fig. 5, the successor of $(\langle x_{\varphi,1} \geq 0\rangle, \{\})$ for instance). Otherwise, we will iteratively compute a new set of constraints Φ, and a new set of variables IA_3. Initially, $\Phi :=$ **true**, $\mathrm{IA}_3 = \emptyset$. Pick each variable x_{q_j,i_j}, with j ranging from 1 to m in some order, and consider all the atoms in the conjunction C_j:

1. if q is an atom of C_j: when $1 \leq j \leq k$ (active variable), add $x'_{q,\ell} = x_{q_j,i_j}$ to Φ, where $\ell \geq 2$ is the smallest index (greater than 2) such that $x'_{q,\ell}$ is not used in Φ; when $k+1 \leq j \leq m$ (inactive variable), add $x'_{q,0}$ to IA_3,
2. if $x.q$ is an atom of C_j, add $x'_{q,1} = 0$ to Φ,
3. if $\bar{x}.q$ is an atom of C_j, add $x'_{q,0}$ to IA_3

Define $Z_3 = Z_2 \wedge \Phi$.

Remove old variables. Tighten all the constraints of Z_3 by the canonicalization procedure. After canonicalization, remove all the old unprimed variables, and remove the primes from the newly introduced variables, i.e., $x'_{q,1}$ becomes $x_{q,1}$ and so on. This new node is the required (Z', IA').

Zone Graph of a 1-ATA. The initial node is (Z_0, I_0) where $Z_0 := (x_{q_0,1} = 0)$ (with q_0 being the initial state) and $\mathrm{IA}_0 = \emptyset$. There is a special node $(Z_\emptyset, \{\})$

denoting the empty configuration, with $[\![Z_\emptyset]\!] = \emptyset$. Successors are systematically computed by enumerating over all the outgoing targets, and performing the successor computation as explained earlier. The resulting graph that is computed is called the zone graph of the 1-ATA. A node (Z, IA) is said to be accepting if for every $x \in Var(Z) \cup \mathrm{IA}$, we have $\mathrm{loc}(x)$ to be accepting. In particular, the special node $(Z_\emptyset, \{\})$ is accepting. We can prove the soundness and completeness of this zone graph – the language of a 1-ATA is non-empty iff there is a path in the zone graph starting from the initial node to an accepting node.

5 The Entailment Relation

Zone enumeration suffers from two sources of infinity: (1) the width of the zone (number of active states) can increase in an unbounded manner, and (2) the constants appearing in the zone constraints can be unbounded too. For timed automata, the number of clocks (and hence the width of the zones) is fixed. However, the second challenge does manifest and there is a long line of work coming up with better termination mechanisms that tackle the unbounded growth of constants. The termination mechanism is essentially a subsumption relation between zones, which allows to prune the search.

For zone-based universality in 1-clock timed automata, an *entailment* relation between zones was used as a termination mechanism [1]. We provide an algorithm for this check that makes use of zone operations from the timed automata literature. We then prove that the entailment is NP-hard. As the entailment check is an important operation in the zone graph, done each time a new zone is added, the NP-hardness illuminates the difficulty caused due to point (1) above.

Finally, for 1-ATA with bounded width, where (1) is not an issue any more, we provide a slight modification to the test, which makes it polynomial-time checkable, and yet ensures termination.

The entailment check is based on an equivalence between configurations $\gamma \simeq_M \gamma'$, first proposed in [31]. The equivalence can be adapted to our setting by considering the $\perp$ states by adding that there is a one-to-one correspondence between inactive states in γ and γ'.

Definition 3. *Let $M \in \mathbb{N}$ be the largest constant appearing in $\mathcal{A}$. For two configurations γ and γ', we say that γ is region equivalent to γ', or that $\gamma \simeq_M \gamma'$, if we can define a bijection $h : \gamma \to \gamma'$ such that for every $(q, v), (q_1, v_1), (q_2, v_2) \in \gamma$:*

- $\mathrm{loc}(h(q, v)) = q$,
- $val(h(q, v)) = \perp$ *iff* $v = \perp$, *and* $0 \leq val(h(q, v)) \leq M$ *iff* $0 \leq v \leq M$,
- *if* $0 \leq v \leq M$, *then* $\lfloor v \rfloor = \lfloor val(h(q, v)) \rfloor$, *and* $\mathrm{fract}(v) = 0$ *iff* $\mathrm{fract}(val(h(q, v))) = 0$,
- *if* $0 \leq v_1, v_2 \leq M$, *then* $\mathrm{fract}(v_1) \leq \mathrm{fract}(v_2)$ *iff* $\mathrm{fract}(val(h(q_1, v_1))) \leq \mathrm{fract}(val(h(q_2, v_2)))$.

For the rest of the document, it is sufficient to observe that $\simeq_M$ is a time-abstract bisimulation on the configurations. The equivalence $\simeq_M$ is defined on

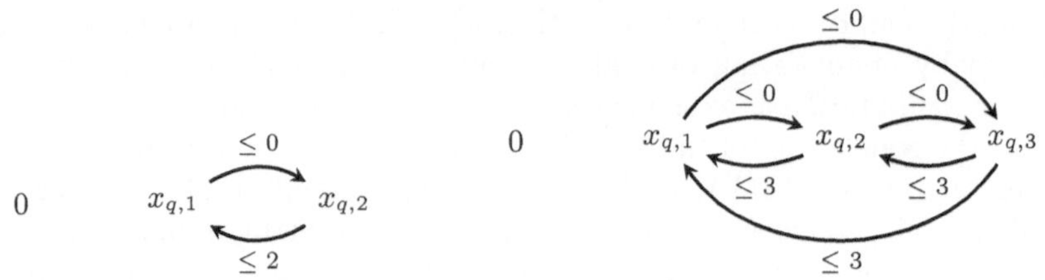

Fig. 6: Zone Z_1 on the left, and Z_2 on the right for Example 4 represented in a graphical notation. An edge $x \xrightarrow{\triangleleft c} y$ stands for constraint $y - x \triangleleft c$. Edges $x \xrightarrow{\leq 0} 0$ are all omitted, as well as edges $0 \xrightarrow{<\infty} x$.

configurations with the same size. However, as noticed in Fig. 4, the size of the configurations could be unbounded, and we need a way to relate configurations of different sizes. We adapt an entailment relation proposed in [1] to our setting:

Definition 4 (Entailment relation). *For configurations γ, γ', we say that γ is entailed by γ', or $\gamma \sqsubseteq_M \gamma'$, if there exists a subset $\gamma'' \subseteq \gamma'$ such that $\gamma \simeq_M \gamma''$.*

For nodes $(Z, \mathrm{IA}), (Z', \mathrm{IA}')$, we say $(Z, \mathrm{IA}) \sqsubseteq_M (Z', \mathrm{IA}')$, or (Z, IA) is entailed by (Z', IA'), if for all $\gamma' \in (Z', \mathrm{IA}')$, there exists $\gamma \in (Z, \mathrm{IA})$ s.t. $\gamma \sqsubseteq_M \gamma'$.

When M is clear from the context, we simply write $\sqsubseteq$ instead of $\sqsubseteq_M$.

The idea is that when $\gamma \sqsubseteq \gamma'$, if γ' reaches an accepting configuration, then so does γ. In other words, if γ' (the bigger configuration) is able to discharge all its obligations, the small configuration γ (with fewer obligations) is also able to discharge its own obligations. The same idea holds in the case of nodes. If $(Z, \mathrm{IA}) \sqsubseteq (Z', \mathrm{IA}')$, then if (Z', IA') reaches an accepting node, then so does (Z, IA). Therefore, if we reach (Z', IA') during the zone enumeration, while (Z, IA) has already been visited, we can stop exploring (Z', IA'). Moreover, by definition of $\simeq_M$, it follows that $(Z, IA) \sqsubseteq (Z', IA')$ iff $IA \subseteq IA'$ and $Z \sqsubseteq Z'$ (relation restricted to active states).

We still need to show that this pruning results in a finite zone graph. To do so, we will use the following: a relation $\prec\ \subseteq A \times A$ for a set A is a well-quasi order (WQO) if for every infinite sequence $a_1 a_2 \ldots$ where $a_1, a_2, \cdots \in A$, there exists some i and j such that $a_i \prec a_j$. It was shown in [1] that the entailment restricted to zones $Z \sqsubseteq Z'$ is a well-quasi order. Since the IA sets are finite, $IA \subseteq IA'$ is also a well-quasi order. Finite cartesian products of WQOs are WQOs [27]. Now, we can observe since $\sqsubseteq$ is a WQO on nodes, every sequence of nodes will hit a node which is bigger than an existing node w.r.t $\sqsubseteq$ and hence the computation will terminate at this node, implying that every path in the zone graph pruned using $\sqsubseteq$, is finite.

Example 4. Let nodes $(Z_1, \emptyset)$ and $(Z_2, \emptyset)$ be defined by the constraints represented as a graph Fig. 6, where an edge $x \xrightarrow{\triangleleft\, c_{xy}} y$ represents the constraint $y - x \triangleleft c_{xy}$. We take $M = 3$. With this value for M, we can replace $\gamma \simeq_M \gamma'$ with

$\gamma = \gamma'$ for the configurations used in this example. Then, we claim that $(Z_1, \emptyset) \sqsubseteq (Z_2, \emptyset)$. We now explain why – alongside, we illustrate a source of difficulty in the check. Firstly, $\emptyset \subseteq \emptyset$, and so the test on the inactive component trivially passes. Next, pick an arbitary $\gamma_2 \in Z_2$ of the form $\gamma_2 = \{(q, v_1), (q, v_2), (q, v_3)\}$.

Suppose $v_1 - v_2 \leq 2$. Consider the projection $\gamma_2' = \{(q, v_1), (q, v_2)\}$. By examining the constraints of Z_1, we conclude that $\gamma_2' \in Z_1$. Setting $\gamma_1 = \gamma_2'$, we have $\gamma_1 \in Z_1$, such that $\gamma_1 \sqsubseteq \gamma_2$. Suppose $v_1 - v_2 > 2$. Then, γ_2' as above, does not belong to Z_1. However, if $v_1 - v_2 > 2$ (equivalently $v_2 - v_1 < -2$), and since $\gamma_2 \in Z_2$, by the constraint $v_1 - v_3 \leq 3$, we infer that $v_2 - v_3 < -1$. Now, we can take the subset $\gamma_2'' = \{(q, v_2), (q, v_3)\}$ and set $\gamma_1 = \gamma_2''$ to get $\gamma_1 \in Z_1$ (mapping (q, v_2) to $x_{q,1}$ and (q, v_3) to $x_{q,2}$) such that $\gamma_1 \sqsubseteq \gamma_2$.

Notice that different configurations in Z_2 require different projections as witnesses for the entailment. As we will see, this makes the problem NP-hard.

Algorithm for the Entailment Check. To get to an algorithm to check entailment on two zones, we look at the cases when the entailment will not hold, i.e. the conditions when $(Z, \mathrm{IA}) \not\sqsubseteq (Z', \mathrm{IA}')$. We observe that $(Z, \mathrm{IA}) \not\sqsubseteq (Z', \mathrm{IA}')$ if either $\mathrm{IA} \not\subseteq \mathrm{IA}'$, or

$$\text{there is some } \gamma' \in Z' \text{ s.t. for every } \gamma \in Z \text{ and every } \gamma'' \subseteq \gamma',\ \gamma \not\simeq_M \gamma'' \quad (1)$$

As the first case is easy to check, we assume $\mathrm{IA} \subseteq \mathrm{IA}'$ and focus on the second case, which means that for this particular $\gamma' \in Z'$, for each of its subsets γ'', there is no configuration that is region equivalent to it in Z. We see that for $\gamma'' \subseteq \gamma'$ such that loc-sign$(\gamma'') \neq$ loc-sign(Z), there will trivially be no region equivalent configuration for γ'' in Z. Thus we look at the subsets γ'' such that loc-sign(γ'') = loc-sign(Z) and investigate the entailment check (1).

Consider a one-to-one mapping $r : Var(Z) \to Var(Z')$ that preserves locations: that is, for every $x \in Var(Z)$, we have $\mathrm{loc}(x) = \mathrm{loc}(r(x))$. Let range$(r)$ be the set of variables of $Var(Z')$ that are in the range of r. Rename variables $Var(Z)$ to $y_1, y_2, \ldots, y_n$, making Z a zone over these fresh variables. Similarly, project Z' to range(r) and rename $r(y_1), r(y_2), \ldots, r(y_n)$ as $y_1, y_2, \ldots, y_n$. Call the resulting zones Z_r and Z'_r respectively. These are zones on the same fixed set of variables and are hence amenable to techniques from timed automata literature. A valuation $v' \in Z'_r$ associates a real to each variable y_i. Define: $N_r := \{v' \in Z'_r \mid \forall v \in Z_r, v \not\simeq_M v'\}$. Given Z_r, Z'_r, there is a method to check if N_r is non-empty [23]. Using this algorithm, we can in fact describe N_r as a finite union of zones. The complexity of this method is quadratic in the number of variables used in Z_r and Z'_r. Then, we translate N_r back to Z': let $N'_r = \{\gamma' \in Z' \mid \gamma' \text{ restricted to } r \text{ belongs to } N_r\}$.

Lemma 3. *For two zones (Z, IA) and (Z', IA'), $(Z, \mathrm{IA}) \not\sqsubseteq (Z', \mathrm{IA}')$ iff $\mathrm{IA} \not\subseteq \mathrm{IA}'$ or $\bigcap_{r \in R_{loc}} N'_r \neq \emptyset$, where R_{loc} denotes the set of all location preserving one-to-one mappings from $Var(Z)$ to $Var(Z')$*

The above lemma leads to an algorithm: compute N'_r for each location preserving mapping in time $\mathcal{O}(|Var(Z)|^2)$ and check if the intersection is non-empty.

Notice that the number of mappings r could be exponential in $|Var(Z)|$ – it is bounded by the number of subsets of size $|Var(Z)|$ in $Var(Z')$, which is given by $\binom{|Var(Z')|}{|Var(Z)|}$. The overall complexity is $\mathcal{O}(\binom{|Var(Z')|}{|Var(Z)|} \cdot |Var(Z)|^2)$. In practice, one can envisage an algorithm that enumerates each location preserving r, and keeps the intersection of N'_r among all the r enumerated so far. This intersection will be a union of zones. If at some point, the intersection becomes empty, the algorithm can stop. Otherwise, the algorithm continues until all location-preserving maps are considered.

Hardness. Let $V = \{p_1, p_2, \dots, p_n\}$ be a set of propositional variables, taking values either $\top$ (true) or $\bot$ (false). An assignment fixes a value to each of the variables. A positive literal is an element of $L_+ = V$ and a negative literal is an element of $L_- = \{\bar{p}_1, \bar{p}_2, \dots, \bar{p}_n\}$. When $p_i = \top$, the literal $\bar{p}_i = \bot$ and vice-versa. We denote by $L := L_+ \cup L_-$, the set of all literals. A 3-clause (hereafter called simply a clause) C is a disjunction of *three* literals. The clause is said to be monotone if either all its literals are positive, or all its literals are negative: $C \subseteq L_+$ or $C \subseteq L_-$. A boolean formula in 3-CNF form is a conjunction of clauses: $\bigwedge_{i=1}^{i=m} C_i$. A 3-CNF formula is monotone if each of its clauses is monotone. The MONOTONE-3-SAT problem: given a monotone 3-CNF formula, is it satisfiable? This problem is known to be NP-complete [20].

Let 1-ATA-ZONE-NON-ENTAILMENT be the decision problem which takes as input two nodes (Z, IA), (Z', IA'), and a constant M, and checks if $(Z, \mathrm{IA}) \not\sqsubseteq (Z', \mathrm{IA}')$, where $\sqsubseteq$ makes use of the region equivalence w.r.t M.

Theorem 1. *1-ATA-ZONE-NON-ENTAILMENT is* NP*-hard.*

To prove the theorem, we present a reduction from MONOTONE-3-SAT to 1-ATA-ZONE-NON-ENTAILMENT. Given a monotone 3-CNF formula φ, we construct zones $(Z_\varphi, \mathrm{IA}_\varphi)$, $(Z'_\varphi, \mathrm{IA}'_\varphi)$ together with the constant M_φ such that φ is satisfiable iff $(Z_\varphi, \mathrm{IA}_\varphi) \not\sqsubseteq (Z'_\varphi, \mathrm{IA}'_\varphi)$. To start off with, we assume $\mathrm{IA}_\varphi = \mathrm{IA}'_\varphi = \emptyset$. This means $(Z_\varphi, \mathrm{IA}_\varphi) \not\sqsubseteq (Z'_\varphi, \mathrm{IA}'_\varphi)$ iff $(Z_\varphi, \emptyset) \not\sqsubseteq (Z'_\varphi, \emptyset)$. So for simplicity we will use the notation $Z_\varphi \not\sqsubseteq Z'_\varphi$ in the rest of this section. We now define the zones constructed for the monotone 3-CNF formula $\varphi = C_1 \wedge C_2 \wedge \dots \wedge C_n$, where we assume that $C_1, \dots, C_k$ have all positive literals and $C_{k+1}, \dots, C_n$ have all negative literals.

The Idea: We see that φ is satisfied iff *there exists some assignment* of the variables of φ such that *for every clause* C_i for $1 \leq i \leq n$, C_i is true. Comparing this with the definition of non-entailment: $Z_\varphi \not\sqsubseteq Z'_\varphi$ iff *there exists* $\gamma' \in Z'_\varphi$ such that *for every* $\gamma'' \subseteq \gamma'$ *and* $\gamma \in Z_\varphi$, we have $\gamma'' \not\simeq_M \gamma$. In our construction, we will pick a large enough constant M_φ so that $\gamma'' \simeq_M \gamma$ boils down to simply $\gamma'' = \gamma$. So, we see that we need to construct zones Z_φ and Z'_φ such that each $\gamma' \in Z'_\varphi$ corresponds to an assignment, and having no subset $\gamma'' \subseteq \gamma'$ with $\gamma'' \in Z_\varphi$ corresponds to every clause being true. To encode assignments through our zones, we use two variables, x and y and use the value of their difference, $y-x$, to decide the assignment of the variable they encode: $[0, 1]$ as false and $(1, 2]$ as true. Due to space constraints, we present an overview of our construction

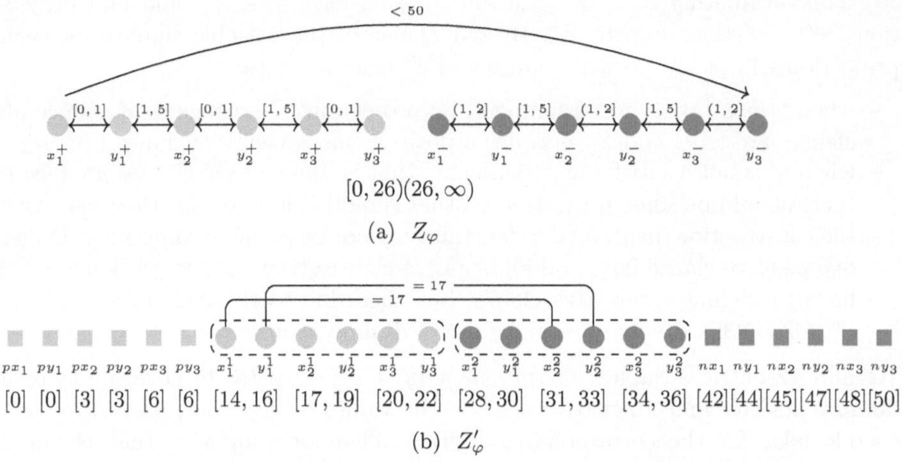

(a) Z_φ

(b) Z'_φ

Fig. 7: Zones Z_φ, Z'_φ for formula $\varphi = (p_1 \vee p_2 \vee p_3) \wedge (\bar{p}_4 \vee \bar{p}_1 \vee \bar{p}_5)$.

though an example $\varphi = (p_1 \vee p_2 \vee p_3) \wedge (\bar{p}_4 \vee \bar{p}_1 \vee \bar{p}_5)$. Fig. 7 illustrates the construction for φ.

Zone Z'_φ. We want to construct this zone to encode the possible assignments to the variables of φ. To do this, we consider variables x_j^i, y_j^i for $1 \leq j \leq 3$ and $1 \leq i \leq m$ with constraints $0 \leq y_j^i - x_j^i \leq 2$. Each valuation to the x, y variables assigns a value between 0 to 2 for the difference. For it to encode an assignment, the assigned values should be consistent across all occurrences of the same variable: for example, in Fig. 7, variable p_1 occurs in C_1 positively and in C_2 negatively. We add two constraints $x_2^2 - x_1^1 = 17$ and $y_2^2 - y_1^1 = 17$. These constraints ensure that $y_1^1 - x_1^1 = y_2^2 - x_2^2$. Therefore, in all occurrences of the variable, we have the difference $y - x$ to be the same, and hence we can extract an assignment of the variables of φ from a valuation of the x, y variables in Z'_φ. Fig. 7 also shows other variables px_j, py_j and nx_j, ny_j, which we will explain later, after describing Z_φ.

Zone Z_φ*:* In the entailment check $Z_\varphi \sqsubseteq Z'_\varphi$, we pick a $\gamma' \in Z'_\varphi$, select a subset $\gamma'' \subseteq \gamma'$ and check whether $\gamma'' \notin Z_\varphi$. The subset selection γ'' should correspond to a clause and Z_φ intuitively encodes one clause and checks if all literals are false. Since negative and positive clauses need to be dealt with differently, we instead encode two clauses, one positive (the green part of Z_φ in Fig. 7) and one negative (the red part in the figure). The constraints of Z_φ ensure that all literals are false, hence the two selected clauses are false. The extra constraints $[1, 5]$ ensure that all three literals are picked from the same clause. Moreover, for the mapping to work correctly, we set $\mathrm{loc}(x_j^i) = q_x$ and $\mathrm{loc}(x_i^+) = \mathrm{loc}(x_i^-) = q_x$, and $\mathrm{loc}(y_j^i) = q_y$, $\mathrm{loc}(y_i^+) = \mathrm{loc}(y_i^-) = q_y$ for all relevant indices.

While evaluating $Z_\varphi \not\sqsubseteq Z'_\varphi$, we will consider each $\gamma' \in Z'_\varphi$, and find projections $\gamma'' \subseteq \gamma'$ that map to Z_φ. By our choice of the variable signatures, each projection will pick a positive clause and a negative clause.

- when γ' is a satisfying assignment, no projection γ'' belongs to Z_φ, since all clauses are true, and Z_φ encodes a positive and a negative clause.
- when γ' is not a satisfying assignment (that is, some of the clauses are false), there should be some projection γ'' that should belong to Z_φ. However, with the construction mentioned so far, this may not be possible: suppose γ' makes one positive clause false and all negative clauses true; any projection γ'' will have to include a negative clause, but then due to the constraints on Z_φ, $\gamma'' \notin Z_\varphi$. This is where we make use of dummy clauses.

We add 12 extra variables px_j, py_j, nx_j, ny_j for $j = 1, 2, 3$ to correspond to a dummy positive and a negative clause. The values of $py_j - px_j$ and $ny_j - nx_j$ encode false for the corresponding literals. Therefore, for a γ' that is not a satisfying assignment, we can pick a projection γ'' that chooses a clause that is made false by γ' and a dummy clause of the opposite polarity. This takes care of the second point above. However, due to this addition, there is a projection γ'' consisting of both dummy clauses. One can always choose this pair of dummy clauses to belong to Z_φ, even when γ' is a satisfying assignment. To eliminate this, we add a final constraint to Z_φ which bounds the distance between the two extremes ($y_3^- - x_1^+ < 50$ in Fig. 7) preventing choosing both the dummy clauses. The full details of the construction can be found in [12]

Entailment for Bounded Width 1-ATAs. Suppose we start with a 1-ATA $\mathcal{A}$ with width k. Then in the zone graph computation as explained in Section 4, every variable name $x_{q,i}$ will have index $i \leq k$. For each state q, there are at most k variables in the zone (along with potentially an $x_{q,0}$ in the inactive part). In such a scenario, we can use a modified entailment check that only compares zones with the same set of variable names.

Let Z and Z' be two zones such that $Var(Z) = Var(Z')$. Let $\iota : Var(Z) \to Var(Z')$ be the identity mapping. Let $\simeq_M^\iota$ be the equivalence of Definition 3 with h being the identity function ι. So, $\simeq_M^\iota$ is simply the classical region equivalence of [4]. This can be naturally lifted to zones: $(Z, \mathrm{IA}) \sqsubseteq_M^b (Z', \mathrm{IA}')$ if $\mathrm{IA} \subseteq \mathrm{IA}'$ and for every $\gamma' \in Z'$ there is some $\gamma \in Z$ such that $\gamma \simeq_M^\iota \gamma'$. he check $\sqsubseteq^b$ is identical to a simulation test $\preccurlyeq_M$ used in the timed automata literature [23], but with the direction reversed: $(Z, \mathrm{IA}) \sqsubseteq_M^b (Z', \mathrm{IA}')$ iff $\mathrm{IA} \subseteq \mathrm{IA}'$ and $Z' \preccurlyeq_M Z$. This test can be done with quadratic complexity and is known to be a well-quasi-order on zones with a fixed number of variables.

Notice that $\sqsubseteq^b$ implies $\sqsubseteq$, but not the other way around. Therefore, the zone graph obtained by $\sqsubseteq^b$ is correct but it could be an over-approximation, with potentially more nodes. However, the entailment check $\sqsubseteq^b$ is efficient. Moreover, since there are at most $|Q| \times k$ number of variables generated, and since we know that for each $j \leq |Q| \times k$, the relation $\sqsubseteq^b$ induces a well-quasi-order, the overall computation using $\sqsubseteq^b$ terminates: if not, there is an infinite path in the zone graph with an infinite subsequence of zones over the same set of variables, which is non-increasing w.r.t $\sqsubseteq_M^b$. A contradiction.

6 Conclusion

In timed automata literature, 1-ATAs have played the role of an excellent technical device for logic-to-automata translations. Perhaps, owing to the high general complexity, there have been no attempts at aligning the analysis of 1-ATAs to the well developed timed automata algorithms. Typically, 1-ATAs are converted to equivalent network of timed automata, with a blowup in the control states. Our aim in this paper is to pass on the message that these conversions of 1-ATAs to TAs may not really be needed and instead one could embed the analysis of 1-ATAs into the current timed automata algorithms. To substantiate this claim, we have given a zone-based emptiness algorithm for 1-ATAs. In [12], we show how we can lift it to a model-checking algorithm, by providing a zone graph for the product of a timed automaton with a 1-ATA (the $A \times B_{\neg\varphi}$ as described in the introduction). We have also demonstrated a logical fragment which induces bounded width 1-ATAs, for which the operations in the zone enumeration are as good as the counterparts used in timed automata reachability. One important idea that we wish to highlight is the addition of an explicit deactivation operation for 1-ATAs. This is a simple trick that can have a substantial impact on the analysis.

We believe that the theoretical foundations we lay here pave the way for studying further optimizations, implementing the ideas and understanding the practical impact. As part of future work, it would be interesting to study whether the best known simulation techniques for timed automata [10] can be incorporated into the 1-ATA zone graphs. What about ATAs with multiple clocks? They are undecidable, of course. But, are there good restrictions that make the modeling more succinct and still enable a zone-based analysis? It would be interesting to extend the notion of boundedness on multi-clock ATAs and find out if there are fragments of MTL (MITL for instance) that are expressible by it. This study also motivates the question of understanding zone-based liveness algorithms for ATAs, in particular, bounded-width ATAs. Another direction is to explore the extensive literature based on antichain algorithms studied for alternating finite automata emptiness, in the real-time setting. Here is another natural problem that we leave open: given a 1-ATA $\mathcal{A}$ and a constant k, decide whether $\mathcal{A}$ has width k?

References

1. Abdulla, P.A., Ouaknine, J., Quaas, K., Worrell, J.: Zone-based universality analysis for single-clock timed automata. In: International Conference on Fundamentals of Software Engineering. pp. 98–112. Springer (2007)
2. Akshay, S., Gastin, P., Govind, R., Joshi, A.R., Srivathsan, B.: A unified model for real-time systems: Symbolic techniques and implementation. In: Enea, C., Lal, A. (eds.) Computer Aided Verification - 35th International Conference, CAV 2023, Paris, France, July 17-22, 2023, Proceedings, Part I. Lecture Notes in Computer Science, vol. 13964, pp. 266–288. Springer (2023). https://doi.org/10.1007/978-3-031-37706-8_14, https://doi.org/10.1007/978-3-031-37706-8_14

3. Akshay, S., Gastin, P., Govind, R., Srivathsan, B.: MITL model checking via generalized timed automata and a new liveness algorithm. In: Majumdar, R., Silva, A. (eds.) 35th International Conference on Concurrency Theory, CONCUR 2024, September 9-13, 2024, Calgary, Canada. LIPIcs, vol. 311, pp. 5:1–5:19. Schloss Dagstuhl - Leibniz-Zentrum für Informatik (2024). https://doi.org/10.4230/LIPICS.CONCUR.2024.5, https://doi.org/10.4230/LIPIcs.CONCUR.2024.5
4. Alur, R., Dill, D.L.: A theory of timed automata. Theoretical computer science **126**(2), 183–235 (1994)
5. Alur, R., Feder, T., Henzinger, T.A.: The benefits of relaxing punctuality. Journal of the ACM (JACM) **43**(1), 116–146 (1996)
6. Alur, R., Fix, L., Henzinger, T.A.: Event-clock automata: A determinizable class of timed automata. Theoretical Computer Science **211**(1-2), 253–273 (1999)
7. Behrmann, G., Bouyer, P., Larsen, K.G., Pelánek, R.: Lower and upper bounds in zone-based abstractions of timed automata. Int. J. Softw. Tools Technol. Transf. **8**(3), 204–215 (2006). https://doi.org/10.1007/S10009-005-0190-0, https://doi.org/10.1007/s10009-005-0190-0
8. Behrmann, G., David, A., Larsen, K.G.: A tutorial on uppaal. In: Bernardo, M., Corradini, F. (eds.) Formal Methods for the Design of Real-Time Systems, International School on Formal Methods for the Design of Computer, Communication and Software Systems, SFM-RT 2004, Bertinoro, Italy, September 13-18, 2004, Revised Lectures. Lecture Notes in Computer Science, vol. 3185, pp. 200–236. Springer (2004). https://doi.org/10.1007/978-3-540-30080-9_7, https://doi.org/10.1007/978-3-540-30080-9_7
9. Bouyer, P.: Forward analysis of updatable timed automata. Formal Methods Syst. Des. **24**(3), 281–320 (2004). https://doi.org/10.1023/B:FORM.0000026093.21513.31, https://doi.org/10.1023/B:FORM.0000026093.21513.31
10. Bouyer, P., Gastin, P., Herbreteau, F., Sankur, O., Srivathsan, B.: Zone-based verification of timed automata: Extrapolations, simulations and what next? In: Bogomolov, S., Parker, D. (eds.) Formal Modeling and Analysis of Timed Systems - 20th International Conference, FORMATS 2022, Warsaw, Poland, September 13-15, 2022, Proceedings. Lecture Notes in Computer Science, vol. 13465, pp. 16–42. Springer (2022). https://doi.org/10.1007/978-3-031-15839-1_2, https://doi.org/10.1007/978-3-031-15839-1_2
11. Bouyer, P., Markey, N., Ouaknine, J., Worrell, J.: The cost of punctuality. In: 22nd Annual IEEE Symposium on Logic in Computer Science (LICS 2007). pp. 109–120. IEEE (2007)
12. Bouyer, P., Srivathsan, B., Vishwanath, V.: Model-checking real-time systems: revisiting the alternating automaton route (2025), https://arxiv.org/abs/2501.17576
13. Brihaye, T., Estiévenart, M., Geeraerts, G.: On mitl and alternating timed automata. In: Formal Modeling and Analysis of Timed Systems: 11th International Conference, FORMATS 2013, Buenos Aires, Argentina, August 29-31, 2013. Proceedings 11. pp. 47–61. Springer (2013)
14. Brihaye, T., Geeraerts, G., Ho, H., Monmege, B.: Mightyl: A compositional translation from MITL to timed automata. In: Majumdar, R., Kuncak, V. (eds.) Computer Aided Verification - 29th International Conference, CAV 2017, Heidelberg, Germany, July 24-28, 2017, Proceedings, Part I. Lecture Notes in Computer Science, vol. 10426, pp. 421–440. Springer (2017). https://doi.org/10.1007/978-3-319-63387-9_21, https://doi.org/10.1007/978-3-319-63387-9_21

15. Daws, C., Tripakis, S.: Model checking of real-time reachability properties using abstractions. In: Steffen, B. (ed.) Tools and Algorithms for Construction and Analysis of Systems, 4th International Conference, TACAS '98, Held as Part of the European Joint Conferences on the Theory and Practice of Software, ETAPS'98, Lisbon, Portugal, March 28 - April 4, 1998, Proceedings. Lecture Notes in Computer Science, vol. 1384, pp. 313–329. Springer (1998). https://doi.org/10.1007/BFB0054180, https://doi.org/10.1007/BFb0054180
16. Dill, D.L.: Timing assumptions and verification of finite-state concurrent systems. In: Sifakis, J. (ed.) Automatic Verification Methods for Finite State Systems, International Workshop, Grenoble, France, June 12-14, 1989, Proceedings. Lecture Notes in Computer Science, vol. 407, pp. 197–212. Springer (1989). https://doi.org/10.1007/3-540-52148-8_17, https://doi.org/10.1007/3-540-52148-8_17
17. Doyen, L., Raskin, J.: Antichains for the automata-based approach to model-checking. Log. Methods Comput. Sci. **5**(1) (2009), http://arxiv.org/abs/0902.3958
18. Doyen, L., Raskin, J.: Antichain algorithms for finite automata. In: Esparza, J., Majumdar, R. (eds.) Tools and Algorithms for the Construction and Analysis of Systems, 16th International Conference, TACAS 2010, Held as Part of the Joint European Conferences on Theory and Practice of Software, ETAPS 2010, Paphos, Cyprus, March 20-28, 2010. Proceedings. Lecture Notes in Computer Science, vol. 6015, pp. 2–22. Springer (2010). https://doi.org/10.1007/978-3-642-12002-2_2, https://doi.org/10.1007/978-3-642-12002-2_2
19. Fiedor, T., Holík, L., Hruska, M., Rogalewicz, A., Síc, J., Vargovcík, P.: Reasoning about regular properties: A comparative study. In: Pientka, B., Tinelli, C. (eds.) Automated Deduction - CADE 29 - 29th International Conference on Automated Deduction, Rome, Italy, July 1-4, 2023, Proceedings. Lecture Notes in Computer Science, vol. 14132, pp. 286–306. Springer (2023). https://doi.org/10.1007/978-3-031-38499-8_17, https://doi.org/10.1007/978-3-031-38499-8_17
20. Garey, M.R., Johnson, D.S.: Computers and Intractability: A Guide to the Theory of NP-Completeness. W. H. Freeman (1979)
21. Gastin, P., Mukherjee, S., Srivathsan, B.: Reachability in Timed Automata with Diagonal Constraints. In: Schewe, S., Zhang, L. (eds.) 29th International Conference on Concurrency Theory (CONCUR 2018). Leibniz International Proceedings in Informatics (LIPIcs), vol. 118, pp. 28:1–28:17. Schloss Dagstuhl–Leibniz-Zentrum fuer Informatik, Dagstuhl, Germany (2018). https://doi.org/10.4230/LIPIcs.CONCUR.2018.28, http://drops.dagstuhl.de/opus/volltexte/2018/9566
22. Herbreteau, F., Point, G.: TChecker. https://github.com/fredher/tchecker (v02 - April 2019)
23. Herbreteau, F., Srivathsan, B., Walukiewicz, I.: Better abstractions for timed automata. Information and Computation **251**, 67–90 (2016)
24. Holík, L., Vargovcík, P.: Antichain with SAT and tries. In: Chakraborty, S., Jiang, J.R. (eds.) 27th International Conference on Theory and Applications of Satisfiability Testing, SAT 2024, August 21-24, 2024, Pune, India. LIPIcs, vol. 305, pp. 15:1–15:24. Schloss Dagstuhl - Leibniz-Zentrum für Informatik (2024). https://doi.org/10.4230/LIPICS.SAT.2024.15, https://doi.org/10.4230/LIPIcs.SAT.2024.15
25. Kant, G., Laarman, A., Meijer, J., van de Pol, J., Blom, S., van Dijk, T.: LTSmin: High-performance language-independent model checking. In: TACAS. Lecture Notes in Computer Science, vol. 9035, pp. 692–707. Springer (2015)

26. Krishna, S.N., Madnani, K., Pandya, P.K.: Logics meet 1-clock alternating timed automata. In: Schewe, S., Zhang, L. (eds.) 29th International Conference on Concurrency Theory, CONCUR 2018, September 4-7, 2018, Beijing, China. LIPIcs, vol. 118, pp. 39:1–39:17. Schloss Dagstuhl - Leibniz-Zentrum für Informatik (2018). https://doi.org/10.4230/LIPICS.CONCUR.2018.39, https://doi.org/10.4230/LIPIcs.CONCUR.2018.39
27. Kruskal, J.B.: The theory of well-quasi-ordering: A frequently discovered concept. Journal of Combinatorial Theory, Series A **13**(3), 297–305 (1972). https://doi.org/https://doi.org/10.1016/0097-3165(72)90063-5, https://www.sciencedirect.com/science/article/pii/0097316572900635
28. Larsen, K.G., Pettersson, P., Yi, W.: UPPAAL in a nutshell. STTT **1**(1-2), 134–152 (1997)
29. Lasota, S., Walukiewicz, I.: Alternating timed automata. In: International Conference on Foundations of Software Science and Computation Structures. pp. 250–265. Springer (2005)
30. Lasota, S., Walukiewicz, I.: Alternating timed automata. ACM Trans. Comput. Log. **9**(2), 10:1–10:27 (2008). https://doi.org/10.1145/1342991.1342994, https://doi.org/10.1145/1342991.1342994
31. Ouaknine, J., Worrell, J.: On the language inclusion problem for timed automata: Closing a decidability gap. In: Proceedings of the 19th Annual IEEE Symposium on Logic in Computer Science, 2004. pp. 54–63. IEEE (2004)
32. Ouaknine, J., Worrell, J.: On the decidability of metric temporal logic. In: 20th IEEE Symposium on Logic in Computer Science (LICS 2005), 26-29 June 2005, Chicago, IL, USA, Proceedings. pp. 188–197. IEEE Computer Society (2005). https://doi.org/10.1109/LICS.2005.33, https://doi.org/10.1109/LICS.2005.33
33. Ouaknine, J., Worrell, J.: On the decidability and complexity of metric temporal logic over finite words. Logical Methods in Computer Science **3** (2007)
34. Srivathsan, B.: Reachability in timed automata. ACM SIGLOG News **9**(3), 6–28 (2022). https://doi.org/10.1145/3559736.3559738, https://doi.org/10.1145/3559736.3559738
35. Sun, J., Liu, Y., Dong, J.S., Pang, J.: PAT: Towards flexible verification under fairness. Lecture Notes in Computer Science, vol. 5643, pp. 709–714. Springer (2009)
36. Tóth, T., Hajdu, A., Vörös, A., Micskei, Z., Majzik, I.: Theta: a framework for abstraction refinement-based model checking. In: Stewart, D., Weissenbacher, G. (eds.) Proceedings of the 17th Conference on Formal Methods in Computer-Aided Design. pp. 176–179 (2017). https://doi.org/10.23919/FMCAD.2017.8102257
37. Vargovcík, P., Holík, L.: Simplifying alternating automata for emptiness testing. In: Oh, H. (ed.) Programming Languages and Systems - 19th Asian Symposium, APLAS 2021, Chicago, IL, USA, October 17-18, 2021, Proceedings. Lecture Notes in Computer Science, vol. 13008, pp. 243–264. Springer (2021). https://doi.org/10.1007/978-3-030-89051-3_14, https://doi.org/10.1007/978-3-030-89051-3_14
38. Wulf, M.D., Doyen, L., Henzinger, T.A., Raskin, J.: Antichains: A new algorithm for checking universality of finite automata. In: Ball, T., Jones, R.B. (eds.) Computer Aided Verification, 18th International Conference, CAV 2006, Seattle, WA, USA, August 17-20, 2006, Proceedings. Lecture Notes in Computer Science, vol. 4144, pp. 17–30. Springer (2006). https://doi.org/10.1007/11817963_5, https://doi.org/10.1007/11817963_5

Author Index

P. A. Abdulla and D. Kesner (Eds.): FoSSaCS 2025, LNCS 15691, pp. 423–424, 2025.
https://doi.org/10.1007/978-3-031-90897-2